Lecture Notes in Computer Science 13904

Founding Editors

Gerhard Goos
Juris Hartmanis

Editorial Board Members

The series Lecture Notes in Computer Science (LNCS), including its subseries Lecture Notes in Artificial Intelligence (LNAI) and Lecture Notes in Bioinformatics (LNBI), has established itself as a medium for the publication of new developments in computer science and information technology research, teaching, and education.

LNCS enjoys close cooperation with the computer science R & D community, the series counts many renowned academics among its volume editors and paper authors, and collaborates with prestigious societies. Its mission is to serve this international community by providing an invaluable service, mainly focused on the publication of conference and workshop proceedings and postproceedings. LNCS commenced publication in 1973.

Alberto Del Pia · Volker Kaibel

Editors

Integer Programming and Combinatorial Optimization

24th International Conference, IPCO 2023
Madison, WI, USA, June 21–23, 2023
Proceedings

 Springer

Editors
Alberto Del Pia 🆔
University of Wisconsin-Madison
Madison, WI, USA

Volker Kaibel 🆔
Otto-von-Guericke-Universität
Magdeburg, Sachsen-Anhalt, Germany

ISSN 0302-9743　　　　　　　　ISSN 1611-3349 (electronic)
Lecture Notes in Computer Science
ISBN 978-3-031-32725-4　　　　ISBN 978-3-031-32726-1 (eBook)
https://doi.org/10.1007/978-3-031-32726-1

This Springer imprint is published by the registered company Springer Nature Switzerland AG
The registered company address is: Gewerbestrasse 11, 6330 Cham, Switzerland

Preface

This volume contains extended abstracts of the papers presented at IPCO 2023, the 24th Conference on Integer Programming and Combinatorial Optimization, held on June 21–23, 2023 in Madison, Wisconsin, USA.

IPCO is under the auspices of the Mathematical Optimization Society. Since its first edition, held at the University of Waterloo, Canada in May 1990, it has become a most important forum for presenting the latest results on the theory and practice of the various aspects of discrete optimization. For this year's 24th edition the conference had a Program Committee consisting of 16 members. In response to the Call for Papers, we received 119 submissions and accepted 33 papers, with an acceptance ratio of 28%. In a single-blind review process, each submission was reviewed by at least three Program Committee members, and 246 additional reviews were provided by external experts. Because of the limited number of time slots for presentations, many excellent submissions could not be accepted. The page limit for contributions to these proceedings was set to 15. We expect the full versions of the extended abstracts appearing in this Lecture Notes in Computer Science volume to be submitted for publication in refereed journals. A special issue of Mathematical Programming, Series B containing such versions is in process.

As has become a good tradition, IPCO 2023 had a Best Paper Award, which was given to Daniel Dadush, Friedrich Eisenbrand, and Thomas Rothvoss for their paper "From approximate to exact integer programming." This year, IPCO was preceded by a Summer School held during June 19–20, 2023, with lectures by Amitabh Basu (Johns Hopkins University, USA), Fatma Kilinç-Karzan (Carnegie Mellon University, USA), and Domenico Salvagnin (University of Padova, Italy). We thank them warmly for their contributions. We would also like to thank

- the authors who submitted their research to IPCO;
- the members of the Program Committee;
- the expert additional reviewers;
- the members of the Local Organizing Committee;
- the Mathematical Optimization Society, in particular the members of its IPCO Steering Committee, Karen Aardal, Oktay Günlük, Jochen Könemann, and Giacomo Zambelli;
- EasyChair for making the paper management simple and effective; and
- Springer for their efficient cooperation in producing this volume and for financial support for the Best Paper Award.

We would further like to thank the following sponsors for their financial support: the Wisconsin Institute for Discovery and the Department of Industrial & Systems Engineering at the University of Wisconsin-Madison, the Air Force Office of Scientific Research, the Office of Naval Research, FICO, Google, Gurobi Optimization, and The Optimization Firm.

March 2023

Alberto Del Pia
Volker Kaibel

Organization

Program Committee

Merve Bodur	University of Toronto, Canada
Jose Correa	Universidad de Chile, Chile
Alberto Del Pia	University of Wisconsin-Madison, USA
Yuri Faenza	Columbia University, USA
Volker Kaibel (Chair)	OVGU Magdeburg, Germany
Simge Kucukyavuz	Northwestern University, USA
Andrea Lodi	Cornell Tech, USA
Diego Moran	Universidad Adolfo Ibañez, Chile
Giacomo Nannicini	University of Southern California, USA
Britta Peis	RWTH Aachen, Germany
Mohit Singh	Georgia Institute of Technology, USA
Martin Skutella	TU Berlin, Germany
Juan Pablo Vielma	Massachusetts Institute of Technology, USA
Jens Vygen	University of Bonn, Germany
Stefan Weltge	TU München, Germany
Giacomo Zambelli	London School of Economics and Political Science, UK

Sponsors

WISCONSIN
INSTITUTE FOR DISCOVERY
AT THE UNIVERSITY OF WISCONSIN-MADISON

**Department of Industrial
and Systems Engineering**
UNIVERSITY OF WISCONSIN-MADISON

FICO®

Google

 GUROBI
OPTIMIZATION

 THE
OPTIMIZATION
FIRM

Contents

Information Complexity of Mixed-Integer Convex Optimization

Amitabh Basu[1], Hongyi Jiang[2], Phillip Kerger[1], and Marco Molinaro[3,4(✉)]

[1] Department of Applied Mathematics and Statistics, Johns Hopkins University,
Baltimore, USA
{abasu9,pkerger}@jhu.edu

[2] School of Civil and Environmental Engineering, Cornell University, Ithaca, USA
hj348@cornell.edu

[3] Microsoft Research, Redmond, USA
mmolinaro@microsoft.com

[4] Computer Science Department, PUC-Rio, Rio de Janeiro, Brazil

Abstract. We investigate the information complexity of mixed-integer convex optimization under different kinds of oracles. We establish new lower bounds for the standard first-order oracle, improving upon the previous best known lower bound. This leaves only a lower order linear term (in the dimension) as the gap between the lower and upper bounds. Further, we prove the first set of results in the literature (to the best of our knowledge) on information complexity with respect to oracles based on first-order information but restricted to binary queries, and discuss various special cases of interest thereof.

Keywords: Mixed-integer optimization · Information complexity

1 First-order Information Complexity

We consider the problem class of *mixed-integer convex optimization*:

$$\inf\{f(\mathbf{x},\mathbf{y}) : (\mathbf{x},\mathbf{y}) \in C, (\mathbf{x},\mathbf{y}) \in \mathbb{Z}^n \times \mathbb{R}^d\}, \tag{1}$$

where $f : \mathbb{R}^n \times \mathbb{R}^d \to \mathbb{R} \cup \{+\infty\}$ is a convex (possibly nondifferentiable) function and $C \subseteq \mathbb{R}^n \times \mathbb{R}^d$ is a closed, convex set. Given $\varepsilon > 0$, we wish to report a point in

$$S((f,C),\varepsilon) := \{(\mathbf{x},\mathbf{y}) \in C \cap \mathrm{dom}(f) \cap (\mathbb{Z}^n \times \mathbb{R}^d) : f(\mathbf{x},\mathbf{y}) \le f(\mathbf{x}',\mathbf{y}') + \varepsilon,$$
$$\forall(\mathbf{x}',\mathbf{y}') \in C \cap \mathrm{dom}(f) \cap (\mathbb{Z}^n \times \mathbb{R}^d)\}.$$

The first and third authors gratefully acknowledge support from Air Force Office of Scientific Research (AFOSR) grant FA95502010341 and National Science Foundation (NSF) grant CCF2006587. The fourth author was supported in part by the Coordenação de Aperfeiçoamento de Pessoal de Nível Superior (CAPES, Brasil) - Finance Code 001, by Bolsa de Produtividade em Pesquisa #312751/2021-4 from CNPq, and FAPERJ grant "Jovem Cientista do Nosso Estado".

A. Del Pia and V. Kaibel (Eds.): IPCO 2023, LNCS 13904, pp. 1–13, 2023.
https://doi.org/10.1007/978-3-031-32726-1_1

A point in $S((f, C), \varepsilon)$ will be called an ε-*approximate solution* and points in $C \cap \mathrm{dom}(f) \cap (\mathbb{Z}^n \times \mathbb{R}^d)$ will be called *feasible solutions*. We say that $\mathbf{x}_1, \ldots, \mathbf{x}_n$ are the *integer-valued decision variables* or simply the *integer variables* of the problem, and $\mathbf{y}_1, \ldots, \mathbf{y}_d$ are called the *continuous variables*.

The notion of *information complexity* (a.k.a *oracle complexity* or *analytical complexity*) goes back to foundational work by Nemirovski and Yudin [8] on convex optimization (without integer variables) and is based on the following. An algorithm for reporting an ε-approximate solution to an instance (f, C) must be "given" the instance somehow. Allowing only instances with explicit, algebraic descriptions (e.g., the case of linear programming) can be restrictive. To work with more general, nonlinear instances, the algorithm is allowed to make queries to an oracle to collect information about the instance. The standard oracle that has been studied over the past several decades is the so-called *first-order oracle*, which consists of two parts: i) a *separation* oracle that receives a point $\mathbf{z} \in \mathbb{R}^{n+d}$ and reports "YES" if $\mathbf{z} \in C$ and otherwise reports a separating hyperplane for $\mathbf{z}$ and C, ii) a *subgradient* oracle that receives a point $\mathbf{z} \in \mathbb{R}^{n+d}$ and reports $f(\mathbf{z})$ and a subgradient for f at $\mathbf{z}$. The goal is to design a query strategy that can report an ε-approximate solution after making the smallest number of queries. Tight lower and upper bounds (differing by only a small constant factor) on the number of queries were obtained by Nemirovski and Yudin in their seminal work [8] (the case with no integer variables); roughly speaking, the bound is $\Theta\left(d \log\left(\frac{1}{\varepsilon}\right)\right)$. These insights were extended to the mixed-integer setting in [2,3, 9], with the best known lower and upper bounds stated in [2].

Observe that the response to any separation/subgradient query is a vector in $\mathbb{R}^{n+d}$. Thus, each query reveals at least $n+d$ bits of information about the instance. A more careful accounting that measures the "amount of information" accrued would track the total number of bits obtained as opposed to just the total number of oracle queries made. A natural question, posed in [2], is whether the bounds from the classical analysis would change if one uses this new measure of the total number of bits, as opposed to the number of queries. The intuition, roughly, is that one should need a factor $(n + d) \log\left(\frac{1}{\varepsilon}\right)$ larger than the number of first-order queries, because one should need to probe at least $\log\left(\frac{1}{\varepsilon}\right)$ bits in $n+d$ coordinates to recover the full subgradient/separating hyperplane (up to desired approximations). We attempt to make some progress on this question in this paper.

The above discussion suggests that one should consider oracles that return a desired bit of a desired coordinate of the separating hyperplane vector or subgradient. However, one can imagine making other binary queries on the instance; for example, one can pick a direction and ask for the sign of the inner product of the subgradient and this direction. In fact, one can consider more general binary queries that have nothing to do with subgradients/separating hyperplanes. If one allows *all* possible binary queries, i.e., one can use any function from the space of instances to $\{0, 1\}$ as a query, then one can simply ask for the appropriate bits of the true minimizer and in $O((n + d) \log(1/\varepsilon))$ queries, one can get an ε-approximate solution. A matching lower bound follows from a fairly straightforward counting argument. Thus, allowing for all possible binary queries gives the same information complexity bound as the original Nemirovski-Yudin bound with subgradient queries in the $n = 0$ (no integer variables) case, but is an exponential improvement when $n \geq 1$ (see [2] and the discussion below). What this shows is that the bounds on information complexity can be quite different under different oracles. With all possible binary queries, while each query reveals only a single bit of information, the queries themselves are a much richer class and this compensates to give the same bound in the continuous case and exponentially better bounds in the presence of integer variables. Thus, to get a better understanding of this trade-off, we restrict to queries that still extract information from the subgradient or separating hyperplane at a point, and are thus "local" in a sense. ($\mathcal{I}_{n,d,R,\rho,M}$ is the set of instances we focus on throughout the paper, see Definition 2 for a formal definition.)

Definition 1. *An oracle using first-order information consists of two parts:*

1. *For every $\mathbf{z} \in [-R, R]^{n+d}$, there exist two maps $g_{\mathbf{z}}^{\mathrm{sep}} : \mathcal{I}_{n,d,R,\rho,M} \to \mathbb{R}^{n+d}$ and $g_{\mathbf{z}}^{\mathrm{sub}} : \mathcal{I}_{n,d,R,\rho,M} \to \mathbb{R} \times \mathbb{R}^{n+d}$ such that for all $(f, C) \in \mathcal{I}_{n,d,R,\rho,M}$ the following properties hold.*
   (a) *$C \subseteq \{\mathbf{z}' \in \mathbb{R}^{n+d} : \langle g_{\mathbf{z}}^{\mathrm{sep}}(C), \mathbf{z}' \rangle < \langle g_{\mathbf{z}}^{\mathrm{sep}}(C), \mathbf{z} \rangle\}$ if $\mathbf{z} \notin C$ and $g_{\mathbf{z}}^{\mathrm{sep}}(C) = \mathbf{0}$ if $\mathbf{z} \in C$. In other words, $g_{\mathbf{z}}^{\mathrm{sub}}(f)$ returns a (normal vector to a) separating hyperplane if $\mathbf{z} \notin C$. We will assume that a nonzero response $g_{\mathbf{z}}^{\mathrm{sep}}(C)$ has norm 1, since scalings do not change the separation property.*
   (b) *$g_{\mathbf{z}}^{\mathrm{sub}}(f) \in \{f(\mathbf{z})\} \times \partial f(\mathbf{z})$, where $\partial f(\mathbf{z})$ denotes the subdifferential (the set of all subgradients) of f at $\mathbf{z}$. In other words, $g_{\mathbf{z}}^{\mathrm{sub}}(f)$ returns the function value and a subgradient for f at $\mathbf{z}$. If $f(\mathbf{z}) = +\infty$, $g_{\mathbf{z}}^{\mathrm{sub}}(f)$ returns a separating hyperplane for $\mathbf{z}$ and the domain of f.[1]*
   Such maps will be called first-order maps. *A collection of first-order maps, one for every $\mathbf{z}$, is called a (complete)* first-order chart *and will be denoted by $\mathcal{G}$.*
2. *There are two sets of functions $\mathcal{H}^{\mathrm{sep}}$ and $\mathcal{H}^{\mathrm{sub}}$ with domains $\mathbb{R}^{n+d}$ and $\mathbb{R} \times \mathbb{R}^{n+d}$ respectively. We will use the notation $\mathcal{H} = \mathcal{H}^{\mathrm{sep}} \cup \mathcal{H}^{\mathrm{sub}}$. $\mathcal{H}$ will be called the collection of* permissible queries *of the oracle.*

The algorithm, at any iteration, can choose a point $\mathbf{z}$ and a function $h \in \mathcal{H}$ and it receives the response $h(g_{\mathbf{z}}^{\mathrm{sep}}(\widehat{f}))$ or $h(g_{\mathbf{z}}^{\mathrm{sub}}(\widehat{C}))$, depending on whether $h \in \mathcal{H}^{sep}$ or $h \in \mathcal{H}^{sub}$, where $(\widehat{f}, \widehat{C})$ is the unknown instance.

In particular, when $\mathcal{H}^{\mathrm{sep}}$ and $\mathcal{H}^{\mathrm{sub}}$ consist only of the identity function, we recover a standard first-order oracle. We will also study the cases where $\mathcal{H}^{\mathrm{sep}}$ and $\mathcal{H}^{\mathrm{sub}}$ consist of functions that map a vector to a particular bit of a particular coordinate, or the sign of the inner product with a particular direction, or the set of all possible binary functions. In the last case, the oracle will be called the *general binary oracle based on $\mathcal{G}$*.

---

[1] When the function value is $+\infty$, we will count this as a separation query. In other words, below we will assume without further comment that every functional query returns a finite real value for the function and a subgradient at the queried point.

1.1   Our Results

It is not hard to see that if we consider all possible instances of (1), then any adaptive query strategy has infinite information complexity because a finite number of queries cannot distinguish between all possible instances. Thus, bounds on the information complexity must be based on appropriate parameterizations of the problem. We will focus on the following standard parameterization.

Definition 2. $\mathcal{I}_{n,d,R,\rho,M}$ is the set of all instances of (1) such that

(i) $C \cap \mathrm{dom}(f)$ is contained in the box $\{\mathbf{z} \in \mathbb{R}^n \times \mathbb{R}^d : \|\mathbf{z}\|_\infty \leq R\}$.

(ii) If $(\mathbf{x}^\star, \mathbf{y}^\star)$ is an optimal solution of the instance, then there exists $\hat{\mathbf{y}} \in \mathbb{R}^d$ satisfying $\{(\mathbf{x}^\star, \mathbf{y}) : \|\mathbf{y} - \hat{\mathbf{y}}\|_\infty \leq \rho\} \subseteq C$. In other words, there is a "strictly feasible" point $(\mathbf{x}^\star, \hat{\mathbf{y}})$ in the same fiber as the optimum $(\mathbf{x}^\star, \mathbf{y}^\star)$.

(iii) f is Lipschitz continuous with respect to the $\|\cdot\|_\infty$-norm with Lipschitz constant M on $\{\mathbf{x}\} \times [-R, R]^d$ for all $\mathbf{x} \in [-R, R]^n \cap \mathbb{Z}^n$, where we use the convention that if f is identically $+\infty$ on $\{\mathbf{x}\} \times [-R, R]^d$, then any M works on this fiber. In other words, for any $(\mathbf{x}, \mathbf{y}), (\mathbf{x}, \mathbf{y}') \in (\mathbb{Z}^n \times \mathbb{R}^d) \cap [-R, R]^{n+d}$ with $\|\mathbf{y} - \mathbf{y}'\|_\infty \leq R$, $|f(\mathbf{x}, \mathbf{y}) - f(\mathbf{x}, \mathbf{y}')| \leq M\|\mathbf{y} - \mathbf{y}'\|_\infty$ with the convention that $\infty - \infty = 0$.

We obtain the following results in this paper.

Results for $n \geq 1$ (allowing integer variables).

1. In the classical setting where $\mathcal{H}^{sep}$ and $\mathcal{H}^{sub}$ consist only of the identity function (i.e. each query receives the entire subgradient/separating hyperplane), we improve the best known lower bound (from [2]) on the number of queries needed for the general mixed-integer case. In particular, we show that one needs at least $\Omega\left(2^n d \log\left(\frac{MR}{\min\{\rho,1\}\varepsilon}\right)\right)$ queries, improving upon the previous bound of $\Omega\left(2^n d \log\left(\frac{R}{\rho}\right)\right)$. We mention here that the first lower bound on information complexity with integer constrained variables was established in [3], for a specific class of algorithms/query strategies called *cutting-plane schemes*. The lower bounds stated here (and in [2]) do not make any assumptions on the algorithms/query strategies.

2. This lower bound of $\Omega\left(2^n d \log\left(\frac{MR}{\min\{\rho,1\}\varepsilon}\right)\right)$ for the classical setting is complemented by an upper bound of $O\left(2^n d(n + d) \log\left(\frac{MR}{\min\{\rho,1\}\varepsilon}\right)\right)$ in the literature. This was first obtained in [9]; see [2] for a self-contained exposition. As mentioned before, we expect the information complexity in the setting of binary oracles using first-order information to be at most a factor $(n + d) \log\left(\frac{MR}{\min\{\rho,1\}\varepsilon}\right)$ larger than the classical setting. We rigorously prove this, under the additional assumption that the fiber containing the optimal solution contains a point that is ρ-deep inside the feasible region C, i.e., a ball of radius ρ centered at this point is contained in the set C. Note that this is a stronger assumption compared to item (ii) in the definition of $\mathcal{I}_{n,d,R,\rho,M}$.

Results for $n = 0$ (continuous case with no integer variables).

1. When the separation and subgradient oracles can only be accessed through (all) binary queries (i.e., $\mathcal{H}^{\text{sep}}$ and $\mathcal{H}^{\text{sub}}$ consist of all possible binary functions on $\mathbb{R}^d$), we show that strictly more queries are needed compared to the classic setting of full oracle access. More precisely, compared to the classic bound of $\Theta\left(d\log\left(\frac{MR}{\min\{\rho,1\}\varepsilon}\right)\right)$, we show that one needs at least $\tilde{\Omega}\left(\max\left\{d^{\frac{8}{7}}, d\log\left(\frac{MR}{\min\{\rho,1\}\varepsilon}\right)\right\}\right)$ binary queries (where $\tilde{\Omega}$ hides polylogarithmic factors in d). This is obtained using recent lower bounds on "memory-constrained" algorithms for convex optimization from [6].

2. We establish an upper bound of $O\left(d^2\log\left(\frac{MR}{\min\{\rho,1\}\varepsilon}\right)^2\right)$ for binary queries using first-order oracles. This is an extension of a result from [11] that considered the unconstrained optimization setting.

3. We establish an upper bound of $O\left(\log|\mathcal{I}| + d\log\left(\frac{MR}{\min\{\rho,1\}\varepsilon}\right)\right)$ for any finite subclass of instances $\mathcal{I} \subseteq \mathcal{I}_{n,d,R,\rho,M}$. Note that this can beat the lower bound from Item 1 above, e.g., if $|\mathcal{I}| = 2^{O(d)}$.

1.2   Formal Definitions and Statement of Results

For any set X, we will use X^* to denote the set of all finite sequences of elements from X (e.g., $\{0,1\}^*$ denotes the set of all finite binary strings).

Definition 3. *Given an oracle using first-order information $(\mathcal{G}, \mathcal{H})$, let $\mathcal{Q}$ be the set of possible queries that can be made under this oracle (i.e., pairs $(\mathbf{z}, h)$ where $\mathbf{z} \in \mathbb{R}^{n+d}$ is a query point and $h \in \mathcal{H}$). Let H denote the response set of the functions in $\mathcal{H}$ (e.g., vectors for standard first-order oracles, or $\{0,1\}$ for binary queries).*

A query strategy is a function $D : (\mathcal{Q} \times H)^* \to \mathcal{Q}$. The transcript $\Pi(D, I)$ of a strategy D on an instance $I = (f, C)$ is the sequence of query and response pairs $(q_i, q_i(I))$, $i = 1, 2, \ldots$ obtained when one applies D on I, i.e., $q_1 = D(\emptyset)$ and $q_i = D((q_1, q_1(I)), \ldots, (q_{i-1}, q_{i-1}(I)))$ for $i \geq 2$.

The ε-information complexity $\text{icomp}_\varepsilon(D, I, \mathcal{G}, \mathcal{H})$ of an instance I for a query strategy D, with access to an oracle using first-order information $(\mathcal{G}, \mathcal{H})$, is defined as the minimum natural number k such that the set of all instances which return the same responses as the instance I to the first k queries of D have a common ε-approximate solution. The ε-information complexity of the problem class $\mathcal{I}_{n,d,R,\rho,M}$, with access to an oracle using first-order information $(\mathcal{G}, \mathcal{H})$, is defined as

$$\text{icomp}_\varepsilon(n, d, R, \rho, M, \mathcal{G}, \mathcal{H}) := \inf_D \sup_{I \in \mathcal{I}_{n,d,R,\rho,M}} \text{icomp}_\varepsilon(D, I, \mathcal{G}, \mathcal{H})$$

where the infimum is taken over all query strategies.

We can now formally state our main results. Let $\mathcal{H}^{\text{bit}}$ be the set of binary queries that return a desired bit (of a desired coordinate) of a subgradient/separating hyperplane/function value. Let $\mathcal{H}^{\text{dir}}$ be the set of binary queries that returns the sign of the inner product of the subgradient/separating hyperplane with a desired direction, or a desired bit of the function value.

Results for $n \geq 1$ (allowing integer variables).

Theorem 1. *There exists a complete first-order chart $\mathcal{G}$ such that for the standard first-order oracle based on $\mathcal{G}$ (i.e., $\mathcal{H}$ consists of the identity functions), we have*

$$\mathrm{icomp}_\varepsilon(n, d, R, \rho, M, \mathcal{G}, \mathcal{H}) = \Omega\left(2^n\left(1 + d\log\left(\frac{MR}{\min\{\rho, 1\}\varepsilon}\right)\right)\right).$$

Theorem 2. *Assume $d \geq 1$. For $U > 0$, consider the subclass of instances of $\mathcal{I}_{n,d,R,\rho,M}$ whose objective function values lie in $[-U, U]$, and the fiber over the optimal solution contains a $\mathbf{z}$ such that the $(n + d)$-dim ρ-radius ball centered at $\mathbf{z}$ is contained in C. There exists a query strategy for this subclass that uses $(\mathcal{G}, \mathcal{H})$, where $\mathcal{G}$ is any complete first-order chart and $\mathcal{H}$ is either $\mathcal{H}^{\mathrm{bit}}$ or $\mathcal{H}^{\mathrm{dir}}$, that reports an ε-approximate solution by making at most*

$$O\left(2^n d\,(n+d)\log\left(\frac{dMR}{\min\{\rho, 1\}\varepsilon}\right)\right) \cdot \left((n+d)\log\left(\frac{(n+d)MR}{\varepsilon}\right) + \log\frac{U}{\varepsilon}\right)$$

queries.

Prescribing an *a priori* range for objective function values is not a serious restriction for two reasons: i) The difference between the maximum and the minimum values of an objective function in $\mathcal{I}_{n,d,R,\rho,M}$ is at most $2MR$, and ii) All optimization problems whose objective functions differ by a constant are equivalent. We also comment that while we assume $d \geq 1$ in Theorem 2, similar bounds can be established for the $d = 0$ case. We omit this here because a unified expression for the $d = 0$ and $d \geq 1$ cases becomes unwieldy and difficult to parse.

We remark that we can obtain the same result when the objective function can only be accessed through comparisons of the form "Is $f(\mathbf{z}) \leq f(\mathbf{z}')$?", i.e., no access to the subgradients ∂f. Such algorithms are particularly useful in learning from users' behaviors, since while a user typically cannot accurately report its (dis)utility value $f(\mathbf{z})$ for an option $\mathbf{z}$, it can more reliably compare the values $f(\mathbf{z})$ and $f(\mathbf{z}')$ of two options; see [5,10] and references therein for discussions and algorithms in the continuous case. To the best of our knowledge, no such algorithm for the mixed-integer case has appeared explicitly in the literature.

We also remark that the additional assumption that a $(n + d)$-dim ρ-radius is contained in C can be weakened by using a Lenstra-style algorithm, but this yields a much worse dependence in d and n.

Results for $n = 0$ (continuous case with no integer variables).

Theorem 3. *There exists a complete first-order chart $\mathcal{G}$ such that for the general binary oracle based on $\mathcal{G}$ (i.e., $\mathcal{H}$ consists of all possible binary functions on $\mathbb{R}^d$), we have*

$$\mathrm{icomp}_\varepsilon(d, R, \rho, M, \mathcal{G}, \mathcal{H}) = \tilde{\Omega}\left(\max\left\{d^{\frac{8}{7}}, d\log\left(\frac{MR}{\min\{\rho, 1\}\varepsilon}\right)\right\}\right),$$

where $\tilde{\Omega}$ hides polylogarithmic factors in d.

Theorem 4. *For $U > 0$, consider the subclass of instances of $\mathcal{I}_{d,R,\rho,M}$ whose objective function values lie in $[-U, U]$. There exists a query strategy for this subclass that uses $(\mathcal{G}, \mathcal{H})$, where $\mathcal{G}$ is any complete first-order chart and $\mathcal{H}$ is either $\mathcal{H}^{\mathrm{bit}}$ or $\mathcal{H}^{\mathrm{dir}}$, that reports an ε-approximate solution by making at most*

$$O\left(d \log\left(\frac{dMR}{\min\{\rho, 1\}\varepsilon}\right)\right) \cdot \left(d \log\left(\frac{dMR}{\varepsilon}\right) + \log\frac{U}{\varepsilon}\right)$$

queries.

Theorem 5. *Given any subclass of finitely many instances $\mathcal{I} \subset \mathcal{I}_{n,d,R,\rho,M}$ and any complete first-order chart $\mathcal{G}$, there exists a query strategy for this subclass using the general binary oracle based on $\mathcal{G}$ that reports an ε-approximate solution by making at most $O\left(\log|\mathcal{I}| + d\log\left(\frac{MR}{\min\{\rho,1\}\varepsilon}\right)\right)$ queries.*

1.3   Discussion and Future Avenues

The concept of information complexity in continuous convex optimization and its study go back several decades, and it is considered a fundamental question in convex optimization. In comparison, much less work on information complexity has been carried out in the presence of integer constrained variables. Nevertheless, we believe there are important and challenging questions that come up in that domain that are worth studying. Further, even within the context of continuous convex optimization, the notion of information complexity has almost exclusively focused on the number of first-order queries. As we hope to illustrate with the results of this paper, considering other kinds of oracles lead to very interesting questions at the intersection of mathematical optimization and information theory. In particular, the study of binary oracles promises to give a more refined understanding of the fundamental question "How much information about an optimization instance do we need to be able to solve it with provable guarantees?". For instance, establishing *any superlinear (in the dimension)* lower bound for the continuous problem with binary oracles, like the one in Theorem 3, seems to be nontrivial. In fact, the results from [6], on which Theorem 3 is based, were considered a breakthrough in establishing superlinear lower bounds on space complexity of convex optimization. Even so, the right bound is conjectured to be quadratic in the dimension (see Theorem 4) and our Theorem 3 is far from that at this point. We thus view the results of this paper as expanding our understanding of information complexity of optimization in two different dimensions: what role does the presence of integer variables play and what role does the nature of the oracle play (with or without integer variables)? For integer variables, our first result brings the lower bound closer to the best known upper bound on information complexity based on the classical subgradient oracle. The remaining gap is now simply a factor linear in the dimension. A conjecture in convex geometry first articulated in [9, Conjecture 4.1.20] and elaborated upon in [2,3] would resolve this and would show that the right upper bound is essentially equal to the lower bound we prove in this paper. Therefore, we have reasons to believe that the right bound is the one we obtain in this paper.

Beyond this, we believe the following additional conjectures to be good catalysts for future research, especially in regard to understanding the interplay of integer variables and other oracles.

Conjecture 1. *Given an oracle* $(\mathcal{G}, \mathcal{H})$ *based on first-order information, suppose there is a family of instances that establishes a lower bound* $\ell(d, R, \rho, M, \mathcal{G}, \mathcal{H})$ *for the* $n = 0$ *(continuous) case. Then there exists a family of instances that establishes a lower bound of* $2^n \cdot \ell(d, R, \rho, M, \mathcal{G}, \mathcal{H})$ *for the* $n \geq 1$ *case, i.e., the general mixed-integer case.*

Conjecture 2. *If there exists a query strategy with worst case information complexity* $u(n, d, R, \rho, M, \mathcal{G})$ *under the standard first-order oracle based on a complete first-order chart* $\mathcal{G}$, *then there exists a query strategy with worst case information complexity* $u(n, d, R, \rho, M, \mathcal{G}) \cdot O\left((n + d) \log\left(\frac{MR}{\rho\varepsilon}\right)\right)$ *under the general binary oracle based* $\mathcal{G}$.

Both of the above results, if true, would be useful "transfer" theorems: the first one for lower bounds, the second one for upper bounds. Conjecture 1 takes a lower bound result for the continuous problem and lifts it to the general mixed-integer case with a factor of 2^n. This would be a general tool that can then give Theorem 1 as a special case and also give a mixed-integer version of Theorem 3 as a corollary. Further, if future research on the information complexity of continuous convex optimization results in better/different lower bounds, these would immediately imply new lower bounds for the mixed-integer case as well. For instance, we believe the following conjecture to be true for the continuous convex optimization problem.

Conjecture 3. *There exists a complete first-order chart* $\mathcal{G}$ *such that the general binary oracle based on* $\mathcal{G}$ *has information complexity* $\Omega\left(d^2 \log\left(\frac{MR}{\rho\varepsilon}\right)^2\right)$.

Another version of Conjecture 3 is also stated in the language of "memory-constrained" algorithms (see Sect. 2.2 below) in [6,11].

Conjecture 2 can be used to take upper bound results proved in the standard first-order oracle setting and get upper bound results in the general binary oracle setting. For instance, if the upper bound for the general mixed-integer problem is improved by resolving the convex geometry conjecture mentioned above (and we believe the lower bound is correct and the upper bound is indeed loose), then this would also give better upper bounds for the general binary oracle setting.

2 Proof Sketches

2.1 Proof Sketch of Theorem 1

The general strategy to prove Theorem 1 is the following: Given any query strategy D, we will construct two instances $(f_1, C_1), (f_2, C_2) \in \mathcal{I}_{n,d,R,\rho,M}$ such that the transcripts $\Pi(D, (f_1, C_1))$ and $\Pi(D, (f_2, C_2))$ are equal for the first k terms if k is less than the lower bound, but $S((f_1, C_1), \varepsilon) \cap S((f_2, C_2), \varepsilon) = \emptyset$.

The lower bound $\mathrm{icomp}_\varepsilon(n, d, R, \rho, M, \mathcal{G}, \mathcal{H}) \geq \left(2^n \cdot d \log \left(\frac{2R}{3\rho} \right) \right)$ was established in [2, Theorem 4.2]. Thus, it suffices to show $\mathrm{icomp}_\varepsilon(n, d, R, \rho, M, \mathcal{G}, \mathcal{H}) \geq d 2^n \left\lfloor \log_8 \left(\frac{MR}{2\varepsilon} \right) \right\rfloor$.

The idea is to use a family $\mathcal{F}$ of convex functions defined over $[-R, R]^d$ described in [7,8] such that one needs at least $d \left\lfloor \log_8 \left(\frac{MR}{2\varepsilon} \right) \right\rfloor$ subgradient queries to report an ε-approximate solution. In fact, for any query strategy in $\mathbb{R}^d$, one can adversarially construct subgradient responses such that if less than $d \left\lfloor \log_8 \left(\frac{MR}{2\varepsilon} \right) \right\rfloor$ are made, one can report two functions from $\mathcal{F}$ that would have provided the same responses as given by the adversary and yet have disjoint sets of ε-approximate solutions. We now mimic this by putting the family $\mathcal{F}$ over the fibers $\{\mathbf{x}\} \times [-R, R]^d$ for $\mathbf{x} \in \{0, 1\}^n$. Our constraint set C is going to be simply $[0, 1]^n \times [-R, R]^d$; thus, the separation oracle queries will provide no information.

We will create a nested sequence of polyhedra $Y_0 \supseteq Y_1 \supseteq \ldots \supseteq Y_k$ contained in $[0, 1]^n \times [-R, R]^d \times \mathbb{R}$, where Y_i corresponds to query i. This sequence will depend on the queries made and will determine our responses. The set Y_k will be used to construct the epigraphs of two functions f_1 and f_2, whose ε-approximate minimizers in $\mathbb{Z}^n \times \mathbb{R}^d$ will be disjoint. We now enumerate different cases:

1. If the query point is outside $[0, 1]^n \times [-R, R]^d$, then the function value is reported to be $+\infty$ and a separating hyperplane is reported. Y_i is not updated.
2. If the query point $\mathbf{z} = (\mathbf{x}, \mathbf{y})$ is inside $[0, 1]^n \times [-R, R]^d$, but $\mathbf{x} \notin \{0, 1\}^n$, then we simply report the function value from the current Y_i set interpreted as an epigraph and any subgradient at this point on Y_i. Y_i is not updated.
3. If the query point $\mathbf{z} = (\mathbf{x}, \mathbf{y}) \in \{0, 1\}^n \times [-R, R]^d$, we look at what the response from the adversary would have been for the family $\mathcal{F}$ at the query point $\mathbf{y}$, and we rotate the corresponding subgradient hyperplane so that it is valid for all other fibers $\{\mathbf{x}'\} \times [-R, R]^d$ for $\mathbf{x}' \in \{0, 1\}^n \setminus \{\mathbf{x}\}$, as well as all points queried inside the hypercube, but not the fibers. This can be done because the fibers are compact, and only a finite number of queries have been made. We then update Y_i by intersecting with this rotated halfspace, and report this rotated halfspace as the response for this query.

If $k < 2^n d \left\lfloor \log_8 \left(\frac{MR}{2\varepsilon} \right) \right\rfloor$ queries have been made, there must exist a fiber – corresponding to say $\mathbf{x}^\star \in \{0, 1\}^n$ – on which less than $d \left\lfloor \log_8 \left(\frac{MR}{2\varepsilon} \right) \right\rfloor$ queries were made. We now take the two functions $\tilde{f}_1, \tilde{f}_2$ from $\mathcal{F}$ that would have given the same responses on that fiber, with disjoint ε-approximate solutions. On the other fibers corresponding to $\mathbf{x} \neq \mathbf{x}^\star$, we consider any function $\tilde{f}^{\mathbf{x}}$ from $\mathcal{F}$ that would have returned the same responses on that fiber (this can be ensured to exist given the structure of the family $\mathcal{F}$), and define $\tilde{Y}^{\mathbf{x}} := \{\mathbf{x}\} \times \mathrm{epi}(\tilde{f}^{\mathbf{x}})$. For query points $\mathbf{z}$ not on the fibers, we consider the sets $\tilde{Y}^{\mathbf{z}} := Y_k \cap (\{\mathbf{z}\} \times \mathbb{R})$. For $i = 1, 2$, we define E_i as the convex hull of $\{\mathbf{x}^\star\} \times \mathrm{epi}(\tilde{f}_i)$ and all the $\tilde{Y}^{\mathbf{x}}$, $\mathbf{x} \neq \mathbf{x}^\star$ and $\tilde{Y}^{\mathbf{z}}$ for query points $\mathbf{z}$ not on the fibers. These convex hulls are the epigraphs of two functions f_1, f_2 whose ε-approximate minimizers can be shown to be exactly the points of the form $(\mathbf{x}^\star, \mathbf{y})$, where $\mathbf{y}$ is an ε-approximate minimizer of $\tilde{f}_1, \tilde{f}_2$, respectively. These are disjoint sets and we are done.

2.2   Proof of Theorem 3

We need to introduce the idea of *information memory* of any query strategy/algorithm.

Definition 4. *A first-order query strategy with* information memory *comprises three functions:*

1. $\phi_{\mathrm{query}} : \{0,1\}^* \to [-R, R]^n \times [-R, R]^d$.
2. $\phi_{\mathrm{update}}^{\mathrm{sep}} : (\mathbb{R}^n \times \mathbb{R}^d) \times \{0,1\}^* \to \{0,1\}^*$.
3. $\phi_{\mathrm{update}}^{\mathrm{sub}} : (\mathbb{R} \times (\mathbb{R}^n \times \mathbb{R}^d)) \times \{0,1\}^* \to \{0,1\}^*$.

Given access to a (complete) first-order chart $\mathcal{G}$, the query strategy maintains an information memory, *which is a finite length binary string in $\{0,1\}^*$, initialized as the empty string. At every iteration $k = 1, 2, \ldots$, the query strategy computes $\mathbf{z}_k := \phi_{query}(r_{k-1})$ and updates its memory using either $r_k = \phi_{\mathrm{update}}^{\mathrm{sep}} \left(g_{\mathbf{z}_k}^{\mathrm{sep}}(\widehat{C}), r_{k-1} \right)$ or $r_k = \phi_{\mathrm{update}}^{\mathrm{sep}} \left(g_{\mathbf{z}_k}^{\mathrm{sub}}(\widehat{f}), r_{k-1} \right)$, where $(\widehat{f}, \widehat{C})$ is the unknown true instance. After finitely many iterations, the query strategy does a final computation based on its information memory and reports an ε-approximate solution, i.e., there is a final function $\phi_{fin} : \{0,1\}^* \to \mathbb{Z}^n \times \mathbb{R}^d$.*

The information memory complexity *of an algorithm for an instance is the maximum length of its information memory r_k over all iterations k during the processing of this instance.*

Proposition 6. *Let $\mathcal{G}$ be a (complete) first-order chart. For any first-order query strategy $\mathcal{A}$ with information memory that uses $\mathcal{G}$, there exists a query strategy $\mathcal{A}'$ using the general binary oracle based on $\mathcal{G}$, such that for any instance (f, C), if $\mathcal{A}$ stops after T iterations with information memory complexity Q, $\mathcal{A}'$ stops after making at most $Q \cdot T$ oracle queries.*

Conversely, for any query strategy $\mathcal{A}'$ using the general binary oracle based on $\mathcal{G}$, there exists a first-order query strategy $\mathcal{A}$ with information memory such that for any instance (f, C), if $\mathcal{A}'$ stops after T iterations, $\mathcal{A}$ stops after making at most T iterations with information memory complexity at most T.

Proof. Let $\mathcal{A}$ be a first-order query strategy with information memory. We can simulate $\mathcal{A}$ by the query strategy whose queries are precisely the bits of the information memory state r_k at each iteration k of $\mathcal{A}$. More formally, the query is $\mathbf{z} = \phi_{\text{query}}(r_{k-1})$ and $h(\cdot) = (\phi_{\text{update}}^{\text{sep}}(\cdot, r_{k-1}))_i$ or $h(\cdot) = (\phi_{\text{update}}^{\text{sub}}(\cdot, r_{k-1}))_i$, depending on which type of query was made.

Conversely, given a query strategy $\mathcal{A}'$ based on the general binary oracle, we can simulate it with a first-order query strategy with information memory where in each iteration, we simply append the new bit queried by $\mathcal{A}'$ to the current state of the memory. □

The following is (a rephrasing of) a result from [6].

Theorem 7. [6, Theorem 1] *For every $\delta \in [0, 1/4]$, there is a class of instances $\mathcal{I} \subseteq \mathcal{I}_{n,d,R,\rho,M}$ and a (complete) first-order chart $\mathcal{G}$ such that any first-order query strategy with information memory must have either $d^{1.25-\delta}$ information memory complexity (in the worst case) or make at least $\tilde{\Omega}(d^{1+\frac{4}{3}\delta})$ iterations (in the worst case).*

Proof of Theorem 3. Setting $\delta = \frac{3}{28}$ in Theorem 7, we obtain that any first-order query strategy uses either $d^{8/7}$ information memory or makes at least $\tilde{\Omega}(d^{8/7})$ iterations. Using the second part of Proposition 6, we obtain the desired lower bound of $\tilde{\Omega}(d^{8/7})$ on the number of queries made by any query strategy using the general binary oracle based on $\mathcal{G}$. □

2.3   Proof Sketch of Theorem 5

We will sketch the proof for solving the feasibility problem, optimization being handled in a similar way by incorporating subgradients. Thus, we have a finite set of instance $\mathcal{I} \subseteq \mathcal{I}_{d,R,\rho}$ with only continuous variables, a true (unknown) instance $C \in \mathcal{I}$, and our goal is to report a point in C using few *binary queries* to a separation oracle. For that, we design a procedure that maintains a family $\mathcal{U} \subseteq \mathcal{I}$ of the instances that are still possible (which always includes the true instance C), along with a polyhedron P containing C. We start with $\mathcal{U} = \mathcal{I}$ and $P = [-R, R]^d$. We will show that we can always either reduce $|\mathcal{U}|$ or vol(P) by a constant fraction with each query. For that, while $|\mathcal{U}| > 1$ do the following:

- Set $\mathbf{p}$ equal to be the centroid of P. If the separation oracle at $\mathbf{p}$ reports that $\mathbf{p} \in C$, then we return $\mathbf{p}$. Otherwise:

- **Case 1:** For every possible answer $\mathbf{v} \in \mathbb{R}^d$ to the separation oracle, at most half of the sets C' in $\mathcal{U}$ give that answer for the point $\mathbf{p}$, namely $g_{\mathbf{p}}^{\text{sep}}(C') = \mathbf{v}$. Then, there is a set of answers $V \subseteq \mathbb{R}^d$ such that the number of sets $C' \in \mathcal{U}$ with $g_{\mathbf{p}}^{\text{sep}}(C') \in V$ is between $\frac{1}{4}|\mathcal{U}|$ and $\frac{3}{4}|\mathcal{U}|$. Then querying whether the true instance has $g_{\mathbf{p}}^{\text{sep}}(C) \in V$ (using the binary query h where $h(\mathbf{v}) = 1$ iff $\mathbf{v} \in V$) we can eliminate at least a quarter of the instances of $\mathcal{U}$ as not possible. So update $\mathcal{U}$ by deleting those instances from it.

- **Case 2:** There exists $\bar{\mathbf{v}} \in \mathbb{R}^d$ such that more than half of the instances C' in $\mathcal{U}$ have $g_{\mathbf{p}}^{\text{sep}}(C') = \bar{\mathbf{v}}$. Query whether the true instance has $g_{\mathbf{p}}^{\text{sep}}(C) = \bar{\mathbf{v}}$ (using the binary query h that takes value 1 at $\bar{\mathbf{v}}$ and 0 everywhere else). If $g_{\mathbf{p}}^{\text{sep}}(C) \neq \bar{\mathbf{v}}$, then remove from $\mathcal{U}$ all instances C' such that $g_{\mathbf{p}}^{\text{sep}}(C') = \bar{\mathbf{v}}$, reducing the size of $\mathcal{U}$ by at least half. Otherwise, we then know the exact separating hyperplane for the true instance C, namely $g_{\mathbf{P}}^{\mathrm{sep}}(C) = \bar{\mathbf{v}}$, and so employ it to update the relaxation as $P \leftarrow P \cap \{\mathbf{x} \in \mathbb{R}^d : \langle \bar{\mathbf{v}}, \mathbf{x} \rangle \leq \langle \bar{\mathbf{v}}, \mathbf{p} \rangle\}$.

In each step, either the size of $\mathcal{U}$ decreases by at least $1/4$, or the volume of P decreases by a factor of at least $\frac{1}{e}$ (by Grünbaum's Theorem [4]). The former can only happen $O(\log |\mathcal{I}|)$ times until $\mathcal{U}$ becomes a singleton (in which case we know the true instance), whereas the latter can happen at most $O\left(d \log \left(\frac{R}{\rho}\right)\right)$ times, since C is always a subset of P and C contains an ℓ_∞-ball of radius ρ; thus, $\mathrm{vol}(P) \geq (2\rho)^d$. So the procedure succeeds in $O\left(\log |\mathcal{I}| + d \log \left(\frac{R}{\rho}\right)\right)$ queries.

2.4   Proof Sketch of Theorems 2 and 4

Here we consider all instances $\mathcal{I}_{n,d,R,\rho,M}$ and again the goal is to solve convex mixed-integer (Theorem 2) and continuous (Theorem 4) instances using few binary queries (more specifically, bit queries or inner product sign queries) to the separation and subgradient oracles. Our strategy is to: 1) solve the the problems using approximate subgradients/separating hyperplanes; 2) use binary queries to construct such approximations.

For the first item, we use the algorithm of [3] based on the *centerpoint*: this is a point in the convex set where every halfspace supported on it cuts off a significant (mixed-integer) volume of the set. Similar to the previous section, the algorithm keeps an outer relaxation P of the feasible region C, and repeatedly applies separation or subgradient-based cuts through the centerpoint of P; the assumption that the feasible region contains a ball (in the optimal fiber) establishes a volume lower bound that essentially limits the number of iterations of the algorithm. While the original algorithm of [3] uses exact separation/subgradient oracles, we show, not surprisingly, that approximate ones suffice.

The next item is to construct approximate separation/subgradient oracles by using few binary queries to the exact ones. In case of bit queries $\mathcal{H}^{\mathrm{bit}}$ this is can be easily done by querying enough bits of the latter. The case of inner product sign queries $\mathcal{H}^{\mathrm{dir}}$, namely that given a vector $\mathbf{g}$ we can pick a direction $\mathbf{a}$ and ask "Is $\langle \mathbf{a}, \mathbf{g} \rangle \geq 0$?", is more interesting. It boils down to approximating the vector $\mathbf{g}$ (subgradient/separating hyperplane) using few such queries.[2]

Lemma 1. *For any vector $\mathbf{g} \in \mathbb{R}^d$, using $O(d \log \frac{d}{\varepsilon'})$ inner product sign queries we can obtain a unit-length vector $\hat{\mathbf{g}} \in \mathbb{R}^d$ such that $\left\| \hat{\mathbf{g}} - \frac{\mathbf{g}}{\|\mathbf{g}\|} \right\| \leq \varepsilon'$.*

Proof. sketch. We prove by induction on d that with $d \log \frac{8}{\delta}$ queries we can obtain a unit-length vector $\hat{\mathbf{g}}$ with $\left\| \hat{\mathbf{g}} - \frac{\mathbf{g}}{\|\mathbf{g}\|} \right\| \leq 2d\delta$; the lemma follows by setting $\delta = \frac{\varepsilon'}{2d}$.

In the 2-dimensional case can use $\log \frac{8}{\delta}$ queries to perform binary search and obtain a cone of angle $\frac{\delta\pi}{4}$ that contains $\mathbf{g}$; any unit-length vector $\hat{\mathbf{g}}$ in this cone has $\left\| \hat{\mathbf{g}} - \frac{\mathbf{g}}{\|\mathbf{g}\|} \right\| \leq \delta \leq 4\delta$ as desired.

---

[2] This is related to (actively) learning the linear classifier whose normal is given by $\mathbf{g}$ [1]. These methods can perhaps be adapted to our setting, but we present a different and self-contained statement and proof.

For the general case $d > 2$, we consider any 2-dim subspace A of $\mathbb{R}^d$ and apply the previous case to obtain in $\log \frac{d}{\delta}$ queries an approximation $\tilde{\mathbf{g}} \in A$ to the projection $\Pi_A \mathbf{g}$ of $\mathbf{g}$ onto A with guarantee $\| \|\Pi_A \mathbf{g}\| \cdot \tilde{\mathbf{g}} - \Pi_A \mathbf{g}\| \leq \delta \|\Pi_A \mathbf{g}\| \leq \delta\|\mathbf{g}\|$. Then we consider the $(d-1)$-dim subspace $B := \mathrm{span}\{\tilde{\mathbf{g}}, A^\perp\}$ and by induction, with $2(d-1)\log \frac{8}{\delta}$ queries we obtain a vector $\hat{\mathbf{g}} \in B$ that approximates the projection $\Pi_B \mathbf{g}$ of $\mathbf{g}$ onto B with guarantee $\| \|\Pi_B \mathbf{g}\| \cdot \hat{\mathbf{g}} - \Pi_B \mathbf{g}\| \leq 2(d-1)\delta\|\Pi_B \mathbf{g}\| \leq 2(d-1)\delta\|\mathbf{g}\|$ (a total of less than $2d \log \frac{8}{\delta}$ queries was then used). One can then show that $\hat{\mathbf{g}}$ is the desired approximation of $\mathbf{g}$, namely (letting $\lambda_A := \|\Pi_A \mathbf{g}\|$ and $\lambda_B := \|\Pi_B \mathbf{g}\|$)

$$\big\|\mathbf{g} - \|\mathbf{g}\| \cdot \hat{\mathbf{g}}\big\| \leq \|\mathbf{g} - \Pi_B \mathbf{g}\| + \|\Pi_B \mathbf{g} - \lambda_B \cdot \hat{\mathbf{g}}\| + \|\lambda_B \cdot \hat{\mathbf{g}} - \|\mathbf{g}\| \cdot \hat{\mathbf{g}}\| \leq 2d\delta\|\mathbf{g}\|,$$

the upper bound on the first and third terms in the middle inequality following from the fact $\mathrm{dist}(\mathbf{g}, B) \leq \delta\|\mathbf{g}\|$ (by the guarantee of $\tilde{\mathbf{g}}$) and the upper bound on the second term following from the approximation guarantee of $\hat{\mathbf{g}}$. $\square$

References

1. Balcan, M.-F., Long, P.: Active and passive learning of linear separators under log-concave distributions. In: Shalev-Shwartz, S., Steinwart, I., (eds.) Proceedings of the 26th Annual Conference on Learning Theory, volume 30 of Proceedings of Machine Learning Research, pp. 288–316. Princeton, NJ, USA, 12–14 June 2013. PMLR (2013)
2. Basu, A.: Complexity of optimizing over the integers. to appear in Mathematical Programming, Series A (2022)
3. Basu, A., Oertel, T.: Centerpoints: a link between optimization and convex geometry. SIAM J. Optim. **27**(2), 866–889 (2017)
4. Grünbaum, B.: Partitions of mass-distributions and of convex bodies by hyperplanes. Pacific J. Math. **10**, 1257–1261 (1960)
5. Jamieson, K.G., Nowak, R.D., Recht, B.: Query complexity of derivative-free optimization. In: Proceedings of the 25th International Conference on Neural Information Processing Systems - Volume 2, NIPS2012, pp. 2672–2680, Red Hook, NY, USA. Curran Associates Inc. (2012)
6. Marsden, A., Sharan, V., Sidford, A., Valiant, G.: Efficient convex optimization requires superlinear memory. arXiv preprint arXiv:2203.15260 (2022)
7. Nemirovski, A.: Efficient methods in convex programming. Lecture Notes (1994)
8. Nemirovski, A.S., Yudin, D,B.: Problem complexity and method efficiency in optimization. John Wiley (1983)
9. Oertel, T.: Integer convex minimization in low dimensions, Ph. D. thesis, Diss., Eidgenössische Technische Hochschule ETH Zürich, Nr. 22288 (2014)
10. Protasov, V.Y.: Algorithms for approximate calculation of the minimum of a convex function from its values. Math. Notes **59**(1), 69–74 (1996)
11. Woodworth, B., Srebro, N.: Open problem: the oracle complexity of convex optimization with limited memory. In: Conference on Learning Theory, pp. 3202–3210. PMLR (2019)

Efficient Separation of RLT Cuts for Implicit and Explicit Bilinear Products

Ksenia Bestuzheva[1]([✉])[ID], Ambros Gleixner[1,2][ID], and Tobias Achterberg[3][ID]

[1] Zuse Institute Berlin, Berlin, Germany
{bestuzheva,gleixner}@zib.de
[2] HTW Berlin, Berlin, Germany
[3] Gurobi GmbH, Frankfurt am Main, Germany
achterberg@gurobi.com

Abstract. The reformulation-linearization technique (RLT) is a prominent approach to constructing tight linear relaxations of non-convex continuous and mixed-integer optimization problems. The goal of this paper is to extend the applicability and improve the performance of RLT for bilinear product relations. First, a method for detecting bilinear product relations implicitly contained in mixed-integer linear programs is developed based on analyzing linear constraints with binary variables, thus enabling the application of bilinear RLT to a new class of problems. Our second contribution addresses the high computational cost of RLT cut separation, which presents one of the major difficulties in applying RLT efficiently in practice. We propose a new RLT cutting plane separation algorithm which identifies combinations of linear constraints and bound factors that are expected to yield an inequality that is violated by the current relaxation solution. A detailed computational study based on implementations in two solvers evaluates the performance impact of the proposed methods.

Keywords: Reformulation-linearization technique · Bilinear products · Cutting planes · Mixed-integer programming

1 Introduction

The reformulation-linearization technique (RLT) was first proposed by Adams and Sherali [1–3] for bilinear problems with binary variables, and has been applied to mixed-integer [18–20], general bilinear [21] and polynomial [24] problems. RLT constructs valid polynomial constraints, then linearizes these constraints by using nonlinear relations given in the problem and applying relaxations when such relations are not available. If relations used in the linearization step are violated by a relaxation solution, this procedure may yield violated cuts. By increasing the degree of derived polynomial constraints, hierarchies of

© The Author(s), under exclusive license to Springer Nature Switzerland AG 2023
A. Del Pia and V. Kaibel (Eds.): IPCO 2023, LNCS 13904, pp. 14–28, 2023.
https://doi.org/10.1007/978-3-031-32726-1_2

relaxations can be constructed, which were shown to converge to the convex hull representation of MILPs and mixed-integer polynomial problems where continuous variables appear linearly [18–20].

RLT has been shown to provide strong relaxations [21,23], but this comes at the cost of excessive numbers of cuts. To address this, Sherali and Tuncbilek [25] proposed a technique to add a subset of RLT cuts, depending on signs of coefficients of monomial terms in the original constraints and the RLT constraints. Furthermore, the reduced RLT technique [12–14,22] yields equivalent representations with fewer nonlinear terms for polynomial problems containing linear equality constraints.

We focus on RLT for bilinear products, which is of particular interest due to the numerous applications whose models involve nonconvex quadratic nonlinearities [3,5,7–9,17]. Even in the bilinear case, large numbers of factors to be multiplied and of RLT cuts that are generated as a result remain an issue that can lead to considerable slowdowns, both due to the cost of cut separation and the large sizes of resulting LP relaxations.

The first contribution of this paper is a new approach to applying RLT to MILPs. Unlike the approaches that only introduce multilinear relations via multiplication [18,19], this approach detects and enforces bilinear relations that are already implicitly present in the model. A bilinear product relation where one multiplier is a binary variable and the other multiplier is a variable with finite bounds can be equivalently written as two linear constraints. We identify such pairs of linear constraints that implicitly encode a bilinear product relation, then utilize this relation in the generation of RLT cuts.

The second contribution of this paper addresses the major bottleneck for applying RLT successfully in practice, which stems from prohibitive costs of separating RLT cuts, by proposing an efficient separation algorithm. This algorithm considers the signs of bilinear relation violations in a current LP relaxation solution and the signs of coefficients in linear constraints in order to ignore combinations of factors that will not produce a violated inequality. Furthermore, we propose a technique which projects the linear constraints onto a reduced space and constructs RLT cuts based on the resulting much smaller system.

The rest of the paper is organized as follows. In Sect. 2, RLT for bilinear products is explained. In Sect. 3, we describe the technique for deriving bilinear product relations from MILP constraints. Section 4 presents the new cut separation algorithm, and computational results are presented in Sect. 5.

2 RLT for Bilinear Products

We consider mixed-integer (nonlinear) programs (MI(N)LPs) of the extended form where auxiliary variables w are introduced for all bilinear products:

$$\min \mathbf{c}^{\mathrm{T}}\mathbf{x} \tag{1a}$$

$$\text{s.t. } A\mathbf{x} \leqslant \mathbf{b}, \tag{1b}$$

$$g(\mathbf{x}, \mathbf{w}) \leqslant 0, \tag{1c}$$

$$x_i x_j \lesseqgtr w_{ij} \text{ for all } (i,j) \in \mathcal{I}^w, \tag{1d}$$

$$\underline{\mathbf{x}} \leqslant \mathbf{x} \leqslant \overline{\mathbf{x}}, \quad \underline{\mathbf{w}} \leqslant \mathbf{w} \leqslant \overline{\mathbf{w}}, \tag{1e}$$

$$x_j \in \mathbb{R} \text{ for all } j \in \mathcal{I}^c, \quad x_j \in \{0,1\} \text{ for all } j \in \mathcal{I}^b, \tag{1f}$$

with $\mathcal{I} = \mathcal{I}^c \cup \mathcal{I}^b$ being a disjoint partition of variables $\mathbf{x}$ and $\mathbf{x}$ having dimension $|\mathcal{I}| = n$. In the above formulation, $\underline{\mathbf{x}}, \overline{\mathbf{x}} \in \overline{\mathbb{R}}^n$, $\underline{\mathbf{w}}, \overline{\mathbf{w}} \in \overline{\mathbb{R}}^{|\mathcal{I}^w|}$ ($\overline{\mathbb{R}} = \mathbb{R} \cup \{-\infty, +\infty\}$), $\mathbf{c} \in \mathbb{R}^n$ and $\mathbf{b} \in \mathbb{R}^{m^{(l)}}$ are constant vectors and $A \in \mathbb{R}^{m^{(l)} \times n}$ is a coefficient matrix, and the function g defines the nonlinear constraints. Constraint (1d) defines the bilinear product relations in the problem and allows for inequalities and equations. Let $\mathcal{I}^p$ denote the set of indices of all variables that participate in bilinear product relations (1d).

Solvers typically employ McCormick inequalities [16] to construct an LP relaxation of constraints (1d). These inequalities describe the convex hull of the set given by the relation $x_i x_j \lesseqgtr w_{ij}$:

$$\underline{x}_i x_j + x_i \underline{x}_j - \underline{x}_i \underline{x}_j \leqslant w_{ij}, \quad \overline{x}_i x_j + x_i \overline{x}_j - \overline{x}_i \overline{x}_j \leqslant w_{ij}, \tag{2a}$$

$$\underline{x}_i x_j + x_i \overline{x}_j - \underline{x}_i \overline{x}_j \geqslant w_{ij}, \quad \overline{x}_i x_j + x_i \underline{x}_j - \overline{x}_i \underline{x}_j \geqslant w_{ij}, \tag{2b}$$

where (2a) is a relaxation of $x_i x_j \leqslant w_{ij}$ and (2b) is a relaxation of $x_i x_j \geqslant w_{ij}$.

In the presence of linear constraints (1b), this relaxation can be strengthened by adding RLT cuts. Consider a linear constraint: $\sum_{k=1}^n a_{1k} x_k \leqslant b_1$. Multiplying this constraint by nonnegative bound factors $(x_j - \underline{x}_j)$ and $(\overline{x}_j - x_j)$, where $\underline{x}_j$ and $\overline{x}_j$ are finite, yields valid nonlinear inequalities. We will derive the RLT cut using the lower bound factor. The derivation is analogous for the upper bound factor. The multiplication, referred to as the reformulation step, yields:

$$\sum_{k=1}^n a_{1k} x_k (x_j - \underline{x}_j) \leqslant b_1 (x_j - \underline{x}_j).$$

This nonlinear inequality is then linearized in order to obtain a valid linear inequality. The following linearizations are applied to each nonlinear term $x_k x_j$:

- $x_k x_j$ is replaced by w_{kj} if the relation $x_k x_j \leqslant w_{kj}$ exists in the problem and $a_{1k} \leqslant 0$, or if the relation $x_k x_j \geqslant w_{kj}$ exists and $a_{1k} \geqslant 0$, or if the relation $x_k x_j = w_{kj}$ exists in the problem,
- if $k = j \in \mathcal{I}^b$, then $x_k x_j = x_j^2 = x_j$,
- if $k = j \notin \mathcal{I}^b$, then $x_k x_j = x_j^2$ is outer approximated by a secant from above or by a tangent from below, depending on the sign of the coefficient,
- if $k \neq j$, $k, j \in \mathcal{I}^b$ and one of the four clique constraints is implied by the linear constraints (1b), then: $x_k + x_j \leqslant 1 \Rightarrow x_k x_j = 0$; $x_k - x_j \leqslant 0 \Rightarrow x_k x_j = x_k$; $-x_k + x_j \leqslant 0 \Rightarrow x_k x_j = x_j$; $-x_k - x_j \leqslant -1 \Rightarrow x_k x_j = x_j + x_j - 1$,
- otherwise, $x_k x_j$ is replaced by its McCormick relaxation.

The key step is the replacing of products $x_k x_j$ with the variables w_{kj}. When a bilinear product relation $x_k x_j \lesseqgtr w_{kj}$ does not hold for the current relaxation solution, this substitution may lead to an increase in the violation of the inequality, thus possibly producing a cut that is violated by the relaxation solution.

In the case that we have a linear equation constraint $\sum_{k=1}^{n} a_{1k} x_k = b_1$ and all nonlinear terms can be replaced using equality relations, then RLT produces an equation cut. Otherwise, the equation constraint is treated as two inequalities $\sum_{k=1}^{n} a_{1k} x_k \leq b_1$ and $\sum_{k=1}^{n} a_{1k} x_k \geq b_1$ to produce inequality cuts.

3   Detection of Implicit Products

Consider a product relation $w_{ij} = x_i x_j$, where x_i is binary. It can be equivalently rewritten as two implications: $x_i = 0 \Rightarrow w_{ij} = 0$ and $x_i = 1 \Rightarrow w_{ij} = x_j$. With the use of the big-M technique, these implications can be represented as linear constraints, provided that the bounds of x_j are finite:

$$w_{ij} - \overline{x}_j x_i \leq 0, \quad w_{ij} - x_j - \underline{x}_j x_i \leq -\underline{x}_j \tag{3a}$$
$$-w_{ij} + \underline{x}_j x_i \leq 0, \quad -w_{ij} + x_j + \overline{x}_j x_i \leq \overline{x}_j. \tag{3b}$$

Linear constraints with binary variables can be analyzed in order to detect constraint pairs of the forms (3). The method can be generalized to allow for bilinear relations of the following form, with $A, B, C, D \in \mathbb{R}$:

$$Ax_i + Bw_{ij} + Cx_j + D \lesseqgtr x_i x_j \tag{4}$$

Theorem 1. *Consider two linear constraints depending on the same three variables x_i, x_j and w_{ij}, where x_i is binary:*

$$a_1 x_i + b_1 w_{ij} + c_1 x_j \leq d_1, \tag{5a}$$
$$a_2 x_i + b_2 w_{ij} + c_2 x_j \leq d_2. \tag{5b}$$

If $b_1 b_2 > 0$ and $\gamma = c_2 b_1 - b_2 c_1 \neq 0$, then these constraints imply the following product relation:

$$(1/\gamma)((b_2(a_1 - d_1) + b_1 d_2)x_i + b_1 b_2 w_{ij} + b_1 c_2 x_j - b_1 d_2) \leq x_i x_j \quad if\ b_1/\gamma \geq 0,$$

$$(1/\gamma)((b_2(a_1 - d_1) + b_1 d_2)x_i + b_2 b_2 w_{ij} + b_1 c_2 x_j - b_1 d_2) \geq x_i x_j \quad if\ b_1/\gamma \leq 0.$$

Proof. We begin by writing the bilinear relation (4), treating its coefficients and inequality sign as unknown, and reformulating it as two implications:

$$x_i = 1 \qquad \Rightarrow Bw_{ij} + (C-1)x_j \lesseqgtr -D - A, \tag{6a}$$
$$x_i = 0 \qquad \Rightarrow Bw_{ij} + Cx_j \lesseqgtr -D, \tag{6b}$$

where the inequality sign must be identical in both implied inequalities. Similarly, we rewrite constraints (5) with scaling parameters α and β:

$$x_i = 1 \qquad \Rightarrow \alpha b_1 w_{ij} + \alpha c_1 x_j \lesseqgtr \alpha(d_1 - a_1), \qquad (7a)$$
$$x_i = 0 \qquad \Rightarrow \beta b_2 w_{ij} + \beta c_2 x_j \lesseqgtr \beta d_2, \qquad (7b)$$

where the inequality signs depend on the signs of α and β.

The goal is to find the coefficients A, B, C and D and the inequality sign. We require that coefficients and inequality signs in implications (6) and (7) match. Solving the resulting system yields:

$$b_1 b_2 > 0, \ A = (1/\gamma)(b_2(a_1 - d_1) + b_1 d_2)$$
$$B = b_1 b_2/\gamma, \ C = b_1 c_2/\gamma, \ D = -b_1 d_2/\gamma, \ \gamma \neq 0,$$

where $\gamma = c_2 b_1 - b_2 c_1$ and the inequality sign is '$\leq$' if $b1/\gamma \geq 0$, and '$\geq$' if $b1/\gamma \leq 0$. Thus, the bilinear relation stated in this theorem is obtained. $\qquad\square$

Although the conditions of the theorem are sufficient for the bilinear product relation to be implied by the linear constraints, in practice more conditions are checked before deriving such a relation. In particular:

- At least one of the coefficients a_1 and a_2 must be nonzero. Otherwise, the product relation is always implied by the linear constraints, including when $0 < x_i < 1$.
- The signs of the coefficients of the binary variable x_i must be different, that is, one linear relation is more restrictive when $x_i = 1$ and the other when $x_i = 0$. While this is not necessary for the non-redundancy of the derived product relation, by requiring this we focus on stronger implications (for instance, for a linear relation $a_1 x_i + b_1 w_{ij} + c_1 x_j \leq d_1$ with $a_1 > 0$, we use the more restrictive implication $x_i = 1 \ \Rightarrow b_1 w_{ij} + c_1 x_j \leq d_1 - a_1$ rather than the less restrictive implication $x_i = 0 \ \Rightarrow b_1 w_{ij} + c_1 x_j \leq d_1$).

In separation, the product relation (4) is treated similarly to product relations $w_{ij} \lesseqgtr x_i x_j$, with the linear left-hand side $A x_i + B w_{ij} + C x_j + D$ being used instead of the individual auxiliary variable w_{ij}.

The detection algorithm searches for suitable pairs of linear relations and derives product relations from them. Let x_i, as before, be a binary variable. The following relation types are considered as candidates for the first relation in such a pair: implied relations of the form $x_i = \xi \ \Rightarrow \tilde{b}_1 w_{ij} + \tilde{c}_1 x_j \leq \tilde{d}_1$, where $\xi = 0$ or $\xi = 1$; and implied bounds of the form $x_i = \xi \ \Rightarrow w_{ij} \leq \tilde{d}_1$.

The second relation in a pair can be: an implied relation of the form $x_i = \bar{\xi} \ \Rightarrow \tilde{b}_2 w_{ij} + \tilde{c}_2 x_j \leq \tilde{d}_2$, where $\bar{\xi}$ is the complement of ξ; if w_{ij} is non-binary, an implied bound of the form $x_i = \bar{\xi} \ \Rightarrow w_{ij} \leq \tilde{d}_2$; if w_{ij} is binary, a clique containing the complement of x_i if $\xi = 1$ or x_i if $\xi = 0$, and w_{ij} or its complement; a constraint on x_j and w_{ij}; or a global bound on w_{ij}. Cliques are constraints of the form: $\sum_{k \in \mathcal{J}} x_k + \sum_{k \in \overline{\mathcal{J}}}(1 - x_k) \leq 1$, where $\mathcal{J} \subseteq \mathcal{I}^b$, $\overline{\mathcal{J}} \subseteq \mathcal{I}^b$ and $\mathcal{J} \cap \overline{\mathcal{J}} = \emptyset$.

4 Separation Algorithm

We present a new algorithm for separating RLT cuts within an LP-based branch-and-bound solver. The branch-and-bound algorithm builds LP relaxations of problem (1) by constructing linear underestimators of functions g in the constraint $g(\mathbf{x}, \mathbf{w}) \leqslant 0$ and McCormick inequalities for constraints (1d).

Let $(\mathbf{x}^*, \mathbf{w}^*)$ be the solution of an LP relaxation at a node of the branch-and-bound tree, and suppose that $(\mathbf{x}^*, \mathbf{w}^*)$ violates the relation $x_i x_j \lesseqgtr w_{ij}$ for some $i, j \in \mathcal{I}^w$. Separation algorithms generate cuts that separate $(\mathbf{x}^*, \mathbf{w}^*)$ from the feasible region, and add those cuts to the solver's cut storage.

The standard separation algorithm, which will serve as a baseline for comparisons, iterates over all linear constraints. For each constraint, it iterates over all variables x_j that participate in bilinear relations and generates RLT cuts using bound factors of x_j. Violated cuts are added to the MINLP solver's cut storage.

4.1 Row Marking

Let the bound factors be denoted as $f_j^{(\ell)}(\mathbf{x}) = x_j - \underline{x}_j$ and $f_j^{(u)}(\mathbf{x}) = \overline{x}_j - x_j$. Consider a linear constraint multiplied by a bound factor:

$$f_j^{(\cdot)}(\mathbf{x}) \mathbf{a}_r \mathbf{x} \leqslant f_j^{(\cdot)}(\mathbf{x}) b_r. \tag{8}$$

The ith nonlinear term is $a'_{ri} x_i x_j$, where $a'_{ri} = a_{ri}$ when multiplying by $(x_j - \underline{x}_j)$ and $a'_{ri} = -a_{ri}$ when multiplying by $(\overline{x}_j - x_j)$. Following the procedure described in Sect. 2, RLT may replace the product $x_i x_j$ with w_{ij}. The product can also be replaced with a linear expression, but this does not change the reasoning, and we will only use w_{ij} in this section.

If $w_{ij}^* \neq x_i^* x_j^*$, then such a replacement will change the violation of (8). The terms whose replacement will increase the violation are of interest, that is, the terms where:

$$a'_{ri} x_i^* x_j^* \leqslant a'_{ri} w_{ij}^*.$$

This determines the choice of bound factors to multiply with:

$$x_i^* x_j^* < w_{ij}^* \Rightarrow \begin{array}{l} \text{multiply by } (x_j - \underline{x}_j) \text{ if } a_{ri} > 0, \\ \text{multiply by } (\overline{x}_j - x_j) \text{ if } a_{ri} < 0, \end{array}$$

$$x_i^* x_j^* > w_{ij}^* \Rightarrow \begin{array}{l} \text{multiply by } (\overline{x}_j - x_j) \text{ if } a_{ri} > 0, \\ \text{multiply by } (x_j - \underline{x}_j) \text{ if } a_{ri} < 0. \end{array}$$

The separation algorithm is initialized by creating data structures to enable efficient access to 1) all variables appearing in bilinear products together with a given variable and 2) the bilinear product relation involving two given variables.

For each variable x_i, linear rows are marked in order to inform the separation algorithms which bound factors of x_i they should be multiplied with, if any. The algorithm can work with inequality rows in both '$\leqslant$' and '$\geqslant$' forms as well as equation rows. For each bilinear product $x_i x_j$, the row marking algorithm iterates over all linear rows that contain x_j with a nonzero coefficient. These rows are stored in a sparse array and have one of the following marks:

– MARK_LT: the row contains a term $a_{rj}x_j$ such that $a_{rj}x_i^* x_j^* < a_{rj}w_{ij}^*$;
– MARK_GT: the row contains a term $a_{rj}x_j$ such that $a_{rj}x_i^* x_j^* > a_{rj}w_{ij}^*$;
– MARK_BOTH: the row contains terms fitting both cases above.

Row marks are represented by integer values 1, 2 and 3, respectively, and are stored in two sparse arrays, *row_idcs* and *row_marks*, the first storing sorted row indices and the second storing the corresponding marks. In the algorithm below, we use the notation $mark(r)$ to denote accessing the mark of row r by performing a search in *row_idcs* and retrieving the corresponding entry in *row_marks*. We also define a sparse matrix W with entries w_{ij}.

Input: x^*, w^*, W
1 $marks := \varnothing$
2 **for** $i \in \mathcal{I}^p, j \in nnz(\boldsymbol{w}_i)$ **do**
3   **for** r *such that* $j \in nnz(\boldsymbol{a}_r)$ **do**
4   **if** $r \notin marks$ **then**
5           $marks \leftarrow r$
6           $mark(r) := 0$
7   **if** $a_{rj}x_i^* x_j^* < a_{rj}w_{ij}^*$ **then**
8           $mark(r) \mathrel{|}= $ MARK_LT
9   **else**
10          $mark(r) \mathrel{|}= $ MARK_GT

The algorithm iterates over the sparse array of marked rows and generates RLT cuts for the following combinations of linear rows and bound factors:

– If $mark = $ MARK_LT, then "$\leq$" constraints are multiplied with the lower bound factor and "$\geq$" constraints are multiplied with the upper bound factor;
– If $mark = $ MARK_GT, then "$\leq$" constraints are multiplied with the upper bound factor and "$\geq$" constraints are multiplied with the lower bound factor;
– If $mark = $ MARK_BOTH, then both "$\leq$" and "$\geq$" constraints are multiplied with both the lower and the upper bound factors;
– Marked equality constraints are always multiplied with x_i itself.

4.2 Projection Filtering

If at least one of the variables x_i and x_j has a value equal to one of its bounds, then the McCormick relaxation (2) is tight for the relation $w_{ij} = x_i x_j$. Therefore, if x_i or x_j is at a bound and the McCormick inequalities are satisfied, then the product relation is also satisfied. We describe the equality case here, and the reasoning is analogous for the inequality case of $x_i x_j \lessgtr w_{ij}$.

Consider the linear system $A\mathbf{x} \leqslant \mathbf{b}$ projected onto the set of variables whose values are not equal to either of their bounds.

$$\sum_{k \in \mathcal{J}^1} a_{rk} x_k \leqslant b_r - \sum_{k \in \mathcal{J}^2} a_{rk} x_k^*, \ \forall r \in 1, \ldots, m^{(l)},$$

where $\mathcal{J}^1 \subseteq \mathcal{I}$ is the set of all problem variables whose values in the solution $\mathbf{x}^*$ of the current LP relaxation are not equal to one of their bounds, and $\mathcal{J}^2 = \mathcal{I} \setminus \mathcal{J}^1$.

Violation is then first checked for RLT cuts generated based on the projected linear system. Only if such a cut, which we will refer to as a projected RLT cut, is violated, then the RLT cut for the same bound factor and the corresponding constraint in the original linear system will be constructed. Since $\mathbf{x}^*$ is a basic LP solution, in practice either $x_k^* = \underline{x}_k$ or $x_k^* = \overline{x}_k$ holds for many of the variables, and the projected system often has a considerably smaller size than the original system.

In the projected system multiplied with a bound factor $f_j^{(\cdot)}(\mathbf{x})$:

$$f_j^{(\cdot)}(\mathbf{x}) \cdot \sum_{k \in \mathcal{J}^1} a_{rk} x_k \leqslant f_j^{(\cdot)}(\mathbf{x})(b_r - \sum_{k \in \mathcal{J}^2} a_{rk} x_k^*), \ \forall r \in 1, \ldots, m^{(l)},$$

the only nonlinear terms are $x_j x_k$ with $k \in \mathcal{J}^1$, and therefore, no substitution $x_i x_k \to w_{ik}$ is performed for $k \in \mathcal{J}^2$. If the McCormick inequalities for x_i, x_k and w_{ik} hold, then $x_i^* x_k^* = w_{ik}^*$ for $k \in \mathcal{J}^2$, and checking the violation of a projected RLT cut is equivalent to checking the violation of a full RLT cut.

Depending on the solver, McCormick inequalities may not be satisfied at $(\mathbf{x}^*, \mathbf{w}^*)$. Thus, it is possible that $x_i^* x_k^* \neq w_{ik}^*$ for some $k \in \mathcal{J}^2$, but these violations will not contribute to the violation of the projected RLT cut. In this case, projection filtering has an additional effect: for violated bilinear products involving variables whose values in $\mathbf{x}^*$ are at bound, the violation of the product will be disregarded when checking the violation of RLT cuts. Thus, adding McCormick cuts will be prioritized over adding RLT cuts.

5   Computational Results

5.1   Setup

We tested the proposed methods on the MINLPLib[1] [6] test set and a test set comprised of instances from MIPLIB3, MIPLIB 2003, 2010 and 2017 [10], and Cor@l [15]. These test sets consist of 1846 MINLP instances and 666 MILP instances, respectively. After structure detection experiments, only those instances were chosen for performance evaluations that either contain bilinear products in the problem formulation, or where our algorithm derived bilinear products. This resulted in test sets of 1357 MINLP instances and 195 MILP instances.

---

[1] https://www.minlplib.org.

The algorithms were implemented in the MINLP solver SCIP [4]. We used a development branch of SCIP (githash dd6c54a9d7) compiled with SoPlex 5.0.2.4, CppAD 20180000.0, PaPILO 1.0.0.1, bliss 0.73p and Ipopt 3.12.11. The experiments were carried out on a cluster of Dell Poweredge M620 blades with 2.50GHz Intel Xeon CPU E5-2670 v2 CPUs, with 2 CPUs and 64GB memory per node. The time limit was set to one hour, the optimality gap tolerance to 10^{-4} for MINLP instances and to 10^{-6} for MILP instances, and the following settings were used for all runs, where applicable:

- The maximum number of unknown bilinear terms that a product of a row and a bound factor can have in order to be used was 20. Unknown bilinear terms are those terms $x_i x_j$ for which no w_{ij} variable exists in the problem, or its extended formulation which SCIP constructs for the purposes of creating an LP relaxation of an MINLP.
- RLT cut separation was called every 10 nodes of the branch-and-bound tree.
- In every non-root node where separation was called, 1 round of separation was performed. In the root node, 10 separation rounds were performed.
- Unless specified otherwise, implicit product detection and projection filtering were enabled and the new separation algorithm was used.

5.2   Impact of RLT Cuts

In this subsection we evaluate the performance impact of RLT cuts. The following settings were used: *Off* - RLT cuts are disabled; *ERLT* - RLT cuts are added for products that exist explicitly in the problem; *IERLT* - RLT cuts are added for both implicit and explicit products. The setting *ERLT* was used for the MINLP test set only, since MILP instances contain no explicitly defined bilinear products.

We report overall numbers of instances, numbers of solved instances, shifted geometric means of the runtime (shift 1 s), and the number of nodes in the branch-and-bound tree (shift 100 nodes), and relative differences between settings. Additionally, we report results on subsets of instances. Affected instances are instances where a change of setting leads to a difference in the solving process, indicated by a difference in the number of LP iterations. [x,timelim] denotes the subset of instances which took the solver at least x seconds to solve with at least one setting, and were solved to optimality with at least one setting. All-optimal is the subset of instances which were solved to optimality with both settings.

Table 1 shows the impact of RLT cuts on MILP performance. We observe a slight increase in time when RLT cuts are enabled, and a slight decrease in number of nodes. The difference is more pronounced on 'difficult' instances: a 9% decrease in number of nodes on subset [100,timelim] and 28% on subset [1000,timelim], and a decrease of 21% in the mean time on subset [1000,timelim].

Table 2 reports the impact of RLT cuts derived from explicitly defined bilinear products. A substantial decrease in running times and tree sizes is observed across all subsets, with a 15% decrease in the mean time and a 19% decrease in

Table 1. Impact of RLT cuts: MILP instances

Subset	instances	Off			IERLT			IERLT/Off	
		solved	time	nodes	solved	time	nodes	time	nodes
All	971	905	45.2	1339	909	46.7	1310	1.03	0.98
Affected	581	571	48.8	1936	575	51.2	1877	1.05	0.97
[0,tilim]	915	905	34.4	1127	909	35.6	1104	1.04	0.98
[1,tilim]	832	822	47.2	1451	826	49.0	1420	1.04	0.98
[10,tilim]	590	580	126.8	3604	584	133.9	3495	1.06	0.97
[100,tilim]	329	319	439.1	9121	323	430.7	8333	0.98	0.91
[1000,tilim]	96	88	1436.7	43060	92	1140.9	31104	0.79	0.72
All-optimal	899	899	31.9	1033	899	34.1	1053	1.07	1.02

the number of nodes on all instances, and a 87% decrease in the mean time and a 88% decrease in the number of nodes on the subset [1000,timelim]. 223 more instances are solved with *ERLT* than with *Off*.

Table 3 evaluates the impact of RLT cuts derived from implicit bilinear products. Similarly to MILP instances, the mean time slightly increases and the mean number of nodes slightly decreases when additional RLT cuts are enabled, but on MINLP instances, the increase in the mean time persists across different instance subsets and is most pronounced (9%) on the subset [100,timelim], and the number of nodes increases by $6 - 7\%$ on subsets [100,timelim] and [1000,timelim].

Table 2. Impact of RLT cuts derived from explicit products: MINLP instances

Subset	instances	Off			ERLT			ERLT/Off	
		solved	time	nodes	solved	time	nodes	time	nodes
All	6622	4434	67.5	3375	4557	57.5	2719	0.85	0.81
Affected	2018	1884	18.5	1534	2007	10.6	3375	0.57	0.51
[0,timelim]	4568	4434	10.5	778	4557	8.2	569	0.78	0.73
[1,timelim]	3124	2990	28.3	2081	3113	20.0	1383	0.71	0.67
[10,timelim]	1871	1737	108.3	6729	1860	63.6	3745	0.59	0.56
[100,tilim]	861	727	519.7	35991	850	196.1	12873	0.38	0.36
[1000,tilim]	284	150	2354.8	196466	273	297.6	23541	0.13	0.12
All-optimal	4423	4423	8.6	627	4423	7.5	518	0.87	0.83

Table 3. Impact of RLT cuts derived from implicit products: MINLP instances

		ERLT			IERLT			ERLT/IERLT	
Subset	instances	solved	time	nodes	solved	time	nodes	time	nodes
All	6622	4565	57.0	2686	4568	57.4	2638	1.01	0.98
Affected	1738	1702	24.2	1567	1705	24.8	1494	1.02	0.95
[0,timelim]	4601	4565	8.5	587	4568	8.6	576	1.01	0.98
[1,timelim]	3141	3105	21.1	1436	3108	21.4	1398	1.01	0.97
[10,timelim]	1828	1792	74.1	4157	1795	75.4	4012	1.02	0.97
[100,tilim]	706	670	359.9	22875	673	390.4	24339	1.09	1.06
[1000,tilim]	192	156	1493.3	99996	159	1544.7	107006	1.03	1.07
All-optimal	4532	4532	7.7	540	4532	7.8	529	1.02	0.98

Table 4 reports numbers of instances for which a change in the root node dual bound was observed, where the relative difference is quantified as $\frac{\gamma_2 - \gamma_1}{\gamma_1}$, where γ_1 and γ_2 are root node dual bounds obtained with the first and second settings, respectively. The range of the change is specified in the column 'Difference', and each column shows numbers of instances for which one or the other setting provided a better dual bound, within given range.

The results of comparisons *Off/IERLT* for MILP instances and *Off/ERLT* for MINLP instances are consistent with the effect of RLT cuts on performance observed in Tables 1 and 2. Interestingly, *IERLT* performs better than *ERLT* in terms of root node dual bound quality. Thus, RLT cuts derived from implicit products in MINLP instances tend to improve root node relaxations.

5.3   Separation

In Table 5, the setting *Marking-off* employs the standard separation algorithm, and *Marking-on* enables the row marking and projection filtering algorithms described in Sect. 4. Row marking reduces the running time by 63% on MILP instances, by 70% on affected MILP instances, by 12% on MINLP instances and by 22% on affected MINLP instances. The number of nodes increases when row marking is enabled because, due to the decreased separation time, the solver can

Table 4. Root node dual bound differences

	MILP	MINLP	
Difference	Off / IERLT	Off / ERLT	ERLT / IERLT
0.01-0.2	54 / 62	224 / 505	379 / 441
0.2-0.5	2 / 4	23 / 114	44 / 48
0.5-1.0	0 / 3	40 / 150	19 / 30
>1.0	0 / 2	4 / 182	4 / 23

explore more nodes before reaching the time limit: this is confirmed by the fact that on the subset All-optimal, the number of nodes remains nearly unchanged.

Table 5. Separation algorithm comparison

Test set	subset	instances	Marking-off			Marking-on			M-on/M-off	
			solved	time	nodes	solved	time	nodes	time	nodes
MILP	All	949	780	124.0	952	890	45.2	1297	0.37	1.37
	Affected	728	612	156.6	1118	722	46.4	1467	0.30	1.31
	All-optimal	774	774	58.4	823	774	21.2	829	0.36	1.01
MINLP	All	6546	4491	64.5	2317	4530	56.4	2589	0.88	1.12
	Affected	3031	2949	18.5	1062	2988	14.3	1116	0.78	1.05
	All-optimal	4448	4448	9.1	494	4448	7.4	502	0.81	1.02

Table 6 analyzes the percentage of time that RLT cut separation takes out of overall running time, showing the arithmetic mean and maximum over all instances, numbers of instances for which the percentage was within a given interval, and numbers of failures. The average percentage is reduced from 54.2% to 2.8% for MILP instances and from 15.1% to 2.4% for MINLP instances, and the maximum percentage is reduced from 99.6% to 71.6% for MILP instances, but remains at 100% for MINLP instances. The numbers of failures are reduced with *Marking-on*, mainly due to avoiding failures that occur when the solver runs out of memory.

Table 6. Separation times

Test set	Setting	avg %	max %	N(< 5%)	N(5-20%)	N(20-50%)	N(50-100%)	fail
MILP	*Marking-off*	54.2	99.6	121	117	169	552	16
	Marking-on	2.8	71.6	853	87	31	4	0
MINLP	*Marking-off*	15.1	100.0	3647	1265	1111	685	77
	Marking-on	2.4	100.0	6140	376	204	49	16

Projection filtering has a minor impact on performance. When comparing the runs where projection filtering is disabled and enabled, the relative difference in time and nodes does not exceed 1% on both MILP and MINLP instances, except for affected MILP instances where projection filtering decreases the number of nodes by 4%. This is possibly occurring due to the effect of prioritizing McCormick inequalities to RLT cuts when enforcing derived product relations. The number of solved instances remains almost unchanged, with one less instance being solved on both MILP and MINLP test sets when projection filtering is enabled.

5.4   Experiments with Gurobi

In this subsection we present results obtained by running the mixed-integer quadratically-constrained programming solver Gurobi 10.0 beta [11]. The algorithms for implicit product detection and RLT cut separation are the same as in SCIP, although implementation details may differ between the solvers.

The internal Gurobi test set was used, comprised of models sent by Gurobi customers and models from public benchmarks, chosen in a way that avoids overrepresenting any particular problem class. Whenever RLT cuts were enabled, so was implicit product detection, row marking and projection filtering. The time limit was set to 10000 s.

Table 7 shows, for both MILP and MINLP test sets, the numbers of instances in the test sets and their subsets, and the ratios of shifted geometric means of running time and number of nodes of the runs with RLT cuts enabled, to the same means obtained with RLT cuts disabled. The last row shows the numbers of instances solved with one setting and unsolved with the other, that is, for example, "RLT off: +41" means that 41 instances were solved with the setting "off" that were not solved with the setting "on".

While the results cannot be directly compared to those obtained with SCIP due to the differences in the experimental setup, we observe the same tendencies. In particular, RLT cuts yield small improvements on MILP instances which become more pronounced on subsets [100,timelim] and [1000,timelim], and larger improvements are observed on MINLP instances both in terms of geometric means and numbers of solved instances. Relative differences are comparable to those observed with SCIP, but the impact of RLT cuts is larger in Gurobi, and no slowdown is observed with Gurobi on any subset of MILP instances.

Table 7. Results obtained with Gurobi 10.0 beta

Subset	MILP			MINLP		
	instances	timeR	nodeR	instances	timeR	nodeR
All	5011	0.99	0.97	806	0.73	0.57
[0,timelim]	4830	0.99	0.96	505	0.57	0.44
[1,timelim]	3332	0.98	0.96	280	0.40	0.29
[10,timelim]	2410	0.97	0.93	188	0.29	0.20
[100,timelim]	1391	0.95	0.91	114	0.17	0.11
[1000,timelim]	512	0.89	0.83	79	0.12	0.08
Solved	RLT off: +41; RLT on: +37			RLT off: +2; RLT on: +35		

5.5   Summary

RLT cuts yield a considerable performance improvement for MINLP problems and a small performance improvement for MILP problems which becomes more pronounced for challenging instances. The new separation algorithm drastically reduces the computational burden of RLT cut separation and is essential to an efficient implementation of RLT cuts, enabling the speedups we observed when activating RLT.

Acknowledgements. The work for this article has been conducted within the Research Campus Modal funded by the German Federal Ministry of Education and Research (BMBF grant numbers 05M14ZAM, 05M20ZBM).

References

1. Adams, W.P., Sherali, H.D.: A tight linearization and an algorithm for zero-one quadratic programming problems. Manage. Sci. **32**(10), 1274–1290 (1986)
2. Adams, W.P., Sherali, H.D.: Linearization strategies for a class of zero-one mixed integer programming problems. Oper. Res. **38**(2), 217–226 (1990)
3. Adams, W.P., Sherali, H.D.: Mixed-integer bilinear programming problems. Math. Program. **59**(1), 279–305 (1993)
4. Bestuzheva, K., et al.: Enabling research through the SCIP optimization suite 8.0. ACM Trans. Math. Softw. (2023). https://doi.org/10.1145/3585516
5. Buchheim, C., Wiegele, A., Zheng, L.: Exact algorithms for the quadratic linear ordering problem. INFORMS J. Comput. **22**(1), 168–177 (2010)
6. Bussieck, M.R., Drud, A.S., Meeraus, A.: MINLPLib - a collection of test models for mixed-integer nonlinear programming. INFORMS J. Comput. **15**(1), 114–119 (2003). https://doi.org/10.1287/ijoc.15.1.114.15159
7. Castillo, I., Westerlund, J., Emet, S., Westerlund, T.: Optimization of block layout design problems with unequal areas: a comparison of MILP and MINLP optimization methods. Comput. Chem. Eng. **30**(1), 54–69 (2005)
8. Frank, S., Steponavice, I., Rebennack, S.: Optimal power flow: a bibliographic survey I. Energy syst. **3**(3), 221–258 (2012)
9. Frank, S., Steponavice, I., Rebennack, S.: Optimal power flow: a bibliographic survey II. Energy Syst. **3**(3), 259–289 (2012)
10. Gleixner, A., et al.: MIPLIB 2017: data-driven compilation of the 6th mixed-integer programming library. Math. Program. Comput. **13**(3), 443–490 (2021). https://doi.org/10.1007/s12532-020-00194-3
11. Gurobi Optimization, LLC: Gurobi Optimizer Reference Manual (2022). https://www.gurobi.com
12. Liberti, L.: Reduction constraints for the global optimization of NLPs. Int. Trans. Oper. Res. **11**(1), 33–41 (2004)
13. Liberti, L.: Reformulation and convex relaxation techniques for global optimization. Ph.D. thesis. Springer (2004)
14. Liberti, L.: Linearity embedded in nonconvex programs. J. Global Optim. **33**(2), 157–196 (2005)
15. Linderoth, J.T., Ralphs, T.K.: Noncommercial software for mixed-integer linear programming. Integer Programm. Theory Practice **3**, 253–303 (2005)
16. McCormick, G.P.: Computability of global solutions to factorable nonconvex programs: Part I - convex underestimating problems. Math. Program. **10**(1), 147–175 (1976)
17. Misener, R., Floudas, C.A.: Advances for the pooling problem: modeling, global optimization, and computational studies. Appl. Comput. Math. **8**(1), 3–22 (2009)

18. Sherali, H.D., Adams, W.P.: A hierarchy of relaxations between the continuous and convex hull representations for zero-one programming problems. SIAM J. Discret. Math. **3**(3), 411–430 (1990)
19. Sherali, H.D., Adams, W.P.: A hierarchy of relaxations and convex hull characterizations for mixed-integer zero-one programming problems. Discret. Appl. Math. **52**(1), 83–106 (1994)
20. Sherali, H.D., Adams, W.P.: A reformulation-linearization technique (RLT) for semi-infinite and convex programs under mixed 0–1 and general discrete restrictions. Discret. Appl. Math. **157**(6), 1319–1333 (2009)
21. Sherali, H.D., Alameddine, A.: A new reformulation-linearization technique for bilinear programming problems. J. Global Optim. **2**(4), 379–410 (1992)
22. Sherali, H.D., Dalkiran, E., Liberti, L.: Reduced RLT representations for nonconvex polynomial programming problems. J. Global Optim. **52**(3), 447–469 (2012)
23. Sherali, H.D., Smith, J.C., Adams, W.P.: Reduced first-level representations via the reformulation-linearization technique: results, counterexamples, and computations. Discret. Appl. Math. **101**(1–3), 247–267 (2000)
24. Sherali, H.D., Tuncbilek, C.H.: A global optimization algorithm for polynomial programming problems using a reformulation-linearization technique. J. Global Optim. **2**(1), 101–112 (1992)
25. Sherali, H.D., Tuncbilek, C.H.: New reformulation linearization/convexification relaxations for univariate and multivariate polynomial programming problems. Oper. Res. Lett. **21**(1), 1–9 (1997)

A Nearly Optimal Randomized Algorithm for Explorable Heap Selection

Sander Borst[1]([envelope])[iD], Daniel Dadush[1]([envelope])[iD], Sophie Huiberts[2]([envelope])[iD], and Danish Kashaev[1]([envelope])[iD]

[1] Centrum Wiskunde & Informatica (CWI), Amsterdam, The Netherlands
{sander.borst,dadush,danish.kashaev}@cwi.nl
[2] Columbia University, New York, USA
sophie@huiberts.me

Abstract. Explorable heap selection is the problem of selecting the nth smallest value in a binary heap. The key values can only be accessed by traversing through the underlying infinite binary tree, and the complexity of the algorithm is measured by the total distance traveled in the tree (each edge has unit cost). This problem was originally proposed as a model to study search strategies for the branch-and-bound algorithm with storage restrictions by Karp, Saks and Widgerson (FOCS '86), who gave deterministic and randomized $n \cdot \exp(O(\sqrt{\log n}))$ time algorithms using $O(\log(n)^{2.5})$ and $O(\sqrt{\log n})$ space respectively. We present a new randomized algorithm with running time $O(n \log(n)^3)$ against an oblivious adversary using $O(\log n)$ space, substantially improving the previous best randomized running time at the expense of slightly increased space usage. We also show an $\Omega(\log(n)n/\log(\log(n)))$ lower bound for any algorithm that solves the problem in the same amount of space, indicating that our algorithm is nearly optimal.

1 Introduction

Many important problems in theoretical computer science are fundamentally search problems. The objective of these problems is to find a certain solution from the search space. In this paper we analyze a search problem that we call *explorable heap selection*. The problem is related to the famous branch-and-bound algorithm and was originally proposed by Karp, Widgerson and Saks [13] to model node selection for branch-and-bound with low space-complexity. Furthermore, as we will explain later, the problem remains practically relevant to branch-and-bound even in the full space setting.

Due to space limitations, we have omitted several proofs. These can be found in [7]. This project has received funding from the European Research Council (ERC) under the European Union's Horizon 2020 research and innovation programme (grant agreement QIP–805241).

The explorable heap selection problem[1] is an online graph exploration problem for an agent on a rooted (possibly infinite) binary tree. The nodes of the tree are labeled by distinct real numbers (the key values) that increase along every path starting from the root. The tree can thus be thought of as a min-heap. Starting at the root, the agent's objective is to select the nth smallest value in the tree while minimizing the distance traveled, where each edge of the tree has unit travel cost. The key value of a node is only revealed when the agent visits it, and the problem thus has an online nature. When the agent learns the key value of a node, it still does not know the rank of this value.

A simple selection strategy is to use the best-first rule, which repeatedly explores the unexplored node whose parent has the smallest key value. While this rule is optimal in terms of the number of nodes that it explores, namely $\Theta(n)$, the distance traveled by the agent can be far from optimal. In the worst-case, an agent using this rule will need to travel a distance of $\Theta(n^2)$ to find the nth smallest value. A simple bad example for this rule is to consider a rooted tree consisting of two paths (which one can extend to a binary tree), where the two paths are consecutively labeled by all positive even and odd integers respectively.

Improving on the best-first strategy, Karp, Saks and Wigderson [13] gave a randomized algorithm with expected cost $n \cdot \exp(O(\sqrt{\log(n)}))$ using $O(\sqrt{\log(n)})$ working space. They also showed how to make the algorithm deterministic using $O(\log(n)^{2.5})$ space. In this work, our main contribution is an improved randomized algorithm with expected cost $O(n \log(n)^3)$ using $O(\log(n))$ space. Given the $\Omega(n)$ lower bound, our travel cost is optimal up to logarithmic factors. Furthermore we show that any algorithm for explorable heap selection that only uses s units of memory, must take at least $n \cdot \log_s(n)$ time in expectation. An interesting open problem is the question whether a superlinear lower bound also holds without any restriction on the memory usage.

To clarify the memory model, it is assumed that any key value and $O(\log n)$ bit integer can be stored using $O(1)$ space. We also assume that maintaining the current position in the tree does not take up memory. Furthermore, we assume that key value comparisons and moving across an edge of the tree require $O(1)$ time. Under these assumptions, the running times of the above algorithms are in fact proportional to their travel cost. Throughout the paper, we will thus use travel cost and running time interchangeably.

Motivation. The motivation to look at this problem comes from the branch-and-bound algorithm. This is a well-known algorithm that can be used for solving many types of problems. In particular, it is often used to solve integer linear programs (ILP), which are of the form $\arg\min\{c^\top x : x \in \mathbb{Z}^n, Ax \leq b\}$. In that setting, branch-and-bound works by first solving the linear programming (LP) relaxation, which does not have integrality constraints. The value of the solution to the relaxation forms a lower bound on the objective value of the original problem. Moreover, if this solution only has integral components, it is also optimal for the original problem. Otherwise, the algorithm chooses a component x_i for which the solution value $\hat{x}_i$ is not integral. It then creates two new subproblems, by either adding the constraint $x_i \leq \lfloor \hat{x}_i \rfloor$ or $x_i \geq \lceil \hat{x}_i \rceil$. This

---

[1] [13] did not name the problem, so we have given a descriptive name here.

operation is called *branching*. The tree of subproblems, in which the children of a problem are created by the branching operation, is called the branch-and-bound tree. Because a subproblem contains more constraints than its parent, its objective value is greater or equal to the one of its parent.

At the core, the algorithm consists of two important components: the branching rule and the node selection rule. The branching rule determines how to split up a problem into subproblems, by choosing a variable to branch on. Substantial research has been done on branching rules, see, e.g., [2,4,14,15].

The node selection rule decides which subproblem to solve next. Not much theoretical research has been done on the choice of the node selection rule. Traditionally, the best-first strategy is thought to be optimal from a theoretical perspective because this rule minimizes the number of nodes that need to be visited. However, to efficiently implement this rule the solver needs space proportional to the number of explored nodes, because all of them need to be kept in memory. In contrast to this, a simple strategy like depth-first search only needs to store the current solution. Unfortunately, performing a depth-first search can lead to an arbitrarily bad running time. This was the original motivation for introducing the explorable heap selection problem [13]. By guessing the number N of branch-and-bound nodes whose LP values are at most that of the optimal IP solution (which can be done via successive doubling), a search strategy for this problem can be directly interpreted as a node selection rule. The algorithm that they introduced can therefore be used to implement branch-and-bound efficiently in only $O\left(\sqrt{\log(N)}\right)$ space.

In practice, computers are usually able to store all explored nodes of the branch-and-bound tree in memory. However, many MIP-solvers still make use of a hybrid method that consists of both depth-first and best-first searches. This is not only done because depth-first search uses less memory, but also because it is often faster. Experimental studies have confirmed that the depth-first strategy is in many cases faster than best-first one [8]. This seems contradictory, because the running time of best-first search is often thought to be theoretically optimal.

In part, this contradiction can be explained by the fact that actual IP-solvers often employ complementary techniques and heuristics on top of branch-and-bound, which might benefit from depth-first searches. Additionally, a best-first search can hop between different parts of the tree, while a depth first search subsequently explores nodes that are very close to each other. In the latter case, the LP-solver can start from a very similar state, which is known as warm starting. This is faster for a variety of technical reasons [1]. For example, this can be the case when the LP-solver makes use of the LU-factorization of the optimal basis matrix [16]. Through the use of dynamic algorithms, computing this can be done faster if a factorization for a similar LP-basis is known [19]. Because of its large size, MIP-solvers will often not store the LU-factorization for all nodes in the tree. This makes it beneficial to move between similar nodes in the branch-and-bound tree. Furthermore, moving from one part of the tree to another means that the solver needs to undo and redo many bound changes, which also takes up time. Hence, the amount of distance traveled between nodes in the tree is a metric that influences the running time. This can also be observed when running the academic MIP-solver SCIP [12].

The explorable heap selection problem captures these benefits of locality by measuring the running time in terms of the amount of travel through the tree. Therefore, we argue that this problem is still relevant for the choice of a node selection rule, even if all nodes can be stored in memory.

Related Work. The explorable heap selection problem was first introduced in [13]. Their result was later applied to prove an upper bound on the parallel running time of branch-and-bound [18].

When random access to the heap is provided at constant cost, selecting the nth value in the heap can be done by a deterministic algorithm in $O(n)$ time by using an additional $O(n)$ memory for auxilliary data structures [11].

The explorable heap selection problem can be thought of as a *search game* [3] and bears some similarity to the *cow path problem*. In the cow path problem, an agent explores an unweighted unlabeled graph in search of a target node. The location of the target node is unknown, but when the agent visits a node they are told whether or not that node is the target. The performance of an algorithm is judged by the ratio of the number of visited nodes to the distance of the target from the agent's starting point. In both the cow path problem and the explorable heap selection problem, the cost of backtracking and retracing paths is an important consideration. The cow path problem on infinite b-ary trees was studied in [9] under the assumption that when present at a node the agent can obtain an estimate on that node's distance to the target.

Other explorable graph problems exist without a target, where typically the graph itself is unknown at the outset. There is an extensive literature on exploration both in graphs and in the plane. Models have been studied, in which one tried to minimize either the distance traveled or the amount of used memory. For more information we refer to [6,20] and the references therein.

Outline. In Sect. 2 we formally introduce the explorable heap selection problem and any notation we will use. In Sect. 3 we introduce a new algorithm for solving this problem and provide a running time analysis. In Sect. 4 we give a lower bound on the complexity of solving explorable heap selection using a limited amount of memory.

2   The Explorable Heap Selection Problem

In this section we introduce the formal model for the explorable heap selection problem. The input to the algorithm is an infinite binary tree $T = (V, E)$, where each node $v \in V$ has an associated real value, denoted by $\text{val}(v) \in \mathbb{R}$. We assume that all the values are distinct and that for each node in the tree, the values of its children are larger than its own value. The binary tree T is thus a heap.

We want to find the nth smallest value in this tree. This may be seen as an online graph exploration problem where an agent can move in the tree and learns the value of a node each time he explores it. At each time step, the agent resides at a vertex $v \in V$ and may decide to move to either the left child, the right child or the parent of v (if it exists, i.e. if v is not the root of the tree). Each traversal of an edge costs one unit of time, and the complexity of an algorithm for this problem is thus measured by the total traveled distance in the binary tree. The algorithm is also allowed to store values in memory.

For a node $v \in V$, also per abuse of notation written $v \in T$, we denote by $T^{(v)}$ the subtree of T rooted at v. For a tree T and a value $\mathcal{L} \in \mathbb{R}$, we define the subtree $T_{\mathcal{L}} := \{v \in T \mid \text{val}(v) \leq \mathcal{L}\}$. We denote the nth smallest value in T by $\mathsf{SELECT}^T(n)$. This is the quantity that we are interested in finding algorithmically. We say that a value $\mathcal{V} \in \mathbb{R}$ is *good* for a tree T if $\mathcal{V} \leq \mathsf{SELECT}^T(n)$ and *bad* otherwise. Similarly, we call a node $v \in T$ *good* if $\text{val}(v) \leq \mathsf{SELECT}^T(n)$ and *bad* otherwise. We use $[k]$ to refer to the set $\{1, \ldots, k\}$. When we write $\log(n)$, we assume the base of the logarithm to be 2.

We will often instruct the agent to move to an already discovered good vertex $v \in V$. The way this is done algorithmically is by saving $\text{val}(v)$ in memory and starting a depth first search at the root, turning back every time a value strictly bigger than $\text{val}(v)$ is encountered until finally finding $\text{val}(v)$. This takes at most $O(n)$ time, since we assume v to be a good node. If we instruct the agent to go back to the root from a certain vertex $v \in V$, this is simply done by traveling back in the tree, choosing to go to the parent of the current node at each step.

In later sections, we will often say that a subroutine takes a subtree $T^{(v)}$ as input. This implicitly means that we in fact pass it $\text{val}(v)$ as input, make the agent travel to $v \in T$ using the previously described procedure, call the subroutine from that position in the tree, and travel back to the original position at the end of the execution. Because the subroutine knows the value $\text{val}(v)$ of the root of $T^{(v)}$, it can ensure it never leaves the subtree $T^{(v)}$, thus making it possible to recurse on a subtree as if it were a rooted tree by itself.

We will sometimes want to pick a value uniformly at random from a set of values $\{\mathcal{V}_1, \ldots, \mathcal{V}_k\}$ of unknown size that arrives in a streaming fashion, for instance when we traverse a part of the tree T by doing a depth first search. That is, we see the value $\mathcal{V}_i$ at the ith time step, but do not longer have access to it in memory once we move on to $\mathcal{V}_{i+1}$. This can be done by generating random values $\{X_1, \ldots, X_k\}$ where, at the ith time step, $X_i = \mathcal{V}_i$ with probability $1/i$, and $X_i = X_{i-1}$ otherwise. It is easy to check that X_k is a uniformly distributed sample from $\{\mathcal{V}_1, \ldots, \mathcal{V}_k\}$.

3 A New Algorithm

The authors of [13] presented a deterministic algorithm that solves the explorable heap selection problem in $n \cdot \exp(O(\sqrt{\log(n)}))$ time and $O(n\sqrt{\log(n)})$ space. By replacing the binary search that is used in the algorithm by a randomized variant, they can decrease the space requirements. This way, they get a randomized algorithm with expected running time $n \cdot \exp(O(\sqrt{\log(n)}))$ and space complexity $O(\sqrt{\log(n)})$. Alternatively, the binary search can be implemented in a deterministic way by [17] to get the same running time with $O(\log(n)^{2.5})$ space.

We present a randomized algorithm with a running time $O(n \log(n)^3)$ and space complexity $O(\log(n))$. Unlike the algorithms mentioned before, our algorithm fundamentally relies on randomness to bound its running time. This bound only holds when the algorithm is run on a tree with labels that are fixed before the execution of the algorithm. That is, the tree must be generated by an adversary that is oblivious to the choices made by the algorithm. This is a stronger assumption than is needed for the algorithm that is given in [13], which also works against adaptive adversaries. An adaptive adversary is able to defer the decision of the node label to the time that the node is explored. Note that this distinction does not really matter for the application of the algorithm as a node selection rule in branch-and-bound, since there the node labels are fixed because they are derived from the integer program.

Theorem 1. *There exists a randomized algorithm that solves the explorable heap selection problem, with expected running time $O(n \log(n)^3)$ and $O(\log(n))$ space.*

The explorable heap selection problem can be seen as the problem of finding all n good nodes. Both our method and that of [13] function by first identifying a subtree consisting of only good nodes. The children of the leaves of this subtree are called "roots" and the subtree is extended by finding a number of new good nodes under these roots in multiple rounds.

In [13] this is done by running $O(c^{\sqrt{2\log(n)}})$ different rounds, for some constant $c > 1$. In each round, the algorithm finds $n/c^{\sqrt{2\log(n)}}$ new good nodes. These nodes are found by recursively exploring each active root and using binary search on the observed values to discover which of these values are good. Which active roots are recursively explored further depends on which values are good. The recursion in the algorithm is at most $O(\sqrt{\log(n)})$ levels deep, which is where the space complexity bound comes from.

In our algorithm, we take a different approach. We will call our algorithm consecutively with $n = 1, 2, 4, 8, \ldots$. Hence, for a call to the algorithm, we can assume that we have already found at least $n/2$ good nodes. These nodes form a subtree of the original tree T. In each round, our algorithm chooses a random root under this subtree and finds every good node under it. It does so by doing recursive subcalls to the main algorithm on this root with values $n = 1, 2, 4, 8, \ldots$. As soon as the recursively obtained node is a bad node, the algorithm stops searching the subtree of this root, since it is guaranteed that all the good nodes there have been found. The largest good value that is found can then be used to find additional good nodes under the other roots without recursive calls, through a simple depth-first search. Assuming that the node values were fixed in advance, we expect this largest good value to be greater than half of the other roots' largest good values. Similarly, we expect its smallest bad value to be smaller than half of the other roots' smallest bad values. By this principle, a sizeable fraction of the roots can, in expectation, be ruled out from getting a recursive call. Each round a new random root is selected until all good nodes have been found. This algorithm allows us to effectively perform binary search on the list of roots, ordered by the largest good value contained in each of their subtrees in $O(\log n)$ rounds, and the same list ordered by the smallest bad values (Lemma 2). Bounding the expected number of good nodes found using recursive subcalls requires a subtle induction on two parameters (Lemma 1): both n and the number of good nodes that have been identified so far.

3.1   The Algorithm

The EXTEND procedure is the core of our algorithm. It finds the nth smallest value in the tree, under the condition that the kth smallest value $\mathcal{L}_0$ is provided to the algorithm for some $k \geq n/2$. Using this procedure, SELECT(n) can be solved by consecutively calling EXTEND$(T, n_i, k_i, \mathcal{L}_i)$ with parameters $(n_i, k_i) = (2^i, 2^{i-1})$ for $i \in \{1, \ldots, \lceil \log(n) \rceil\}$.

Algorithm 1. The EXTEND procedure

1: **Input:** T: tree which is to be explored.
2:   $n \in \mathbb{N}$: total number of good values to be found in the tree T, satisfying $n \geq 2$.
3:   $k \in \mathbb{N}$: number of good values already found in the tree T, satisfying $k \geq n/2$.
4:   $\mathcal{L}_0 \in \mathbb{R}$: value satisfying DFS$(T, \mathcal{L}_0, n) = k$.
5: **Output:** the nth smallest value in T.
6: **procedure** EXTEND$(T, n, k, \mathcal{L}_0)$
7:     $\mathcal{L} \leftarrow \mathcal{L}_0, \mathcal{U} \leftarrow \infty$
8:     **while** $k < n$ **do**
9:         $r \leftarrow$ random element from ROOTS$(T, \mathcal{L}_0, \mathcal{L}, \mathcal{U})$
10:        $\mathcal{L}' \leftarrow \max(\mathcal{L}, \text{val}(r))$
11:        $k' \leftarrow$ DFS$(T, \mathcal{L}', n)$ // count the number of values $\leq \mathcal{L}'$ in T
12:        $c \leftarrow$ DFS$(T^{(r)}, \mathcal{L}', n)$ // count the number of values $\leq \mathcal{L}'$ in $T^{(r)}$
13:        $c' \leftarrow \min(n - k' + c, 2c)$ // increase the number of values to be found in $T^{(r)}$
14:        **while** $k' < n$ **do** // loop until it is certified that SELECT$^T(n) \leq \mathcal{L}'$
15:            $\mathcal{L}' \leftarrow$ EXTEND$(T^{(r)}, c', c, \mathcal{L}')$
16:            $k' \leftarrow$ DFS$(T, \mathcal{L}', n)$
17:            $c \leftarrow c'$
18:            $c' \leftarrow \min(n - k' + c, 2c)$
19:        **end while**
20:        $\tilde{\mathcal{L}}, \tilde{\mathcal{U}} \leftarrow$ GOODVALUES$(T, T^{(r)}, \mathcal{L}', n)$ // find the good values in $T^{(r)}$
21:        $\mathcal{L} \leftarrow \max(\mathcal{L}, \tilde{\mathcal{L}})$, $\mathcal{U} \leftarrow \min(\mathcal{U}, \tilde{\mathcal{U}})$
22:        $k \leftarrow$ DFS$(T, \mathcal{L}, n)$ // compute the number of good values found in T
23:    **end while**
24:    **return** $\mathcal{L}$
25: **end procedure**

Let us describe a few invariants from the EXTEND procedure.

- $\mathcal{L}$ and $\mathcal{U}$ are respectively lower and upper bounds on SELECT$^T(n)$ during the whole execution of the procedure.
- The integer k counts the number of values $\leq \mathcal{L}$ in the full tree T.
- After an iteration of the inner while loop, $\mathcal{L}'$ is set to the cth smallest value in $T^{(r)}$. The variable c' then corresponds to the next value we would like to find

in $T^{(r)}$ if we were to continue the search. Note that $c' \leq 2c$, enforcing that the recursive call to EXTEND satisfies its precondition, and that $c' \leq n - (k' - c)$ implies that $(k' - c) + c' \leq n$, which implies that the recursive subcall will not spend time searching for a value that is known in advance to be bad.

– k' always counts the number of values $\leq \mathcal{L}'$ in the full tree T. It is important to observe that this is a global parameter, and does not only count values below the current root. Moreover, $k' \geq n$ implies that we can stop searching below the current root, since it is guaranteed that all good values in $T^{(r)}$ have been found, i.e., $\mathcal{L}'$ is larger than all the good values in $T^{(r)}$.

We now describe the subroutines used in the EXTEND procedure.

The Procedure DFS. The procedure DFS is a variant of depth first search. The input to the procedure is T, a cutoff value $\mathcal{L} \in \mathbb{R}$ and an integer $n \in \mathbb{N}$. The procedure returns the number of vertices in T whose value is at most $\mathcal{L}$.

It achieves that by exploring the tree T in a depth first search manner, starting at the root and turning back as soon as a node $w \in T$ such that $\mathrm{val}(w) > \mathcal{L}$ is encountered. Moreover, if the number of nodes whose value is at most $\mathcal{L}$ exceeds n during the search, the algorithm stops and returns $n + 1$.

The algorithm output is the following integer. whose value is at most $\mathcal{L}$:

$$\mathrm{DFS}(T, \mathcal{L}, n) := \min \{|T_\mathcal{L}|, n + 1\}.$$

Observe that the DFS procedure allows us to check whether a node $w \in T$ is a good node, i.e. whether $\mathrm{val}(w) \leq \mathsf{SELECT}^T(n)$. Indeed, w is good if and only if $\mathrm{DFS}(T, \mathrm{val}(w), n) \leq n$.

This procedure visits only nodes in $T_\mathcal{L}$ or its direct descendants and its running time is thus $O(n)$. The space complexity is $O(1)$, since the only values needed to be stored in memory are $\mathcal{L}$, $\mathrm{val}(v)$, where v is the root of the tree T, and a counter for the number of good values found so far.

The Procedure Roots. The procedure ROOTS takes as input a tree T as well as a lower bound $\mathcal{L}_0 \in \mathbb{R}$ on the value of $\mathsf{SELECT}^T(n)$. We assume that the main algorithm has already found all the nodes $w \in T$ satisfying $\mathrm{val}(w) \leq \mathcal{L}_0$. This means that the remaining values the main algorithm needs to find in T are all lying in the subtrees of the following nodes, that we call the $\mathcal{L}_0$-*roots of T* (Fig. 1):

$$R(T, \mathcal{L}_0) := \{r \in T \setminus T_{\mathcal{L}_0} \mid r \text{ is a child of a node in } T_{\mathcal{L}_0}\}$$

In words, these are all the vertices in T one level deeper in the tree than $T_{\mathcal{L}_0}$. In addition to that, the procedure takes two other parameters $\mathcal{L}, \mathcal{U} \in \mathbb{R}$ as input, which correspond to (another) lower and upper bound on the value of $\mathsf{SELECT}^T(n)$. These lower and upper bounds will be updated during the execution of the main algorithm. A key observation is that these bounds can allow us to remove certain roots in $R(T, \mathcal{L}_0)$ from consideration, in the sense that all the good values in that root's subtree will be certified to have already been found (Fig. 2):

$$\mathrm{ROOTS}(T, \mathcal{L}_0, \mathcal{L}, \mathcal{U}) := \{r \in R(T, \mathcal{L}_0) \mid \exists w \in T^{(r)} \text{ with } \mathrm{val}(w) \in (\mathcal{L}, \mathcal{U})\}$$

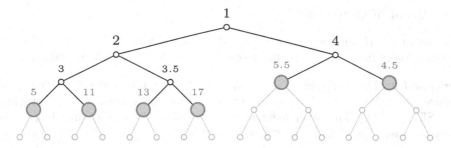

Fig. 1. An illustration of $R(T, \mathcal{L}_0)$ with $\mathcal{L}_0 = 4$. The number above each vertex is its value, the blue nodes are $R(T, \mathcal{L}_0)$, whereas the subtree above is $T_{\mathcal{L}_0}$. (Color figure online)

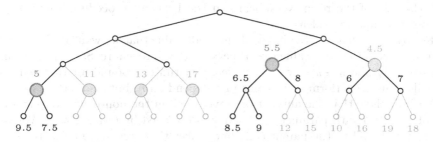

Fig. 2. An illustration of the ROOTS procedure with $\mathcal{L}_0 = 4, \mathcal{L} = 7$ and $\mathcal{U} = 10$. Only two active roots remain, and are both colored in blue. The other roots are considered killed since all the good values have been found in their subtrees. (Color figure online)

This subroutine can be implemented by running a depth first search starting at the root of T and exploring $T_{\mathcal{L}}$ with its direct descendants. Since $\mathcal{L}$ is known to be good, the running time is bounded by $O(|T_{\mathcal{L}}|) = O(n)$. In the main algorithm, we will only need this procedure in order to select a root from ROOTS$(T, \mathcal{L}_0, \mathcal{L}, \mathcal{U})$ uniformly at random, without having to store the whole list in memory. This can then be achieved in $O(1)$ space, since one then only needs to store val$(v), \mathcal{L}_0, \mathcal{L}$ and $\mathcal{U}$ in memory, where v is the root of the tree T.

The Procedure GoodValues. The procedure GOODVALUES takes as input a tree T, a subtree $T^{(r)}$, a value $\mathcal{L}' \in \mathbb{R}_{\geq 0}$ and an integer $n \in \mathbb{N}$. The procedure then analyzes the set $S := \{\text{val}(w) \mid w \in T^{(r)}, \text{val}(w) \leq \mathcal{L}'\}$ and outputs both the largest good value and the smallest bad value in that set, that we respectively call $\mathcal{L}$ and $\mathcal{U}$. If no bad values exist in S, the algorithm sets $\mathcal{U} = \infty$.

The procedure can be implemented using a randomized binary search on the values in S, where the procedure DFS is used to check whether a value is good. This makes the procedure have a running time of $O(n \log n)$.

The procedure only needs $O(1)$ space, since the only values necessary to be kept in memory are val(v) (where v is the root of the tree T), val(r), $\mathcal{L}$, $\mathcal{U}$ and $\mathcal{L}'$, as well as the fact that every call to DFS also requires $O(1)$ space.

3.2   Proof of Correctness

Theorem 2. *At the end of the execution of Algorithm 1, $\mathcal{L}$ is set to the nth smallest value in T. Moreover, the algorithm is guaranteed to terminate.*

Proof sketch. The variable $\mathcal{L}$ is always set to the first output of the procedure GOODVALUES, which is always the value of a good node, implying $\mathcal{L} \leq \mathsf{SELECT}^T(n)$. The other inequality follows since the outer while loop ends when at least n good nodes have been found in T.

3.3   Running Time Analysis

The main challenge in analyzing the running time of the algorithm is dealing with the cost of the recursive subcalls in the EXTEND procedure. For this we rely on two important ideas.

Firstly, note that n is the index of the node value that we want to find, while k is the index of the node value that is passed to the procedure. So, the procedure needs to only find $n - k$ new good nodes. Our runtime bound for the recursive subcalls that are performed does not just depend on n, but also on $n - k$.

We will show that the amount of travel done in the non-recursive part of a call of EXTEND with parameters n and k is bounded by $O(n \log(n)^2)$. We will charge this travel to the parent call that makes these recursive calls. Hence, a parent call that does z recursive calls with parameters $(n_1, k_1), \ldots, (n_z, k_z)$ will be charged a cost of $\sum_{i=1}^{z} n_i \log(n_i)^2$. In our analysis, we will show that this sum can be upper bounded by $(n - k) \log(n)^2$. So, for every recursive call with parameters n and k, a cost of at most $(n - k) \log(n)^2$ is incurred by the caller.

Now we just need to bound the sum over $(n - k) \log(n)^2$ for all calls with parameters n and k that are done. We do this by first considering a single algorithm call with parameters n and k that makes z recursive subcalls with parameters $(n_1, k_1), \ldots, (n_z, k_z)$. For such a subcall, we would like to bound the sum $\sum_{i=1}^{z} (n_i - k_i) \log(n_i)^2$ by $(n - k) \log(n)^2$. However, this bound does not hold deterministically. Instead, we show that this bound does hold in expectation.

Now we know that every layer of recursion incurs an expected cost of at most $(n - k) \log(n)^2$. Because the parameter n will decrease by at least a constant factor in each layer of recursion, there can be at most $O(\log(n))$ layers. An upper bound of $O((n - k) \log(n)^3)$ on the expected running time of the EXTEND then follows for the recursive part.

Combining this with the upper bound of $O(n \log(n)^2)$ on the non-recursive part, we get a total running time of $O(n \log(n)^2) + O((n - k) \log(n)^3)$ for the EXTEND procedure, which then implies a running time of $O(n \log(n)^3)$ for the SELECT procedure.

Let us now prove these claims. We first show that the expectation of $\sum_{i=1}^{z} (n_i - k_i)$ is bounded.

Lemma 1. *Let z be the number of recursive calls that are done in the main loop of* EXTEND $(T, n^\star, k^\star, \mathcal{L})$ *with parameter $k \geq 1$. For $i \in [z]$, let n_i and k_i be the* *values of n and k that are given as parameters to the ith such subcall. Then:*

$$\mathbb{E}\left[\sum_{i=1}^{z} n_i - k_i\right] \leq n^\star - k^\star.$$

Proof. Assume we have m roots, whose order is fixed. For $i \in [z]$, let $r_i \in [m]$ be such that the ith recursive subcall is done on the root with index r_i. For $t \in [m]$, let $s_t = \sum_{i=1}^{z} 1_{r_i=t}(n_i - k_i)$. From the algorithm we see that when $r_i = t$, all successive recursive calls will also be on root t, until all good nodes under this root have been found. The updated values of $\mathcal{L}$ and $\mathcal{U}$ ensure this root is never selected again after this, hence all iterations i with $r_i = t$ are consecutive. Now let a_t, b_t be variables that respectively denote the first and last indices i with $r_i = t$. When there is no iteration i with $r_i = t$, then $a_t = b_t = \infty$.

For two calls i and $i+1$ with $r_i = t = r_{i+1}$, observe that after call i already n_i good nodes under root t have been found. On line 15, c' corresponds to n_i and c corresponds to k_i, hence $k_{i+1} = n_i$. Therefore, the definition of s_t is a telescoping series and can be rewritten as $s_t = n_{b_t} - k_{a_t}$, when we define $k_\infty = n_\infty = 0$.

Let $p = n^\star - k^\star$ and let $W = \{w_1, \ldots, w_p\}$ denote the p smallest values under T that are larger than $\mathcal{L}_0$, in increasing order. Now each of these values in W will be part of a subtree generated by one of the roots. For the $j \in [p]$, let $d_j \in [m]$ be such that value w_j is part of the subtree of root d_j. Let $S_t = \{j \in [p] : d_j = t\}$. We will now show that for each root r_t, we have $\mathbb{E}[s_t] \leq |S_t|$. This will imply that $\mathbb{E}[\sum_{i=1}^{z} n_i - k_i] = \sum_{t=1}^{m} \mathbb{E}[s_t] \leq \sum_{t=1}^{m} |S_t| = n^\star - k^\star$.

First, consider a root t with $t \neq d_p$. On line 9, each iteration a random root is chosen. In every iteration root d_p will be among the active roots. So the probability that this root is chosen before root t is at least a half. In that case, after the iteration of root d_p, $\mathcal{L}$ will be set to w_p. Then DFS$(T, \mathcal{L}, n)$ returns n, and the algorithm terminates. Since no subcalls are done on root t, $s_i = 0$.

If the algorithm does do subcalls i with $r_i = t$, then consider iteration b_t, the last iteration i that has $r_i = t$. Before this iteration, already k_{b_t} good nodes under the root have been found by the algorithm. It can be seen in the algorithm on lines 13 and 18 that $n_{b_t} \leq 2k_{b_t}$. Hence $s_t = n_{b_t} - k_{a_t} \leq n_{b_t} \leq 2k_{b_t} \leq 2|S_t|$. We therefore have $\mathbb{E}[s_t] \leq \frac{1}{2} \cdot 0 + \frac{1}{2} \cdot 2|S_t| = |S_t|$.

Now consider the root d_p. If $S_{d_p} = [p]$, then $s_p = n_{b_{d_p}} - k_{a_{d_p}} \leq n^\star - k^\star = |S_{d_p}|$, because $n_{b_{d_p}} \leq n^\star$ and $k_{a_{d_p}} \geq k^\star$. If $S_{d_p} \subsetneq [p]$, then there exists a j with $d_j \neq d_p$. Thus, we can define $j^\star = \max\{j \in [p] : d_j \neq d_p\}$. With probability a half, root $d_{j^\star}$ is considered before root d_p. If this happens, $\mathcal{L}$ will be equal to $w_{j^\star}$ when root d_p is selected by the algorithm. In particular, this means that $k_{a_{d_p}}$ will be equal to $j^\star$. Recall the stated invariant that $c' \leq n^\star - k^\star = p$, and hence $n_{b_{d_p}} = c' \leq p$. Now we can see that $s_{d_p} = n_{b_{d_p}} - k_{a_{d_p}} \leq p - j^\star$.

If root d_p is chosen before root $d_{j^\star}$, then consider the last recursive call b_{d_p} to EXTEND that we do on root d_p. Define $A = [k' - k^\star] \cap S_{d_p}$, i.e. the set of all good values under root d_p that have been found so far. We distinguish two cases.

If $k' - k^\star \geq j^\star$, i.e., when all good values under $d_{j^\star}$ have been found, then by definition of $j^\star$, $[p] \setminus [k' - k^\star] \subseteq [p] \setminus [j^\star] \subseteq S_{d_p}$. Because A and $[p] \setminus [k' - k^\star]$

are disjoint, we have $|A| + (n^\star - k') = |A| + |[p] \setminus [k' - k^\star]| \leq |S_{d_p}|$. Hence, we have $c' \leq n^\star - k' + c = n^\star - k' + |A| \leq |S_{d_p}|$. Therefore, $s_{d_p} \leq n_{b_{d_p}} = c' \leq |S_{d_p}|$.

If $k' - k^\star < j^*$ at the time of subcall b_{d_p}, then the last good value under d_{j^*} has yet to be found, implying that $A \subseteq [j^*]$. From the definition of $j^\star$ we get $[p] \setminus [j^\star] \subseteq S_{d_p}$. Hence, $|A| \leq |S_{d_p}| - |[p] \setminus [j^\star]| = |S_{d_p}| - (p - j^\star)$. Thus $c' \leq 2c = 2|A| \leq 2(|S_{d_p}| - (p - j^*))$. So, in this case we have $s_{d_p} \leq n_{b_{d_p}} = c' \leq 2(|S_{d_p}| - (p - j^*))$.

Collecting the three cases above, we find that

$$\mathbb{E}[s_{d_p}] \leq \frac{1}{2} \cdot (p - j^*) + \frac{1}{2} \cdot \max\left(|S_{d_p}|, 2(|S_{d_p}| - (p - j^*))\right)$$

$$\leq \max\left(\frac{1}{2}|S_{d_p}| + \frac{1}{2}(p - j^*), |S_{d_p}| - \frac{1}{2}(p - j^*)\right).$$

Lastly, by definition of j^* we have $[p] \setminus [j^*] \subseteq S_{d_p}$, from which it follows that $p - j^* \leq |S_{d_p}|$. We finish the proof by observing that this implies

$$\max\left(\frac{1}{2}|S_{d_p}| + \frac{1}{2}(p - j^*), |S_{d_p}| - \frac{1}{2}(p - j^*)\right) \leq |S_{d_p}|,$$

which finishes the proof.

Lemma 2. *The expected number of times that the outermost while-loop (at line 8) is executed by the procedure* EXTEND *is at most* $O(\log(n))$.

Proof sketch. Let $A_\ell(\mathcal{L}) := \{r_j : \ell_j > \mathcal{L}\}$ and $A_u(\mathcal{U}) := \{r_j : u_j < \mathcal{U}\}$. Observe that $\text{ROOTS}(T, \mathcal{L}_0, \mathcal{L}, \mathcal{U}) = A_\ell(\mathcal{L}) \cup A_u(\mathcal{U})$ for any $\mathcal{L} \leq \mathcal{U}$. One can show that in each iteration the size of either $A_\ell(\mathcal{L})$ or $A_u(\mathcal{U})$ halves in expectation. Hence, in expectation at most $O(\log R)$ iterations are needed, where $\log R$ is the initial number of roots. Since $R \leq n$, the lemma follows.

By an elementary analysis of the algorithm and applying Lemma 2 we can prove the following lemma.

Lemma 3. *The expected running time of the non-recursive part of every call to* EXTEND *is* $O(n \log(n)^2)$.

Finally we are able to prove the running time bound.

Lemma 4. *Let* $R(T, n, k)$ *denote the running time of a call to* EXTEND$(T, n, k, \mathcal{L}_0)$. *Then there exists* $C > 0$ *such that*

$$\mathbb{E}[R(T, n, k)] \leq 5C(n - k)\log(n)^3 + Cn\log(n)^2.$$

Proof. We will prove this with induction on $r := \lceil \log(n) \rceil$. For $r = 1$, we have $n \leq 2$. In this case R is constant, proving our induction base.

Now consider a call EXTEND$(T, n, k, \mathcal{L}_0)$ and assume the induction claim is true when $\lceil \log(n) \rceil \leq r - 1$. By Lemma 3, we can choose C such that this running time is bounded by $C \cdot n \log(n)$.

Now we move on to the recursive part of the algorithm. All calls to EXTEND(T, n, k, $\mathcal{L}_0$) with $k = 0$ will have $n = 1$, so each of these calls takes only $O(1)$ time. Hence we can safely ignore these calls.

Let z be the number of recursive calls to EXTEND(T, n, k, $\mathcal{L}_0$) that are done from the base call with $k \geq 1$. Let T_i, k_i, n_i for $i \in [z]$ be the arguments of these function calls. Note that $n/2 \geq n_i \geq 2$ for all i. By the induction hypothesis the expectation of the recursive part of the running time is:

$$
\mathbb{E}\left[\sum_{i=1}^{z} R(T_i, n_i, k_i)\right] \leq \mathbb{E}\left[\sum_{i=1}^{z} 5C \log(n_i)(n_i - k_i)\log(n_i)^2 + Cn_i\log(n_i)^2\right]
$$

$$
\leq 5C\log(n/2)\,\mathbb{E}\left[\sum_{i=1}^{r} n_i - k_i\right]\log(n)^2 + C\log(n)^2\sum_{i=1}^{r} n_i
$$

$$
\leq 5C(\log(n) - 1)(n - k)\log(n)^2 + 5C\log(n)^2(n - k)
$$

$$
\leq 5C(n - k)\log(n)^3.
$$

Here we used Lemma 1 as well as the fact that $\sum_{i=1}^{r} n_i \leq 4(n - k)$. To see this, consider an arbitrary root q with s good values under it. Now $\sum_{i=1}^{z} \mathbf{1}_{T_i = T^{(q)}} n_i \leq \sum_{i=2}^{\lceil \log(s+1) \rceil} 2^i \leq 2^{\lceil \log(s+1) \rceil + 1} \leq 4s$. In total there are $n - k$ good values under the roots, and hence $\sum_{i=1}^{z} n_i \leq 4(n - k)$.

Adding the expected running time of the recursive and the non-recursive part, we see that

$$
\mathbb{E}[R(T, n, k)] \leq 5C(n - k)\log(n)^3 + Cn\log(n)^2.
$$

3.4   Space Complexity Analysis

We prove in this section the space complexity of our algorithm.

Theorem 3. *The procedure* EXTEND *runs in* $O(\log(n))$ *space.*

Proof. Observe that the subroutines DFS, ROOTS and GOODVALUES all require $O(1)$ memory, as argued in their respective analyses. Hence the space complexity of the non-recursive part of the EXTEND is $O(1)$. Any recursive subcall EXTEND(T_i, n_i, k_i, $\mathcal{L}_i$) resulting from a call to EXTEND(T, n, k, $\mathcal{L}$), will have $n_i \leq n/2$. Hence, the depth of recursion is at most $O(\log(n))$, which implies that the same is true for the space complexity.

4   Lower Bound

In general, no lower bound is known for the running time of the selection problem. However, we will show that any algorithm with space complexity at most s, has a running time of at least $\Omega(n\log_s(n))$. The tree that is used for the lower bound construction is very simple: a root with two trails of length $O(n)$ attached to it.

We will make use of a variant of the communication complexity model. In this model there are two agents A and B, that both have access to their own sets of values in S_A and S_B respectively. These sets are the input. We have $|S_A| = n+1$ and $|S_B| = n$. Assume that all values S_A and S_B are different. Now consider the problem where player A wants to compute the median of $S_A \cup S_B$.

Because the players only have access to their own values, they need to communicate. They use a protocol, that can consist of multiple rounds. In every odd round, player A can do computations and send units of information to player B. In every even round, player B does computations and sends information to player A. We assume that sending one value from S_A or S_B takes up one *unit of information*. Furthermore, we assume that, except for comparisons, no operations can be performed on the values.

We can reduce median computation to the explorable heap selection problem.

Lemma 5. *If there is a algorithm that solves* SELECT$(3n)$ *in* $f(n)n$ *time and* g *space, then there is a protocol for median computation that uses* $f(n)/2$ *rounds in each of which at most* g *units of information are sent.*

By showing a lower bound on the number of necessary rounds for median computation we can now prove the lower bound.

Theorem 4. *The time complexity of any randomized algorithm for* SELECT(n) *with at most* g *units of storage is* $\Omega(n \log_{g+1}(n))$.

References

1. Achterberg, T.: Constraint Integer Programming. Ph.D. thesis, TU Berlin (2009)
2. Achterberg, T., Koch, T., Martin, A.: Branching rules revisited. Oper. Res. Lett. **33**(1), 42–54 (2005). https://doi.org/10.1016/j.orl.2004.04.002
3. Alpern, S., Gal, S.: The Theory of Search Games and Rendezvous, vol. 55. Springer, New York (2006). https://doi.org/10.1007/b100809
4. Balcan, M.F., Dick, T., Sandholm, T., Vitercik, E.: Learning to branch. In: ICML (2018)
5. Banerjee, S., Cohen-Addad, V., Gupta, A., Li, Z.: Graph searching with predictions, December 2022
6. Berman, P.: On-line searching and navigation. In: Fiat, A., Woeginger, G.J. (eds.) Online Algorithms. LNCS, vol. 1442, pp. 232–241. Springer, Heidelberg (1998). https://doi.org/10.1007/BFb0029571
7. Borst, S., Dadush, D., Huiberts, S., Kashaev, D.: A nearly optimal randomized algorithm for explorable heap selection, October 2022. https://doi.org/10.48550/arXiv.2210.05982
8. Clausen, J., Perregaard, M.: On the best search strategy in parallel branch-and-bound: best-first search versus lazy depth-first search. Ann. Oper. Res. **90**, 1–17 (1999)
9. Dasgupta, P., Chakrabarti, P.P., DeSarkar, S.C.: A near optimal algorithm for the extended cow-path problem in the presence of relative errors. In: Thiagarajan, P.S. (ed.) FSTTCS 1995. LNCS, vol. 1026, pp. 22–36. Springer, Heidelberg (1995). https://doi.org/10.1007/3-540-60692-0_38
10. Diks, K., Fraigniaud, P., Kranakis, E., Pelc, A.: Tree exploration with little memory. J. Algorithms **51**(1), 38–63 (2004). https://doi.org/10.1016/j.jalgor.2003.10.002

11. Frederickson, G.: An optimal algorithm for selection in a min-heap. Inf. Comput. **104**(2), 197–214 (1993). https://doi.org/10.1006/inco.1993.1030
12. Gleixner, A.M.: Personal communication, November 2022
13. Karp, R.M., Saks, M.E., Wigderson, A.: On a search problem related to branch-and-bound procedures. In: FOCS, pp. 19–28 (1986)
14. Linderoth, J.T., Savelsbergh, M.W.P.: A computational study of search strategies for mixed integer programming. INFORMS J. Comput. **11**(2), 173–187 (1999). https://doi.org/10.1287/ijoc.11.2.173
15. Lodi, A., Zarpellon, G.: On learning and branching: a survey. TOP **25**(2), 207–236 (2017). https://doi.org/10.1007/s11750-017-0451-6
16. Morrison, D.R., Jacobson, S.H., Sauppe, J.J., Sewell, E.C.: Branch-and-bound algorithms: a survey of recent advances in searching, branching, and pruning. Discret. Optim. **19**, 79–102 (2016). https://doi.org/10.1016/j.disopt.2016.01.005
17. Munro, J., Paterson, M.: Selection and sorting with limited storage. Theoret. Comput. Sci. **12**(3), 315–323 (1980). https://doi.org/10.1016/0304-3975(80)90061-4
18. Pietracaprina, A., Pucci, G., Silvestri, F., Vandin, F.: Space-efficient parallel algorithms for combinatorial search problems. J. Parallel Distrib. Comput. **76**, 58–65 (2015)
19. Suhl, L.M., Suhl, U.H.: A fast LU update for linear programming. Ann. Oper. Res. **43**(1), 33–47 (1993). https://doi.org/10.1007/BF02025534
20. Kamphans, T.: Models and algorithms for online exploration and search. Ph.D. thesis, Rheinische Friedrich-Wilhelms-Universität Bonn (2006). https://hdl.handle.net/20.500.11811/2622

Sparse Approximation over the Cube

Sabrina Bruckmeier[1]([📧]) [ID], Christoph Hunkenschröder[2] [ID],
and Robert Weismantel[1]

[1] ETH Zürich, Zürich, Switzerland
{sabrina.bruckmeier,robert.weismantel}@ifor.math.ethz.ch
[2] TU Berlin, Berlin, Germany
hunkenschroeder@tu-berlin.de

Abstract. This paper presents an analysis of the NP-hard minimization problem $\min\{\|b - Ax\|_2 : x \in [0,1]^n, |\mathrm{supp}(x)| \le \sigma\}$, where $\mathrm{supp}(x) := \{i \in [n] : x_i \ne 0\}$ and σ is a positive integer. The object of investigation is a natural relaxation where we replace $|\mathrm{supp}(x)| \le \sigma$ by $\sum_i x_i \le \sigma$. Our analysis includes a probabilistic view on when the relaxation is exact. We also consider the problem from a deterministic point of view and provide a bound on the distance between the images of optimal solutions of the original problem and its relaxation under A. This leads to an algorithm for generic matrices $A \in \mathbb{Z}^{m \times n}$ and achieves a polynomial running time provided that m and $\|A\|_\infty$ are fixed.

Keywords: Sparse Approximation · Subset Selection · Signal Recovery

1 Introduction and Literature Review

Due to the recent development of machine learning, data science and signal processing, more and more data is generated, but only a part of it might be necessary in order to already make predictions in a sufficiently good manner. Therefore, the question arises to best approximate a signal b by linear combinations of no more than σ vectors A_i from a suitable dictionary $A = (A_1, \ldots, A_n) \in \mathbb{R}^{m \times n}$:

$$\min \|Ax - b\|_2 \text{ subject to } \|x\|_0 \le \sigma, \tag{1}$$

where $\|x\|_0 := |\{i \in [n] : x_i \ne 0\}|$. Additionally, many areas of application – as for example portfolio selection theory, sparse linear discriminant analysis, general linear complementarity problems or pattern recognition – require the solution x to satisfy certain polyhedral constraints. For instance motivated by computer tomography, lower and upper bounds on the variables are considered in [31]. While there exists a large variety of ideas how to tackle this problem, the majority of them relies on the matrix A satisfying conditions such as being sampled in a specific way or being close to behaving like an orthogonal system, that might be hard to verify. Additionally, these algorithms commonly yield results only with a certain probability or within an approximation factor that again highly depends on the properties of A. A discussion of these ideas and

A. Del Pia and V. Kaibel (Eds.): IPCO 2023, LNCS 13904, pp. 44–57, 2023.
https://doi.org/10.1007/978-3-031-32726-1_4

different names and variants of this problem is postponed to the end of the introduction. In this work, we develop an exact algorithm that, without these limitations on A, solves the *Sparse Approximation problem* in $[0,1]$-variables,

$$\min_x \|Ax - b\|_2 \text{ subject to } x \in [0,1]^n \text{ and } \|x\|_0 \le \sigma. \qquad (P_0)$$

Theorem 1. *Given $A \in \mathbb{Z}^{m \times n}, b \in \mathbb{Z}^m$ and $\sigma \in \mathbb{Z}_{\ge 1}$, we can find an optimal solution x to Problem (P_0) in $(m\|A\|_\infty)^{\mathcal{O}(m^2)} \cdot \text{poly}(n, \ln(\|b\|_1))$ arithmetic operations.*

Relaxing the pseudonorm $\| \cdot \|_0$ by $\| \cdot \|_1$ is a commonly used technique in the literature. In contrast to previous results we are able to bound the distance between the images of these solutions under A without any further assumptions on the input data and therefore derive a *proximity result* that – to the best of our knowledge – has not been known before.

Theorem 2. *Let $\hat{x}$ be an optimal solution to the following relaxation of (P_0):*

$$\min_x \|Ax - b\|_2 \text{ subject to } x \in [0,1]^n \text{ and } \|x\|_1 \le \sigma.$$

Every optimal solution $x^\star$ of (P_0) satisfies

$$\|Ax^\star - A\hat{x}\|_2 \le 2\|\hat{x} - \lfloor\hat{x}\rfloor\|_1 \max_{i=1,\dots,n} \|A_i\|_2 \le 2\,m^{3/2}\|A\|_\infty,$$

where $\lfloor\hat{x}\rfloor$ denotes the vector $\hat{x}$ rounded down component-wise.

We also illuminate our approach from a probabilistic point of view. Specifically, the hard instances are those where b is relatively close to the boundary of the polytope $Q := \{Ax : x \in [0,1]^n, \|x\|_1 \le \sigma\}$. Conversely, if b is deep inside Q or far outside of Q, then with high probability, an optimal solution to the relaxation solves the initial problem (P_0).

The paper is organized as follows. We conclude the introduction by providing an overview on related literature. Section 2 discusses preliminaries. The probabilistic analysis of a target vector b is carried out in Sect. 3. We then discuss a worst-case proximity bound between optimal solutions of (P_0) and a natural relaxation in Sect. 4. This will allow us to formalize a deterministic algorithm in Sect. 5.

In the literature, Problem (1) can be found under various modifications and names, see e.g. [4,5,8,27]. A common variant in the context of random measurements is often called *Sparse Recovery*, cf. [20], or *Subset Selection for (linear) regression*, cf. [10], while the name *(Best) Subset Selection* is generally used without further interpretation cf. [8,11,34], in contrast to *Signal Recovery* or *Signal Reconstruction* as in [3]. If the vector b can be represented exactly, the problem is called *Exact Sparse Approximation* or *Atomic Decomposition*, cf. [7,21,32,35]. Since the differences are marginal and the names in the literature not well-defined, we restrain ourselves to the name *Sparse Approximation* for simplicity.

In general, there are two common strategies used to tackle *Sparse Approximation*: Greedy algorithms and algorithms based on relaxations. A detailed discussion of those is beyond the scope of this paper. Let us rather put these approaches into context below. The algorithms either recover the optimal support only under certain conditions (compare [1,8,9,32]), recover it with high probability (see for example [13,34]) or approximate the solution (for instance [10,14,21]). Unfortunately, because of their high computational cost most common greedy algorithms are not sufficient for large systems, though experiments suggest that there still exist applicable greedy approaches, such as the Dropping Forward-Backward Scheme, introduced by Nguyen [26]. While the idea of relaxing the pseudonorm $\|\cdot\|_0$ by the norm $\|\cdot\|_1$, as done for example in Basis Pursuit by Chen, Donoho and Saunders [7], might seem intuitive, for a long time the success of this method was not quite understood. This changed as Candes, Romberg and Tao [4,6] discovered and improved the Uniform Uncertainty Principle. For the usually problematic case of having not enough data points, the Dantzig Selector presented by Candes and Tao [5] yields a sophisticated estimator with high probability. Similarly, LASSO based methods, see for instance [27], either recover the support with high probability exactly under certain conditions, or fail with high probability if the conditions are not met, cf. [33]. Finally, Garmanik and Zadik [17] revealed interesting structural results, that explain the above mentioned all-or-nothing behavior. There also exists a series of papers in a similar line of thought that relaxes $\|\cdot\|_0$ by smooth, non-decreasing, concave functions, see [12,15,16,19,22,24,29,30]. It can be shown that these relaxations converge towards the optimal solution of (P_0). Qian et al. [28] and Çivril [35] proved that, unless $P = NP$, for a general matrix A Pareto Optimization and the two greedy algorithms, Forward Selection and Orthogonal Matching Pursuit, are almost the best we can hope for. This motivated a search for more efficiently solvable classes of A, cf. [3,11,18,20]. Finally, it should be mentioned that there exists a variety of Branch-and-Bound algorithms whose success though is in general only tested experimentally, see [2,23].

2   Preliminaries

Let $A \in \mathbb{Z}^{m \times n}$ and $b \in \mathbb{Z}^m$. Moreover, let $\mathrm{supp}(x)$ denote the support of x, i.e. $\mathrm{supp}(x) := \{i \in [n] : x_i \neq 0\}$ and set $\|x\|_0 := |\mathrm{supp}(x)|$. For the rest of the paper, $x^\star$ denotes an optimal solution for (P_0) for a given integer $\sigma \in \mathbb{Z}_{\geq 1}$. A natural convex relaxation of (P_0) is given by

$$\min_x \|Ax - b\|_2 \text{ subject to } x \in [0,1]^n \text{ and } \|x\|_1 \leq \sigma. \qquad (P_1)$$

An optimal solution to (P_1) will be denoted by $\hat{x}$ throughout the paper. When $m = 1$, there exists an optimal solution $\hat{x}$ for (P_1) that has at most one fractional variable (see Lemma 4). This solution is also feasible for (P_0), and hence optimal. The idea of our approach is to establish a proximity result for $A\hat{x}$ and $Ax^\star$ respectively, that we can exploit algorithmically. This proximity bound depends on m which comes as no surprise, given that the problem is NP-hard even for fixed values of m. The latter statement can be verified by reducing the NP-hard partition problem to an instance of (P_0).

Theorem 3. *The problem (P_0) is NP-hard, even if $m = 2$.*

A simple but important ingredient of our proximity theorem is the following fact that can be derived from elementary linear programming theory.

Lemma 4 (Few fractional entries).

1. *Let x be a feasible point for (P_0). There exists a solution x' such that $Ax = Ax'$ with at most m fractional entries.*
2. *Let x be a feasible point for (P_1). There exists a solution x' such that $Ax = Ax'$ with at most m fractional entries.*

Proof. 1. Let x be a solution of (P_0) and denote $S = \mathrm{supp}(x)$. Let A_S denote the submatrix of A comprising the columns with indices in S. The set

$$P_S(x) := \{y \in \mathbb{R}^{|S|} : \ Ax = A_S y, \ 0 \le y \le 1\}$$

is a polytope. It is non-empty since $x \in P$, hence it has at least one vertex v. By standard LP theory, at least $|S| - m$ inequalities of the form $0 \le y \le 1$ are tight at v. It follows that v has at most m fractional entries. The vertex v can easily be extended to a solution x' of (P_0) by adding zero-entries.

2. Given the solution x to (P_1), consider the optimization problem

$$\min\{\sum_{i=1}^{n} y_i : \ y \in P_{\{1,\dots,n\}}(x)\}.$$

Let v be an optimal vertex solution. From Part 1. v has at most m fractional entries. Since x is feasible for the above problem, $\sum_{i=1}^{n} v_i \le \sum_{i=1}^{n} x_i \le \sigma$. □

3 The ℓ_1-Relaxation for Random Targets b

In order to shed some light on Problem (P_0) and its natural convex relaxation (P_1) we first provide a probabilistic analysis to what extend optimal solutions of (P_1) already solve (P_0). Let $Q := \{Ax \in \mathbb{R}^m : \ x \in [0,1]^n, \|x\|_1 \le \sigma\}$ be the set of all points we can represent with the ℓ_1-relaxation. This section deals with the question which vectors b are "easy" target vectors. It turns out that if b is "deep" inside Q or far outside of Q, then the corresponding instances of (P_0) are easy with very high probability. In fact, there almost always exist optimal solutions of (P_1) that are already feasible for (P_0) and hence optimal. Conversely, if b is close to the boundary of Q, then the probability that an optimal solution of (P_1) solves (P_0) is almost 0.

Theorem 5. *Let $A \in \mathbb{R}^{m \times n}$ and $\sigma \ge m$ be an integer. If $b \in \frac{\sigma - m + 1}{\sigma} Q$, then there exists $x^\star \in [0,1]^n$ with $\|x^\star\|_0 \le \sigma$ and $Ax^\star = b$.*

Proof. If $b \in \frac{\sigma - m + 1}{\sigma} Q$, there exists a vector $\hat{x} \in [0, \frac{\sigma - m + 1}{\sigma}]^n$ such that $b = A\hat{x}$ and $\|\hat{x}\|_1 \leq \sigma - m + 1$. Let v be a vertex of $\{x \in [0,1]^n : Ax = b, \|x\|_1 \leq \sigma - m + 1\}$, which contains $\hat{x}$. According to the constraint $\|x\|_1 \leq \sigma - m + 1$, v has at most $\sigma - m + 1$ integral non-zero entries. By Lemma 4, v has at most m fractional entries. However, if there are fractional entries present, we can only have $\sigma - m$ integral entries, thus, $\|v\|_0 \leq \sigma$. □

Intuitively, the following theorem states that if b is sampled far away from Q, then (P_1) provides a solution to (P_0) as well. Here $B := \{x \in \mathbb{R}^m : \|x\|_2 \leq 1\}$ denotes the Euclidean unit ball.

Theorem 6. *Let $A \in \mathbb{R}^{m \times n}$, $\lambda \geq 0$ and $\sigma \geq 1$ be an integer. If b is sampled uniformly at random from the convex set $Q + \lambda B$, then with probability at least*

$$\left(\frac{\lambda}{\lambda + \sigma \sqrt{m} \|A\|_\infty} \right)^m$$

there exists $x \in \{0,1\}^n$ that is optimal for (P_1) and (P_0).

Proof. Define $P := \{x \in [0,1]^n : \|x\|_1 \leq \sigma\}$, hence we have $Q = \{Ax : x \in P\}$. Observe that all vertices of P are in $\{0,1\}^n$, and as a consequence any vertex v of Q can be written as

$$v = Ax \text{ with } x \text{ a vertex in } P \text{ that is integral.} \qquad (2)$$

Hence, whenever an optimal solution to $\min\{\|b - x\|_2 : x \in Q\}$ is attained by a vertex of Q, the problem (P_1) has an optimal integral vertex solution v. Since an integral solution to (P_1) is also feasible for (P_0), the vector v is also optimal for (P_0).

Let V be the vertex set of Q. For $v \in V$, denote the normal cone of v by

$$C_v := \{c \in \mathbb{R}^m : c^\mathsf{T}(w - v) \leq 0 \; \forall w \in Q\}.$$

Fix a vertex v and assume $b \in v + C_v$. We next show that v is an optimal solution to $\min\{\|b - x\|_2^2 : x \in Q\}$. Since $b = v + c$ with $c^\mathsf{T}(v - w) \geq 0$ for all $w \in Q$, we obtain

$$\|b - w\|_2^2 = \|v - w + c\|_2^2 = \|v - w\|_2^2 + \|c\|_2^2 + 2c^\mathsf{T}(v - w) \geq \|c\|_2^2 = \|b - v\|_2^2,$$

showing that v is optimal. By Eq. (2) there exists an integral $x \in P$ such that $v = Ax$ and hence x is optimal for (P_0). It remains to calculate the probability that $b \in v + C_v$ for some vertex v of Q. We obtain

$$\mathrm{vol}\left(\bigcup_{v \in V} (v + C_v) \cap (Q + \lambda B) \right) = \mathrm{vol}\left(\bigcup_{v \in V} C_v \cap \lambda B \right)$$

$$= \mathrm{vol}\left(\left(\bigcup_{v \in V} C_v \right) \cap \lambda B \right)$$

$$= \mathrm{vol}(\lambda B) = \lambda^m \, \mathrm{vol}(B).$$

In the second to last equality we used that the normal cones C_v tile the space $\mathbb{R}^m$.

Let $\mu > 0$ be a constant s.t. $Q \subseteq \mu B$, e.g. $\mu = \sigma\sqrt{m}\|A\|_\infty$. We have the containment $Q + \lambda B \subseteq \mu B + \lambda B = (\mu + \lambda)B$ that allows us to estimate $\mathrm{vol}(Q + \lambda B) \leq (\lambda + \mu)^m \, \mathrm{vol}(B)$. The probability that b is sampled in one of the normal cones is therefore

$$\frac{\mathrm{vol}(\lambda B)}{\mathrm{vol}(Q + \lambda B)} \geq \frac{\lambda^m}{(\lambda + \mu)^m} \geq \left(\frac{\lambda}{\lambda + \sigma\sqrt{m}\|A\|_\infty}\right)^m.$$

$\square$

Let us briefly comment on the probability quantity $\rho := (\lambda/(\lambda + \sigma\sqrt{m}\|A\|_\infty))^m$. If we choose $\lambda = 2m^{3/2}\sigma\|A\|_\infty$ in Theorem 6, then $\rho \geq 1/2$, as one can verify with Bernoulli's inequality. Figure 1 depicts the geometry underlying the proof of Theorem 6. The vector b_1 is sampled from the dotted area and hence, an optimal solution of (P_1) may use 2 fractional entries, and thus have support $\sigma + 1$. On the other hand, the vector b_2 is sampled from the dashed area, which leads to the solution of (P_1) corresponding to a vertex of Q. In the second case (P_1) has an integral solution, which automatically solves (P_0).

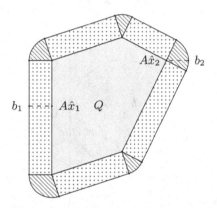

Fig. 1. The sampling of the vector b from $Q + \lambda B$

4   Proximity Between Optimal Solutions of (P_0) and (P_1)

In this section we illuminate the Problems (P_0) and (P_1) from a deterministic point of view and develop worst-case bounds for the distance of the images of corresponding optimal solutions under A. Our point of departure is an optimal solution $\hat{x}$ of (P_1). The target is to show that there exists an optimal solution $x^\star$ of (P_0) satisfying $\|A(\hat{x} - x^\star)\|_2 \leq 2m^{3/2}\|A\|_\infty$. Our strategy is to define a hyperplane containing $A\hat{x}$ in the space of target vectors b that separates b from all vectors Ax with x feasible for (P_0). The next step is to show that if we perturb $\hat{x}$ along the fractional variables, we will remain in this hyperplane. This has the side-effect that we can find a feasible solution for (P_0) whose image is in the vicinity of $A\hat{x}$. The triangle inequality and basic geometry then come into play to establish the claimed bound.

We introduce the hyperplane tangent to the ball $B(b, \|b - A\hat{x}\|_2)$ in $A\hat{x}$,

$$H := \{y \in \mathbb{R}^m : (b - A\hat{x})^{\mathsf{T}} y = (b - A\hat{x})^{\mathsf{T}} A\hat{x}\}.$$

Lemma 7. *We have $(b - A\hat{x})^{\mathsf{T}}(Ax - A\hat{x}) \leq 0$ for any point x feasible for (P_1).*

Proof. Assume that there exists a point x feasible for (P_1) for which the inequality $(b - A\hat{x})^{\mathsf{T}}(Ax - A\hat{x}) > 0$ holds. As a convex combination the point $p := A(\hat{x} + \varepsilon(x - \hat{x}))$ is feasible for (P_1) for each $\varepsilon \in [0, 1]$, and we can estimate the objective value as

$$\|b - p\|_2^2 = \|b - A\hat{x}\|_2^2 + \varepsilon^2 \|A(x - \hat{x})\|_2^2 - 2\varepsilon(b - A\hat{x})^{\mathsf{T}}(Ax - A\hat{x}) < \|b - A\hat{x}\|_2^2$$

for ε small enough. This contradicts the optimality of $\hat{x}$. We illustrate the argument geometrically in Fig. 2. □

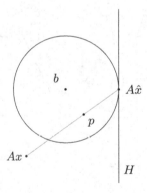

Fig. 2. If H does not separate b from Ax, there is a point p closer to b than $A\hat{x}$.

An important property is that H contains many points that we can easily generate from $\hat{x}$. This is made precise below. (Recall that A_i denotes the i-th column of A).

Lemma 8. *Define $\mathcal{F} := \{i \in [n] : \hat{x}_i \notin \mathbb{Z}\}$ where $\hat{x}$ is an optimal solution to (P_1). We have*

$$A\hat{x} + \left\{ \sum_{i \in \mathcal{F}} \lambda_i A_i : \sum_{i \in \mathcal{F}} \lambda_i = 0 \right\} \subseteq H.$$

Proof. Let $v := b - A\hat{x} \in \mathbb{R}^m$ be the normal vector of H and let

$$y = \sum_{i \in \mathcal{F}} \lambda_i e_i \quad \text{for some } \lambda_i \in \mathbb{R} \text{ with } \sum_{i \in \mathcal{F}} \lambda_i = 0.$$

Since $\hat{x}_i \in (0,1)$ for all $i \in \mathcal{F}$, there exists $\varepsilon > 0$ such that both points $\hat{x} + \varepsilon y$ and $\hat{x} - \varepsilon y$ are feasible for (P_1). By Lemma 7, we must have $v^\mathsf{T} A(\hat{x} + \varepsilon y - \hat{x}) = v^\mathsf{T} A \varepsilon y \leq 0$ and $-v^\mathsf{T} A \varepsilon y \leq 0$, resulting in $v^\mathsf{T} A y = 0$. Thus, $A(\hat{x} + y) \in H$. $\square$

With these results we are now able to show a proximity result (Theorem 2) between $A\hat{x}$ and $Ax^\star$. Here, $\lfloor \hat{x} \rfloor$ denotes the vector $\hat{x}$ rounded down component-wise.

Proof (Theorem 2). Given an optimal solution $\hat{x}$ of (P_1), let $\mathcal{F} = \{i \in [n] : \hat{x}_i \in (0,1)\}$. Without loss of generality, we may assume that $|\mathcal{F}| \leq m$ and $\mathcal{F} = \{1, 2, \ldots, |\mathcal{F}|\}$. Let $k := \sum_{i \in \mathcal{F}} \hat{x}_i$, and construct a feasible solution y for (P_0) from $\hat{x}$ as follows:

$$y_i := \begin{cases} 1, & 1 \leq i \leq \lfloor k \rfloor \\ k - \lfloor k \rfloor, & i = \lceil k \rceil \\ 0, & \lceil k \rceil + 1 \leq i \leq |\mathcal{F}| \\ \hat{x}_i, & i \notin \mathcal{F}. \end{cases}$$

The point y satisfies $0 \leq y_i \leq 1$ for all $i \in [n]$ and $\|y\|_0 = \lceil \|\hat{x}\|_1 \rceil \leq \sigma$. Since

$$\sum_{i \in \mathcal{F}} y_i = \sum_{i \in \mathcal{F}} \hat{x}_i, \tag{3}$$

Lemma 8 implies that $Ay \in H$ and hence

$$\|b - Ay\|_2^2 = \|b - A\hat{x}\|_2^2 + \|A\hat{x} - Ay\|_2^2.$$

Assume $\|Ax^\star - A\hat{x}\|_2 > \|Ay - A\hat{x}\|_2$ holds for some optimal solution $x^\star$ for (P_0). Since y is feasible for (P_0), we also know $\|b - Ay\|_2 \geq \|b - Ax^\star\|_2$. We are now prepared to estimate (using Lemma 7 in the third line)

$$\begin{aligned} \|b - Ax^\star\|_2^2 &= \|b - A\hat{x} + A\hat{x} - Ax^\star\|_2^2 \\ &= \|b - A\hat{x}\|_2^2 + \|A\hat{x} - Ax^\star\|_2^2 + 2(b - A\hat{x})^\mathsf{T}(A\hat{x} - Ax^\star) \\ &\geq \|b - A\hat{x}\|_2^2 + \|A\hat{x} - Ax^\star\|_2^2 \\ &> \|b - A\hat{x}\|_2^2 + \|A\hat{x} - Ay\|_2^2 \\ &= \|b - Ay\|_2^2, \end{aligned}$$

showing that $x^\star$ is not optimal. The proof is finished by observing that Eq. (3) also implies $\|\hat{x} - y\|_1 \leq 2\|\hat{x} - \lfloor \hat{x} \rfloor\|_1$, and consequently

$$\|Ay - A\hat{x}\|_2 \leq 2\|\hat{x} - \lfloor \hat{x} \rfloor\|_1 \max_{i=1,\ldots,n} \|A_i\|_2 \leq 2m^{3/2}\|A\|_\infty.$$

$\square$

5   A Deterministic Algorithm

The results presented so far give rise to a conceptually simple algorithm. Compute an optimal solution $\hat{x}$ to (P_1). According to the proximity Theorem 2, we can limit our search for an optimal right-hand side vector $b^\star = Ax^\star$ in the vicinity of $A\hat{x}$. Since $b^\star$ might be fractional, we cannot enumerate all possible right-hand sides. Instead, we refine our approach by decomposing $x^\star = z^\star + f^\star$ into its integral part $z^\star$ and its fractional part $f^\star$. We first guess the support $\mathcal{F}$ of the fractional entries, which satisfies $|\mathcal{F}| \leq m$ by Lemma 4. For the remaining variables, we next establish a candidate set $Z^\star$ comprising the potential vectors $z^\star$ in the decomposition of $x^\star$. It will be essential to determine a bound on $|Z^\star|$. This is where the proximity theorem comes into play. We now enumerate the elements of $Z^\star$ and extend each of them by a vector $f^\star$ whose support is in the index set $\mathcal{F}$ that we guessed upfront. A composition of these two solutions will provide $x^\star$.

This section is devoted to analyze this conceptually simple algorithm and this way shed some light on some of the details required.

Before we describe the decomposition $x^\star = z^\star + f^\star$ in more detail, we discuss the standard obstacle in convex optimization that $\hat{x}$ can only be approximated. To be more precise, we call a solution $\bar{x}$ to (P_1) ε-close, if

$$\|b - A\bar{x}\|_2^2 - \|b - A\hat{x}\|_2^2 \leq \varepsilon^2. \tag{4}$$

We obtain a canonical corollary from the proximity Theorem 2.

Corollary 9. *Let $\bar{x}$ be an ε-close solution of (P_1).*

1. Every optimal solution $x^\star$ of (P_0) satisfies

$$\|Ax^\star - A\bar{x}\|_\infty \leq 2\, m^{3/2} \|A\|_\infty + \varepsilon.$$

2. The integral part $z^\star$ of $x^\star$ satisfies

$$\|Az^\star - A\bar{x}\|_\infty \leq 3\, m^{3/2} \|A\|_\infty + \varepsilon.$$

Proof. We start by estimating the distance from $A\bar{x}$ to $A\hat{x}$ for an optimal solution $\hat{x}$ of (P_1). We have

$$\|b - A\bar{x}\|_2^2 = \|b - A\hat{x} + A\hat{x} - A\bar{x}\|_2^2$$
$$= \|b - A\hat{x}\|_2^2 + \|A\hat{x} - A\bar{x}\|_2^2 + 2(b - A\hat{x})^\mathsf{T}(A\hat{x} - A\bar{x}),$$

where the last term is non-negative by Lemma 7. Rearranging terms, we obtain

$$\|A\hat{x} - A\bar{x}\|_2^2 = \|b - A\bar{x}\|_2^2 - \|b - A\hat{x}\|_2^2 - 2(b - A\hat{x})^\mathsf{T}(A\hat{x} - A\bar{x}) \leq \varepsilon^2.$$

Applying the triangle inequality and combining the above estimate with Theorem 2, we have

$$\|Ax^\star - A\bar{x}\|_\infty \leq \|Ax^\star - A\bar{x}\|_2 \leq \|Ax^\star - A\hat{x}\|_2 + \|A\hat{x} - A\bar{x}\|_2$$
$$\leq 2\, m^{3/2} \|A\|_\infty + \varepsilon.$$

For Part 2, recall that $x^\star - z^\star = f^\star$ with $\|f^\star\|_0 \le m$, implying the inequality $\|A(x^\star - z^\star)\|_\infty \le m\|A\|_\infty$. We obtain

$$
\begin{aligned}
\|Az^\star - A\bar{x}\|_\infty &= \|Az^\star - Ax^\star + Ax^\star - A\bar{x}\|_\infty \\
&\le \|A(z^\star - x^\star)\|_\infty + \|Ax^\star - A\bar{x}\|_\infty \\
&\le m\|A\|_\infty + 2\,m^{3/2}\|A\|_\infty + \varepsilon \le 3\,m^{3/2}\|A\|_\infty + \varepsilon.
\end{aligned}
$$

$\square$

We next outline the decomposition $x^\star = z^\star + f^\star$. As a first step, we guess the support of a minimal index set of fractional entries.

Lemma 10. *There are $(2\|A\|_\infty + 1)^{m^2}$ potentially different index sets $\mathrm{supp}(f^\star)$.*

Proof. We notice that a minimal index set of fractional entries uses distinct columns from the matrix A. There are at most $(2\|A\|_\infty + 1)^m$ distinct columns of A. Since the cardinality of a minimal index set of fractional entries is bounded by m, there are at most $((2\|A\|_\infty + 1)^m)^m = (2\|A\|_\infty + 1)^{m^2}$ potentially different index sets. $\square$

A canonical approach would be to search for the vector $f^\star$. Then we run into the problem that our objective is nonlinear, and hence $f^\star$ depends on $z^\star$. This requires us to first search for an optimal $z^\star$ and then use continuous optimization techniques to compute $f^\star$. In order to avoid to determine a minimal index set of fractional entries, we also allow entries with index in $\mathcal{F}$ to be integral. Then we need to guess only sets $\mathcal{F} \subseteq [n]$ with $|\mathcal{F}| = m$.

We denote by $A_{\backslash f^\star}$ the matrix A without the columns with index in $\mathrm{supp}(f^\star)$. The next theorem shows that we can compute a small set $Z^\star$ of possible vectors for $z^\star$.

Theorem 11. *Let $\bar{x}$ be an ε-close solution to (P_1). If $\mathrm{supp}(f^\star)$ is fixed, we can compute a set $Z^\star \subseteq \{0,1\}^n$ of candidate vectors such that $x^\star = z^\star + f^\star$ with $z^\star \in Z^\star$. This requires us to solve at most $(6m^{3/2}\|A\|_\infty + 2\varepsilon + 1)^m$ linear integer programming problems.*

Proof. We have $Az^\star \in A\bar{x} + [-D_\varepsilon, D_\varepsilon]^m \cap \mathbb{Z}^m$, where $D_\varepsilon = 3m^{3/2}\|A\|_\infty + \varepsilon$ by Corollary 9. For every $b^\star \in A\bar{x} + [-D_\varepsilon, D_\varepsilon]^m \cap \mathbb{Z}^m$ we solve the integer feasibility problem

$$
A_{\backslash f^\star} y = b^\star, \quad \sum_{i=1}^{n-m} y_i \le \sigma - m, \quad y \in \{0,1\}^{n-m}.
$$

If it has a feasible solution y, we can insert zero entries according to $\mathrm{supp}(f^\star)$ and obtain a vector $z \in \{0,1\}^n$ that qualifies as the vector $z^\star$. The set $Z^\star$ is the set of all extended vectors z. $\square$

It remains to compose each $z^\star \in Z^\star$ with a vector $f^\star$. This is accomplished by solving a series of least-square problems. The reason why we proceed in this way is that it allows us to compute the exact vector $f^\star$ as opposed to an ε-close solution.

Lemma 12 (Extension lemma). *For each $z \in Z^\star$ an optimal solution f to* $\min\{\|b - Az - Af\|_2 : \text{supp}(f) \subseteq \text{supp}(f^\star), 0 \leq f \leq 1\}$ *can be computed in* $\mathcal{O}(3^m m^3)$ *arithmetic operations.*

Proof. As $f_i = 0$ for $i \notin \text{supp}(f^\star)$, we can restrict to the matrix $A_{f^\star} \in \mathbb{Z}^{m \times m}$ and solve the equivalent problem $\min\{\|b' - A_{f^\star}g\|_2 : g \in [0,1]^m\}$ for $b' := b - Az$. Without the variable bounds this is a least-square problem that can be solved in $\mathcal{O}(m^3)$ arithmetic operations. Let $g^\star$ be an optimal solution. We guess the sets $S_0 := \{i : g_i^\star = 0\}$ and $S_1 := \{i : g_i^\star = 1\}$, and afterwards solve the modified least-square problem $\min\{\|b - A_{f^\star}g\|_2 : g_i = 0 \; \forall i \in S_0, g_i = 1 \; \forall i \in S_1\}$. If the solution g is in $[0,1]^n$, its extension $f \in [0,1]^n$ qualifies as $f^\star$. In the end, we pick the best among all feasible extensions. As there are 3^m guesses, this finishes the proof. $\qquad\square$

This completes the presentation of the main steps to prove Theorem 1. In fact, in order to obtain an optimal solution to (P_0) one proceeds as follows. We first guess the set $\text{supp}(f^\star)$, determine the set $Z^\star$ and compute for every $z^\star \in Z^\star$ an optimal vector $f^\star$. The best of all those solutions solves (P_0). As a last technicality, we have to show how to find an ε-close solution $\bar{x}$ for which we fall back on [25, Chap. 8].

Lemma 13 ([25, Chap. 8]). *We can find a $\sqrt{m}\|A\|_\infty$-close solution for (P_1) in $\mathcal{O}\left(n^{7/2} \ln\left(n^2\sigma\|b\|_1\right)\right)$ arithmetic operations.*

Proof. We apply the results presented in [25, Chap. 8] that depend on several parameters. Let $P := \{x \in [0,1]^n : \|x\|_1 \leq \sigma\}$ denote the feasible region of (P_1) and $\hat{x}$ an optimal solution. We first need to estimate

$$\mathcal{D} := \max\{\|b - Ay\|_2^2 - \|b - A\hat{x}\|_2^2 : y \in P\}.$$

For any $y \in P$ we can estimate

$$\begin{aligned}\|b - Ay\|_2^2 - \|b - A\hat{x}\|_2^2 &= \|Ay\|_2^2 - \|A\hat{x}\|_2^2 + 2b^\mathsf{T}A(\hat{x} - y) \\ &\leq \sigma^2 m\|A\|_\infty^2 + 4\|b\|_1\sigma\|A\|_\infty \\ &\leq 4\sigma^2 m\|A\|_\infty^2(\|b\|_1 + 1),\end{aligned}$$

resulting in $\mathcal{D} \leq 4\sigma^2 m\|A\|_\infty^2(\|b\|_1 + 1)$. As the initial point in the interior of P that is required in [25, Chap. 8] we choose $w := \frac{\sigma}{n+\sigma} \cdot \mathbf{1}$ where $\mathbf{1}$ denotes the all-ones vector. Next we estimate the *asymmetry coefficient*

$$\alpha(P : w) := \max\{t : w + t(w - P) \subseteq P\}.$$

Since $[0, \frac{\sigma}{n}]^n \subseteq P \subseteq [0,1]^n$, for $t = \frac{\sigma}{n}$ we obtain

$$w + t(w - P) \subseteq w + t(w - [0,1]^n) = \left[0, \tfrac{\sigma}{n}\right]^n \subseteq P,$$

thus $\alpha(P : w) \geq \frac{\sigma}{n}$. By [25, Chap. 8, Eq. 8.1.5] we can compute a feasible solution $\bar{x}$ of (P_1) satisfying $\|b - A\bar{x}\|_2^2 - \|b - A\hat{x}\|_2^2 \leq \delta\mathcal{D}$ in $O(1)(2n+1)^{1.5}n^2 \ln\left(\frac{2n+1}{\alpha(P:w)\delta}\right)$ arithmetic operations. Finally, by choosing $\delta = \frac{1}{4\sigma^2(\|b\|_1+1)}$ finding a $\sqrt{m}\|A\|_\infty$-close solution takes $\mathcal{O}\left(n^{7/2} \ln\left(n^2\sigma\|b\|_1\right)\right)$ arithmetic operations. $\qquad\square$

6 Extension

A natural generalization of our problem is to consider arbitrary upper bounds $u_i > 0$, i.e.

$$\min_x \|Ax - b\|_2 \text{ subject to } \|x\|_0 \leq \sigma \text{ and } 0 \leq x_i \leq u_i \text{ for all } i \in [n]. \quad (P_0')$$

The natural convex relaxation of (P_0') is given by:

$$\min_x \|Ax - b\|_2 \text{ subject to } \sum_{i=1}^{n} \frac{x_i}{u_i} \leq \sigma \text{ and } 0 \leq x_i \leq u_i \text{ for all } i \in [n]. \quad (P_1')$$

The results of Sects. 4 and 5 extend to this generalization in a straight-forward manner. For the algorithm it implies that the number of arithmetic operations increases by an additional factor of $\|u\|_\infty^m$. The reason is the core of our approach: The proximity bound between optimal solutions for (P_0') and (P_1') respectively, increases by this factor. The proximity bound must however depend on $\|u\|_\infty$ as the following example shows:

Let n and u be even, non-negative integers. Set $A := \mathbb{1}, \sigma := \frac{n}{2}$ and $b = \frac{u}{2}\mathbb{1}$ where $\mathbb{1}$ denotes the all-ones vector. It can easily be checked that $\hat{x} = \frac{u}{2}\mathbb{1}$ is optimal for (P_1') while

$$x_i^\star = \begin{cases} \frac{u}{2}, i \in [\sigma] \\ 0, i \in [n] \setminus [\sigma] \end{cases}$$

is optimal for (P_0'). This shows that any approach aiming for a logarithmic dependency on $\|u\|_\infty$ requires techniques that are different from the ideas presented in this paper.

Acknowledgements. The second and third author acknowledge support by the Einstein Foundation Berlin.

References

1. Ament, S., Gomes, C.: On the optimality of backward regression: Sparse recovery and subset selection. In: ICASSP 2021–2021 IEEE International Conference on Acoustics, Speech and Signal Processing (ICASSP), June 2021. https://doi.org/10.1109/icassp39728.2021.9415082
2. Beale, E.M.L., Kendall, M.G., Mann, D.W.: The discarding of variables in multivariate analysis. Biometrika **54**(3–4), 357–366 (1967). https://doi.org/10.1093/biomet/54.3-4.357
3. Candes, E., Romberg, J., Tao, T.: Robust uncertainty principles: exact signal reconstruction from highly incomplete frequency information. IEEE Trans. Inf. Theory **52**(2), 489–509 (2006). https://doi.org/10.1109/TIT.2005.862083

4. Candes, E., Tao, T.: Decoding by linear programming. IEEE Trans. Inf. Theory **51**(12), 4203–4215 (2005). https://doi.org/10.1109/TIT.2005.858979

5. Candes, E., Tao, T.: The Dantzig selector: statistical estimation when p is much larger than n. Ann. Stat. **35**(6), 2313–2351 (2007). https://doi.org/10.1214/009053606000001523

6. Candes, E.J., Romberg, J.K., Tao, T.: Stable signal recovery from incomplete and inaccurate measurements. Commun. Pure Appl. Math. **59**(8), 1207–1223 (2006). https://doi.org/10.1002/cpa.20124

7. Chen, S.S., Donoho, D.L., Saunders, M.A.: Atomic decomposition by basis pursuit. SIAM Rev. **43**(1), 129–159 (2001). https://doi.org/10.1137/S1064827596304010

8. Couvreur, C., Bresler, Y.: On the optimality of the backward greedy algorithm for the subset selection problem. SIAM J. Matrix Anal. Appl. **21**(3), 797–808 (2000). https://doi.org/10.1137/S0895479898332928

9. Das, A., Kempe, D.: Algorithms for subset selection in linear regression. In: Proceedings of the Fortieth Annual ACM Symposium on Theory of Computing. STOC 2008, pp. 45–54. Association for Computing Machinery, New York (2008). https://doi.org/10.1145/1374376.1374384

10. Das, A., Kempe, D.: Submodular meets spectral: greedy algorithms for subset selection, sparse approximation and dictionary selection. In: Proceedings of the 28th International Conference on International Conference on Machine Learning. ICML 2011, pp. 1057–1064. Omnipress, Madison (2011). https://doi.org/10.5555/3104482.3104615

11. Del Pia, A., Dey, S.S., Weismantel, R.: Subset selection in sparse matrices. SIAM J. Optim. **30**(2), 1173–1190 (2020). https://doi.org/10.1137/18M1219266

12. Di Lorenzo, D., Liuzzi, G., Rinaldi, F., Schoen, F., Sciandrone, M.: A concave optimization-based approach for sparse portfolio selection. Optim. Methods Softw. **27**(6), 983–1000 (2012). https://doi.org/10.1080/10556788.2011.577773

13. Donoho, D.: Compressed sensing. IEEE Trans. Inf. Theory **52**(4), 1289–1306 (2006). https://doi.org/10.1109/TIT.2006.871582

14. Elenberg, E.R., Khanna, R., Dimakis, A.G., Negahban, S.: Restricted strong convexity implies weak submodularity. Ann. Stat. **46**(6B), 3539–3568 (2018). https://doi.org/10.1214/17-AOS1679

15. Feng, M., Mitchell, J.J., Pang, J.S., Shen, X., Waechter, A.: Complementarity formulations of ℓ_0-norm optimization. Pac. J. Optim. **14**(2), 273–305 (2018)

16. Fung, G.M., Mangasarian, O.L.: Equivalence of minimal ℓ_0- and ℓ_1-norm solutions of linear equalities, inequalities and linear programs for sufficiently small p. J. Optim. Theory Appl. **151**(1), 1–10 (2011). https://doi.org/10.1007/s10957-011-9871-x

17. Gamarnik, D., Zadik, I.: High dimensional regression with binary coefficients. estimating squared error and a phase transition. In: Proceedings of the 2017 Conference on Learning Theory. Proceedings of Machine Learning Research, vol. 65, pp. 948–953. PMLR, 07–10 July 2017

18. Gao, J., Li, D.: A polynomial case of the cardinality-constrained quadratic optimization problem. J. Glob. Optim. **56**(4), 1441–1455 (2013). https://doi.org/10.1007/s10898-012-9853-z

19. Ge, D., Jiang, X., Ye, Y.: A note on the complexity of LP minimization. Math. Program. **129**, 285–299 (2011). https://doi.org/10.1007/s10107-011-0470-2

20. Gilbert, A., Indyk, P.: Sparse recovery using sparse matrices. Proc. IEEE **98**(6), 937–947 (2010). https://doi.org/10.1109/JPROC.2010.2045092

21. Gilbert, A.C., Muthukrishnan, S., Strauss, M.J.: Approximation of functions over redundant dictionaries using coherence. In: SODA, pp. 243–252. Citeseer (2003). https://doi.org/10.5555/644108.644149
22. Migot, T., Haddou, M.: A smoothing method for sparse optimization over polyhedral sets. In: Le Thi, H.A., Pham Dinh, T., Nguyen, N.T. (eds.) Modelling, Computation and Optimization in Information Systems and Management Sciences. AISC, vol. 359, pp. 369–379. Springer, Cham (2015). https://doi.org/10.1007/978-3-319-18161-5_31
23. Hocking, R.R., Leslie, R.N.: Selection of the best subset in regression analysis. Technometrics 9(4), 531–540 (1967). https://doi.org/10.1080/00401706.1967.10490502
24. Mangasarian, O.: Minimum-support solutions of polyhedral concave programs. Optimization 45(1–4), 149–162 (1999). https://doi.org/10.1080/02331939908844431
25. Nesterov, Y., Nemirovski, A.: Interior-point polynomial algorithms in convex programming. In: SIAM Studies in Applied Mathematics (1994)
26. Nguyen, T.: Dropping forward-backward algorithms for feature selection. CoRR abs/1910.08007 (2019)
27. Oymak, S., Thrampoulidis, C., Hassibi, B.: The squared-error of generalized lasso: a precise analysis. In: 2013 51st Annual Allerton Conference on Communication, Control, and Computing (Allerton), pp. 1002–1009 (2013). https://doi.org/10.1109/Allerton.2013.6736635
28. Qian, C., Yu, Y., Zhou, Z.H.: Subset Selection by Pareto Optimization. NIPS 2015, pp. 1774–1782. MIT Press, Cambridge (2015). https://doi.org/10.5555/2969239.2969437
29. Rinaldi, F.: Concave programming for finding sparse solutions to problems with convex constraints. Optim. Methods Softw. 26(6), 971–992 (2011). https://doi.org/10.1080/10556788.2010.511668
30. Rinaldi, F., Schoen, F.: Concave programming for minimizing the zero-norm over polyhedral sets. Comput. Optim. Appl. 46, 467–486 (07 2010). https://doi.org/10.1007/s10589-008-9202-9
31. Teng, Y., Qi, S., Xiao, D., Xu, L., Li, J., Kang, Y.: A general solution to least squares problems with box constraints and its applications. Math. Probl. Eng. 2016 (2016)
32. Tropp, J.: Greed is good: algorithmic results for sparse approximation. IEEE Trans. Inf. Theory 50(10), 2231–2242 (2004). https://doi.org/10.1109/TIT.2004.834793
33. Wainwright, M.J.: Sharp thresholds for high-dimensional and noisy sparsity recovery using ℓ_1-constrained quadratic programming (lasso). IEEE Trans. Inf. Theor. 55(5), 2183–2202 (2009). https://doi.org/10.1109/TIT.2009.2016018
34. Zhu, J., Wen, C., Zhu, J., Zhang, H., Wang, X.: A polynomial algorithm for best-subset selection problem. Proc. Natl. Acad. Sci. 117(52), 33117–33123 (2020). https://doi.org/10.1073/pnas.2014241117
35. Çivril, A.: A note on the hardness of sparse approximation. Inf. Process. Lett. 113(14), 543–545 (2013). https://doi.org/10.1016/j.ipl.2013.04.014

Recycling Inequalities for Robust Combinatorial Optimization with Budget Uncertainty

Christina Büsing[1] , Timo Gersing[1]([✉]) , and Arie M.C.A. Koster[2]

[1] Combinatorial Optimization, RWTH Aachen University, Aachen, Germany
{buesing,gersing}@combi.rwth-aachen.de
[2] Discrete Optimization, RWTH Aachen University, Aachen, Germany
koster@math2.rwth-aachen.de

Abstract. Robust combinatorial optimization with budget uncertainty is one of the most popular approaches for integrating uncertainty in optimization problems. The existence of a compact reformulation for (mixed-integer) linear programs and positive complexity results give the impression that these problems are relatively easy to solve. However, the practical performance of the reformulation is actually quite poor when solving robust integer problems due to its weak linear relaxation.

To overcome the problems arising from the weak formulation, we propose a procedure to derive new classes of valid inequalities for robust binary optimization problems. For this, we recycle valid inequalities of the underlying deterministic problem such that the additional variables from the robust formulation are incorporated. The valid inequalities to be recycled may either be readily available model constraints or actual cutting planes, where we can benefit from decades of research on valid inequalities for classical optimization problems.

We first demonstrate the strength of the inequalities theoretically, by proving that recycling yields a facet-defining inequality in surprisingly many cases, even if the original valid inequality was not facet-defining. Afterwards, we show in a computational study that using recycled inequalities leads to a significant improvement of the computation time when solving robust optimization problems.

Keywords: Robust Optimization · Combinatorial Optimization · Integer Programming · Polyhedral Combinatorics

1 Introduction

Robust optimization is a widely used approach for integrating uncertainties into optimization models. The concept of budgeted uncertainty by Bertsimas and Sim [6] has received particular attention. However, despite its popularity and the amount of research devoted to solving these kind of robust optimization problems, instances of practical size often still pose a considerable challenge for

A. Del Pia and V. Kaibel (Eds.): IPCO 2023, LNCS 13904, pp. 58–71, 2023.
https://doi.org/10.1007/978-3-031-32726-1_5

MILP solvers [7]. In this context, we propose a new class of valid inequalities for robust combinatorial optimization problems that are easy to compute and can lead to a significant reduction of the computation time.

Without uncertainties, the so called *nominal* combinatorial optimization problem NOM is defined as $\min\{\sum_{i\in[n]} c_i x_i | Ax \leq b, x \in \{0,1\}^n\}$, with $c \in \mathbb{R}^n$, $A \in \mathbb{R}^{m\times n}$, and $b \in \mathbb{R}^m$. Here, $[n] = \{1,\ldots,n\}$. In the case of uncertainty in the objective, the coefficients c_i are replaced by uncertain coefficients c_i' from an interval $[c_i, c_i + \hat{c}_i]$. We say that c_i' can *deviate* from its *nominal value* c_i by up to the *deviation* $\hat{c}_i$. Since the worst-case, in which all coefficients c_i' deviate to $c_i + \hat{c}_i$, is unlikely, Bertsimas and Sim [6] define an uncertainty budget $\Gamma \in [0, n]$ and only consider *scenarios* where at most $\lfloor \Gamma \rfloor$ coefficients c_i' deviate to $c_i + \hat{c}_i$ and one coefficient may deviate to $c_i + (\Gamma - \lfloor \Gamma \rfloor)\hat{c}_i$. The robust counterpart, in which we optimize against the worst-case, can be stated as

$$\min \sum_{i\in[n]} c_i x_i + \max_{\substack{S\cup\{t\}\subseteq[n]:\\ |S|\leq\lfloor\Gamma\rfloor, t\notin S}} \left((\Gamma - \lfloor\Gamma\rfloor)\hat{c}_t x_t + \sum_{i\in S} \hat{c}_i x_i \right)$$

$$\text{s.t. } Ax \leq b, x \in \{0,1\}^n.$$

Dualizing the inner maximization problem [6] yields the compact robust problem

$$\text{ROB} \qquad \min \Gamma z + \sum_{i\in[n]} (c_i x_i + p_i)$$

$$\text{s.t. } (x, p, z) \in \mathcal{P}^{\text{ROB}}, x \in \{0,1\}^n$$

with

$$\mathcal{P}^{\text{ROB}} = \left\{ (x, p, z) \middle| \begin{array}{ll} Ax \leq b \\ p_i + z \geq \hat{c}_i x_i & \forall i \in [n] \\ x \in [0,1]^n, p \in \mathbb{R}^n_{\geq 0}, z \in \mathbb{R}_{\geq 0} \end{array} \right\}.$$

Unfortunately, the formulation $\mathcal{P}^{\text{ROB}}$ is quite weak, often leading to much higher computation times for solving ROB compared to NOM. In fact, the relative integrality gap of the formulation $\mathcal{P}^{\text{ROB}}$ may be arbitrarily large, even if the integrality gap of the corresponding nominal problem is zero. This is shown in the following example from [7].

Example 1. Consider the easy problem of selecting the cheapest of n elements $\min\left\{\sum_{i\in[n]} c_i x_i \middle| \sum_{i\in[n]} x_i = 1, x \in \{0,1\}^n\right\}$. The integrality gap is zero for all $c \in \mathbb{R}^n$. However, if we consider an instance of the uncertain counterpart ROB with $c \equiv 0$, $\hat{c} \equiv 1$, and $\Gamma = 1$

$$\min\left\{ z + \sum_{i\in[n]} p_i \middle| \begin{array}{ll} \sum_{i\in[n]} x_i = 1 \\ p_i + z \geq x_i & \forall i \in [n] \\ x \in \{0,1\}^n, p \in \mathbb{R}^n_{\geq 0}, z \in \mathbb{R}_{\geq 0} \end{array} \right\}$$

then $(x, p, z) = \left(\frac{1}{n}, \ldots, \frac{1}{n}, 0, \ldots, 0, \frac{1}{n}\right)$ is the unique optimal fractional solution of value $\frac{1}{n}$, while the objective value of an optimal integer solution is 1. Hence, the integrality gap is $\frac{1}{1/n} = n$, and thus unbounded.

The above example shows that optimal continuous solutions for ROB tend to be highly fractional, as small values of x_i allow for covering all right-hand sides $\hat{c}_i x_i$ in the constraints $p_i + z \geq \hat{c}_i x_i$ with a small value of z, while choosing $p \equiv 0$. On the one hand, such solutions are exactly what we aim for when striving for robustness, as we distribute the risk as much as possible. On the other hand, highly fractional optimal solutions for the linear relaxation imply the need for much branching, and thus a high computational effort when solving ROB.

Bertsimas et al. [4] as well as Fischetti and Monaci [9] tested the practical performance of the compact reformulation $\mathcal{P}^{\mathrm{ROB}}$ compared to a separation approach using an alternative formulation with exponentially many inequalities, each one modeling a scenario from the uncertainty set. Unfortunately, the alternative formulation is, despite its size, as weak as $\mathcal{P}^{\mathrm{ROB}}$ and performs worse for robust integer problems (but better for continuous problems) [4,9]. Joung and Park [16] propose cuts that dominate the classic scenario inequalities and can be separated by considering the robustness term as a submodular function and greedily solving a maximization problem over the corresponding polymatroid. Atamtürk [3] addresses this issue by proposing four different strong formulations. The strongest of these preserves the integrality gap of the nominal problem, but all four formulations are very large and hence, are outperformed by $\mathcal{P}^{\mathrm{ROB}}$ [7].

The weak relaxation can be avoided by tentatively fixing the variable z to different values, resulting in a series of nominal problems NOM to be solved instead [2,5,19]. Lee and Kwon [17] showed that at most $\left\lceil \frac{n-\Gamma}{2} \right\rceil + 1$ nominal problems have to be solved. However, the computational effort is usually still higher compared to solving ROB directly if n is large [7]. Hansknecht et al. [14] improve on this with their divide & conquer approach, in which one prunes many non-optimal values for z. In [7], non-optimal values for z are pruned even more efficiently by exploiting structural insights and strong linearizations derived from the following bilinear formulation

$$\mathcal{P}^{\mathrm{BIL}} = \left\{ (x, p, z) \left| \begin{array}{ll} Ax \leq b \\ p_i + x_i z \geq \hat{c}_i x_i & \forall i \in [n] \\ x \in [0, 1]^n, p \in \mathbb{R}_{\geq 0}^n, z \in \mathbb{R}_{\geq 0} \end{array} \right. \right\}.$$

This bilinear formulation strengthens the robustness constraints $p_i + z \geq \hat{c}_i x_i$ by multiplying z with x_i, which is valid due to $x_i \in \{0, 1\}$. While the bilinearity is rather hindering for practical purposes, $\mathcal{P}^{\mathrm{BIL}}$ is theoretically very strong. In fact, there exists no polyhedral formulation $\mathcal{P}$ for ROB with $\mathcal{P} \subsetneq \mathcal{P}^{\mathrm{BIL}}$.

Contribution. In this paper, we use the bilinear formulation $\mathcal{P}^{\mathrm{BIL}}$ as a foundation for the new class of *recycled inequalities*. To obtain these, we combine the strength of the bilinear inequalities with the structural properties provided by inequalities for the nominal problem NOM. By doing so, we can use inequalities for NOM a second time to improve the formulation $\mathcal{P}^{\text{ROB}}$. We show that in many cases they even define facets of the convex hull of integer-feasible solutions

$$\mathcal{C}^{\text{ROB}} = \text{conv}\left(\left\{(x,p,z) \in \mathcal{P}^{\text{ROB}} \middle| x \in \{0,1\}^n\right\}\right).$$

A preliminary computational study reveals that separating recycled inequalities can lead to a drastic improvement of both integrality gap and solving times. First experiments with adapted MIPLIB [12] instances (to be presented in the journal version) confirm these results for a broad set of robust problems.

All implemented algorithms and generated test instances are published together with a package of algorithms for solving robust combinatorial optimization problems [10] and benchmark instances [11] for those very problems.

Outline. In Sect. 2, we show how to derive recycled inequalities from valid inequalities for NOM. In Sect. 3, we characterize valid inequalities for which the respective recycled inequality is facet-defining. We also provide examples indicating that this applies for many well-known valid inequalities for classical optimization problems. In Sect. 4, we test recycled inequalities in a computational study, highlighting their practical value.

2   Recycling Valid Inequalities

As already mentioned, the bilinear inequalities $p_i + x_i z \geq \hat{c}_i x_i$ play a crucial role for our recycled inequalities. To understand their strength intuitively, we recall our observations from Example 1. There, we noticed that choosing fractional values for x_i is tempting, as we are then able to meet the inequalities $p_i + z \geq \hat{c}_i x_i$ with a small value of z and $p \equiv 0$. However, this advantage vanishes for the bilinear inequalities $p_i + x_i z \geq \hat{c}_i x_i$, as we always have $z \geq \hat{c}_i$ for $x_i \neq 0$ and $p_i = 0$. To make use of this in practice, it would be beneficial to carry over the strength of the bilinear inequalities to a linear formulation.

Multiplying linear inequalities with variables as an intermediate step in order to achieve a stronger linear formulation is not a new approach. For the Reformulation-Linearization-Technique by Sherali and Adams [20], one multiplies constraints with variables and linearizes the resulting products afterwards via substitution with auxiliary variables. Our approach is different in the sense that we don't directly linearize the bilinear inequalities, and thus don't create auxiliary variables. Instead, we combine several of the bilinear inequalities in order to estimate the non-linear terms against a linear term, using a valid inequality for the corresponding nominal problem. From now on, let

$$\mathcal{C}^{\text{NOM}} = \text{conv}\left(\{x \in \{0,1\}^n | Ax \leq b\}\right)$$

be the convex hull of all integer nominal solutions. Then we combine the bilinear inequalities and valid inequalities for $\mathcal{C}^{\text{NOM}}$ as follows.

Theorem 1. *Let $\sum_{i\in[n]} \pi_i x_i \leq \pi_0$ be a valid inequality for $\mathcal{C}^{NOM}$ with $\pi \geq 0$. Then the inequality*

$$\pi_0 z + \sum_{i\in[n]} \pi_i p_i \geq \sum_{i\in[n]} \pi_i \hat{c}_i x_i \tag{1}$$

is valid for $\mathcal{C}^{ROB}$.

Proof. Summing the bilinear constraints $p_i + x_i z \geq \hat{c}_i x_i$, each with a factor of π_i, we obtain

$$\sum_{i\in[n]} \pi_i p_i + \sum_{i\in[n]} \pi_i x_i z \geq \sum_{i\in[n]} \pi_i \hat{c}_i x_i,$$

which is a valid inequality for $\mathcal{C}^{ROB}$ due to $\pi \geq 0$. Now, since $z \geq 0$ holds, we have $\sum_{i\in[n]} \pi_i x_i z \leq \pi_0 z$, which proves the statement. □

As we reuse the valid inequality $\sum_{i\in[n]} \pi_i x_i \leq \pi_0$ to strengthen the formulation $\mathcal{P}^{ROB}$, we call (1) the *recycled inequality* of $\sum_{i\in[n]} \pi_i x_i \leq \pi_0$. In accordance with the requirements of Theorem 1, we call $\sum_{i\in[n]} \pi_i x_i \leq \pi_0$ *recyclable* if it is valid for $\mathcal{C}^{NOM}$ and $\pi \geq 0$.

Note that we could also derive the concept of recycled inequalities on the basis of the even stronger bilinear inequalities $x_i(p_i + z) \geq \hat{c}_i x_i$, resulting from multiplying both p_i and z with x_i. However, after summing the bilinear inequalities with factors π_i, this would yield the term $\sum_{i\in[n]} \pi_i x_i p_i$, which we can only estimate against $\sum_{i\in[n]} \pi_i p_i$, yielding the same result as above.

To get a better understanding for recycled inequalities, let us recognize how they compare to the bilinear inequalities over the course of their construction. First, note that the sum of the bilinear inequalities is weaker than the bilinear inequalities themselves. Hence, when separating a recycled inequality to cut-off a fractional solution $(\tilde{x}, \tilde{p}, \tilde{z}) \in \mathcal{P}^{NOM}$, the inequality to be recycled should only support indices $i \in [n]$ with $\pi_i > 0$ for which the bilinear inequality $\tilde{p}_i + \tilde{x}_i \tilde{z} \geq \hat{c}_i \tilde{x}_i$ is violated or tight. A second potential weakening occurs when applying the estimation $\sum_{i\in[n]} \pi_i x_i z \leq \pi_0 z$. This implies that recycling $\sum_{i\in[n]} \pi_i x_i \leq \pi_0$ is especially interesting if it is binding for $\tilde{x}$.

Revisit Example 1, where we can recycle the valid inequality $\sum_{i\in[n]} x_i \leq 1$ implied by the constraint $\sum_{i\in[n]} x_i = 1$. The corresponding recycled inequality $z + \sum_{i\in[n]} p_i \geq \sum_{i\in[n]} x_i$ yields $z + \sum_{i\in[n]} p_i \geq 1$, and thus the optimal objective value of the linear relaxation is now equal to the optimal integer objective value. This intuitively highlights the strength of the recycled inequalities in the case where both properties, a binding recyclable valid inequality and the violation of supported bilinear inequalities, coincide.

3   Facet-Defining Recycled Inequalities

In this section, we show that recycled inequalities often define facets of the convex hull of the robust problem $\mathcal{C}^{ROB}$. To this end, we first determine the dimension of $\mathcal{C}^{ROB}$ and assume for the sake of simplicity that the sets of solutions to our problems are non-empty.

Lemma 1. *We have* $\dim\left(\mathcal{C}^{ROB}\right) = \dim\left(\mathcal{C}^{NOM}\right) + n + 1$.

Proof. For a polytope $P \subseteq \mathbb{R}^n$, the number $n - \dim(P)$ equals the maximum number of linearly independent equations that are met by all $x \in P$. Let $\sum_{i\in[n]}\left(\omega_i x_i + \omega_{n+i}p_i\right) + \omega_{2n+1}z = \omega_0$ be an equation that is satisfied by all $(x, p, z) \in \mathcal{C}^{ROB}$. Since p and z can be raised arbitrarily and $\mathcal{C}^{ROB} \neq \emptyset$, we have $\omega_{n+1} = \cdots = \omega_{2n+1} = 0$ and thus $\sum_{i\in[n]}\omega_i x_i = \omega_0$. Hence, the equations that are met by all $(x, p, z) \in \mathcal{C}^{ROB}$ are exactly the equations that are met by all $x \in \mathcal{C}^{NOM}$, which implies

$$\dim\left(\mathcal{C}^{ROB}\right) = 2n + 1 - \left(n - \dim\left(\mathcal{C}^{NOM}\right)\right) = \dim\left(\mathcal{C}^{NOM}\right) + n + 1.$$

$\square$

Knowing the dimension of $\mathcal{C}^{ROB}$, we are now able to study facet-defining recycled inequalities. For this, we only consider inequalities $\sum_{i\in[n]}\pi_i x_i \leq \pi_0$ consisting of variables with uncertain objective coefficients, i.e., we have $\pi_i = 0$ for all $i \in [n]$ with $\hat{c}_i = 0$. We call inequalities with this property *uncertainty-exclusive inequalities*. Note that these are the only interesting inequalities for recycling, because we can always drop variables x_i with $\hat{c}_i = 0$. This strengthens the corresponding recycled inequality by removing $\pi_i p_i$ from the left-hand side, while the right-hand side doesn't change due to $\pi_i \hat{c}_i x_i = 0$. The following theorem characterizes exactly under which conditions recyclable, uncertainty-exclusive inequalities $\sum_{i\in[n]}\pi_i x \leq \pi_0$ yield facet-defining recycled inequalities, based on the *face* $F(\pi) = \left\{x \in \mathcal{C}^{NOM}\middle|\sum_{i\in[n]}\pi_i x = \pi_0\right\}$. The statement may seem very technical at first glance, but we will see afterwards that it is quite powerful and has some surprising implications.

Theorem 2. *Let $\sum_{i\in[n]}\pi_i x_i \leq \pi_0$ be a recyclable, uncertainty-exclusive inequality and $e^j \in \mathbb{R}^{n+1}$ be the unit-vector for $j \in S = \{i \in [n]|\pi_i = 0\}$. Then the recycled inequality (1) is facet-defining for $\mathcal{C}^{ROB}$ if and only if there exist vectors $\left\{\tilde{x}^1, \ldots, \tilde{x}^{n-|S|}\right\} \subseteq F(\pi)$ such that $\left\{e^j|j \in S\right\} \cup \left\{\left(\tilde{x}^1, 1\right), \ldots, \left(\tilde{x}^{n-|S|}, 1\right)\right\}$ are linearly independent.*

Proof. First, note that the face of the recycled inequality is not equal to $\mathcal{C}^{ROB}$, since p and z can be raised arbitrarily. Thus, it is facet-defining if and only if there exist $\dim\left(\mathcal{C}^{ROB}\right) = \dim\left(\mathcal{C}^{NOM}\right) + n + 1$ affinely independent vectors $(x, p, z) \in \mathcal{C}^{ROB}$ that satisfy it with equality.

Regardless of π, there are $\dim\left(\mathcal{C}^{NOM}\right) + 1 + |S|$ affinely independent $(x, p, z) \in \mathcal{C}^{ROB}$ satisfying (1) with equality. For this, let $\left\{x^0, \ldots, x^{\dim(\mathcal{C}^{NOM})}\right\} \subseteq \mathcal{C}^{NOM}$ be affinely independent. We choose $\left(x^j, \hat{c} \odot x^j, 0\right)$ for each $j \in \left\{0, \ldots, \dim\left(\mathcal{C}^{NOM}\right)\right\}$, where $\hat{c} \odot x^j$ refers to the component-wise multiplication, i.e., $\left(\hat{c} \odot x^j\right)_i = \hat{c}_i x_i^j$. By definition, $\left(x^j, \hat{c} \odot x^j, 0\right)$ is within $\mathcal{C}^{ROB}$ and satisfies (1) with equality. Additionally, we choose $\left(x^0, \hat{c} \odot x^0 + e^j, 0\right)$ for each $j \in S$. Here, $e^j \in \mathbb{R}^n$ with some abuse of notation. Again, this vector is within $\mathcal{C}^{ROB}$ and satisfies (1) with equality due to $\pi_j = 0$.

Now, the recycled inequality (1) is facet-defining if and only if there exists a *suitable extension* of the vectors above, consisting of additional vectors $\left(\tilde{x}^j, \tilde{p}^j, \tilde{z}^j\right)_{j \in [n-|S|]}$ that satisfy (1) with equality and are affinely independent to the vectors above. Such vectors need to satisfy the property

$$\tilde{p}_i^j = \left(\hat{c}_i - \tilde{z}^j\right) \tilde{x}_i^j \text{ for all } i \in [n] \setminus S \text{ and } j \in [n-|S|], \tag{2}$$

as otherwise

$$\pi_0 \tilde{z}^j + \sum_{i \in [n]} \pi_i \tilde{p}_i^j > \pi_0 \tilde{z}^j - \sum_{i \in [n]} \pi_i \tilde{z}^j \tilde{x}_i^j + \sum_{i \in [n]} \pi_i \hat{c}_i \tilde{x}_i^j \geq \sum_{i \in [n]} \pi_i \hat{c}_i \tilde{x}_i^j.$$

One can show that any vectors $\left(\tilde{x}^j, \tilde{p}^j, \tilde{z}^j\right)_{j \in [n-|S|]}$ with property (2) are affinely independent to the ones above if and only if $\tilde{z}^j > 0$ for all $j \in [n-|S|]$ and $\left\{e^j \mid j \in S\right\} \cup \left\{\left(\tilde{x}^1, 1\right), \ldots, \left(\tilde{x}^{n-|S|}, 1\right)\right\}$ are linearly independent. To show this, one subtracts $\left(x^0, \hat{c} \odot x^0, 0\right)$ from all other vectors, yielding vectors that are linearly independent if and only if the desired affine independency holds. Writing the vectors in a matrix and performing basic column and row transformations implies the result. We omit this step here due to space limitations.

With $\tilde{z}^j > 0$ and $\pi_i \tilde{p}_i^j = \pi_i \left(\hat{c}_i - \tilde{z}^j\right) \tilde{x}_i^j$, we also have

$$\pi_0 \tilde{z}^j + \sum_{i \in [n]} \pi_i \tilde{p}_i^j = \sum_{i \in [n]} \pi_i \hat{c}_i \tilde{x}_i^j \Leftrightarrow \pi_0 \tilde{z}^j = \sum_{i \in [n]} \pi_i \tilde{z}^j \tilde{x}_i^j \Leftrightarrow \pi_0 = \sum_{i \in [n]} \pi_i \tilde{x}_i^j,$$

and thus $\left\{\tilde{x}^1, \ldots, \tilde{x}^{n-|S|}\right\} \subseteq F(\pi)$ if and only if $\left(\tilde{x}^j, \tilde{p}^j, \tilde{z}^j\right)_{j \in [n-|S|]}$ fulfill the recycled inequality (1) with equality. This shows the necessity of the condition.

Now, let $\left\{\tilde{x}^1, \ldots, \tilde{x}^{n-|S|}\right\} \subseteq F(\pi)$ be as specified in the theorem. To show sufficiency of the condition, we only need to construct vectors $\left(\tilde{x}^j, \tilde{p}^j, \tilde{z}^j\right)_{j \in [n-|S|]}$ satisfying property (2) and $\tilde{z}^j > 0$ for all $j \in [n-|S|]$. For each $j \in [n-|S|]$, we choose $\left(\tilde{x}^j, \tilde{p}^j, \tilde{z}\right)$ with $\tilde{z} = \min\left\{\hat{c}_i \mid i \in [n], \hat{c}_i > 0\right\}$ and $\tilde{p}_i^j = \max\left\{0, \hat{c}_i - \tilde{z}\right\} \tilde{x}_i^j$ for all $i \in [n]$. Then $\left(\tilde{x}^j, \tilde{p}^j, \tilde{z}\right)$ is by definition within $\mathcal{C}^{\text{ROB}}$ and satisfies $\tilde{z} > 0$. Since $\sum_{i \in [n]} \pi_i x_i \leq \pi_0$ is uncertainty-exclusive, we have $\pi_i = 0$ for all $\hat{c}_i < \tilde{z}$, and thus $\tilde{p}_i^j = (\hat{c}_i - \tilde{z}) \tilde{x}_i^j$ for all $i \in [n] \setminus S$. Therefore, $\left(\tilde{x}^j, \tilde{p}^j, \tilde{z}\right)$ also satisfies property (2), which completes the proof.   □

A straightforward, but powerful implication of Theorem 2 is that recycling a uncertainty-exclusive inequality yields always a facet-defining inequality if $\dim\left(F(\pi)\right) = n - 1$ holds. This is because there already exist n affinely independent vectors satisfying $\sum_{i \in [n]} \pi_i x = \pi_0$, which implies that there exist appropriate vectors $\left\{x^1, \ldots, x^{n-|S|}\right\}$. Note that $\dim\left(F(\pi)\right) = n - 1$ holds if $F(\pi)$ is either a facet of a full-dimensional polytope $\mathcal{C}^{\text{NOM}}$ or if $\sum_{i \in [n]} \pi_i x \leq \pi_0$ is actually an equation with $F(\pi) = \mathcal{C}^{\text{NOM}}$ and $\dim\left(\mathcal{C}^{\text{NOM}}\right) = n - 1$. This is summarized in the following corollary.

Corollary 1. *Let $\sum_{i \in [n]} \pi_i x_i \leq \pi_0$ be a recyclable, uncertainty-exclusive inequality. The recycled inequality (1) is facet-defining for $\mathcal{C}^{\text{ROB}}$ if one of the following holds:*

- C^{NOM} is full-dimensional and $F(\pi)$ is a facet of C^{NOM},
- $\dim(C^{NOM}) = n - 1$ and $F(\pi) = C^{NOM}$.

Contrary to first intuition, it is also possible to obtain facet-defining inequalities by recycling weaker inequalities that are neither facet-defining nor equations. This is because Theorem 2 suggests that an inequality defining a low-dimensional face can also be recycled to a facet-defining inequality if we have $\pi_i = 0$ for many $i \in [n]$. For example, consider an independent set problem on a graph with vertices $V = [n]$ and let $Q \subseteq V$ be a clique. Then the clique inequality $\sum_{i \in Q} x_i \leq 1$ dominates all inequalities $\sum_{i \in Q'} x_i \leq 1$ with $Q' \subsetneq Q$ and is facet-defining if and only if Q is a maximal clique with respect to inclusion [8]. However, the recycled inequality $z + \sum_{i \in Q'} p_i \geq \sum_{i \in Q'} \hat{c}_i x_i$ is facet-defining for all cliques $Q' \subseteq Q$. This is because the set $\{\tilde{x}^1, \ldots, \tilde{x}^{n-|S|}\} = \{e^j | j \in Q'\}$ meets the criteria of Theorem 2 with $S = V \setminus Q'$. Other examples include odd-hole inequalities for the independent set problem [18] and minimal cover inequalities for the knapsack problem [8]. These are in general not facet-defining for their respective polytope, but yield facet-defining recycled inequality for the robust counterpart. All these examples are covered by the following corollary.

Corollary 2. *Let C^{NOM} be a full-dimensional polyhedron such that $x \in C^{NOM}$ and $0 \leq x' \leq x$ implies $x' \in C^{NOM}$. Furthermore, let $\sum_{i \in [n]} \pi_i x_i \leq \pi_0$ be a recyclable, uncertainty-exclusive inequality. The recycled inequality (1) is facet-defining for C^{ROB} if $\sum_{i \in [n]} \pi_i x_i \leq \pi_0$ is facet-defining for the restricted solution space $\{x \in C^{NOM} | x_i = 0 \text{ for all } \pi_i = 0\}$.*

Note that the additional requirements on C^{NOM} imply that the restricted solution space is of dimension $n - |S|$, which guarantees that we find appropriate vectors $\{\tilde{x}^1, \ldots, \tilde{x}^{n-|S|}\}$.

One now might raise the question whether inequalities recycled from dominated inequalities are actually of practical interest or whether they do not really matter due to the special structure of the objective function. The following example demonstrates that it can be beneficial to weaken an inequality before it is recycled.

Example 2. Consider the robust problem

$$
\min \left\{ 2z + \sum_{i \in [5]} -x_i + p_i \left| \begin{array}{l} 3x_5 + \sum_{i \in [4]} x_i \leq 3 \\ z + p_i \geq x_i \\ x \in \{0,1\}^5, p \in \mathbb{R}^5_{\geq 0}, z \in \mathbb{R}_{\geq 0} \end{array} \right. \forall i \in [5] \right\}.
$$

Choosing $x = (\frac{3}{4}, \ldots, \frac{3}{4}, 0)$, $p \equiv 0$, and $z = \frac{3}{4}$ yields an optimal solution for the linear relaxation of value $-\frac{3}{2}$. Recycling constraint $3x_5 + \sum_{i \in [4]} x_i \leq 3$ yields $3z + 3p_5 + \sum_{i \in [4]} p_i \geq 3x_5 + \sum_{i \in [4]} x_i$. After adding the recycled inequality, an optimal choice is given by $x = (\frac{3}{4}, \ldots, \frac{3}{4}, 0)$, $p = (0, \ldots, 0, \frac{1}{4})$, and $z = \frac{3}{4}$, with an objective value of $-\frac{5}{4}$. Note that we now choose $p_5 > 0$ even though $x_5 = 0$ holds. This is because raising p_5 has the same effect on the recycled inequality as raising z, but is cheaper in the objective function. Since the bilinear inequality $p_5 + x_5 z \geq \hat{c}_5 x_5$ now has a slack of $\frac{1}{4}$, our observation from the last section suggests that it may be beneficial to drop x_5 from the valid inequality for recycling. In fact, when recycling the dominated inequality $\sum_{i \in [4]} x_i \leq 3$ instead, we obtain $3z + \sum_{i \in [4]} p_i \geq \sum_{i \in [4]} x_i$ and an optimal choice is now given by $x = (1,1,1,0,0)$, $p \equiv 0$, and $z = 1$, which yields an objective value of -1.

We can benefit from this insight on dominated inequalities when recycling within a separation procedure to cut-off a fractional solution $(\tilde{x}, \tilde{p}, \tilde{z}) \in \mathcal{P}^{\mathrm{ROB}}$. The violation of a recycled inequality is given by $\sum_{i \in [n]} \pi_i (\hat{c}_i \tilde{x}_i - \tilde{p}_i) - \pi_0 \tilde{z}$. In order to maximize the violation, we can drop all variables x_i from the recyclable inequality $\sum_{i \in [n]} \pi_i x_i \leq \pi_0$ with $\hat{c}_i \tilde{x}_i - \tilde{p}_i < 0$. We use this in our computational study in the next section, where we show that recycled inequalities are not only interesting from a theoretical point of view, but also computationally relevant.

4   Computational Study

Due to space limitations, we present a preliminary computational study, in which we test recycled inequalities for robust counterparts of two classical combinatorial problems, namely the weighted independent set problem and the weighted bipartite matching problem.

In the following, we compare (i) computation times to asses an algorithm's performance and (ii) integrality gaps to evaluate the strength of a formulation. Since displaying these for all algorithms and instances is impractical, we give aggregated values using the *shifted geometric mean*, as proposed by Achterberg [1]. This is defined as $\left(\Pi_{i=1}^{k} (v_i + s)^{1/k} \right) - s$ for values $v_1, \ldots, v_k \in \mathbb{R}_{\geq 0}$ and a *shifting parameter* $s \in \mathbb{R}_{\geq 0}$. We always use $s = 1$ second for aggregating computation times and $s = 1\%$ for aggregating integrality gaps. Furthermore, we use a time limit of 3600 seconds for each algorithm and instance and set the computation time to this value if an algorithm reaches the limit. Note that this is a bias in favor of algorithms that reach the time limit for many instances.

All experiments have been implemented in Java 11 and are performed on a single core of a Linux machine with an Intel® Core™ i7-5930K CPU @ 3.50GHz, with 4 GB RAM reserved for each calculation. All LPs and MILPs are solved using Gurobi version 9.5.0 [13] in single thread mode and all other settings at default, if not stated otherwise.

All implemented algorithms [10] and generated test instances [11] are freely available online.

4.1   Robust Independent Set

To show the effect of recycling a class of well-known valid inequalities in a separation procedure, we consider the robust maximum weighted independent set problem on a graph with nodes V and edges E

$$\max\left\{\sum_{v\in V}c_v x_v - \Gamma z - \sum_{v\in V}p_v \;\middle|\; \begin{array}{ll} x_v + x_w \leq 1 & \forall\,\{v,w\}\in E \\ p_v + z \geq \hat{c}_v x_v & \forall\,v\in V \\ x\in\{0,1\}^V, p\in\mathbb{R}^V_{\geq 0}, z\in\mathbb{R}_{\geq 0} \end{array}\right\}.$$

As seen in Sect. 3, recycling a clique inequality $\sum_{v\in Q} x_v \leq 1$ yields a facet-defining inequality for all cliques $Q \subseteq V$. We compare the separation of recycled clique inequalities in the root node of the branching tree against the robust default formulation $\mathcal{P}^{\mathrm{ROB}}$, which solely uses the constraints $p_i + z \geq \hat{c}_i x_i$. For this, we use Gurobi's callback to add the recycled inequalities as user cuts [13]. Every time Gurobi invokes the callback in the root node and reports a current optimal fractional solution $(\tilde{x}, \tilde{p}, \tilde{z}) \in \mathcal{P}^{\mathrm{ROB}}$, we try to compute cliques $Q \subseteq V$ for which the recycled inequality $z + \sum_{v\in Q} p_v \geq \sum_{v\in Q} \hat{c}_v x_v$ is violated. Since a node $v \in V$ positively contributes to the violation if $\hat{c}_v \tilde{x}_v - \tilde{p}_v > 0$ holds, we essentially need to solve a maximum weighted clique problem with weights $\hat{c}_v \tilde{x}_v - \tilde{p}_v$. To separate many recycled inequalities at once, we extend each node $v \in V$ with $\hat{c}_v \tilde{x}_v - \tilde{p}_v > 0$ greedily to a clique $Q_v \subseteq V$ with $v \in Q_v$. For this, we start with $Q_v = \{v\}$ and then iteratively add $v' \in N(Q_v)$ such that $\hat{c}_{v'} \tilde{x}_{v'} - \tilde{p}_{v'}$ is maximal and non-negative. Finally, we return the corresponding recycled inequality to Gurobi as a user cut if its violation is positive.

As a basis for our test instances, we use the graphs of the second DIMACS implementation challenge on the clique problem [15]. Of the 66 DIMACS graphs, we choose the 46 graphs that have at most 500 nodes, as otherwise the nominal problem is already very hard. For each node $v \in V$, we generate independent and uniformly distributed values $c_v \in \{900, \ldots, 1000\}$ and correlated deviations $\hat{c}_v = \lceil \xi_v c_v \rceil$, with $\xi_v \in [0.45, 0.55]$ being an independent and uniformly distributed random variable. Since robust problems tend to be hard for Γ being somewhere around half the number of variables with $x_i = 1$ [7], we greedily compute an independent set $S \subseteq V$ and define $\Gamma = \left\lfloor \frac{|S|}{2} \right\rfloor$. For this, we start with $S = \emptyset$ and then iteratively add nodes $v \in V \setminus N[S]$ such that $|V \setminus N[S \cup \{v\}]|$ is maximal, with $N[S']$ being the closed neighborhood of S'. Using this procedure, we randomly generate five robust independent set problems for each of the 46 DIMACS graphs, leaving us with 230 robust instances.

We show computational results for the robust default formulation $\mathcal{P}^{\mathrm{ROB}}$ and the recycling of clique inequalities in Table 1. Here, we see that the shifted geometric mean of the integrality gaps is reduced absolutely by 220% from 1427.59% to 1207.59% when using recycled clique inequalities. For computing these gaps, we use the dual bound obtained by heuristically separating recycled clique inequalities for subsequent linear relaxations until no violated cuts are found. While the absolute reduction of the integrality gap is quite impressive, the relative reduction does not adequately reflect the strength of the recycled inequalities. This is due to the large integrality gap of the nominal problem, which constitutes a major part of the total gap. Therefore, we also test a stronger formulation for the nominal problem, in which we replace every

Table 1. Computational results for 230 instances of the robust maximum weighted independent set problem.

		robust default formulation			separate recycled clique inequalities		
nominal formulation	Gurobi's cuts	tilim	time	int. gap	tilim	time	int. gap
edge	enabled	24	26.15	1427.59%	22	31.03	1207.59%
	disabled	40	51.01		20	20.34	
clique	enabled	61	133.62	135.30%	63	141.97	56.25%
	disabled	78	187.21		54	89.24	

constraint $x_v + x_w \leq 1$ with $\sum_{v \in Q} x_v \leq 1$ for a clique $Q \subseteq V$ with $\{v, w\} \subseteq Q$. This *clique formulation* has a much tighter linear relaxation compared to the previous *edge formulation*, and thus reduces the contribution of the nominal problem to the integrality gap. Indeed, Table 1 shows that separating recycled clique inequalities reduces the integrality gap by more than one half when using the clique formulation. Apart from this observation, the clique formulation is not of practical interest, as the solver performs better on the edge formulation.

Using the edge formulation, we are able to solve 2 more instances when recycling clique inequalities, but observe an increase of the computation time. This seems to be due to some interference with Gurobi's own cutting planes. When disabling Gurobi's cutting planes, recycling is much better than using the default formulation. In fact, disabling Gurobi's cuts and using recycled clique inequalities is the overall best performing approach, solving the most instances in the least amount of computation time. This is true for both nominal formulations and indicates that, given a careful implementation, recycling clique inequalities yields a significant speedup compared to the robust default formulation.

4.2 Robust Bipartite Matching

We now consider the robust maximum weighted bipartite matching problem

$$\max \left\{ \sum_{e \in E} c_e x_e - \Gamma z - \sum_{e \in E} p_e \middle| \begin{array}{ll} \sum_{e \in \delta(v)} x_e \leq 1 & \forall v \in V \\ p_e + z \geq \hat{c}_e x_e & \forall e \in E \\ x \in \{0, 1\}^E, p \in \mathbb{R}^E_{\geq 0}, z \in \mathbb{R}_{\geq 0} \end{array} \right\}$$

on a bipartite graph with nodes V and edges E. In contrast to the independent set problem, for which the standard nominal formulation is quite weak, we have $\mathcal{P}^{\mathrm{NOM}} = \mathcal{C}^{\mathrm{NOM}}$ for the bipartite matching problem [8]. That is, the integrality gap of the robust counterpart is only due to the robust substructure, which allows us to test the strength of recycled inequalities to their limit.

We randomly generate instances by first dividing a given set of nodes $V = [n]$ into two partitions $U = [\lceil \frac{n}{2} \rceil]$ and $W = \{\lceil \frac{n}{2} \rceil + 1, \ldots, n\}$. Afterwards, we

Table 2. Computational results for the robust maximum weighted bipartite matching problem.

nodes	Gurobi's cuts	robust default formulation			recycle constraints			recycle constraints and separate dominated		
		tilim	time	int. gap	tilim	time	int. gap	tilim	time	int. gap
50	enabled	0	0.78	19.532%	0	0.53	0.326%	0	0.64	0.319%
	disabled	10	3600.00		0	0.25		0	0.28	
100	enabled	5	603.37	22.82%	0	4.76	0.319%	0	5.59	0.316%
	disabled	10	3600.00		0	14.83		0	15.62	
150	enabled	4	1405.15	23.66%	0	122.11	0.269%	0	158.81	0.265%
	disabled	10	3600.00		6	1809.62		7	1873.43	

sample for each node $u \in U$ a random number $\phi_u \in [0,1]$, modeling the probability with which an edge incident to u exists. Then for every $w \in W$, we add the edge $\{u, w\}$ with probability ϕ_u. Analogously to the independent set problem, every weight is a random number $c_e \in \{900, \ldots, 1000\}$ and the correlated deviations are $\hat{c}_e = \lceil \xi_e c_e \rceil$ with $\xi_e \in [0.45, 0.55]$. Finally, as the number of edges in a solution will most likely be near to $\frac{n}{2}$, we set $\Gamma = \lfloor \frac{n}{4} \rfloor$. We use this procedure to generate ten instances for different numbers of nodes $n \in \{50, 100, 150\}$.

Table 2 shows computational results for the robust default formulation and two different approaches for using recycled inequalities. The first approach recycles all constraints $\sum_{e \in \delta(v)} x_e \leq 1$ for $v \in V$. The second approach additionally separates violated recycled inequalities corresponding to dominated inequalities $\sum_{e \in E'} x_e \leq 1$ with $E' \subseteq \delta(v)$ for $v \in V$ in the root node of the branching tree.

It is evident that recycling inequalities is significantly better than solely using the default formulation. We observe a significant strengthening of the formulation, leading to a reduction of the integrality gap to nearly one-hundredth for $n = 150$ nodes. This strength also translates to a higher number of instances solved and much lower computation times. For $n = 150$, recycling constraints leads to a speedup of more than 1000% with Gurobi's cuts enabled.

The reduced integrality gap obtained by recycling dominated constraints compared to the sole recycling of constraints $\sum_{e \in \delta(v)} x_e \leq 1$ shows that recycling dominated inequalities can improve the strength of the linear relaxation in practice. However, as the recycled constraints already perform very well for these instances, the improvement in the linear relaxation is very small. In fact, the minor strengthening of the linear relaxation cannot compensate for the computational load imposed by the additional inequalities, which leads to higher computation times.

In any case, recycling valid inequalities yields a significant speed-up compared to the default formulation. First experiments with adapted MIPLIB [12] instances, which will be part of the full paper, confirm that this is also true for a broad set of different robust problems. Here, we even observe that recycling dominated inequalities can have a strong positive effect on the strength of the linear relaxation.

5   Conclusion

In this paper, we proposed and analyzed recycled inequalities for robust combinatorial optimization problems with budget uncertainty. These can be derived in linear time from valid inequalities for the nominal problem, which gives the possibility to easily reuse model constraints and well known classical valid inequalities in order to strengthen the linear relaxation of the robust problem. We highlighted the theoretical strength of recycled inequalities by proving that they often define facets of the convex hull of the robust problem, even when the underlying valid inequality is dominated.

Our preliminary computational experiments reveal that recycled inequalities are not only interesting from a theoretical point of view, but can also yield a substantial speed-up in the optimization process. They thus extend the boundaries of computational tractability for one of the most popular approach for integrating uncertainties into optimization problems.

Acknowledgements. This work was partially supported by the German Federal Ministry of Education and Research (grants no. 05M16PAA) within the project "Health-FaCT - Health: Facility Location, Covering and Transport", the Freigeist-Fellowship of the Volkswagen Stiftung, and the German research council (DFG) Research Training Group 2236 UnRAVeL.

Code Availability. All tested algorithms have been implemented in Java and are available on GitHub, see [10].

Data Availability. All test instances used in our computational study are published and available for download, sharing, and reuse, see [11].

References

1. Achterberg, T.: Constraint integer programming. Ph.D. Thesis, Technische Universitat Berlin (2007)
2. Álvarez-Miranda, E., Ljubić, I., Toth, P.: A note on the Bertsimas & Sim algorithm for robust combinatorial optimization problems. 4OR **11**(4), 349–360 (2013)
3. Atamtürk, A.: Strong formulations of robust mixed 0–1 programming. Math. Program. **108**(2–3), 235–250 (2006)
4. Bertsimas, D., Dunning, I., Lubin, M.: Reformulation versus cutting-planes for robust optimization. CMS **13**(2), 195–217 (2016)
5. Bertsimas, D., Sim, M.: Robust discrete optimization and network flows. Math. Program. **98**(1–3), 49–71 (2003)
6. Bertsimas, D., Sim, M.: The price of robustness. Oper. Res. **52**(1), 35–53 (2004)
7. Büsing, C., Gersing, T., Koster, A.M.: A branch and bound algorithm for robust binary optimization with budget uncertainty. Math. Program. Comput. (2023). https://doi.org/10.1007/s12532-022-00232-2
8. Conforti, M., Cornuéjols, G., Zambelli, G., et al.: Integer Programming, vol. 271. Springer, Cham (2014). https://doi.org/10.1007/978-3-319-11008-0
9. Fischetti, M., Monaci, M.: Cutting plane versus compact formulations for uncertain (integer) linear programs. Math. Program. Comput. **4**(3), 239–273 (2012)

10. Gersing, T.: Algorithms for robust binary optimization, December 2022. https://doi.org/10.5281/zenodo.7463371
11. Gersing, T., Büsing, C., Koster, A.: Benchmark Instances for Robust Combinatorial Optimization with Budgeted Uncertainty, December 2022. https://doi.org/10.5281/zenodo.7419028
12. Gleixner, A., et al.: MIPLIB 2017: data-driven compilation of the 6th mixed-integer programming library. Math. Program. Comput. **13**(3), 443–490 (2021). https://doi.org/10.1007/s12532-020-00194-3
13. Gurobi Optimization, LLC: Gurobi optimizer reference manual, version 9.5 (2022). http://www.gurobi.com
14. Hansknecht, C., Richter, A., Stiller, S.: Fast robust shortest path computations. In: 18th Workshop on Algorithmic Approaches for Transportation Modelling, Optimization, and Systems (ATMOS 2018). Schloss Dagstuhl-Leibniz-Zentrum fuer Informatik (2018)
15. Johnson, D.S., Trick, M.A.: Cliques, coloring, and satisfiability: second DIMACS implementation challenge, 11–13 October 1993, vol. 26. American Mathematical Society (1996)
16. Joung, S., Park, S.: Robust mixed 0–1 programming and submodularity. INFORMS J. Optim. **3**(2), 183–199 (2021). https://doi.org/10.1287/ijoo.2019.0042
17. Lee, T., Kwon, C.: A short note on the robust combinatorial optimization problems with cardinality constrained uncertainty. 4OR **12**(4), 373–378 (2014)
18. Padberg, M.W.: On the facial structure of set packing polyhedra. Math. Program. **5**(1), 199–215 (1973)
19. Park, K., Lee, K.: A note on robust combinatorial optimization problem. Manag. Sci. Financ. Eng. **13**(1), 115–119 (2007)
20. Sherali, H.D., Adams, W.P.: A Reformulation-Linearization Technique for Solving Discrete and Continuous Nonconvex Problems, vol. 31. Springer, New York (2013). https://doi.org/10.1007/978-1-4757-4388-3

Inapproximability of Shortest Paths on Perfect Matching Polytopes

Jean Cardinal[1] and Raphael Steiner[2]([✉])

[1] Université Libre de Bruxelles (ULB), Brussels, Belgium
`jean.cardinal@ulb.be`
[2] ETH Zurich, Zürich, Switzerland
`raphaelmario.steiner@inf.ethz.ch`

Abstract. We consider the computational problem of finding short paths in the skeleton of the perfect matching polytope of a bipartite graph. We prove that unless P = NP, there is no polynomial-time algorithm that computes a path of constant length between two vertices at distance two of the perfect matching polytope of a bipartite graph. Conditioned on P $\neq$ NP, this disproves a conjecture by Ito, Kakimura, Kamiyama, Kobayashi and Okamoto [SIAM Journal on Discrete Mathematics, 36(2), pp. 1102-1123 (2022)]. Assuming the Exponential Time Hypothesis we prove the stronger result that there exists no polynomial-time algorithm computing a path of length at most $\left(\frac{1}{4} - o(1)\right) \frac{\log N}{\log \log N}$ between two vertices at distance two of the perfect matching polytope of an N-vertex bipartite graph. These results remain true if the bipartite graph is restricted to be of maximum degree three.

The above has the following interesting implication for the performance of pivot rules for the simplex algorithm on simply-structured combinatorial polytopes: If P $\neq$ NP, then for every simplex pivot rule executable in polynomial time and every constant $k \in \mathbb{N}$ there exists a linear program on a perfect matching polytope and a starting vertex of the polytope such that the optimal solution can be reached using only two monotone non-degenerate steps from the starting vertex, yet the pivot rule will require at least k non-degenerate steps to reach the optimal solution. This result remains true in the more general setting of pivot rules for so-called *circuit-augmentation algorithms*.

Keywords: Perfect matching polytopes · Simplex method · Pivot rules · Circuit augmentations · Combinatorial reconfiguration

1 Introduction

The history of linear programming is intimately intertwined with that of Dantzig's simplex algorithm. While the simplex and its many variants are among

R. Steiner–Supported by an ETH Postdoctoral Fellowship.
A full version of this article can be found at https://arxiv.org/abs/2210.14608. Proofs of statements marked with ⋆ are deferred to the full version.

the most studied algorithms ever, a number of fundamental questions remain open. It is not known, for instance, whether there exists a pivot rule that makes the simplex method run in strongly polynomial time. Since the publication of the first examples of linear programs that make the original simplex algorithm run in exponential time, many alternative pivot rules have been proposed, fostering a tremendous amount of work in the past 75 years, both from the combinatorial and complexity-theoretic point of views.

The simplex algorithm follows a monotone path on the skeleton of the polytope defining the linear program. The following natural question was recently raised by De Loera, Kafer, and Sanità [16]:

"Can one hope to find a pivot rule that makes the simplex method use a shortest monotone path?".

As an answer, they proved that given an initial solution to a linear program, it is NP-hard to find a $(2-\varepsilon)$-approximate shortest monotone path to an optimal solution. It implies that unless P = NP, no polynomial-time pivot rule for the simplex can be guaranteed to reach an optimal solution in a minimum number of (non-degenerate) steps.

A similar result can also be deduced from two independent contributions, by Aichholzer, Cardinal, Huynh, Knauer, Mütze, Steiner, and Vogtenhuber [2] on the one hand, and by Ito, Kakimura, Kamiyama, Kobayashi, and Okamoto [29] on the other hand. They proved that the above result holds for perfect matching polytopes of planar and bipartite graphs, albeit with a slightly weaker inapproximability factor of 3/2 instead of 2. Ito et al. [29] conjecture that there exists a constant-factor approximation algorithm for the problem of finding a shortest path between two perfect matchings on the perfect matching polytope.

Our main result is a disproof of this conjecture under the P $\neq$ NP assumption: Strengthening the previous inapproximability results mentioned above, we show that unless P = NP no C-approximation for a shortest path between two vertices at distance 2 of a bipartite perfect matching polytope can be found in polynomial time, for any (arbitrarily large) choice of $C > 0$. We also give an even stronger inapproximability result under the *Exponential Time Hypothesis* (ETH). The latter states that the 3-SAT problem cannot be solved in worst-case subexponential time, and is one of the main computational assumptions of the fine-grained complexity program [39]. As a consequence, there is not much hope of finding a pivot rule for the simplex algorithm yielding good approximations of the shortest path towards an optimal solution, even when the linear program is integer and its associated matrix totally unimodular.

1.1   Our Result

We consider the complexity of computing short paths on the 0/1 polytope associated with perfect matchings of a bipartite graph. Given a balanced bipartite graph $G = (V, E)$, where V is partitioned into two equal-size independent sets A and B, we define the *perfect matching polytope* $P_G \subseteq \mathbb{R}^E$ of G as the convex hull of the 0/1 incidence vectors of perfect matchings of G.

It is well-known (see e.g. Chapter 18 in [38]) that for *bipartite* graphs G, there is a nice halfspace representation of P_G. An edge-vector $(x_e)_{e \in E} \in \mathbb{R}^E$ is in P_G if and only if the following hold.

$$\sum_{e \ni v} x_e = 1, \qquad (\forall v \in V) \tag{1}$$

$$x_e \geq 0, \qquad (\forall e \in E). \tag{2}$$

The above is a compact encoding of P_G, with a number of constraints and variables of size polynomial in G. The assumption that G is bipartite is crucial here: For non-bipartite G the polytope defined by the above constraints has non-integral vertices and is thus not a representation of P_G [38]. The matrix of this representation of a perfect matching polytope of a bipartite graph G is simply the vertex-edge-incidence matrix of G, which is totally unimodular. The problem of maximizing a linear functional $w^T x$ subject to constraints (1) and (2) corresponds exactly to the problem of finding a perfect matching M of G whose weight $\sum_{e \in M} w_e$ is maximal.

Given that the simplex algorithm moves along the edges of a polytope, it is crucial for our considerations to understand adjacency of vertices on P_G. The following result is well-known [13, 30].

Lemma 1. *For a bipartite graph G, two vertices of P_G corresponding to two perfect matchings M_1 and M_2 are adjacent in the skeleton of P_G if and only if the symmetric difference $M_1 \triangle M_2$ is a cycle in G.*

This cycle is said to be *alternating* in both matchings, and one matching can be obtained from the other by *flipping* this alternating cycle. In general, we will say that two perfect matchings are *at distance at most k* from each other on P_G, for some positive integer k, if one can be obtained from the other by successively flipping at most k alternating cycles.

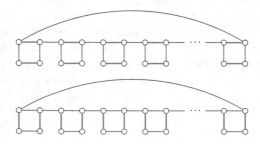

Fig. 1. Two perfect matchings at distance two on the perfect matching polytope, but whose symmetric difference consists of an arbitrarily large number of even cycles.

Note that given any two perfect matchings M_1 and M_2 of a bipartite graph G, it is always the case that $M_1 \triangle M_2$ is a collection of vertex-disjoint even cycles that are alternating in both matchings. The number of such cycles is therefore an upper bound on the distance between M_1 and M_2 on P_G. Interestingly, this upper bound can be arbitrarily larger than the actual distance. Figure 1 shows a construction of a graph G with two matchings at distance two on P_G, whose symmetric difference consists of an arbitrary number of cycles.

Our main result is the following.

Theorem 1. *Let $k \geq 2$ be any fixed integer. Unless $\mathsf{P} = \mathsf{NP}$, there does not exist any polynomial-time algorithm solving the following problem:*

Input: *A bipartite graph G of maximum degree 3 and a pair of perfect matchings M_1, M_2 of G at distance at most 2 on the polytope P_G.*
Output: *A path from M_1 to M_2 in the skeleton of P_G, of length at most k.*

More strongly, for every absolute constant $\delta > 0$, unless the Exponential Time Hypothesis fails, no polynomial-time algorithm can solve the above problem when k is allowed to grow with the number N of vertices of G as $k(N) = \left\lfloor \left(\frac{1}{4} - \delta\right) \frac{\log N}{\log \log N} \right\rfloor$.

A path on the perfect matching polytope of a bipartite graph G is said to be *monotone* with respect to some weight vector $w = (w_e)_{e \in E} \in \mathbb{R}^E$ on the edges of G if the perfect matchings along the path have monotonically increasing total weights. Given two perfect matchings M_1 and M_2 at distance two on the perfect matching polytope, one can assign weights to edges so that (i) the path of length two between them is strictly monotone, and (ii) M_2 is the unique matching of maximal weight (this will be formally proven later in Lemma 4). This allows us to formulate our result as one about the hardness of reaching an optimal solution from a given feasible solution of a linear program on the perfect matching polytope.

Corollary 1. *Unless $\mathsf{P} = \mathsf{NP}$, there does not exist any polynomial-time constant-factor approximation algorithm for the following optimization problem:*

Input: *A bipartite graph $G = (V, E)$ of maximum degree 3, a weight function $E \to \mathbb{R}^+$, and a perfect matching M of G.*
Output: *A shortest monotone path on P_G from M to a maximum-weight perfect matching of G.*

Furthermore, assuming ETH, for an arbitrary but fixed $\delta > 0$ no polynomial-time algorithm can achieve an approximation ratio of less than $\left(\frac{1}{8} - \delta\right) \frac{\log N}{\log \log N}$, where $N := |V(G)|$.

This corollary can be further interpreted as a statement on the existence of a polynomial-time pivot rule that would make the simplex method use an approximately shortest monotone path to a solution. Any such pivot rule could be used as an approximation algorithm for the above problem, contradicting the computational hypotheses. While making this statement, it is important to point out that here (and throughout this paper) we always measure the number of *non-degenerate* steps during an execution of a pivot rule for the simplex algorithm, and not the natural alternative, which would be the number of all steps (including both degenerate and non-degenerate steps). A *degenerate* step in a simplex algorithm here means a step that changes the basis of active variables, but not the value of the current feasible solution.

1.2   Pivot Rules for Circuit-Augmentation Algorithms

Our work on distances in the skeleton of P_G for bipartite graphs G was originally motivated by questions regarding so-called *circuit moves (or circuit augmentations)*, that have been recently studied in linear programming [11,15,16] as well as in the context of relaxations of the Hirsch conjecture concerning the diameter of polytopes [10,32]. A *circuit move* extends the simplex-paradigm of moving along an incident edge of the constraint-polyhedron, by additionally allowing to move along certain non-edge directions, called *circuits*. Given a linear program, the circuits in a well-defined sense represent all possible edge-directions that could occur after changing the right-hand side of the LP. The following is a formal definition.

Definition 1 (cf. Definition 1 in [16]). *Given a polyhedron of the form*

$$\mathcal{P} = \{\mathbf{x} \in \mathbb{R}^n | A\mathbf{x} = \mathbf{b}, B\mathbf{x} \leq \mathbf{d}\},$$

a circuit is a vector $\mathbf{g} \in \mathbb{R}^n \setminus \{\mathbf{0}\}$ *such that*

1. $A\mathbf{g} = 0$, *and*
2. $B\mathbf{g}$ *is inclusion-wise support-minimal in the collection* $\{By|Ay = \mathbf{0}, y \neq \mathbf{0}\}$.

Given an LP $\{\max \mathbf{c}^T \mathbf{x} | \mathbf{x} \in \mathcal{P}\}$ for a polyhedron $\mathcal{P}$, a current feasible solution $\mathbf{x} \in \mathcal{P}$ and a circuit $\mathbf{g}$ with $\mathbf{c}^T \mathbf{g} > 0$, a *circuit move* then consists of moving to a new feasible solution $\mathbf{x}' = \mathbf{x} + t^*\mathbf{g}$, where $t^* \geq 0$ is maximal w.r.t. $\mathbf{x} + t^*\mathbf{g} \in \mathcal{P}$. Note that in general, an optimization algorithm based on a pivot rule for circuit moves may traverse several non-vertices of the polyhedron before reaching an optimal solution.

Our interest in the perfect matching polytope for understanding the complexity of circuit-pivot algorithms came from the following statement (see the full version for a proof).

Lemma 2 (⋆). *Let G be a bipartite graph. Then if x is a vertex of P_G, and $x' \neq x$ is obtained from x by a circuit move, then x' is also a vertex of P_G and adjacent to x on the skeleton of P_G.*

This lemma implies that any sequence of circuit moves, applied starting from a vertex of P_G, will follow a monotone path on the skeleton of P_G from vertex to vertex. Consequently, Corollary 1 also yields an inapproximability result for polynomial pivot rules for circuit augmentation, as follows.

Corollary 2. *Unless* P = NP, *there does not exist a polynomial-time constant-factor approximation algorithm for the following problem.*

Input: *A bipartite graph G of maximum degree 3, a vertex $\mathbf{x} \in P_G$ and a linear objective function.*
Output: *A shortest sequence of circuit moves on P_G from $\mathbf{x}$ to an optimal solution.*

Assuming ETH, no polynomial-time algorithm can achieve an approximation ratio of less than $\left(\frac{1}{8} - \delta\right) \frac{\log N}{\log \log N}$, where $N := |V(G)|$ and $\delta > 0$ is a constant.

A related inapproximability result (but for the largest improvement of the objective function via a *single* circuit step) was obtained by Borgwardt, Brand, Feldmann and Koutecký [9].

1.3 Related Works

Our work relates to two main threads of research in combinatorics and computer science: one obviously related to the complexity of the simplex method and linear programming in general, and another more recent one, aiming at building a thorough understanding of the computational complexity of so-called *combinatorial reconfiguration problems*.

Complexity of the Simplex Method. In 1972, Klee and Minty showed that the original simplex method had an exponential worst-case behavior on what came to be known as Klee-Minty cubes [34]. Since then, many other variants have been shown to have exponential or superpolynomial lower bounds [4,7,19,24,31], although subexponential rules are known [26]. More dramatic complexity results have been obtained recently [1,20]. In particular, it was shown by Fearney and Savani [21] that Dantzig's original simplex method can solve PSPACE-complete problems: Given an initial vertex, deciding whether some variable will ever be chosen by the algorithm to enter the basis is PSPACE-complete. The simplex method is also a key motivation for studying the diameter of polytopes, in particular the Hirsch conjecture, refuted in 2012 by Santos [37].

The hardness result on approximating monotone paths given by De Loera, Kafer, and Sanità [16] is in fact a corollary of the NP-hardness of the following problem: Given a feasible extreme point solution of the bipartite matching polytope and an objective function, decide whether there is a neighbor extreme point that is optimal. A related result for circulation polytopes was proved by Barahona and Tardos [5]. These two results, as well as the hardness results from Aichholzer et al. [2] and Ito et al. [29] rely on the NP-hardness of the Hamiltonian cycle problem. In order to deal with the approximability of the shortest path, we have to resort to more recent inapproximability results on the longest cycle problem [6].

Reconfiguration of Matchings. The field of *combinatorial reconfiguration* deals with the problems of transforming a given discrete structure, typically a feasible solution of a combinatorial optimization problem, into another one using elementary combinatorial moves [23,27,28,36]. The reachability problem, for instance, asks whether there exists such a transformation, while the shortest reconfiguration path problem asks for the minimum number of elementary moves.

A number of recent works in this vein deal with reconfiguration of matchings in graphs [8,12,25,28,33]. Ito et al. [28] proved that the reachability problem between matchings of size at least some input number k and under single edge addition or removal was solvable in polynomial time. This was extended to an adjacency relation involving two edges by Kamiński, Medvedev and Milanič [33]. The problem of finding the shortest reconfiguration path under this model was shown to be NP-hard [12,25]. Another line of work involves flip graphs on perfect matchings in which the adjacency relation corresponds to flips of alternating cycles of length exactly four [8,14,17,18,35]. Note that for bipartite graphs, this flip graph is precisely the subgraph of the skeleton of the perfect matching polytope that consists of edges of length two. Bonamy, Bousquet, Heinrich, Ito, Kobayashi, Mary, Mühlentaler and Wasa [8] proved that the reachability problem in these flip graphs is PSPACE-complete.

2   Proof of Theorem 1

2.1   Preliminaries

First note that perfect matchings of a bipartite graph $G = (A \cup B, E)$ with $|A| = |B|$ can also be represented by orientations of G in which every vertex in A has outdegree one and every vertex in B has indegree one. The edges of the matching are those oriented from A to B. Alternating cycles in a perfect matching are one-to-one with directed cycles in this orientation, and flipping the cycle amounts to reverting the orientations of all its arcs. We will switch from one representation to another when convenient.

We prove Theorem 1 by reducing from the problem of approximating the longest directed cycle in a digraph. We rely on the following two results from Björklund, Husfeldt, and Khanna given as Theorems 1 and 2 in [6].

Theorem 2 (Björklund, Husfeldt, Khanna [6]). *Consider the problem of computing a long directed cycle in a given Hamiltonian digraph D on n vertices.*

1. *For every fixed $0 < \varepsilon < 1$, unless $\mathsf{P} = \mathsf{NP}$, there does not exist any polynomial-time algorithm that returns a directed cycle of length at least n^ε in D.*
2. *For every polynomial-time computable increasing function $f : \mathbb{N} \to \mathbb{N}$ in $\omega(1)$, unless the Exponential Time Hypothesis fails, there does not exist any polynomial-time algorithm that returns a directed cycle of length at least $f(n) \log n$ in D.*

Note that in the two problems, the input graph is guaranteed to be Hamiltonian, yet it remains hard to explicitly *construct* a directed cycle of some guaranteed length. Characterising the approximability of the longest cycle problem in undirected graphs is a longstanding open question [3,22].

The second ingredient of our proof is the following lemma, perhaps of independent interest, that bounds the increase in length of a longest directed cycle after a number of cycle flips in a digraph.

Lemma 3. *Let G be an undirected graph, and let $C_1, \ldots, C_t$ be a sequence of (not necessarily distinct) cycles in G. Let $D_0, D_1, \ldots, D_t$ be a sequence of orientations of G such that for each $i \in [t]$ the cycle C_i is directed in D_{i-1} and such that D_i is obtained from D_{i-1} by flipping C_i.*

There exists a polynomial-time algorithm that, given as input a number ℓ, the orientations $D_0, \ldots, D_t$ and a directed cycle C in D_t of length $|C| > \ell^{t+1}$, computes a directed cycle in D_0 of length at least ℓ.

The bound of Lemma 3 can be shown to be essentially tight. We refer to the full version of this paper for an explicit description of a directed graph whose maximum directed cycle is of length ℓ, but after a sequence of at most t cycle flips, it contains a directed cycle of length at least $(\ell/2)^{t+1}$.

2.2 Reduction

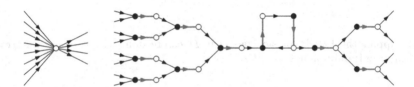

Fig. 2. Illustration of the reduction in the proof of Theorem 1. Every vertex in the given Hamiltonian digraph D (left) is replaced by the depicted gadget (right), yielding a maximum degree-three bipartite graph with a perfect matching.

We now give a proof of Theorem 1, assuming Lemma 3.

Proof (Theorem 1). We consider the first problem in Theorem 2: For a fixed $\varepsilon > 0$, given a Hamiltonian digraph D on n vertices, return a directed cycle of length at least n^ε. We first construct a digraph D' from D by replacing every vertex v of D by the gadget illustrated on Fig. 2. The gadgets are obtained by applying the following transformations[1] to every vertex v of D:

---

[1] We note that the sole prupose of splitting vertices into binary trees is to restrict the maximum degree of the graph, the remainder of the proof is only based on the 4-cycles in the middle of the gadgets.

1. The set of incoming arcs of v is decomposed into a balanced binary tree with $\deg_D^-(v)$ leaves and a degree-one root identified to v. Each internal node of this binary tree (that is, all nodes except for the leaves and the roots of degree 1) is further split into an arc. All arcs of the tree are oriented towards the root.
2. The set of outgoing arcs are split into a tree with $\deg_D^+(v)$ leaves in a similar fashion, with all arcs oriented away from the root. The roots of the in- and out-trees are both identified with v and thus equal to each other.
3. Finally, the vertex v itself is replaced by a directed 4-cycle, such that the single incoming arc from the first tree and the single outgoing arc from the second tree have adjacent endpoints on the cycle.

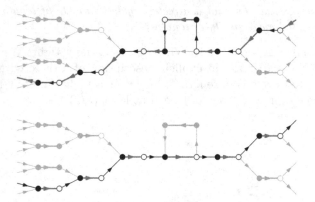

Fig. 3. Flipping the 4-cycles of each gadget in D' can be done with two successive cycle flips, using the Hamiltonian cycle of D.

The digraph D' thus obtained is bipartite and subcubic.

Furthermore, it is easy to see by construction that for every vertex $v \in V(D)$, the corresponding gadget in D' has at most

$$4\deg_D^-(v) + 4 + 4\deg_D^+(v) \leq 8(n-1) + 4 < 8n$$

vertices, such that $N := |V(D')| < n \cdot 8n = 8n^2$, and D' is of polynomial size.

Furthermore, the orientation of D' is such that every vertex in one side of the bipartition has outdegree one, and every vertex in the other has indegree one, hence it corresponds to a perfect matching M_1. By flipping the alternating 4-cycle in each gadget, we obtain another perfect matching M_2. We observe that M_2 can be obtained from M_1 in two cycle flips, by using the Hamiltonian cycle of D twice (see Fig. 3). Hence, while $M_1 \triangle M_2$ consists of n disjoint 4-cycles, M_2 is in fact at distance two from M_1 on the perfect matching polytope of D'. The underlying undirected graph of D' together with the two perfect matchings M_1 and M_2 therefore constitute an instance of the problem described in Theorem 1. We now show that any sequence of length at most k of alternating cycle flips transforming M_1 into M_2 can be turned in polynomial time into a cycle of length at least n^ε in D, for some $\varepsilon > 0$ depending solely on k.

Consider a sequence of k cycles $C'_1, C'_2, \ldots, C'_k$ in the underlying graph of D', such that C'_1 is alternating with respect to the matching M_1 in D' (and thus a directed cycle in D'); for every $i = 2, \ldots, k$ the cycle C'_i is alternating with respect to the perfect matching $M_1 \Delta C_1 \Delta \cdots \Delta C_{i-1}$ (and thus a directed cycle in the digraph obtained from D' after flipping the cycles $C'_1, C'_2, \ldots, C'_{i-1}$, in this order); and such that flipping all k alternating cycles in sequence transforms M_1 into M_2. Observe that the sum of the lengths of the cycles in this sequence must be at least n, since all the orientations of the 4-cycles in the n different gadgets in D' have to be flipped, and since every single cycle C'_i can intersect at most $|C'_i|$ different gadget-4-cycles. Let $\ell = \lceil n^{1/(k+2)} \rceil$. We have

$$\sum_{i=1}^{k} |C'_i| \geq n = (1 - o(1))\ell^{k+2} > \sum_{i=1}^{k} \ell^{i+1},$$

hence from the pigeonhole principle, at least one cycle C'_i in the sequence has length $|C'_i| > \ell^{i+1}$. From Lemma 3, applied with this value of i, we can now compute in polynomial time a directed cycle in D' of length at least $n^{\varepsilon'}$ for $\varepsilon' = 1/(k+2)$. Let us call this cycle C'. Note that for every gadget in D' corresponding to a vertex v of D, either C' is vertex-disjoint from this gadget, or it traverses it via exactly one directed path, consisting of a leaf-to-root path in the in-tree, a directed path of length 7 touching the 4-cycle of the gadget, and then a root-to-leaf path in the out-tree. From this it follows that by contracting the edges of the gadgets, the cycle C' in D' can be mapped to a cycle C in D. Note that the in- and out-degree of a vertex in D is at most $n - 1$, thus all the in- and out-trees in D' corresponding to the gadgets have depth at most $2\lceil \log_2 n \rceil$. Consequently, the length of C' can be shrinked by at most a factor of $4\lceil \log_2 n \rceil + 7$ by contracting the gadgets. In other words, we obtain a directed cycle C in D of length at least $n^{\varepsilon'}/(4\lceil \log_2 n \rceil + 7) = n^\varepsilon$, for $\varepsilon = \varepsilon' - o(1)$. Hence if we can obtain in polynomial time a sequence of at most $k = O(1)$ flips transforming M_1 into M_2, we can also find a cycle of length at least n^ε in D for some fixed $\varepsilon > 0$. This establishes the first statement of Theorem 1.

It remains to prove the second statement. We consider the second problem in Theorem 2, in which we seek a path of length at least $f(n) \log n$, for some computable function $f(n) = \omega(1)$. Suppose that for some $\delta > 0$ there is a polynomial-time algorithm that can find a sequence of at most $k = k(N) = \left\lfloor \left(\frac{1}{4} - \delta\right) \frac{\log N}{\log \log N} \right\rfloor$ flips transforming M_1 into M_2. Note that for n large enough

$$k + 2 \leq \left(\frac{1}{4} - \delta\right) \frac{\log 8n^2}{\log \log 8n^2} + 2 < \frac{1}{2}(1 - \delta)\frac{\log n}{\log \log n}.$$

Now, from the same reasoning as above, we can turn such an algorithm into a polynomial-time algorithm that finds a directed cycle in D of length at least

$$\frac{n^{1/(k+2)}}{4\lceil \log_2 n \rceil + 7} > \frac{n^{2 \cdot \log \log n / ((1-\delta)\log n)}}{O(\log n)} = \Omega\left(\frac{\log^{2/(1-\delta)} n}{\log n}\right) = f(n)\log n,$$

for a computable function $f(n) = \Omega(\log^{2\delta/(1-\delta)} n) = \omega(1)$. This, from Theorem 2, is impossible unless the Exponential Time Hypothesis fails.   □

In order to deduce Corollary 1 from Theorem 1, we need the following lemma.

Lemma 4 ($\star$).   *Given a bipartite graph $G = (V, E)$, let M_1 and M_2 be two perfect matchings in G at distance two on the perfect matching polytope, hence such that $M_2 = (M_1 \Delta C_1)\Delta C_2$ for some pair C_1, C_2 of cycles in G, and such that $M_1\Delta C_1 =: M'$ is also a perfect matching. Then there exists a weight function $w : E \to \mathbb{R}^+$ such that*

1. *M_2 is the unique maximum-weight perfect matching of G,*
2. *$w(M_1) < w(M') < w(M_2)$ (where $w(M) = \sum_{e\in M} w(e)$).*

In other words, there exists a linear program over the perfect matching polytope of G such that the path $M_1, M_1\Delta C_1 = M' = M_2\Delta C_2, M_2$ is a strictly monotone path and M_2 is the unique optimum.

2.3   Proof of Lemma 3

Proof (Lemma 3). Let the orientations $D_0, D_1, \ldots, D_t$ of G be given as input, together with a directed cycle C in D_t and a number $\ell \in \mathbb{N}$ such that $|C| > \ell^{t+1}$.

Our algorithm starts by computing the sequence of cycles $C_1, \ldots, C_t$ by determining for each $i \in [t]$ the set of edges with different orientation in D_{i-1} and D_i. Next we compute in polynomial time the subgraph H of G which is the union of the cycles $C_1, \ldots, C_t$ in G. We in particular compute a list of the vertex sets of its connected components, which we call $Z_1, \ldots, Z_c$ for some number $c \geq 1$.

We need the following fact, proved in the full version:

Claim ✚ ($\star$). *For each $r \in [c]$ the induced subdigraph $D_0[Z_r]$ of D_0 is strongly connected.*

Let $(x_0, x_1, \ldots, x_{k-1}, x_k = x_0)$ be the cyclic list of vertices on the directed cycle C in D_t, with edges oriented from x_i to x_{i+1} for all $i \in [k-1]$. By assumption on the input, we have $k = |C| > \ell^{t+1}$.

We first check if C is vertex-disjoint from H, in which case we may return C, which is then also a directed cycle in D_0 of length $k > \ell^{t+1} \geq \ell$, as desired.

Otherwise, C intersects some of the components of H. We then for each vertex $x_i \in V(C)$ compute a label $\text{lab}(x_i) \in [c+1]$, defined as $\text{lab}(x_i) := r$ if $x_i \in Z_r$ lies in the r-th component of H, and $\text{lab}(x_i) := c+1$ if x_i is not a vertex of H. We next compute an auxiliary weighted directed multigraph M on the vertex set $[c]$ as follows: For every maximal subsequence of C, of length at least two, of the form $x_i, x_{i+1}, \ldots, x_j$ (addition to be understood modulo k) such that $\mathrm{lab}(x_s) = c+1$ for all $s = i+1, \ldots, j-1$ (if any), we add an additional arc from $\mathrm{lab}(x_i)$ to $\mathrm{lab}(x_j)$ and give it weight $j - i$, the corresponding number of arcs in C. Note that the total arc weight in M is exactly $|C|$, while the total number of arcs is exactly $|V(C) \cap V(H)| \le |V(H)|$. The construction of M is illustrated in Fig. 4.

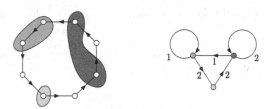

Fig. 4. Construction of the auxiliary directed multigraph M in the proof of Lemma 3. The cycle C is shown on the left, together with the connected components of H that it intersects. The resulting weighted directed multigraph M is shown on the right.

Furthermore, by definition every vertex in M has the same number of incoming and outgoing arcs. Hence, we may compute in polynomial time an edge-disjoint decomposition of M into directed cycles (including possible loops) in M. Let $W_1, \ldots, W_p$ for some $p \in \mathbb{N}$ be the list of edge-disjoint directed cycles in this decomposition of M. We now create, for each W_i, a directed cycle K_i in D_0 of length $|K_i| \ge \mathrm{weight}(W_i)$, where $\mathrm{weight}(W_i)$ is the total arc weight of W_i, as follows: Let $(l_0, l_1, \ldots, l_s = l_0)$ be the cyclic vertex-sequence of W_i.

For each arc (l_j, l_{j+1}) in W_i, we consider the corresponding subsequence $P(l_j, l_{j+1})$ of C which starts in Z_{l_j}, ends in $Z_{l_{j+1}}$, and all whose internal vertices are not contained in H. We note that since arcs outside H have the same orientation in D_0 and D_t, the subsequence $P(l_j, l_{j+1})$ is a directed path or a directed cycle also in D_0 which starts in Z_{l_j} and ends in $Z_{l_{j+1}}$.

We first check whether there exists an index j such that $P(l_j, l_{j+1})$ is a directed cycle. In this case, necessarily W_i is a loop (i.e. $s = 0$) and $l_j = l_{j+1} = l_0$. We thus may simply put $K_i := P(l_j, l_{j+1})$, with $\mathrm{weight}(W_i) = |K_i|$ satisfied by definition of the weights in M.

Otherwise, each of the $P(l_j, l_{j+1})$ is a directed path in D_0. We now make use of Claim ✦, which tells us that $D_0[Z_{l_j}]$ is strongly connected for every $j = 0, 1, \ldots, s - 1$. We may therefore compute in polynomial time for each $j = 0, 1, \ldots, s-1$ a directed path Q_j in $D_0[Z_{l_j}]$ (possibly consisting of a single vertex) which connects the endpoint of $P(l_{j-1}, l_j)$ to the starting point of $P(l_j, l_{j+1})$ (index-addition modulo s). Crucially, note that any two directed paths in the collection $\{P(l_j, l_{j+1}), Q_j | j = 0, 1, \ldots, s-1\}$ are vertex-disjoint except for shared common endpoints. We now compute the directed cycle K_i in D_0, which is the union of the directed paths $P(l_j, l_{j+1})$ and the directed paths Q_j for $j = 0, \ldots, s - 1$. It is clear that its length $|K_i|$ is lower-bounded by the sum of the lengths of the $P(l_j, l_{j+1})$, which by definition of M equals the sum of arc-weights on W_i, i.e., we indeed have $|K_i| \geq \text{weight}(W_i)$ also in this case.

After having computed the directed cycles $K_1, \ldots, K_p$ in D_0, the algorithm checks whether one of the cycles has length $|K_i| \geq \ell$. If so, it returns the cycle K_i and the algorithm stops with the desired output. Otherwise, we have $|K_i| < \ell$ for $i = 1, \ldots, p$, which implies that

$$\ell^{t+1} < |C| = \text{weight}(M) = \sum_{i=1}^{p} \text{weight}(W_i) \leq \sum_{i=1}^{p} |K_i| \leq p(\ell - 1).$$

Note that p is at most as large as the number of arcs in M, which in turn is bounded by $|V(H)|$. We thus obtain

$$\ell^{t+1} < |V(H)| \cdot (\ell - 1) \leq \sum_{i=1}^{t} |C_i| \cdot (\ell - 1).$$

This yields that

$$\sum_{i=1}^{t} |C_i| > \frac{\ell^{t+1}}{\ell - 1} > \sum_{i=1}^{t} \ell^i.$$

Therefore there exists $i \in \{1, \ldots, t\}$ such that $|C_i| > \ell^i$. The algorithm proceeds by finding one cycle C_i with this property. Note that C_i is a directed cycle in the orientation D_{i-1} of G. Hence a recursive call of the algorithm to the input $D_0, D_1, \ldots, D_{i-1}$ and the cycle C_i will yield a directed cycle of length at least ℓ in D_0, as desired. This proves the correctness of the described algorithm. As all steps between two recursive calls are executable in polytime in the size of G and t, and since there will clearly be at most $t - 1$ recursive calls in any execution of the algorithm, the whole algorithm runs in polynomial time, as desired. $\square$

References

1. Adler, I., Papadimitriou, C., Rubinstein, A.: On simplex pivoting rules and complexity theory. In: Lee, J., Vygen, J. (eds.) IPCO 2014. LNCS, vol. 8494, pp. 13–24. Springer, Cham (2014). https://doi.org/10.1007/978-3-319-07557-0_2
2. Aichholzer, O., et al.: Flip distances between graph orientations. Algorithmica **83**(1), 116–143 (2021)
3. Alon, N., Yuster, R., Zwick, U.: Color-coding. J. ACM **42**(4), 844–856 (1995)
4. Avis, D., Friedmann, O.: An exponential lower bound for Cunningham's rule. Math. Program. **161**(1–2), 271–305 (2017)
5. Barahona, F., Tardos, É.: Note on Weintraub's minimum-cost circulation algorithm. SIAM J. Comput. **18**(3), 579–583 (1989)
6. Björklund, A., Husfeldt, T., Khanna, S.: Approximating longest directed paths and cycles. In: Díaz, J., Karhumäki, J., Lepistö, A., Sannella, D. (eds.) ICALP 2004. LNCS, vol. 3142, pp. 222–233. Springer, Heidelberg (2004). https://doi.org/10.1007/978-3-540-27836-8_21

7. Bland, R.G.: New finite pivoting rules for the simplex method. Math. Oper. Res. **2**(2), 103–107 (1977)
8. Bonamy, M., et al.: The perfect matching reconfiguration problem. In: Rossmanith, P., Heggernes, P., Katoen, J. (eds.) 44th International Symposium on Mathematical Foundations of Computer Science, MFCS 2019, August 26–30, 2019, Aachen, Germany. LIPIcs, vol. 138, pp. 80:1–80:14. Schloss Dagstuhl - Leibniz-Zentrum für Informatik (2019)
9. Borgwardt, S., Brand, C., Feldmann, A.E., Koutecký, M.: A note on the approximability of deepest-descent circuit steps. Oper. Res. Lett. **49**(3), 310–315 (2021)
10. Borgwardt, S., Finhold, E., Hemmecke, R.: On the circuit diameter of dual transportation polyhedra. SIAM J. Discrete Math. **29**(1), 113–121 (2015)
11. Borgwardt, S., Viss, C.: A polyhedral model for enumeration and optimization over the set of circuits. Discret. Appl. Math. **308**, 68–83 (2022)
12. Bousquet, N., Hatanaka, T., Ito, T., Mühlenthaler, M.: Shortest reconfiguration of matchings. In: Sau, I., Thilikos, D.M. (eds.) WG 2019. LNCS, vol. 11789, pp. 162–174. Springer, Cham (2019). https://doi.org/10.1007/978-3-030-30786-8_13
13. Chvátal, V.: On certain polytopes associated with graphs. J. Comb. Theory, Ser. B **18**(2), 138–154 (1975)
14. Cioabă, S.M., Royle, G., Tan, Z.K.: On the flip graphs on perfect matchings of complete graphs and signed reversal graphs. Australas. J. Comb. **81**, 480–497 (2021)
15. De Loera, J.A., Hemmecke, R., Lee, J.: On augmentation algorithms for linear and integer-linear programming: from Edmonds-Karp to Bland and beyond. SIAM J. Optim. **25**(4), 2494–2511 (2015)
16. De Loera, J.A., Kafer, S., Sanità, L.: Pivot rules for circuit-augmentation algorithms in linear optimization. SIAM J. Optim. **32**(3), 2156–2179 (2022)
17. Diaconis, P.W., Holmes, S.P.: Matchings and phylogenetic trees. Proc. Natl. Acad. Sci. USA **95**(25), 14600–14602 (1998)
18. Diaconis, P.W., Holmes, S.P.: Random walks on trees and matchings. Electron. J. Probab. **7**(6), 1–17 (2002)
19. Disser, Y., Friedmann, O., Hopp, A.V.: An exponential lower bound for Zadeh's pivot rule. CoRR abs/1911.01074 (2019). http://arxiv.org/abs/1911.01074
20. Disser, Y., Skutella, M.: The simplex algorithm is NP-mighty. ACM Trans. Algorithms **15**(1), 5:1–5:19 (2019)
21. Fearnley, J., Savani, R.: The complexity of the simplex method. In: Servedio, R.A., Rubinfeld, R. (eds.) Proceedings of the Forty-Seventh Annual ACM on Symposium on Theory of Computing, STOC 2015, Portland, OR, USA, June 14–17, 2015, pp. 201–208. ACM (2015)
22. Gabow, H.N., Nie, S.: Finding a long directed cycle. ACM Trans. Algorithms **4**(1), 7:1–7:21 (2008)
23. Gima, T., Ito, T., Kobayashi, Y., Otachi, Y.: Algorithmic meta-theorems for combinatorial reconfiguration revisited. In: Chechik, S., Navarro, G., Rotenberg, E., Herman, G. (eds.) 30th Annual European Symposium on Algorithms, ESA 2022, September 5–9, 2022, Berlin/Potsdam, Germany. LIPIcs, vol. 244, pp. 61:1–61:15. Schloss Dagstuhl - Leibniz-Zentrum für Informatik (2022)
24. Goldfarb, D., Sit, W.Y.: Worst case behavior of the steepest edge simplex method. Discret. Appl. Math. **1**(4), 277–285 (1979)
25. Gupta, M., Kumar, H., Misra, N.: On the complexity of optimal matching reconfiguration. In: Catania, B., Královič, R., Nawrocki, J., Pighizzini, G. (eds.) SOFSEM 2019. LNCS, vol. 11376, pp. 221–233. Springer, Cham (2019). https://doi.org/10.1007/978-3-030-10801-4_18

26. Hansen, T.D., Zwick, U.: An improved version of the random-facet pivoting rule for the simplex algorithm. In: Servedio, R.A., Rubinfeld, R. (eds.) Proceedings of the Forty-Seventh Annual ACM on Symposium on Theory of Computing, STOC 2015, Portland, OR, USA, June 14–17, 2015, pp. 209–218. ACM (2015)
27. van den Heuvel, J.: The complexity of change. In: Blackburn, S.R., Gerke, S., Wildon, M. (eds.) Surveys in Combinatorics 2013, London Mathematical Society Lecture Note Series, vol. 409, pp. 127–160. Cambridge University Press (2013)
28. Ito, T., et al.: On the complexity of reconfiguration problems. Theor. Comput. Sci. **412**(12–14), 1054–1065 (2011)
29. Ito, T., Kakimura, N., Kamiyama, N., Kobayashi, Y., Okamoto, Y.: Shortest reconfiguration of perfect matchings via alternating cycles. SIAM J. Discret. Math. **36**(2), 1102–1123 (2022)
30. Iwata, S.: On matroid intersection adjacency. Discret. Math. **242**(1–3), 277–281 (2002)
31. Jeroslow, R.G.: The simplex algorithm with the pivot rule of maximizing criterion improvement. Discret. Math. **4**(4), 367–377 (1973)
32. Kafer, S., Pashkovich, K., Sanità, L.: On the circuit diameter of some combinatorial polytopes. SIAM J. Discret. Math. **33**(1), 1–25 (2019)
33. Kaminski, M., Medvedev, P., Milanic, M.: Complexity of independent set reconfigurability problems. Theor. Comput. Sci. **439**, 9–15 (2012)
34. Klee, V., Minty, G.J.: How good is the simplex algorithm? In: Inequalities, III (Proc. Third Sympos., Univ. California, Los Angeles, Calif., 1969; dedicated to the memory of Theodore S. Motzkin), pp. 159–175. Academic Press, New York (1972)
35. Monroy, R.F., Flores-Peñaloza, D., Huemer, C., Hurtado, F., Wood, D.R., Urrutia, J.: On the chromatic number of some flip graphs. Discret. Math. Theor. Comput. Sci. **11**(2), 47–56 (2009)
36. Nishimura, N.: Introduction to reconfiguration. Algorithms **11**(4), 52 (2018)
37. Santos, F.: A counterexample to the Hirsch conjecture. Ann. Math. **176**(1), 383–412 (2012)
38. Schrijver, A.: Combinatorial Optimization: Polyhedra and Efficiency, Algorithms and Combinatorics, vol. 24. Springer (2003)
39. Williams, V.V.: On some fine-grained questions in algorithms and complexity. In: Proceedings of the International Congress of Mathematicians (ICM 2018), pp. 3447–3487. World Scientific (2018)

Monoidal Strengthening and Unique Lifting in MIQCPs

Antonia Chmiela[1], Gonzalo Muñoz[2(✉)], and Felipe Serrano[3]

[1] Zuse Institute Berlin, Berlin, Germany
chmiela@zib.de
[2] Institute of Engineering Sciences, Universidad de O'Higgins, Rancagua, Chile
gonzalo.munoz@uoh.cl
[3] I2DAMO GmbH, Berlin, Germany
serrano@i2damo.de

Abstract. Using the recently proposed maximal quadratic-free sets and the well-known monoidal strengthening procedure, we show how to improve intersection cuts for quadratically-constrained optimization problems by exploiting integrality requirements. We provide an explicit construction that allows an efficient implementation of the strengthened cuts along with computational results showing their improvements over the standard intersection cuts. We also show that, in our setting, there is *unique lifting* which implies that our strengthening procedure is generating the best possible cut coefficients for the integer variables.

Keywords: MIQCP · monoidal strengthening · unique lifting

1 Introduction

In recent years, we have seen multiple efforts in generating valid linear inequalities to quadratically constrained quadratic programs (QCQPs) which, using an epigraph formulation, we can assume have the form

$$\min\{\bar{c}^\mathsf{T} s : s \in S \subseteq \mathbb{R}^p\} \tag{1}$$

where $S = \{s \in \mathbb{R}^p : s^\mathsf{T} Q_i s + b_i^\mathsf{T} s + c_i \leq 0, \ i = 1, \ldots, m\}$. One of the approaches to generate such valid inequalities has been the *intersection cut* paradigm [1,13, 19] which works as follows. We assume we have $f \notin S$, a basic feasible solution of a *linear programming* (LP) relaxation of (1). Additionally, we assume we have a simplicial conic relaxation $K \supseteq S$ with apex f, and an S-free set C—a convex set satisfying $\mathrm{int}(C) \cap S = \emptyset$—such that $f \in \mathrm{int}(C)$. Using these ingredients, we can find a cutting plane separating f from S. In Fig. 1 we show a simple intersection cut in the case when all p rays of K intersect the boundary of the S-free set C. In such case, the intersection cut is simply defined by the hyperplane containing all such intersection points. It is well known that one can assume C to be described as $C = \{s \in \mathbb{R}^p : \phi(s - f) \leq 1\}$ where ϕ is a sublinear function. For instance, ϕ

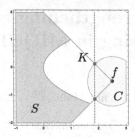

Fig. 1. An intersection cut (red) separating f from S (blue). The cut is computed using the intersection points of an S-free set C (green) and the rays of a simplicial cone $K \supseteq S$ (boundary in orange) with apex $f \notin S$. Figure obtained from [8] (Color figure online).

can be chosen as the gauge of $C - f$ [17]. Further assuming w.l.o.g that the LP relaxation is in standard form, we consider the constraint $f + \sum_{i=1}^{p} r^i s_i \in S$ with $r^i \in \mathbb{R}^p$ (e.g. the extreme rays of K) and $s_i \in \mathbb{R}_+$. Under these considerations, the intersection cut separating f is

$$\sum_{i=1}^{p} \phi(r^i)s_i \geq 1. \tag{2}$$

In [8,16] a method for constructing *maximal* quadratic-free sets (which ensures separation of any basic feasible solution $f \notin S$) and a computational implementation was developed, with positive results in a broad class of problems.

One of the limitations of these cutting planes is that they do not use any integrality information: if we were to add integrality requirements to (1)—thus obtaining an MIQCP—the intersection cuts would be completely oblivious to this.

In this work, we remedy this via the *monoidal strengthening* framework [2]; a strengthening of intersection cuts based on integrality information. Monoidal strengthening leverages the fact that some of the s_i in (2) are integer. The idea is to take the relation $f + \sum_{i=1}^{p} r^i s_i \in S$ and modify it in the following way. Assume that all s_i are restricted to be integer. The above relation implies that $f + \sum_{i=1}^{p}(r^i + m_i)s_i \in S + \sum_{i=1}^{p} m_i s_i$. The points $\sum_{i=1}^{p} m_i s_i$ form a monoid $M = \{m : m = \sum_{i=1}^{p} m_i s_i, s_i \in \mathbb{Z}_+\}$, that is, M satisfies $0 \in M$ and $M + M = M$. Thus, we obtain the new relation: $f + \sum_{i=1}^{p}(r^i + m_i)s_i \in S + M$. If it turns out that C is still $S + M$ free, then we can use the function ϕ to generate a new cut. The above is summarized in the following result by Balas and Jeroslow.

Theorem 1 ([2] **Theorem 1**). *Let M be a monoid such that C is $S + M$-free and let $I = \{i \in [p] : s_i \in \mathbb{Z}\}$ be the index set of the integer variables. Then,*

$$\sum_{i \notin I} \phi(r^i)s_i + \sum_{i \in I} \inf_{m \in M} \phi(r^i + m)s_i \geq 1$$

is valid and dominates the intersection cut.

There are two main challenges in this technique. Firstly, to find a monoid M such that C stays $S + M$-free. Note that, equivalently, we can find a monoid M such that $C - M$ is (possibly non-convex) S-free[1]. Secondly, to efficiently solve the problem $\psi(r) := \inf_{m \in M} \phi(r + m)$ for r the rays associated to integer non-basic variables and thus obtain a stronger cut coefficients. In this work we tackle both tasks.

In most of this article S is defined using a single quadratic inequality. As noted in [16], using linear transformations (diagonalization and homogenization), one can shift the focus from a generic quadratic set, $S = \{s \in \mathbb{R}^p : s^\top Q s + b^\top s + c \leq 0\}$, to one of the following two sets:

$$S^h := \{(x, y) \in \mathbb{R}^{n+m} : \|x\| \leq \|y\|\}, \tag{3}$$

$$S^g := \{(x, y) \in \mathbb{R}^{n+m} : \|x\| \leq \|y\|, a^\top x + d^\top y = -1\}. \tag{4}$$

where $\max\{\|a\|, \|d\|\} = 1$. Whether S gets mapped to S^h or S^g depends on whether the quadratic defining S is homogeneous or not. Thus, one of the goals of this paper will be: using C as maximal S^h- and S^g-free sets of [16], to find a monoid M such that C is $S^h + M$- or $S^g + M$-free and, subsequently, strengthen the corresponding intersection cut.

Monoidal strengthening is also related to lifting [3,9–11]. A nice property of ψ is that it is subadditive[2]. This implies that with ψ we obtain *sequence independent lifting*, i.e. we can apply the strengthening to all integer variables at the same time. However, the monoidal lifting function ψ is, in general, just one possible way of lifting. We can define the best possible coefficient that a particular integer variable can achieve, with the so-called *lifting function* π [4,10].

$$\pi(r) = \sup \left\{ \frac{1 - \phi(s)}{\sigma} : f + s + \sigma r \in S, \sigma \in \mathbb{Z}_{\geq 1} \right\}. \tag{5}$$

In general, π is not subadditive so we do not have sequence independent lifting with it [4]. When it is subadditive, we say that there is *unique lifting*, because π dominates any other lifting. For the case $S = \mathbb{Z}^n \cap P$ it is well understood when we have unique lifting [3]. Our final goal is to show that in our setting there is unique lifting; more specifically, we show that choosing ϕ to be a *minimal representation*[3] of $C - f$, we have that $\pi = \psi$.

Contributions. Our main contributions are: (1) we show that the monoidal strengthening framework does not produce any strengthening when S is defined using a homogeneous quadratic; (2) in the non-homogeneous case, we show a family of cases where monoidal strengthening can be applied and an explicit monoid construction based on a maximal S^g-free set of [16] which can be used for this strengthening; (3) we show an explicit formula for how to efficiently

---

[1] With a slight abuse of notation, we refer to a *non-convex* set $C - M$ as S-free whenever the convex set $C - m$ is S-free for every $m \in M$.

[2] A function ψ is *subadditive* if $\psi(x + y) \leq \psi(x) + \psi(y)$.

[3] This means that if ρ is such that $C = \{s \in \mathbb{R}^p : \rho(s - f) \leq 1\}$ then $\rho(s) \geq \phi(s)$.

compute $\psi(r)$ in practice; (4) we show that in our setting there is unique lifting which, in particular, implies that ψ yields the best coefficients in the strengthening of the intersection cut; and (5) we present extensive computational results that show the impact of this strengthening procedure.

We remark that, even though our constructions are based on the structure of one quadratic, they can also be applied to MIQCPs with multiple quadratic inequalities: using our approach, it suffices to have one quadratic inequality being violated in order to ensure separation.

In the interest of space, we do not present all details in this extended abstract. We refer the reader to our preprint available in [7].

2   Monoidal Strengthening in the Homogeneous Case

In this section, we analyze the case of S^h and show that the monoidal strengthening framework does not produce any improvements when the cuts are created using *maximal* S^h-free sets. The main reason behind this fact is that S^h is a cone, and consequently every maximal S^h-free set is a convex cone [6, Corollary 3][4]; we show below why this is not a good setting for monoidal strengthening. In fact, the results in this section apply to a generic *closed cone* S and are stated with respect to such set.

As mentioned before, for a given S-free set C, we are interested in finding a monoid M such that $C - M$ is S-free. The following result shows that in this case $C - M = M$. We remark that cone($\cdot$) is the cone generated by a set, which may not be convex.

Theorem 2. *Let* $S, C \subseteq \mathbb{R}^n$ *where* S *is a closed cone and* C *is a convex maximal* S-free *set. Let* $M \subseteq \mathbb{R}^n$ *be a monoid such that* $C - M$ *is* S-free, *then* $C - M = C$. *In particular, this implies that the cut obtained from monoidal strengthening would be the same as the standard intersection cut obtained through* C.

Proof (sketch). Since M is a monoid, $0 \in M$ and thus $C \subseteq C - M$. It can be shown that cl cone(M) is a convex cone such that $C -$ cl cone(M) is S-free. Note that $C -$ cl cone(M) is convex, and thus the maximality of C implies that $C -$ cl cone(M) $\subseteq C$. Since $C - M \subseteq C -$ cl cone(M), we conclude $C - M = C$.

This last result shows that in the presence of a maximal S-free set C, there is not much to be gained from the monoidal strengthening framework when S is a cone. This negative property, nonetheless, can be reinterpreted as a way of detecting "non-maximality" of an S-free set: if one could find a monoid M such that $C - M$ is S-free and $C - M \neq C$, then C is not maximal. We formalize this in the next result.

---

[4] This citation deals with a particular set S, but the proof can be easily extended to any conic set S.

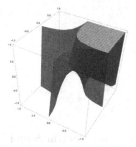   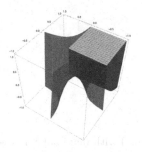

(a) Slices of S (blue) and C_θ (orange)   (b) Slices of S (blue) and C'_θ (orange)

Fig. 2. Three-dimensional slices of S, C_θ and C'_θ in Example 1 given by $a = 1/10$.

Proposition 1. *Let S be a closed cone and let C be a full dimensional closed convex S-free cone. If there exists $r \notin -C$ such that C is $S + \text{cone}(r)$-free, then C is not a maximal S-free set. Furthermore, $C + \text{cone}(-r)$ is S-free and strictly contains C.*

The next example illustrates an application of the last proposition, in a connection with the work of [6].

Example 1. Consider the set $S = \{(a, b, c, d) \in \mathbb{R}^4 : ad = bc, a \geq 0\}$. Although this set does not fall into either the forms S^h or S^g which are our main objects of interest, it is still a closed conic set to which the results of this section apply. The set S is studied in [6], and appears when using lifted variables $X_{i,j}$ representing bilinear terms $x_i x_j$. Let $C_\theta = \{(a, b, c, d) \in \mathbb{R}^4 : \cos(\theta)(a + d) + \sin(\theta)(b - c) \geq \sqrt{(a - d)^2 + (b + c)^2}\}$. In [6, Theorem 7], the authors show that C_θ is maximal S-free for values of θ that satisfy $\cos(\theta) = 0$ or $\sin(\theta) = 0$. Using the results of this section, we can prove that if θ is such that $\cos(\theta) \neq 0$ and $\sin(\theta) \neq 0$, then C_θ is not maximal S-free. More specifically, we can show that $-e_4 \notin -C_\theta$ and that C_θ is $S + \text{cone}(-e_4)$-free, where $e_4 = (0, 0, 0, 1)$. Using Proposition 1, this implies that $C'_\theta = C_\theta + \text{cone}(e_4)$ is S-free and strictly contains C_θ. We leave out the details for the sake of brevity. In Fig. 2 we show a 3-dimensional slice of the 4-dimensional sets S and C_θ, for $\theta = \pi/4$, showing how the S-free was enlarged.

We remark that one can actually show that C'_θ is *maximal* S-free using the maximality criteria of [15].

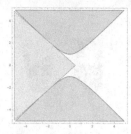 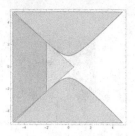

(a) S (blue) with maximal S-free set C (orange). In this case the two inequalities of C intersect S.

(b) Set of points not in S and "to the left of the exposing points" (green). Note that the green region is not contained on the orange region: see the top left and bottom left.

Fig. 3. Constuction of the monoid for a maximal S-free set.

3   Monoidal Strengthening in the Non-homogeneous Case

In this case, the monoidal strengthening framework does produce improvements. The intuition for our construction is as follows. Consider the maximal S-free set C represented in Fig. 3a.

The set is maximal, because all of its defining inequalities $\alpha^\mathsf{T} s \leq \beta$ have *exposing points* [16], that is, there exists $s_0 \in C \cap S$ with $\alpha^\mathsf{T} s_0 = \beta$ such that if $\gamma^\mathsf{T} s \leq \delta$ is any other non-trivial valid inequality for C that is tight at s_0, then there exists a $\mu > 0$ such that $\gamma = \mu\alpha$ and $\beta = \mu\delta$. For example, if C is a polyhedron and $s_0 \in C \cap S$ is an exposing point of an inequality, then that inequality is a facet and s_0 is in its relative interior. Thus, in the example of Fig. 3a, the two exposing points of C are the points of the facets of C that are tangent to S. We see that a way of translate C such that the translation is S-free is by moving the apex of C to a point not in S and to the left of the exposing points (see Fig. 3b). This is the basic idea behind our monoid construction, and below we show how to formalize it.

3.1   A Technical Consideration for S^g

Before motivating the construction the monoid, we need to provide some details on the construction of maximal S^g-free presented in [16]. This construction starts from the maximal S^h-free set

$$C_\lambda = \{(x, y) \in \mathbb{R}^{n+m} : \|y\| \leq \lambda^\mathsf{T} x\}, \tag{6}$$

where λ is a vector in the unit sphere. Note that C_λ can be equivalently described as $C_\lambda = \{(x, y) \in \mathbb{R}^{n+m} : \beta^\mathsf{T} y \leq \lambda^\mathsf{T} x \ \forall \beta \in D_1\}$ where D_1 is the unit sphere of appropriate dimension. The proof that C_λ is maximal S^h-free boils down to noting that for each $\hat{\beta}$, the vector $(\lambda, \hat{\beta}) \in S^h \cap C_\lambda$ is tight for the inequality

$\hat{\beta}^\mathsf{T} y \le \lambda^\mathsf{T} x$ and for no other of C_λ. This means that each inequality of C_λ indexed by β has an *exposing point* in S^h.

Moving to the non-homogeneous case, since $S^g = S^h \cap H$, where $H = \{(x,y) \in \mathbb{R}^{n+m} : a^\mathsf{T} x + d^\mathsf{T} y = -1\}$, the set $C = C_\lambda \cap H$ is clearly S^g-free[5], but it is not necessarily maximal. The maximal S^g-free constructed in [16] first identifies the inequalities of C_λ for which an exposing point can be found in H, keeps them, and relaxes the rest.

The inequalities given by β such that $\|\beta\| = 1$ and $a^\mathsf{T}\lambda + d^\mathsf{T}\beta < 0$ are the ones that have the desired exposing points; these exposing points are

$$-\frac{1}{a^\mathsf{T}\lambda + d^\mathsf{T}\beta}(\lambda, \beta)$$

Maximality of the resulting set is shown using the exposing points above and, for relaxed inequalities, a diverging sequence in S^g that approaches the inequality indefinitely (an exposing sequence). This is due to the fact that these relaxed inequalities may have never intersect S^g.

In our current monoid construction, we require that all inequalities to have exposing points. This requirement translates to $a^\mathsf{T}\lambda + d^\mathsf{T}\beta < 0$ for all β with $\|\beta\| = 1$. This, in turn, reduces to $\|d\| < -a^\mathsf{T}\lambda$. Note that this condition implies that $C = C_\lambda \cap H$ is maximal S^g-free with respect to H [16]. Additionally, this implies that we can assume $\|a\| = \max\{\|a\|, \|d\|\} = 1$. We believe that when these assumptions are not fulfilled, monoidal strengthening cannot be applied. Proving this conjecture is part of future work.

3.2 Monoid Construction

Using the considerations of the previous section, we can formalize the notion of "left of the exposing points": we first consider the halfspace $\{(x,y) \in \mathbb{R}^{n+m} : (a - \lambda^\mathsf{T} a\lambda)^\mathsf{T} x \ge 0\}$ which contains the exposing points and the directions of lineality of $C_\lambda \cap H$. Secondly, when translating C_λ by a vector m we can modify m by a vector in the lineality space of C_λ without changing the translation. Thus, we restrict to vectors m that live in a subspace that contains the exposing points and is orthogonal to the lineality space of $C_\lambda \cap H$. This subspace is given by $\langle\{\lambda, a\}\rangle \times \mathbb{R}^m$ where $\langle\{\lambda, a\}\rangle$ denotes the span of $\{\lambda, a\}$. Thus, we have the following set representing the points "left of the exposing points":

$$L = \{(x,y) \in \langle\{\lambda, a\}\rangle \times \mathbb{R}^m : a^\mathsf{T} x + d^\mathsf{T} y = -1, \|x\| \ge \|y\|, (a - \lambda^\mathsf{T} a\lambda)^\mathsf{T} x \ge 0\}.$$

To obtain the translation we find the apex of $C_\lambda \cap H$ in the space $\langle\{\lambda, a\}\rangle \times \mathbb{R}^m$. This point is given by

$$\nu = (x_0, 0) := \left(\frac{-1}{1 - (\lambda^\mathsf{T} a)^2}a + \frac{\lambda^\mathsf{T} a}{1 - (\lambda^\mathsf{T} a)^2}\lambda, 0\right). \tag{7}$$

---

[5] Note that S^g is contained on a halfspace, so S^g-freeness is with respect to the induced topology in H.

Thus, $L - \nu$ is a candidate to represent the translations of C that would result in an S^g-free set. Note that the assumptions $\|d\| < -\lambda^{\mathsf{T}} a$ and $\|d\| \le 1$ imply $(\lambda^{\mathsf{T}} a)^2 < 1$, thus ν is well-defined. Recall that the translations we consider for C are given by "minus the monoid" and that a monoid must contain the origin, therefore our candidate for a monoid is

$$M = \{(x, y) \in \langle\{\lambda, a\}\rangle \times \mathbb{R}^m : a^{\mathsf{T}} x + d^{\mathsf{T}} y = 0, \|x - x_0\| \ge \|y\|, \qquad (8)$$
$$(a - \lambda^{\mathsf{T}} a\lambda)^{\mathsf{T}} x \le -1\} \cup \{(0,0)\}.$$

Theorem 3. *Let M be defined as in* (8) *with $\|d\| < -\lambda^{\mathsf{T}} a$ and $\|a\| = \|\lambda\| = 1$. The set M is a monoid.*

Proof (sketch). This proof is highly technical, so we just present the high-level strategy for obtaining the desired result. See [7] for the details.

We equivalently show that $-M$ is a monoid. Thus, we take two vector $(x_i, y_i) \in -M$, $i = 1, 2$, and show that their sum is in $-M$. This is trivial whenever one of the vectors is $(0,0)$. The linear constraints in the definition of $-M$ are satisfied trivially, hence the main argument is to show that $\|x_1 + x_2 + x_0\| \ge \|y_1 + y_2\|$. This is equivalent to showing that the value of the following optimization problem is non-negative.

$$\min_{x_i, y_i} \{\|x_1 + x_2 + x_0\|^2 - \|y_1 + y_2\|^2 : (x_i, y_i) \in -M \setminus \{(0,0)\}, i = 1, 2\}$$

Using that $\|x_i + x_0\| \ge \|y_i\|$ and $(a - \lambda^{\mathsf{T}} a\lambda)^{\mathsf{T}} x_i \ge 1 \Leftrightarrow -x_0^{\mathsf{T}} x_i \ge \|x_0\|^2$, we can lower bound the objective function by $\|x_1 + x_2 + x_0\|^2 - \|y_1 + y_2\|^2 \ge 2x_1^{\mathsf{T}} x_2 - 2y_1^{\mathsf{T}} y_2 - \|x_0\|^2$. Hence, to show that

$$\min_{x_i, y_i} \{x_1^{\mathsf{T}} x_2 - y_1^{\mathsf{T}} y_2 : (x_i, y_i) \in -M \setminus \{(0,0)\}, i = 1, 2\} \qquad (P)$$

is lower bounded by $\frac{1}{2}\|x_0\|^2$ suffices. Note that we can decompose $y_i = \omega_i d + \rho_i$ where ρ_i is orthogonal to d. Furthermore, since $x_i \in \langle\{a, \lambda\}\rangle$, we can write $x_i = \theta_i a + \eta_i \lambda$. Using this together with the fact that $\lambda^{\mathsf{T}} x_0 = 0$ and $a^{\mathsf{T}} x_0 = -1$, the hyperplane in $-M$ becomes $0 = a^{\mathsf{T}} x_i + d^{\mathsf{T}} y_i = \theta_i + \eta_i \lambda^{\mathsf{T}} a + \omega_i \|d\|^2$. Furthermore, we get $-x_0^{\mathsf{T}} x_i \ge \|x_0\|^2 \Leftrightarrow \theta_i \ge \|x_0\|^2$, and expanding the nonlinear constraint we reformulate problem (P) as

$$\min_{\theta_i, \eta_i, \omega_i, \rho_i} \quad \theta_1\theta_2 + \eta_1\eta_2 + \theta_1\eta_2\lambda^{\mathsf{T}} a + \theta_2\eta_1\lambda^{\mathsf{T}} a - w_1 w_2 \|d\|^2 - \rho_1^{\mathsf{T}} \rho_2$$
$$\text{s.t.} \quad 0 \le \theta_i^2 + \eta_i^2 + 2\theta_i\eta_i\lambda^{\mathsf{T}} a - 2\theta_i + \|x_0\|^2 - w_i^2\|d\|^2 - \|\rho_i\|^2$$
$$\|x_0\|^2 \le \theta_i \qquad\qquad (P_{exp})$$
$$\|d\|^2 \omega_i = -\theta_i + \eta_i \lambda^{\mathsf{T}} a$$

The remainder of the proof focuses on showing the desired lower bound for this problem. The key elements of the proof involve: first showing that the problem is simply bounded, then showing that constraint $\|x_0\|^2 \le \theta_i$ can be assumed to be tight. This is shown leveraging results from [18]. Using this we show the desired lower bound $(P_{exp}) \ge \frac{1}{2}\|x_0\|^2$.

The last required result for monoidal strengthening is the following.

Theorem 4. *Let S^g and C_λ be defined as in (4) and (6) respectively, and $H = \{(x,y) \in \mathbb{R}^{n+m} : a^\mathsf{T}x + d^\mathsf{T}y = -1\}$. Let M be defined as in (8) with $\|d\| < -\lambda^\mathsf{T}a$ and $\|a\| = \|\lambda\| = 1$. The set $C_\lambda \cap H - M$ is S^g-free.*

4   Solving the Monoidal Strengthening Problem

In order to strengthen the cut using Theorem 1 and the construction in Sect. 3, we need to solve $\psi(r) = \inf_{m \in M} \phi(r+m)$, where ϕ is such that $C_\lambda \cap H = \{s : \phi(s-f) \le 1\}$. From now on, $\lambda = \frac{f_x}{\|f_x\|}$, where f is the point we want to separate, i.e., $f \notin S^g$. Furthermore, we restrict $C_\lambda \cap H$ to $\langle\{\lambda, a\}\rangle \times \mathbb{R}^m$ because any representation of $C_\lambda \cap H$ is invariant in the directions of the lineality space of $C_\lambda \cap H$, namely, $\langle\{\lambda, a\}\rangle^\perp \times \{0\}$. Thus, we define $C = C_\lambda \cap H \cap \langle\{\lambda, a\}\rangle \times \mathbb{R}^m$. Likewise, we restrict all rays to $\langle\{\lambda, a\}\rangle \times \mathbb{R}^m$.

We work with the minimal representation of $C - f$; we can prove it is given by

$$\phi(s) = \begin{cases} \sup_{\|\beta\|=1} \frac{\beta^\mathsf{T}s_y - \lambda^\mathsf{T}s_x}{\lambda^\mathsf{T}f_x - \beta^\mathsf{T}f_y}, & \text{if } s \in H \text{ and } s_x \in \langle\{\lambda, a\}\rangle \\ +\infty, & \text{otherwise.} \end{cases} \tag{9}$$

The monoidal problem is equivalent to $\psi(r) = \inf\{\tau : \phi(r+m) \le \tau, m \in M\}$. In order to understand this problem better, we need to understand the set $\{s : \phi(s) \le \tau\}$.

Lemma 1. *Let ϕ be the minimal representation of $C - f$ given in (9). Then $\{s : \phi(s) \le \tau\} = C - \nu - \tau(f - \nu)$, where ν is defined in (7) (the apex of $C_\lambda \cap H$ in the space $\langle\{\lambda, a\}\rangle \times \mathbb{R}^m$).*

From this lemma, we can show that

$$\psi(r) = \inf\{\tau : r + \nu + \tau(f - \nu) \in C - M\}. \tag{10}$$

In other words, solving the monoidal strengthening problem reformulates to finding the first intersection point between the line $l(\tau) = r + \nu + \tau(f - \nu)$, and the set $C - M$.

It can be shown that $C - M = L \cup C$, thus, $\psi(r) = \inf\{\tau : l(\tau) \in L \cup C\} = \min\{\tau_1, \tau_2\}$, where $\tau_1 = \inf\{\tau : l(\tau) \in L\}$ and $\tau_2 = \inf\{\tau : l(\tau) \in C\}$. Note that τ_2 corresponds to the normal intersection cut coefficient $\phi(r)$. The following proposition shows how to evaluate $\psi(r)$.

Proposition 2. *Let $\bar{\tau}$ be the largest root of the univariate quadratic equation $\|l_x(\tau)\|^2 = \|l_y(\tau)\|^2$. If the root exists and $l(\bar{\tau}) \in L$, then $\psi(r) = \bar{\tau}$. Otherwise, $\psi(r) = \phi(r)$.*

To finalize this section, we show how to use this result starting from a general quadratic constraint. Consider S to be defined by a general quadratic constraint, i.e., $S = \{s \in \mathbb{R}^p : s^\mathsf{T}Qs + b^\mathsf{T}s + c \le 0\}$ with $Q \in \mathbb{R}^{p \times p}$, $b \in \mathbb{R}^p$ and $c \in \mathbb{R}$. Let

$\bar{s} \notin S$ be the point we want to separate. In [8] the authors transform S using the eigenvalue decomposition $Q = V\Theta V^\mathsf{T}$. Let θ_i, $i \in [p]$, be the eigenvalues of Q, and let $I_+ = \{i : \theta_i > 0\}$, $I_- = \{i : \theta_i < 0\}$ and $I_0 = \{i : \theta_i = 0\}$. Furthermore, denote by v_i the i-th eigenvector of Q, that is, the i-th column of V. We avoid showing the full transformation here, but an important fact is that the conditions $\|d\| < -\lambda^\mathsf{T} a < 1$ we need for applying monoidal strengthening in S^g become

$$(V^\mathsf{T} b)_{I_0} = 0 \quad \wedge \quad c - \frac{1}{4} \sum_{i \in I_+ \cup I_-} \frac{(v_i^\mathsf{T} b)^2}{\theta_i} > 0 \tag{11}$$

The following result summarizes the necessary computations.

Proposition 3. *Suppose conditions* (11) *are met. The computation of* $\psi(r)$ *for a given ray* r *reduces to computing the largest root of* $A\tau^2 + B\tau + D = 0$ *where*

$$A = \sum_{i \in I_+ \cup I_-} \theta_i \left(v_i^\mathsf{T}(\bar{s} - \nu) \right)^2, \quad B = 2 \sum_{i \in I_+ \cup I_-} \theta_i \left(v_i^\mathsf{T}(\bar{s} - \nu) \right) \left(v_i^\mathsf{T}(r + \nu + \frac{b}{2\theta_i}) \right),$$

$$D = \sum_{i \in I_+ \cup I_-} \theta_i \left(v_i^\mathsf{T}(r + \nu + \frac{b}{2\theta_i}) \right)^2 + \kappa$$

$$\nu = -\frac{\kappa}{\sum_{j \in I_+} \theta_j (v_j^\mathsf{T}(\bar{s} + \frac{b}{2\theta_j}))^2} \sum_{j \in I_+} v_{ij}(v_j^\mathsf{T}(\bar{s} + \frac{b}{2\theta_j})) - \sum_{j \in I_+ \cup I_-} v_{ij} \frac{v_j^\mathsf{T} b}{2\theta_j}$$

$$\kappa = c - \frac{1}{4} \sum_{i \in I_+ \cup I_-} \frac{(v_i^\mathsf{T} b)^2}{\theta_i}$$

We note that we also need to compute the cut coefficient $\phi(r)$, *but this can also be done efficiently as shown in [8].*

The expressions on the previous proposition may not provide too much insight themselves, as they are accumulating a series of transformations to bring S to S^g. However, we believe their value relies in that, given an eigenvalue decomposition for a general quadratic inequality, one can simply plug-in the desired parameters and obtain a univariate quadratic that yields the strengthened coefficients of an intersection cut.

5   Unique Lifting

As mentioned in the introduction, monoidal strengthening is just one way of improving the cut coefficients of integer variables. The best possible coefficient that a particular integer variable can achieve is given by the lifting function π defined in (5). If π is subadditive, then there is unique lifting [4]. This means that the lifting using π can be applied simultaneously to all rays r^i corresponding to integer variables and dominates any other lifting. In this section, we show that if we use ϕ the minimal representation of $C - f$ shown in (9), we have $\pi = \psi$;

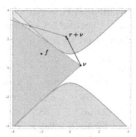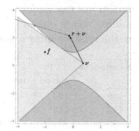

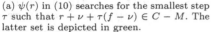

(a) $\psi(r)$ in (10) searches for the smallest step τ such that $r + \nu + \tau(f - \nu) \in C - M$. The latter set is depicted in green.

(b) $\pi_1(r)$ in (13) searches for the largest step τ such that $r + \nu + \tau(f - \nu) \in S^g - \mathrm{rec}(C)$. The latter set is depicted in green

Fig. 4. Comparison of the definitions of ψ and π_1 showing why they are equal. In both figures, S^g is depicted in blue and C in orange. (Color figure online)

since ψ is a subadditive function we obtain unique lifting and, moreover, that the procedure of Sect. 4 yields the best possible lifting coefficients.

Let π_1 be the restriction of π to $\sigma = 1$. Slightly reformulating the optimization problem, we see that

$$\pi_1(r) = \sup\left\{\tau \;:\; f + s + r \in S^g, \phi(s) \leq 1 - \tau\right\} \tag{12}$$

Using Lemma 1, we can show that $f + s + r \in S^g$ and $\phi(s) \leq 1 - \tau$ reformulate to $r + \nu + \tau(f - \nu) \in S^g - (C - \nu)$. Since $C - \nu = \mathrm{rec}(C)$, (12) becomes

$$\pi_1(r) = \sup\left\{\tau \;:\; r + \nu + \tau(f - \nu) \in S^g - \mathrm{rec}(C)\right\}. \tag{13}$$

Notice that problem (13) is very similar to the monoidal problem (10). Moreover, we can use this—plus structural results we leave out for the sake of space—to show that $\psi(r) = \pi_1(r)$. This is almost what we want. In Fig. 4 we illustrate the definitions of both π_1 and ψ to provide some intuition on why this result holds. To make the connection with π we prove the following lemma.

Lemma 2. *Let M be a monoid such that C is $S + M$-free and $\pi_1(r) = \sup\{1 - \phi(s) \;:\; f + s + r \in S\}$. If π_1 is subadditive, then $\pi = \pi_1$ and we have unique lifting.*

Since ψ is subadditive and $\psi = \pi_1$, we directly obtain the following theorem.

Theorem 5. *Consider ψ the monoidal strengthening function and π the lifting function, both defined using ϕ as in (9). Then $\pi = \psi$, in particular, there is unique lifting.*

6   Computational Results

In this section, we show results of computational experiments testing the efficacy of the monoidal strengthening procedure we propose. We embedded the

Table 1. Summary of results for branch-and-bound experiments. Rows labeled $[t, 7200]$ consider instances where one of the settings took at least t seconds. Columns labeled *relative* show the relative improvement of MONOIDAL compared to ICUTS.

subset	instances	ICUTS			MONOIDAL			relative	
		solved	time	nodes	solved	time	nodes	time	nodes
,all	189	113	221.87	5282	115	214.63	5321	0.97	0.97
$[0, 7200]$	115	113	22.81	936	115	21.56	883	0.95	0.94
$[1, 7200]$	83	81	67.62	2377	83	62.40	2184	0.92	0.92
$[10, 7200]$	81	79	72.54	2574	81	66.56	2341	0.92	0.91
$[100, 7200]$	23	21	724.66	186545	23	565.24	144747	0.78	0.78
$[1000, 7200]$	10	8	2475.04	631764	10	1252.96	307639	0.51	0.49

computation of the monoidal strengthening cut coefficients in SCIP 8.0 [5] as a subroutine of the already implemented intersection cut generator. As the underlying LP solver, we used CPLEX 12.10.0.0. For testing, we used a Linux cluster of Intel Xeon CPU E5-2680 0 2.70 GHz with 20MB cache and 64GB main memory. The time limit in all experiments was set to two hours. The test set we consider consists of the publicly available instances of the MINLPLib [14] and QPLib [12]. We selected all non-convex instances with (mixed)-integer constriants and at least one quadratic constraint of the correct case, leaving us with 95 instances. Furthermore, we filtered out all instances that are either infeasible, where no dual bound was found or where monoidal strengthening could not been applied. This leaves us with a heterogeneous test set of 63 instances with 8–23826 variables and 12–24971 constraints. All experiments are run with three different permutations for each instance. We treat every instance-permutation pair as an individual instance, since permuting the constraints and variables of a problem formulation may considerably change the solving process.

We consider two different settings that are both based on SCIP's default settings: ICUTS additionally generates the original intersection cuts, whereas MONOIDAL uses the strengthened cutting planes if possible. Furthermore, we restrict ICUTS and MONOIDAL to add at most 20 intersection cuts per quadratic constraint. We found this to be the best performing setting compared to default SCIP.

Summarized results can be found in Table 1. MONOIDAL consistently outperforms ICUTS with respect to solving time as well as number of nodes needed. On the whole test set, the strengthened intersection cuts reduce both metrics by around 3% while solving two more instances. This improvement increases when looking at harder instances: On the hardest test set $[1000, 7200]$ containing only instances for which at least one setting needs 1000 seconds or more, MONOIDAL uses 49% less time and 51% less nodes.

These results show that the proposed monoidal strengthening procedure significantly improves the standard intersection cuts, which highlights the importance of exploiting integrality whenever possible. Nonetheless, our cuts are currently not able to improve the overall performance of default SCIP. One of the main reasons is that our cuts, while helping in obtaining better dual bounds, are negatively affecting the performance of SCIP's primal heuristics. Improving this behavior is subject of ongoing work.

References

1. Balas, E.: Intersection cuts–a new type of cutting planes for integer programming. Oper. Res. **19**(1), 19–39 (1971)
2. Balas, E., Jeroslow, R.G.: Strengthening cuts for mixed integer programs. Eur. J. Oper. Res. **4**(4), 224–234 (1980)
3. Basu, A., Campelo, M., Conforti, M., Cornuéjols, G., Zambelli, G.: Unique lifting of integer variables in minimal inequalities. Math. Program. **141**(1–2), 561–576 (2012)
4. Basu, A., Dey, S.S., Paat, J.: Nonunique lifting of integer variables in minimal inequalities. SIAM J. Discret. Math. **33**(2), 755–783 (2019)
5. Bestuzheva, K., et al.: The SCIP Optimization Suite 8.0. ZIB-Report 21–41, Zuse Institute Berlin, December 2021
6. Bienstock, D., Chen, C., Munoz, G.: Outer-product-free sets for polynomial optimization and oracle-based cuts. Math. Program. 1–44 (2020)
7. Chmiela, A., Muñoz, G., Serrano, F.: Monoidal strengthening and unique lifting in MIQCPs (2022). https://www.gonzalomunoz.org/publications/
8. Chmiela, A., Muñoz, G., Serrano, F.: On the implementation and strengthening of intersection cuts for QCQPs. Math. Program. pp. 1–38 (2022)
9. Conforti, M., Cornuéjols, G., Zambelli, G.: A geometric perspective on lifting. Oper. Res. **59**(3), 569–577 (2011)
10. Dey, S.S., Wolsey, L.A.: Two row mixed-integer cuts via lifting. Math. Program. **124**(1–2), 143–174 (2010)
11. Fukasawa, R., Poirrier, L., Xavier, Á.S.: The (not so) trivial lifting in two dimensions. Math. Program. Comput. **11**(2), 211–235 (2018). https://doi.org/10.1007/s12532-018-0146-5
12. Furini, F., et al.: A library of quadratic programming instances. Programming Computation, QPLIB (2018)
13. Glover, F.: Convexity cuts and cut search. Oper. Res. **21**(1), 123–134 (1973)
14. MINLP library (2010). http://www.minlplib.org/
15. Muñoz, G., Serrano, F.: Maximal quadratic-free sets. In: Bienstock, D., Zambelli, G. (eds.) IPCO 2020. LNCS, vol. 12125, pp. 307–321. Springer, Cham (2020). https://doi.org/10.1007/978-3-030-45771-6_24
16. Muñoz, G., Serrano, F.: Maximal quadratic-free sets. Math. Program. 1–42 (2021)
17. Rockafellar, R.T.: Convex Analysis. Princeton University Press, Princeton (1970)
18. Santana, A., Dey, S.S.: The convex hull of a quadratic constraint over a polytope. SIAM J. Optim. **30**(4), 2983–2997 (2020)
19. Tuy, H.: Concave programming with linear constraints. In: Doklady Akademii Nauk, vol. 159, pp. 32–35. Russian Academy of Sciences (1964)

From Approximate to Exact Integer Programming

Daniel Dadush[1], Friedrich Eisenbrand[2], and Thomas Rothvoss[3(✉)]

[1] CWI, Amsterdam, The Netherlands
dadush@cwi.nl
[2] EPFL, Lausanne, Switzerland
friedrich.eisenbrand@epfl.ch
[3] University of Washington, Seattle, USA
rothvoss@uw.edu

Abstract. Approximate integer programming is the following: For a given convex body $K \subseteq \mathbb{R}^n$, either determine whether $K \cap \mathbb{Z}^n$ is empty, or find an integer point in the convex body $2 \cdot (K - c) + c$ which is K, scaled by 2 from its center of gravity c. Approximate integer programming can be solved in time $2^{O(n)}$ while the fastest known methods for exact integer programming run in time $2^{O(n)} \cdot n^n$. So far, there are no efficient methods for integer programming known that are based on approximate integer programming. Our main contribution are two such methods, each yielding novel complexity results.

First, we show that an integer point $x^* \in (K \cap \mathbb{Z}^n)$ can be found in time $2^{O(n)}$, provided that the *remainders* of each component x_i^* mod ℓ for some arbitrarily fixed $\ell \geq 5(n+1)$ of x^* are given. The algorithm is based on a *cutting-plane technique*, iteratively halving the volume of the feasible set. The cutting planes are determined via approximate integer programming. Enumeration of the possible remainders gives a $2^{O(n)} n^n$ algorithm for general integer programming. This matches the current best bound of an algorithm by Dadush (2012) that is considerably more involved. Our algorithm also relies on a new *asymmetric approximate Carathéodory theorem* that might be of interest on its own.

Our second method concerns integer programming problems in standard equation form $Ax = b, 0 \leq x \leq u, x \in \mathbb{Z}^n$. Such a problem can be reduced to the solution of $\prod_i O(\log u_i + 1)$ approximate integer programming problems. This implies, for example that *knapsack* or *subset-sum* problems with *polynomial variable range* $0 \leq x_i \leq p(n)$ can be solved in time $(\log n)^{O(n)}$. For these problems, the best running time so far was $n^n \cdot 2^{O(n)}$.

A full version of this paper can be found under https://arxiv.org/abs/2211.03859.

D. Dadush—Supported by ERC Starting Grant no. 805241-QIP.

F. Eisenbrand—Supported by the Swiss National Science Foundation (SNSF) grant 185030 and 207365.

T. Rothvoss—Supported by NSF CAREER grant 1651861 and a David & Lucile Packard Foundation Fellowship.

A. Del Pia and V. Kaibel (Eds.): IPCO 2023, LNCS 13904, pp. 100–114, 2023.
https://doi.org/10.1007/978-3-031-32726-1_8

1   Introduction

Many *combinatorial optimization problems* as well as many problems from the *algorithmic geometry of numbers* can be formulated as an *integer linear program*

$$\max\{\langle c, x \rangle \mid Ax \le b, x \in \mathbb{Z}^n\} \qquad (1)$$

where $A \in \mathbb{Z}^{m \times n}, b \in \mathbb{Z}^m$ and $c \in \mathbb{Z}^n$, see, e.g. [16,27,30]. Lenstra [23] has shown that integer programming can be solved in polynomial time, if the number of variables is fixed. A careful analysis of his algorithm yields a running time of $2^{O(n^2)}$ times a polynomial in the binary encoding length of the input of the integer program. Kannan [19] has improved this to $n^{O(n)}$, where, from now on we ignore the extra factor that depends polynomially on the input length. The current best algorithm is the one of Dadush [10] with a running time of $2^{O(n)} \cdot n^n$.

The question whether there exists a *singly exponential time*, i.e., a $2^{O(n)}$-algorithm for integer programming is one of the most prominent open problems in the area of algorithms and complexity. Integer programming can be described in the following more general form. Here, a *convex body* is synonymous for a full-dimensional compact and convex set.

Integer Programming (IP)

Given a *convex body* $K \subseteq \mathbb{R}^n$, find an integer solution $x^* \in K \cap \mathbb{Z}^n$ or assert that $K \cap \mathbb{Z}^n = \emptyset$.

The convex body K must be well described in the sense that there is access to a *separation oracle*, see [16]. Furthermore, one assumes that K contains a ball of radius $r > 0$ and that it is contained in some ball of radius R. In this setting, the current best running times hold as well. The additional polynomial factor in the input encoding length becomes a polynomial factor in $\log(R/r)$ and the dimension n. Central to this paper is *Approximate integer programming* which is as follows.

Approximate Integer Programming (Approx-IP)

Given a *convex body* $K \subseteq \mathbb{R}^n$, let $c \in \mathbb{R}^n$ be its center of gravity. Either find an integer vector $x^* \in (2 \cdot (K - c) + c) \cap \mathbb{Z}^n$, or assert that $K \cap \mathbb{Z}^n = \emptyset$.

The convex body $2 \cdot (K - c) + c$ is K scaled by a factor of 2 from its center of gravity. The algorithm of Dadush [11] solves approximate integer programming in singly exponential time $2^{O(n)}$. Despite its clear relation to exact integer programming, there is no reduction from exact to approximate known so far. Our guiding question is the following: Can approximate integer programming be used to solve the exact version of (specific) integer programming problems?

1.1   Contributions of This Paper

We present two different algorithms to reduce the exact integer programming problem (IP) to the approximate version (APPROX-IP).

a) Our first method is a randomized cutting-plane algorithm that, in time $2^{O(n)}$ and for any $\ell \geq 5(n+1)$ finds a point in $K \cap (\mathbb{Z}^n / \ell)$ with high probability, if K contains an integer point. This algorithm uses an oracle for (APPROX-IP) on K intersected with one side of a hyperplane that is close to the center of gravity. Thereby, the algorithm collects ℓ integer points close to K. The collection is such that the convex combination with uniform weights $1/\ell$ of these points lies in K. If, during an iteration, no point is found, the volume of K is roughly halved and eventually K lies on a lower-dimensional subspace on which one can recurse.

b) If equipped with the component-wise remainders $v \equiv x^*$ (mod ℓ) of a solution x^* of (IP), one can use the algorithm to find a point in $(K - v) \cap \mathbb{Z}^n$ and combine it with the remainders to a full solution of (IP), using that $(K - v) \cap \ell \mathbb{Z}^n \neq \emptyset$. This runs in singly exponential randomized time $2^{O(n)}$. Via enumeration of all remainders, one obtains an algorithm for (IP) that runs in time $2^{O(n)} \cdot n^n$. This matches the best-known running time for general integer programming [11], which is considerably involved.

c) Our analysis depends on a new *approximate Carathéodory theorem* that we develop in Sect. 4. While approximate Carathéodory theorems are known for centrally symmetric convex bodies [4,26,28], our version is for general convex sets and might be of interest on its own.

d) Our second method is for integer programming problems $Ax = b, x \in \mathbb{Z}^n, 0 \leq x \leq u$ in equation standard form. We show that such a problem can be reduced to $2^{O(n)} \cdot (\prod_i \log(u_i + 1))$ instances of (APPROX-IP). This yields a running time of $(\log n)^{O(n)}$ for such IPs, in which the variables are bounded by a polynomial in the dimension. The so-far best running time for such instances $2^{O(n)} \cdot n^n$. Well known benchmark problems in this setting are *knapsack* and *subset-sum* with polynomial upper bounds on the variables, see Sect. 5.

1.2 Related Work

If the convex body K is an ellipsoid, then the integer programming problem (IP) is the well known *closest vector problem (CVP)* which can be solved in time $2^{O(n)}$ with an algorithm by Micciancio and Voulgaris [25]. Blömer and Naewe [7] previously observed that the sampling technique of Ajtai et al. [1] can be modified in such a way as to solve the closest vector approximately. More precisely, they showed that a $(1 + \epsilon)$-approximation of the closest vector problem can be found in time $O(2 + 1/\epsilon)^n$ time. This was later generalized to arbitrary convex sets by Dadush [11]. This algorithm either asserts that the convex body K does not contain any integer points, or it finds an integer point in the body stemming from K is scaled by $(1+\epsilon)$ from its center of gravity. Also the running time of this randomized algorithm is $O(2 + 1/\epsilon)^n$. In our paper, we restrict to the case $\epsilon = 1$ which can be solved in singly exponential time. The technique of reflection sets was also used by Eisenbrand et al. [13] to solve (CVP) in the ℓ_∞-norm approximately in time $O(2 + \log(1/\epsilon))^n$.

In the setting in which integer programming can be attacked with dynamic programming, tight upper and lower bounds on the complexity are known [14,17,20]. Our $n^n \cdot 2^{O(n)}$ algorithm could be made more efficient by constraining the possible remainders of a solution (mod ℓ) efficiently. This barrier is different than the one in classical integer-programming methods that are based on branching on flat directions [16,23] as they result in a branching tree of size $n^{O(n)}$.

The *subset-sum problem* is as follows. Given a set $Z \subseteq \mathbb{N}$ of n positive integers and a *target value* $t \in \mathbb{N}$, determine whether there exists a subset $S \subseteq Z$ with $\sum_{s \in S} s = t$. Subset sum is a classical NP-complete problem that serves as a benchmark in algorithm design. The problem can be solved in pseudopolynomial time [5] by dynamic programming. The current fastest pseudopolynomial-time algorithm is the one of Bringmann [8] that runs in time $O(n + t)$ up to polylogarithmic factors. There exist instances of subset-sum whose set of feasible solutions, interpreted as $0/1$ incidence vectors, require numbers of value n^n in the input, see [2]. Lagarias and Odlyzko [21] have shown that instances of subset sum in which each number of the input Z is drawn uniformly at random from $\{1, \ldots, 2^{O(n^2)}\}$ can be solved in polynomial time with high probability. The algorithm of Lagarias and Odlyzko is based on the LLL-algorithm [22] for lattice basis reduction.

2 Preliminaries

A *lattice* Λ is the set of integer combinations of linearly independent vectors, i.e. $\Lambda := \Lambda(B) := \{Bx \mid x \in \mathbb{Z}^r\}$ where $B \in \mathbb{R}^{n \times r}$ has linearly independent columns. The *determinant* is the volume of the r-dimensional parallelepiped spanned by the columns of the basis B, i.e. $\det(\Lambda) := \sqrt{\det_r(B^T B)}$. We say that Λ has *full rank* if $n = r$. In that case the determinant is simply $\det(\Lambda) = |\det_n(B)|$. For a full rank lattice Λ, we denote the dual lattice by $\Lambda^* = \{y \in \mathbb{R}^n \mid \langle x, y \rangle \in \mathbb{Z} \ \forall x \in \Lambda\}$. Note that $\det(\Lambda^*) \cdot \det(\Lambda) = 1$. For an introduction to lattices, we refer to [24].

A set $Q \subseteq \mathbb{R}^n$ is called a *convex body* if it is convex, compact and has a non-empty interior. A set Q is *symmetric* if $Q = -Q$. Recall that any symmetric convex body Q naturally induces a norm $\| \cdot \|_Q$ of the form $\|x\|_Q = \min\{s \geq 0 \mid x \in sQ\}$. For a full rank lattice $\Lambda \subseteq \mathbb{R}^n$ and a symmetric convex body $Q \subseteq \mathbb{R}^n$ we denote $\lambda_1(\Lambda, Q) := \min\{\|x\|_Q \mid x \in \Lambda \setminus \{0\}\}$ as the length of the shortest vector with respect to the norm induced by Q. We denote the Euclidean ball by $B_2^n := \{x \in \mathbb{R}^n \mid \|x\|_2 \leq 1\}$ and the ℓ_∞-ball by $B_\infty^n := [-1,1]^n$. An (origin centered) *ellipsoid* is of the form $\mathcal{E} = A(B_2^n)$ where $A : \mathbb{R}^n \to \mathbb{R}^n$ is an invertible linear map. For any such ellipsoid $\mathcal{E}$ there is a unique positive definite matrix $M \in \mathbb{R}^{n \times n}$ so that $\|x\|_\mathcal{E} = \sqrt{x^T M x}$. The *barycenter* (or *centroid*) of a convex body Q is the point $\frac{1}{\text{Vol}_n(Q)} \int_Q x \, dx$. We will use the following version of (APPROX-IP) that runs in time $2^{O(n)}$, provided that the symmetrizer for the used center c is large enough. This is the case for c being the center of gravity, see Theorem 3. Note that the center of gravity of a convex body can be (approximately) computed in randomized polynomial time [6,12].

Theorem 1 (Dadush [11]). *There is a $2^{O(n)}$-time algorithm* APXIP(K, c, Λ) *that takes as input a convex set $K \subseteq \mathbb{R}^n$, a point $c \in K$ and a lattice $\Lambda \subseteq \mathbb{R}^n$. Assuming that $\text{Vol}_n((K - c) \cap (c - K)) \geq 2^{-\Theta(n)} \text{Vol}_n(K)$ the algorithm either returns a point $x \in (c + 2(K - c)) \cap \Lambda$ or returns* EMPTY *if $K \cap \Lambda = \emptyset$.*

One of the classical results in the geometry of numbers is Minkowski's Theorem which we will use in the following form:

Theorem 2 (Minkowski's Theorem). *For a full rank lattice $\Lambda \subseteq \mathbb{R}^n$ and a symmetric convex body $Q \subseteq \mathbb{R}^n$ one has*

$$\lambda_1(\Lambda, Q) \le 2 \cdot \left(\frac{\det(\Lambda)}{Vol_n(Q)} \right)^{1/n}$$

We will use the following bound on the density of sublattices which is an immediate consequence of Minkowski's Second Theorem. Here we abbreviate $\lambda_1(\Lambda) := \lambda_1(\Lambda, B_2^n)$.

Lemma 1. *Let $\Lambda \subseteq \mathbb{R}^n$ be a full rank lattice. Then for any k-dimensional sublattice $\tilde{\Lambda} \subseteq \Lambda$ one has $\det(\tilde{\Lambda}) \ge (\frac{\lambda_1(\Lambda)}{\sqrt{k}})^k$.*

Finally, we revisit a few facts from *convex geometry*. Details and proofs can be found in the excellent textbook by Artstein-Avidan, Giannopoulos and Milman [3].

Lemma 2 (Grünbaum's Lemma). *Let $K \subseteq \mathbb{R}^n$ be any convex body and let $\langle a, x \rangle = \beta$ be any hyperplane through the barycenter of K. Then $\frac{1}{e} Vol_n(K) \le Vol_n(\{x \in K \mid \langle a, x \rangle \le \beta\}) \le (1 - \frac{1}{e}) Vol_n(K)$.*

For a convex body K, there are two natural symmetric convex bodies that approximate K in many ways: the "inner symmetrizer" $K \cap (-K)$ (provided $\mathbf{0} \in K$) and the "outer symmetrizer" in form of the *difference body* $K - K$. The following is a consequence of a more general inequality of Milman and Pajor.

Theorem 3. *Let $K \subseteq \mathbb{R}^n$ be any convex body with barycenter $\mathbf{0}$. Then $Vol_n(K \cap (-K)) \ge 2^{-n} Vol_n(K)$.*

In particular Theorem 3 implies that choosing c as the barycenter of K in Theorem 1 results in a $2^{O(n)}$ running time—however this will not be the choice that we will later make for c. Also the size of the difference body can be bounded:

Theorem 4 (Inequality of Rogers and Shephard). *For any convex body $K \subseteq \mathbb{R}^n$ one has $Vol_n(K - K) \le 4^n Vol_n(K)$.*

Recall that for a convex body Q with $\mathbf{0} \in int(Q)$, the *polar* is $Q^\circ = \{y \in \mathbb{R}^n \mid \langle x, y \rangle \le 1 \ \forall x \in Q\}$. We will use the following relation between volume of a symmetric convex body and the volume of the polar; to be precise we will use the lower bound (which is due to Bourgain and Milman).

Theorem 5 (Blaschke-Santaló-Bourgain-Milman). *For any symmetric convex body $Q \subseteq \mathbb{R}^n$ one has*

$$C^n \le \frac{Vol_n(Q) \cdot Vol_n(Q^\circ)}{Vol_n(B_2^n)^2} \le 1$$

where $C > 0$ is a universal constant.

We will also rely on the result of Frank and Tardos to reduce the bit complexity of constraints:

Theorem 6 (Frank, Tardos [15]). *There is a polynomial time algorithm that takes $(a, b) \in \mathbb{Q}^{n+1}$ and $\Delta \in \mathbb{N}_+$ as input and produces a pair $(\tilde{a}, \tilde{b}) \in \mathbb{Z}^{n+1}$ with $\|\tilde{a}\|_\infty, |\tilde{b}| \le 2^{O(n^3)} \cdot \Delta^{O(n^2)}$ so that $\langle a, x \rangle = b \Leftrightarrow \langle \tilde{a}, x \rangle = \tilde{b}$ and $\langle a, x \rangle \le b \Leftrightarrow \langle \tilde{a}, x \rangle \le \tilde{b}$ for all $x \in \{-\Delta, \dots, \Delta\}^n$.*

3   The Cut-Or-Average Algorithm

First, we discuss our CUT-OR-AVERAGE algorithm that on input of a convex set K, a lattice Λ and integer $\ell \geq 5(n+1)$, either finds a point $x \in \frac{\Lambda}{\ell} \cap K$ or decides that $K \cap \Lambda = \emptyset$ in time $2^{O(n)}$. Note that for any polyhedron $K = \{x \in \mathbb{R}^n \mid Ax \leq b\}$ with rational A, b and lattice Λ with basis B one can compute a value of Δ so that $\log(\Delta)$ is polynomial in the encoding length of A, b and B and $K \cap \Lambda \neq \emptyset$ if and only if $K \cap [-\Delta, \Delta]^n \cap \Lambda \neq \emptyset$. See Schrijver [31] for details. In other words, w.l.o.g. we may assume that our convex set is bounded. The pseudo code of the algorithm can be found in Fig. 1. An intuitive description of the algorithm is as follows: we compute the barycenter c of K and an ellipsoid $\mathcal{E}$ that approximates K up to a factor of $R = n + 1$. Then we iteratively use the oracle for approximate integer programming from Theorem 1 to find a convex combination z of lattice points in a 3-scaling of K until z is close to the barycenter c. If this succeeds, then we can directly use an asymmetric version of the *Approximate Carathéodory Theorem* (Lemma 9) to find an unweighted average of ℓ lattice points that lies in K; this would be a vector of the form $x \in \frac{\Lambda}{\ell} \cap K$. If the algorithm fails to approximately express c as a convex combination of lattice points, then we will have found a hyperplane H going almost through the barycenter c so that $K \cap H_{\geq}$ does not contain a lattice point. Then the algorithm continues searching in $K \cap H_{\leq}$ (Fig. 2). This case might happen repeatedly, but after polynomial number of times, the volume of K will have dropped below a threshold so that we may recurse on a *single* $(n-1)$-dimensional subproblem. We will now give the detailed analysis. Note that in order to obtain a clean exposition we did not aim to optimize any constant. However by merely tweaking the parameters one could make the choice of $\ell = (1 + \varepsilon)n$ work for any constant $\varepsilon > 0$.

3.1   Bounding the Number of Iterations

We begin the analysis with a few estimates that will help us to bound the number of iterations.

Lemma 3. *Any point x found in line (7) lies in a 3-scaling of K around c, i.e. $x \in c + 3(K - c)$ assuming $0 < \rho \leq 1$.*

Proof. We verify that

$$x \in (c - \rho d) + 2(K - (c - \rho d)) = c + 2(K - c) + \rho d \subseteq c + 3(K - c)$$

using that $\|\rho d\|_{\mathcal{E}} = \rho \leq 1$.

Next we bound the distance of z to the barycenter:

Lemma 4. *At the beginning of the kth iterations of the WHILE loop on line (5), one has $\|c - z\|_{\mathcal{E}}^2 \leq \frac{9R^2}{k}$.*

Proof. We prove the statement by induction on k. At $k = 1$, by construction on line (4), $z \in c + 2(K - c) \subseteq c + 2R\mathcal{E}$. Thus $\|c - z\|_{\mathcal{E}}^2 \leq (2R)^2 \leq 9R^2$, as needed.

Input: Convex set $K \subseteq \mathbb{R}^n$, lattice Λ, parameter $\ell \geq 5(n+1)$
Output: Either a point $x \in K \cap \frac{\Lambda}{\ell}$ or conclusion that $K \cap \Lambda = \emptyset$

(1) WHILE $\lambda_1(\Lambda^*, (K-K)^\circ) > \frac{1}{2}$ DO
   (2) Compute barycenter c of K.
   (3) Let $\mathcal{E} := \{x \in \mathbb{R}^n \mid x^T M x \leq 1\}$ be **0**-centered ellipsoid with $c + \mathcal{E} \subseteq K \subseteq c + R \cdot \mathcal{E}$ for $R := n+1$, let $\rho := \frac{1}{4n}$.
   (4) Let $z := \text{ApxIP}(K, c, \Lambda)$, $X = \{z\}$. If $z = \text{EMPTY}$, Return EMPTY.
   (5) WHILE $\|c - z\|_{\mathcal{E}} > \frac{1}{4}$ DO
      (6) Let $d := \frac{-(z-c)}{\|z-c\|_{\mathcal{E}}}$, $a := -M(z-c)$.
      (7) Compute $x := \text{ApxIP}(K \cap \{x \in \mathbb{R}^n \mid \langle a, x \rangle \geq \langle a, c + \rho d/2 \rangle\}, c + \rho d, \Lambda)$.
      (8) IF $x = \text{EMPTY}$ THEN replace K by $K' := K \cap \{x \in \mathbb{R}^n \mid \langle a, x \rangle \leq \langle a, c + \rho d/2 \rangle\}$ and GOTO (1).
      (9) ELSE $X := X \cup \{x\}$, $z := (1 - \frac{1}{|X|})z + \frac{x}{|X|}$.
   (10) Compute $\mu \in \frac{\mathbb{Z}_{\geq 0}^X}{\ell}$ with $\sum_{x \in X} \mu_x = 1$ and $\sum_{x \in X} \mu_x x \in K$ using Asymmetric Approximate Carathéodory.
   (11) Return $\sum_{x \in X} \mu_x x$.
(12) Compute $y \in \Lambda^* \setminus \{\mathbf{0}\}$ with $\|y\|_{(K-K)^\circ} \leq \frac{1}{2}$.
(13) Find $\beta \in \mathbb{Z}$ so that $K \cap \Lambda \subseteq U$ with $U = \{x \in \mathbb{R}^n \mid \langle y, x \rangle = \beta\}$.
(14) IF $n = 1$ THEN $U = \{x^*\}$; Return x^* if $x^* \in \Lambda$ and return "$K \cap \Lambda = \emptyset$" otherwise.
(15) Recurse on $(n-1)$-dim. instance CUT-OR-AVERAGE$(K \cap U, \Lambda \cap U, \ell)$.

Fig. 1. The Cut-Or-Average algorithm.

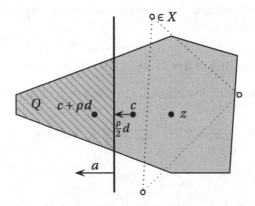

Fig. 2. Visualization of the inner WHILE loop where $Q := K \cap \{x \in \mathbb{R}^n \mid \langle a, x \rangle \geq \langle a, c + \frac{\rho}{2}d \rangle\}$.

Now assume $k \geq 2$. Let z, z' denote the values of z during iteration $k-1$ before and after the execution of line (9) respectively, and let x be the vector found on line (7) during iteration $k-1$. Note that $z' = (1 - \frac{1}{k})z + \frac{1}{k}x$. By the induction hypothesis, we have that $\|z - c\|_{\mathcal{E}}^2 \leq 9R^2/(k-1)$. Our goal is to show that $\|z' - c\|_{\mathcal{E}}^2 \leq 9R^2/k$. Letting d denote the normalized version of $z - c$, we see that $\|d\|_{\mathcal{E}} = 1$ and hence $d \in K - c$. By construction $\langle a, x - c \rangle \geq 0$ and from Lemma 3 we have $x \in c + 3(K - c)$ which implies

$\|x - c\|_{\mathcal{E}} \le 3R$. The desired bound on the $\mathcal{E}$-norm of $z' - c$ follows from the following calculation:

$$\|z' - c\|_{\mathcal{E}}^2 = \left\|\left(1 - \frac{1}{k}\right)(z - c) + \frac{1}{k}(x - c)\right\|_{\mathcal{E}}^2$$

$$= \left(1 - \frac{1}{k}\right)^2 \|z - c\|_{\mathcal{E}}^2 - 2\left(1 - \frac{1}{k}\right)\frac{1}{k}\langle a, x - c\rangle + \frac{1}{k^2}\|x - c\|_{\mathcal{E}}^2$$

$$\le \left(1 - \frac{1}{k}\right)^2 \|z - c\|_{\mathcal{E}}^2 + \frac{1}{k^2}\|x - c\|_{\mathcal{E}}^2$$

$$\le \left(\left(1 - \frac{1}{k}\right)^2 \frac{1}{k-1} + \frac{1}{k^2}\right) \cdot 9R^2 = \frac{9R^2}{k}.$$

In particular Lemma 4 implies an upper bound on the number of iterations of the inner WHILE loop:

Corollary 1. *The WHILE loop on line (5) never takes more than $36R^2$ iterations.*

Proof. By Lemma 4, for $k := 36R^2$ one has $\|c - z\|_{\mathcal{E}}^2 \le \frac{9R^2}{k} \le \frac{1}{4}$. $\square$

Next, we prove that every time we replace K by $K' \subset K$ in line (8), its volume drops by a constant factor.

Lemma 5. *In step (8) one has $Vol_n(K') \le (1 - \frac{1}{e}) \cdot (1 + \frac{\rho}{2})^n \cdot Vol_n(K)$ for any $\rho \ge 0$. In particular for $0 \le \rho \le \frac{1}{4n}$ one has $Vol_n(K') \le \frac{3}{4} Vol_n(K)$.*

Proof. The claim is invariant under affine linear transformations, hence we may assume w.l.o.g. that $\mathcal{E} = B_2^n$, $M = I_n$ and $c = \mathbf{0}$. Note that then $B_2^n \subseteq K \subseteq RB_2^n$. Let us abbreviate $K_{\le t} := \{x \in K \mid \langle d, x\rangle \le t\}$. In this notation $K' = K_{\le \rho/2}$. Recall that Grünbaum's Lemma (Lemma 2) guarantees that $\frac{1}{e} \le \frac{Vol_n(K_{\le 0})}{Vol_n(K)} \le 1 - \frac{1}{e}$. Moreover, it is well known that the function $t \mapsto Vol_n(K_{\le t})^{1/n}$ is concave on its support, see again [3]. Then

$$Vol_n(K_{\le 0})^{1/n} \ge \left(\frac{1}{1 + \rho/2}\right) \cdot Vol_n(K_{\le \rho/2})^{1/n} + \left(\frac{\rho/2}{1 + \rho/2}\right) \cdot \underbrace{Vol_n(K_{\le -1})^{1/n}}_{\ge 0}$$

$$\ge \left(\frac{1}{1 + \rho/2}\right) \cdot Vol_n(K_{\le \rho/2})^{1/n}$$

and so

$$\left(1 - \frac{1}{e}\right) \cdot Vol_n(K) \ge Vol_n(K_{\le 0}) \ge \left(\frac{1}{1 + \rho/2}\right)^n \cdot Vol_n(K_{\le \rho/2})$$

Rearranging gives the first claim in the form $Vol_n(K_{\le \rho/2}) \le (1 - \frac{1}{e}) \cdot (1 + \frac{\rho}{2})^n \cdot Vol_n(K)$. For the 2nd part we verify that for $\rho \le \frac{1}{4n}$ one has $(1 - \frac{1}{e}) \cdot (1 + \frac{\rho}{2})^n \le (1 - \frac{1}{e}) \cdot \exp(\frac{\rho}{2}) \le \frac{3}{4}$. $\square$

Lemma 6. *Consider a call of CUT-OR-AVERAGE on (K, Λ) where $K \subseteq rB_2^n$ for some $r > 0$. Then the total number of iterations of the outer WHILE loop over all recursion levels is bounded by $O(n^2 \log(\frac{nr}{\lambda_1(\Lambda)}))$.*

Proof. Consider any recursive run of the algorithm. The convex set will be of the form $\tilde{K} := K \cap U$ and the lattice will be of the form $\tilde{\Lambda} := \Lambda \cap U$ where U is a subspace and we denote $\tilde{n} := \dim(U)$. We think of $\tilde{K}$ and $\tilde{\Lambda}$ as $\tilde{n}$-dimensional objects. Let $\tilde{K}_t \subseteq \tilde{K}$ be the convex body after t iterations of the outer WHILE loop. Recall that $\mathrm{Vol}_{\tilde{n}}(\tilde{K}_t) \leq (\frac{3}{4})^t \cdot \mathrm{Vol}_{\tilde{n}}(\tilde{K})$ by Lemma 5 and $\mathrm{Vol}_{\tilde{n}}(\tilde{K}) \leq r^{\tilde{n}}\mathrm{Vol}_{\tilde{n}}(B_2^{\tilde{n}})$. Our goal is to show that for t large enough, there is a non-zero lattice vector $y \in \tilde{\Lambda}^*$ with $\|y\|_{(\tilde{K}_t - \tilde{K}_t)^\circ} \leq \frac{1}{2}$ which then causes the algorithm to recurse. To prove existence of such a vector y, we use Minkowski's Theorem (Theorem 2) followed by the Blaschke-Santaló-Bourgain-Milman Theorem (Theorem 5) to obtain

$$\lambda_1(\tilde{\Lambda}^*, (\tilde{K}_t - \tilde{K}_t)^\circ) \overset{\text{Thm 2}}{\leq} 2 \cdot \left(\frac{\det(\tilde{\Lambda}^*)}{\mathrm{Vol}_{\tilde{n}}((\tilde{K}_t - \tilde{K}_t)^\circ)}\right)^{1/\tilde{n}}$$

$$\overset{\text{Thm 5}}{\leq} 2C \cdot \left(\frac{\mathrm{Vol}_{\tilde{n}}(\tilde{K}_t - \tilde{K}_t)}{\det(\tilde{\Lambda}) \cdot \mathrm{Vol}_{\tilde{n}}(B_2^{\tilde{n}})^2}\right)^{1/\tilde{n}}$$

$$\overset{\text{Thm 4}}{\leq} 2 \cdot 4 \cdot \frac{\sqrt{\tilde{n}}}{2} \cdot C\left(\frac{\mathrm{Vol}_{\tilde{n}}(\tilde{K}_t)}{\det(\tilde{\Lambda}) \cdot \mathrm{Vol}_{\tilde{n}}(B_2^{\tilde{n}})}\right)^{1/\tilde{n}}$$

$$\leq 4C\sqrt{\tilde{n}} \cdot r \cdot \frac{(3/4)^{t/\tilde{n}}}{\det(\tilde{\Lambda})^{1/\tilde{n}}} \leq 4C \cdot \frac{\tilde{n} \cdot r}{\lambda_1(\Lambda)} \cdot (3/4)^{t/\tilde{n}}$$

Here we use the convenient estimate of $\mathrm{Vol}_{\tilde{n}}(B_2^{\tilde{n}}) \geq \mathrm{Vol}_{\tilde{n}}(\frac{1}{\sqrt{\tilde{n}}}B_\infty^{\tilde{n}}) = (\frac{2}{\sqrt{\tilde{n}}})^{\tilde{n}}$. Moreover, we have used that by Lemma 1 one has $\det(\tilde{\Lambda}) \geq (\frac{\lambda_1(\Lambda)}{\sqrt{\tilde{n}}})^{\tilde{n}}$. Then $t = \Theta(\tilde{n}\log(\frac{\tilde{n}r}{\lambda_1(\Lambda)}))$ iterations suffice until $\lambda_1(\tilde{\Lambda}^*, (\tilde{K}_t - \tilde{K}_t)^\circ) \leq \frac{1}{2}$ and the algorithm recurses. Hence the total number of iterations of the outer WHILE loop over all recursion levels can be bounded by $O(n^2\log(\frac{nr}{\lambda_1(\Lambda)}))$.

The iteration bound of Lemma 6 can be improved by amortizing the volume reduction over the different recursion levels following the approach of Jiang [18]. We refrain from that to keep our approach simple.

3.2 Correctness and Efficiency of Subroutines

Next, we verify that the subroutines are used correctly. The proofs in this section are deferred to the full version of this paper.

Lemma 7. *For any convex body $K \subseteq \mathbb{R}^n$ one can compute the barycenter c and a 0-centered ellipsoid $\mathcal{E}$ in randomized polynomial time so that $c + \mathcal{E} \subseteq K \subseteq c + (n+1)\mathcal{E}$.*

In order for the call of APXIP in step (7) to be efficient, we need that the symmetrizer of the set is large enough volume-wise, see Theorem 1. In particular for any parameters $2^{-\Theta(n)} \leq \rho \leq 0.99$ and $R \leq 2^{O(n)}$ we will have $\mathrm{Vol}_n((Q - \tilde{c}) \cap (\tilde{c} - Q)) \geq 2^{-\Theta(n)}\mathrm{Vol}_n(Q)$ which suffices for our purpose.

Lemma 8. *In step (7), the set $Q := \{x \in K \mid \langle a, x \rangle \geq \langle a, c + \frac{\rho}{2}d \rangle\}$ and the point $\tilde{c} := c + \rho d$ satisfy $\mathrm{Vol}_n((Q - \tilde{c}) \cap (\tilde{c} - Q)) \geq (1 - \rho)^n \cdot \frac{\rho}{2R} \cdot 2^{-n} \cdot \mathrm{Vol}_n(Q)$.*

3.3   Conclusion on the Cut-Or-Average Algorithm

From the discussion above, we can summarize the performance of the algorithm in Fig. 1 as follows:

Theorem 7. *Given a full rank matrix $B \in \mathbb{Q}^{n \times n}$ and parameters $r > 0$ and $\ell \geq 5(n+1)$ with $\ell \in \mathbb{N}$ and a separation oracle for a closed convex set $K \subseteq rB_2^n$, there is a randomized algorithm that with high probability finds a point $x \in K \cap \frac{1}{\ell}\Lambda(B)$ or decides that $K \cap \Lambda(B) = \varnothing$. Here the running time is $2^{O(n)}$ times a polynomial in $\log(r)$ and the encoding length of B.*

This can be easily turned into an algorithm to solve integer linear programming:

Theorem 8. *Given a full rank matrix $B \in \mathbb{Q}^{n \times n}$, a parameter $r > 0$ and a separation oracle for a closed convex set $K \subseteq rB_2^n$, there is a randomized algorithm that with high probability finds a point $x \in K \cap \Lambda(B)$ or decides that there is none. The running time is $2^{O(n)} n^n$ times a polynomial in $\log(r)$ and the encoding length of B.*

Proof. Suppose that $K \cap \Lambda \neq \varnothing$ and fix an (unknown) solution $x^* \in K \cap \Lambda$. We set $\ell := \lceil 5(n+1) \rceil$. We iterate through all $v \in \{0, \ldots, \ell-1\}^n$ and run Theorem 7 on the set K and the shifted lattice $v + \ell\Lambda$. For the outcome of v with $x^* \equiv v \mod \ell$ one has $K \cap (v + \ell\Lambda) \neq \varnothing$ and so the algorithm will discover a point $x \in K \cap (v + \Lambda)$.

4   An Asymmetric Approximate Carathéodory Theorem

The *Approximate Carathéodory Theorem* states the following.

Given any point-set $X \subseteq B_2^n$ in the unit ball with $\mathbf{0} \in \mathrm{conv}(X)$ and a parameter $k \in \mathbb{N}$, there exist $u_1, \ldots, u_k \in X$ (possibly with repetition) such that

$$\left\| \frac{1}{k} \sum_{i=1}^{k} u_i \right\|_2 \leq O(1/\sqrt{k}).$$

The theorem is proved, for example, by Novikoff [28] in the context of the *perceptron algorithm*. An ℓ_p-version was provided by Barman [4] to find Nash equilibria. Deterministic and nearly-linear time methods to find the convex combination were recently described in [26]. In the following, we provide a generalization to asymmetric convex bodies and the dependence on k will be weaker but sufficient for our analysis of our CUT-OR-AVERAGE algorithm from Sect. 3.

Recall that with a symmetric convex body K, we one can associate the *Minkowski norm* $\|\cdot\|_K$ with $\|x\|_K = \inf\{s \geq 0 \mid x \in sK\}$. In the following we will use the same definition also for an arbitrary convex set K with $\mathbf{0} \in K$. Symmetry is not given but one still has $\|x + y\|_K \leq \|x\|_K + \|y\|_K$ for all $x, y \in \mathbb{R}^n$ and $\|\alpha x\|_K = \alpha \|x\|_K$ for $\alpha \in \mathbb{R}_{\geq 0}$. Using this notation we can prove the main result of this section.

Lemma 9. *Given a point-set $X \subseteq K$ contained in a convex set $K \subseteq \mathbb{R}^n$ with $\mathbf{0} \in \mathrm{conv}(X)$ and a parameter $k \in \mathbb{N}$, there exist $u_1, \ldots, u_k \in X$ (possibly with repetition) so that*

$$\left\| \frac{1}{k} \sum_{i=1}^{k} u_i \right\|_K \leq \min\{|X|, n+1\}/k.$$

Moreover, given X as input, the points $u_1, \ldots, u_k$ can be found in time polynomial in $|X|$, k and n.

Proof. Let $\ell = \min\{|X|, n+1\}$. The claim is true whenever $k \leq \ell$ since then we may simply pick an arbitrary point in X. Hence from now on we assume $k > \ell$.

By Carathéodory's theorem, there exists a convex combination of zero, using ℓ elements of X. We write $\mathbf{0} = \sum_{i=1}^{\ell} \lambda_i v_i$ where $v_i \in X$, $\lambda_i \geq 0$ for $i \in [\ell]$ and $\sum_{i=1}^{\ell} \lambda_i = 1$. Consider the numbers $L_i = (k - \ell)\lambda_i + 1$. Clearly, $\sum_{i=1}^{\ell} L_i = k$. This implies that there exists an integer vector $\mu \in \mathbb{N}^{\ell}$ with $\mu \geq (k - \ell)\lambda$ and $\sum_{i=1}^{\ell} \mu_i = k$. It remains to show that we have

$$\left\| \frac{1}{k} \sum_{i=1}^{\ell} \mu_i v_i \right\|_K \leq \ell / k.$$

In fact, one has

$$\left\| \sum_{i=1}^{\ell} \mu_i v_i \right\|_K = \left\| \sum_{i=1}^{\ell} \underbrace{(\mu_i - (k-\ell)\lambda_i)}_{\geq 0} v_i + \underbrace{(k-\ell)}_{\geq 0} \sum_{i=1}^{\ell} \lambda_i v_i \right\|_K$$

$$\leq \sum_{i=1}^{\ell} (\mu_i - (k-\ell)\lambda_i) \underbrace{\| v_i \|_K}_{\leq 1} + (k-\ell) \underbrace{\left\| \sum_{i=1}^{\ell} \lambda_i v_i \right\|_K}_{=0} \leq \ell.$$

For the moreover part, note that the coefficients $\lambda_1, \ldots, \lambda_\ell$ are the extreme points of a linear program which can be found in polynomial time. Finally, the linear system $\mu \geq \lceil (k-\ell)\lambda \rceil, \sum_{i=1}^{\ell} \mu_i = k$ has a totally unimodular constraint matrix and the right hand side is integral, hence any extreme point solution is integral as well, see e.g. [31]. ∎

Lemma 10. *For any integer $\ell \geq 5(n+1)$, the convex combination μ computed in line (10) satisfies $\sum_{x \in X} \mu_x x \in K$.*

Proof. We may translate the sets X and K so that $c = \mathbf{0}$ without affecting the claim. Recall that $z \in \text{conv}(X)$. By Carathéodory's Theorem there are $v_1, \ldots, v_m \in X$ with $m \leq n+1$ so that $z \in \text{conv}\{v_1, \ldots, v_m\}$ and so $\mathbf{0} \in \text{conv}\{v_1 - z, \ldots, v_m - z\}$. We have $v_i \in 3K$ by Lemma 3 and $-z \in \frac{1}{4}\mathcal{E} \subseteq \frac{1}{4}K$ as well as $z \in \frac{1}{4}K$. Hence $\| v_i - z \|_K \leq \| v_i \|_K + \| -z \|_K \leq \frac{13}{4}$. We apply Lemma 9 and obtain a convex combination $\mu \in \frac{\mathbb{Z}_{\geq 0}^m}{\ell}$ with $\| \sum_{i=1}^{m} \mu_i (v_i - z) \|_{\frac{13}{4}K} \leq \frac{m}{\ell}$. Then

$$\left\| \sum_{i=1}^{m} \mu_i v_i \right\|_K \leq \left\| \sum_{i=1}^{m} \mu_i (v_i - z) \right\|_K + \underbrace{\| z \|_K}_{\leq 1/4} \leq \frac{13}{4} \frac{m}{\ell} + \frac{1}{4} \leq 1$$

if $\ell \geq \frac{13}{3} m$. This is satisfies if $\ell \geq 5(n+1)$. ∎

5 IPs with Polynomial Variable Range

Now we come to our second method that reduces (IP) to (APPROX-IP) that applies to integer programming in *standard equation form*

$$Ax = b, \, x \in \mathbb{Z}^n, \, 0 \leq x_i \leq u_i, \, i = 1, \ldots, n, \tag{2}$$

Here, $A \in \mathbb{Z}^{m \times n}$, $b \in \mathbb{Z}^m$, and the $u_i \in \mathbb{N}_+$ are positive integers that bound the variables from above. Our main goal is to prove the following theorem.

Theorem 9. *The integer feasibility problem in standard equation form (see (2)) can be solved in time* $2^{O(n)} \prod_{i=1}^{n} \log_2(u_i + 1)$.

We now describe the algorithm. It is again based on the approximate integer programming technique of Dadush [11]. We exploit it to solve integer programming exactly via the technique of *reflection sets* developed by Cook et al. [9]. For each $i = 1, \ldots, n$ we consider the two families of hyperplanes that slice the feasible region with the shifted lower and upper bounds respectively

$$x_i = 2^{j-1} \text{ and } x_i = u_i - 2^{j-1}, 0 \le j \le \lceil \log_2(u_i) \rceil. \tag{3}$$

Following [9], we consider two points w, v that lie in the region between two consecutive planes $x_i = 2^{j-1}$ and $x_i = 2^j$ for some j. Suppose that $w_i \le v_i$ holds. Let s be the point such that $w = 1/2(s + v)$. The line-segment s, v is the line segment w, v scaled by a factor of 2 from v. Let us consider what can be said about the i-th component of s. Clearly $s_i \ge 2^{j-1} - (2^j - 2^{j-1}) = 0$. Similarly, if w and v lie in the region in-between $x_i = 0$ and $x_i = 1/2$, then $s_i \ge -1/2$. We conclude with the following observation.

Lemma 11. *Consider the hyperplane arrangement defined by the equations (3) as well as by $x_i = 0$ and $x_i = u_i$ for $1 \le i \le n$. Let $K \subseteq \mathbb{R}^n$ a cell of this hyperplane arrangement and $v \in K$. If K' is the result of scaling K by a factor of 2 from v, i.e.*

$$K' = \{v + 2(w - v) \mid w \in K\},$$

then K' satisfies the inequalities $-1/2 \le x_i \le u_i + 1/2$ for all $1 \le i \le n$.

We use this observation to prove Theorem 9:

Proof (Proof of Theorem 9). The task of (2) is to find an integer point in the affine subspace defined by the system of equations $Ax = b$ that satisfies the bound constraints $0 \le x_i \le u_i$. We first partition the feasible region with the hyperplanes (3) as well as $x_i = 0$ and $x_i = u_i$ for each i. We then apply the approximate integer programming algorithm with approximation factor 2 on each convex set $P_K = \{x \in \mathbb{R}^n \mid Ax = b\} \cap K$ where K ranges over all cells of the arrangement. In $2^{O(n)}$ time, the algorithm either finds an integer point in the convex set C_K that results from P_K by scaling it with a factor of 2 from its center of gravity, or it asserts that P_K does not contain an integer point. Clearly, $C_K \subseteq \{x \in \mathbb{R}^n \mid Ax = b\}$ and if the algorithm returns an integer point x^*, then, by Lemma 11, this integer point also satisfies the bounds $0 \le x_i \le u_i$. The running time of the algorithm is equal to the number of cells times $2^{O(n)}$ which is $2^{O(n)} \prod_{i=1}^{n} \log_2(u_i + 1)$.

IPs in Inequality Form

We can also use Theorem 9 to solve integer linear programs in *inequality form*. Here the efficiency is strongly dependent on the number of inequalities.

Theorem 10. *Let $A \in \mathbb{Q}^{m \times n}$, $b \in \mathbb{Q}^m$, $c \in \mathbb{Q}^n$ and $u \in \mathbb{N}_+^n$. Then the integer linear program*

$$\max\{\langle c, x \rangle \mid Ax \leq b, 0 \leq x \leq u, x \in \mathbb{Z}^n\}$$

can be solved in time $n^{O(m)} \cdot (2\log(1 + \Delta))^{O(n+m)}$ where $\Delta := \max\{u_i \mid i = 1, \ldots, n\}$.

Proof. Via binary search it suffices to solve the feasibility problem

$$\langle c, x \rangle \geq \gamma, \; Ax \leq b, 0 \leq x \leq u, x \in \mathbb{Z}^n \tag{4}$$

in the same claimed running time. We apply the result of Frank and Tardos (Theorem 6) and replace c, γ, A, b by integer-valued objects of bounded $\|\cdot\|_\infty$-norm so that the feasible region of (4) remains the same. Hence we may indeed assume that $c \in \mathbb{Z}^n$, $\gamma \in \mathbb{Z}$, $A \in \mathbb{Z}^{m \times n}$ and $b \in \mathbb{Z}^m$ with $\|c\|_\infty, |\gamma|, \|A\|_\infty, \|b\|_\infty \leq 2^{O(n^3)} \cdot \Delta^{O(n^2)}$. Any feasible solution x to (4) has a slack bounded by $\gamma - \langle c, x \rangle \leq |\gamma| + \|c\|_\infty \cdot n \cdot \Delta \leq N$ where we may choose $N := 2^{O(n^3)}\Delta^{O(n^2)}$. Similarly $b_i - A_i x \leq N$ for all $i \in [n]$. We can then introduce slack variables $y \in \mathbb{Z}_{\geq 0}$ and $z \in \mathbb{Z}_{\geq 0}^m$ and consider the system

$$\begin{aligned} \langle c, x \rangle + y = \gamma, &\quad Ax + z = b, \\ 0 \leq x \leq u, &\quad 0 \leq y \leq N, \; 0 \leq z_j \leq N \, \forall j \in \lfloor m \rfloor, \\ (x, y, z) \in \mathbb{Z}^{n+1+m} \end{aligned} \tag{5}$$

in equality form which is feasible if and only if (4) is feasible. Then Theorem 9 shows that such an integer linear program can be solved in time

$$2^{O(n+m)} \cdot \left(\prod_{i=1}^n \ln(1 + u_i) \right) \cdot (\ln(1 + N))^{m+1} \leq n^{O(m)} \cdot (2\log(1 + \Delta))^{O(n+m)}.$$

Subset Sum and Knapsack

The *subset-sum problem (with multiplicities)* is an integer program of the form (2) with one linear constraint. Polak and Rohwedder [29] have shown that subset-sum with multiplicities—that means $\sum_{i=1}^n x_i z_i = t, 0 \leq x_i \leq u_i \, \forall i \in [n], x \in \mathbb{Z}^n$—can be solved in time $O(n + z_{max}^{5/3})$ times a polylogarithmic factor where $z_{max} := \max_{i=1,\ldots,n} z_i$. The algorithm of Frank and Tardos [15] (Theorem 6) finds an equivalent instance in which z_{max} is bounded by $2^{O(n^3)} u_{max}^{O(n^2)}$. All-together, if each multiplicity is bounded by a polynomial $p(n)$, then the state-of-the-art for subset-sum with multiplicities is straightforward enumeration resulting in a running time $n^{O(n)}$ which is the current best running time for integer programming. We can significantly improve the running time in this regime. This is a direct consequence of Theorem 10.

Corollary 2. *The subset sum problem with multiplicities of the form $\sum_{i=1}^n x_i z_i = t, 0 \leq x \leq u, x \in \mathbb{Z}^n$ can be solved in time $2^{O(n)} \cdot (\log(1 + \|u\|_\infty))^n$. In particular if each multiplicity is bounded by a polynomial $p(n)$, then it can be solved in time $(\log n)^{O(n)}$.*

Knapsack with multiplicities is the following integer programming problem

$$\max\{\langle c, x \rangle \mid x \in \mathbb{Z}_{\geq 0}^n, \langle a, x \rangle \leq \beta, 0 \leq x \leq u\}, \tag{6}$$

where $c, a, u \in \mathbb{Z}_{\geq 0}^n$ are integer vectors. Again, via the preprocessing algorithm of Frank and Tardos [15] (Theorem 6) one can assume that $\|c\|_\infty$ as well as $\|a\|_\infty$ are bounded by $2^{O(n^3)} u_{\max}^{O(n^2)}$. If each u_i is bounded by a polynomial in the dimension, then the state-of-the-art for this problem is again straightforward enumeration which leads to a running time of $n^{O(n)}$. Also in this regime, we can significantly improve the running time which is an immediate consequence of Theorem 10.

Corollary 3. *A knapsack problem* (6) *can be solved in time* $2^{O(n)} \cdot (\log(1 + \|u\|_\infty))^n$. *In particular if* $\|u\|_\infty$ *is bounded by a polynomial* $p(n)$ *in the dimension, it can be solved in time* $(\log n)^{O(n)}$.

References

1. Miklós Ajtai, R.K., Sivakumar, D.: A sieve algorithm for the shortest lattice vector problem. In: Proceedings of the Thirty-Third Annual ACM Symposium on Theory of Computing, pp. 601–610 (2001)
2. Alon, N., Vũ, V.H.: Anti-hadamard matrices, coin weighing, threshold gates, and indecomposable hypergraphs. J. Comb. Theory Ser. A **79**(1), 133–160 (1997)
3. Artstein-Avidan, S., Giannopoulos, A., Milman, V.D.: Asymptotic geometric analysis. Part I, Volume 202 of Mathematical Surveys and Monographs. American Mathematical Society, Providence, RI (2015)
4. Barman, S.: Approximating nash equilibria and dense bipartite subgraphs via an approximate version of Caratheodory's theorem. In: Proceedings of the Forty-Seventh Annual ACM Symposium on Theory of Computing, pp. 361–369 (2015)
5. Bellman, R.: Dynamic programming. Science **153**(3731), 34–37 (1966)
6. Bertsimas, D., Vempala, S.: Solving convex programs by random walks. J. ACM (JACM) **51**(4), 540–556 (2004)
7. Blömer, J., Naewe, S.: Sampling methods for shortest vectors, closest vectors and successive minima. Theor. Comput. Sci. **410**(18), 1648–1665 (2009)
8. Bringmann, K.: A near-linear pseudopolynomial time algorithm for subset sum. In: Proceedings of the Twenty-Eighth Annual ACM-SIAM Symposium on Discrete Algorithms, pp. 1073–1084. SIAM (2017)
9. Cook, W., Hartmann, M., Kannan, R., McDiarmid, C.: On integer points in polyhedra. Combinatorica **12**(1), 27–37 (1992)
10. Dadush, D.: Integer Programming, Lattice Algorithms, and Deterministic Volume Estimation. Georgia Institute of Technology (2012)
11. Dadush, D.: A randomized sieving algorithm for approximate integer programming. Algorithmica **70**(2), 208–244 (2014)
12. Dyer, M., Frieze, A., Kannan, R.: A random polynomial-time algorithm for approximating the volume of convex bodies. J. ACM **38**(1), 1–17 (1991)
13. Eisenbrand, F., Hähnle, N., Niemeier, M.: Covering cubes and the closest vector problem. In: Proceedings of the Twenty-Seventh Annual Symposium on Computational Geometry, pp. 417–423 (2011)
14. Eisenbrand, F., Weismantel, R.: Proximity results and faster algorithms for integer programming using the Steinitz lemma. ACM Trans. Algorithms (TALG) **16**(1), 1–14 (2019)
15. Frank, A., Tardos, É.: An application of simultaneous diophantine approximation in combinatorial optimization. Combinatorica **7**(1), 49–65 (1987)
16. Grötschel, M., Lovász, L., Schrijver, A.: Geometric Algorithms and Combinatorial Optimization. Algorithms and Combinatorics, vol. 2. Springer, Heidelberg (1988). https://doi.org/10.1007/978-3-642-78240-4

17. Jansen, K., Rohwedder, L.: On integer programming and convolution. In: 10th Innovations in Theoretical Computer Science Conference (ITCS 2019). Schloss Dagstuhl-Leibniz-Zentrum fuer Informatik (2018)
18. Jiang, H.: Minimizing convex functions with integral minimizers. In: SODA, pp. 976–985. SIAM (2021)
19. Kannan, R.: Minkowski's convex body theorem and integer programming. Math. Oper. Res. **12**(3), 415–440 (1987)
20. Knop, D., Pilipczuk, M., Wrochna, M.: Tight complexity lower bounds for integer linear programming with few constraints. ACM Trans. Comput. Theory (TOCT) **12**(3), 1–19 (2020)
21. Lagarias, J.C., Odlyzko, A.M.: Solving low-density subset sum problems. J. ACM (JACM) **32**(1):229–246 (1985)
22. Lenstra, A.K., Lenstra, H.W., Lovász, L.: Factoring polynomials with rational coefficients. Mathematische annalen **261**(ARTICLE), 515–534 (1982)
23. Lenstra Jr., H.W.: Integer programming with a fixed number of variables. Math. Oper. Res. **8**(4), 538–548 (1983)
24. Micciancio, D., Goldwasser, S.: Complexity of lattice problems - a cryptograhic perspective. The Kluwer International Series in Engineering and Computer Science, vol. 671. Springer, New York (2002). https://doi.org/10.1007/978-1-4615-0897-7
25. Micciancio, D., Voulgaris, P.: A deterministic single exponential time algorithm for most lattice problems based on voronoi cell computations. In: Proceedings of the Forty-Second ACM Symposium on Theory of Computing, pp. 351–358 (2010)
26. Mirrokni, V., Leme, R.P., Vladu, A., Wong, S.C.: Tight bounds for approximate Carathéodory and beyond. In: International Conference on Machine Learning, pp. 2440–2448. PMLR (2017)
27. Nemhauser, G.L., Wolsey, L.A.: Integer programming. In: Nemhauser, G.L., et al. (eds.) Optimization. Handbooks in Operations Research and Management Science, chapter VI, vol. 1, pp. 447–527. Elsevier (1989)
28. Novikoff, A.B.: On convergence proofs for perceptrons. Technical report, Office of Naval Research, Washington, D.C. (1963)
29. Polak, A., Rohwedder, L., Węgrzycki, K.: Knapsack and subset sum with small items. In: 48th International Colloquium on Automata, Languages, and Programming (ICALP 2021), number CONF, pp. 106–1. Schloss Dagstuhl-Leibniz-Zentrum für Informatik (2021)
30. Schrijver, A.: Polyhedral combinatorics. In: Graham, R., Grötschel, M., Lovász, L. (eds.) Handbook of Combinatorics, chapter 30, vol. 2, pp. 1649–1704. Elsevier (1995)
31. Schrijver, A.: Theory of Linear and Integer Programming. Wiley-Interscience Series in Discrete Mathematics and Optimization. Wiley, Hoboken (1999)

Optimizing Low Dimensional Functions over the Integers

Daniel Dadush[1], Arthur Léonard[2], Lars Rohwedder[3(✉)], and José Verschae[4]

[1] CWI, Amsterdam, Netherlands
dadush@cwi.nl
[2] ENS, Paris, France
arthur.leonard@ens.psl.eu
[3] Maastricht University, Maastricht, Netherlands
l.rohwedder@maastrichtuniversity.nl
[4] Pontificia Universidad Católica de Chile, Santiago, Chile
jverschae@uc.cl

Abstract. We consider box-constrained integer programs with objective $g(Wx) + c^\mathsf{T} x$, where g is a "complicated" function with an m dimensional domain. Here we assume we have $n \gg m$ variables and that $W \in \mathbb{Z}^{m \times n}$ is an integer matrix with coefficients of absolute value at most Δ. We design an algorithm for this problem using only the mild assumption that the objective can be optimized efficiently when all but m variables are fixed, yielding a running time of $n^m (m\Delta)^{O(m^2)}$. Moreover, we can avoid the term n^m in several special cases, in particular when $c = 0$.

Our approach can be applied in a variety of settings, generalizing several recent results. An important application are convex objectives of low domain dimension, where we imply a recent result by Hunkenschröder et al. [SIOPT'22] for the 0-1-hypercube and sharp or separable convex g, assuming W is given explicitly. By avoiding the direct use of proximity results, which only holds when g is separable or sharp, we match their running time and generalize it for arbitrary convex functions. In the case where the objective is only accessible by an oracle and W is unknown, we further show that their proximity framework can be implemented in $n(m\Delta)^{O(m^2)}$-time instead of $n(m\Delta)^{O(m^3)}$. Lastly, we extend the result by Eisenbrand and Weismantel [SODA'17, TALG'20] for integer programs with few constraints to a mixed-integer linear program setting where integer variables appear in only a small number of different constraints.

1 Introduction

Integer programming has played a crucial role in many areas of computer science, operations research, and more recently, data science. Its modelling power allows

The first author has received funding from the European Research Council (ERC) under the European Unions Horizon 2020 research and innovation programme: grant agreement QIP805241.

The fourth author was partially funded by Fondecyt grant Nr. 1221460 and Centro de Modelamiento Matemático (CMM), FB210005, BASAL funds, ANID-Chile.

A. Del Pia and V. Kaibel (Eds.): IPCO 2023, LNCS 13904, pp. 115–126, 2023.
https://doi.org/10.1007/978-3-031-32726-1_9

to capture a large diversity of settings. However, its general intractability makes it challenging to derive a general algorithmic theory, and hence the focus has been to consider meaningful special cases. The main theoretical result in this area has been the algorithm by Lenstra [10], and the improvement by Kannan [7], which show that integer programs are tractable as long as the dimension is constant. In recent years, a surge of interest appeared regarding efficient algorithms for integer programs under other assumptions. More recently, the seminal work by Eisenbrand and Weismantel [4] for integer programs with a constant number of constraints and bounded matrix coefficients sparked a new trend of improved algorithms and lower bounds; see, e.g., [2,6,9].

In this paper, we study a new general framework that encompasses and further extends many of the settings found in the literature. Consider the problem of optimizing a low dimensional objective function over a high dimensional space $\mathbb{Z}^n$. Formally, the problem is defined as

$$
\begin{aligned}
\min \ & c^\mathsf{T} x + g(Wx) \\
& \ell_i \leq x_i \leq u_i && \text{for all } i \in \{1, 2, \ldots, n\}, && (1) \\
& x \in \mathbb{Z}^n.
\end{aligned}
$$

We assume that $W \in \mathbb{Z}^{m \times n}$ has entries of absolute value at most Δ. Here, W can be interpreted as a projection matrix to a space of low dimension $m \ll n$, where then the function $g : \mathbb{R}^m \to \mathbb{R} \cup \{\infty\}$ is applied to the projection. We can think of W as extracting a relatively small set of features from x. The vectors $\ell \in (\mathbb{Z} \cup \{-\infty\})^n$ and $u \in (\mathbb{Z} \cup \{\infty\})^n$ are arbitrary variable bounds and c represents a linear cost function.

Crucially, we only make a very mild assumption on g, namely that we can solve (1) when all but m of the variables are fixed: given any $I \dot\cup J = [n]$ with $|I| = m$ and any fixing $z \in \mathbb{Z}^J$ of the J-variables, we require that

$$
\begin{aligned}
\min \ & c_I x + g(W_I x + W_J z) \\
\text{s.t.} \ & \ell_i \leq x_i \leq u_i && \text{for all } i \in I, && (2) \\
& x \in \mathbb{Z}^I,
\end{aligned}
$$

can be solved efficiently. Here $c_I = (c_i)_{i \in I}$ is the vector c restricted to indices in I and similarly W_I (resp. W_J) is the matrix W restricted to columns indexed by I (resp. J). The requirement is intuitively necessary, because the only plausible approach to efficiently solve the very general setting (1) is to exploit that the function g is low dimensional. If we cannot even optimize over it in a low dimensional coordinate subspace of $\mathbb{Z}^n$, then there is no hope to optimize over it on all of $\mathbb{Z}^n$. Perhaps the most natural such setting is when g is convex and can be accessed through gradient and function evaluation queries. Then (2) can be solved in time that is exponential only in m, but polynomial in the other input parameters, by the Lenstra-Kannan algorithm [7,10].

If g is indeed convex, the Lenstra-Kannan algorithm can also be used to directly solve (1) in time $\Delta^{\Delta^{O(m)}} \cdot \langle \text{input} \rangle^{O(1)}$, where $\langle \text{input} \rangle$ denotes the encoding size of the input. Indeed, we can merge variables with the same columns in

W, which reduces the dimension of the problem to $n' = (2\Delta + 1)^m$. Notice that the linear part of the objective may not remain linear, but it does remain convex. Thus, we can apply Lenstra-Kannan to solve the problem in the claimed running time.

This indicates that the problem is tractable for small values of Δ and m. Our main result is an algorithm that avoids the double exponential running time.

Theorem 1. *For any function g, problem (1) can be solved in time*

$$n^m \cdot (m\Delta)^{O(m^2)} \cdot Q \ ,$$

where Q is the query time of the oracle for (2). In particular, for a convex function g, the term Q can be replaced by $\langle input \rangle^{O(1)}$.

We notice that in this theorem the bound of Q for the convex case follows by using the Lenstra-Kannan algorithm to solve the small dimensional subproblem (2). In this case, the $m^{O(m)}$ factor in the running time of the Lenstra-Kannan algorithm can be omitted as it is upper bounded by $(m\Delta)^{O(m^2)}$. Regarding the n^m term, as we explain below, it can be made lower order in interesting concrete settings. We also remark that a term of the form Δ^m cannot be avoided due to reductions from integer linear programming (see Sect. 1.1) and lower bounds for that problem [6].

1.1   Applications

Low dimensional convex functions. The main inspiration for this work is a recent study by Hunkenschröder, Pokutta, and Weismantel [5], who consider the problem

$$\min_{x \in \{0,1\}^n} g(Wx) \ , \tag{3}$$

where $W \in \mathbb{Z}^{m \times n}$ with entries of absolute value at most Δ, $g : \mathbb{R}^m \to \mathbb{R}$ is a "nice" *sharp* or separable convex function, and the algorithm can make function and gradient evaluations to the objective $g(Wx)$. They further distinguish between the case where W is given explicitly and where W is unknown to the algorithm. Assuming g is separable, they provide an $n(m\Delta)^{O(m^2)}$-time algorithm when W is known, and an $n(mL\Delta)^{O(m^3)}$-time algorithm when W is unknown and g is assumed to have L-Lipschitz gradients[1]. They show similar results when g is suitably "sharp", though we omit the statements for concision.

As a direct application of Theorem 1, we extend the result of [5] to arbitrary convex functions when W is known.

Corollary 1. *When W is known and g is an arbitrary convex function, problem (3) can be solved in time*

$$O(nm) + (m\Delta)^{O(m^2)} \ .$$

---

[1] They further require g to have an integer valued gradient on integer inputs.

The reduction to Theorem 1 is as follows: we first abandon the restriction of $x \in \{0,1\}^n$ in favor of the general bounded integer variables. Then, since any two variables with the same column in the projection matrix W can be merged to one (by adapting the box-constraints), we may assume without loss of generality that $n \leq (2\Delta + 1)^m$.

One of the main motivations in the work by Hunkenschröder et al. [5] is to solve certain types of regression problems. For example, they examine an integer *compressed sensing* problem, where one receives a small number m of linear measurements of a high dimensional integral signal $x^* \in \{0,1\}^n$ which one would like to (approximately) reconstruct. The received measurements are of the form $b = Wx^*$, where $W \in \mathbb{Z}^{m \times n}$ is an unknown matrix with coefficients of size at most Δ. As an approximation to x^*, they compute the minimizer of $\min\{\|b - Wx\|^2 : x \in \{0,1\}^n\}$, under the assumption that one can only access W indirectly via gradient and function evaluation queries to $f(x) = \|b - Wx\|^2$.

As we will explain later, in the compressed sensing and related settings, one can essentially avoid any overhead from not knowing W. While we focus above on the case where W is known, using orthogonal techniques, we can also improve the running times in the unknown W setting by modifying the Hunkenschröder et al. framework. We defer further discussion of their framework and our related improvements to Sect. 4.

Mixed-Integer Linear Programming. Eisenbrand and Weismantel [4] studied the complexity of integer programs of the form

$$
\begin{aligned}
&\min \ c^\mathsf{T} x \\
&\text{s.t. } Ax = b, \\
&\quad \ell_i \leq x_i \leq u_i \qquad \text{for all } i \in \{1, \ldots, n\}, \\
&\quad x \in \mathbb{Z}^n.
\end{aligned}
\tag{4}
$$

Specifically, they considered the setting where A has few rows and then used the Steinitz Lemma to obtain an algorithm with running time $(m\Delta)^{O(m^2)} \cdot n$, where Δ is the maximum absolute value in A. This has inspired a line of work for similar settings, see for example [3,6,8,9]. Our setting is a generalization of theirs: take $A = W$ and let

$$
g(Ax) = \begin{cases} 0 & \text{if } Ax = b, \\ \infty & \text{otherwise.} \end{cases}
$$

Here subproblem (2) corresponds to solving integer programming in m dimensions, which can be done using Lenstra-Kannan. Alternatively, one could model the problem as minimizing the convex function $g(Ax) = \|Ax - b\|$ for some suitable norm. Moreover, our model generalizes beyond the scope of Eisenbrand and

Weismantel's work to mixed-integer linear programming. Consider the problem

$$\begin{aligned}
\min \; & c^\mathsf{T} x + d^\mathsf{T} y \\
\text{s.t. } & Ax + By = b, \\
& \ell_i \le x_i \le u_i, && \text{for all } i \in \{1, \ldots, n\} \qquad (5) \\
& x \in \mathbb{Z}^n, \\
& y \in P \subseteq \mathbb{R}^h.
\end{aligned}$$

Here P is some polytope that can impose additional constraints on the continuous variables. We can encode this problem in (1) by setting $W = A$ and

$$g(Ax) = \begin{cases} \min\{d^\mathsf{T} y : By = b - Ax, y \in P\} & \text{if this minimum exists,} \\ \infty & \text{otherwise.} \end{cases}$$

Notice that the oracle problem (2) in this case forms a mixed-integer linear program itself, but with only m many integer variables; hence it can be solved efficiently with the algorithm by Lenstra-Kannan.

Corollary 2. *Assuming P can be efficiently separated over, problem (5) can be solved in time*

$$n^m \cdot (m\varDelta)^{O(m^2)} \cdot \langle \text{input} \rangle^{O(1)} .$$

We emphasize here that $\varDelta$ is only a bound on the entries of A, but not necessarily on those of B. Compared to the algorithm for the pure integer setting in [4], our running time has an extra factor of n^m, which however vanishes in some settings: for example, when $c = 0$ or $u_i = \infty$ for all i. In those cases we can again merge variables that share the same column in W. The only other example we are aware of that extends Eisenbrand and Weismantel's setting to mixed-integer linear programming is the work by Brand, Koutecký, and Ordyniak [2]. Their setting can be considered orthogonal to ours. On the one hand, they study a much more general structure of bounded treedepth programs, of which integer programs with a bounded number of constraints are the simplest special case. On the other hand, they impose these structural restrictions also on the continuous variables (and additionally bounds on their coefficients), whereas we impose essentially no restrictions on the structure of continuous variables or their coefficients.

To appreciate this, let us remark a pleasing aspect of the (straight-forward) extension of Lenstra-Kannan to mixed-integer linear programs: it combines the tractability of integer programs in fixed dimension with the tractability of linear programs in any dimension, achieving essentially a generalization of both. Eisenbrand and Weismantel's algorithm, on the other hand, concerns the tractability of integer programs with a fixed number of constraints (adding the necessary assumption that $\varDelta$ is bounded). In a similar spirit to the aforementioned generalization, our algorithm combines this with the tractability of (arbitrary) linear programs.

Integer linear programming with few complex variables. Recall the integer programming setting (4) studied by Eisenbrand and Weismantel, for which they gave an algorithm with running time $(m\Delta)^{O(m^2)} \cdot n$ (with Δ being the maximum absolute value in A). The interesting parameter regime for this algorithm is therefore when m and Δ are very small. Already for $m = 1$ this formulation easily captures the Knapsack problem, which is weakly NP-hard and therefore we cannot hope to reduce the dependency on Δ to, say, $\log(\Delta)$ while still maintaining polynomial dependency on n. In Lenstra-Kannan, on the other hand, the dependency on the coefficients of the matrix is polynomial in the encoding size, i.e., in $\log(\Delta)$, but the dependency on n is exponential. These two rather orthogonal results can be combined using Theorem 1.

Corollary 3. *Consider the integer programming problem in (4) and partition the columns of A into "simple" columns where the entries are bounded by Δ in absolute value and "complex" columns where they are arbitrary. Suppose that there are only k many complex columns. Then we can solve (4) in time*

$$n^m \cdot (m\Delta)^{m^2} \cdot k^{O(k)} \cdot \langle \text{input} \rangle^{O(1)} .$$

For this we proceed as follows. Let S and C be the index sets of the simple and complex columns and accordingly let A_S and A_C be the matrix A restricted to these column sets. We define Problem (1) only on x_S, the variables for the simple columns. Then let

$$g(A_S x_S) = \begin{cases} \min\{c_C^\mathsf{T} x_C : A_C x_C = b - A_S x_S\} & \text{if this minimum exists,} \\ \infty & \text{otherwise.} \end{cases}$$

The resulting subproblem (2) is then an integer program with $m + k$ variables that can be solved using Lenstra-Kannan. We note that one could even add to (4) arbitrary additional constraints on the complex columns and still solve the problem in the same way.

Variable-sized Knapsack. Antoniadis et al. [1] introduce a variant of the Knapsack problem with a non-linear cost function associated with the used capacity. They show that the case where this function is concave is polynomial time solvable and describe an FPTAS for the convex case. Our result can be used to devise a pseudopolynomial time algorithm for the convex case: the problem can be expressed as

$$\max \left\{ \sum_{i=1}^n p_i x_i - g\left(\sum_{i=1}^n w_i x_i \right) : x_i \in \{0, 1, \ldots, u_i\} \text{ for all } i \right\} . \tag{6}$$

where p_i is the profit of item i, w_i the weight, and $u_i \in \mathbb{Z}_{\geq 0} \cup \{\infty\}$ is a bound on the number of items of this type. Straightforward generalizations to multi-dimensional knapsack follow in a similar way.

Corollary 4. *For a convex function g, problem (6) can be solved in time*

$$(n + w_{\max})^{O(1)}.$$

Here, the oracle problem (2) reduces to a simple binary search. In general our result fits well to problems with a similar spirit, where the constraints are not hard, but they induce some penalty.

1.2   Overview of Techniques

The related results for more restrictive cases in [4] and [5] are based on proximity: the continuous relaxation of the problem, where the integer requirement is omitted, is solved and if one can show that the solution for the relaxation and the actual solution differ only slightly, then this can be exploited in reducing the search space. The precise proximity theorem in [4] is as follows.

Theorem 2 (Eisenbrand and Weismantel *[4]*). *Let z be an optimal vertex solution to the linear program*

$$\max \left\{ c^{\mathsf{T}} x : Ax = b \text{ and } \ell_i \le x_i \le u_i \text{ for all } i \right\},$$

where $A \in \mathbb{Z}^{m \times n}$ has entries of size at most Δ. If there exists an integer solution, then there is also an optimal integer solution x^* with

$$\|x^* - z\|_1 \le m(2m\Delta + 1)^m .$$

Hunkenschröder et al. [5] consider the optimal solution to the continuous relaxation of (3). In the special cases of separable convex and strict convex functions they show that a similar proximity holds, which is a crucial ingredient in their algorithm.

Already for general convex functions, however, the proximity bound can be very large, as shown in an example in [5]. This forms a serious obstacle towards our main result. We manage to circumvent this and still rely on proximity by applying it in a different way. Consider for sake of illustration that we were able to determine the value of $b^* = Wx^*$, where x^* is the optimal solution of (1). Then it would be easy to recover x^* (solving our problem) by applying the integer linear programming algorithm by Eisenbrand and Weismantel [4]. The algorithm works by computing the continuous solution z to $Wx = b^*$ and then using that $\|z - x^*\|_1$ is bounded by Theorem 2. Indeed, this bound still holds in our case when fixing b^*. However, it is not clear how to compute or guess b^*, nor how to compute z without knowing b^*.

Let us now consider the case that the domain of each variable is $\mathbb{Z}_{\ge 0}$, which is slightly simpler than the bounded case. Here we may assume that z has only m non-zero components, which we can guess from n^m candidates. We still do not know b^* or z, but we trivially know z on the $n - m$ zero components. Intuitively, this is enough to apply proximity to recover x^* on the zero components of z. Moreover, recovering x^* on the non-zero components of z is only an m-dimensional problem, where we can apply the oracle problem (2).

For our general result with arbitrarily bounded variable domains, there is another obstacle: if we try to generalize the previous line of arguments, it is still true that there are only m "special" variables in z, namely variables that are not tight on either of their bounds. For the remaining variables, however, it is not immediately obvious whether they equal the lower bound or the upper bound and if we do not know this, it is unclear how to determine x^* on these tight variables. We overcome this by guessing enough information about the dual so that we can use complementary slackness to infer which bound that the tight variables attain.

2   Non-negative Variables

For simplicity, in this section we first prove our main result for the variable domain $\mathbb{Z}_{\geq 0}$, that is, $\ell_i = 0$ and $u_i = \infty$ for all i. Consider the optimal solution x^* to (1) and define $b^* = Wx^*$. Furthermore, let z be an optimal vertex solution to $\min\{c^\mathsf{T}x : Wx = b^*, x \in \mathbb{R}^n_{\geq 0}\}$. We emphasize that z is not necessarily integral. By Theorem 2 there is an optimal integer solution x' to $\min\{c^\mathsf{T}x : Wx = b^*, x \in \mathbb{Z}^n_{\geq 0}\}$ with $\|x' - z\|_1 \leq O(m\Delta)^m$. We can assume without loss of generality that $x' = x^*$. Since z is a vertex solution, it has at least $n - m$ zero components T. It follows that

$$\|x^*_T\|_1 = \|x^*_T - z_T\|_1 \leq \|x^* - z\|_1 \leq O(m\Delta)^m \ .$$

Thus,

$$\|W_T x^*_T\|_1 \leq m\Delta \cdot \|x^*_T\|_1 \leq O(m\Delta)^{m+1} \ .$$

We now guess the indices of variables in T from the n^m many candidates and we guess the value of $b^{(T)} := W_T x^*_T$ from the $O(m\Delta)^{(m+1)m}$ many candidates. It is now easy to recover x^*_T (or an equivalent solution) by solving

$$\min \left\{ c^\mathsf{T}_T x_T : W_T x_T = b^{(T)} \text{ and } x_i \in \mathbb{Z}_{\geq 0} \text{ for all } i \in T \right\}.$$

Here we use the algorithm by Eisenbrand and Weismantel [4] or the improvement in [6]. This requires time $(m\Delta)^{O(m)} \cdot n$, which is insignificant compared to the number of guesses above. The algorithm assumes a solution to the LP relaxation is given, which, however, only serves the purpose of having a vector close to the optimal solution (in ℓ_1-norm). For this purpose we can also simply take z_T (the zero vector). Let L be the set of indices not in T. To recover x^*_L we need to solve

$$\min \{c^\mathsf{T}_L x^*_L + g(W_L x^*_L + W_T x^*_T) : x^*_i \in \mathbb{Z}_{\geq 0} \text{ for all } i \in L\}.$$

This corresponds to an oracle query of the form (2). For each guess of T and $b^{(T)}$ we compute a solution in this way and return the best among them. The running time, which is dominated by the number of guesses, is therefore

$$n^m \cdot (m\Delta)^{O(m^2)} \cdot Q \ . \tag{7}$$

In fact, the n^m term here can be omitted, since one may assume without loss of generality that no two columns of W are equal and therefore $n \leq (2\Delta + 1)^m$.

3   Bounded Variables

Let again x^* denote an optimal solution to (1) and $b^* = Wx^*$. Let z be an optimal solution to

$$\min\left\{c^\mathsf{T}x : Wx = b^* \text{ and } \ell_i \le x_i \le u_i \text{ for all } i\right\}.$$

We assume that c is augmented slightly by adding ε^i to the ith component for all i for some very small ε, which essentially implements a lexicographic tie-breaking rule between solutions. Here ε can be treated symbolically. We note that the dual of this linear program is

$$\max\left\{b^{*\mathsf{T}}y + \ell^\mathsf{T}s^\ell - u^\mathsf{T}s^u : c - W^\mathsf{T}y = s^\ell - s^u \text{ and } s_i^\ell, s_i^u \in \mathbb{R}_{\ge 0}^n, y \in \mathbb{R}^m\right\}.$$

Let y, s^ℓ, s^u be an optimal vertex solution to the dual. Then there are m linearly independent rows $(W^\mathsf{T})_i$ with $s_i^\ell = s_i^u = 0$ (otherwise it would not be a vertex solution). We guess these rows among the n^m candidates, which fully determines y and in particular $c - W^\mathsf{T}y$. We may assume that $(c - W^\mathsf{T}y)_i \neq 0$ for the other $n - m$ rows, which follows from the perturbation with ε. If $(c - W^\mathsf{T}y)_i > 0$ for some i we know that $s_i^u > 0$ and likewise if $(c - W^\mathsf{T}y)_i < 0$, then $s_i^\ell > 0$. By complementary slackness we can determine for these rows that $z_i = u_i$ (respectively, $z_i = \ell_i$). It follows that for $n - m$ variables T we now determined its value in z. Let L denote the m other variables. We now proceed similar to the previous section. We again have that

$$\|x_T^* - z_T\|_1 \le O(m\Delta)^m.$$

This implies that

$$\|W_T x_T^* - W_T z_T\|_1 \le m\Delta \cdot \|x_T^*\| \le O(m\Delta)^{m+1}.$$

Since we know the value of $W_T z_T$, we can guess $b^{(T)} = W_T x_T^*$ among the $O(m\Delta)^{(m+1)m}$ many candidates. Then we recover x_T^* using the algorithm by Eisenbrand and Weismantel [4] (where we can use z_T instead of an LP solution) and x_L^* by applying (2) to

$$\min\{c_L^\mathsf{T}x_L^* + g(W_L x_L^* + W_T x_T^*) : x_i^* \in \{\ell_i, \ell_i + 1, \ldots, u_i\} \text{ for all } i \in L\}.$$

Finally, we return the best solution computed for any guess.

4   Overview of Hunkenschröder Et Al. [5] and Related Improvements

We now explain the high-level algorithm of Hunkenschröder et al. [5] in more detail, as well as some improvements to their framework in the unknown W case.

Their algorithm starts with an optimal solution z to the continuous relaxation $\min\{g(Wx) : x \in [0,1]^n\}$ having at most m fractional components (which is easy to show to always exists). Here, z is assumed to be given by an oracle. For the cases they consider, e.g., the separable case, they prove that there is a "nearby" optimal integral solution x^* satisfying $\|x^* - z\|_1 \leq (m\Delta)^{O(m)}$. Function g being separable means that it can be decomposed into a sum of functions each depending only on a single dimension, that is, $g(Wx) = g_1((Wx)_1) + \ldots + g_m((Wx)_m)$. Using the proximity result, they guess $b^* = Wx^* \in \mathbb{Z}^m$, where the number of guesses is bounded by $(m\Delta)^{O(m^2)}$ (modulo an n factor, this is the dominant term in the complexity), noting that $\|W(x^* - z)\|_\infty \leq (m\Delta)\|x^* - z\|_1$. They then recover an optimal solution by solving the integer program $Wx = b^*, x \in \{0,1\}^n$. Note that this version of the algorithm requires W to be known.

When W is unknown, they show that one can replace W by a proxy matrix W', whose rows correspond to linearly independent gradients of $f(x) := g(Wx)$ seen so far by the algorithm. Their first observation is that the gradients of $\nabla f(x) = W^\mathsf{T} \nabla g(Wx)$ are linear combinations of the rows of W. Their second crucial observation is that for $b^* = W'x^*$, any integer solution to $W'x = b^*, x \in \{0,1\}^n$, is either optimal or has a gradient $\nabla f(x)$ outside the row span of W', in which case we can add an extra row to W'. Thus, one can iterate the guessing procedure with W replaced by W' at most m times before finding an optimal solution. The blowup in complexity in this setting comes from a lack of control over the coefficients appearing in W'. Indeed, this is precisely why they require that g has an L-Lipschitz gradient and integral gradients on integral inputs.

We remark that this idea can be implemented more efficiently without suffering from the worse parameters of W'. First, we observe that the cardinality of the set

$$B_N = \{W'x : x \in \{0,1\}^n, \|\lfloor z \rfloor - x\|_1 \leq N\}$$

can be bounded solely in N and the parameters of W. This is because each row of W' is a linear combination of rows of W. Hence, $Wx = Wx'$ implies $W'x = W'x'$ and therefore $|B_N| \leq O(N\Delta)^m$. Next, notice that B_N can be enumerated in time polynomial in n and $|B_N|$: this follows from an induction over n. To this end, for all $n' \leq n+1$ we define

$$B_N^{(n')} = \{W'x : x \in \{0,1\}^n, \|\lfloor z \rfloor - x\|_1 \leq N, \text{ and } x_i = \lfloor z_i \rfloor \text{ for all } i \geq n'\},$$

where $B_N^{(n+1)} = B_N$. We now iteratively generate the sets $B_{N'}^{(n')}$ by using $B_{N'}^1 = \{W'\lfloor z \rfloor\}$ and the recurrence

$$B_{N'}^{(n'+1)} = \begin{cases} B_{N'}^{(n')} \cup (B_{N'-1}^{(n')} + W_{n'}') & \text{if } \lfloor z_{n'} \rfloor = 0, \\ B_{N'}^{(n')} \cup (B_{N'-1}^{(n')} - W_{n'}') & \text{if } \lfloor z_{n'} \rfloor = 1. \end{cases}$$

Here $W_{n'}'$ is the n'th column of W'. We note that every vector in $B_{N'-1}^{(n')}$ is generated from some x with $x_{n'} = 0$ iff $\lfloor z_{n'} \rfloor = 0$. Hence, when adding (resp. removing) $W_{n'}'$ there is again a legal x generating this vector (where $\|\lfloor z \rfloor - x\|_1$ has increased by one).

As in the algorithm of Hunkenschröder et al., we now start with W' having only the single row $\nabla f(\lfloor z \rfloor)$. Then for every element of B_N we consider a corresponding integer solution x (note that such an x can easily be recovered in the above recurrence) and check if $\nabla f(x)$ and the rows of W' are linearly independent. If so, we add the gradient as a new row to W'. We repeat for at most m iterations until no new row is added. Then we return the best solution x^* seen during this process.

Theorem 3. *Let $g : \mathbb{R}^m \to R$ be a convex function, let $f(x) := g(Wx)$ be accessible via a function value and gradient oracle, where $W \in \mathbb{Z}^{m \times n}$ is an unknown matrix with entries of absolute value at most Δ. Then given an optimal solution z to the continuous relaxation $\min\{f(x) : x \in [0,1]^n\}$ with at most m fractional entries, one can compute an optimal integral solution in time $n(N\Delta)^{O(m)}$. Here N is the minimum $\|z - x^*\|_1$ over all optimal integer solutions x^*. In particular, when g is separable convex, the running time becomes $n(m\Delta)^{O(m^2)}$.*

5   Conclusion and Open Questions

In this paper we have demonstrated that the task of optimizing low dimensional functions over a projection as introduced by Hunkenschröder et al. [5] remains tractable even in much more general settings than originally considered. This creates a bridge also to other lines of work in integer optimization, such as integer programs with few constraints [4].

Our main result leaves open a few questions about the complexity of algorithms for problem (1) or the central case of g being a convex function. As mentioned before, one cannot hope to avoid a term of Δ^m in the running time because of known conditional lower bounds. The necessity of the n^m term or the m^2 exponent, however, appears less clear.

The algorithm for integer programming by Eisenbrand and Weismantel [4], a special case of our setting (see applications), does not require the n^m term and in many cases we can avoid it as well by merging duplicate columns of W. It would be nice if this term could be removed in general, or at least in the convex case.

Related to the m^2 exponent, there is already a notorious question arising from [4]. There, Eisenbrand and Weismantel gave an improved algorithm with exponent $O(m)$ instead of $O(m^2)$ for the case that there are no upper variable bounds, but with bounds they require $O(m^2)$. It remains unclear whether this is necessary. In our case even without upper bounds our algorithm need the exponent $O(m^2)$. In fact, this exponent arises in several places: when guessing the support (assuming $n \approx \Delta^m$) and when guessing the projection of the tight variables $b^{(T)}$.

References

1. Antoniadis, A., Huang, C.-C., Ott, S., Verschae, J.: How to pack your items when you have to buy your knapsack. In: Proceedings of MFCS, pp. 62–73 (2013)
2. Brand, C., Koutecký, M., Ordyniak, S.: Parameterized algorithms for MILPs with small treedepth. In: Proceedings of AAAI, pp. 12249–12257 (2021)

3. Cslovjecsek, J., Eisenbrand, F., Hunkenschröder, C., Rohwedder, L., Weismantel, R.: Block-structured integer and linear programming in strongly polynomial and near linear time. In: Proceedings of SODA, pp. 1666–1681 (2021)
4. Eisenbrand, F., Weismantel, R.: Proximity results and faster algorithms for integer programming using the Steinitz lemma. ACM Trans. Algorithms **16**(1), 5:1-5:14 (2020)
5. Hunkenschröder, C., Pokutta, S., Weismantel, R.: Optimizing a low-dimensional convex function over a high-dimensional cube. SIAM J. Optim. 2022, to appear
6. Jansen, K., Rohwedder, L.: On integer programming, discrepancy, and convolution. Math. Oper. Res. (2022, to appear)
7. Kannan, R.: Improved algorithms for integer programming and related lattice problems. In: Proceedings of STOC, pp. 193–206 (1983)
8. Klein, K.-M.: About the complexity of two-stage stochastic IPs. Math. Program. **192**(1), 319–337 (2022)
9. Knop, D., Pilipczuk, M., Wrochna, M.: Tight complexity lower bounds for integer linear programming with few constraints. ACM Trans. Comput. Theory **12**(3), 191–1919 (2020)
10. Lenstra, H.W., Jr.: Integer programming with a fixed number of variables. Math. Oper. Res. **8**(4), 538–548 (1983)

Configuration Balancing for Stochastic Requests

Franziska Eberle[1]($\boxtimes$), Anupam Gupta[2]($\boxtimes$), Nicole Megow[3]($\boxtimes$),
Benjamin Moseley[2]($\boxtimes$), and Rudy Zhou[2]($\boxtimes$)

[1] London School of Economics and Political Science, London, UK
f.eberle@lse.ac.uk
[2] Carnegie Mellon University, Pittsburgh, PA, USA
anupamg@cs.cmu.edu, {moseleyb,rbz}@andrew.cmu.edu
[3] University of Bremen, Bremen, Germany
nicole.megow@uni-bremen.de

Abstract. The configuration balancing problem with stochastic requests generalizes well-studied resource allocation problems such as load balancing and virtual circuit routing. There are given m resources and n requests; each request has multiple possible *configurations*, each of which increases the load of each resource by some amount. The goal is to select one configuration for each request to minimize the *makespan*: the load of the most-loaded resource. In the stochastic setting, the amount by which a configuration increases the resource load is uncertain until the configuration is chosen, but we are given a probability distribution.

We develop both offline and online algorithms for configuration balancing with stochastic requests. When the requests are known offline, we give a non-adaptive policy for configuration balancing with stochastic requests that $O(\frac{\log m}{\log \log m})$-approximates the optimal adaptive policy, which matches a known lower bound for the special case of load balancing on identical machines. When requests arrive online in a list, we give a non-adaptive policy that is $O(\log m)$ competitive. Again, this result is asymptotically tight due to information-theoretic lower bounds for special cases (e.g., for load balancing on unrelated machines). Finally, we show how to leverage adaptivity in the special case of load balancing on *related* machines to obtain a constant-factor approximation offline and an $O(\log \log m)$-approximation online. A crucial technical ingredient in all of our results is a new structural characterization of the optimal adaptive policy that allows us to limit the correlations between its decisions.

Keywords: stochastic scheduling · stochastic routing · load balancing

1 Introduction

This paper considers the *configuration balancing* problem: there are m resources and n requests. Request j has q_j configurations $x_j(1), \ldots, x_j(q_j) \in \mathbb{R}^m_{\geq 0}$. We

F. Eberle—Supported by the Dutch Research Council (NWO), Netherlands Vidi grant 016.Vidi.189.087.

must choose one configuration $c_j \in [q_j]$ per request, which adds $x_j(c_j)$ to the load vector on the resources. The goal is to minimize the makespan, i.e., the load of the most-loaded resource. Configuration balancing captures many natural resource allocation problems where requests compete for a finite pool of resources and the task is to find a "fair" allocation in which no resource is over-burdened. Two well-studied problems of this form arise in scheduling and routing.

(i) In *load balancing* a.k.a. *makespan minimization*, there are m (unrelated) machines and n jobs. Scheduling job j on machine i increases the load of i by $p_{ij} \geq 0$. The goal is to schedule each job on some machine to minimize the makespan, i.e., the load of the most-loaded machine.

(ii) In *virtual circuit routing* or *congestion minimization*, there is a directed graph $G = (V, E)$ on m edges with edge capacities $c_e > 0$ for $e \in E$, and n requests, each request consisting of a source-sink pair (s_j, t_j) in G and a demand $d_j \geq 0$. The goal is to route each request j from s_j to t_j via some directed path, increasing the load/congestion of each edge e on the path by d_j/c_e, while the objective is to minimize the load of the most-loaded edge.

Configuration balancing captures both problems by taking the m resources to be the m machines or edges, respectively; each configuration now corresponds to assigning a job to some machine or routing a request along some path.

Typically, job sizes or request demands are not known exactly when solving resource allocation problems in practice. This motivates the study of algorithms under uncertainty, where an algorithm must make decisions given only partial/uncertain information about the input. Uncertainty can be modeled in different ways. In exceptional cases, a *non-clairvoyant* algorithm that has *no* knowledge about the loads of requests may perform surprisingly well; an example is Graham's greedy list scheduling for load balancing on identical machines [15]. In general, a non-clairvoyant algorithm cannot perform well. Hence, we consider a stochastic model, where the unknown input follows some known distribution but the actual realization is a priori unknown. Such a model is natural when there is historical data available from which such distributions can be deduced.

In the *configuration balancing with stochastic requests* problem, we assume that each configuration c of request j is a random vector $X_j(c)$ with known distribution $\mathcal{D}_j(c)$ supported on $\mathbb{R}_{\geq 0}^m$ such that the $X_j(c)$'s are independent across different requests j. The actual realized vector of a configuration c of request j is only observed after *irrevocably* selecting this particular configuration for request j. The objective is to minimize the expected maximum load (makespan) $\mathbb{E}\left[\max_i \sum_{j=1}^n X_{ij}(c_j)\right]$, where c_j is the configuration chosen for request j. We assume that we have oracle access to the $\mathcal{D}_j(c)$'s; in particular we assume that in constant time, we can compute any needed statistic of the distribution $\mathcal{D}_j(c)$.

Further, we distinguish whether there is an additional dimension of uncertainty or not, namely the knowledge about the request set. In the *offline* setting, the set of requests and the distributions of the configurations of each request are known up-front, and they can be selected and assigned to the resources irrevocably in any order. In the *online* setting, requests are not known in advance and they are revealed one-by-one (online-list model). The algorithm learns the stochastic information on configurations of a request upon its arrival, and must select one of them without knowledge of future arrivals. After a configuration is chosen irrevocably, the next request arrives.

In general, we allow an algorithm to base the next decision on knowledge about the realized vectors of all previously selected request configurations. We call such policies *adaptive*. Conversely, a *non-adaptive* policy is one that fixes the particular configuration chosen for a request without using any knowledge of the realized configuration vectors.

The goal of this paper is to investigate the power of adaptive and non-adaptive policies for online and offline configuration balancing with stochastic requests. We quantify the performance of an algorithm by bounding the worst-case ratio of the achieved expected makespan and the minimal expected makespan achieved by an optimal offline adaptive policy. We say that an algorithm ALG α-*approximates* an algorithm ALG' if, for any input instance, the expected makespan of ALG is at most a factor α larger than the expected makespan of ALG'; we refer to α also as *approximation ratio*. For online algorithms, the term *competitive ratio* refers to their approximation ratio.

1.1   Our Results

Main Result. As our first main result, we present non-adaptive algorithms for offline and online configuration balancing with stochastic requests.

Theorem 1. *For configuration balancing with stochastic requests there is a randomized offline algorithm that computes a non-adaptive policy that is a $\Theta\left(\frac{\log m}{\log \log m}\right)$-approximation and an efficient deterministic online algorithm that is a $\Theta(\log m)$-approximation when comparing to the optimal offline adaptive policy. Both algorithms run in polynomial time in the number of resources and the total number of configurations over all requests.*

The offline analysis relies on a linear programming (LP) relaxation of configuration balancing, which has a known integrality gap of $\Theta\left(\frac{\log m}{\log \log m}\right)$, even for virtual circuit routing [25], implying that the analysis is tight. In the online setting, our analysis employs a potential function to greedily determine which configuration to choose for each request. In particular, we generalize the idea by [3] to the setting of configuration balancing with stochastic requests and match a known lower bound for online deterministic load balancing on unrelated machines by [5].

If the configurations are not given explicitly as part of the input or the number of configurations is large, then efficiently solving the problem requires us to be able to optimize over configurations in polynomial time.

Applications. These results would hold for both load balancing on unrelated machines and virtual circuit routing if we could guarantee that either the configurations are given explicitly or the respective subproblems can be solved efficiently. We can ensure this in both cases.

For stochastic load balancing on unrelated machines, the resources are the m machines, and each job has m possible configurations – one corresponding to assigning that job to each machine. Thus, we can efficiently represent all configurations. Further, here the LP relaxation of configuration balancing used in Theorem 1 is equivalent to the LP relaxation of the generalized assignment problem (GAP) solved in [33], which gives a deterministic rounding algorithm. Hence, Theorem 1 implies the following theorem. We omit the proof in this extended abstract; see [14] for proof.

Theorem 2. *There exist efficient deterministic algorithms that compute a non-adaptive policy for load balancing on unrelated machines with stochastic jobs that achieve an $\Theta\left(\frac{\log m}{\log \log m}\right)$-approximation offline and an $\Theta(\log m)$-approximation online when comparing to the optimal offline adaptive policy.*

These results are asymptotically tight due to the lower bound of $\Omega\left(\frac{\log m}{\log \log m}\right)$ on the adaptivity gap [17] and the lower bound of $\Omega(\log m)$ on the competitive ratio of any deterministic online algorithm, even for deterministic requests [5]. This implies that the adaptivity gap for stochastic load balancing is $\Theta\left(\frac{\log m}{\log \log m}\right)$.

For virtual circuit routing, the resources are the m edges and each request has a configuration for each possible routing path. Thus, efficiently solving the subproblems requires more work as the configurations are only given *implicitly* and there can be exponentially many. For the offline setting, since the LP relaxation has (possibly) exponentially many variables, we design an efficient separation oracle for the dual LP in order to efficiently solve the primal. For the online setting, we carefully select a subset of polynomially many configurations that contain the configuration chosen by the greedy algorithm, even when presented with all configurations. Thus, Theorem 1 implies that stochastic requests are not harder to approximate than deterministic requests. We omit the proof in this extended abstract; see [14] for proof.

Theorem 3. *For routing with stochastic requests, there exist an efficient randomized offline algorithm computing a non-adaptive policy that is a $\Theta\left(\frac{\log m}{\log \log m}\right)$-approximation and an efficient deterministic online algorithm that computes an $\Theta(\log m)$-approximation when comparing to the optimal offline adaptive policy.*

Adaptive Policies for Related Machines. When each request j has m configurations and configuration $c \in [m]$ can be written as $X_j(c) = \frac{X_j}{s_i} e_c$, where $e_c \in \mathbb{R}^m$ is the cth standard unit vector, the problem is also known as *load balancing on related machines*. We say that X_j is the size of request (or job) j and s_i is the speed of resource (or machine) i. In this special case, we show how to leverage adaptivity to overcome the $\Omega\left(\frac{\log m}{\log \log m}\right)$ lower bound on the adaptivity gap. Interestingly, our adaptive algorithms begin with a similar *non-adaptive* assignment of jobs to machines, but we deviate from the assignment adaptively to obtain our improved algorithms.

Theorem 4. *For load balancing on related machines with stochastic jobs, there exist efficient deterministic algorithms that compute an adaptive offline $O(1)$-approximation and an adaptive online $O(\log\log m)$-approximation when comparing to the optimal offline adaptive policy.*

It remains an interesting open question whether the online setting admits an $O(1)$-competitive algorithm.

1.2   Technical Overview

We illustrate the main idea behind our non-adaptive policies, which compare to the optimal offline adaptive policy. Throughout this paper, we let OPT denote the optimal adaptive policy as well as its makespan. As in many other stochastic optimization problems, our goal is to give a good deterministic proxy for the makespan of a policy. Then, our algorithm will optimize over this deterministic proxy to obtain a good solution. First, we observe that if all configurations were bounded with respect to $\mathbb{E}[\text{OPT}]$ in every entry, then selecting configurations such that each resource has expected load $O(\mathbb{E}[\text{OPT}])$ gives the desired $O(\frac{\log m}{\log\log m})$-approximation by standard concentration inequalities for independent sums with bounded increments. Thus, in this case the expected load on each resource is a good proxy. However, in general, we have no upper bound on $X_{ij}(c)$, so we cannot argue as above. We turn these unbounded random variables (RVs) into bounded ones in a standard way by splitting each request into *truncated* and *exceptional* parts.

Definition 1 (Truncated and Exceptional Parts). *Fix $\tau \geq 0$ as threshold. For a RV X, its truncated part (w.r.t. threshold τ) is $X^T := X \cdot \mathbb{1}_{X<\tau}$ and its exceptional part is $X^E := X \cdot \mathbb{1}_{X \geq \tau}$. Note that $X = X^T + X^E$.*

It is immediate that the truncated parts $X_{ij}^T(c)$ are bounded in $[0, \tau]$. Taking $\tau = O(\mathbb{E}[\text{OPT}])$, we can control their contribution to the makespan using concentration. It remains to find a good proxy for the contribution of exceptional parts to the makespan. This is one of the main technical challenges of our work as we aim to compare against the optimal adaptive policy: adaptive policies have much better control over the exceptional parts than non-adaptive ones.

Concretely, let c_j be the configuration chosen by some fixed policy for request j. Note that c_j itself can be a random variable in $\{1, \dots, q_j\}$. We want to control the quantity $\mathbb{E}\left[\max_i \sum_{j=1}^n X_{ij}^E(c_j)\right]$. Since we have no reasonable bound on the $X_{ij}^E(c_j)$'s, for non-adaptive policies, we can only upper bound the expected maximum by the following sum

$$\mathbb{E}\left[\max_{1\leq i\leq m}\sum_{j=1}^n X_{ij}^E(c_j)\right] \leq \sum_{j=1}^n \mathbb{E}\left[\max_{1\leq i\leq m} X_{ij}^E(c_j)\right]. \tag{1}$$

We call the right hand side *total (expected) exceptional load*. The above inequality is tight up to constants for non-adaptive policies, so it seems like the total expected exceptional load is a good proxy to use for our algorithm. However, it is far from tight for adaptive policies as the following example shows.

Example 1. Recall that in load balancing on related machines, each request j has m configurations and configuration $c \in [m]$ has the special form of $X_j(c) = \frac{X_j}{s_i} e_c$, where X_j is the processing time of job j and s_i is the speed of machine i. We assume that there is one fast machine with speed $s_1 = 1$ and $m - 1$ slow machines with speed $s_2 = \ldots = s_m = \frac{1}{\tau m}$, where $\tau > 0$ is the truncation threshold. There are m jobs: a stochastic one with processing time $X_j \sim \tau \cdot \text{Ber}\left(\frac{1}{\tau}\right)$ and $m - 1$ deterministic jobs with processing time $X_j \equiv \frac{1}{m}$. The optimal adaptive policy first schedules the stochastic job on the fast machine. If its realized size is 0, then it schedules all deterministic jobs on the fast machine as well. Otherwise the realized size is τ and it schedules one deterministic job on each slow machine, implying $\mathbb{E}[\text{OPT}] = \left(1 - \frac{1}{\tau}\right)\left(\frac{m-1}{m}\right) + \frac{1}{\tau} \cdot \tau = \Theta(1)$. However, the total expected exceptional load (w.r.t. τ) is $\sum_{i,j} \mathbb{E}\left[X_{ij}^E \cdot \mathbb{1}_{j \to i}\right] = \frac{1}{\tau}(m\tau) = m$, where $j \to i$ denotes that job j is assigned to machine i, i.e., configuration i is chosen for j.

In the example, the optimal adaptive policy accrues a lot of exceptional load, but this does not have a large effect on the makespan. Concretely, (1) can be loose by a $\Omega(m)$-factor for adaptive policies. Thus, it seems that the total exceptional load is a bad proxy in terms of lower-bounding OPT. However, we show that, by comparing our algorithm to a *near-optimal* adaptive policy rather than the optimal one, the total exceptional load becomes a good proxy in the following sense. This is the main technical contribution of our work, and it underlies all of our algorithmic techniques.

Theorem 5. *For configuration balancing with stochastic requests, there exists an adaptive policy with expected maximum load and total expected exceptional load at most $2 \cdot \mathbb{E}[\text{OPT}]$ with respect to any truncation threshold $\tau \geq 2 \cdot \mathbb{E}[\text{OPT}]$. Further, any configuration c selected by this policy satisfies $\mathbb{E}\left[\max_i X_i(c)\right] \leq \tau$.*

The proof of the above relies on carefully modifying the "decision tree" representing the optimal adaptive policy; see [14] for proof. In light of Theorem 5, the deterministic proxies we consider are the expected truncated load on each resource and the total expected exceptional load. All of our algorithms then proceed by ensuring that both quantities are bounded with respect to $\mathbb{E}[\text{OPT}]$. In the offline case, we round a natural assignment-type linear program (LP), and in the online case, we use a potential-function argument. All of these algorithms actually output non-adaptive policies. For the special case of related-machines load balancing, we also compute a non-adaptive assignment but instead of following it exactly, we deviate using adaptivity and give improved solutions.

1.3   Related Work

While stochastic optimization problems have long been studied [6,11], approximation algorithms for them are more recent [13,29]. By now, multi-stage stochastic problems (where uncertain information is revealed in stages) are well-understood [9,19,34]. In contrast, more dynamic models, where the exact value of an unknown parameter becomes known at times depending on the algorithms decisions (serving a request) still remain poorly understood. Some exceptions come from stochastic knapsack [8,12,16,28] as well as stochastic scheduling and routing which we discuss below.

Scheduling. For load balancing with deterministic sizes, a 2-approximation in the most general unrelated-machines offline setting [26] is known. For identical machines $(p_{ij} = p_j$ for all jobs $j)$, the greedy algorithm (called *list scheduling*) is a $(2 - \frac{1}{m})$-approximation algorithm [15]. This guarantee holds even when the jobs arrive online and *nothing* is known about job sizes. This implies a $(2 - \frac{1}{m})$-approximate *adaptive* policy for stochastic load balancing on identical machines.

Apart from this, prior work on stochastic scheduling has focused on approximating the optimal *non-adaptive* policy. There are non-adaptive $O(1)$-approximations known for identical machines [24], unrelated machines [17], the ℓ_q-norm objective [30], and monotone, symmetric norms [21].

In contrast, our work focuses on approximating the stronger optimal *adaptive* policy. The *adaptivity gap* (the ratio between the expected makespan of the optimal adaptive and non-adaptive policies) can be $\Omega\left(\frac{\log m}{\log \log m}\right)$ even for the simplest case of identical machines [17]. Thus, previous work on approximating the optimal non-adaptive policy does not immediately give any non-trivial approximation guarantees for our setting. The only previous work on adaptive stochastic policies for load-balancing (beyond the highly-adaptive list scheduling) is by [32]. They propose scheduling policies whose degree of adaptivity can be controlled by parameters and show an approximation factor of $O(\log \log m)$ for scheduling on identical machines.

Online load balancing with deterministic jobs is also well studied [4]. On identical machines, the aforementioned list scheduling algorithm [15] is $(2 - \frac{1}{m})$-competitive. For unrelated machines, there is a deterministic $O(\log m)$-competitive algorithm [3] and this is best possible [5]. When the machines are uniformly related, [7] design an $O(1)$-competitive algorithm for minimizing the makespan. [22,23] study the multi-dimensional generalization to vector scheduling under the makespan and the ℓ_q-norm objective.

To the best of our knowledge, configuration balancing has not been explicitly defined before. The techniques of [3] give an $O(\log m)$-competitive algorithm for deterministic requests. It is also studied for packing integer programs [1,2,18].

Routing. For oblivious routing with stochastic demands, [20] give an algorithm which is an $O(\log^2 n)$-approximation with high probability. Here, "oblivious" refers to the requirement that the chosen path between a source-sink pair must not depend on the current congestion of the network. In particular, after specifying a set of paths for each possible source-sink pair, a demand matrix is drawn from an a-priori known distribution and each demand needs to be routed along one of the predefined paths. The obliviousness requirement is very different from our setting and makes the two models essentially incomparable.

When $d_j = 1$ for each source-sink pair, there is an $O\left(\frac{\log m}{\log \log m}\right)$-approximation algorithm by [31], which is best possible, unless NP $\subseteq$ ZPTIME$(n^{\log \log n})$ [10].

In the online setting, when the source-sink pairs arrive online in a list and have to be routed before the next pair arrives, [3] give a lower bound of $\Omega(\log n)$ on the competitive ratio of any deterministic online algorithm in directed graphs, where n is the number of vertices. They also give a matching upper bound. For more details on online routing we refer to the survey [27].

2   Configuration Balancing with Stochastic Requests

In this section, we prove our main results for the most general problem we consider: configuration balancing. We give an $O\left(\frac{\log m}{\log\log m}\right)$-approximation offline and an $O(\log m)$-approximation online; both algorithms are non-adaptive. Before describing the algorithms, we give our main structural theorem that enables all of our results. Roughly, we show that instead of comparing to the optimal adaptive policy, by losing only a constant factor in the approximation ratio, we can compare to a near-optimal policy that behaves like a non-adaptive one (w.r.t. the proxy objectives we consider, namely, the total expected exceptional load).

2.1   Structural Theorem

In this section, we show that there exists a near-optimal policy as guaranteed by Theorem 5. To this end, we modify the optimal policy by "restarting" whenever an exceptional request is encountered. Additionally, we ensure that this modified policy never selects a configuration c for a request j with $\mathbb{E}\left[\max_i X_{ij}(c)\right] > \tau$.

We let J denote the set of requests. For any subset $J' \subseteq J$, we let $\mathrm{OPT}(J')$ denote the optimal adaptive policy (and its maximum load) on the set of requests J'. Note that $\mathrm{OPT}(\emptyset) = 0$. Our (existential) algorithm to construct such a policy will begin by running the optimal policy $\mathrm{OPT}(J)$ on all requests. However, once an exceptional request is encountered or the next decision will choose a configuration with too large expected maximum, we cancel $\mathrm{OPT}(J)$ and instead recurse on all remaining requests, ignoring all previously-accrued loads; see Algorithm 1. The idea of our analysis is that we recurse with small probability; see [14].

Theorem 5. *For configuration balancing with stochastic requests, there exists an adaptive policy with expected maximum load and total expected exceptional load at most $2 \cdot \mathbb{E}[\mathrm{OPT}]$ with respect to any truncation threshold $\tau \geq 2 \cdot \mathbb{E}[\mathrm{OPT}]$. Further, any configuration c selected by this policy satisfies $\mathbb{E}\left[\max_i X_i(c)\right] \leq \tau$.*

Having this near-optimal policy at hand, the upshot is that we can bound our subsequent algorithms with respect to the following LP relaxation ($\mathrm{LP_C}$) for configuration balancing with stochastic requests. The variable y_{cj} denotes selecting configuration c for request j. We take our threshold between the truncated and exceptional parts to be τ. Using the natural setting of the y-variables as the probabilities of the policy from Theorem 5, it is straight-forward to show that

---

Algorithm 1: Policy $S(J)$

$R \leftarrow J$ // remaining requests

if $R = \emptyset$ **then**

    return empty policy   // finish

while $R \neq \emptyset$ **do**

    $j \leftarrow$ first / next request considered by $\mathrm{OPT}(J)$

    $c_j \leftarrow$ configuration chosen for request j by $\mathrm{OPT}(J)$

1   **if** $\mathbb{E}\big[\max_i X_{ij}(c_j)\big] > \tau$ **then**   // maximum too large

       break

    else

       choose c_j for request j // $S(J)$ follows $\mathrm{OPT}(J)$

       $R \leftarrow R \setminus \{j\}$ // update remaining requests

2   **if** $\max_i X_{ij}(c_j) \geq \tau$ **then**   // exceptional configuration observed

          break

run $S(R)$ // recurse with remaining requests

---

the following LP relaxation has a feasible solution, formalized in Lemma 1.

$$\sum_{c=1}^{q_j} y_{cj} = 1 \quad \forall\, j \in [n]$$
$$\sum_{j=1}^{n} \sum_{c=1}^{q_j} \mathbb{E}[X_{ij}^{T}(c)] \cdot y_{cj} \leq \tau \quad \forall\, i \in [m]$$
$$\sum_{j=1}^{n} \sum_{c=1}^{q_j} \mathbb{E}[\max_i X_{ij}^{E}(c)] \cdot y_{cj} \leq \tau$$
$$y_{cj} = 0 \quad \forall\, j \in [n], \forall\, c \in [q_j] : \mathbb{E}[\max_i X_{ij}(c)] > \tau$$
$$y_{cj} \geq 0 \quad \forall\, j \in [n], \forall\, c \in [q_j]$$

$$(\mathsf{LP_C})$$

Lemma 1. $(\mathsf{LP_C})$ *has a feasible solution for any* $\tau \geq 2 \cdot \mathbb{E}[\mathrm{OPT}]$.

2.2   Offline Setting

Our offline algorithm is the natural randomized rounding of $(\mathsf{LP_C})$. For the truncated parts, the following inequality bounds their contribution to the makespan.

Lemma 2. *Let* $S_1, \ldots, S_m$ *be sums of independent RVs bounded in* $[0, \tau]$ *for some* $\tau > 0$ *such that* $\mathbb{E}[S_i] \leq \tau$ *for all* $i \in [m]$. *Then,* $\mathbb{E}[\max_i S_i] = O\big(\frac{\log m}{\log \log m}\big)\tau$.

To bound the contribution of the exceptional parts, we use (1), i.e., the total expected exceptional load. Using binary search for the correct choice of τ and re-scaling the instance down by the current value of τ, it suffices to give an efficient algorithm that either

- outputs a non-adaptive policy with expected makespan $O\big(\frac{\log m}{\log \log m}\big)$, or
- certifies that $\mathbb{E}[\mathrm{OPT}] > 1$.

This is because for $\tau \in (\mathbb{E}[\mathrm{OPT}], 2 \cdot \mathbb{E}[\mathrm{OPT}]]$, the re-scaling guarantees $\mathbb{E}[\mathrm{OPT}] \in [\frac{1}{2}, 1)$ on the scaled instance, in which case the algorithm achieves expected makespan $O\big(\frac{\log m}{\log \log m}\big) = O\big(\frac{\log m}{\log \log m}\big) \cdot \mathbb{E}[\mathrm{OPT}]$.

Algorithm 2: Offline Configuration Balancing with Stochastic Requests

---

try to solve (LP$_C$) with $\tau = 2$
if (LP$_C$) *is feasible* **then**
    | let y^* be the outputted feasible solution
    | **for** *each request j* **do**
    | | independently sample $c \in [q_j]$ with probability y^*_{cj}
    | | choose sampled c as c_j
else
    | return "$\mathbb{E}[\text{OPT}] > 1$"

To that end, we use the natural independent randomized rounding of (LP$_C$). That is, if (LP$_C$) has a feasible solution y^*, for request j, we choose configuration c as configuration c_j independently with probability y^*_{cj}; see Algorithm 2.

If the configurations are given explicitly as part of the input, then (LP$_C$) can be solved in polynomial time and, thus, Algorithm 2 runs in polynomial time. Hence, the $O\left(\frac{\log m}{\log \log m}\right)$-approximate non-adaptive policy for configuration balancing with stochastic requests (Theorem 1) follows from the next lemma.

Lemma 3. *If* (LP$_C$) *can be solved in polynomial time, Algorithm 2 is a polynomial-time randomized algorithm that either outputs a non-adaptive policy with expected makespan $O\left(\frac{\log m}{\log \log m}\right)$, or certifies correctly that $\mathbb{E}[\text{OPT}] > 1$.*

2.3 Online Setting

We now consider online configuration balancing where n stochastic requests arrive online one-by-one, and for each request, one configuration has to be irrevocably selected before the next request appears. We present a non-adaptive online algorithm that achieves a competitive ratio of $O(\log m)$, which is best possible due to the lower bound of $\Omega(\log m)$ [5].

By a standard guess-and-double scheme, we may assume that we have a good guess of $\mathbb{E}[\text{OPT}]$. We omit the proof, which is analogous to its virtual-circuit-routing counterpart in [3].

Lemma 4. *Given an instance of online configuration balancing with stochastic requests, suppose there exists an online algorithm that, given parameter $\lambda > 0$, never creates an expected makespan more than $\alpha \cdot \lambda$, possibly terminating before handling all requests. Further, if the algorithm terminates prematurely, then it certifies that $\mathbb{E}[\text{OPT}] > \lambda$. Then, there exists an $O(\alpha)$-competitive algorithm for online configuration balancing with stochastic requests. Further, the resulting algorithm preserves non-adaptivity.*

We will build on the same technical tools as in the offline case. In particular, we wish to compute a non-adaptive assignment online with small expected truncated load on each resource and small total expected exceptional load. To achieve this, we generalize the greedy potential function approach of [3]. Our two new ingredients are to treat the exceptional parts of a request's configuration as a resource requirement for an additional, artificial resource and to compare the potential of our solution directly with a *fractional* solution to (LP$_C$).

Now we describe our potential function, which is based on an exponential/soft-max function. Let λ denote the current guess of the optimum as required by Lemma 4. We take $\tau = 2\lambda$ as our truncation threshold. Given load vector $L \in \mathbb{R}^{m+1}$, our potential function is

$$\phi(L) = \sum_{i=0}^{m} (3/2)^{L_i/\tau}.$$

For $i \in [m]$, we ensure the ith entry of L is the *expected* truncated load on resource i and use the 0th entry as a virtual resource that is the total expected exceptional load. For any request j, let L_j be the expected load vector after handling the first j requests, with L_{ij} denoting its ith entry. Let $L_{i0} := 0$ for all i. Upon arrival of request j, our algorithm tries to choose the configuration $c_j \in [q_j]$ that minimizes the increase in potential; see Algorithm 3.

Algorithm 3: Online Configuration Balancing with Stochastic Requests

$\ell \leftarrow \log_{3/2}(2m + 2)$
$\lambda \leftarrow$ current guess of $\mathbb{E}[\text{OPT}]$
$\tau \leftarrow 2\lambda$ truncation threshold
upon *arrival of request j* **do**

1    $c_j \leftarrow \arg\min_{c \in [q_j]} \left((3/2)^{(L_{0j-1} + \mathbb{E}[\max_{i \in [m]} X_{ij}^E(c)])/\tau} + \right.$

$\left. \sum_{i=1}^{m} (3/2)^{(L_{ij-1} + \mathbb{E}[X_{ij}^T(c)])/\tau} \right) - \phi(L_{j-1})$

   if $L_{ij-1} + \mathbb{E}[X_{ij}^T(c_j)] \leq \ell\tau$ for all $i \in [m]$ **and** $L_{0j-1} + \mathbb{E}[\max_{i \in [m]} X_{ij}^E(c_j)] \leq \ell\tau$
   then
   | choose c_j for j
   | $L_{ij} \leftarrow L_{ij-1} + \mathbb{E}[X_{ij}^T(c_j)]$ for all $i \in [m]$
   | $L_{0j} \leftarrow L_{0j-1} + \mathbb{E}[\max_{i \in [m]} X_{ij}^E(c_j)]$
   else
   | return "$\mathbb{E}[\text{OPT}] > \lambda$"

To analyze this algorithm, we compare its makespan with a solution to (LP$_C$). This LP has an integrality gap of $\Omega\left(\frac{\log m}{\log \log m}\right)$, which follows immediately from the path assignment LP for virtual circuit routing [25]. Hence, a straightforward analysis of Algorithm 3 comparing to a rounded solution to (LP$_C$) gives an assignment with expected truncated load per machine and total expected exceptional load $O\left(\log m \cdot \frac{\log m}{\log \log m}\right) \cdot \mathbb{E}[\text{OPT}]$. To get a tight competitive ratio of $O(\log m)$, we avoid the integrality gap by comparing to a *fractional* solution to (LP$_C$), and we use a slightly different inequality than Lemma 2 for the regime where the mean of the sums is larger than the increments by at most a $O(\log m)$-factor.

Lemma 5. *Let $S_1, \ldots, S_m$ be sums of independent RVs bounded in $[0, \tau]$ for $\tau > 0$ such that $\mathbb{E}[S_i] \leq O(\log m)\tau$ for all $1 \leq i \leq m$. Then, $\mathbb{E}[\max_i S_i] \leq O(\log m)\tau$.*

We give the guarantee for Algorithm 3, which implies the $O(\log m)$-competitive algorithm for online configuration balancing with stochastic requests.

Lemma 6. *Suppose the minimizing configuration in Line 1 can be found in polynomial time. Then Algorithm 3 runs in polynomial time; it is deterministic, non-adaptive and correctly solves the subproblem of Lemma 4 for $\alpha = O(\log m)$.*

3   Load Balancing on Related Machines

In this section, we improve on Theorem 2 in the special case of related machines, where each machine i has a speed parameter $s_i > 0$ and each job j an independent size X_j such that $X_{ij} = \frac{X_j}{s_i}$. Recall that we gave a non-adaptive $O\left(\frac{\log m}{\log \log m}\right)$-approximation for unrelated machines. However, the adaptivity gap is $\Omega\left(\frac{\log m}{\log \log m}\right)$ even for load balancing on identical machines where every machine has the same speed. Thus, to improve on Theorem 2, we need to use adaptivity.

The starting point of our improved algorithms is the same non-adaptive assignment for unrelated-machines load balancing. However, instead of non-adaptively assigning job j to the specified machine i, we adaptively assign j to the least loaded machine with similar speed to i. We formalize this idea and briefly explain the algorithms for offline and online load balancing on related machines.

Machine Smoothing. In this part, we define a notion of *smoothed machines*. We show that by losing a constant factor in the approximation ratio, we may assume that the machines are partitioned into at most $O(\log m)$ groups such that machines within a group have the same speed and the size of the groups shrinks geometrically. Thus, by "machines with similar speed to i," we mean machines in the same group.

Formally, we transform an instance $\mathcal{I}$ of load balancing on m related machines with stochastic jobs into an instance $\mathcal{I}_s$ with so-called "smoothed machines" and the same set of jobs with the following three properties:

(i)   The machines are partitioned into $m' = O(\log m)$ groups such that group k consists of m_k machines with speed exactly s_k such that $s_1 < s_2 < \cdots < s_{m'}$.

(ii) For all groups $1 \le k < m'$, we have $m_k \ge \frac{3}{2} m_{k+1}$.

(iii) $\text{OPT}(\mathcal{I}_s) = O(\text{OPT}(\mathcal{I}))$.

To this end, we suitably decrease machine speeds and delete machines from the original instance $\mathcal{I}$; see [14] for the algorithm and the technical details.

Lemma 7. *There is an efficient algorithm that, given an instance $\mathcal{I}$ of load balancing with m related machines and stochastic jobs, computes an instance $\mathcal{I}_s$ of smoothed machines with the same set of jobs satisfying Properties (i) to (iii).*

A similar idea for machine smoothing has been employed by Im et al. [23] for deterministic load balancing on related machines. In their approach, they ensure that the *total processing power* of the machines in a group decreases geometrically rather than the number of machines.

Offline Setting. We run Algorithm 2 on the configuration balancing instance defined by the load balancing instance with smoothed machines. Given a job-to-machine assignment, we list schedule the jobs assigned to a particular group on the machines of this group. In the proof of Theorem 4, we rely on the following strong bound on the expected maximum of the truncated load; see [14].

Lemma 8. *Let $c_1, \ldots, c_m \in \mathbb{N}_{\geq 1}$ be constants such that $c_i \geq \frac{3}{2}c_{i+1}$ for all $1 \leq i \leq m$. Let $S_1, \ldots, S_m$ be sums of independent random variables bounded in $[0, \tau]$ such that $\mathbb{E}[S_i] \leq c_i \tau$ for all $1 \leq i \leq m$. Then, $\mathbb{E}\left[\max_i \frac{S_i}{c_i}\right] \leq O(\tau)$.*

Online Setting. We apply a similar framework as above. Note that our online configuration balancing algorithm loses a logarithmic factor in the number of resources, so to obtain a $O(\log \log m)$-approximation, we aggregate each group (in the smoothed-machines instance) as a single resource. Intuitively, this definition captures the fact that we will average all jobs assigned to a group over the machines in this group. Thus, our configuration balancing instance will have only $O(\log m)$ resources and applying Theorem 1 proves Theorem 4; see [14].

Conclusion

We considered the configuration balancing problem under uncertainty. In contrast to the (often overly optimistic) clairvoyant settings and the (often overly pessimistic) non-clairvoyant settings, we consider the stochastic setting where each request j presents a set of random vectors, and we need to (adaptively) pick one of these vectors, to minimize the *expected* maximum load over the m resources. We give logarithmic bounds for several general settings (which are existentially tight), and a much better $O(1)$ offline and $O(\log \log m)$ online bound for the related machines setting. Closing the gap for online related-machines load balancing remains an intriguing open problem. More generally, getting a better understanding of both adaptive and non-adaptive algorithms for stochastic packing and scheduling problems remains an exciting direction for research.

References

1. Agrawal, S., Devanur, N.R.: Fast algorithms for online stochastic convex programming. In: Proceedings of SODA, pp. 1405–1424 (2015)
2. Agrawal, S., Wang, Z., Ye, Y.: A dynamic near-optimal algorithm for online linear programming. Oper. Res. **62**(4), 876–890 (2014)
3. Aspnes, J., Azar, Y., Fiat, A., Plotkin, S.A., Waarts, O.: On-line routing of virtual circuits with applications to load balancing and machine scheduling. J. ACM **44**(3), 486–504 (1997)

4. Azar, Y.: On-line load balancing. In: Fiat, A., Woeginger, G.J. (eds.) Online Algorithms. LNCS, vol. 1442, pp. 178–195. Springer, Heidelberg (1998). https://doi.org/10.1007/BFb0029569
5. Azar, Y., Naor, J., Rom, R.: The competitiveness of on-line assignments. J. Algorithms **18**(2), 221–237 (1995)
6. Beale, E.M.L.: On minimizing a convex function subject to linear inequalities. J. Roy. Stat. Soc. Ser. B. Methodol. **17**, 173–184; discussion, 194–203 (1955)
7. Berman, P., Charikar, M., Karpinski, M.: On-line load balancing for related machines. J. Algorithms **35**(1), 108–121 (2000)
8. Bhalgat, A., Goel, A., Khanna, S.: Improved approximation results for stochastic knapsack problems. In: Proceedings of SODA, pp. 1647–1665. SIAM (2011)
9. Charikar, M., Chekuri, C., Pál, M.: Sampling bounds for stochastic optimization. In: Chekuri, C., Jansen, K., Rolim, J.D.P., Trevisan, L. (eds.) APPROX/RANDOM -2005. LNCS, vol. 3624, pp. 257–269. Springer, Heidelberg (2005). https://doi.org/10.1007/11538462_22
10. Chuzhoy, J., Guruswami, V., Khanna, S., Talwar, K.: Hardness of routing with congestion in directed graphs. In: Proceedings of STOC, pp. 165–178. ACM (2007)
11. Dantzig, G.B.: Linear programming under uncertainty. Manag. Sci. **1**, 197–206 (1955)
12. Dean, B.C., Goemans, M.X., Vondrák, J.: Approximating the stochastic knapsack problem: the benefit of adaptivity. Math. Oper. Res. **33**(4), 945–964 (2008)
13. Dye, S., Stougie, L., Tomasgard, A.: The stochastic single resource service-provision problem. Naval Res. Logist. **50**(8), 869–887 (2003)
14. Eberle, F., Gupta, A., Megow, N., Moseley, B., Zhou, R.: Configuration balancing for stochastic requests. CoRR, abs/2208.13702 (2022)
15. Graham, R.L.: Bounds on multiprocessing timing anomalies. SIAM J. Appl. Math. **17**(2), 416–429 (1969)
16. Gupta, A., Krishnaswamy, R., Molinaro, M., Ravi, R.: Approximation algorithms for correlated knapsacks and non-martingale bandits. In: Ostrovsky, R. (ed.) Proceedings of FOCS, pp. 827–836. IEEE Computer Society (2011)
17. Gupta, A., Kumar, A., Nagarajan, V., Shen, X.: Stochastic load balancing on unrelated machines. Math. Oper. Res. **46**(1), 115–133 (2021)
18. Gupta, A., Molinaro, M.: How the experts algorithm can help solve LPS online. Math. Oper. Res. **41**(4), 1404–1431 (2016)
19. Gupta, A., Pál, M., Ravi, R., Sinha, A.: Sampling and cost-sharing: approximation algorithms for stochastic optimization problems. SIAM J. Comput. **40**(5), 1361–1401 (2011)
20. Hajiaghayi, M.T., Kim, J.H., Leighton, T., Räcke, H.: Oblivious routing in directed graphs with random demands. In: Proceedings of STOC, pp. 193–201. ACM (2005)
21. Ibrahimpur, S., Swamy, C.: Approximation algorithms for stochastic minimum-norm combinatorial optimization. In: Proceedings of FOCS, pp. 966–977. IEEE (2020)
22. Im, S., Kell, N., Kulkarni, J., Panigrahi, D.: Tight bounds for online vector scheduling. SIAM J. Comput. **48**(1), 93–121 (2019)
23. Im, S., Kell, N., Panigrahi, D., Shadloo, M.: Online load balancing on related machines. In: Proceedings of STOC, pp. 30–43. ACM (2018)
24. Kleinberg, J.M., Rabani, Y., Tardos, É.: Allocating bandwidth for bursty connections. SIAM J. Comput. **30**(1), 191–217 (2000)
25. Leighton, T., Rao, S., Srinivasan, A.: Multicommodity flow and circuit switching. In: HICSS (7), pp. 459–465. IEEE Computer Society (1998)

26. Lenstra, J.K., Shmoys, D.B., Tardos, É.: Approximation algorithms for scheduling unrelated parallel machines. Math. Program. **46**, 259–271 (1990)
27. Leonardi, S.: On-line network routing. In: Fiat, A., Woeginger, G.J. (eds.) Online Algorithms. LNCS, vol. 1442, pp. 242–267. Springer, Heidelberg (1998). https://doi.org/10.1007/BFb0029572
28. Ma, W.: Improvements and generalizations of stochastic knapsack and Markovian bandits approximation algorithms. Math. Oper. Res. **43**(3), 789–812 (2018)
29. Möhring, R.H., Schulz, A.S., Uetz, M.: Approximation in stochastic scheduling: the power of LP-based priority policies. J. ACM **46**(6), 924–942 (1999)
30. Molinaro, M.: Stochastic p load balancing and moment problems via the l-function method. In: Proceedings of SODA, pp. 343–354. SIAM (2019)
31. Raghavan, P., Thompson, C.D.: Randomized rounding: a technique for provably good algorithms and algorithmic proofs. Combinatorica **7**(4), 365–374 (1987)
32. Sagnol, G., Schmidt, D., Waldschmidt, G.: Restricted adaptivity in stochastic scheduling. In: Proceedings of ESA. LIPIcs, vol. 204, pp. 79:1–79:14. Schloss Dagstuhl - Leibniz-Zentrum für Informatik (2021)
33. Shmoys, D.B., Tardos, É.: An approximation algorithm for the generalized assignment problem. Math. Program. **62**, 461–474 (1993)
34. Swamy, C., Shmoys, D.B.: Sampling-based approximation algorithms for multi-stage stochastic optimization. SIAM J. Comput. **41**(4), 975–1004 (2012)

An Update-and-Stabilize Framework for the Minimum-Norm-Point Problem

Satoru Fujishige[1] , Tomonari Kitahara[2] , and László A. Végh[3]([✉])

[1] Research Institute for Mathematical Sciences, Kyoto University,
Kyoto 606-8502, Japan
fujishig@kurims.kyoto-u.ac.jp
[2] Faculty of Economics, Kyushu University, Fukuoka 819-0395, Japan
tomonari.kitahara@econ.kyushu-u.ac.jp
[3] Department of Mathematics, London School of Economics and Political Science,
London WC2A 2AE, UK
L.Vegh@lse.ac.uk

Abstract. We consider the minimum-norm-point (MNP) problem of polyhedra, a well-studied problem that encompasses linear programming. Inspired by Wolfe's classical MNP algorithm, we present a general algorithmic framework that performs first order update steps, combined with iterations that aim to 'stabilize' the current iterate with additional projections, i.e., finding a locally optimal solution whilst keeping the current tight inequalities. We bound the number of iterations polynomially in the dimension and in the associated circuit imbalance measure. In particular, the algorithm is strongly polynomial for network flow instances. The conic version of Wolfe's algorithm is a special instantiation of our framework; as a consequence, we obtain convergence bounds for this algorithm. Our preliminary computational experiments show a significant improvement over standard first-order methods.

1 Introduction

We study the minimum-norm-point (MNP) problem

$$\text{Minimize } \tfrac{1}{2}||Ax - b||^2 \text{ subject to } \mathbf{0} \le x \le u,\, x \in \mathbb{R}^N, \qquad (P)$$

where m and n are positive integers, $M = \{1, \cdots, m\}$ and $N = \{1, \cdots, n\}$, $A \in \mathbb{R}^{M \times N}$ is a matrix with rank $\text{rk}(A) = m$, $b \in \mathbb{R}^M$, and $u \in (\mathbb{R} \cup \{\infty\})^N$.

This is an extended abstract. The full version including all omitted proofs is available on arXiv:2211.02560.

SF's research is supported by JSPS KAKENHI Grant Numbers JP19K11839 and 22K11922 and by the Research Institute for Mathematical Sciences, an International Joint Usage/Research Center located in Kyoto University. TK is supported by JSPS KAKENHI Grant Number JP19K11830. LAV's research is supported by the European Research Council (ERC) under the European Union's Horizon 2020 research and innovation programme (grant agreement no. 757481–ScaleOpt).

A. Del Pia and V. Kaibel (Eds.): IPCO 2023, LNCS 13904, pp. 142–156, 2023.
https://doi.org/10.1007/978-3-031-32726-1_11

We will use the notation $\mathbf{B}(u) := \{x \in \mathbb{R}^N \mid \mathbf{0} \le x \le u\}$ for the feasible set. The problem (P) generalizes the linear programming (LP) feasibility problem: the optimum value is 0 if and only if $Ax = b$, $x \in \mathbf{B}(u)$ is feasible. We say that (P) is an *uncapacitated instance* if $u(i) = \infty$ for all $i \in N$.

The formulation (P) belongs to a family of problems for which Necoara, Nesterov, and Glineur [16] showed linear convergence bounds of first order methods. That is, the number of iterations needed to find an ε-approximate solution depends linearly on $\log(1/\varepsilon)$. Such convergence has been known for strongly convex functions, but this property does not hold for (P). However, [16] shows that restricted variants of strong convexity also suffice. For problems of the form (P), the required property follows using Hoffman-proximity bounds [13]; see [19] and the references therein for recent results on Hoffman-proximity.

We propose a new algorithmic framework for the minimum-norm-point problem (P) that uses *stabilizing steps* between first order updates. Our algorithm terminates with an exact optimal solution in a finite number of iterations. Moreover, we show $\text{poly}(n, \kappa)$ running time bounds for multiple instantiations of the framework, where κ is the *circuit imbalance measure* associated with the matrix $(A \mid I_m)$ (see Sect. 2). This gives strongly polynomial bounds whenever κ is constant; in particular, $\kappa = 1$ for network flow feasibility. We note that if $A \in \mathbb{Z}^{M \times N}$, then $\kappa \le \Delta(A)$ for the maximum subdeterminant $\Delta(A)$. Still, κ can be exponential in the encoding length of the matrix.

The stabilizing step is inspired by Wolfe's classical minimum-norm-point algorithm [24]. This considers the variant of (P) where the box constraint $x \in \mathbf{B}(u)$ is replaced by $\sum_{i \in N} x_i = 1$, $x \ge \mathbf{0}$. Wolfe's algorithm is reminiscent of the simplex method. It comprises major and minor cycles, and at the end of every major cycle, the algorithm maintains a *corral* solution: for a linearly independent set of columns, the current point is the nearest point to b in the affine hull of these columns, while it also falls inside their convex hull. Wolfe's algorithm has been successfully employed as a subroutine in various optimization problems, e.g., submodular function minimization [11], see also [1,8,10]. Beyond the trivial 2^n bound, the convergence analysis remained elusive; the first bound with $1/\varepsilon$-dependence was given by Chakrabarty et al. [2] in 2014. Lacoste-Julien and Jaggi [14] gave a $\log(1/\varepsilon)$ bound, parametrized by the *pyramidal width* of the polyhedron. Recently, De Loera et al. [5] showed an example of exponential time behaviour of Wolfe's algorithm for the *min-norm insertion rule* (the analogue of a pivot rule); no exponential example for other insertion rules such as the *linopt* rule used in the application for submodular minimization.

Wolfe's algorithm works with a polytope in V-representation. Concurrently with Wolfe's work, Wilhelmsen [23] proposed an equivalent algorithm for uncapacitated (conic) instances of (P), i.e., for a polytope in H-representation. This algorithm can be seen as a special instantiation of our framework, and we show an $O(n^4 \kappa^2 \|A\|^2 \log(n + \kappa))$ iteration bound.

A significant difference compared to the Wolfe and Wolfe–Wilhelmsen algorithms is that the supports of our iterates are not required to be independent. This provides much additional flexibility: our algorithm can be combined with a variety of first order methods. This feature also yields a significant advantage in our computational experiments.

Overview of the Algorithm. A key concept in our algorithm is the *centroid mapping*, defined as follows. For disjoint subsets $I_0, I_1 \subseteq N$, we let $\mathbf{L}(I_0, I_1)$ denote the affine subspace of $\mathbb{R}^N$ where $x(i) = 0$ for $i \in I_0$ and $x(i) = u(i)$ for $i \in I_1$. For $x \in \mathbf{B}(u)$, let $I_0(x)$ and $I_1(x)$ denote the subsets of coordinates i with $x(i) = 0$ and $x(i) = u(i)$, respectively. A *centroid mapping* $\Psi : \mathbf{B}(u) \to \mathbb{R}^N$ is a mapping with the property that $\Psi(x) \in \arg\min_y\{\frac{1}{2}\|Ay - b\|^2 \mid y \in \mathbf{L}(I_0(x), I_1(x))\}$. This mapping may not be unique, since the columns of A corresponding to $\{i \in N \mid 0 < x(i) < u(i)\} = N \setminus (I_0(x) \cup I_1(x))$ may not be independent: the optimal *centroid set* is itself an affine subspace. The point $x \in \mathbf{B}(u)$ is *stable* if $\Psi(x) = x$. Stable points can be seen as the analogues of corral solutions in Wolfe's algorithm.

Every major cycle starts with an update step and ends with a stable point. The update step could be any first-order step satisfying some natural requirements, such as variants of Frank–Wolfe, projected gradient, or Wolfe updates. As long as the current iterate is not optimal, this update strictly improves the objective. Finite convergence follows by the fact that there can be at most 3^n stable points.

After the update step, we start a sequence of minor cycles. From the current iterate $x \in \mathbf{B}(u)$, we move to $\Psi(x)$ in case $\Psi(x) \in \mathbf{B}(u)$, or to the intersection of the boundary of $\mathbf{B}(u)$ and the line segment $[x, \Psi(x)]$ otherwise. The minor cycles finish once $x = \Psi(x)$ is a stable point. The objective $\frac{1}{2}\|Ax - b\|^2$ is decreasing in every minor cycle, and at least one new coordinate $i \in N$ is set to 0 or to $u(i)$. Thus, the number of minor cycles in any major cycle is at most n. One can use various centroid mappings, with only a mild requirement on Ψ (see Sect. 2.2).

We present a $\mathrm{poly}(n, \kappa)$ convergence analysis in the uncapacitated case for projected gradient and Wolfe updates. We expect that similar arguments extend to the capacitated case. The proof has two key ingredients. First, we show linear convergence of the first-order update steps (Theorem 3). Such a bound follows already from [16]; we present a simple self-contained proof exploiting properties of stable points and the uncapacitated setting. The second step of the analysis shows that in every $\mathrm{poly}(n, \kappa)$ iterations, we can identify a new variable that will never become zero in subsequent iterations (Theorem 2). The proof relies on proximity arguments: we show that for any iterate x and any subsequent iterate x', the distance $\|x - x'\|$ can be upper bounded in terms of n, κ, and the optimality gap at x.

In Sect. 5, we present preliminary computational experiments using randomly generated problem instances of various sizes. We compare the performance of different variants of our algorithm to standard gradient methods. The algorithm performs much better with projected gradient updates than with Wolfe updates. We compare an 'oblivious' centroid mapping and one that chooses $\Psi(x)$ as the a nearest point to x in the centroid set in the *'local norm'* (see Sect. 2.2). The latter one appears to be significantly better. For choices of parameters $n \geq 2m$, our method with projected gradient updates and local norm mapping outperforms the accelerated gradient method—the best among classical methods—by a factor 10 or more in computational time.

Related Work. Arguments that show strongly polynomial convergence by gradually revealing the support of an optimal solution are prevalent in combinatorial optimization. These date back to Tardos's [21] groundbreaking work giving the first strongly polynomial algorithm for minimum-cost flows. Our proof is closer to the dual 'abundant arc' algorithms in [9,17]. Tardos generalized the above result for general LP's, giving a running time dependence $\text{poly}(n, \log \Delta(A))$, where $\Delta(A)$ is the largest subdeterminant of A. This framework was recently strengthened in [4] to $\text{poly}(n, \log \kappa(A))$ running time for the circuit imbalance measure $\kappa(A)$. We note that the above algorithms—along with many other strongly polynomial algorithms in combinatorial optimization—modify the problem directly once new information is learned about the optimal support. In contrast, our algorithm does not require any such modifications, nor a knowledge or estimate on the condition number κ.

Strongly polynomial algorithms with $\text{poly}(n, \log \kappa(A))$ running time bounds can also be obtained using layered least squares interior point methods. This line of work was initiated by Vavasis and Ye [22] using a related condition measure $\bar{\chi}(A)$. An improved version that also established the relation between $\bar{\chi}(A)$ and $\kappa(A)$ was recently given by Dadush et al. [3]. We refer the reader to the survey [6] for properties and further applications of circuit imbalances.

Further Related Work. There are similarities between our algorithm and the Iteratively Reweighted Least Squares (IRLS) method that has been intensively studied since the 1960's [15,18]. For some $p \in [0, \infty]$, $A \in \mathbb{R}^{M \times N}$, $b \in \mathbb{R}^M$, the goal is to approximately solve $\min\{\|x\|_p : Ax = b\}$. At each iteration, a weighted minimum-norm point $\min\{\langle w^{(t)}, x \rangle : Ax = b\}$ is solved, where the weights $w^{(t)}$ are iteratively updated. The LP-feasibility problem $Ax = b$, $0 \leq x \leq 1$ for finite upper bounds $u = 1$ can be phrased as an ℓ_∞-minimization problem $\min\{\|x\|_\infty : Ax = b - A1/2\}$. Ene and Vladu [7] gave an efficient variant of IRLS for ℓ_1 and ℓ_∞-minimization; see their paper for further references. Some variants of our algorithm solve a weighted least squares problem with changing weights in the stabilizing steps. There are however significant differences between IRLS and our method. The underlying optimization problems are different, and IRLS does not find an exact optimal solution in finite time. Applied to LP in the ℓ_∞ formulation, IRLS satisfies $Ax = b$ throughout while violating the box constraints $0 \leq x \leq u$. In contrast, iterates of our algorithm violate $Ax = b$ but maintain $0 \leq x \leq u$. The role of the least squares subroutines is also rather different in the two settings.

2 Preliminaries

Notation. We use $N \oplus M$ for disjoint union (or direct sum) of the copies of the two sets. For a matrix $A \in \mathbb{R}^{M \times N}$, $i \in M$ and $j \in N$, we denote the ith row of A by A_i and jth column by A^j. Also for any matrix X denote by $X^\top$ the matrix transpose of X. We let $\| \cdot \|_p$ denote the ℓ_p vector norm; we use $\| \cdot \|$ to denote the Euclidean norm $\| \cdot \|_2$. For a matrix $A \in \mathbb{R}^{M \times N}$, we let $\|A\|$ denote the spectral norm, that is, the $\ell_2 \to \ell_2$ operator norm.

For any $x, y \in \mathbb{R}^M$ we define $\langle x, y \rangle = \sum_{i \in M} x(i)y(i)$. We will use this notation also in other dimensions. We let $[x, y] := \{\lambda x + (1 - \lambda)y \mid \lambda \in [0, 1]\}$ denote the line segment between the vectors x and y.

Elementary Vectors and Circuits. For a linear space $W \subsetneq \mathbb{R}^N$, $g \in W$ is an *elementary vector* if g is a support minimal nonzero vector in W, that is, no $h \in W \setminus \{\mathbf{0}\}$ exists such that $\mathrm{supp}(h) \subsetneq \mathrm{supp}(g)$, where supp denotes the support of a vector. We let $\mathcal{F}(W) \subseteq W$ denote the set of elementary vectors. A *circuit* in W is the support of some elementary vector; these are precisely the circuits in the associated linear matroid $\mathcal{M}(W)$.

The subspaces $W = \{\mathbf{0}\}$ and $W = \mathbb{R}^N$ are called trivial subspaces, all other subspaces are nontrivial. We define the *circuit imbalance measure*

$$\kappa(W) := \max \left\{ \left| \frac{g(j)}{g(i)} \right| \,\middle|\, g \in \mathcal{F}(W), i, j \in \mathrm{supp}(g) \right\}$$

for nontrivial subspaces and $\kappa(W) = 1$ for trivial subspaces. For a matrix $A \in \mathbb{R}^{M \times N}$, we use $\kappa(A)$ to denote $\kappa(\ker(A))$.

Recall that a matrix is *totally unimodular (TU)* if the determinant of every square submatrix is 0, +1, or −1. A result by Cederbaum from 1957 shows that $\kappa(W) = 1$ if and only if there exists a TU matrix $A \in \mathbb{R}^{M \times N}$ such that $W = \ker(A)$. We also note that if $A \in \mathbb{Z}^{M \times N}$, then $\kappa(A) \le \Delta(A)$.

We say that the vector $y \in \mathbb{R}^N$ *conforms* to $x \in \mathbb{R}^N$ if $x(i)y(i) > 0$ whenever $y(i) \ne 0$. Given a subspace $W \subseteq \mathbb{R}^N$, a *conformal circuit decomposition* of a vector $v \in W$ is a decomposition

$$v = \sum_{k=1}^{\ell} h^k,$$

where $\ell \le n$ and $h^1, h^2, \ldots, h^\ell \in \mathcal{F}(W)$ are elementary vectors that conform to v. A fundamental result on elementary vectors asserts that for every subspace $W \subseteq \mathbb{R}^N$, every $v \in W$ admits a conformal circuit decomposition, see e.g. [12, 20]. Note that there may be multiple conformal circuit decompositions of a vector.

Given $A \in \mathbb{R}^{M \times N}$, we define the extended subspace $\mathcal{X}_A \subset \mathbb{R}^{N \oplus M}$ as $\mathcal{X}_A := \ker(A \mid -I_M)$. Hence, for every $v \in \mathbb{R}^N$, $(v, Av) \in \mathcal{X}_A$. For $v \in \mathbb{R}^N$, the *generalized path-circuit decomposition of v with respect to A* is a decomposition $v = \sum_{k=1}^{\ell} h^k$, where $\ell \le n$, and for each $1 \le k \le \ell$, $(h^k, Ah^k) \in \mathbb{R}^{N \oplus M}$ is an elementary vector in $\mathcal{X}_A$ that conforms to (v, Av). Moreover, h^k is an *inner vector* in the decomposition if $Ah^k = \mathbf{0}$ and an *outer vector* otherwise.

We say that $v \in \mathbb{R}^N$ is *cycle-free with respect to A*, if all generalized path-circuit decompositions of v contain outer vectors only. The following lemma will play a key role in analyzing our algorithms.

Lemma 1. *For any $A \in \mathbb{R}^{M \times N}$, let $v \in \mathbb{R}^N$ be cycle-free with respect to A. Then, $\|v\|_\infty \leq \kappa(\mathcal{X}_A)\|Av\|_1$ and $\|v\|_2 \leq n\kappa(\mathcal{X}_A)\|Av\|_2$.*

Remark 1. We note that a similar argument shows that $\|A\| \leq \min\{n\kappa(\mathcal{X}_A), \sqrt{n}\tau(A)\kappa(\mathcal{X}_A)\}$, where $\tau(A)$ is the maximum size of $\mathrm{supp}(Ah)$ for an elementary vector $(h, Ah) \in \mathcal{X}_A$.

Example 1. If $A \in \mathbb{R}^{M \times N}$ is the node-arc incidence matrix of a directed graph $D = (M, N)$. The system $Ax = b$, $x \in \mathbf{B}(u)$ corresponds to a network flow feasibility problem. Here, $b(i)$ is the demand of node $i \in M$, i.e., the inflow minus the outflow at i is required to be $b(i)$. Recall that A is a TU matrix; consequently, $(A| - I_M)$ is also TU, and $\kappa(\mathcal{X}_A) = 1$. Our algorithm is strongly polynomial in this setting. Note that inner vectors correspond to cycles and outer vectors to paths; this motivates the term 'generalized path-circuit decomposition'. We also note $\tau(A) = 2$, and thus $\|A\| \leq 2\sqrt{n}$ in this case.

2.1   Optimal Solutions and Proximity

We define the set
$$Z(A, u) := \{Ax \mid x \in \mathbf{B}(u)\}.$$

Thus, Problem (P) is to find the point in $Z(A, u)$ that is nearest to b with respect to the Euclidean norm. We note that if the upper bounds u are finite, $Z(A, u)$ is called a *zonotope*.

Throughout, we let p^* denote the optimum value of (P). Note that whereas the optimal solution x^* may not be unique, the vector $b^* := Ax^*$ is unique by strong convexity; we have $p^* = \frac{1}{2}\|b - b^*\|^2$. We use

$$\eta(x) := \frac{1}{2}\|Ax - b\|^2 - p^*$$

to denote the optimality gap for $x \in \mathbf{B}(u)$. The point $x \in \mathbf{B}(u)$ is an ε-*approximate solution* if $\eta(x) \leq \varepsilon$.

For a point $x \in \mathbf{B}(u)$, let $I_0(x) := \{i \in N : x(i) = 0\}$, $I_1(x) := \{i \in N : x(i) = u(i)\}$, and $J(x) := N \setminus (I_0(x) \cup I_1(x))$. The gradient of the objective $\frac{1}{2}\|Ax - b\|^2$ in (P) can be written as

$$g^x := A^\top (Ax - b).$$

We recall the first order optimality conditions: $x \in \mathbf{B}(u)$ is an optimal solution to (P) if and only if $g^x(i) = 0$ for all $i \in J(x)$, $g^x(i) \geq 0$ for all $i \in I_0(x)$, and $g^x(i) \leq 0$ for all $i \in I_1(x)$. Using Lemma 1, we can show:

Lemma 2. *For any $x \in \mathbf{B}(u)$, there exists an optimal solution x^* to (P) such that $\|x - x^*\|_\infty \leq \kappa(\mathcal{X}_A)\|Ax - b^*\|_1$, and hence, $\|x - x^*\|_2 \leq n\kappa(\mathcal{X}_A)\|Ax - b^*\|_2$.*

2.2 The Centroid Mapping

Let us denote by 3^N the set of all ordered pairs (I_0, I_1) of disjoint subsets $I_0, I_1 \subseteq N$, and let $I_* := \{i \in N \mid u(i) < \infty\}$. For any $(I_0, I_1) \in 3^N$ with $I_1 \subseteq I_*$, we let

$$\mathbf{L}(I_0, I_1) := \{x \in \mathbb{R}^N \mid \forall i \in I_0 : x(i) = 0, \ \forall i \in I_1 : x(i) = u(i) \}.$$

We call $\{Ax \mid x \in \mathbf{B}(u) \cap \mathbf{L}(I_0, I_1)\} \subseteq Z(A, u)$ a *pseudoface* of the $Z(A, u)$. We note that every face of $Z(A, u)$ is a pseudoface, but there might be pseudofaces that do not correspond to any face. We define a *centroid set* for (I_0, I_1) as

$$\mathcal{C}(I_0, I_1) := \arg\min_y \{\|Ay - b\| \mid y \in \mathbf{L}(I_0, I_1))\}.$$

Proposition 1. *For $(I_0, I_1) \in 3^N$ with $I_1 \subseteq I_*$, $\mathcal{C}(I_0, I_1)$ is an affine subspace of $\mathbb{R}^N$, and there exists $w \in \mathbb{R}^M$ such that $Ay = w$ for every $y \in \mathcal{C}(I_0, I_1)$.*

The *centroid mapping* $\Psi : \mathbf{B}(u) \to \mathbb{R}^N$ is a mapping that satisfies

$$\Psi(\Psi(x)) = \Psi(x) \quad \text{and} \quad \Psi(x) \in \mathcal{C}(I_0(x), I_1(x)), \ \forall x \in \mathbf{B}(u)$$

We say that $x \in \mathbf{B}(u)$ is a *stable point* if $\Psi(x) = x$. A simple, 'oblivious' centroid mapping arises by taking the minimum-norm point of the centroid set:

$$\Psi(x) := \arg\min\{\|y\| \mid y \in \mathcal{C}(I_0(x), I_1(x))\}. \tag{1}$$

However, this mapping has some undesirable properties. For example, we may have an iterate x that is already in $\mathcal{C}(I_0(x), I_1(x))$, but $\Psi(x) \neq x$. Instead, we aim for centroid mappings that move the current point 'as little as possible'. The centroid mapping Ψ is called *cycle-free*, if the vector $\Psi(x) - x$ is cycle-free w.r.t. A for every $x \in \mathbf{B}(u)$.

Lemma 3. *For every $x \in \mathbf{B}(u)$, let $D(x) \in \mathbb{R}_{>0}^{N \times N}$ be a positive diagonal matrix. Then, the following $\Psi(x)$ defines a cycle-free centroid mapping:*

$$\Psi(x) := \arg\min\{\|D(x)(y - x)\| \mid y \in \mathcal{C}(I_0(x), I_1(x))\}. \tag{2}$$

We emphasize that $D(x)$ in the statement is a function of x and can be any positive diagonal matrix. Note also that the diagonal entries for indices in $I_0(x) \cup I_1(x)$ do not matter. In our experiments, defining $D(x)$ with diagonal entries $1/x(i) + 1/(u(i) - x(i))$ for $i \in J(x)$ performs particularly well. Intuitively, this choice aims to move less the coordinates close to the boundary. The next proposition follows from Lagrangian duality. We note that $\Psi(x)$ as in (1) or (2) can be computed by solving a system of linear equations.

Proposition 2. *For a partition $N = I_0 \cup I_1 \cup J$, the centroid set can be written as $\mathcal{C}(I_0, I_1) = \{y \in \mathbf{L}(I_0, I_1) \mid (A^J)^\top (Ay - b) = \mathbf{0}\}$.*

3   The Update-and-Stabilize Framework

Now we describe a general algorithmic framework $\text{MNPZ}(A, b, u)$ for solving (P), shown in Algorithm 1. Similarly to Wolfe's MNP algorithm, the algorithm comprises major and minor cycles. We maintain a point $x \in \mathbf{B}(u)$, and x is stable at the end of every major cycle. Each major cycle starts by calling the subroutine $\text{Update}(x)$; the only general requirement on this subroutine is:

(U1) for $y = \text{Update}(x)$, $y = x$ if and only if x is optimal to (P), and $\|Ay - b\| < \|Ax - b\|$ otherwise, and

(U2) if $y \neq x$, then for any $\lambda \in [0, 1)$, $z = \lambda y + (1 - \lambda)x$ satisfies $\|Ay - b\| < \|Az - b\|$.

Property (U1) can be obtained from any first order algorithm; we introduce some important examples below. Property (U2) might be violated if using a fixed step-length, which is a common choice. In order to guarantee (U2), we can post-process the first order update that returns y' by choosing y as the optimal point on the line segment $[x, y']$.

The algorithm terminates once $x = \text{Update}(x)$. In the minor cycles, as long as $w := \Psi(x) \neq x$, i.e., x is not stable, we set $x := w$ if $w \in \mathbf{B}(u)$; otherwise, we set the next x as the intersection of the line segment $[x, w]$ and the boundary of $\mathbf{B}(u)$. The requirement (U1) is already sufficient to show finite termination.

Algorithm 1: $\text{MNPZ}(A, b, u)$

   Input : $A \in \mathbb{R}^{M \times N}$, $b \in \mathbb{R}^M$, $u \in (\mathbb{R} \cup \{\infty\})^N$
   Output: An optimal solution x to (P)
1 $x \leftarrow$ initial point from $\mathbf{B}(u)$;
2 **repeat**
3      $x \leftarrow \text{Update}(x)$;   // Major cycle
4      $w \leftarrow \Psi(x)$;
5   **while** $\Psi(x) \neq x$ // Minor cycle
6   **do**
7          $\alpha^* \leftarrow \arg\max\{\alpha \in [0, 1] \mid x + \alpha(w - x) \in \mathbf{B}(u)\}$;
8          $x \leftarrow x + \alpha^*(w - x)$;
9          $w \leftarrow \Psi(x)$;
10    $x \leftarrow w$;
11 **until** $x = \text{Update}(x)$
12 **return** x

Theorem 1. *Consider any* $\text{Update}(x)$ *subroutine that satisfies (U1) and any centroid mapping* Ψ. *The algorithm* $\text{MNPZ}(A, b, u)$ *finds an optimal solution to (P) within* 3^n *major cycles. Every major cycle contains at most* n *minor cycles.*

We can implement the Update(x) subroutine satisfying (U1) and (U2) using various first order methods for constrained optimization. Recall the notation g^x for the gradient g^x; we use $g = g^x$ when x is clear from the context. The following property of stable points can be compared to the optimality conditions:

Lemma 4. *If $x(= \Psi(x))$ is a stable point, then $g(j) = 0$ for all $j \in J(x)$.*

We now describe three classical options. We stress that the choice of the centroid mapping Ψ can be chosen independently of the update step.

The Frank–Wolfe Update. The Frank–Wolfe or *conditional gradient* method is applicable only in the case when $u(i)$ is finite for every $i \in N$. In every update step, we start by computing $\bar{y}$ as the minimizer of the linear objective $\langle g, y \rangle$ over $\mathbf{B}(u)$, that is,

$$\bar{y} \in \arg\min\{\langle g, y \rangle \mid y \in \mathbf{B}(u)\}.$$

We set Update$(x) := x$ if $\langle g, \bar{y} \rangle = \langle g, x \rangle$, or $y = $ Update(x) is selected so that y minimizes $\frac{1}{2}\|Ay - b\|^2$ on the line segment $[x, \bar{y}]$. Clearly, $\bar{y}(i) = 0$ if $g(i) > 0$, and $\bar{y}(i) = u(i)$ if $g(i) < 0$. But, $\bar{y}(i)$ can be chosen arbitrarily if $g(i) = 0$. In this case, we keep $\bar{y}(i) = x(i)$; this will be significant to guarantee stability of solutions in the analysis.

The Projected Gradient Update. The projected gradient update moves in the opposite gradient direction to $\bar{y} := x - \lambda g$ for some step-length $\lambda > 0$, and obtains the output $y = $ Update(x) as the projection y of $\bar{y}$ to the box $\mathbf{B}(u)$. This projection simply changes every negative coordinate to 0 and every $\bar{y}(i) > u(i)$ to $y(i) = u(i)$. To ensure (U2), we can perform an additional step that replaces y by the point $y' \in [x, y]$ that minimizes $\frac{1}{2}\|Ay' - b\|^2$.

Consider now an uncapacitated instance (i.e., $u(i) = \infty$ for all $i \in N$), and let x be a stable point. Recall $I_1(x) = \emptyset$ in the uncapacitated setting. Lemma 4 allows us to write the projected gradient update in the following simple form that also enables to use optimal line search. Define

$$z^x(i) := \max\{-g^x(i), 0\},$$

and use $z = z^x$ when clear from the context. Note that x is optimal to (P) if and only if $z = \mathbf{0}$. We use the optimal line search

$$y := \arg\min_y \left\{\tfrac{1}{2}\|Ay - b\|^2 \mid y = x + \lambda z, \lambda \geq 0\right\}.$$

If $z \neq \mathbf{0}$, this can be written explicitly as

$$y := x + \frac{\|z\|^2}{\|Az\|^2} z.$$

To verify this formula, note that $\|z\|^2 = -\langle g, z \rangle$, since either $z(i) = 0$ or $z(i) = -g(i)$.

The Wolfe Update. Our third update resembles Wolfe's algorithm. In the uncapacitated case, it corresponds to the Wolfe–Wilhelmsen algorithm. Given a stable point $x \in \mathbf{B}(u)$, we select a coordinate $j \in N$ where either $j \in I_0(x)$ and $g(j) < 0$ or $j \in I_1(x)$ and $g(j) > 0$, and set y such that $y(i) = x(i)$ if $i \neq j$, and $y(j)$ is chosen in $[0, u(j)]$ so that $\frac{1}{2}\|Ay - b\|^2$ is minimized.

Analogously to Wolfe's algorithm, we can maintain basic solutions throughout. Namely, if A^J is linearly independent for $J = J(x)$, then one can show that $A^{J'}$ is also linearly independent for $J' = J(y) = J \cup \{j\}$, where $y = \mathtt{Update}(x)$. Assume we start with $x = \mathbf{0}$, i.e., $J(x) = I_1(x) = \emptyset$, $I_0(x) = N$. Then, $A^{J(x)}$ remains linearly independent throughout. Hence, every stable solution x is a basic solution to (P). Note that whenever $A^{J(x)}$ is linearly independent, $\mathcal{C}(I_0(x), I_1(x))$ contains a single point, hence, $\Psi(x)$ is uniquely defined.

Consider now the uncapacitated setting. For $z = z^x$, let us return $y = x$ if $z = \mathbf{0}$. Otherwise, let $j \in \arg\max_k z(k)$; note that $j \in I_0(x)$. Let

$$y(i) := \begin{cases} x(i) & \text{if } i \in N \setminus \{j\}, \\ \frac{z(i)}{\|A^*\|^2} & \text{if } i = j. \end{cases}$$

It is easy to verify that the Frank–Wolfe, projected gradient, and Wolfe update rules all satisfy (U1) and (U2). For projected gradient, for the updates in the uncapacitated form as described above, (U2) is guaranteed. For the general form with upper bounds, we can perform a post-processing as noted above to ensure (U2). We say that $\mathtt{Update}(x)$ is a *cycle-free update rule*, if for every $x \in \mathbf{B}(u)$ and $y = \mathtt{Update}(x)$, $x - y$ is cycle-free w.r.t. A. One can show that the Frank–Wolfe, projected gradient, and Wolfe updates are all cycle-free.

4   Analysis

Theorem 2. *Consider an uncapacitated instance of (P), and assume we use a cycle-free centroid mapping. Algorithm 1 terminates with an optimal solution in $O(n^3 \kappa^2(\mathcal{X}_A)\|A\|^2 \log(n + \kappa(\mathcal{X}_A)))$ major cycles using projected gradient updates, and in $O(n^4 \kappa^2(\mathcal{X}_A)\|A\|^2 \log(n + \kappa(\mathcal{X}_A)))$ major cycles using Wolfe updates. In both cases, the total number of minor cycles is $O(n^4 \kappa^2(\mathcal{X}_A)\|A\|^2 \log(n + \kappa(\mathcal{X}_A)))$.*

Proximity Bounds. We show that if using a cycle-free update rule and a cycle-free centroid mapping, the movement of the iterates in Algorithm 1 can be bounded by the change in the objective value. First, a nice property of the centroid set is that the movement of Ax directly relates to the decrease in the objective value. Namely,

Lemma 5. *For $x \in \mathbf{B}(u)$, let $y \in \mathcal{C}(I_0(x), I_1(x))$. Then, $\|Ax - Ay\|^2 = \|Ax - b\|^2 - \|Ay - b\|^2$. Consequently, if Ψ is a cycle-free centroid mapping and $y = \Psi(x)$, then*

$$\|x - y\|^2 \leq n^2 \kappa^2(\mathcal{X}_A) \left(\|Ax - b\|^2 - \|Ay - b\|^2 \right).$$

Next, let us consider the movement of x during a call to $\mathtt{Update}(x)$.

Lemma 6. *Let $x \in \mathbf{B}(u)$ and $y = \mathtt{Update}(x)$. Then, $\|Ax - Ay\|^2 \leq \|Ax - b\|^2 - \|Ay - b\|^2$. If using a cycle-free update rule, we also have*

$$\|x - y\|^2 \leq n^2 \kappa^2(\mathcal{X}_A) \left(\|Ax - b\|^2 - \|Ay - b\|^2 \right).$$

Lemma 7. *Let $x \in \mathbf{B}(u)$, and let x' be an iterate obtained by consecutive t major or minor updates of Algorithm 1 using a cycle-free update rule and a cycle-free centroid mapping, starting from x. Then,*

$$\|x - x'\| \leq n\kappa(\mathcal{X}_A)\sqrt{2t} \cdot \sqrt{\tfrac{1}{2}\|Ax - b\|^2 - \tfrac{1}{2}\|Ax' - b\|^2}.$$

Geometric Convergence of the Projected Gradient and Wolfe Updates. Recall that $\eta(x)$ denotes the optimality gap at x.

Theorem 3. *Consider an uncapacitated instance of (P), and let $x \geq 0$ be a stable point. Then for $y = \mathtt{Update}(x)$ using the projected gradient update, we have $\eta(y) \leq \left(1 - 1/(2n^2\kappa^2(\mathcal{X}_A)\|A\|^2)\right)\eta(x)$. Using the Wolfe updates, we have $\eta(y) \leq \left(1 - 1/(2n^3\kappa^2(\mathcal{X}_A)\|A\|^2)\right)\eta(x)$.*

The theorem follows easily from the next two lemmas. First, we formulate the update progress using optimal line search, and next, we use Lemma 2 to bound $\|z\|$.

Lemma 8. *For a stable point $x \geq 0$, the projected gradient update satisfies $\|Ax - b\|^2 - \|Ay - b\|^2 \geq \|z\|^2/\|A\|^2$, and the Wolfe update satisfies $\|Ax - b\|^2 - \|Ay - b\|^2 = z(j)^2/\|A^j\|^2$.*

Lemma 9. *For a stable point $x \geq 0$ and the update direction $z = z^x$, we have $\|z\| \geq \sqrt{\eta(x)}/(\sqrt{2}n\kappa(\mathcal{X}_A))$.*

Proof. Let $x^* \geq 0$ be an optimal solution to (P) as in Lemma 2, and $b^* = Ax^*$. Using convexity of $f(x) := \tfrac{1}{2}\|Ax - b\|^2$, we have $p^* = f(x^*) \geq f(x) + \langle g, x^* - x \rangle \geq f(x) - \langle z, x^* - x \rangle$, where the second inequality follows by noting that for each $i \in N$, either $z(i) = -g(i)$, or $z(i) = 0$ and $g(i)(x^*(i) - x(i)) \geq 0$. From the Cauchy-Schwarz inequality and Lemma 2, we get $p^* \geq f(x) - \|z\| \cdot \|x^* - x\| \geq f(x) - n\kappa(\mathcal{X}_A)\|Ax - b^*\| \cdot \|z\|$, that is, $\|z\| \geq \eta(x)/n\kappa(\mathcal{X}_A)\|Ax - b^*\|$. The proof is complete by showing $2\eta(x) \geq \|Ax - b^*\|^2$.

Recalling that $\eta(x) = \tfrac{1}{2}\|Ax - b\|^2 - \tfrac{1}{2}\|Ax^* - b\|^2$ and that $b^* = Ax^*$, this is equivalent to $\langle Ax - Ax^*, Ax^* - b \rangle \geq 0$. This can be further written as $\langle x - x^*, g^{x^*} \rangle \geq 0$, which is implied by the first order optimality condition at x^*. This proves $2\eta(x) \geq \|Ax - b^*\|^2$, and hence the lemma follows.

Overall Convergence Bounds. We now prove Theorem 2. Using Lemma 7 and Theorem 3, we can derive the following stronger proximity bound:

Lemma 10. *Consider an uncapacitated instance of (P). Let $x \geq 0$ be an iterate of Algorithm 1 using projected gradient updates, and let $x' \geq 0$ be any later iterate. Then, for a value $\Theta := O(n^{2.5}\kappa^2(\mathcal{X}_A)\|A\|)$, we have $\|x - x'\| \leq \Theta\sqrt{\eta(x)}$.*

We need one more auxiliary lemma.

Lemma 11. *Consider an uncapacitated instance of (P), and let $x \geq \mathbf{0}$ be a stable point. Let $\hat{x} \geq \mathbf{0}$ such that for each $i \in N$, either $\hat{x}(i) = x(i)$, or $\hat{x}(i) = 0 < x(i)$. Then, $\|A\hat{x} - b\|^2 = \|Ax - b\|^2 + \|A\hat{x} - Ax\|^2$.*

For the threshold Θ as in Lemma 10 and for any $x \geq \mathbf{0}$, let us define

$$J^{\star}(x) := \left\{ i \mid x(i) > \Theta\sqrt{\eta(x)} \right\}.$$

The following is immediate from Lemma 10.

Lemma 12. *Consider an uncapacitated instance of (P). Let $x \geq \mathbf{0}$ be an iterate of Algorithm 1 using projected gradient updates, and $x' \geq \mathbf{0}$ be any later iterate. Then, $J^{\star}(x) \subseteq J(x')$.*

Proof (Proof of Theorem 2). At any point of the algorithm, let $J^{\star}$ denote the union of the sets $J^{\star}(x)$ for all iterations thus far. Consider a stable iterate x at the beginning of any major cycle, and let $\varepsilon := \sqrt{\eta(x)}/4n\Theta\|A\|$. Theorem 3 for projected gradient updates guarantees that within $O(n^2\kappa^2(\mathcal{X}_A)\|A\|^2 \log(n + \kappa(\mathcal{X}_A)))$ major cycles we arrive at an iterate x' such that $\sqrt{\eta(x')} < \varepsilon$. The bound is $O(n^3\kappa^2(\mathcal{X}_A)\|A\|^2 \log(n + \kappa(\mathcal{X}_A)))$ for Wolfe updates. We note that $\log(n + \kappa(\mathcal{X}_A) + \|A\|) = O(\log(n + \kappa(\mathcal{X}_A)))$ according to Remark 1. We show that

$$J^{\star}(x') \cap I_0(x) \neq \emptyset. \tag{3}$$

From here, we can conclude that $J^{\star}$ was extended between iterates x and x'. This may happen at most n times, and so we get the claimed bounds on the number of major cycles. The bound on the minor cycles for projected gradient updates follows since every major cycle contains at most n minor cycles. For Wolfe updates, it follows since every major cycle adds on one component to $J(x)$ whereas every minor cycle removes at least one. Hence, the total number of minor cycles is at most m plus the total number of major cycles.

For a contradiction, assume that (3) does not hold. Thus, for every $i \in I_0(x)$, we have $x'(i) \leq \Theta\varepsilon$. Let us define $\hat{x} \in \mathbb{R}^N$ as $\hat{x}(i) := 0$ if $i \in I_0(x)$, and $\hat{x}(i) := x'(i)$ if $i \in J(x)$. By the above assumption, $\|\hat{x} - x'\|_\infty \leq \Theta\varepsilon$, and therefore $\|A\hat{x} - Ax'\| \leq \sqrt{n}\Theta\|A\|\varepsilon$. From Lemma 11, we can bound $\|A\hat{x} - b\|^2 = \|Ax' - b\|^2 + \|A\hat{x} - Ax'\|^2 \leq 2p^* + (n\Theta^2\|A\|^2 + 2)\varepsilon^2$. Recall that since x is a stable solution, $\|Ax - b\| = \min\{\|Ay - b\| : y \in \mathbf{L}(I_0(x), \emptyset)\}$. Since $\hat{x}$ is a feasible solution to this program, it follows that $\|A\hat{x} - b\|^2 \geq \|Ax - b\|^2$. We get that $2\eta(x) = \|Ax - b\|^2 - 2p^* \leq \|A\hat{x} - b\|^2 - 2p^* \leq (n\Theta^2\|A\|^2 + 2)\varepsilon^2$, in contradiction with the choice of ε.

5 Computational Experiments

We give preliminary computational experiments of different versions of our algorithm, and compare them to standard gradient methods. The experiments were programmed and executed by MATLAB version R2021b on a personal computer having 11th Gen Intel(R) Core(TM) i7-11370H @ 3.30GHz and 16GB of memory. We present results on randomly generated uncapacitated instances. The full version contains more experiments, also on capacitated instances, and also including the Frank–Wolfe method.

The entries of the $m \times n$ matrix A and the m dimensional vector b were chosen independently uniformly at random from the interval $[-0.5, 0.5]$. Thus, the underlying LP $Ax = b$, $x \geq 0$ may or may not be feasible. For the case $m = 1000$, $n = 1050$, this leads to infeasible instance with high probability. In this case, we also generated feasible instances by sampling coefficients $w(i) \in [0, 1]$ uniformly at random, and setting $b = Aw$.

We test each combination of two update methods: Projected Gradient (PG) and Wolfe (W); and two centroid mappings, the 'oblivious' mapping (1) and the 'local norm' mapping (2) with diagonal entries $1/x(i) + 1/(u(i) - x(i))$. Recall that for Wolfe updates, there is a unique centroid mapping. We benchmark against standard constrained first order methods: the projected gradient (PG), and the projected fast (accelerated) gradient method (PFG). In contrast to our algorithm, these do not finitely terminate. We stopped these algorithms once they found a near-optimal solution within a certain accuracy threshold.

We stopped each algorithm when the computation time reached 180 s. For each (m, n), we test all the algorithms 10 times and the results shown below are the 10-run averaged figures. Table 1 shows the overall computational times; values in brackets show the number of trials whose computation time exceeded 180 s. For the 'near-square' case $m = 1000$, $n = 1050$, status 'I' denotes infeasible and 'F' feasible instances. Table 2 shows the number of major cycles and the total number of minor cycles.

In our framework, projected gradient updates perform significantly better than Wolfe updates, except for infeasible 'near-square' instances. For Wolfe updates, the number of major and minor cycles is similar; projected gradient performs much fewer major cycles. Among the two centroid mappings, the 'local-norm' update (2) performs significantly better than the 'oblivious' update (1).

There is a marked difference between infeasible and feasible 'near-square' instances. Our algorithms perform well on feasible instances. For infeasible instances, the running time is much longer, with an excessive number of minor cycles.

As one may expect, projected fast gradient is significantly better than project gradient. For 'rectangular' $(n \geq 2m)$ instances, our method with projected gradient updates together with the centroid mapping (2), outperforms fast gradient by a factor 10 or more. This is despite the fact that centroid mappings are computationally more expensive than first order methods. For feasible near-square instances, the performance in these two cases is similar. However, for infeasible near-square instances, our algorithms are outperformed by projected gradient and projected fast gradient methods.

Table 1. Computation time (in sec) for random uncapacitated instances

m	100	200	300	500	500	1000	1000
n	200	400	600	1000	3000	1050	1050
Status						I	F
PG+(1)	0.37	3.19	11.07	79.17	142.67 (4)	58.57	5.68
PG+(2)	0.04	0.31	0.81	1.76	2.37	57.72	0.86
W	0.20	1.90	6.69	33.75	47.83	22.80	149.73
PG	3.49	87.07 (3)	118.91 (5)	112.85 (5)	5.31	6.26	180.00 (10)
PFG	0.17	4.13	17.48	47.33 (1)	6.17	5.15	92.59

Table 2. # of major cycles (first number) and total # of minor cycles (second number) for random uncapacitated instances

m	100	200	300	500	500	1000	1000
n	200	400	600	1000	3000	1050	1050
Status						I	F
PG+(1)	6.0	9.5	11.6	11.9	1.0	4.1	1.0
	199.7	462.6	759.6	1505.2	1513.9	599.7	31.6
PG+(2)	2.2	2.8	2.6	2.2	1.0	4.1	1.0
	23.5	30.9	36.9	29.8	1.0	570.2	3.1
W	121	265.4	401.1	653.5	501.4	526.7	1091.6
	144.6	333.4	506.1	810.2	508.2	530.8	1182.2

Acknowledgments. The third author would like to thank Richard Cole, Daniel Dadush, Christoph Hertrich, Bento Natura, and Yixin Tao for discussions on first order methods and circuit imbalances.

References

1. Bach, F.: Learning with submodular functions: a convex optimization perspective. Found. Trends Mach. Learn. **6**(2–3), 145–373 (2013)
2. Chakrabarty, D., Jain, P., Kothari, P.: Provable submodular minimization using Wolfe's algorithm. In: Advances in Neural Information Processing Systems, vol. 27 (2014)
3. Dadush, D., Huiberts, S., Natura, B., Végh, L.A.: A scaling-invariant algorithm for linear programming whose running time depends only on the constraint matrix. In: Proceedings of the 52nd Annual ACM Symposium on Theory of Computing (STOC), pp. 761–774 (2020)
4. Dadush, D., Natura, B., Végh, L.A.: Revisiting Tardos's framework for linear programming: faster exact solutions using approximate solvers. In: Proceedings of the 61st Annual IEEE Symposium on Foundations of Computer Science (FOCS), pp. 931–942 (2020)

5. De Loera, J.A., Haddock, J., Rademacher, L.: The minimum Euclidean-norm point in a convex polytope: Wolfe's combinatorial algorithm is exponential. SIAM J. Comput. **49**(1), 138–169 (2020)
6. Ekbatani, F., Natura, B., Végh, A.L.: Circuit imbalance measures and linear programming. In: Surveys in Combinatorics 2022. London Mathematical Society Lecture Note Series, pp. 64–114. Cambridge University Press, Cambridge (2022)
7. Ene, A., Vladu, A.: Improved convergence for ℓ_1 and ℓ_∞ regression via iteratively reweighted least squares. In: International Conference on Machine Learning, pp. 1794–1801. PMLR (2019)
8. Fujishige, S.: Lexicographically optimal base of a polymatroid with respect to a weight vector. Math. Oper. Res. **5**(2), 186–196 (1980)
9. Fujishige, S.: A capacity-rounding algorithm for the minimum-cost circulation problem: a dual framework of the Tardos algorithm **35**(3), 298–308 (1986)
10. Fujishige, S., Hayashi, T., Yamashita, K., Zimmermann, U.: Zonotopes and the LP-Newton method. Optim. Eng. **10**(2), 193–205 (2009)
11. Fujishige, S., Isotani, S.: A submodular function minimization algorithm based on the minimum-norm base. Pac. J. Optim. **7**(1), 3–17 (2011)
12. Fulkerson, D.: Networks, frames, blocking systems. Math. Decis. Sci. Part I, Lect. Appl. Math. **2**, 303–334 (1968)
13. Hoffman, A.J.: On approximate solutions of systems of linear inequalities. J. Res. Natl. Bur. Stand. **49**(4), 263–265 (1952)
14. Lacoste-Julien, S., Jaggi, M.: On the global linear convergence of Frank-Wolfe optimization variants. In: Advances in Neural Information Processing Systems, vol. 28 (2015)
15. Lawson, C.L.: Contribution to the theory of linear least maximum approximation. Ph.D. thesis (1961)
16. Necoara, I., Nesterov, Y., Glineur, F.: Linear convergence of first order methods for non-strongly convex optimization. Math. Program. **175**(1), 69–107 (2019)
17. Orlin, J.B.: A faster strongly polynomial minimum cost flow algorithm. Oper. Res. **41**(2), 338–350 (1993)
18. Osborne, M.R.: Finite Algorithms in Optimization and Data Analysis. Wiley, Hoboken (1985)
19. Peña, J., Vera, J.C., Zuluaga, L.F.: New characterizations of Hoffman constants for systems of linear constraints. Math. Program. 1–31 (2020)
20. Rockafellar, R.T.: The elementary vectors of a subspace of R^N. In: Combinatorial Mathematics and Its Applications: Proceedings North Carolina Conference, Chapel Hill, 1967, pp. 104–127. The University of North Carolina Press (1969)
21. Tardos, É.: A strongly polynomial minimum cost circulation algorithm. Combinatorica **5**(3), 247–255 (1985)
22. Vavasis, S.A., Ye, Y.: A primal-dual interior point method whose running time depends only on the constraint matrix **74**(1), 79–120 (1996)
23. Wilhelmsen, D.R.: A nearest point algorithm for convex polyhedral cones and applications to positive linear approximation. Math. Comput. **30**(133), 48–57 (1976)
24. Wolfe, P.: Finding the nearest point in a polytope. Math. Program. **11**(1), 128–149 (1976)

Stabilization of Capacitated Matching Games

Matthew Gerstbrein[1], Laura Sanità[2], and Lucy Verberk[3(✉)]

[1] University of Waterloo, Waterloo, Canada
`mlgerstbrein@uwaterloo.ca`
[2] Bocconi University, Milan, Italy
`laura.sanita@unibocconi.it`
[3] Eindhoven University of Technology, Eindhoven, The Netherlands
`l.p.a.verberk@tue.nl`

Abstract. An edge-weighted, vertex-capacitated graph G is called *stable* if the value of a maximum-weight capacity-matching equals the value of a maximum-weight *fractional* capacity-matching. Stable graphs play a key role in characterizing the existence of stable solutions for popular combinatorial games that involve the structure of matchings in graphs, such as network bargaining games and cooperative matching games.

The vertex-stabilizer problem asks to compute a minimum number of players to block (i.e., vertices of G to remove) in order to ensure stability for such games. The problem has been shown to be solvable in polynomial-time, for unit-capacity graphs. This stays true also if we impose the restriction that the set of players to block must not intersect with a given specified maximum matching of G.

In this work, we investigate these algorithmic problems in the more general setting of arbitrary capacities. We show that the vertex-stabilizer problem with the additional restriction of avoiding a given maximum matching remains polynomial-time solvable. Differently, without this restriction, the vertex-stabilizer problem becomes NP-hard and even hard to approximate, in contrast to the unit-capacity case.

Finally, in unit-capacity graphs there is an equivalence between the stability of a graph, existence of a stable solution for network bargaining games, and existence of a stable solution for cooperative matching games. We show that this equivalence does not extend to the capacitated case.

Keywords: Matching · Game theory · Network bargaining

1 Introduction

Network Bargaining Games (NBG) and *Cooperative Matching Games* (CMG) are popular combinatorial games involving the structure of matchings in graphs. CMG were introduced in the seminal paper of Shapley and Shubik 50 years ago [18], and have been widely studied since then. NBG are relatively more recent, and were defined by Kleinberg and Tardos [14] as a generalization of Nash's 2-player bargaining solution [17].

© The Author(s), under exclusive license to Springer Nature Switzerland AG 2023
A. Del Pia and V. Kaibel (Eds.): IPCO 2023, LNCS 13904, pp. 157–171, 2023.
https://doi.org/10.1007/978-3-031-32726-1_12

Instances of these games are described by a graph $G = (V, E)$ with weights $w \in \mathbb{R}^E_{\geq 0}$, where the vertices and the edges model the players and their potential interactions, respectively. The value of a *maximum-weight matching*, denoted as $\nu(G)$, is the total value that players can collectively accumulate. The goal, roughly speaking, is to assign values to players in such a way that players have no incentive to deviate from the current allocation.

Formally, in an instance of a NBG, players want to enter in a deal with one of their neighbours, and agree on how to split the value of the deal given by the weight of the corresponding edge. Hence, an outcome is naturally associated with a matching M of G representing the deals, and allocation vector $y \in \mathbb{R}^V_{\geq 0}$ with $y_u + y_v = w_{uv}$ if $uv \in M$, and $y_v = 0$ if v is not matched. An outcome (M, y) is *stable* if each player's allocation y_u is at least as large as their *outside option*, formally defined as $\max_{v: uv \in E \setminus M} \{w_{uv} - y_v\}$.

In an instance of a CMG, one wants to find an allocation of total value $\nu(G)$, given by a vector $y \in \mathbb{R}^V_{\geq 0}$ in which no subset of players can gain more by forming a coalition. This condition is enforced by the constraint $\sum_{v \in S} y_v \geq \nu(G[S])$ for all $S \subseteq V$, where $G[S]$ indicates the subgraph of G induced by the vertices in S. Such allocation is called *stable*, and the set of stable allocations constitutes the *core* of the game.

Despite having been defined in different contexts, there is a tight link between stable solutions of these types of games. In particular, if each game is played on the same graph G, then it has been shown that either a stable solution exists for both games, or for neither game. This follows as both games admit the same polyhedral characterization of instances with stable solutions [7,14]. Specifically, a stable solution exists if and only if $\nu(G)$ equals the value of the standard linear programming (LP) relaxation of the maximum matching problem defined as

$$\nu_f(G) := \max \left\{ w^\top x : \sum_{u: uv \in E} x_{uv} \leq 1 \; \forall v \in V, \; x \geq 0 \right\}. \tag{1}$$

A graph G for which $\nu(G) = \nu_f(G)$ is called *stable*. As a result of this characterization, it is easy to see that there are graphs which do not admit stable solutions (to either type of game), such as odd cycles. Given that not all graphs are stable, naturally arises the *stabilization* problem of how to minimally modify a graph to turn it into a stable one. Stabilization problems attracted a lot of attention in the literature in the past years (see e.g. [1,3–6,13,15,16]).

In this context, very natural operations to stabilize graphs are edge- and vertex-removal operations. Those have an interesting interpretation: they correspond to blocking interactions and players, respectively, in order to ensure a stable outcome. While removing a minimum number of edges to stabilize a graph is NP-hard already for unit weight graphs [4], and even hard-to-approximate with a constant factor [11,15], stabilizing the graph via vertex-removal operations turned out to be solvable in polynomial-time. Specifically, [1,13] showed that computing a minimum-cardinality set of players to block in order to stabilize an unweighted graph (called the *vertex-stabilizer problem*) can be done in polynomial time. Furthermore, [1] showed that computing a minimum set of players to block in order to make a given maximum matching realizable as a stable outcome (called the *M-vertex-stabilizer problem*) is also efficiently solvable. The authors of [15] showed that both results generalize to weighted graphs.

This paper focuses on *Capacitated NBG*, introduced by Bateni et al [2] as a generalization of NBG, to capture the more realistic scenario where players are allowed to enter in more than one deal. This generalization can be modeled by allowing for vertex capacities $c \in \mathbb{Z}_{\geq 0}^V$. The notion of a matching is therefore generalized to a *c-matching*, where each vertex $v \in V$ is matched with at most c_v vertices. In this case, the value of a maximum-weight *c*-matching of a graph G is denoted as $\nu^c(G)$, and the standard LP relaxation is given by

$$\nu_f^c(G) := \max \left\{ w^\top x : \sum_{u:uv \in E} x_{uv} \leq c_v \; \forall v \in V, \; 0 \leq x \leq 1 \right\}. \qquad (2)$$

An outcome to the NBG is associated with a *c*-matching M and a vector $a \in \mathbb{R}_{\geq 0}^{2E}$ that satisfies $a_{uv} + a_{vu} = w_{uv}$ if $uv \in M$, and $a_{uv} = a_{vu} = 0$ otherwise. The concepts of outside option and stable outcome can be defined similarly as in the unit-capacity case, see [2].

The authors of [2] proved that the LP characterization of stable solutions generalize, i.e., there exist a stable outcome for the capacitated NBG on G if and only if $\nu^c(G) = \nu_f^c(G)$ (i.e., G is *stable*). Farczadi et al [9] show that some other important properties of NBG extend to this capacitated generalization, such as the possibility to efficiently compute a so-called *balanced* solution (we refer to [9] for details).

The goal of this paper is to investigate whether the other two significant features of NBG mentioned before generalize to the capacitated setting. Namely:

(i) *Can one still efficiently stabilize instances via vertex-removal operations?*
(ii) *Does the equivalence between existence of stable allocations for capacitated CMG and existence of stable solutions for capacitated NBG still hold?*

Our Results. In this paper we provide an answer to the above questions.

We investigate the *M*-vertex-stabilizer problem and the vertex-stabilizer problem in the capacitated setting in Sects. 3 and 4, respectively. While for unit-capacity graphs both problems are efficiently solvable, we show that adding capacities makes the complexity status of the vertex-stabilizer problem diverge. In particular, we prove that the vertex-stabilizer problem is NP-complete, and no $n^{1-\varepsilon}$-approximation is possible, for any $\varepsilon > 0$, unless P=NP. Note that a trivial n-approximation algorithm can be easily developed.

In contrast, we show that the *M*-vertex-stabilizer problem is still polynomial-time solvable in the capacitated setting. Our results here extend those of [15] for unit-capacity graphs, and builds upon an auxiliary construction of [9].

Finally, in Sect. 5 we show that the equivalence between stability of a graph, existence of a stable allocation for CMG and existence of a stable outcome for NBG does *not* extend in the capacitated setting. In particular, we provide an unstable graph which does attain a stable allocation for the capacitated CMG.[1]

---

[1] It is stated in [8] (Theorem 2.3.9) that a stable allocation for capacitated CMG exists iff G is stable, but our example shows this statement is not correct.

2 Preliminaries and Notation

Problem Definition. A set $S \subseteq V$ is called a **vertex-stabilizer** if $G \setminus S$ is stable, where $G \setminus S$ is the subgraph induced by the vertices $V \setminus S$. We say that a vertex-stabilizer S *preserves* a matching M of G if M is a matching in $G \setminus S$.

We now formally define the stabilization problems considered in this paper.

Vertex-stabilizer Problem: given $G = (V, E)$ with edge weights $w \in \mathbb{R}_{\geq 0}^E$ and vertex capacities $c \in \mathbb{Z}_{\geq 0}^V$, find a vertex-stabilizer of minimum cardinality.

M-vertex-stabilizer Problem: given $G = (V, E)$ with edge weights $w \in \mathbb{R}_{\geq 0}^E$, vertex capacities $c \in \mathbb{Z}_{\geq 0}^V$, and a maximum-weight c-matching M, find a vertex-stabilizer of minimum cardinality among the ones preserving M.

An instance of the vertex-stabilizer problem will be denoted as (G, w, c). An instance of the M-vertex-stabilizer problem will be denoted as $[(G, w, c), M]$. We say that an instance is stable if G is stable. Without loss of generality, we can assume that c_v is bounded by the degree of $v \in V$.

Notation. For a vertex v, we let $\delta(v)$ be the set of edges of G incident into it, we let $N(v)$ be the set of its neighbours, and $N^+(v) = N(v) \cup \{v\}$. For $F \subseteq E$, we denote by d_v^F the degree of v in G with respect to the edges in F. We define $w(F) := \sum_{e \in F} w_e$. Given a c-matching M, we say that $v \in V$ is *exposed* if $d_v^M = 0$, and *covered* if $d_v^M > 0$. We also use these terms for feasible solutions x of (2), called *fractional c-matchings*, e.g., v is exposed if $\sum_{e \in \delta(v)} x_e = 0$. We let $n := |V|$, and $\triangle$ denote the symmetric difference operator.

We denote a $(uv\text{-})$walk W by listing its edges and endpoints sequentially, i.e., by $W = (u; e_1, \ldots, e_k; v)$. We define its inverse as $W^{-1} = (v; e_k, \ldots, e_1; u)$. Note that a path is a walk in which edges do not repeat, and internal vertices do not repeat. A cycle is a path which starts and ends at the same vertex. If we refer to the edge set of a walk W, we just write W. Note that this can be a multi-set.

Duality and Augmenting Structures. The dual of (2) is given by

$$\tau_f^c(G) := \min \left\{ c^\top y + \mathbf{1}^\top z : y_u + y_v + z_{uv} \geq w_{uv} \ \forall uv \in E, y \geq 0, z \geq 0 \right\}. \quad (3)$$

A solution (y, z) feasible for (3) is called a *fractional vertex cover*. By LP theory, we have $\nu^c(G) \leq \nu_f^c(G) = \tau_f^c(G)$.

Definition 1. *We say that a walk W is M-alternating (w.r.t. a matching M) if it alternates edges in M and edges not in M. We say W is M-augmenting if it is M-alternating and $w(W \setminus M) > w(W \cap M)$. An M-alternating uv-walk W is proper if $d_u^{W \triangle M} \leq c_u$ and $d_v^{W \triangle M} \leq c_v$.*

Definition 2. *Given an M-alternating walk $W = (u; e_1, \ldots, e_k; v)$ and an $\varepsilon > 0$, the ε-augmentation of W is the vector $x^{M/W}(\varepsilon) \in \mathbb{R}^E$ given by*

$$x_e^{M/W}(\varepsilon) = \begin{cases} 1 - \kappa(e)\varepsilon & \text{if } e \in M, \\ \kappa(e)\varepsilon & \text{if } e \notin M, \end{cases} \quad (4)$$

where $\kappa(e) = |\{i \in [k] : e_i = e\}|$. We say that W is feasible if there exists an $\varepsilon > 0$ such that the corresponding ε-augmentation of W is a fractional c-matching.

Remark 1. A feasible M-alternating walk with distinct endpoints is proper.

Definition 3. An odd cycle $C = (v; e_1, \ldots, e_k; v)$ is called an M-blossom if it is M-alternating such that either e_1 and e_k are both in M, or are both not in M. The vertex v is called the base of the blossom.

Definition 4. An M-flower $C \cup P$ consists of an M-blossom C with base u and an M-alternating path $P = (u; e_1, \ldots, e_k; v)$ such that (P, C, P^{-1}) is M-alternating and feasible. The vertex v is called the root of the flower. The flower is M-augmenting if

$$w(C \setminus M) + 2w(P \setminus M) > w(C \cap M) + 2w(P \cap M). \tag{5}$$

Definition 5. An M-bi-cycle $C \cup P \cup D$ consists of two M-blossoms C and D with bases u and v, respectively, and an M-alternating path $P = (u; e_1, \ldots, e_k; v)$ such that (P, D, P^{-1}, C) is M-alternating. The bi-cycle is M-augmenting if

$$w(C \setminus M) + 2w(P \setminus M) + w(D \setminus M) > w(C \cap M) + 2w(P \cap M) + w(D \cap M). \tag{6}$$

Note that, in the last two definitions, it may happen that P has no edges.

Auxiliary Construction. We will use a construction given in [9], to transform an M-vertex-stabilizer instance $[(G, w, c), M]$ into another one $([(G', w', \mathbf{1}), M'])$ defined on an auxiliary graph with unit capacities.
    Construction: $[(G, w, c), M] \rightarrow [(G', w', \mathbf{1}), M']$

1. For each $v \in V$, create the set $C_v = \{v_1, \ldots, v_{c_v}\}$ of c_v copies of v, add C_v to $V(G')$, and initialize $J(v) = \{1, \ldots, c_v\}$.
2. For each $uv \in M$, add a single edge $u_i v_j$ to both $E(G')$ and M' with edge-weight w_{uv}, where $i \in J(u)$ and $j \in J(v)$ are chosen arbitrarily. Remove i and j from $J(u)$ and $J(v)$, respectively.
3. For each edge $uv \in E \setminus M$, add an edge $u_i v_j$ to $E(G')$ with edge-weight w_{uv}, for all $u_i \in C_u$ and $v_j \in C_v$.

See Fig. 1 for an example. In this figure it is easy to see that the matching M' in G' is not maximum, even though M is maximum in G.[2]

Remark 2. If $[(G, w, c), M]$ has auxiliary $[(G', w', \mathbf{1}), M']$, and $X \subseteq V$ is any set of vertices which avoids M, then $(G \setminus X)' = G' \setminus X'$, where $X' = \cup_{v \in X} C_v$.

We define a map η to go back from the auxiliary graph G' to the original graph G. Specifically, if $u_i \in V(G') \cap C_u$ for some $u \in V$, then $\eta(u_i) := u$, and if $u_i v_j \in E(G')$ such that $u_i \in C_u$, $v_j \in C_v$ for some $u, v \in V$, then $\eta(u_i v_j) := uv$. This extends in the obvious way to paths, cycles, walks, and so on.
    We will need the following theorem.

---

[2] It was stated in [9, corollary 1] that M is maximum if and only if M' is maximum, but this example shows this to be false.

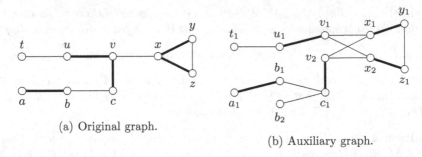

(a) Original graph.

(b) Auxiliary graph.

Fig. 1. Example of the auxiliary construction on an instance $[(G, w, c), M]$. Capacities are all 1 except for $c_v = c_x = c_b = 2$. Weights are all 1 except for $w_{bc} = 0.5$. The matching is displayed as bold edges.

Theorem 1. $[(G, w, c), M]$ *is not stable if and only if the graph* G' *in the auxiliary instance* $[(G', w', 1), M']$ *contains at least one of the following: (i) an* M'*-augmenting flower; (ii) an* M'*-augmenting bi-cycle; (iii) a proper* M'*-augmenting path; (iv) an* M'*-augmenting cycle.*

Proof. It was proven in [9, Theorem 2] that $[(G, w, c), M]$ is not stable if and only if $[(G', w', 1), M']$ is not stable. We distinguish two scenarios for when the latter condition occurs. If M' is maximum-weight, then G' contains an M'-augmenting flower or bi-cycle, see [15, Theorem 1]. If M' is not maximum-weight, G' must contain a proper M'-augmenting path or cycle, by standard matching theory. $\square$

We will refer to an augmenting structure of type $(i) - (iv)$ in Theorem 1 as a *basic* augmenting structure. The next lemma follows from [15].

Lemma 1. *Let* $[(G', w', 1), M']$ *be an unstable instance of NBG.*

(a) For any M'*-exposed vertex* u, *one can compute a feasible* M'*-augmenting walk starting at* u *of length at most* $3|V(G')|$, *or determine that none exists, in polynomial time.*

(b) A feasible M'*-augmenting* uv*-walk contains a feasible* M'*-augmenting* uv*-path (proper if* $u \neq v$), *an* M'*-augmenting cycle, an* M'*-augmenting flower rooted at* u *or* v, *or an* M'*-augmenting bi-cycle. Furthermore, this augmenting structure can be computed in polynomial time.*

Proof. (a) When given a graph G', a matching M', a vertex u, and an integer k, algorithm 3 in [15] computes a feasible M'-augmenting uv-walk of length at most k, or determines none exist, for all $v \in V(G')$. Correctness is shown in Lemma 7 and 8 in [15]. The algorithm is polynomial time in k, $|V(G')|$, and $|E(G')|$. We use this algorithm and select an arbitrary v for which a uv-walk is returned, or determine that no such walk starting at u exists. Since we set $k = 3|V(G')|$, this procedure terminates in polynomial time.

(b) Lemma 9 in [15] states that a feasible M'-augmenting uv-walk contains a feasible M'-augmenting uv-path, an M'-augmenting cycle, an M'-augmenting flower rooted at u or v, or an M'-augmenting bi-cycle. By remark 1 the path is proper if $u \neq v$. Lemma 9 in [15] is proven in a constructive way, hence it also gives a way to compute the augmenting structure in polynomial time.   □

The following easy lemma will be useful.

Lemma 2. *Given* $[(G, w, c), M]$ *and auxiliary* $[(G', w', \mathbf{1}), M']$, *let P be a feasible M'-augmenting walk. Then,* $\eta(P)$ *is a feasible M-augmenting walk.*

Proof. Let $e_1 = uv$ and $e_2 = vw$ be two consecutive edges on P. Then $\eta(e_1)$ and $\eta(e_2)$ are the corresponding edges on $\eta(P)$, and they are both incident with $\eta(v)$. Hence, $\eta(P)$ is a walk. For any edge e on P, we have $e \in M'$ if and only if $\eta(e) \in M$. In addition, $w'_e = w_{\eta(e)}$. So, $\eta(P)$ is an M-augmenting walk. Suppose $P = (u; e_1, \ldots, e_k; v)$. Feasibility of P means that either $e_1 \in M'$, or u is M'-exposed. Likewise for e_k and v. It follows that either $\eta(e_1) \in M$, or $\eta(u)$ is M-unsaturated. Likewise for $\eta(e_k)$ and $\eta(v)$. This means $\eta(P)$ is feasible.   □

The next theorem is standard.

Theorem 2. $[(G, w, c), M]$ *is stable if and only if G does not contain a feasible M-augmenting walk.*

Proof. ($\Rightarrow$) Assume there exists a feasible M-augmenting walk W. Since W is augmenting, $w(W \setminus M) > w(W \cap M)$, and since W is feasible, $x^{M/W}(\varepsilon)$ is a fractional c-matching for some fixed $\varepsilon > 0$. Together they imply

$$\nu_f^c(G) \geq w^\top x^{M/W}(\varepsilon) = w(M) - \varepsilon w(W \cap M) + \varepsilon w(W \setminus M) > w(M), \quad (7)$$

i.e., the instance $[(G, w, c), M]$ is not stable.

($\Leftarrow$) Assume the instance is not stable. Then by Theorem 1, the graph G' from the auxiliary $[(G', w', \mathbf{1}), M']$ contains a basic augmenting structure, which clearly is a feasible M'-augmenting walk P. Then $\eta(P)$ is a feasible M-augmenting walk, by Lemma 2.   □

3 M-vertex-stabilizer

The goal of this section is to prove the following theorem.

Theorem 3. *The M-vertex-stabilizer problem on weighted, capacitated graphs can be solved in polynomial time.*

Overview of the Strategy. A natural strategy would be to first apply the auxiliary construction described in Sect. 2 to reduce to unit-capacity instances, and then apply the algorithm proposed in [15] which solves the problem exactly. However, there is a critical issue with this strategy. Namely, the auxiliary construction applied to unstable instances does *not* always preserve maximality of the corresponding matchings, as shown in Fig. 1. In that example, the matching M' is not maximum in G'. The algorithm of [15], if applied to an instance where the given matching is not maximum, is not guaranteed to find an optimal solution, but only a 2-approximate one (see Theorem 12 in [15]). In addition, since the auxiliary construction splits a vertex into multiple ones, we may even get infeasible solutions. As a concrete example of this, the algorithm of [15] applied to the instance of Fig. 1b will include b_2 in its proposed solution. Mapping this solution to our capacitated instance would imply to remove b, which is clearly not allowed as b is M-covered.

To overtake this issue, we do not apply the algorithm of [15] as a black-box, but use parts of it (highlighted in Lemma 1) in a careful way. In particular, we use it to compute a sequence of feasible augmenting walks in G'. We actually show that the walks in G' which might create the issue described before when mapped backed to G, are the walks in which at least one edge of G is traversed more than once in opposite directions, and that have two distinct endpoints. When this happens, we prove that we can modify the walk and get one where the endpoints coincide, which will still be feasible and augmenting. In this latter case, we can then either correctly identify a vertex to remove (the unique endpoint), or determine that the instance cannot be stabilized.

A More Detailed Description. We start by defining the operation of *traceback*, which we will use to modify the feasible augmenting walks, when needed.

Definition 6. *Given $[(G, w, c), M]$ and an M-alternating walk $P = (u; e_1, \ldots, e_k; v)$ which repeats an edge in opposite directions, let t be the least index such that $e_t = e_s$ for some $s < t$, and e_s and e_t are traversed in opposite directions by P. Then the u-traceback and v-traceback of P are defined as the walks $tb(P, u) = (e_1, \ldots, e_t, e_{s-1}, e_{s-2}, \ldots, e_1)$ and $tb(P, v) = (e_k, e_{k-1} \ldots, e_s, e_{t+1}, e_{t+2}, \ldots, e_k)$.*

The next lemma explains how to use the traceback operation. Due to space constraint, the proof is deferred to the full version of this extended abstract [10].

Lemma 3. *Given $[(G, w, c), M]$ and auxiliary $[(G', w', 1), M']$, let $P' = (u_i; e'_1, \ldots, e'_k; v_j)$ be a proper M'-augmenting path such that both u_i and v_j are M'-exposed and $\eta(u_i) \neq \eta(v_j)$. Then $tb(\eta(P'), \eta(u_i))$ and $tb(\eta(P'), \eta(v_j))$ are well-defined, feasible M-alternating walks, and at least one of them is M-augmenting.*

Proof (Proof of Theorem 3). Let $[(G, w, c), M]$ be the input for the M-vertex-stabilizer problem, with auxiliary $[(G', w', 1), M']$. Algorithm 1 iteratively considers an M'-exposed vertex u_i, and computes a feasible M'-augmenting walk U starting at u_i, if one exists. Lemma 2 implies that $\eta(U)$ is a feasible M-augmenting walk in G. Theorem 2 implies that we need to remove at least one vertex of the walk $\eta(U)$ to stabilize the graph. Note that every vertex $a \neq u_i, v_j$ of U is M'-covered, and hence, $\eta(a)$ is M-covered. Therefore, the only vertices we can potentially remove are $\eta(u_i)$ or $\eta(v_j)$. Hence, if both $\eta(u_i)$ and $\eta(v_j)$ are M-covered, the graph cannot be stabilized and Algorithm 1 checks this in line 9. If only one among $\eta(u_i)$ and $\eta(v_j)$ is M-covered, then necessarily we have to remove the M-exposed vertex among the two. Algorithm 1 checks this in

---

Algorithm 1: finding an M-vertex-stabilizer

---

input: $[(G, w, c), M]$

1 compute the auxiliary $[(G', w', \mathbf{1}), M']$

2 initialize $S \leftarrow \emptyset$, $L \leftarrow M'$-exposed vertices

3 **while** $L \neq \emptyset$ **do**

4   select $u_i \in L$ and compute a feasible M'-augmenting walk starting at u_i using lemma 1(a)

5   **if** *no such walk exists* **then**

6           $L \leftarrow L \setminus \{u_i\}$

7   **else**

8   consider the computed feasible M'-augmenting $u_i v_j$-walk

9   **if** *both $\eta(u_i)$ and $\eta(v_j)$ are M-covered* **then**

10   **return** infeasible

11   **else if** *$\eta(u_i)$ is M-covered and $\eta(v_j)$ is not* **then**

12               $S \leftarrow S \cup \eta(v_j)$, $G \leftarrow G \setminus \eta(v_j)$, $G' \leftarrow G' \setminus C_{\eta(v_j)}$, $L \leftarrow L \setminus C_{\eta(v_j)}$

13   **else if** *$\eta(v_j)$ is M-covered and $\eta(u_i)$ is not* **then**

14               $S \leftarrow S \cup \eta(u_i)$, $G \leftarrow G \setminus \eta(u_i)$, $G' \leftarrow G' \setminus C_{\eta(u_i)}$, $L \leftarrow L \setminus C_{\eta(u_i)}$

15   **else**

16   **if** $\eta(u_i) = \eta(v_j)$ **then**

17                   $S \leftarrow S \cup \eta(u_i)$, $G \leftarrow G \setminus \eta(u_i)$, $G' \leftarrow G' \setminus C_{\eta(u_i)}$, $L \leftarrow L \setminus C_{\eta(u_i)}$

18   **else**

19   find a basic M'-augmenting structure W contained in the $u_i v_j$-walk using lemma 1(b)

20   **if** *W is an M'-augmenting cycle or bi-cycle* **then**

21   **return** infeasible

22   **if** *W is an M'-augmenting flower rooted at u_i* **then**

23                       $S \leftarrow S \cup \eta(u_i)$, $G \leftarrow G \setminus \eta(u_i)$, $G' \leftarrow G' \setminus C_{\eta(u_i)}$, $L \leftarrow L \setminus C_{\eta(u_i)}$

24   **if** *W is an M'-augmenting flower rooted at v_j* **then**

25                       $S \leftarrow S \cup \eta(v_j)$, $G \leftarrow G \setminus \eta(v_j)$, $G' \leftarrow G' \setminus C_{\eta(v_j)}$, $L \leftarrow L \setminus C_{\eta(v_j)}$

26   **if** *W is a proper M'-augmenting $u_i v_j$-path* **then**

27   compute $\mathrm{tb}(\eta(W), \eta(u_i))$ and $\mathrm{tb}(\eta(W), \eta(v_j))$

28   **if** *$\mathrm{tb}(\eta(W), \eta(u_i))$ is M-augmenting* **then**

29                           $S \leftarrow S \cup \eta(u_i)$, $G \leftarrow G \setminus \eta(u_i)$, $G' \leftarrow G' \setminus C_{\eta(u_i)}$, $L \leftarrow L \setminus C_{\eta(u_i)}$

30   **if** *$\mathrm{tb}(\eta(W), \eta(v_j))$ is M-augmenting* **then**

31                           $S \leftarrow S \cup \eta(v_j)$, $G \leftarrow G \setminus \eta(v_j)$, $G' \leftarrow G' \setminus C_{\eta(v_j)}$, $L \leftarrow L \setminus C_{\eta(v_j)}$

32 **if** $w(M) < \nu_f^c(G)$ **then**

33   **return** infeasible

34 **else**

35   **return** S

---

line 11 and 13. Note that, by remark 2, instead of computing a new auxiliary for the modified G, we can just remove $C_{\eta(u_i)}$ (resp. $C_{\eta(v_j)}$) from G'. Similarly, if $\eta(u_i) = \eta(v_j)$ and $\eta(u_i)$ is M-exposed, we necessarily have to remove $\eta(u_i)$. Algorithm 1 checks this in line 16. If instead $\eta(u_i) \neq \eta(v_j)$, and both are M'-exposed, we apply Lemma 1(b) to find a basic augmenting structure W contained in U. Once again, we know by Lemma 2 and Theorem 2 that we need to remove a vertex in $\eta(W)$. In case W is a cycle or bi-cycle, all vertices of $\eta(W)$ are M-covered so the graph cannot be stabilized and Algorithm 1 checks this in line 20. In case W is a M'-augmenting flower with base u_i or v_j, Algorithm 1 accordingly removes $\eta(u_i)$ or $\eta(v_j)$ as all other vertices in $\eta(W)$ are M-covered, in line 23 and 25. Finally, if W is a proper (because $\eta(u_i) \neq \eta(v_j)$) M'-augmenting path, by Lemma 3 we know that we can find a feasible M-augmenting walk, where the only M-exposed vertex is either $\eta(u_i)$ or $\eta(v_j)$. Once again, this implies that this vertex must be removed. Algorithm 1 does so in lines 29 and 31.

From the discussion so far, it follows that when we exit the while loop each vertex in S is a necessary vertex to be removed from G, in order to stabilize the instance. We now argue that either removing all vertices in S is also sufficient, or G cannot be stabilized. Suppose that $G \setminus S$ is not stable. Theorem 1 implies that $(G \setminus S)'$ contains a basic augmenting structure Q. Note that Q cannot be an M'-augmenting flower with exposed root, or a proper M'-augmenting path with at least one exposed endpoint. To see this, observe that a flower and path are feasible M'-augmenting walks of length at most $3|V(G')|$ and $|V(G')|$, respectively. Hence, they would have been found by Algorithm 1 in line 4, contradicting that Q exists in $(G \setminus S)'$. It follows that Q is a basic augmenting structure where all vertices are M'-covered. By Lemma 2 $\eta(Q)$ is a feasible M-augmenting walk where all vertices are M-covered. This implies that G cannot be stabilized. Furthermore, using the ε-augmentation of $\eta(Q)$ we can obtain a fractional c-matching whose value is strictly greater than $w(M)$. Hence, $w(M) < \nu_f^c(G \setminus S)$. Algorithm 1 correctly determines this in line 32. This proves correctness of our algorithm.

Finally, we argue about the running time of the algorithm. Note that each operation that the algorithm performs can be done in polynomial time. Furthermore, after each iteration of the while loop, we either determine that the instance cannot be stabilized, or remove a vertex from G. Therefore, the while loop can be executed at most n times. The result follows. □

We close this section with a remark. The authors in [15] have also considered the following problem: given a weighted graph G and a (non necessarily maximum-weight) matching M, find a minimum-cardinality $S \subseteq V$ such that $G \setminus S$ is stable, and M is a maximum-weight matching in $G \setminus S$. This is a generalization of our definition of the M-vertex-stabilizer problem, which essentially allows M to be not maximum-weight.[3] The authors show that this problem is

---

[3] In fact, this is the way the M-vertex-stabilizer problem is defined in [15]. We instead use the original definition in [1,6] which assumes M to be maximum.

NP-hard, but admits a 2-approximation algorithm (we mentioned this in the strategy overview), which is best possible assuming Unique Game Conjecture.

With a minor modification of Algorithm 1, we can generalize this result to the capacitated setting. We state here the result, and refer to the full version of this extended abstract paper for details [10].

Theorem 4. *Given a weighted, capacitated graph $G = (V, E)$ and a c-matching M, the problem of computing a minimum-cardinality $S \subseteq V$ such that $G \setminus S$ is stable, and M is a maximum-weight c-matching in $G \setminus S$, admits an efficient 2-approximation algorithm.*

4 Vertex-Stabilizer

The goal of this section is to prove the following theorem.

Theorem 5. *The vertex-stabilizer problem on capacitated graphs is NP-complete, even if all edges have unit-weight. Furthermore, no efficient $n^{1-\varepsilon}$-approximation exists for any $\varepsilon > 0$, unless $P = NP$.*

Note that, given an unstable instance (G, w, c), removing all vertices (but two) trivially yields a stable graph. This gives a (trivial) n-approximation algorithm for the vertex-stabilizer problem. The theorem above essentially implies that one cannot hope for a much better approximation. To prove it, we will use:

Minimum Independent Dominating Set (MIDS) Problem. Given a graph $G = (V, E)$, compute a minimum-cardinality subset $S \subseteq V$ that is independent (for all $uv \in E$ at most one of u and v is in S) and dominating (for all $v \in V$ at least one $u \in N^+(v)$ is in S).

There is no efficient $n^{1-\varepsilon}$-approximation for any $\varepsilon > 0$ for the MIDS problem, unless $P = NP$. [12, corollary 3]

Proof (Proof of Theorem 5). The decision variant of the problem asks to find a vertex-stabilizer of size at most k. This problem is in NP, since if a vertex set S is given, it can be verified in polynomial time if $|S| \le k$ and if $\nu^c(G \setminus S) = \nu_f^c(G \setminus S)$. We prove the NP-hardness and the inapproximability result by giving an approximation-preserving reduction from the MIDS problem.

Let $G = (V, E)$ be an instance of the MIDS problem. For $v \in V$, we define the gadget Γ_v by

$$V(\Gamma_v) = N^+(v) \cup \{v_1, v_2, v_3, v_4\}, \tag{8}$$

$$E(\Gamma_v) = \{uv_1 : u \in N^+(v)\} \cup \{v_1v_2, v_2v_3, v_3v_4, v_2v_4\}. \tag{9}$$

For $e = uv \in E$ and $i \in \{1, \dots, n\}$, we define the gadget Γ_{uv}^i by

$$V(\Gamma_{uv}^i) = \{u, v, e_1^i, e_2^i, e_3^i, e_4^i, e_5^i\}, \tag{10}$$

$$E(\Gamma_{uv}^i) = \{ue_1^i, ve_1^i, e_1^i e_2^i, e_1^i e_3^i, e_3^i e_4^i, e_4^i e_5^i, e_3^i e_5^i\}. \tag{11}$$

168   M. Gerstbrein et al.

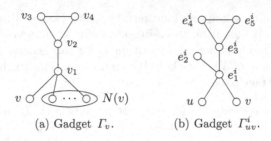

(a) Gadget Γ_v. (b) Gadget Γ_{uv}^i.

Fig. 2. Examples of gadgets.

See Fig. 2 for an example of these gadgets. Now let G' be defined as the union of all Γ_v and all Γ_{uv}^i, such that vertices from V overlap. We set the capacity as follows: $c_v = d_v^{E(G')}$ for all $v \in V$, $c_{v_1} = d_v^E + 1$ for all $v \in V$, $c_{e_1^i} = c_{e_3^i} = 2$ for $e_1^i, e_3^i \in V(\Gamma_{uv}^i)$ for all $e = uv \in E$ and $i \in \{1, \ldots, n\}$, and $c_v = 1$ for all remaining $v \in V(G')$. All edges are set to have unit-weight. The key point is:

Claim. G has an independent dominating set of size at most k if and only if $(G', \mathbf{1}, c)$ has a vertex-stabilizer of size at most k.

Proof. ($\Rightarrow$) Let S be an independent dominating set of G of size k. The vertices in S naturally correspond with vertices in G'. We show that S is a vertex-stabilizer of $(G', \mathbf{1}, c)$.

We define a c-matching M and fractional vertex cover (y, z) on $G' \setminus S$ as follows. First, set $y_v = 0$ for all $v \in V \setminus S$.

Next, for all $v \in V$, consider Γ_v. Add $\{uv_1 : u \in N^+(v) \setminus S\} \cup \{v_1v_2, v_3v_4\}$ to M. Note that at least one vertex from $N^+(v)$ is in S, since S is dominating. Set $y_{v_1} = 0$, $y_{v_2} = 1$, $y_{v_3} = y_{v_4} = 0.5$, $z_e = 1$ for all $e \in \{uv_1 : u \in N^+(v) \setminus S\}$ and $z_e = 0$ for the remaining edges in the gadget.

Finally, for all $e = uv \in E$ and $i \in \{1, \ldots, n\}$, consider Γ_{uv}^i. Since S is independent, at most one of u and v is in S. If neither are in S, add both ue_1^i and ve_1^i to M. If one of them is in S, without loss of generality let it be u, then add ve_1^i and $e_1^ie_2^i$ to M. Furthermore, add $e_3^ie_4^i$ and $e_3^ie_5^i$ to M. Set $y_{e_1^i} = 1$, $y_{e_2^i} = 0$, $y_{e_3^i} = y_{e_4^i} = y_{e_5^i} = 0.5$, and $z_f = 0$ for all edges f in the gadget.

Let x be the indicator vector of M. One can verify that x and (y, z) satisfy the complementary slackness conditions for $\nu_f^c(G' \setminus S)$ and $\tau_f^c(G' \setminus S)$. Since x is integral, this implies that $G' \setminus S$ is stable.

($\Leftarrow$) Let S be a vertex-stabilizer of $(G', \mathbf{1}, c)$ of size k. We show that: (i) S contains at least one vertex of each gadget Γ_v; (ii) without loss of generality, one can assume that at most one of u and v is in S for each edge $uv \in E$.

(i) Suppose for the sake of contradiction that there is some $v \in V$ such that S contains no vertices of Γ_v. Since $G' \setminus S$ is stable, there is a maximum-cardinality fractional c-matching x^*, that is integral. Define for each $e \in E(G' \setminus S)$

$$
x_e = \begin{cases}
x_e^* & \text{if } e \in E(G' \setminus S) \setminus E[\Gamma_v], \\
1 & \text{if } e \in \{uv_1 : u \in N^+(v)\}, \\
0 & \text{if } e = v_1 v_2, \\
0.5 & \text{if } e \in \{v_2 v_3, v_3 v_4, v_2 v_4\}.
\end{cases} \tag{12}
$$

Note that x is a fractional c-matching in $G' \setminus S$, since x^* is. However, $\sum_{e \in E[\Gamma_v]} x_e = d_v + 2.5 > \sum_{e \in E[\Gamma_u]} x_e^*$, since x^* is integral. Hence, $\mathbf{1}^\top x > \mathbf{1}^\top x^*$, contradicting the optimality of x^*.

(ii) Suppose there is some $e = uv \in E$ such that S contains both u and v. All gadgets Γ_{uv}^i are then components in $G' \setminus S$. If u and v are the only vertices in S from some component Γ_{uv}^i, then a maximum-cardinality fractional c-matching in this components is given by $x_{e_1^i e_2^i} = x_{e_1^i e_3^i} = 1$ and $x_{e_3^i e_4^i} = x_{e_4^i e_5^i} = x_{e_3^i e_5^i} = 0.5$. Which means this component is not stable, and thus $G' \setminus S$ is not stable, a contradiction. Hence, S must contain at least one vertex of each Γ_{uv}^i that is neither u nor v. Consequently, $k = |S| \geq n + 2$. Since G has only n vertices, it obviously has an independent dominating set of size at most n, and hence of size at most k. Such a set can for example be obtained by a greedy approach. Hence, for the remainder of the proof we can assume that at most one of u and v is in S for each $uv \in E$.

We now create a set $S' \subseteq V$ from S, that is an independent dominating set of G of size at most k, as follows. Iterate over $v \in V$. Let $S_v = S \cap V(\Gamma_v)$. Note that $S_v \neq \emptyset$ by (i). Define

$$
S_v' = \begin{cases}
(S_v \cup S') \cap N^+(v) & \text{if this is nonempty}, \\
v & \text{otherwise}.
\end{cases} \tag{13}
$$

Set $S' = S' \cup S_v'$, and repeat for the next vertex.

Clearly, all S_v''s are nonempty, which means that S' contains at least one vertex from $N^+(v)$ for all $v \in V$, which means S' is dominating.

Suppose for the sake of contradiction that S' contains both u and v for some edge $uv \in E$. We know S did not contain both of them, by (ii). If S contained exactly one of them, without loss of generality let it be u. Then, when v is considered by the iterative process, $(S_v \cup S') \cap N^+(v)$ contains u, but not v. In particular, this means that we did not add v to S_v' and consequently also not to S', a contradiction. If S contained neither of them, then because we do the process iteratively, one of them will be added first to S'. Without loss of generality let it be u. Then again, when v is considered by the iterative process, $(S_v \cup S') \cap N^+(v)$ contains u but not v, so we reach a contradiction in the same way. In conclusion, S' is independent.

For all $v \in V$, before we added S_v' to S', we had $|S_v' \setminus S'| \leq |S_v|$. Consequently, $|S'| \leq \cup_{v \in V} |S_v| \leq |S| = k$. $\qquad\square$

By this claim, any minimum-cardinality vertex-stabilizer of $(G', 1, c)$ is of the same size as any minimum independent dominating set of G. Further, any efficient α-approximation algorithm for the vertex-stabilizer problem translates into an efficient α-approximation algorithm for the MIDS problem. Hence, the result follows from the inapproximability of the MIDS problem.   □

5   Capacitated Cooperative Matching Games

Cooperative matching games in unit-capacity graphs, defined in the introduction, extend quite easily to capacitated graphs, by replacing each v with v^c. In unit-capacity graphs G the following statements are equivalent [7,14]:

 (i) G is stable,
 (ii) there exists an allocation in the core of the CMG on G,
(iii) there exists a stable outcome for the NBG on G.

We here note that the equivalence does not extend to capacitated graphs.

In particular, as mentioned in the introduction, we still have $(i) \iff (iii)$ proven in [2, corollary 3.3]. The implication $(i) \implies (ii)$ still holds, and follows from [2, lemma 3.4][4]. However, the graph G given in Fig. 3 shows that $(ii) \not\implies (i)$ (and hence $(ii) \not\implies (iii)$).

Assuming all the edges of G in Fig. 3 have unit weight, it is quite easy to see that $v^c(G) = 3$ and $v_f^c(G) = 3.5$, thus G is not stable. One can check that $y = (1, 1, 1, 0)$ is in the core.

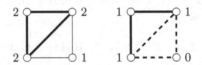

Fig. 3. On the left: the graph G where the values close to the vertices indicate the capacities. Bold edges indicate a maximum c-matching. On the right: the graph G where the values close to the vertices indicate the allocation y. A maximum fractional c-matching is given by $x_e = \frac{1}{2}$ for dashed edges, $x_e = 1$ otherwise.

Acknowledgements. The second and third authors are supported by the NWO VIDI grant VI.Vidi.193.087. The second author thanks the 2021 Hausdorff Research Institute for Mathematics Program Discrete Optimization, during which part of this work was developed.

References

1. Ahmadian, S., Hosseinzadeh, H., Sanità, L.: Stabilizing network bargaining games by blocking players. Math. Program. **172**, 249–275 (2018)
2. Bateni, M., Hajiaghayi, M., Immorlica, N., Mahini, H.: The cooperative game theory foundations of network bargaining games (2010)

---

[4] [2] assumes that the graph is bipartite, but bipartiteness is not needed in their proof.

3. Biró, P., Kern, W., Paulusma, D.: On solution concepts for matching games. In: Kratochvíl, J., Li, A., Fiala, J., Kolman, P. (eds.) Theory Appl. Models Comput., pp. 117–127. Springer, Berlin Heidelberg, Berlin, Heidelberg (2010)
4. Bock, A., Chandrasekaran, K., Könemann, J., Peis, B., Sanità, L.: Finding small stabilizers for unstable graphs. Math. Program. **154**, 173–196 (2015)
5. Chandrasekaran, K.: Graph stabilization: a survey. In: Fukunaga, T., Kawarabayashi, K. (eds.) Combinatorial Optimization and Graph Algorithms, pp. 21–41. Springer, Singapore (2017). https://doi.org/10.1007/978-981-10-6147-9_2
6. Chandrasekaran, K., Gottschalk, C., Könemann, J., Peis, B., Schmand, D., Wierz, A.: Additive stabilizers for unstable graphs. Discret. Optim. **31**, 56–78 (2019)
7. Deng, X., Ibaraki, T., Nagamochi, H.: Algorithmic aspects of the core of combinatorial optimization games. Math. Oper. Res. **24**(3), 751–766 (1999)
8. Farczadi, L.: Matchings and games on networks, Ph. D. thesis, University of Waterloo (2015)
9. Farczadi, L., Georgiou, K., Könemann, J.: Network bargaining with general capacities. arXiv preprint arXiv:1306.4302 (2013)
10. Gerstbrein, M., Sanità, L., Verberk, L.: Stabilization of capacitated matching games. arXiv preprint (2022)
11. Gottschalk, C.: Personal communication (2018)
12. Halldórsson, M.M.: Approximating the minimum maximal independence number. Inf. Process. Lett. **46**(4), 169–172 (1993)
13. Ito, T., Kakimura, N., Kamiyama, N., Kobayashi, Y., Okamoto, Y.: Efficient stabilization of cooperative matching games. Theoret. Comput. Sci. **677**, 69–82 (2017)
14. Kleinberg, J.M., Tardos, É.: Balanced outcomes in social exchange networks. In: Proceedings of the 40th STOC, pp. 295–304 (2008)
15. Koh, Z.K., Sanità, L.: Stabilizing weighted graphs. Math. Oper. Res. **45**(4), 1318–1341 (2020)
16. Könemann, J., Larson, K., Steiner, D.: Network bargaining: using approximate blocking sets to stabilize unstable instances. In: Serna, M. (ed.) SAGT 2012. LNCS, pp. 216–226. Springer, Heidelberg (2012). https://doi.org/10.1007/978-3-642-33996-7_19
17. Nash, J.F.: The bargaining problem. Econometrica **18**, 155–162 (1950)
18. Shapley, L., Shubik, M.: The assignment game i: The core. Internat. J. Game Theory **1**(1), 111–130 (1971)

Designing Optimization Problems
with Diverse Solutions

Oussama Hanguir[1]([✉]), Will Ma[2], and Christopher Thomas Ryan[3]

[1] Industrial Engineering and Operations Research, Columbia University,
New York, NY 10027, USA
oh2204@columbia.edu
[2] Graduate School of Business, Columbia University, New York, NY 10027, USA
wm2428@gsb.columbia.edu
[3] UBC Sauder School of Business, University of British Columbia,
Vancouver, BC V6T 1Z2, Canada
chris.ryan@sauder.ubc.ca

Abstract. We consider the problem of designing a linear program that
has diverse solutions as the right-hand side varies. This problem arises
in video game settings where designers aim to have players use different
"weapons" or "tactics" as they progress. We model this design question
as a choice over the constraint matrix A and cost vector c to maximize
the number of possible *supports* of unique optimal solutions (what we
call "loadouts") of Linear Programs $\max\{c^\top x \mid Ax \leq b, x \geq 0\}$ with
nonnegative data considered over all resource vectors b. We provide an
upper bound on the optimal number of loadouts and provide a family
of constructions that have an asymptotically optimal number of loadouts. The upper bound is based on a connection between our problem
and the study of triangulations of point sets arising from polyhedral
combinatorics, and specifically the combinatorics of the cyclic polytope.
Our asymptotically optimal construction also draws inspiration from the
properties of the cyclic polytope.

Keywords: linear programming · triangulations · diversity

1 Introduction

In this paper, we formulate the problem of *designing* linear programs that allow
for *diversity* in their optimal solutions. This setting is motivated by video games,
in particular, the design of competitive games where players optimize their
strategies to improve their in-game status. For such games, a desideratum for
game designers is for optimizing players to play different strategies at different
stages of the game. Informally, we interpret the player's problem as solving a Linear Program of the form $\max\{c^\top x \mid Ax \leq b, x \geq 0\}$. Players at different stages
of the game have different resource vectors b. The columns of A correspond to
the tools that the player can use in the game. We call a subset of these tools
(represented by subsets of the columns of A) a *loadout* (which literally means
the equipment carried into battle by a soldier), if they correspond to the support

© The Author(s), under exclusive license to Springer Nature Switzerland AG 2023
A. Del Pia and V. Kaibel (Eds.): IPCO 2023, LNCS 13904, pp. 172–186, 2023.
https://doi.org/10.1007/978-3-031-32726-1_13

of an optimal solution x^* to the linear program $\max\{c^\top x \mid Ax \le b, x \ge 0\}$ for some resource vector b (In fact, we require x^* to be the unique optimal solution of this linear program, for reasons that will become clear later). The support of a vector corresponds to a selection of the available tools, forming a strategy for how the player approaches the game given available resources. We assume that the game designer is able to choose A and c. We refer to this choice as the *design* of the game. We measure the *diversity* of a design as the number of possible loadouts that arise as the resource vector b changes.

For a fixed design (A, c) and resource vector b, players solve the linear program

$$LP(A, c, b): \qquad \max\{c^\top x \mid Ax \le b, x \ge 0\}, \tag{1}$$

where c, A, and b all have nonnegative data. If x is the optimal solution of $LP(A, c, b)$, we define the support of x as $\operatorname{supp}(x) \triangleq \{i \in \{1, \ldots, n\} \mid x_i > 0\}$. If x is the *unique* optimal solution of $LP(A, c, b)$, then we call $\operatorname{supp}(x)$ an *optimal loadout* (or simply, a loadout) of design (A, c).

For fixed n and m, the *loadout maximization problem* is to choose c and A that maximize the total number of loadouts of the design (A, c). That is, the goal is to design benefits for each tool (the vector c) and limitations on investing in tools (the matrix A) so that the linear programs $LP(A, c, b)$ have as many possible supports of unique optimal solutions as possible, as b varies in $\mathbb{R}^m$.

Our Contributions: Our first contribution consists in introducing the loadout maximization problem and establishing a link between the loadout maximization problem and the theory of polyhedral subdivisions and triangulations. In particular, for a fixed design (A, c), the theory of triangulations offers a nice decomposition of the cone generated by the columns of the constraint matrix A. This decomposition depends on the objective vector c. We show that for a fixed design (A, c), the loadouts can be seen as elements of this decomposition. This allows us to use a set of powerful tools from the theory of triangulations to prove structural results on the loadouts of a design.

Our second contribution is to show a non-trivial upper bound on the number of loadouts of any design. The upper bound involves an interesting connection to the faces of the so-called cyclic polytope, a compelling object central to the theory of polyhedral combinatorics. We also show that this upper bound holds when the constraints of the linear program are equality constraints.

The third contribution of this paper is to present a construction of a design (A, c) with a number of loadouts that asymptotically matches the above upper bound. Furthermore, for cases with few constraints, we present optimal constructions that *exactly* match the upper bound. Due to space constraints, we defer the proofs to the full version[1].

Related Work. Our work is closely related to parametric linear programming, which is the study of how optimal properties depend on parameterizations of the data. The study of parametric linear programming dates back to the work of [12,14], [19], and [18] in the 1950s and 1960s. In parametric programming,

---

[1] Full version: https://arxiv.org/abs/2106.11538.

the objective is to understand the dependence of optimal solutions on one or more parameters; that is, on the entries of A, b, and c. Our work is novel in the sense that the objective is to understand the structure of the supports of optimal solutions by fixing A and c and having b vary in $\mathbb{R}_{\geq 0}^m$. To the best of our knowledge, this question has not previously been studied in the literature.

There have been several studies on the interface of optimization and video games. [8,15,17]. Guo *et al.* [7] study the impact of selling virtual currency on players' gameplay behavior, game provider's strategies, and social welfare. Another significant research direction concentrates on studying "loot boxes" in video games. Chen *et al.* [2] study the design and pricing of loot boxes, while Ryan *et al.* [13] study the pricing and deployment of enhancements that increase the player's chance of completing the game. Chen *et al.* [3] and Huang *et al.* [9] study the problem of in-game matchmaking to maximize a player's engagement in a video game.

2   Statement of Main Results

In this section, we state our main results. To make these statements precise, we require some preliminary definitions. Let $[k]$ denote the set $\{1, \ldots, k\}$ for any positive integer k. Using this notation, we can define the support of $x \in \mathbb{R}_{\geq 0}^n$ as $\mathrm{supp}(x) = \{j \in [n] \mid x_j > 0\}$. For a matrix $A \in \mathbb{R}_{\geq 0}^{m \times n}$, the (i,j)th entry is denoted a_{ij} for $i \in [n]$ and $j \in [m]$, the jth column is denoted A_j for $j \in [n]$, and the ith row is denoted a_i (where a_i is a column vector) for $i \in [m]$. For a column vector $y \in \mathbb{R}^m$, $y^\top A_j$ denote the scalar product of y and column A_j, i.e., $y^\top A_j = \sum_{i=1}^m y_i a_{i,j}$.

Recall the definition of the linear program $LP(A, b, c)$ in (1). As mentioned in the introduction, we are interested in the unique optimal solutions of the design (A, c). For simplicity, we simply call these the *loadouts* of design (A, c); that is, $L \subseteq [n]$ is a loadout of design (A, c) if there exists a nonnegative resource vector $b \in \mathbb{R}_{\geq 0}^m$ such that $LP(A, c, b)$ has a unique optimal solution x^* with $\mathrm{supp}(x^*) = L$. We say that loadout L is *supported by* resource vector b. If $|L| = k$ then we say L is a k-loadout. Given a design (A, c) and an integer $k \in [m]$, let $\mathcal{L}^k(A, c)$ denote the set of all k-loadouts of design (A, c). The set of all loadouts of any size is $\mathcal{L}(A, c) \triangleq \cup_{k=1}^n \mathcal{L}^k(A, c)$.

Using this notation, we can restate the loadout optimization problem. Given dimensions n and m and integer $k \leq n$, the k-*loadout optimization problem* is

$$\max\{|\mathcal{L}^k(A, c)| \mid A \in \mathbb{R}^{m \times n}, c \in \mathbb{R}^n, A \text{ and } c \text{ are nonnegative}\}. \qquad (\mathrm{L}_k)$$

We can assume without loss of generality that the linear programs $LP(A, c, b)$ are bounded and thus possess an optimal solution because otherwise there is no optimal solution and, therefore, no loadout.

Given that a loadout corresponds to the support of a unique solution of a linear program, any optimal solution with support size greater than m cannot be unique. Therefore, the number of k-loadouts when $k > m$ is always equal to zero. This leads us to consider the optimization problems (L_k) only for $k \in$

$\{1, \ldots, \min(m,n)\}$. For convenience, we will avoid the trivial case of $k = 1$ where the optimal number of loadouts is $\min(m,n)$. A final case we eliminate immediately is when $\min(m,n) = n$, i.e. $m \geq n$. In this case, a trivial design is optimal. By setting $A = I_n$ to be the identity matrix of size n, and $c = (1, \ldots, 1)$, we ensure that for $k \in [1, n]$, every one of the $\binom{n}{k}$ subsets is a loadout. In summary, we proceed without loss under the assumption that $n > m \geq k \geq 2$.

2.1 The Cyclic Polytope

All of our bounds are intimately related to the number of faces on the *cyclic polytope*, which is formally defined in Sect. 3. A remarkable aspect of the cyclic polytope is that for $n > m \geq 2$, the cyclic polytope $\mathcal{C}(n,m)$ *simultaneously* maximizes the number of k-dimensional faces for all $k = 0, \ldots, m-1$ among m-dimensional polytopes over n vertices, a property known as McMullen's Upper Bound Theorem [11]. The number of k-dimensional faces on $\mathcal{C}(n,m)$ is given by the formula

$$f_k(\mathcal{C}(n,m)) = \sum_{\ell=0}^{\lfloor m/2 \rfloor} \binom{\ell}{m-k-1} \binom{n-m+\ell-1}{\ell}$$
$$+ \sum_{\ell=\lfloor m/2 \rfloor+1}^{m} \binom{\ell}{m-k-1} \binom{n-\ell-1}{m-\ell}.$$

When $k = m - 1$, through the "hockey stick" identity on Pascal's triangle, this simplifies to

$$f_{m-1}(\mathcal{C}(n,m)) = \binom{n - \lceil m/2 \rceil}{\lfloor m/2 \rfloor} + \binom{n - \lfloor m/2 \rfloor - 1}{\lceil m/2 \rceil - 1}.$$

2.2 Results and Techniques

Theorem 1. *Fix positive integers n, m, k with $n > m \geq k \geq 2$. Then the number of k-loadouts for any design (A, c) with $A \in \mathbb{R}^{m \times n}$ and $c \in \mathbb{R}^n$ satisfies*

$$|\mathcal{L}^k(A,c)| \leq f_{k-1}(\mathcal{C}(n+1, m)) - \binom{m}{k-1}. \tag{2}$$

We note that the trivial upper bound on the number of k-loadouts in a design with n tools is $\binom{n}{k}$. When $m < n$, the RHS of (2) will always be smaller than this trivial upper bound, which shows that having a limited number of resource types in the game does indeed prevent all subsets of tools from being viable.

Theorem 2. *Fix positive integers n, m, k with $n > m \geq k \geq 2$. Then there exists an explicit design (A, c) with $A \in \mathbb{R}_{\geq 0}^{m \times n}$ and $c \in \mathbb{R}_{\geq 0}^n$ that satisfy*

$$|\mathcal{L}^k(A,c)| \geq \begin{cases} f_{k-1}(\mathcal{C}(n,m)) & \text{if } k < m/2 \\ f_{k-1}(\mathcal{C}(n,m))/2 & \text{if } k \geq m/2 \text{ and } m \text{ is odd,} \\ & \text{or } k = m/2 \text{ and } m \text{ is even} \\ f_{k-1}(\mathcal{C}(n,m))/4 & \text{if } k > m/2 \text{ and } m \text{ is even.} \end{cases}$$

The constructions from Theorem 2 are always within a 1/4-factor of being optimal asymptotically as $n \to \infty$ because it is known that

$$\lim_{n \to \infty} \frac{f_{k-1}(\mathcal{C}(n,m))}{f_{k-1}(\mathcal{C}(n+1,m))} = 1.$$

Theorem 3. *For $n > m = 3$, there exists an explicit design (A, c) with $A \in \mathbb{R}_{\geq 0}^{m \times n}$ and $c \in \mathbb{R}_{\geq 0}^{n}$ that satisfy $|\mathcal{L}^3(A, c)| \geq 2n - 5$ and $|\mathcal{L}^2(A, c)| \geq 3n - 6$.*

Theorem 4. *For $n > m = 2$, there exists an explicit design (A, c) with $A \in \mathbb{R}^{m \times n}$ and $c \in \mathbb{R}^n$ that satisfy $|\mathcal{L}^2(A, c)| \geq n - 1$.*

The constructions from Theorem 3 and Theorem 4 are *exactly tight*; it can be checked that they match the upper bound expression from Theorem 1 when evaluated at $m = 3$ and $m = 2$.

Example of Construction from Theorem 2 and Intuition. Table 1 shows an example of the asymptotically optimal construction for $m = 4$ and $n = 6$. The fact that the cost vector is $(1, 1, \ldots, 1)$ is simply a normalization and can be assumed without loss. Our construction provides a pattern that game designers can follow to diversify loadouts on a set of tools $1, \ldots, n$, by having two types of constraints. The first type of constraints (rows 1 and 3) accords more importance to tools with big indices (because these tools have lower costs to rows 1 and 3) while the second type of constraints (rows 2 and 4) give more advantage to tools with a small index (because these tools have lower costs to rows 2 and 4). This "tension" between the two types of constraints ensures that a given combination of tools cannot be optimal for too many resource vectors. This captures the rough intuition that a game with an overpowered tool (meaning one that is more useful than the others but also not significantly "cumbersome" to limit its use) leads to uniform strategies among players. In other words, for diversity, all tools should have strengths and weaknesses.

Table 1. Example of our construction with $m = 4$, $n = 6$, and $M = 6^4 + 1$.

c	1	1	1	1	1	1
A	1	2	3	4	5	6
	$M - 1^2$	$M - 2^2$	$M - 3^2$	$M - 4^2$	$M - 5^2$	$M - 6^2$
	1^3	2^3	3^3	4^3	5^3	6^3
	$M - 1^4$	$M - 2^4$	$M - 3^4$	$M - 4^4$	$M - 5^4$	$M - 6^4$

We end this section with a high-level overview of our approach for establishing our upper and lower bounds. All the undefined terminology used here will be defined in more detail in later sections. We prove our upper bound Theorem 1 using a sequence of transformations.

We first introduce the intermediate concept of an *equality loadout* problem that replaces the inequality constraint $Ax \leq b$ with an equality $Ax = b$. We show that for a fixed design (A, c) and for every dimension k, the number of k-loadouts is less than the number of k-equality loadouts (Lemma 1). This allows us to focus on proving an upper bound on the number of equality loadouts. Here, we can exploit the dual structure of the equality LP and prove that equality loadouts belong to a cell complex $\Delta_c(A)$ that is characterized by A and c. Importantly, we show that loadouts correspond to *simplicial* cells in this cell complex (Lemma 2). In turn, this allows us to, without loss of generality, assume that $\Delta_c(A)$ is a *triangulation* (as opposed to an arbitrary subdivision), of a cone in the positive orthant of $\mathbb{R}^m$ (Lemma 3). We show that triangulations of cones in the positive orthant of $\mathbb{R}^m$ correspond to triangulations of points in the lower dimension $\mathbb{R}^{m-1}$ (Lemma 4). Finally, we show that the simplices in this triangulation can be embedded into faces of a simplicial polytope in $\mathbb{R}^m$. Therefore, any upper bound on the number of faces of polytopes in $\mathbb{R}^m$ implies an upper bound on the number of loadouts. This allows us to invoke the "maximality" of the cyclic polytope with respect to its number of faces mentioned in Sect. 2.1. Therefore, the number faces of the cyclic polytope of dimension m bounds the number of faces in a polytope of dimension m, and implies a bound on the number of equality loadouts. We also carefully count the number of extraneous faces added through our transformations, by invoking a bound on the minimal number of faces a polytope can have, which allows us to derive tight bounds for small values of m (Lemma 5).

To prove our complementing lower bound Theorem 2, we first explicitly provide our design (A, c) in Sect. 5.1, which is also inspired by the cyclic polytope. Compared to the cyclic polytope, every even row of the matrix A has been "flipped", as we show in the proof, which we now outline. First, we focus on the dual program of $LP(A, c, b)$ and present a sufficient condition (Definition 4) for loadouts in terms of dual variables (Lemma 7). We show that by taking hyperplanes corresponding to the facets of the cyclic polytope in dimension m, one can attempt to construct dual variables that satisfy the sufficient condition (Lemma 8). Our aforementioned "flipping" of the even rows in A is crucial to this construction of the dual variables. We show that as long as the facet of the cyclic polytope is of the "odd" parity, the constructed dual variables will indeed be sufficient (Lemma 9), and hence such a facet and all of the faces contained within it correspond to loadouts. To be more precise, we require odd parity when m is even, and even parity when m is odd. What we mean by the parity of a facet will be made clear later. Therefore, to count the number of k-loadouts, we need to count the number of $(k-1)$-dimensional faces on a cyclic polytope in dimension m that are contained within at least one odd facet. To the best of our knowledge, this is an unsolved problem in the literature. Nonetheless, using Gale's evenness criterion we can map this to a purely combinatorial problem on binary strings. Through some combinatorial bijections, we show that at worst a factor of 4 is lost when one adds the requirement that the $(k-1)$-dimensional face must be contained within at least one odd facet, with the factor improving to 2 if m is odd, and improving to 1 if k is small. These arguments form the cases in Theorem 2. We should note that generally, a cyclic polytope does not have an equal number of odd and even facets. Therefore, one should not expect this factor to always be 2.

3   Preliminaries

We present terminology we use in the proofs of both Theorems 1 and 2. Additional terminology needed in the proof of only one of these results is found in the relevant sections.

A *d-simplex* is a d-dimensional polytope that is the convex hull of $d + 1$ affinely independent points. For instance, a 0-simplex is a point, a 1-simplex is a line segment and 2-simplex is a triangle. For a matrix $A = (A_1, \ldots, A_n)$ of rank m, let $\mathrm{cone}(A) = \mathrm{cone}(\{A_1, \ldots, A_n\})$ represent the closed convex polyhedral cone $\{Ax \mid x \in \mathbb{R}^n_{\geq 0}\}$. We use the notation $\mathrm{cone}(C)$ to denote the cone generated by the columns indexed by $C \subseteq [n]$. If $C \subseteq [n]$ is a subset of indices, the *relative interior* of C is the relatively open (i.e., open in its affine hull) convex set

$$\mathrm{relint}_A(C) \triangleq \Big\{ \sum_{j \in C} \lambda A_j \mid \lambda_j > 0 \text{ for all } j \in C, \text{ and } \sum_{j \in C} \lambda_j = 1 \Big\}.$$

A subset F of polytope P is a *face* if there exists $\alpha \in \mathbb{R}^n$ and $\beta \in \mathbb{R}$ such that $\alpha^\top x + \beta \leq 0$ for all $x \in P$ and $F = \{x \in P \mid \alpha^\top x + \beta = 0\}$. If $\dim(F) = k$ then F is called a *k-dimensional face* or *k-face*. The faces of dimensions 0, 1, and $\dim(P) - 1$ are called vertices, edges, and facets, respectively. Furthermore, we say that F is face of C, where $F, C \subseteq [n]$, when $\mathrm{cone}(F)$ is a face of $\mathrm{cone}(C)$. We define a polyhedral subdivision of $\mathrm{cone}(A)$ as follows.

Definition 1. *Let $A = (A_1, \ldots, A_n)$ be a matrix of rank m. A collection $\mathscr{S}$ of subsets of $[n]$ is a polyhedral subdivision of $\mathrm{cone}(A)$ if it satisfies the following conditions:*

- *(CP): If $C \in \mathscr{S}$ and F is a face of C, then $F \in \mathscr{S}$. (Closure Property)*
- *(UP): $\mathrm{cone}(\{1, \ldots, n\}) \subset \bigcup_{C \in \mathscr{S}} \mathrm{cone}(C)$. (Union Property)*
- *(IP): If $C, C' \in \mathscr{S}$ with $C \neq C'$, then $\mathrm{relint}_A(C) \cap \mathrm{relint}_A(C') = \emptyset$. (Intersection Property)*

If the set of indices $\{j_1, \ldots, j_k\}$ belongs to a subdivision of $\mathrm{cone}(A)$, then it is called a *cell* of the subdivision, and if the cone is of dimension k, it is called a k-cell. We note that a polyhedral cone subdivision is completely specified by listing its maximal cells.

Next, we define a special subdivision of $\mathrm{cone}(A)$ as a function of the cost vector c. The cells of this subdivision map to the loadouts of the design (A, c). For $A \in \mathbb{R}^{m \times n}_{\geq 0}$ and $c \in \mathbb{R}^n_{\geq 0}$, we define the polyhedral subdivision $\Delta_c(A)$ of $\mathrm{cone}(A)$ as the family of subsets of $\{1, \ldots, n\}$ such that $C \in \Delta_c(A)$ if and only if there exists a column vector $y \in \mathbb{R}^m$ such that $y^\top A_j = c_j$ if $j \in C$ and $y^\top A_j > c_j$ if $j \in \{1, \ldots, n\} \backslash C$. In such a case, we say C is a cell of $\Delta_c(A)$ and that $\Delta_c(A)$ is a *cell complex*. A cell $C \in \Delta_c(A)$ is simplicial if the column vectors $(A_j)_{j \in C}$ are linearly independent. If all the cells of $\Delta_c(A)$ are simplicial, then we say $\Delta_c(A)$ is a *triangulation*. The maximum size of a simplicial cell is m. The next results shows that $\Delta_c(A)$ is indeed a polyhedral subdivision of $\mathrm{cone}(A)$.

Proposition 1. $\Delta_c(A)$ *is a polyherdal subdivision of* $\mathrm{cone}(A)$.

Intuitively, we can think of the subdivision $\Delta_c(A)$ as follows: take the cost vector c, and use it to lift the columns of A to $\mathbb{R}^{n+1}$ then look at the projection of the upper faces (those faces you would see if you "look from above"). This is illustrated in Example 1.

Example 1. Consider the following matrix and cost vectors

$$A = \begin{pmatrix} 1/4 & 1/2 & 3/4 \\ 1 & 1 & 1 \end{pmatrix}, \quad c_1 = (2, 2.125 + \epsilon, 2.25) \quad \text{and} \quad c_2 = (2, 2.125 - \epsilon, 2.25),$$

where $\epsilon > 0$ is a small constant. The corresponding subdivisions of $\mathrm{cone}(A)$ are

$$\Delta_{c_1}(A) = \{\{1,2\}, \{2,3\}, \{1\}, \{2\}, \{3\}, \emptyset\} \quad \text{and} \quad \Delta_{c_2}(A) = \{\{1,2,3\}, \{1\}, \{3\}, \emptyset\}.$$

For example, to see that $\{1,2\}$ is a cell of $\Delta_{c_1}(A)$, we consider $y = (0.5 + 4\epsilon, 1.875 - \epsilon)$. One can verify that $y^\top A_1 = c_1$ and $y^\top A_2 = c_2$, while $y^\top A_3 > c_3$. We observe that for the cost vector c_1, the cell $\{1,2\}$ is simplicial, while for c_2, the cell $\{1,2,3\}$ is not simplicial.

In our definition of simplicial cell, we mentioned that if all the cells in the subdivision $\Delta_c(A)$ are simplicial, then $\Delta_c(A)$ is called a triangulation. More generally, a triangulation of cones is a cone subdivision where all the cells are simplicial (the columns of every cell are linearly independent). We will also define the notion of triangulations of point configurations, that is sets of points whose convex hull is subdivided into simplices. The formal definition mirrors that of polyhedral subdivisions and can be found in the full version of the paper.

Definition 2 (Cyclic Polytope). *The cyclic polytope* $\mathcal{C}(n, d)$ *is defined as the convex hull of* n *distinct vertices on the moment curve* $t \mapsto (t, t^2, \ldots, t^d)$. *The precise choice of which* n *points on this curve are selected is irrelevant for the combinatorial structure of this polytope.*

Definition 3 (f-vector). *The* f*-vector of a* d*-dimensional polytope* P *is given by* $(f_0(P), \ldots, f_{d-1}(P))$, *where* $f_i(P)$ *is the number of* i*-dimensional faces in the* d*-dimensional polytope for all* $i = 0, \ldots, d - 1$. *For instance, a 3-dimensional cube has eight vertices, twelve edges, and six facets, so its* f*-vector is* $(f_0(P), f_1(P), f_2(P)) = (8, 12, 6)$.

4 Upper Bound (Proof of Theorem 1)

Throughout this section we fix positive integers $n > m \geq 2$ and $A \in \mathbb{R}^{m \times n}_{\geq 0}, c \in \mathbb{R}^n_{\geq 0}$. We start by formally introducing the *equality* loadout problem. We consider the parametric family of linear programming problems with equality constraints

$$LP_=(A, c, b): \quad \max\{c^\top x \mid Ax = b, x \geq 0\},$$

By analogy to the definition of loadouts in Sect. 2, an *equality loadout* is defined as a subset of indices $L \subseteq \{1, \ldots, n\}$ such that there exists a resource vector b for which $LP_=(A, c, b)$ has a unique optimal solution x^* such that $\operatorname{supp}(x^*) = L$. If $|L| = k$ then we say that L is a k-equality loadout. Given A and c and an integer $k \in [m]$, let $\mathcal{L}_=^k(A, c)$ denote the family of all equality loadouts L of dimension k. Finally, $\mathcal{L}_=(A, c)$ denotes the family of equality loadouts of all dimensions given A and c. Namely, $\mathcal{L}_=(A, c) \triangleq \cup_{k=1}^m \mathcal{L}_=^k(A, c)$. The following proposition bounds the number of loadouts by the number of equality loadouts, for fixed A and c.

Lemma 1. *For every* $A \in \mathbb{R}_{\geq 0}^{m \times n}$, $c \in \mathbb{R}_{\geq 0}^n$ *and* $k \in [m]$, $\mathcal{L}^k(A, c) \subseteq \mathcal{L}_=^k(A, c)$.

In the rest of this section, assume without loss of generality that A is a full-row rank matrix. We present, for all $k \in [m]$, an upper bound for the number $|\mathcal{L}_=^k(A, c)|$ of equality loadouts of size k with respect to the design (A, c).

Some of the results of this section are known in the literature (an excellent reference is the textbook [4]), but we present them using our notation and adapted to the loadout terminology. We provide proofs for clarity and of our a desire to be as self-contained as possible. The proofs are also suggestive of some aspects of our later constructions in Sect. 5.

From Equality Loadouts to Triangulations. The following result links the optimal solutions of $LP_=(A, c, b)$ to the cells of subdivision $\Delta_c(A)$.

Proposition 2. *([16], Lemma 1.4) The optimal solutions x to $LP_=(A, c, b)$ are the solutions to the problem*

$$\text{Find } x \in \mathbb{R}^n \text{ s.t. } Ax = b, x \geq 0 \text{ and } \operatorname{supp}(x) \text{ lies in a cell of } \Delta_c(A). \quad (3)$$

Lemma 2. *A subset $L \subseteq [n]$ is a loadout of (A, c) if and only if it is a simplicial cell in the subdivision $\Delta_c(A)$.*

The lemma above implies that we can focus on the simplicial cells of the subdivision $\Delta_c(A)$. We next show that we can consider without loss of generality choices of c where all the cells of $\Delta_c(A)$ are simplicial. The idea is that if $\Delta_c(A)$ has some non-simplicial cells, then we can "perturb" the cost vector c to some c' and transform at least one non-simplicial cell into one or more simplicial cells. This perturbation conserves all the simplicial cells of $\Delta_c(A)$ and thus the number of equality loadouts for the design (A, c') cannot be less than the number of equality loadouts for the design (A, c). Without loss of optimality, we can ignore cost vectors c that give rise to non-simplicial cells. We first define the notion of refinement that formalizes the "perturbation" of c.

Given two cell complexes $\mathcal{C}_1$ and $\mathcal{C}_2$, we say that $\mathcal{C}_1$ refines $\mathcal{C}_2$ if every cell of $\mathcal{C}_1$ is contained in a cell of $\mathcal{C}_2$. [4, Lemma 2.3.15] shows that if $c' = c + \epsilon \cdot e$ is perturbation of c with $\epsilon > 0$ sufficiently small and $e = (1, \ldots, 1)$, then the new subdivision $\Delta_{c'}(A)$ refines $\Delta_c(A)$. Since $\Delta_{c'}(A)$ refines $\Delta_c(A)$, then $\Delta_{c'}(A)$ will have more cells. However, it is not clear if $\Delta_{c'}(A)$ will have more simplicial cells than $\Delta_c(A)$. We show in the following lemma that this is the case.

Lemma 3. *A refinement of Δ_c adds to the number of simplicial cells in Δ_c.*

In [4, Corollary 2.3.18], it is shown $\Delta_c(A)$ can be refined to a triangulation within a finite number of refinements (suffices for c' to be generic). Therefore, the lemma above implies that in order to maximize the number of loadouts for any dimension $k \leq m$, we can restrict attention to designs (A, c) such that $\Delta_c(A)$ is a triangulation without loss of generality.

We observe that since the matrix $A \in \mathbb{R}_{\geq 0}^{m \times n}$ has all nonnegative entries, $\mathrm{cone}(A)$ is contained entirely in the positive orthant and therefore cannot contain a line. Cones that do not contain lines are called *pointed*. The following lemma shows that triangulations of pointed cones in dimension m are equivalent to triangulations of a non-restricted set of points (columns) in dimension $m - 1$. This implies that equality loadouts can be seen as cells of a triangulation of a point configuration.

Lemma 4 ([1], Theorem 3.2). *Every triangulation $\mathcal{T}$ of a pointed cone of dimension m can be considered as a triangulation $\mathcal{T}'$ of a point configuration of dimension $m - 1$ such that for $1 \leq k \leq m$, the k-simplices of $\mathcal{T}$ map to $(k - 1)$-simplices of $\mathcal{T}'$.*

Lemma 4 implies that equality loadouts of dimension k correspond to $(k-1)$-simplices in a triangulation of a point configuration in dimension $m - 1$.

From Cells of a Triangulation to Faces of a Polytope ([4], Corollary 2.6.5). We now show that any n-point triangulation in $\mathbb{R}^{m-1}$ can be embedded onto the boundary of an $(n+1)$-vertex polytope in $\mathbb{R}^m$, in a way such that $(k-1)$-simplices in the triangulation correspond to $(k - 1)$-faces on the polytope. We then apply the cyclic polytope upper bound on the number of $(k - 1)$-faces on any $(n+1)$-vertex polytope in $\mathbb{R}^m$ to establish our result. To get a tighter bound, we carefully subtract the "extraneous" faces added from the embedding that did not correspond to $(k-1)$-simplices in the original triangulation. We lower bound the number of such extraneous faces using the lower bound theorem of [10].

Let $\mathcal{T}$ denote the original n-point triangulation in $\mathbb{R}^{m-1}$. We will use $\mathrm{conv}\,\mathcal{T}$ to refer to the polytope obtained by taking the convex hull of all the faces in $\mathcal{T}$. Let $g_{k-1}(\mathcal{T})$ denote the number of $(k - 1)$-simplices in the triangulation $\mathcal{T}$. We embed $\mathrm{conv}\,\mathcal{T}$ into a polytope P in $\mathbb{R}^m$ as follows. Let $z^1, \ldots, z^n \in \mathbb{R}^{m-1}$ denote the vertices in triangulation $\mathcal{T}$. We now define the following lifted points in $\mathbb{R}^m$. For all $i = 1, \ldots, n$, let $\underline{z}^i$ denote the point $(z_1^i, \ldots, z_{m-1}^i, 0)$. For all $i = 1, \ldots, n$, let $\bar{z}^i$ denote the point $(z_1^i, \ldots, z_{m-1}^i, \epsilon)$, for some fixed $\epsilon > 0$. Let $\epsilon > 0$, and replace each point $\underline{z}^i$ that is in the interior of $\mathrm{conv}(\{\underline{z}^1, \ldots, \underline{z}^n\})$ by the "lifted" point $\bar{z}^i = (z_1^i, \ldots, z_{m-1}^i, \epsilon)$. The points on the boundary of $\mathrm{conv}(\{\underline{z}^1, \ldots, \underline{z}^n\})$ are not lifted. Let S be the set of the n points in $\mathbb{R}^m$ after lifting. Let $\mathcal{S}_m$ be the unit sphere of $\mathbb{R}^m$ with center at the origin, and S' be the projection of S onto $\mathcal{S}_m$, where every point is projected along the line connecting the point to the center of the sphere. The set S' has the property that all the points that are on the "equator" hyperplane $z_m = 0$ are exactly the projections of the points of S on the boundary of $\mathrm{conv}(S)$ (the points that were not lifted). The other points of S' are in the "northern hemisphere" (the half space $x_m > 0$). The final step is to adjoin the boundary points to the "south pole", $(0, \ldots, 0, -1) \in \mathbb{R}^m$. Let P be the resulting polytope, i.e., $P = \mathrm{conv}(S')$.

The next lemma shows that for $2 \leq k \leq m$, the $(k-1)$-dimensional faces of P are either $(k-1)$-simplices of $\mathcal{T}$, or $(k-2)$-faces of $\mathcal{T}$ that were adjoined to the south pole.

Lemma 5. *For $2 \leq k \leq m$, we have $f_{k-1}(P) = g_{k-1}(\mathcal{T}) + f_{k-2}(\mathcal{T})$.*

The previous lemma implies that $g_{k-1}(\mathcal{T}) = f_{k-1}(P) - f_{k-2}(\mathcal{T})$. Since P has $n+1$ points, we know from the upper bound theorem that $f_{k-1}(P) \leq f_{k-1}(\mathcal{C}(n+1,m))$. Therefore, $g_{k-1}(\mathcal{T}) \leq f_{k-1}(\mathcal{C}(n+1,m)) - f_{k-2}(\mathcal{T})$, and all we need is a lower bound on $f_{k-2}(\mathcal{T})$. The following lemma uses the lower bound theorem (Theorem 1.1, [10]) to establish a lower bound on $f_{k-2}(\mathcal{T})$. The lower bound theorem presents a lower bound on the number of faces in every dimension among all polytopes of dimension d over p points, for $d \geq 2$ and $p \geq 2$.

Lemma 6. *For $2 \leq k \leq m$, we have $g_{k-1}(\mathcal{T}) \leq f_{k-1}(\mathcal{C}(n+1,m)) - \binom{m}{k-1}$.*

See the full version of the paper to see how these lemmas come together to prove Theorem 1.

5   General Lower Bound (Proof of Theorem 2)

Throughout this section, we fix positive integers $n > m \geq 4$, and explicitly present designs (A, c) that have the number of k-loadouts promised in Theorem 2 for all $k \leq m$. For $m = 2$ and $m = 3$, the exactly optimal designs are presented in the full version. All of the designs constructed in this paper will satisfy the property that A has linearly independent rows, hence we assume in the rest of this section that A is a full row rank matrix.

5.1   Construction Based on Moment Curve

Let $t_1, \ldots, t_n$ be arbitrary real numbers satisfying $0 < t_1 < t_2 < \ldots < t_n$. Let M be an arbitrary constant satisfying $M \geq t^m$. We define the design (A, c) so that $c = (1, \ldots, 1) \in \mathbb{R}^n$ and $A = [v'_m(t_1), \ldots v'_m(t_n)]$. where

$$t \mapsto v'_m(t) = \left(t, M - t^2, t^3, M - t^4, \ldots, \frac{(-1)^m + 1}{2} M - (-1)^m t^m \right)^\top \in \mathbb{R}^m.$$

Note that the final row equals $M - t^m$ if m is even, or t^m if m is odd.

For any such values $t_1, \ldots, t_n$ and M, we will get a design that satisfies our Theorem 2. We set all the entries of the cost vector c to 1 to simplify computations. It is not a requirement and the construction would still hold by setting c_j to be any positive number and scaling the column A_j by a factor of c_j. We will also later show that any of these constructions satisfy our assumption of A having full row rank.

Motivation Behind the Construction. Let P be the convex hull of $\{v'_m(t_1), \ldots v'_m(t_n)\}$. Let $t \mapsto v_m(t) = (t, t^2, t^3, \ldots, t^m)^T \in \mathbb{R}^m$ denote the m-dimensional original moment curve the defines the cyclic polytope.

The choice of the curve v'_m is motivated by role the cyclic polytope plays in our corresponding upper bound Theorem 1. In fact, Theorem 1 shows that the number of k-dimensional loadouts is less than the number of $(k-1)$-dimensional faces of the cyclic polytope $\mathcal{C}(n+1, m)$ (for $2 \le k \le m$). An ideal lower bound proof would connect the number of loadouts to the number of faces of the cyclic polytope. However, simply setting the columns of the constraint matrix A to be points on the moment curve of the cyclic polytope does not guarantee the existence of loadouts. We therefore, introduce the curve v'_m that describes a "rotated" cyclic polytope and show that it is rotated to ensure that the supporting normals of "half" of the facets are nonnegative. We use these rotated facets to construct a number of loadouts that asymptotically matches the upper bound. The rotation is performed by multiplying the even coordinates of the moments curve by -1, and we use a sufficiently big constant M to ensure the positivity of the new constraint matrix.

5.2   Dual Certificate for Loadouts

Using LP duality, we derive a sufficient condition for subsets of $[n]$ to be loadouts.

Definition 4. *A set $C \subseteq [n]$ is an inequality cell of the design (A, c) if there exists a variable $y \in \mathbb{R}^m$ such that*

$$y_i > 0, \quad \forall\, i \in [m];$$
$$y^\top A_j = c_j, \quad \forall\, j \in C;$$
$$y^\top A_j > c_j, \quad \forall\, j \notin C.$$

Here, y can be interpreted as a dual variable. However, in contrast to the definition of a cell that features in Proposition 2, here we require $y > 0$. This is because non-negativity is needed for y to be feasible in the dual when the LP has an inequality constraint $Ax \le b$ instead of an equality constraint as considered in Proposition 2.

Lemma 7. *Suppose $C \subseteq [n]$ is an inequality cell with $|C| = m$. Then every non-empty subset of C is a loadout.*

To establish Lemma 7, we show that for every subset $L \subseteq C$, y will verify the complementary slackness constraints with a primal variable x that has support equal to L. This establishes the optimality of x, and to show its uniqueness, we use the assumption that A has a full row rank equal to m.

Note that this lemma only works in one direction. If L is a loadout, it is not clear that we can find a corresponding dual certificate that satisfies Definition 4. However, for our construction, we only need the direction proved in the lemma.

In order to prove Theorem 2, we consider our design from Sect. 5.1, and show that there are many inequality cells of cardinality m. To do so, we take an arbitrary $C \subseteq [n]$ with $|C| = m$ and consider the hyperplane that goes through the m points $\{v'_m(t_j) \mid j \in C\}$. We show in Lemma 8 that the coefficients of the equation for this hyperplane have the same sign. We then use these coefficients to construct a candidate dual vector y. The last step (Lemma 9) is to show that when the hyperplane satisfies a *gap parity* combinatorial condition, this dual vector will indeed satisfy Definition 4, certifying that C is an inequality cell.

Lemma 8. *Let* $C = \{j_1, \ldots, j_m\} \subseteq [n]$ *be a subset of m indices with $j_1 < \cdots < j_m$. The equation*

$$\det \begin{pmatrix} 1 & \cdots & 1 & 1 \\ v'_m(t_{j_1}) & \cdots & v'_m(t_{j_m}) & y \end{pmatrix} = 0 \tag{4}$$

defines a hyperplane in variable $y \in \mathbb{R}^m$ *that passes through the points* $v'_m(t_{j_1}), \ldots, v'_m(t_{j_m})$. *Furthermore if equation* (4) *is written in the form $\alpha_1 y_1 + \ldots \alpha_m y_m - \beta = 0$, then we have $\alpha_1 \neq 0, \ldots \alpha_m \neq 0, \beta \neq 0$, and*

$$sign(\alpha_1) = \ldots = sign(\alpha_m) = sign(\beta) = (-1)^{\lfloor \frac{m}{2} \rfloor + m + 1},$$

where $sign(\alpha_j)$ is equal to 1 if $\alpha_j > 0$ and equal to -1 otherwise.

We now consider a subset $C = \{j_1, \ldots, j_m\} \subseteq [n]$ with $j_1 < \cdots < j_m$, such that the corresponding hyperplane has equation $\alpha_1 y_1 + \ldots \alpha_m y_m - \beta = 0$, as defined above. The previous lemma shows that the dual variable $y = \alpha/\beta$ satisfies $y_i > 0$ for all $i \in [m]$. We now proceed towards a *gap parity condition* on the subset C under which setting $y = \alpha/\beta$ also satisfies Definition 4.

Definition 5. *(Gaps). For a set $C \subset [n]$, a gap of C refers to an index $i \in [n] \setminus C$. A gap i of C is an even gap if the number of elements in C larger than i is even, and i is an odd gap otherwise.*

Definition 6. *(Facets and Gap Parity). A subset $C \subseteq [n]$ is called a facet if $|C| = m$ and either: (i) all of its gaps are even; or (ii) all of its gaps are odd. If all of its gaps are even, then we call C an even facet and define $g(C) = 2$. On the other hand, if all of its gaps are odd, then we call C an odd facet and define $g(C) = 1$. We let $g(C) \in \{1, 2\}$ denote the gap parity of a facet C, with $g(C)$ being undefined if C is not a facet.*

Lemma 9. *Every facet C with $g(C) \not\equiv m \pmod{2}$ is an inequality cell.*

The proofs of Lemmas 8 and 9 require some technical developments on the sub-determinants of A. The outline of the proof of Lemma 9 is as follows. To show that $C = \{j_1, \ldots, j_m\}$ is an inequality cell, we consider the dual certificate $y = \frac{\alpha}{\beta}$ where $\alpha_1 y_1 + \ldots \alpha_m y_m - \beta = 0$ is the equation of C. By Lemma 8, β and α have the same signs, and that $\beta \neq 0$ and $\alpha_i \neq 0$ for $i \in [m]$. Therefore, $y_i > 0, \quad \forall i \in [m]$. For $j \in C$,

$$y^\top v'_m(t_j) = \frac{\alpha^\top v'_m(t_j)}{\beta} = \frac{\beta}{\beta} = 1 = c_j.$$

The last step is to show $y^\top v'_m(t_j) > c_j$ for $j \notin C$.

5.3 Counting the Number of k-Loadouts

The preceding duality certificates combine to provide a purely combinatorial lower bound on the number of k-loadouts in our construction. Indeed, Lemma 7 shows a subset $L \subseteq [n]$ with $|L| = k$ is a k-loadout as long as L is contained within some inequality cell C. In turn, Lemma 9 shows that C is an inequality cell as long as it is a facet with gap parity opposite to m.

The last step is to count the number of k-subsets that are contained within at least one facet with gap parity opposite to m, for all $k = 1, \ldots, m$. The challenge is not to over-count these subsets because such a subset can be contained in different facets. We map the counting of these subsets to a purely combinatorial problem on binary strings. Through some combinatorial bijections, we show that at worst a factor of 4 is lost when one adds the requirement that the $(k-1)$-dimensional face must be contained within at least one odd facet, with the factor improving to 2 if m is odd, and improving to 1 if k is small. The full proofs are deferred to the full version.

6 Conclusion

We study the novel problem of diversity maximization. This problem can be motivated by the video game design context where designing for diversity is one of its core design philosophies. We model this diversity optimization problem as a parametric linear programming problem where we are interested in the diversity of supports of optimal solutions. Using this model, we establish upper bounds and construct designs that match this upper bound asymptotically.

To our knowledge, this is the first paper to systematically study the question of "diversity maximization" as we have defined it here. The goal here is "diverse-in diverse-out", if two players have right-hand resource vectors, they will optimally play different strategies. We believe there could be other applications for "diverse-in diverse-out" optimization problems. Consider, for example, a diet problem where a variety of ingredients are used in the making of meals, depending on different availability in resources. We leave this exploration for future work. There are also natural extensions to our model and analysis that could be pursued. For instance, we have studied the linear programming version of the problem. An obvious next step is the integer linear setting, which also arises naturally in the design of games.

Just as in our analysis of the linear program, a deep understanding of the parametric nature of the integer optimization problems is necessary to proceed in the integer setting. [16] introduce a theory of reduced Gröbner bases of toric ideals that play a role analogous to triangulations of cones.

Another compelling extension would involve *mixed*-integer decision sets. This will require a deep appreciation of parametric mixed-integer linear programming, a topic that remains of keen interest in the integer programming community (see, for instance, [5,6]). Finally, another direction is to consider multiple objectives for the player. In our setting, we have assumed a single meaningful objective for the player, such as maximizing the damage of a loadout of weapons.

References

1. Beck, M., Robins, S.: Computing the Continuous Discretely: Integer-Point Enumeration in Polyhedra. Springer, NY (2007). https://doi.org/10.1007/978-0-387-46112-0
2. Chen, N., Elmachtoub, A.N., Hamilton, M.L., Lei, X.: Loot box pricing and design. Management Science (forthcoming) (2020)
3. Chen, Z., et al.: EOMM: An engagement optimized matchmaking framework. In: Proceedings of the 26th International Conference on World Wide Web, pp. 1143–1150 (2017)
4. De Loera, J.A., Rambau, J., Santos, F.: Triangulations: Structures for Algorithms and Applications. Springer, Heidelberg (2010). https://doi.org/10.1007/978-3-642-12971-1
5. Eisenbrand, F., Shmonin, G.: Parametric integer programming in fixed dimension. Math. Oper. Res. **33**(4), 839–850 (2008)
6. Gribanov, D., Malyshev, D., Pardalos, P.: Parametric integer programming in the average case: sparsity, proximity, and FPT-algorithms. arXiv preprint arXiv:2002.01307 (2020)
7. Guo, H., Hao, L., Mukhopadhyay, T., Sun, D.: Selling virtual currency in digital games: implications for gameplay and social welfare. Inf. Syst. Res. **30**(2), 430–446 (2019)
8. Guo, H., Zhao, X., Hao, L., Liu, D.: Economic analysis of reward advertising. Prod. Oper. Manag. **28**(10), 2413–2430 (2019)
9. Huang, Y., Jasin, S., Manchanda, P.: "Level up": leveraging skill and engagement to maximize player game-play in online video games. Information Systems Research **30**(3), 927–947 (2019)
10. Kalai, G.: Rigidity and the lower bound theorem 1. Invent. Math. **88**(1), 125–151 (1987)
11. McMullen, P.: The maximum numbers of faces of a convex polytope. Mathematika **17**(2), 179–184 (1970)
12. Mills, H.: Marginal values of matrix games and linear programs. In: Kuhn, H.W., Tucker, A.W. (eds.) Linear Inequalities and Related Systems, pp. 183–194. Princeton University Press (1956)
13. Ryan, C.T., Sheng, L., Zhao, X.: Selling enhanced attempts. Available at SSRN 3751523 (2020)
14. Saaty, T., Gass, S.: Parametric objective function (part 1). J. Oper. Res. Soc. Am. **2**(3), 316–319 (1954)
15. Sheng, L., Ryan, C.T., Nagarajan, M., Cheng, Y., Tong, C.: Incentivized actions in freemium games. Manufacturing & Service Operations Management (2020)
16. Sturmfels, B., Thomas, R.R.: Variation of cost functions in integer programming. Math. Program. **77**(2), 357–387 (1997)
17. Turner, J., Scheller-Wolf, A., Tayur, S.: Scheduling of dynamic in-game advertising. Oper. Res. **59**(1), 1–16 (2011)
18. Walkup, D., Wets, R.: Lifting projections of convex polyhedra. Pac. J. Math. **28**(2), 465–475 (1969)
19. Williams, A.: Marginal values in linear programming. J. Soc. Ind. Appl. Math. **11**(1), 82–94 (1963)

ReLU Neural Networks of Polynomial Size for Exact Maximum Flow Computation

Christoph Hertrich[1]([✉])[iD] and Leon Sering[2][iD]

[1] London School of Economics and Political Science, London, UK
c.hertrich@lse.ac.uk
[2] ETH Zurich, Zurich, Switzerland
sering@math.ethz.ch

Abstract. This paper studies the expressive power of artificial neural networks with rectified linear units. In order to study them as a model of *real-valued* computation, we introduce the concept of *Max-Affine Arithmetic Programs* and show equivalence between them and neural networks concerning natural complexity measures. We then use this result to show that two fundamental combinatorial optimization problems can be solved with polynomial-size neural networks. First, we show that for any undirected graph with n nodes, there is a neural network (with fixed weights and biases) of size $\mathcal{O}(n^3)$ that takes the edge weights as input and computes the value of a minimum spanning tree of the graph. Second, we show that for any directed graph with n nodes and m arcs, there is a neural network of size $\mathcal{O}(m^2 n^2)$ that takes the arc capacities as input and computes a maximum flow. Our results imply that these two problems can be solved with strongly polynomial time algorithms that solely uses affine transformations and maxima computations, but no comparison-based branchings.

Keywords: Neural Network Expressivity · Strongly Polynomial Algorithms · Minimum Spanning Tree Problem · Maximum Flow Problem

1 Introduction

Artificial neural networks (NNs) achieved breakthrough results in various application domains like computer vision, natural language processing, autonomous driving, and many more [40]. Also in the field of combinatorial optimization (CO), promising approaches to utilize NNs for problem solving or improving classical solution methods have been introduced [7]. However, the theoretical understanding of NNs still lags far behind these empirical successes.

All neural networks considered in this paper are *feedforward neural networks with rectified linear unit (ReLU) activations*, one of the most popular models in practice [19]. These NNs are directed, acyclic, computational graphs in which

---

The full version is available on arXiv: https://arxiv.org/abs/2102.06635.

© The Author(s), under exclusive license to Springer Nature Switzerland AG 2023
A. Del Pia and V. Kaibel (Eds.): IPCO 2023, LNCS 13904, pp. 187–202, 2023.
https://doi.org/10.1007/978-3-031-32726-1_14

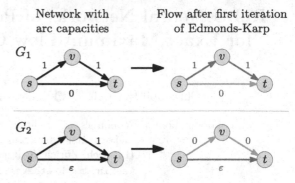

Fig. 2. This example shows that the outcome of one iteration of the Edmonds-Karp algorithm for computing a maximum flow depends discontinuously on the arc capacities. Here, a small adjustment of the capacity of arc st leads to a drastic change of the flow after the first iteration.

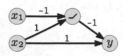

Fig. 1. A small NN with two input neurons x_1 and x_2, a single ReLU neuron labelled with the shape of the ReLU function, and one output neuron y. It computes the function

$$x \mapsto y$$
$$= x_2 - \max\{0, x_2 - x_1\}$$
$$= -\max\{-x_2, -x_1\}$$
$$= \min\{x_1, x_2\}.$$

each edge is equipped with a fixed weight and each node with a fixed bias. Each node (*neuron*) computes an affine transformation of the outputs of its predecessors and applies the ReLU activation function $x \mapsto \max\{0, x\}$ on top. The full NN then computes a function mapping real-valued inputs to real-valued outputs. A simple example is given in Fig. 1.

The neurons are commonly organized in *layers*. The *depth*, *width*, and *size* of an NN are defined as the number of layers, the maximum number of neurons per layer, and the total number of neurons, respectively. An important theoretical question about these NNs is concerned with their expressivity: which functions can be represented by an NN of a certain depth, width, or size?

Neural network expressivity has been thoroughly investigated from an approximation point of view. For example, so-called *universal approximation theorems* [3,11,31] show that every continuous function on a bounded domain can be arbitrarily well approximated with only a single nonlinear layer. However, for a full theoretical understanding of this fundamental machine learning model it is necessary to understand what functions can be *exactly* expressed with different NN architectures. For instance, insights about exact representability have boosted our understanding of the computational complexity of the task to train an NN with respect to both, algorithms [4,36] and hardness results [9,18,20].

It is known that a function can be expressed with a ReLU NN if and only if it is *continuous and piecewise linear* (CPWL) [4]. However, many surprisingly basic questions remain open. For example, it is not known whether two layers of ReLU units (with any width) are sufficient to compute the function $f : \mathbb{R}^4 \to \mathbb{R}$, $x \mapsto \max\{0, x_1, x_2, x_3, x_4\}$ [24, 28].

In this paper we explore another fundamental question within the research stream of exact representability: what are families of CPWL functions that can be represented with ReLU NNs of polynomial size? In other words, using NNs as a model of computation operating on *real* numbers (in contrast to Turing machines or Boolean circuits, which operate on binary encodings), which problems do have polynomial complexity in this model?

Our motivation to study this model stems from a variety of different perspectives, including strongly polynomial time algorithms, arithmetic circuit complexity, parallel computation, and learning theory. We believe that classical combinatorial optimization problems are a natural example to study this model of computation because their algorithmic properties are well understood in each of these areas.

Clearly, if there are polynomial-size NNs to solve a certain problem, and assuming that the weights of these NNs are computable in polynomial time[1], then there exists a strongly polynomial time algorithm for that problem, simply by executing the NN. However, the converse might be false. This is due to the fact that ReLU NNs only allow a very limited set of possible operations, namely affine combinations and maxima computations. In particular, every function computed by such NNs is continuous, making it impossible to realize instructions like a simple **if**-branching based on a comparison of real numbers. In fact, there are related models of computation for which the use of branchings is exponentially powerful [32].

For some CO problems, classical algorithms do not involve comparison-based branchings and, thus, can easily be implemented as an NN. This is, for example, true for many dynamic programs. In these cases, the existence of efficient NNs follows immediately. We refer to Hertrich and Skutella [29] for some examples of this kind. In particular, polynomial-size NNs to compute the length of a shortest path in a network from given arc lengths are possible.

For other problems, like the Minimum Spanning Tree Problem or the Maximum Flow Problem, all classical algorithms use comparison-based branchings. For example, many maximum flow algorithms use them to decide whether an arc is part of the *residual network*. More specifically, in the Edmonds-Karp algorithm a slight perturbation (from 0 to ε) in the capacities can lead to different augmenting path and therefore to a completely different intermediate flow; see Fig. 2. Such a discontinuous behavior can never be represented by a ReLU NN.

---

[1] In circuit complexity language, one would say "if there is a uniform neural network family to solve a certain problem".

1.1   Our Main Results

In order to make it easier to think about NNs in an algorithmic way, we introduce the pseudo-code language *Max-Affine Arithmetic Programs* (MAAPs). We show that MAAPs and NNs are equivalent (up to constant factors) concerning three basic complexity measures corresponding to depth, width, and overall size of NNs. Hence, MAAPs serve as a convenient tool for constructing NNs with bounded size and could be useful for further research about NN expressivity beyond the scope of this paper.

We use this result to prove our two main theorems. The first one shows that computing the value of a minimum spanning tree has polynomial complexity on NNs. The proof is based on a result from subtraction-free circuit complexity [17].

Theorem 1. *For a fixed graph with n vertices, there exists an NN of depth $\mathcal{O}(n \log n)$, width $\mathcal{O}(n^2)$, and size $\mathcal{O}(n^3)$ that correctly maps a vector of edge weights to the value of a minimum spanning tree.*

The second result shows that computing a maximum flow has polynomial complexity on NNs. Since all classical algorithms involve conditional branchings based on the comparison of real numbers, the proof involves the development of a new strongly polynomial maximum flow algorithm which avoids such branchings. While, in terms of standard running times, the algorithm is definitely not competitive with algorithms that exploit comparison-based branchings, it is of independent interest with respect to the structural understanding of flow problems.

Theorem 2. *Let $G = (V, E)$ be a fixed directed graph with $s, t \in V$, $|V| = n$, and $|E| = m$. There exists an NN of depth and size $\mathcal{O}(m^2 n^2)$ and width $\mathcal{O}(1)$ that correctly maps a vector of arc capacities to a vector of flow values in a maximum s-t-flow.*

Let us point out that in case of minimum spanning trees, the NN computes only the objective value, while for maximum flows, the NN computes the actual solution. There is a structural reason for this difference: Due to their continuous nature, ReLU NNs cannot compute a discrete solution vector, like an indicator vector of the optimal spanning tree, because infinitesimal changes of the edge weights would lead to jumps in the output. For the Maximum Flow Problem, however, the optimal flow itself does indeed have a continuous dependence on the arc capacities.

1.2   Discussion of the Results

Before presenting our result in more detail, we discuss the significance and limitations of our results from various perspectives. Due to space constraints, we refer to the full version for a more detailed discussion.

Learning Theory. A standard approach to create a machine learning model usually contains the following two steps. The first step is to fix a particular *hypothesis class.* When using NNs, this means to fix an architecture, that is, the underlying graph of the NN. Then, each possible choice of weights and biases of all affine transformations in the network constitutes one hypothesis in the class. The second step is to run an optimization routine to find a hypothesis in the class that fits given training data as accurately as possible. A core theme in learning theory is to analyse how the choice of the hypothesis class influences different kind of errors made by the machine learning model.

While there exist many attempts to mathematically explain the mysterious success of modern NNs [8], there is still a long way ahead of us. Understanding what CPWL functions are actually contained in the hypothesis classes defined by NNs of a certain size (in particular, polynomial size) is a key insight in this direction. We see our combinatorial, exact perspective as a counterbalance and complement to the usual approximate point of view.

Strongly Polynomial Time Algorithms. As pointed out above, polynomial-size NNs correspond to a subclass of strongly polynomial time algorithms with a very limited set of operations allowed. Given that this subclass stems from one of the most basic machine learning models, our grand vision, to which we contribute with our results, is to understand for different CO problems whether they admit strongly polynomial time algorithms of this type.

Algorithms of this type have not been known before for the two problems considered in this paper. It remains an open question whether such algorithms, and hence, polynomial-size NNs, exist to solve other CO problems for which strongly polynomial time algorithms are known. Can they, for instance, compute the weight of a minimum weight perfect matching in (bipartite) graphs? Can they compute the cost of a minimum cost flow from either node demands or arc costs, while the other of the two quantities is considered to be fixed?

A major open question is also to prove lower bounds on NN sizes. Can we find a family of CPWL functions (corresponding to a CO problem or not) that can be evaluated in strongly polynomial time, but *not* computed by polynomial-size NNs? While proving lower bounds in complexity theory always seems to be a challenging task, we believe that not all hope is lost. For example, in the area of *extended formulations*, it has been shown that there exist problems (in particular, minimum weight perfect matching) which can be solved in strongly polynomial time, but every linear programming formulation to this problem must have exponential size [52]. Possibly, one can show in the same spirit that also polynomial-size NN representations are not achievable.

Boolean Circuits. Even though NNs are naturally a model of real computation, it is worth to have a look at their computational power with respect to Boolean inputs. Interestingly, this makes understanding the computational power of NNs much easier. It is easy to see that ReLU NNs can directly simulate AND-, OR-, and NOT-gates, and thus every Boolean circuit [44]. Hence, in Boolean arith- metics, every problem in P can be solved with polynomial-size NNs.

However, requiring the networks to solve a problem for all possible real- valued inputs seems to be much stronger. Consequently, the class of functions representable with polynomial-size NNs is much less understood than in Boolean arithmetics. Our results suggest that rethinking and forbidding basic algorithmic paradigms (like comparison-based branchings) can help towards improving this understanding.

Arithmetic Circuits. As a circuit model with real-valued computation, ReLU networks are naturally closely related to *arithmetic circuits*. Just like NNs, arithmetic circuits are computational graphs in which each node computes some arithmetic expression (traditionally addition or multiplication) from the outputs of all its predecessors. Arithmetic circuits are well-studied objects in complexity theory [56]. Closer to ReLU NNs, there is a special kind of arithmetic circuits called *tropical circuits* [33]. In contrast to ordinary arithmetic circuits, they contain maximum (or minimum) gates instead of sum gates and sum gates instead of product gates. Thus, they are arithmetic circuits in the max-plus algebra.

A tropical circuit can be simulated by an NN of roughly the same size since NNs can compute maxima and sums. However, neural networks are strictly more powerful than tropical circuits for two reasons: they can realize subtractions (that is, tropical division) by using negative weights and scalar multiplication (tropical exponentiation) with any real number. Thus, lower bounds on the size of tropical circuits do not apply to NNs. A particular example with an exponential gap between NNs and tropical circuits is the computation of the value of a minimum spanning tree. By Jukna and Seiwert [34], no polynomial-size tropical circuit can do this. However, Theorem 1 shows that NNs of cubic size (in the number of nodes of the input graph) are sufficient for this task.

Parallel Computation. Neural networks are naturally a model of parallel computation by performing all operations within one layer at the same time. Without going into detail here, the depth of an NN is related to the running time of a parallel algorithm, its width is related to the required number of processing units, and its size to the total amount of work conducted by the algorithm. One takeaway from this perspective is that, although the result by Arora et al. [4] guarantees that logarithmic depth should be sufficient to compute a maximum flow, this would probably require superpolynomial width and size. The reason is that the Maximum Flow Problem is *P-complete* [22,23], meaning that it probably cannot be efficiently parallelized.

1.3 Further Related Work

Using NNs to solve optimization problems started with so-called *Hopfield networks* in the 1980s [30,35,57], which has also been specialized to the Maximum Flow Problem [2,14,45]. However, the NNs used in these works are conceptually very different from modern feedforward NNs that are considered in this paper.

In recent years interactions between NNs and CO have regained a lot of attention in the literature [7], for example, for boosting MIP solvers [42] and solving specific CO problems [6,16,37,38,47,60]. These approaches usually are of heuristic nature without quality or running time guarantees.

Concerning the expressivity of ReLU neural networks, various trade-offs between depth and width of NNs [4,15,25,27,41,46,51,53,58,59,62] and approaches to count and bound the number of linear regions of a ReLU NN [26,43,50,51,54] have been found. NNs have been studied from a circuit complexity point of view before [5,49,55]. However, these works focus on Boolean circuit complexity of NNs with sigmoid or threshold activation functions. We are not aware of previous work investigating the computational power of ReLU NNs as arithmetic circuits operating on the real numbers.

For an introduction to classical minimum spanning tree and maximum flow algorithms, we refer to textbooks [1,39,61]. The asymptotically fastest known combinatorial maximum flow algorithm due to Orlin [48] runs in $\mathcal{O}(nm)$ time for n nodes and m arcs. Recently, almost linear, weakly polynomial algorithms based on interior point methods have been developed [10]. However, polynomial-size NNs necessarily correspond to strongly polynomial algorithms.

2   Algorithms and Proof Overview

In this section we provide an intuitive overview of how we prove our results. The detailed proofs are deferred to the full version due to space constraints.

Max-Affine Arithmetic Programs. For the purpose of algorithmic investigations of ReLU NNs, we introduce the pseudo-code language *Max-Affine Arithmetic Programs* (MAAPs). A MAAP operates on real-valued variables. The only operations allowed in a MAAP are computing maxima and affine transformations of variables as well as parallel and sequential **for** loops with a *fixed*[2] number of iterations. In particular, no **if** branchings are allowed. With a MAAP A, we associate three complexity measures $d(A)$, $w(A)$, and $s(A)$, which can easily be calculated from a MAAP's description. The intuition behind these measures is that they correspond (up to constant factors) to the depth, width, and size of an NN computing the same function as the MAAP does. We formalize this intuition by proving the following proposition, which is similar to the transformation of circuits into *straight-line programs* in Boolean or arithmetic circuit complexity.

Proposition 3. *For a function $f\colon \mathbb{R}^n \to \mathbb{R}^m$ the following is true.*

(i) *If f can be computed by a MAAP A, then it can also be computed by an NN with depth $d(A) + 1$, width $w(A)$, and size $s(A)$.*
(ii) *If f can be computed by an NN with depth $d + 1$, width w, and size s, then it can also be computed by a MAAP A with $d(A) = d$, $w(A) = 2w$, and $s(A) = 4s$.*

The proof of the proposition works by providing explicit constructions to convert a MAAP into an NN (part (i)), and vice versa (part (ii)) while taking care that the different complexity measures translate respectively.

The takeaway from this exercise is that for proving that NNs of a certain size can compute certain functions, it is sufficient to develop an algorithm in the

---

[2] In this context, *fixed* means that the number of iterations cannot depend on the specific instance. It can still depend on the size of the instance (e.g., the size of the graph in case of the two CO problems considered in this paper).

Algorithm 1: MST_n: Compute the value of a minimum spanning tree for the complete graph on $n \geq 3$ vertices.

Input: Edge weights $(x_{ij})_{1 \leq i < j \leq n}$.

1   $y_n \leftarrow \min_{i \in [n-1]} x_{in}$
2   **for each** $1 \leq i < j \leq n-1$ **do parallel**
3      $x'_{ij} \leftarrow \min \{ x_{ij}, \ x_{in} + x_{jn} - y_n \}$
4   **return** $y_n + \text{MST}_{n-1} \left((x'_{ij})_{1 \leq i < j \leq n-1} \right)$

---

form of a MAAP that computes the same function and to bound its complexity measures $d(A)$, $w(A)$, and $s(A)$.

Minimum Spanning Trees. A spanning tree in an undirected graph is a set of edges that is connected, spans all vertices, and does not contain any cycle. For given edge weights, the Minimum Spanning Tree Problem is to find a spanning tree with the least possible total edge weight.

Classical algorithms for the Minimum Spanning Tree Problem, for example Kruskal's or Prim's algorithm, compare the edge weights and use comparison-based branchings to determine the order in which edges are added to the solution. Thus, they cannot be written as a MAAP or implemented as an NN. Instead, Theorem 1 can be shown by translating an arithmetic circuit with additional division gates by Fomin et al. [17] to a tropical circuit with additional subtraction gates. We refer to the full version for more details.

While this tropicalization is already sufficient to justify the existence of polynomial-size NNs to compute the value of a minimum spanning tree, to unveil the algorithmic ideas behind this construction, we provide an equivalent, completely combinatorial proof of Theorem 1, making use of MAAPs and Proposition 3.

Without loss of generality, we restrict ourselves to complete graphs. Edges missing in the actual input graph can be represented with large weights such that they will never be included in a minimum spanning tree. For $n = 2$ vertices, the MAAP simply returns the weight of the only edge of the graph. For $n \geq 3$, our MAAP is given in Algorithm 1.

Let us mention that the use of recursions is just a technicality because for each fixed n, the recursion can be unrolled and the MAAP can be stated explicitly. In each step, one node of the graph is deleted and all remaining edge weights are updated in such a way that the objective value of the minimum spanning tree problem in the original graph can be calculated from the objective value in the smaller graph. This idea of removing the vertices one by one can be seen as the translation of the so-called *star-mesh transformation* used by Fomin et al. [17] into the combinatorial world.

We prove Theorem 1 in the full version by, firstly, showing that Algorithm 1 indeed computes the correct objective value, and secondly, bounding its complexity measures $d(A)$, $w(A)$, and $s(A)$ and applying Proposition 3.

Maximum Flows. For a given directed graph with a source node s, a sink node t, and nonnegative capacities on each arc, the Maximum Flow Problem asks to find a flow value for each arc such that no capacity is exceeded, the inflow equals the outflow at each node except for s and t, and the outflow at s (or equivalently the inflow at t) is maximised.

Since classical maximum flow algorithms rely on conditional branchings based on the comparison of real numbers (for instance, to check which arcs are contained in the residual network), we develop a new maximum flow algorithm in the form of a MAAP (see Algorithms 2 and 3), which then translates to an NN of the claimed size by Proposition 3. In the description of the algorithm, we assume without loss of generality that for each arc $e = uv \in E$ also its reverse arc vu is contained in E and let $\vec{E}$ denote a subset of all arcs containing exactly one arc for each pair of antiparallel arcs. To point out the ability of neural networks to parallelize well, we sometimes use parallel loops even though this does not significantly reduce asymptotic complexity measures in our case.

To explain our algorithm, let us start by recalling the key ideas of the classical Edmonds-Karp-Dinic algorithm [12,13]. The algorithm repeatedly finds a shortest s-t path in the residual graph $G^* = (V, E^*)$, and sends the maximum possible amount of flow on such a path, that is, saturates at least one arc. The algorithm terminates by returning a minimum cut once t cannot be reached from s in the residual graph. The key insight in the analysis is that the distance from s to t in the residual graph is non-decreasing, and strictly increases within at most m such iterations. Thus, the number of iterations can be bounded by $\mathcal{O}(nm)$.

A shortest path can be characterized by *distance labels*. The vector $d \in \mathbb{R}_+^V$ is a distance labelling if $d(s) = 0$ and $d(v) \leq d(u) + 1$ for every residual arc $uv \in E^*$. If there exists an s-t path P such that $d(v) = d(u) + 1$ for every arc in P, then P is a shortest path. Identifying a shortest path is equivalent to finding distance labels and such a path. We note that the preflow-push algorithm [21] explicitly relies on using distance labels and pushing flow on residual arcs uv with $d(v) = d(u) + 1$. However, finding such a labelling requires **if**-branchings as it needs to identify the arcs in E^*, that is, arcs with positive residual capacity.

At a high level, our algorithm is similar, but it avoids knowing the arcs in the residual graph and the length k of the shortest residual s-t path explicitly. Instead, we guess k in each iteration of the main procedure (Algorithm 2), making sure that we never overestimate the true length. The guess is initialized as $k = 1$ and, in accordance with the Edmonds-Karp-Dinic analysis, we increment k by one in every m iterations. Based on our guess for k, we use a subroutine `FindAugmentingFlow`$_k$ (Algorithm 3) with the following feature: if the actual shortest path length is exactly k, the subroutine will send flow from s to t on (possibly multiple) paths of length exactly k, saturating at least one arc. If the shortest path is longer than k, nothing happens in the current iteration.

Instead of distance labels, the subroutine computes *fattest path values* $a_{i,v}$ (line 7 to 11) that represent the maximum amount of flow that can be sent from v to t on a path of length exactly i. Such values can be obtained by a simple dynamic program that is easy to implement as a MAAP. Thus, a path

---

Algorithm 2: Compute a maximum flow for a fixed graph $G = (V, E)$.

Input: Capacities $(\nu_e)_{e \in E}$.

// Initializing:

1   for each $uv \in \vec{E}$ do parallel

2      $x_{uv} \leftarrow 0$ // flow; negative value correspond to flow on vu

3      $c_{uv} \leftarrow \nu_{uv}$ // residual forward capacities

4      $c_{vu} \leftarrow \nu_{vu}$ // residual backward capacities

// Main part:

5   for $k = 1, \ldots, n - 1$ do

6   for $i = 1, \ldots, m$ do

7        $(y_e)_{e \in \vec{E}} \leftarrow \texttt{FindAugmentingFlow}_k((c_e)_{e \in E})$

         /* Returns an augmenting flow (respecting the residual
         capacities) that only uses paths of length exactly k and
         saturates at least one arc.   */

         // Augmenting:

8   for each $uv \in \vec{E}$ do parallel

9          $x_{uv} \leftarrow x_{uv} + y_{uv}$

10          $c_{uv} \leftarrow c_{uv} - y_{uv}$

11          $c_{vu} \leftarrow c_{vu} + y_{uv}$

12 **return** $(x_e)_{e \in \vec{E}}$

$(s = v_k, v_{k-1}, \ldots, v_1, v_0 = t)$ of length exactly k is contained in the residual network if and only if $a_{i,v_i} > 0$ for all $i = 1, \ldots, k$. Our algorithm makes sure that we only send flow along arcs that are contained in such paths. In particular, the current iteration will send positive flow if and only if $a_{k,s} > 0$. However, we cannot recover the shortest s-t path with capacity $a_{k,s}$. Therefore, in general, flow will not be sent along a single path and the value of the flow output by $\texttt{FindAugmentingFlow}_k$ might be strictly less than $a_{k,s}$.

After computing the $a_{i,v}$ values, $\texttt{FindAugmentingFlow}_k$ greedily pushes flow from s towards t, using a lexicographic selection rule to pick the next arc to push flow on (line 12 to 22). On the high level, this is similar to the preflow-push algorithm, but using the $a_{i,v}$ values that encode the shortest path distance information implicitly. This may leave some nodes with excess flow; a final cleanup phase (line 23 to 29) is needed to send the remaining flow back to the source s.

An example for the $\texttt{FindAugmentingFlow}_k$-subroutine is given in Fig. 3. We emphasize again that, although the description of the subroutine in the example in Fig. 3 seems to rely heavily on the distance of a node to t, this information is calculated and used only in an implicit way via the precomputed $a_{i,v}$ values. This way, we are able to implement the subroutine without the usage of comparison-based branchings.

The proof of correctness for our algorithm consists of two main steps. The first step is the analysis of the subroutine. This involves carefully showing that the returned flow indeed satisfies flow conservation, is feasible with respect to the residual capacities, uses only arcs that lie on a s-t-path of length exactly k in the

---

Algorithm 3: FindAugmentingFlow$_k$ for a fixed graph $G = (V, E)$ and a fixed length k.

Input: Residual capacities $(c_e)_{e \in E}$.

   // Initializing:

1   **for each** $vw \in \vec{E}$ **do parallel**
2      $z_{vw} \leftarrow 0$ // flow in residual network
3      $z_{wv} \leftarrow 0$
4   **for each** $(i, v) \in [k] \times (V \setminus \{t\})$ **do parallel**
5      $Y_v^i \leftarrow 0$ // excessive flow at v in iteration i (from k to 1)
6      $a_{i,v} \leftarrow 0$ // initialize fattest path values

   // Determining the fattest path values:

7   **for each** $v \in N_t^-$ **do parallel**
8      $a_{1,v} \leftarrow c_{vt}$
9   **for** $i = 2, 3, \ldots, k$ **do**
10   **for each** $v \in V \setminus \{t\}$ **do parallel**
11       $a_{i,v} \leftarrow \max_{w \in N_v^+ \setminus \{t\}} \min \{a_{i-1,w}, c_{vw}\}$

   // Pushing flow of value $a_{k,s}$ from s to t:

12 $Y_s^k \leftarrow a_{k,s}$ // excessive flow at s
13 **for** $i = k, k-1, \ldots, 2$ **do**
14   **for** $v \in V \setminus \{t\}$ *in index order* **do**
15   **for** $w \in N_v^+ \setminus \{t\}$ *in index order* **do**
        // Push flow out of v and into w:
16         $f \leftarrow \min \{Y_v^i, c_{vw}, a_{i-1,w} - Y_w^{i-1}\}$ // value we can push over
          vw such that this flow can still arrive at t
17         $z_{vw} \leftarrow z_{vw} + f$
18         $Y_v^i \leftarrow Y_v^i - f$
19         $Y_w^{i-1} \leftarrow Y_w^{i-1} + f$
20 **for each** $v \in N_t^-$ **do parallel**
    // Push flow out of v and into t:
21     $z_{vt} \leftarrow Y_v^1$
22     $Y_v^1 \leftarrow 0$

   // Clean-up by bounding:

23 **for** $i = 2, 3, \ldots, k-1$ **do**
24   **for** $w \in V \setminus \{t\}$ *in reverse index order* **do**
25   **for** $v \in N_w^- \setminus \{t\}$ *in reverse index order* **do**
26         $b \leftarrow \min \{Y_w^i, z_{vw}\}$ // value we can push backwards along vw
27         $z_{vw} \leftarrow z_{vw} - b$
28         $Y_w^i \leftarrow Y_w^i - b$
29         $Y_v^{i+1} \leftarrow Y_v^{i+1} + b$

30 **for each** $uv \in \vec{E}$ **do parallel**
31     $y_{vw} \leftarrow z_{vw} - z_{wv}$
32 **return** $(y_e)_{e \in \vec{E}}$

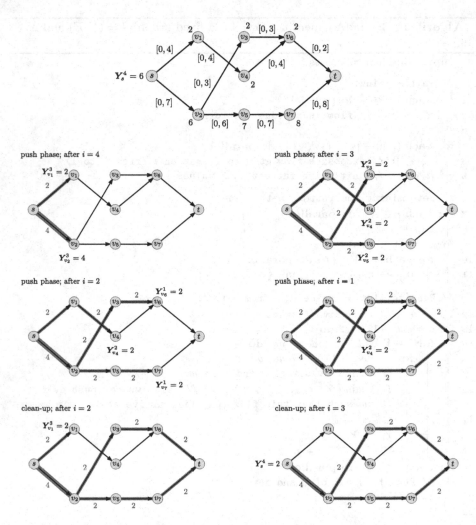

Fig. 3. Example of the `FindAugmentingFlow`$_k$ subroutine for $k = 4$. The edge labels in the top figure are the residual capacity bounds in the current iteration. The first step is to compute the fattest path values $a_{i,v}$, which are depicted as node labels in the top figure. The values Y_v^i always denote the excessive flow of a vertex v with distance i from the sink. All values that are not displayed are zero. At s, we initialize $Y_s^4 = a_{4,s} = 6$. Then, excessive flow is pushed greedily towards the sink, as shown in the four figures in the middle. While doing so, we ensure that at each vertex the arriving flow does not exceed its value $a_{i,v}$. For this reason, flow can get stuck, as it happens at v_4 in this example. Therefore, in a final cleanup phase, depicted in the two bottom figures, we push flow back to the source s. Observe that the result is an s-t-flow that is feasible with respect to the residual capacities, uses only paths of length $k = 4$, and saturates the arc $v_6 t$.

residual network, and most importantly, if such a path exists, it saturates at least one arc. This last property can be shown using the lexicographic selection rule to pick the next arc to push flow on. Note that, in general, the subroutine neither returns a single path (as in the Edmonds-Karp algorithm [13]), nor a blocking flow (as in the Dinic algorithm [12]). The second main step is to show that, nevertheless, the properties of the subroutine are sufficient to ensure that the distance from s to t in the residual network increases at least every m iterations, such that we terminate with a maximum flow after nm iterations.

With the correctness of the whole MAAP at hand, Theorem 2 follows by simply counting the complexity measures $d(A)$, $w(A)$, and $s(A)$, and applying Proposition 3.

Acknowledgements. A large portion of this work was completed while both authors were affiliated with TU Berlin. We thank Max Klimm, Jennifer Manke, Arturo Merino, Martin Skutella, and László Végh for many inspiring and fruitful discussions and valuable comments. Christoph Hertrich acknowledges funding by DFG-GRK 2434 Facets of Complexity and by the European Research Council (ERC) under the European Union's Horizon 2020 research and innovation programme (grant agreement ScaleOpt-757481). Leon Sering acknowledges funding by DFG Excellence Cluster MATH+ (EXC-2046/1, project ID: 390685689).

References

1. Ahuja, R.K., Magnanti, T.L., Orlin, J.B.: Network Flows: Theory, Algorithms, and Applications. Prentice Hall, Upper Saddle River, New Jersey, USA (1993)
2. Ali, M.M., Kamoun, F.: A neural network approach to the maximum flow problem. In: IEEE Global Telecommunications Conference GLOBECOM'91: Countdown to the New Millennium. Conference Record, pp. 130–134 (1991)
3. Anthony, M., Bartlett, P.L.: Neural Network Learning: Theoretical Foundations. Cambridge University Press, Cambridge (1999)
4. Arora, R., Basu, A., Mianjy, P., Mukherjee, A.: Understanding deep neural networks with rectified linear units. In: International Conference on Learning Representations (2018)
5. Beiu, V., Taylor, J.G.: On the circuit complexity of sigmoid feedforward neural networks. Neural Netw. **9**(7), 1155–1171 (1996)
6. Bello, I., Pham, H., Le, Q.V., Norouzi, M., Bengio, S.: Neural combinatorial optimization with reinforcement learning. arXiv:1611.09940 (2016)
7. Bengio, Y., Lodi, A., Prouvost, A.: Machine learning for combinatorial optimization: a methodological tour d'horizon. arXiv:1811.06128 (2018)
8. Berner, J., Grohs, P., Kutyniok, G., Petersen, P.: The modern mathematics of deep learning. arXiv:2105.04026 (2021)
9. Bertschinger, D., Hertrich, C., Jungeblut, P., Miltzow, T., Weber, S.: Training fully connected neural networks is ∃ℝ-complete. arXiv:2204.01368 (2022)
10. Chen, L., Kyng, R., Liu, Y.P., Peng, R., Gutenberg, M.P., Sachdeva, S.: Maximum flow and minimum-cost flow in almost-linear time. arXiv:2203.00671 (2022)
11. Cybenko, G.: Approximation by superpositions of a sigmoidal function. Math. Control Signals Syst. **2**(4), 303–314 (1989)

12. Dinic, E.A.: Algorithm for solution of a problem of maximum flow in a network with power estimation. Soviet Math. Doklady **11**, 1277–1280 (1970)
13. Edmonds, J., Karp, R.M.: Theoretical improvements in algorithmic efficiency for network flow problems. J. ACM **19**(2), 248–264 (1972)
14. Effati, S., Ranjbar, M.: Neural network models for solving the maximum flow problem. Appl. Appl. Math. **3**(3), 149–162 (2008)
15. Eldan, R., Shamir, O.: The power of depth for feedforward neural networks. In: Conference on Learning Theory, pp. 907–940 (2016)
16. Emami, P., Ranka, S.: Learning permutations with Sinkhorn policy gradient. arXiv:1805.07010 (2018)
17. Fomin, S., Grigoriev, D., Koshevoy, G.: Subtraction-free complexity, cluster transformations, and spanning trees. Found. Comput. Math. **16**(1), 1–31 (2016)
18. Froese, V., Hertrich, C., Niedermeier, R.: The computational complexity of ReLU network training parameterized by data dimensionality. arXiv:2105.08675 (2021)
19. Glorot, X., Bordes, A., Bengio, Y.: Deep sparse rectifier neural networks. In: 14th International Conference on Artificial Intelligence and Statistics, pp. 315–323 (2011)
20. Goel, S., Klivans, A.R., Manurangsi, P., Reichman, D.: Tight hardness results for training depth-2 ReLU networks. In: 12th Innovations in Theoretical Computer Science Conference (ITCS '21) (2021)
21. Goldberg, A.V., Tarjan, R.E.: A new approach to the maximum-flow problem. J. ACM (JACM) **35**(4), 921–940 (1988)
22. Goldschlager, L.M., Shaw, R.A., Staples, J.: The maximum flow problem is log space complete for P. Theoretical Comput. Sci. **21**(1), 105–111 (1982)
23. Greenlaw, R., Hoover, H.J., Ruzzo, W.L.: Limits to parallel computation: P-completeness theory. Oxford University Press, Oxford (1995)
24. Haase, C.A., Hertrich, C., Loho, G.: Lower bounds on the depth of integral ReLU neural networks via lattice polytopes. In: International Conference on Learning Representations (ICLR) (2023)
25. Hanin, B.: Universal function approximation by deep neural nets with bounded width and ReLU activations. Mathematics **7**(10), 992 (2019)
26. Hanin, B., Rolnick, D.: Complexity of linear regions in deep networks. In: International Conference on Machine Learning (2019)
27. Hanin, B., Sellke, M.: Approximating continuous functions by ReLU nets of minimal width. arXiv:1710.11278 (2017)
28. Hertrich, C., Basu, A., Di Summa, M., Skutella, M.: Towards lower bounds on the depth of ReLU neural networks. Adv. Neural. Inf. Process. Syst. **34**, 3336–3348 (2021)
29. Hertrich, C., Skutella, M.: Provably good solutions to the knapsack problem via neural networks of bounded size. In: AAAI Conference on Artificial Intelligence (2021)
30. Hopfield, J.J., Tank, D.W.: Neural computation of decisions in optimization problems. Biol. Cybernet. **52**(3), 141–152 (1985)
31. Hornik, K.: Approximation capabilities of multilayer feedforward networks. Neural Netw. **4**(2), 251–257 (1991)
32. Jerrum, M., Snir, M.: Some exact complexity results for straight-line computations over semirings. J. ACM (JACM) **29**(3), 874–897 (1982)
33. Jukna, S.: Lower bounds for tropical circuits and dynamic programs. Theory Comput. Syst. **57**(1), 160–194 (2015)
34. Jukna, S., Seiwert, H.: Greedy can beat pure dynamic programming. Inf. Process. Lett. **142**, 90–95 (2019)

35. Kennedy, M.P., Chua, L.O.: Neural networks for nonlinear programming. IEEE Trans. Circuits Syst. **35**(5), 554–562 (1988)
36. Khalife, S., Basu, A.: Neural networks with linear threshold activations: structure and algorithms. In: International Conference on Integer Programming and Combinatorial Optimization, pp. 347–360. Springer, Cham (2022). https://doi.org/10.1007/978-3-031-06901-7_26
37. Khalil, E., Dai, H., Zhang, Y., Dilkina, B., Song, L.: Learning combinatorial optimization algorithms over graphs. In: Advances in Neural Information Processing Systems 30 (2017)
38. Kool, W., van Hoof, H., Welling, M.: Attention, learn to solve routing problems! In: International Conference on Learning Representations (2019)
39. Korte, B., Vygen, J.: Combinatorial Optimization: Theory and Algorithms, 4th edn. Springer, Heidelberg (2008). https://doi.org/10.1007/3-540-29297-7
40. LeCun, Y., Bengio, Y., Hinton, G.: Deep learning. Nature **521**, 436–444 (2015)
41. Liang, S., Srikant, R.: Why deep neural networks for function approximation? In: International Conference on Learning Representations (2017)
42. Lodi, A., Zarpellon, G.: On learning and branching: a survey. TOP **25**(2), 207–236 (2017). https://doi.org/10.1007/s11750-017-0451-6
43. Montufar, G.F., Pascanu, R., Cho, K., Bengio, Y.: On the number of linear regions of deep neural networks. In: Advances in Neural Information Processing Systems, vol. 27 (2014)
44. Mukherjee, A., Basu, A.: Lower bounds over Boolean inputs for deep neural networks with ReLU gates. arXiv:1711.03073 (2017)
45. Nazemi, A., Omidi, F.: A capable neural network model for solving the maximum flow problem. J. Comput. Appl. Math. **236**(14), 3498–3513 (2012)
46. Nguyen, Q., Mukkamala, M.C., Hein, M.: Neural networks should be wide enough to learn disconnected decision regions. In: International Conference on Machine Learning (2018)
47. Nowak, A., Villar, S., Bandeira, A.S., Bruna, J.: Revised Note on Learning Algorithms for Quadratic Assignment with Graph Neural Networks. arXiv:1706.07450 (2017)
48. Orlin, J.B.: Max flows in O(nm) time, or better. In: Proceedings of the Forty-Fifth Annual ACM Symposium on Theory of Computing (STOC '13), pp. 765–774. Association for Computing Machinery (2013)
49. Parberry, I., Garey, M.R., Meyer, A.: Circuit Complexity and Neural Networks. MIT Press, Cambridge (1994)
50. Pascanu, R., Montufar, G., Bengio, Y.: On the number of inference regions of deep feed forward networks with piece-wise linear activations. In: International Conference on Learning Representations (2014)
51. Raghu, M., Poole, B., Kleinberg, J., Ganguli, S., Dickstein, J.S.: On the expressive power of deep neural networks. In: International Conference on Machine Learning (2017)
52. Rothvoß, T.: The matching polytope has exponential extension complexity. J. ACM (JACM) **64**(6), 1–19 (2017)
53. Safran, I., Shamir, O.: Depth-width tradeoffs in approximating natural functions with neural networks. In: International Conference on Machine Learning (2017)
54. Serra, T., Tjandraatmadja, C., Ramalingam, S.: Bounding and counting linear regions of deep neural networks. In: International Conference on Machine Learning (2018)
55. Shawe-Taylor, J.S., Anthony, M.H., Kern, W.: Classes of feedforward neural networks and their circuit complexity. Neural Netw. **5**(6), 971–977 (1992)

56. Shpilka, A., Yehudayoff, A.: Arithmetic circuits: a survey of recent results and open questions. Now Publishers Inc. (2010)
57. Smith, K.A.: Neural networks for combinatorial optimization: a review of more than a decade of research. INFORMS J. Comput. **11**(1), 15–34 (1999)
58. Telgarsky, M.: Representation benefits of deep feedforward networks. arXiv:1509.08101 (2015)
59. Telgarsky, M.: Benefits of depth in neural networks. In: Conference on Learning Theory, pp. 1517–1539 (2016)
60. Vinyals, O., Fortunato, M., Jaitly, N.: Pointer networks. In: Advances in Neural Information Processing Systems, vol. 28 (2015)
61. Williamson, D.P.: Network Flow Algorithms. Cambridge University Press, Cambridge (2019)
62. Yarotsky, D.: Error bounds for approximations with deep ReLU networks. Neural Netw. **94**, 103–114 (2017)

On the Correlation Gap of Matroids

Edin Husić[1] , Zhuan Khye Koh[2]([✉]) , Georg Loho[3] , and László A. Végh[4]

[1] IDSIA, USI-SUPSI, Lugano, Switzerland
edin.husic@supsi.ch
[2] Centrum Wiskunde & Informatica, Amsterdam, The Netherlands
zhuan.koh@cwi.nl
[3] University of Twente, Enschede, The Netherlands
g.loho@utwente.nl
[4] London School of Economics and Political Science, London, UK
l.vegh@lse.ac.uk

Abstract. A set function can be extended to the unit cube in various ways; the correlation gap measures the ratio between two natural extensions. This quantity has been identified as the performance guarantee in a range of approximation algorithms and mechanism design settings. It is known that the correlation gap of a monotone submodular function is at least $1 - 1/e$, and this is tight for simple matroid rank functions.

We initiate a fine-grained study of the correlation gap of matroid rank functions. In particular, we present an improved lower bound on the correlation gap as parametrized by the rank and girth of the matroid. We also show that for any matroid, the correlation gap of its weighted rank function is minimized under uniform weights. Such improved lower bounds have direct applications for submodular maximization under matroid constraints, mechanism design, and contention resolution schemes.

1 Introduction

A continuous function $h\colon [0,1]^E \to \mathbb{R}_+$ is an *extension* of a set function $f\colon 2^E \to \mathbb{R}_+$ if for every $x \in [0,1]^E$, $h(x) = \mathbb{E}_\lambda[f(S)]$ where λ is a probability distribution over 2^E with marginals x, i.e., $\sum_{S:i\in S} \lambda_S = x_i$ for all $i \in E$. Note that this in particular implies $f(S) = h(\chi_S)$ for every $S \subseteq E$, where χ_S denotes the 0-1 indicator vector of S.

Two natural extensions are the following. The first one corresponds to sampling each $i \in E$ independently with probability x_i, i.e., $\lambda_S = \prod_{i \in S} x_i \prod_{i \notin S} (1 - x_i)$. Thus,

$$F(x) := \sum_{S \subseteq E} f(S) \prod_{i \in S} x_i \prod_{i \notin S} (1 - x_i) \ . \tag{1}$$

This is an extended abstract. The full version of the paper with all proofs is available on arXiv:2209.09896. This project has received funding from the European Research Council (ERC) under the European Union's Horizon 2020 research and innovation programme (grant agreement no. 757481–ScaleOpt).

Z. K. Koh—This work was done while the author was at the London School of Economics.

This is known as the *multilinear extension* in the context of submodular optimization, see [8]. The second extension corresponds to the probability distribution with maximum expectation:

$$\hat{f}(x) := \max_{\lambda} \left\{ \sum_{S \subseteq E} \lambda_S f(S) : \sum_{S \subseteq E : i \in S} \lambda_S = x_i \, \forall i \in E, \sum_{S \subseteq E} \lambda_S = 1, \lambda \geq 0 \right\}. \quad (2)$$

Equivalently, $\hat{f}(x)$ is the upper part of the convex hull of the graph of f; we call it the *concave extension* following terminology of discrete convex analysis [20].

Agrawal, Ding, Saberi and Ye [2] introduced the *correlation gap* as the worst case ratio

$$\mathcal{CG}(f) := \min_{x \in [0,1]^E} \frac{F(x)}{\hat{f}(x)} . \quad (3)$$

It bounds the maximum loss incurred in the expected value of f by ignoring correlations. This quantity plays a fundamental role in stochastic optimization [2, 22], mechanism design [7,18,28], prophet inequalities [11,15,24], and a variety of submodular optimization problems [3,12].

The focus of this paper is on weighted matroid rank functions. For a matroid $\mathcal{M} = (E, \mathcal{I})$ and a weight vector $w \in \mathbb{R}_+^E$, the corresponding *weighted matroid rank function* is given by

$$r_w(S) := \max \{w(T) : T \subseteq S, T \in \mathcal{I}\}.$$

It is monotone nondecreasing and submodular. Recall that a function $f \colon 2^E \to \mathbb{R}$ is *monotone* if $f(X) \leq f(Y)$ for all $X \subseteq Y \subseteq E$, and *submodular* if $f(X) + f(Y) \geq f(X \cap Y) + f(X \cup Y)$ for all $X, Y \subseteq E$.

The correlation gap of a weighted matroid rank function has been identified as the performance guarantee in a range of approximation algorithms and mechanism design settings:

Monotone Submodular Maximization. Calinescu et al. [8] considered the problem of maximizing a sum of weighted matroid rank functions $\sum_{i=1}^m f_i$ subject to a matroid constraint. Using an LP relaxation and pipage rounding [1], they gave a $(1 - 1/e)$-approximation algorithm. This was extended by Shioura [26] to the problem of maximizing a sum of monotone $M^\natural$-concave functions [19]. In [9], a $(1 - 1/e)$-approximation algorithm was obtained for maximizing an arbitrary monotone submodular function subject to a matroid constraint.

A fundamental special case of this model is the *maximum coverage* problem. Given m subsets $E_i \subseteq E$, the corresponding *coverage function* is defined as $f(S) = |\{i \in [m] : E_i \cap S \neq \emptyset\}|$. Note that this is a special case of maximizing a sum of matroid rank functions: $f(S) = \sum_{i=1}^m r_i(S)$ where $r_i(S)$ is the rank function of a rank-1 uniform matroid with support E_i. Even for maximization under a cardinality constraint, there is no better than $(1 - 1/e)$-approximation for this problem unless $P = NP$ (see Feige [16]).

Recently, tight approximations have been established for the special case when the function values $f_i(S)$ are determined by the cardinality of the set S.

Barman et al. [5] studied the *maximum concave coverage* problem: given a monotone concave function $\varphi \colon \mathbb{Z}_+ \to \mathbb{R}_+$ and weights $w \in \mathbb{R}_+^m$, the submodular function is defined as $f(S) = \sum_{i=1}^m w_i \varphi(|S \cap E_i|)$.[1] The maximum coverage problem corresponds to $\varphi(x) = \min\{1, x\}$; on the other extreme, for $\varphi(x) = x$ we get the trivial problem $f(S) = \sum_{j \in S} |\{i \in [m] : j \in E_i\}|$. In [5], they present a tight approximation guarantee for maximizing such an objective subject to a matroid constraint, parametrized by the *Poisson curvature* of the function φ.

This generalizes previous work by Barman et al. [6] which considered $\varphi(x) = \min\{\ell, x\}$ (for $\ell > 1$), motivated by the list decoding problem in coding theory. It also generalizes the work by Dudycz et al. [14] which considered geometrically dominated concave functions φ, motivated by approval voting rules such as Thiele rules, proportional approval voting, and p-geometric rules. In both cases, the obtained approximation guarantees improve over the $1 - 1/e$ factor.

In the full version, we make the observation that the algorithm of Calinescu et al. [8] and Shioura [26] actually has an approximation ratio of $\min_{i \in [m]} CG(f_i)$. We also prove that the Poisson curvature of φ is equal to the correlation gap of the functions $\varphi(|S \cap E_i|)$. Hence, the approximation guarantees in [5,6,14] are in fact correlation gap bounds, and they can be obtained via a single unified algorithm, i.e., the one by Calinescu et al. [8] and Shioura [26]. In particular, the result of Barman et al. [6] which concerned $\varphi(x) = \min\{\ell, x\}$ (for $\ell > 1$) boils down to the analysis of uniform matroid correlation gaps.

Sequential Posted-Price Mechanisms. Following Yan [28], consider a seller with a set of identical services (or goods), and a set E of unit-demand agents. Each agent $i \in E$ has a private valuation v_i for winning the service, and 0 otherwise, where v_i is drawn independently from a known distribution F_i with positive smooth density function over $[0, L]$ for some large L. The seller can only service certain subsets of the agents simultaneously; this is captured by a matroid $\mathcal{M} = (E, \mathcal{I})$ where $\mathcal{I}$ represents the feasible subsets.

Mechanisms like Myerson's mechanism [21] or the VCG mechanism [13,17, 27] have optimal revenue or welfare guarantees, but suffer from complicated formats [4] or high computational overhead [23]. Hence, simple mechanisms are often favoured in practice, such as sequential posted-price mechanisms (SPM), in which the seller makes take-it-or-leave-it price offers to agents one by one. Yan [28] showed that the greedy SPM of Chawla et al. [10] achieves an approximation ratio of $\inf_{w \in \mathbb{R}_+^E} CG(r_w)$, where r_w is the weighted rank function of $\mathcal{M}$ with weights w.

Contention Resolution Schemes. Chekuri et al. [12] introduced contention resolution (CR) schemes as a tool for maximizing a (not necessarily monotone) submodular function f subject to downward-closed constraints, such as matroid constraints, knapsack constraints, and their intersections. Let $\mathcal{M} = (E, \mathcal{I})$ be a matroid imposing one of these constraints. Given a fractional solution x with

---

[1] We note that such functions are exactly the one-dimensional monotone $M^\natural$-concave functions $f_i \colon \mathbb{Z}_+ \to \mathbb{R}_+$.

multilinear extension value $F(x)$, their CR scheme randomly rounds x to an integral solution χ_S where $S \in \mathcal{I}$ such that $\mathbb{E}[\chi_S] \geq \inf_{w \in \mathbb{R}_+^E} C\mathcal{G}(r_w)F(x)$. Here, r_w is again the weighted rank function of $\mathcal{M}$ with weights w.

Motivated by the significance of the correlation gap in algorithmic applications, we study the correlation gap of weighted matroid rank functions. It is well-known that $C\mathcal{G}(f) \geq 1 - 1/e$ for every monotone submodular function f [8]. Moreover, the extreme case $1 - 1/e$ is already achieved by the rank function of a rank-1 uniform matroid as $|E| \to \infty$. More generally, the rank function of a rank-ℓ uniform matroid has correlation gap $1 - e^{-\ell}\ell^\ell/\ell! \geq 1 - 1/e$ [6,28]. Other than for uniform matroids, we are not aware of any previous work that gave better than $1 - 1/e$ bounds on the correlation gap of specific matroids.

First, we show that among all weighted rank functions of a matroid, the smallest correlation gap is realized by its (unweighted) rank function.

Theorem 1. *For any matroid $\mathcal{M} = (E, \mathcal{I})$ with rank function $r = r_1$,*

$$\inf_{w \in \mathbb{R}_+^E} C\mathcal{G}(r_w) = C\mathcal{G}(r).$$

For the purpose of lower bounding $C\mathcal{G}(r_w)$, Theorem 1 allows us to ignore the weights w and just focus on the matroid $\mathcal{M}$. As an application, to bound the approximation ratio of sequential posted-price mechanisms as in [28], it suffices to focus on the underlying matroid. We remark that $\mathcal{M}$ can be assumed to be *connected*, that is, it cannot be written as a direct sum of at least two nonempty matroids. Otherwise, $r = \sum_{i=1}^m r_i$ for matroid rank functions r_i with disjoint supports, and so $C\mathcal{G}(r) = \min_{i \in [m]} C\mathcal{G}(r_i)$. For example, the correlation gap of a partition matroid is equal to the smallest correlation gap of its parts (uniform matroids).

Our goal is to identify the parameters of a matroid which govern its correlation gap. A natural candidate is the rank $r(E)$. However, as pointed out by Yan [28], there exist matroids with arbitrarily high rank whose correlation gap is still $1 - 1/e$, e.g., partition matroids with rank-1 parts. The $1 - e^{-\ell}\ell^\ell/\ell!$ bound for uniform matroids [6,28] is suggestive of girth as another potential candidate. Recall that the *girth* of a matroid is the smallest size of a dependent set. On its own, a large girth does not guarantee improved correlation gap bounds: in the full version, we show that for any $\gamma \in \mathbb{N}$, there exist matroids with girth γ whose correlation gaps are arbitrarily close to $1 - 1/e$.

It turns out that the correlation gap heavily depends on the relative values of the rank and girth of the matroid. Our second result is an improved lower bound on the correlation gap as a function of these two parameters.

Theorem 2. *Let $\mathcal{M} = (E, \mathcal{I})$ be a loopless matroid with rank function r, rank $r(E) = \rho$, and girth γ. Then,*

$$C\mathcal{G}(r) \geq 1 - \frac{1}{e} + \frac{e^{-\rho}}{\rho}\left(\sum_{i=0}^{\gamma-2}(\gamma - 1 - i)\left[\binom{\rho}{i}(e-1)^i - \frac{\rho^i}{i!}\right]\right) \geq 1 - \frac{1}{e}.$$

Furthermore, the last inequality is strict whenever $\gamma > 2$.

Figure 1 illustrates the behaviour of the expression in Theorem 2. For any fixed girth γ, it is monotone decreasing in ρ. On the other hand, for any fixed rank ρ, it is monotone increasing in γ. In the full version, we also give complementing albeit non-tight upper bounds that behave similarly with respect to these parameters. When $\rho = \gamma - 1$, our lower bound simplifies to $1 - e^{-\rho}\rho^\rho/\rho!$, i.e., the correlation gap of a rank-ρ uniform matroid (proven in the full version).

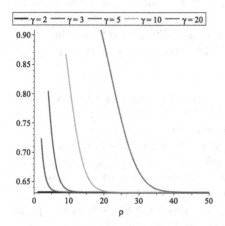 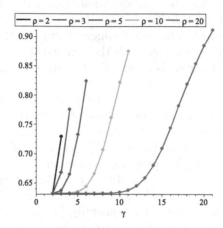

Fig. 1. Our correlation gap bound as a function of the rank ρ and girth γ separately.

The rank and girth have meaningful interpretations in the aforementioned applications. For instance, consider the problem of maximizing a sum of weighted matroid rank functions $\sum_{i=1}^{m} f_i$ under a matroid constraint $(E, \mathcal{J})$. For every $i \in [m]$, let $\mathcal{M}_i$ be the matroid of f_i. In game-theoretic contexts, each f_i usually represents the utility function of agent i. Thus, our goal is to select a bundle of items $S \in \mathcal{J}$ which maximizes the total welfare. If $\mathcal{M}_i$ has girth γ and rank ρ, this means that agent i is interested in $\gamma - 1 \le k \le \rho$ items with positive weights. The special case $\rho = \gamma - 1$ (uniform matroids) has already found applications in list decoding [6] and approval voting [14]. On the other hand, for sequential posted-price mechanisms, if the underlying matroid $\mathcal{M}$ has girth γ and rank ρ, this means that the seller can service $\gamma - 1 \le k \le \rho$ agents simultaneously.

To the best of our knowledge, our results give the first improvement over the $(1 - 1/e)$ bound on the correlation gap of general matroids. We hope that our paper will motivate further studies into more refined correlation gap bounds, exploring the dependence on further matroid parameters, as well as obtaining tight bounds for special matroid classes.

1.1 Our Techniques

We now give a high-level overview of the proofs of Theorem 1 and Theorem 2.

Weighted Rank Functions. The first step in proving both theorems is to deduce structural properties of the points which realize the correlation gap. In Theorem 4, we show that such a point x can be found in the independent set polytope

$\mathcal{P}$. This implies that $\hat{r}_w(x) = w^\top x$ for any weights $w \in \mathbb{R}_+^E$. Moreover, we deduce that $x(E)$ is integral.

To prove Theorem 1, we fix a matroid $\mathcal{M}$ and derive a contradiction for a non-uniform weighting. More precisely, we consider a weighting $w \in \mathbb{R}_+^E$ and a point $x^* \in [0,1]^E$ which give a smaller ratio $R_w(x^*)/\hat{r}_w(x^*) < \mathcal{CG}(r)$. By the above, we can use the simpler form $R_w(x^*)/\hat{r}_w(x^*) = R_w(x^*)/w^\top x^*$. We pick w such that it has the smallest number of different values. If the number of distinct values is at least 2, then we derive a contradiction by showing that a better solution can be obtained by increasing the weights in a carefully chosen value class until they coincide with the next smallest value. The greedy maximization property of matroids is essential for this argument.

Uniform Matroids. Before outlining our proof of Theorem 2, let us revisit the correlation gap of uniform matroids. Let $\mathcal{M} = (E, \mathcal{I})$ be a uniform matroid on n elements with rank $\rho = r(E)$. If $\rho = 1$, then it is easy to verify that the symmetric point $x = (1/n) \cdot \mathbb{1}$ realizes the correlation gap $1 - 1/e$. Since x lies in the independent set polytope, we have $\hat{r}(x) = \mathbb{1}^\top x = 1$. If one samples each $i \in E$ with probability $1/n$, the probability of selecting at least one element is $R(x) = 1 - (1 - 1/n)^n$. Thus, $\mathcal{CG}(r) = 1 - (1 - 1/n)^n$, which converges to $1 - 1/e$ as $n \to \infty$. More generally, for $\rho \geq 1$, Yan [28] showed that the symmetric point $x = (\rho/n) \cdot \mathbb{1}$ similarly realizes the correlation gap $1 - e^{-\rho}\rho^\rho/\rho!$.

Poisson Clock Analysis. To obtain the $(1 - 1/e)$ lower bound on the correlation gap of a monotone submodular function, Calinescu et al. [8] introduced an elegant probabilistic analysis. Instead of sampling each $i \in E$ with probability x_i, they consider n independent *Poisson clocks* of rate x_i that are active during the time interval $[0,1]$. Every clock may send at most one signal from a Poisson process. Let $Q(t)$ be the set of elements whose signal was sent between time 0 and t; the output is $Q(1)$. It is easy to see that $\mathbb{E}[f(Q(1))] \leq F(x)$.

In [8], they show that the derivative of $\mathbb{E}[f(Q(t))]$ can be lower bounded as $f^*(x) - \mathbb{E}[f(Q(t))]$ for every $t \in [0,1]$, where

$$f^*(x) := \min_{S \subseteq E} \left(f(S) + \sum_{i \in E} f_S(i) x_i \right) \qquad (4)$$

is an extension of f such that $f^* \geq \hat{f}$. The bound $\mathbb{E}[f(Q(1))] \geq (1 - 1/e)f^*(x)$ is obtained by solving a differential inequality. Thus, $F(x) \geq \mathbb{E}[f(Q(1))] \geq (1 - 1/e)f^*(x) \geq (1 - 1/e)\hat{f}(x)$ follows.

A Two Stage Approach. If f is a matroid rank function, then $f^* = \hat{f}$ (see Theorem 3). Still, the factor $(1 - 1/e)$ in the analysis of [8] cannot be improved: for an integer $x \in \mathcal{P}$, we lose a factor $(1 - 1/e)$ due to $\mathbb{E}[f(Q(1))] = (1 - 1/e)F(x)$, even though the extensions coincide: $F(x) = \hat{f}(x)$.

Our analysis in Sect. 4 proceeds in two stages. Let $\mathcal{M} = (E, \mathcal{I})$ be a matroid with rank ρ and girth γ. The basic idea is that up to sets of size $\gamma - 1$, our matroid 'looks like' a uniform matroid. Since the correlation gap of uniform matroids is well-understood, we first extract a uniform matroid of rank $\gamma - 1$ from our matroid, and then analyze the contribution from the remaining part separately. More precisely, we decompose the rank function as $r = g + h$, where $g(S) = \min\{|S|, \ell\}$ is the rank function of a uniform matroid of rank $\ell = \gamma - 1$. Note that the residual function $h := f - g$ is not submodular in general, as $h(S) = 0$ for all $|S| \leq \ell$. We will lower bound the multilinear extensions $G(x)$ and $H(x)$ separately. As g is the rank function of a uniform matroid, similarly as above we can derive a tight lower bound on G in terms of its rank $\ell = \gamma - 1$.

Bounding $H(x)$ is based on a Poisson clock analysis as in [8], but is significantly more involved. Due to the monotonicity of h, directly applying the result in [8] would yield $\mathbb{E}[h(Q(1)] \geq (1 - 1/e)h^*(x)$. However, $h^*(x) = 0$ whenever $\mathcal{M}$ is loopless ($\ell \geq 1$), as $h(\emptyset) = 0$ and $h(\{i\}) = 0$ for all $i \in E$. So, the argument of [8] directly only leads to the trivial $\mathbb{E}[h(Q(1))] \geq 0$. Nevertheless, one can still show that, conditioned on the event $|Q(t)| \geq \ell$, the derivative of $\mathbb{E}[H(Q(t))]$ is at least $r^*(x) - \ell - \mathbb{E}[H(Q(t))]$. Let $T \geq 0$ be the earliest time such that $|Q(T)| \geq \ell$, which we call the *activation time* of Q. Then, solving a differential inequality produces $\mathbb{E}[h(Q(1))|T = t] \geq (1 - e^{-(1-t)})(r^*(x) - \ell)$ for all $t \leq 1$.

To lower bound $\mathbb{E}[h(Q(1))]$, it is left to take the expectation over all possible activation times $T \in [0, 1]$. Let $\bar{h}(x) = (r^*(x) - \ell) \int_0^1 \Pr[T = t](1 - e^{-(1-t)})dt$ be the resulting expression. We prove that $\bar{h}(x)$ is concave in each direction $e_i - e_j$ for $i, j \in E$. This allows us to round x to an integer $x' \in [0, 1]^E$ such that $x'(E) = x(E)$ and $\bar{h}(x') \leq \bar{h}(x)$; recall that $x(E) \in \mathbb{Z}$ by Theorem 4. After substantial simplification of $\bar{h}(x')$, we arrive at the formula in Theorem 2, except that ρ is replaced by $x(E)$. So, the rounding procedure effectively shifts the dependency of the lower bound from the value of x to the value of $x(E)$. Since $x(E) \leq \rho$ by Theorem 4, the final step is to prove that the formula in Theorem 2 is monotone decreasing in ρ.

2   Preliminaries

We denote $\mathbb{Z}_+$ and $\mathbb{R}_+$ as the set of nonnegative integers and nonnegative reals respectively. For $n, k \in \mathbb{Z}_+$, $\binom{n}{k} = \frac{n!}{k!(n-k)!}$ if $n \geq k$, and 0 otherwise. For a set S and $i \in S$, $j \notin S$, we use the shorthand $S - i = S \setminus \{i\}$ and $S + j = S \cup \{j\}$. For a function $f : 2^E \to \mathbb{R}$, a set $S \subseteq E$ and an element $i \in E$, let $f_S(i)$ denote the marginal gain of adding i to S, i.e., $f_S(i) := f(S + i) - f(S)$. For $x \in \mathbb{R}^E$ and $S \subseteq E$, we write $x(S) = \sum_{i \in S} x_i$.

Matroids. Let $\mathcal{M} = (E, \mathcal{I})$ be a matroid with rank function $r : 2^E \to \mathbb{Z}_+$. Its *independent set polytope* $\mathcal{P}(r)$ is the convex hull of incidence vectors of independent sets in $\mathcal{I}$. Equivalently, $\mathcal{P}(r) = \{x \in \mathbb{R}_+^E : x(S) \leq r(S) \ \forall S \subseteq E\}$, as shown by Edmonds [25, Theorem 40.2]. We need another classical result by Edmonds [25, Theorem 40.3] on intersecting the independent set polytope with a box.

Theorem 3. *For a matroid rank function* $r : 2^E \to \mathbb{Z}_+$ *and* $x \in \mathbb{R}_+^E$,

$$\max\{y(E) : y \in \mathcal{P}(r), y \leq x\} = \min\{r(T) + x(E \setminus T) : T \subseteq E\}.$$

Probability Distributions. Let $\text{Bin}(n, p)$ denote the binomial distribution with n trials and success probability p. Let $\text{Poi}(\lambda)$ denote the Poisson distribution with rate λ. Recall that $\Pr(\text{Poi}(\lambda) = k) = e^{-\lambda}\lambda^k/k!$ for any $k \in \mathbb{Z}_+$.

Definition 1. *Given random variables X and Y, we say that X is at least Y in the concave order if for every concave function $\varphi : \mathbb{R} \to \mathbb{R}$, we have $\mathbb{E}[\varphi(X)] \geq \mathbb{E}[\varphi(Y)]$ whenever the expectations exist. It is denoted as $X \geq_{cv} Y$.*

Lemma 1 ([6]). *For any $n \in \mathbb{N}$ and $p \in [0, 1]$, we have $\text{Bin}(n, p) \geq_{cv} \text{Poi}(np)$.*

Properties of the Multilinear Extension. For a set function $f : 2^E \to \mathbb{R}$, let $F : [0, 1]^E \to \mathbb{R}$ denote its multilinear extension. We will use the following well-known properties of F, see e.g. [9].

Proposition 1. *If f is monotone, then $F(x) \geq F(y)$ for all $x \geq y$.*

Proposition 2. *If f is submodular, then for any $x \in [0, 1]^E$ and $i, j \in E$, the function $\phi(t) := F(x + t(e_i - e_j))$ is convex.*

3   Locating the Correlation Gap

In this section, given a weighted matroid rank function r_w, we locate a point $x^* \in [0, 1]^E$ on which the correlation gap $\mathcal{CG}(r_w)$ is realized, and derive some structural properties. Using this, we prove Theorem 1, i.e., the smallest correlation gap over all possible weightings is attained by uniform weights. We start with a more convenient characterization of the concave extension of r_w.

Lemma 2. *Let $\mathcal{M} = (E, \mathcal{I})$ be a matroid with rank function r and weights $w \in \mathbb{R}_+^E$. For any $x \in [0, 1]^E$, we have $\hat{r}_w(x) = \max\{w^\top y : y \in \mathcal{P}(r), y \leq x\}$.*

Next, we show that x^* can be chosen to lie in the independent set polytope $\mathcal{P}(r)$; and that $\text{supp}(x^*)$ is a tight set w.r.t. x^*, meaning $x^*(E) = r(\text{supp}(x^*))$.

Theorem 4. *Let $\mathcal{M} = (E, \mathcal{I})$ be a matroid with rank function r. For any weights $w \in \mathbb{R}_+^E \setminus \{0\}$, there exists a point $x^* \in \mathcal{P}(r)$ such that $\mathcal{CG}(r_w) = R_w(x^*)/\hat{r}_w(x^*)$ and $x^*(E) = r(\text{supp}(x^*))$.*

Proof (of Theorem 1). For the purpose of contradiction, suppose that there exist weights $w \in \mathbb{R}_+^E$ and a point $x^* \in [0, 1]^E$ such that $R_w(x^*)/\hat{r}_w(x^*) < \mathcal{CG}(r)$. According to Theorem 4, we may assume that $x^* \in \mathcal{P}(r)$. Thus, $\hat{r}_w(x^*) = w^\top x^*$ by Lemma 2.

Let $w^1 > w^2 > \cdots > w^k \geq 0$ denote the distinct values of w. For each $i \in [k]$, let $E_i \subseteq E$ denote the set of elements with weight w_i. Clearly, $k \geq 2$, as otherwise $R_w(x^*)/\hat{r}_w(x^*) = w^1 R(x^*)/(w^1 x^*(E)) = R(x^*)/x^*(E) \geq \mathcal{CG}(r)$. Let us pick a counterexample with k minimal.

First, we claim that $w_k > 0$. Indeed, if the smallest weight is $w_k = 0$, then $R_w(x^*)$ and $\hat{r}_w(x^*)$ remain unchanged after setting $w_e \leftarrow w^1$ and $x_e^* \leftarrow 0$ for all $e \in E_k$; this contradicts the minimal choice of k.

Let X be the random variable for the set obtained by sampling every element $e \in E$ independently with probability x_e^*. Let $I_X \subseteq X$ denote a maximum weight independent subset of X. Recall the well-known property of matroids that a maximum weight independent set can be selected greedily in decreasing order of the weights w_e. We fix an arbitrary tie-breaking rule inside each set E_i.

The correlation gap of r_w is given by

$$\frac{R_w(x^*)}{\hat{r}_w(x^*)} = \frac{\sum_{S \subseteq E} \Pr(X = S) r_w(S)}{w^\top x^*} = \frac{\sum_{i=1}^{k} w^i \sum_{e \in E_i} \Pr(e \in I_X)}{\sum_{i=1}^{k} w^i x^*(E_i)}.$$

Consider the set

$$J := \underset{i \in [k]}{\arg\min} \, \frac{\sum_{e \in E_i} \Pr(e \in I_X)}{x^*(E_i)}.$$

We claim that $J \setminus \{1\} \neq \emptyset$. Suppose that $J = \{1\}$ for a contradiction. Define the point $x' \in \mathcal{P}(r)$ as $x_e' := x_e^*$ if $e \in E_1$, and $x_e' := 0$ otherwise. Then, we get a contradiction from

$$\mathcal{CG}(r) \leq \frac{R(x')}{\hat{r}(x')} = \frac{w^1 \sum_{e \in E_1} \Pr(e \in I_X)}{w^1 x^*(E_1)} < \frac{\sum_{i=1}^{k} w^i \sum_{e \in E_i} \Pr(e \in I_X)}{\sum_{i=1}^{k} w^i x^*(E_i)} = \frac{R_w(x^*)}{\hat{r}_w(x^*)}.$$

The first equality holds because for every $e \in E_1$, $\Pr(e \in I_X)$ only depends on $x_{E_1}^* = x_{E_1}'$. This is by the greedy choice of I_X: elements in E_1 are selected based only on $X \cap E_1$. The strict inequality is due to $J = \{1\}$, $k \geq 2$ and $w_2 > 0$.

Now, pick any index $j \in J \setminus \{1\}$. Since $w^j > 0$, we have

$$\frac{w^j \sum_{e \in E_j} \Pr(e \in I_X)}{w^j x^*(E_j)} \leq \frac{\sum_{i=1}^{k} w^i \sum_{e \in E_i} \Pr(e \in I_X)}{\sum_{i=1}^{k} w^i x^*(E_i)}.$$

So, we can increase w^j to w^{j-1} without increasing the correlation gap. That is, defining $\bar{w} \in \mathbb{R}_+^E$ as $\bar{w}_e := w^{j-1}$ if $e \in E_j$ and $\bar{w}_e := w_e$ otherwise, we get

$$\frac{R_w(x^*)}{\hat{r}_w(x^*)} \geq \frac{\sum_{i \neq j} w^i \sum_{e \in E_i} \Pr(e \in I_X) + w^{j-1} \sum_{e \in E_j} \Pr(e \in I_X)}{\sum_{i \neq j} w^i x^*(E_i) + w^{j-1} x^*(E_j)}$$

$$= \frac{\sum_{S \subseteq E} \Pr(X = S) r_{\bar{w}}(S)}{\bar{w}^\top x^*} \geq \min_{x \in [0,1]^E} \frac{R_{\bar{w}}(x)}{\hat{r}_{\bar{w}}(x)}.$$

The equality holds because for every $S \subseteq E$, I_S remains a max-weight independent set with the new weights $\bar{w}$. This contradicts the minimal choice of k.   □

4   Lower Bounding the Correlation Gap

This section is dedicated to the proof of Theorem 2. Let $\mathcal{M} = (E, \mathcal{I})$ be a matroid with rank function r, rank $\rho = r(E)$ and girth $\gamma > 1$. By Theorem 4, there exists a point $x^* \in \mathcal{P}(r)$ such that $\mathcal{CG}(r) = R(x^*)/r(x^*)$ and $x^*(E) = r(\text{supp}(x^*))$. For the sake of brevity, we denote $\ell = \gamma - 1$ and $\lambda = x^*(E) \in \mathbb{Z}_+$. Note that if

$\lambda < \ell$, then $\text{supp}(x^*)$ is independent. As $x^*(E) = r(\text{supp}(x^*)) = |\text{supp}(x^*)|$, we have $x_i^* = 1$ for all $i \in \text{supp}(x^*)$. Since x^* is integral, the correlation gap is 1 because $R(x^*) = \hat{r}(x^*)$. Henceforth, we will assume that $\lambda \geq \ell$.

From Lemma 2, we already know that $\hat{r}(x^*) = \mathbb{1}^\top x^* = \lambda$. So, it remains to analyze $R(x^*)$. Let g be the rank function of a rank-ℓ uniform matroid on ground set E, and define the function $h := r - g \geq 0$. By linearity of expectation, $R(x^*) = G(x^*) + H(x^*)$. We lower bound $G(x^*)$ and $H(x^*)$ separately.

4.1 Lower Bounding $G(x^*)$

As g is the rank function of a uniform matroid, the arguments of Yan [28] and Barman et al. [6] apply. In particular, since G is a symmetric polynomial, and convex along $e_i - e_j$ for all $i, j \in E$ by Proposition 2, we have

$$G(x^*) \geq G\left(\frac{\lambda}{n} \cdot \mathbb{1}\right) = \mathbb{E}\left[\min\left\{\text{Bin}\left(n, \frac{\lambda}{n}\right), \ell\right\}\right] \geq \mathbb{E}\left[\min\{\text{Poi}(\lambda), \ell\}\right] . \quad (5)$$

The last inequality follows from Lemma 1. The latter expectation is equal to

$$\sum_{j=1}^{\ell} \Pr(\text{Poi}(\lambda) \geq j) = \sum_{j=1}^{\ell}\left(1 - \sum_{k=0}^{j-1}\frac{\lambda^k e^{-\lambda}}{k!}\right) = \ell - \sum_{k=0}^{\ell-1}(\ell - k)\frac{\lambda^k e^{-\lambda}}{k!}. \quad (6)$$

4.2 Lower Bounding $H(x^*)$

Our analysis of $H(x^*)$ uses the Poisson clock setup of Calinescu et al. [8], which incrementally builds a set $Q(1)$ as follows. Each element $i \in E$ is assigned a Poisson clock of rate x_i^*. We start all the clocks simultaneously at time $t = 0$, and begin with the initial set $Q(0) = \emptyset$. For $t \in [0, 1]$, if the clock on an element i rings at time t, then we add i to our current set $Q(t)$. We stop at time $t = 1$.

Clearly, $\Pr(i \in Q(1)) = 1 - e^{-x_i^*} \leq x_i^*$ for all $i \in E$. Since h is monotone, Proposition 1 yields $H(x^*) \geq H(1 - e^{-x^*}) = \mathbb{E}[h(Q(1))]$, where equality is due to independence of the Poisson clocks. So, it suffices to lower bound $\mathbb{E}[h(Q(1))]$.

Let $t \in [0, 1)$ and consider an infinitesimally small interval $[t, t+dt]$. For each $i \in E$, the probability of adding i during this interval is $\Pr(\text{Poi}(x_i^* dt) \geq 1) = x_i^* dt + O(dt^2)$. Note that the probability of adding two or more elements is also $O(dt^2)$. Since dt is very small, we can effectively neglect all $O(dt^2)$ terms.

Definition 2. *We say that Q is activated at time T if $|Q(t)| < \ell$ for all $t < T$ and $|Q(t)| \geq \ell$ for all $t \geq T$. We call T the activation time of Q.*

Let $S \subseteq E$ where $|S| \geq \ell$ and let $t \geq t' \geq 0$. Conditioning on the events $Q(t) = S$ and $T = t'$, the expected increase of $h(Q(t))$ (up to $O(dt^2)$ terms) is

$$\mathbb{E}[h(Q(t+dt)) - h(Q(t))|Q(t) = S \wedge T = t'] = \sum_{i \in E} r_S(i)x_i^* dt \geq (\lambda - \ell - h(S))dt,$$

where the inequality is due to

$$h(S) + \sum_{i \in E} r_S(i)x_i^* = r(S) - \ell + \sum_{i \in E} r_S(i)x_i^* \geq r^*(x^*) - \ell = \hat{r}(x^*) - \ell = \lambda - \ell.$$

The inequality follows from the definition of r^* in (4), the second equality is by Theorem 3, while the third equality is due to Lemma 2 because $x^* \in \mathcal{P}(r)$. Dividing by dt and taking expectation over S, we obtain for all $t \geq t' \geq 0$,

$$\frac{1}{dt}\mathbb{E}[h(Q(t+dt)) - h(Q(t))|T = t'] \geq \lambda - \ell - \mathbb{E}[h(Q(t))|T = t']. \tag{7}$$

Let $\phi(t) := \mathbb{E}[h(Q(t))|T = t']$. Then, (7) can be written as $\frac{d\phi}{dt} \geq \lambda - \ell - \phi(t)$. To solve this differential inequality, let $\psi(t) := e^t\phi(t)$ and consider $\frac{d\psi}{dt} = e^t(\frac{d\phi}{dt} + \phi(t)) \geq e^t(\lambda - \ell)$. Since $\psi(t') = \phi(t') = 0$, we get

$$\psi(t) = \int_{t'}^t \frac{d\psi}{ds}ds \geq \int_{t'}^t e^s(\lambda - \ell)ds = (e^t - e^{t'})(\lambda - \ell)$$

for all $t \geq t'$. It follows that $\mathbb{E}[h(Q(t))|T = t'] = \phi(t) = e^{-t}\psi(t) \geq (1 - e^{t'-t})(\lambda - \ell)$ for all $t \geq t'$. In particular, at time $t = 1$, we have $\mathbb{E}[h(Q(1))|T = t'] \geq (1 - e^{t'-1})(\lambda - \ell)$ for all $t' \leq 1$. By the law of total expectation,

$$\mathbb{E}[h(Q(1))] \geq (\lambda - \ell)\int_0^1 \Pr(T = t)(1 - e^{t-1})dt. \tag{8}$$

Now, the cumulative distribution function of T is given by

$$\Pr(T \leq t) = 1 - \sum_{\substack{S \subseteq E: \\ |S| < \ell}} \prod_{i \in S}(1 - e^{-x_i^* t}) \prod_{i \notin S} e^{-x_i^* t}$$

$$\overset{\star}{=} 1 - \sum_{S \subseteq E}(-1)^{|S|+\ell-n-1}\binom{|S|-1}{n-\ell}e^{-x^*(S)t}.$$

Any marked equality $\overset{\star}{=}$ indicates that several derivation steps have been skipped, whose details can be found in the full version. Differentiating with respect to t yields the probability density function of T

$$\Pr(T = t) = \frac{d}{dt}\Pr(T \leq t) = \sum_{S \subseteq E}(-1)^{|S|+\ell-n-1}\binom{|S|-1}{n-\ell}x^*(S)e^{-x^*(S)t}.$$

Plugging this back into (8) gives us

$$\mathbb{E}[h(Q(1))] \geq (\lambda - \ell)\sum_{S \subseteq E}(-1)^{|S|+\ell-n-1}\binom{|S|-1}{n-\ell}x^*(S)\int_0^1 e^{-x^*(S)t}(1 - e^{t-1})dt$$

$$= (\lambda - \ell)\sum_{S \subseteq E}(-1)^{|S|+\ell-n-1}\binom{|S|-1}{n-\ell}\left(1 - \frac{1}{e} - \frac{e^{-1} - e^{-x^*(S)}}{x^*(S) - 1}\right)$$

$$= (\lambda - \ell)\left[1 - \frac{1}{e} + \sum_{S \subseteq E}(-1)^{|S|+\ell-n}\binom{|S|-1}{n-\ell}\frac{e^{-1} - e^{-x^*(S)}}{x^*(S) - 1}\right] \tag{9}$$

In the full version, we prove that (9) is concave along $e_i - e_j$ for all $i, j \in E$, when viewed as a function of x^*. This allows us to round x^* to an integral vector $x' \in \{0, 1\}^E$ such that $x'(E) = x^*(E)$ without increasing the value of (9). Note that x' has exactly λ ones and $n - \lambda$ zeroes because $\lambda \in \mathbb{Z}_+$ by Theorem 4. Hence, (9) is lower bounded by

$$(\lambda - \ell)\left[1 - \frac{1}{e} + \sum_{i=0}^{\lambda}\sum_{j=0}^{n-\lambda}\binom{\lambda}{i}\binom{n-\lambda}{j}(-1)^{i+j+\ell-n}\binom{i+j-1}{n-\ell}\frac{e^{-1}-e^{-i}}{i-1}\right]$$

$$\overset{\star}{=} (\lambda - \ell)\left[1 - \frac{1}{e} + \sum_{i=0}^{\ell-1}(-1)^{\ell-i}\binom{\lambda}{i}\binom{\lambda-i-1}{\ell-i-1}\frac{e^{-1}-e^{-(\lambda-i)}}{\lambda-i-1}\right]. \quad (10)$$

Since (10) evaluates to 0 when $\lambda = \ell$, let us assume that $\lambda > \ell$. Then, using $\frac{1}{\lambda-i-1}\binom{\lambda-i-1}{\ell-i-1} = \frac{1}{\lambda-\ell}\binom{\lambda-i-2}{\ell-i-1}$, we can simplify (10) as

$$(\lambda - \ell)\left(1 - \frac{1}{e}\right) + \sum_{i=0}^{\ell-1}(-1)^{\ell-i}\binom{\lambda}{i}\binom{\lambda-i-2}{\ell-i-1}\left(e^{-1}-e^{-(\lambda-i)}\right)$$

$$\overset{\star}{=} \lambda\left(1 - \frac{1}{e}\right) - \ell + e^{-\lambda}\sum_{i=0}^{\ell-1}(-1)^{\ell-i-1}\binom{\lambda}{i}\binom{\lambda-i-2}{\ell-i-1}e^i. \quad (11)$$

The sum in (11) can be viewed as a univariate polynomial of degree $\ell - 1$ in $\alpha \in \mathbb{R}$ for $\alpha = e$. Taking its Taylor expansion at $\alpha = 1$, we can rewrite (11) as

$$\lambda\left(1 - \frac{1}{e}\right) - \ell + e^{-\lambda}\sum_{i=0}^{\ell-1}\binom{\lambda}{i}(\ell-i)(e-1)^i. \quad (12)$$

4.3   Putting Everything Together

We are finally ready to lower bound the correlation gap of the matroid rank function r. Recall that we assumed $\lambda > \ell$ in the previous subsection. Combining the lower bounds (6) and (12) gives us

$$\mathcal{CG}(r) = \frac{G(x^*) + H(x^*)}{\mathbb{1}^\top x^*} = 1 - \frac{1}{e} + \frac{e^{-\lambda}}{\lambda}\sum_{i=0}^{\ell-1}(\ell-i)\left[\binom{\lambda}{i}(e-1)^i - \frac{\lambda^i}{i!}\right]. \quad (13)$$

On the other hand, if $\lambda = \ell$, then $h = 0$. By (6), we obtain

$$\mathcal{CG}(r) = \frac{G(x^*)}{\mathbb{1}^\top x^*} = \frac{G(x^*)}{\ell} \geq 1 - \sum_{k=0}^{\ell-1}\left(1 - \frac{k}{\ell}\right)\frac{\ell^k e^{-\ell}}{k!} = 1 - \frac{\ell^{\ell-1}e^{-\ell}}{(\ell-1)!}, \quad (14)$$

which agrees with (13) when $\lambda = \ell$ (proven in full version).

To finish the proof of Theorem 2, it is left to show that (13) is a decreasing function of λ because $\lambda \leq \rho$. We also need to prove that the final expression is strictly greater than $1 - 1/e$ whenever $\ell \geq 2$. These are done in the full version.

References

1. Ageev, A.A., Sviridenko, M.: Pipage rounding: a new method of constructing algorithms with proven performance guarantee. J. Comb. Optim. **8**(3), 307–328 (2004)
2. Agrawal, S., Ding, Y., Saberi, A., Ye, Y.: Price of correlations in stochastic optimization. Oper. Res. **60**(1), 150–162 (2012)
3. Asadpour, A., Niazadeh, R., Saberi, A., Shameli, A.: Sequential submodular maximization and applications to ranking an assortment of products. In: EC 2022: The 23rd ACM Conference on Economics and Computation, p. 817 (2022)
4. Ausubel, L.M., Milgrom, P.: The lovely but lonely Vickrey auction. In: Cramton, P., Shoham, Y., Steinberg, R. (eds.) Combinatorial Auctions, chap. 1. MIT Press (2006)
5. Barman, S., Fawzi, O., Fermé, P.: Tight approximation guarantees for concave coverage problems. In: 38th International Symposium on Theoretical Aspects of Computer Science (STACS). LIPIcs, vol. 187, pp. 1–17 (2021)
6. Barman, S., Fawzi, O., Ghoshal, S., Gürpinar, E.: Tight approximation bounds for maximum multi-coverage. Math. Program. **192**(1), 443–476 (2022)
7. Bhalgat, A., Chakraborty, T., Khanna, S.: Mechanism design for a risk averse seller. In: Goldberg, P.W. (ed.) WINE 2012. LNCS, vol. 7695, pp. 198–211. Springer, Heidelberg (2012). https://doi.org/10.1007/978-3-642-35311-6_15
8. Calinescu, G., Chekuri, C., Pál, M., Vondrák, J.: Maximizing a submodular set function subject to a matroid constraint (Extended Abstract). In: Fischetti, M., Williamson, D.P. (eds.) IPCO 2007. LNCS, vol. 4513, pp. 182–196. Springer, Heidelberg (2007). https://doi.org/10.1007/978-3-540-72792-7_15
9. Călinescu, G., Chekuri, C., Pál, M., Vondrák, J.: Maximizing a monotone submodular function subject to a matroid constraint. SIAM J. Comput. **40**(6), 1740–1766 (2011)
10. Chawla, S., Hartline, J.D., Malec, D.L., Sivan, B.: Multi-parameter mechanism design and sequential posted pricing. In: Schulman, L.J. (ed.) Proceedings of the 42nd ACM Symposium on Theory of Computing, STOC 2010, Cambridge, Massachusetts, USA, 5–8 June 2010, pp. 311–320. ACM (2010)
11. Chekuri, C., Livanos, V.: On submodular prophet inequalities and correlation gap. In: 14th International Symposium on Algorithmic Game Theory, SAGT. Lecture Notes in Computer Science, vol. 12885, p. 410 (2021)
12. Chekuri, C., Vondrák, J., Zenklusen, R.: Submodular function maximization via the multilinear relaxation and contention resolution schemes. SIAM J. Comput. **43**(6), 1831–1879 (2014)
13. Clarke, E.H.: Multipart pricing of public goods. Public choice, pp. 17–33 (1971)
14. Dudycz, S., Manurangsi, P., Marcinkowski, J., Sornat, K.: Tight approximation for proportional approval voting. In: Bessiere, C. (ed.) Proceedings of the Twenty-Ninth International Joint Conference on Artificial Intelligence, IJCAI 2020, pp. 276–282. ijcai.org (2020)
15. Dughmi, S.: Matroid secretary is equivalent to contention resolution. In: 13th Innovations in Theoretical Computer Science Conference, ITCS. LIPIcs, vol. 215, pp. 1–23 (2022)
16. Feige, U.: A threshold of $\ln n$ for approximating set cover. J. ACM (JACM) **45**(4), 634–652 (1998)
17. Groves, T.: Incentives in teams. Econometrica: J. Econometric Soc. **41**, 617–631 (1973)
18. Hartline, J.D.: Mechanism design and approximation (2013)

19. Leme, R.P.: Gross substitutability: an algorithmic survey. Games Econom. Behav. **106**, 294–316 (2017)
20. Murota, K.: On basic operations related to network induction of discrete convex functions. Optim. Methods Softw. **36**(2–3), 519–559 (2021)
21. Myerson, R.B.: Optimal auction design. Math. Oper. Res. **6**(1), 58–73 (1981)
22. Nikolova, E.: Approximation algorithms for reliable stochastic combinatorial optimization. In: Serna, M., Shaltiel, R., Jansen, K., Rolim, J. (eds.) APPROX/RANDOM -2010. LNCS, vol. 6302, pp. 338–351. Springer, Heidelberg (2010). https://doi.org/10.1007/978-3-642-15369-3_26
23. Nisan, N., Ronen, A.: Computationally feasible VCG mechanisms. J. Arti. Intell. Res. **29**, 19–47 (2007)
24. Rubinstein, A., Singla, S.: Combinatorial prophet inequalities. In: Proceedings of the Twenty-Eighth Annual ACM-SIAM Symposium on Discrete Algorithms, pp. 1671–1687. SIAM (2017)
25. Schrijver, A.: Combinatorial optimization: polyhedra and efficiency, vol. 24. Springer (2003)
26. Shioura, A.: On the Pipage rounding algorithm for submodular function maximization - a view from discrete convex analysis. Discret. Math. Algorithms Appl. **1**(1), 1–24 (2009)
27. Vickrey, W.: Counterspeculation, auctions, and competitive sealed tenders. J. Financ. **16**(1), 8–37 (1961)
28. Yan, Q.: Mechanism design via correlation gap. In: Proceedings of the twenty-second annual ACM-SIAM symposium on Discrete Algorithms, pp. 710–719. SIAM (2011)

A 4/3-Approximation Algorithm
for Half-Integral Cycle Cut Instances
of the TSP

Billy Jin[1]([✉])[iD], Nathan Klein[2][iD], and David P. Williamson[1][iD]

[1] Cornell University, Ithaca, USA
{bzj3,davidpwilliamson}@cornell.edu
[2] University of Washington, Seattle, USA
nwklein@cs.washington.edu

Abstract. A long-standing conjecture for the traveling salesman problem (TSP) states that the integrality gap of the standard linear programming relaxation of the TSP (sometimes called the Subtour LP or the Held-Karp bound) is at most 4/3 for symmetric instances of the TSP obeying the triangle inequality. In this paper we consider the half-integral case, in which a feasible solution to the LP has solution values in $\{0, 1/2, 1\}$. Karlin, Klein, and Oveis Gharan [9], in a breakthrough result, were able to show that in the half-integral case, the integrality gap is at most 1.49993; Gupta et al. [6] showed a slight improvement of this result to 1.4983.

Both of these papers consider a hierarchy of critical tight sets in the support graph of the LP solution, in which some of the sets correspond to *cycle cuts* and the others to *degree cuts*. Here we show that if all the sets in the hierarchy correspond to cycle cuts, then we can find a distribution of tours whose expected cost is at most 4/3 times the value of the half-integral LP solution; sampling from the distribution gives us a randomized 4/3-approximation algorithm. We note that known bad cases for the integrality gap have a gap of 4/3 and have a half-integral LP solution in which all the critical tight sets in the hierarchy are cycle cuts; thus our result is tight.

1   Introduction

In the traveling salesman problem (TSP), we are given a set of n cities and the costs c_{ij} of traveling from city i to city j for all i, j, and the goal of the problem is to find the least expensive tour that visits each city exactly once and returns to its starting point. An instance of the TSP is called *symmetric* if $c_{ij} = c_{ji}$ for all i, j. Costs obey the *triangle inequality* (or are *metric*) if $c_{ij} \leq c_{ik} + c_{kj}$ for all i, j, k. For ease of exposition, we consider the problem input as a complete graph $G = (V, E)$ for the set of cities V, with $c_e = c_{ij}$ for edge $e = (i, j)$. All instances we consider will be symmetric and obey the triangle inequality.

In a breakthrough result, Karlin, Klein, and Oveis Gharan [8] gave the first approximation algorithm with performance ratio better than 3/2, although the

© The Author(s), under exclusive license to Springer Nature Switzerland AG 2023
A. Del Pia and V. Kaibel (Eds.): IPCO 2023, LNCS 13904, pp. 217–230, 2023.
https://doi.org/10.1007/978-3-031-32726-1_16

amount by which the bound was improved is quite small (approximately 10^{-36}). The algorithm follows the Christofides-Serdyukov template by selecting a random spanning tree from the max-entropy distribution, then using a T-join on the odd degree vertices of the tree to create a connected Eulerian subgraph.

One special case of the TSP is known as the *half-integral* case. To understand the half-integral case, we need to introduce a well-known LP relaxation of the TSP, sometimes called the *Subtour LP* or the *Held-Karp bound* [4,7], which is as follows:

$$
\begin{aligned}
\min \quad & \sum_{e \in E} c_e x_e \\
\text{s.t.} \quad & x(\delta(v)) = 2, & \forall\, v \in V, \\
& x(\delta(S)) \geq 2, & \forall\, S \subset V, S \neq \emptyset, \\
& 0 \leq x_e \leq 1, & \forall e \in E,
\end{aligned}
$$

where $\delta(S)$ is the set of all edges with exactly one endpoint in S and we use the shorthand that $x(F) = \sum_{e \in F} x_e$. A half-integral solution to the Subtour LP is one such that $x_e \in \{0, 1/2, 1\}$ for all $e \in E$, and a half-integer instance of the TSP is one whose LP solution is half-integral.

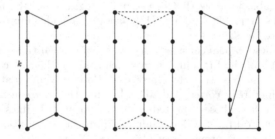

Fig. 1. Illustration of a known worst-case example for the integrality gap for the symmetric TSP with triangle inequality. The figure on the left gives an (unweighted) graph, and costs c_{ij} are the shortest path lengths in the graph. The figure in the center gives the LP solution, in which the dotted edges have value $1/2$, and the solid edges have value 1. The figure on the right gives the optimal tour. The ratio of the cost of the optimal tour to the value of the LP solution tends to $4/3$ as k increases.

The *integrality gap* of an LP relaxation is the worst-case ratio of an optimal integer solution to the linear program to the optimal linear programming solution. Wolsey [13] showed that the analysis of the Christofides-Seryukov algorithm could be used to show that the integrality gap of the Subtour LP is at most $3/2$. It is known that the integrality gap of the Subtour LP is at least $4/3$, due to a set of half-integral graph TSP instances shown in Fig. 1, and another set of half-integral weighted instances due to Boyd and Sebő [2] known as k-donuts. Schalekamp, Williamson, and van Zuylen [11] have conjectured that half-integral instances are the worst-case instances for the integrality gap. It has long been conjectured that the integrality gap is exactly $4/3$, but until the work of Karlin et al. there had been no progress on the conjecture for several decades.

In the case of half-integral instances, some results are known. Mömke and Svensson [10] have shown a $4/3$-approximation algorithm for half-integral graph TSP (in which cost c_{ij} is the number of edges in the shortest i-j path in an input graph), also yielding an integrality gap of $4/3$ for such instances; because of the worst-case examples of Fig. 1, their result is tight. Boyd and Carr [1] give a $4/3$-approximation algorithm (and an integrality gap of $4/3$) for a subclass of half-integer solutions they call triangle points (in which the half-integer edges form disjoint triangles); the examples of Fig. 1 show that their result is tight also. Boyd and Sebő [2] give an upper bound of $10/7$ for a subclass of half-integral solutions they call square points (in which the half-integer edges form disjoint 4-cycles). In a paper released just prior to their general improvement, Karlin, Klein, and Oveis Gharan [9] (KKO) gave a 1.49993-approximation algorithm in the half-integral case; in particular, they show that given a half-integral solution, they can produce a tour of cost at most 1.49993 times the value of the corresponding objective function. Gupta, Lee, Li, Mucha, Newman, and Sarkar [6] improve this factor to 1.4983.

With the improvements on the $3/2$ bound remaining very incremental for weighted instances of the TSP, even in the half-integral case, we turn the question around and look for a large class of weighted half-integral instances for which we can prove that the $4/3$ conjecture is correct, preferably one containing the known worst-case instances.

To define our instances, we turn to some terminology of KKO. The KKO result uses induction on a hierarchy of *critical tight* sets of the half-integral LP solution x. A set $S \subset V$ is *tight* if the corresponding LP constraint is met with equality; that is, $x(\delta(S)) = 2$. A set S is *critical* if it does not cross any other tight set; that is, for any other tight set T, either $S \cap T = \emptyset$ or $S \subseteq T$ or $T \subseteq S$. The critical tight sets then give rise to a natural tree-like hierarchy based on subset inclusion. KKO follow a Christofides-Serdyukov style algorithm that performs induction on the hierarchy. In their analysis, they differentiate between *cycle cuts* (in which the child nodes of a parent are linked by pairs of edges in a chain) and *degree cuts* (in which the child nodes of a parent form a 4-regular graph; more detail is given in subsequent sections).

In this paper, we will consider half-integral instances in which there are only cycle cuts, which we will refer to as half-integral cycle cut instances. Our contribution is to give a randomized $\frac{4}{3}$-approximation algorithm for these instances. More precisely, we give a distribution over connected Eulerian subgraphs such that each edge e is used with expectation at most $\frac{4}{3}x_e$, which implies the result (note that edges are sometimes doubled in the Eulerian graph). Our main theorem is as follows:

Theorem 1. *There is a randomized $4/3$-approximation algorithm for half-integral cycle cut instances of the TSP that produces an Eulerian tour with expected cost at most $\frac{4}{3}\sum_{e\in E}c_e x_e$.*

It is not hard to show that both the bad examples in Fig. 1 and the k-donut instances of Boyd and Sebő [2] are cycle cut instances (Boyd and Carr's result for triangle points works for the examples of Fig. 1, but not for k-donuts). Thus our bound of $4/3$ is tight and cannot be improved.

Our approach to the problem is novel and does not use the same Christofides-Serdyukov framework as employed by KKO and others. Instead, we perform a top-down induction on the hierarchy of critical tight sets. For each set in the hierarchy, we define a set of "patterns" of edges incident on it such that the set has even degree. For each pattern, we give a distribution of edges connecting the chain of child nodes in the cycle cut, which induces a distribution of patterns on each child. Crucially, we then show that there is a *feasible region R* of distributions over patterns, such that if the distribution of patterns on the parent node belongs to R, then the induced distribution on patterns on each child node also belongs to R. Our abstract is structured as follows. We give some needed preliminary definitions in Sect. 2. We then sketch our main result in Sect. 3, and conclude in Sect. 4. Due to space constraints, some proofs are omitted or sketched. The full paper can be accessed at https://arxiv.org/abs/2211.04639.

2   Preliminaries

Given a half-integral LP solution x, we construct a 4-regular 4-edge-connected multigraph $G = (V, E)$ by including a single copy of every edge e for which $x_e = \frac{1}{2}$ and two copies of every edge e for which $x_e = 1$. We state the following for general k-edge-connected multigraphs. In our setting, $k = 4$.

Definition 1. *For a k-edge-connected multigraph $G = (V, E)$, we say:*

- *Any set $S \subseteq V$ such that $|\delta(S)| = k$ (i.e., its boundary is a minimum cut) is a **tight set**.*
- *A set $S \subseteq V$ is **proper** if $2 \le |S| \le n - 2$ and a **singleton** if $|S| = 1$.*
- *Two sets $S, S' \subseteq V$ **cross** if all of $S \smallsetminus S'$, $S' \smallsetminus S$, $S \cap S'$, and $V \smallsetminus (S \cup S') \ne \emptyset$ are non-empty.*

The following are two standard facts about minimum cuts; for proofs see [5].

Lemma 1. *If two tight sets S and S' cross, then each of $S \smallsetminus S'$, $S' \smallsetminus S$, $S \cap S'$ and $\overline{S \cup S'}$ are tight. Moreover, there are no edges from $S \smallsetminus S'$ to $S' \smallsetminus S$, and there are no edges from $S \cap S'$ to $\overline{S \cup S'}$.*

Lemma 2. *Let $G = (V, E)$ be a k-regular k-edge-connected graph. Suppose either $|V| = 3$ or G has at least one proper min cut, and every proper min cut is crossed by some other proper min cut. Then, k is even and G forms a cycle, with $k/2$ parallel edges between each adjacent pair of vertices.*

We now define our class of instances.

Definition 2 (Cycle cut instance). *We say a graph G is a **cycle cut instance** if every non-singleton tight set S can be written as the union of two tight sets $A, B \neq S$.*

As mentioned in the introduction this condition captures the two known integrality gap examples of the subtour LP.

We now show an equivalent definition of cycle cut instances after giving some definitions. First, fix an arbitrary **root vertex** $r \in V$, and for all cuts we consider we will take the side which does not contain r.

Definition 3 (Critical cuts). *A critical cut is any tight set $S \subseteq V \setminus \{r\}$ which does not cross any other tight set.*

Definition 4 (Hierarchy of critical cuts, $\mathcal{H}$). *Let $\mathcal{H} \subseteq 2^{V \setminus r}$ be the set of all critical cuts.*

The hierarchy naturally gives rise to a parent-child relationship between sets as follows:

Definition 5 (Child, parent, $E^{\rightarrow}(S)$). *Let $S \in \mathcal{H}$ such that $|S| \geq 2$. Call the maximal sets $C \in \mathcal{H}$ for which $C \subset S$ the children of S, and call S their parent. Finally, define $E^{\rightarrow}(S)$ to be the set of edges with endpoints in two different children of S.*

Definition 6 (Cycle cut, degree cut). *Let $S \in \mathcal{H}$ with $|S| \geq 2$. Then we call S a **cycle cut** if when $G \setminus S$ and all of the children of S are contracted, the resulting graph forms a cycle of length at least three with two parallel edges between each adjacent node. Otherwise, we call it a **degree cut**.*

While this definition of a cycle cut may sound specialized, due to Lemma 2, cycle cuts arise very naturally from collections of crossing min cuts.

Lemma 3. *If G is a cycle cut instance, then for any choice of r, $\mathcal{H}$ is composed only of cycle cuts (and singletons).*

One can also show that if for some choice of r, $\mathcal{H}$ is composed only of cycle cuts, then G is a cycle cut instance. Thus, in the remainder of the paper, we assume $\mathcal{H}$ is a collection of cycle cuts.

Given $S \in \mathcal{H}$, let $a_0 = G \setminus S$ and let $a_1, \ldots, a_k$ be its children in $\mathcal{H}$ (which are either vertices or cycle cuts). By Lemma 2 $a_0, \ldots, a_k$ can be arranged into a cycle such that two edges go between each adjacent vertex. WLOG let $a_1, \ldots, a_k$ be in counterclockwise order starting from a_0. We call a_1 the leftmost child of S and a_k the rightmost child.

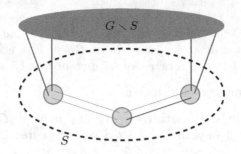

Fig. 2. S is an example of a cycle cut with three children. In blue are contracted critical tight sets. In gray is the rest of the graph with S contracted. As in Lemma 2, we can see that when $G \smallsetminus S$ is contracted into a single vertex, the resulting graph is a cycle with 2 edges between each adjacent vertex. In our recursive proof of our main theorem in Sect. 3, we are given a distribution of Eulerian tours over G/S, so in particular on the red edges here, and will then extend it to G with the blue critical sets contracted by picking a distribution over the black edges. (Color figure online)

Definition 7 (External and internal cycles cuts). *Let* $S \in \mathcal{H}$ *such that* $S \neq V \smallsetminus \{r\}$ *be a cut with parent* S'. *We call* S *external if in the ordering* $a_0, \ldots, a_k$ *of* S' *(as given above),* $S = a_1$ *or* $S = a_k$. *Otherwise, call* S *internal.*

For example, if the blue nodes in Fig. 2 are contracted cycle cuts, the left and right nodes are external, while the middle one is internal. Note that for an cycle cut S with parent S', if S is external then $|\delta(S) \cap \delta(S')| = 2$, and if S is internal then $|\delta(S) \cap \delta(S')| = 0$.

Using the following simple fact, we will now describe our convention for drawing and describing cycle cuts:

Lemma 4. *Let* $A, B, C \in \mathcal{H}$ *be three distinct critical cuts such that* $A \subsetneq B$ *and* $B \cap C = \emptyset$ *or* $B \subseteq C$. *Then* $|\delta(A) \cap \delta(C)| \leq 1$.

Definition 8 $(\delta^L(S), \delta^R(S))$. *Let* $S \in \mathcal{H}$ *be a cycle cut. We will define a partition of* $\delta(S)$ *into two sets* $\delta^L(S), \delta^R(S)$ *each consisting of two edges.*

If $S \neq V \smallsetminus \{r\}$, *then it has a parent* S'. S' *has children* $a_1, \ldots, a_k$ *such that* $S = a_i$ *for* $i \neq 0$. *Let* $\delta^L(S) = \delta(S) \cap \delta(a_{i-1})$ *and* $\delta^R(S) = \delta(S) \cap \delta(a_{i+1 \pmod{k+1}})$. *In other words, we partition the edges of* S *into the two edges going to the left neighbor of* S *in the cycle defined by* S''s *children and the two edges going to the right neighbor.*

Otherwise $S = V \smallsetminus \{r\}$. *Then if* $a_1, \ldots, a_k$ *are the children of* S, *let* $\delta^L(S)$ *consist of an arbitrary edge from* $\delta(a_1) \cap \delta(S)$ *and an arbitrary edge from* $\delta(a_k) \cap \delta(S)$. *Let* $\delta^R(S) = \delta(S) \smallsetminus \delta^L(S)$.

By Lemma 4 and the definition of $\delta^L(S), \delta^R(S)$ for $S = V \smallsetminus \{r\}$, if S' is an external child of a cycle cut S, then $|\delta^L(S) \cap \delta(S')| = |\delta^R(S) \cap \delta(S')| = 1$. This allows us to adopt the following convention for drawing cycle cuts which we will call the **caterpillar drawing** of S: for an example, see Fig. 3. Formally, let $S \in \mathcal{H}$ be a cycle cut with children $a_1, \ldots, a_k \in \mathcal{H}$. Arrange $a_1, \ldots, a_k$ in a horizontal line. First, expand a_1 vertically into its children (if it is not a singleton) such that the unique edge in $\delta^L(S) \cap \delta(a_1)$ is pointing up (if it is a singleton, simply draw this edge pointing up. Then, expand $a_2, \ldots, a_k$ one by one into their respective children (if they exist), placing the children vertically in increasing or decreasing order of their index so that the edges from a_i to a_{i+1} do not cross. If a_k is a singleton, arbitrarily choose which edge to draw pointing up. Otherwise, let a' be the topmost child of a_k. Draw the unique edge in $\delta(S) \cap \delta(a')$ pointing up. There are two types of cycle cuts:

Definition 9 (Straight and twisted cycle cuts). *Let $S \in \mathcal{H}$ be a cycle cut. If $\delta^L(S)$ has both edges pointing up in the caterpillar drawing of S, then call it a straight cycle cut. Otherwise, call it a twisted cycle cut. See Fig. 3 for examples.*

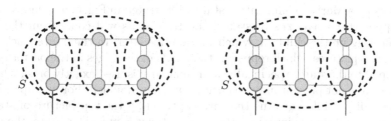

Fig. 3. Caterpillar drawings of two different cycle cuts S. The red edges are in the $\delta^L(S)$ partition, and the blue edges are in the $\delta^R(S)$ partition. The left drawing is a straight cycle cut, and the right is a twisted cycle cut as per Definition 9 (Color figure online).

In the next section, we abbreviate the caterpillar drawing by contracting the non-singleton children of S (see Fig. 4). We do so partially for cleaner pictures but also to emphasize that all the relevant information used by our construction in the following section is contained in the abbreviated pictures.

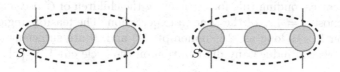

Fig. 4. On the left is a shorthand caterpillar drawing for the straight cycle cut on the left in Fig. 3 obtained by contracting its children. Similarly for the right. We will use this style of picture in future sections.

3 Proof of Theorem 1

We now present a summary of the proof of our main result, a $\frac{4}{3}$-approximation for half-integral cycle-cut instances of the TSP. To prove Theorem 1, we construct a distribution of Eulerian tours such that every edge is used at most $\frac{2}{3}$ of the time. Since $x_e = \frac{1}{2}$ for every edge in the graph, this immediately implies that when we sample a tour from this distribution, its expected cost is at most $\frac{4}{3}$ times the value of the LP. We work on the cycle cut hierarchy from the top down, and inductively specify the distribution of edges that enter every cut.

Figure 4 depicts our convention for visualizing a cycle cut as described in Sect. 2. We say that a cycle cut is *even* if it contains an even number of children, and *odd* otherwise. Fig. 6 illustrates the *patterns* we use, where "pattern" refers to a multiset of edges that enter a cycle cut. For each pattern entering a parent cycle cut, we give (randomized) rules which describe how to connect up its children – this induces a distribution of patterns entering each child. We represent this process using a Markov chain with 4 states, illustrated in Fig. 6. The figure shows the mapping from patterns to states; the transitions will come from the rules for connecting up the children, which we describe later. In the figure, each state contains two pictures, which represent the *parity* of the edges in the patterns that are mapped to the state. Specifically, a present edge is used exactly once, whereas an edge that is not present may be either unused or doubled. For example, Fig. 7 illustrates all possible patterns that are captured by the top picture of state 1. Finally, we maintain the invariant that if a cycle cut is in a given state, then each of the two pictures are equally likely. (When we later give the rules for connecting up the children, we will ensure this invariant is preserved.) Thus, when we say a cycle cut is in a given state with probability p, this means the parity of the pattern entering it follows the top picture in the state with probability $\frac{p}{2}$, and the bottom picture with probability $\frac{p}{2}$. We will use the phrase "the distribution of patterns entering a cycle cut C is (p_1, p_2, p_3, p_4)" to mean that for all $i \in \{1, 2, 3, 4\}$, C is in state i with probability p_i.

To prove our main result, we will give a *feasible region R* of distributions over the states of the Markov chain, such that: 1) If the distribution of patterns entering a cycle cut C belongs to R, there is a way to connect up the children of C such that the distribution on each child also belongs to R, and 2) for each $\mathbf{p} \in R$, the corresponding rule for connecting the children of C uses each edge in $E^{\rightarrow}(C)$ at most $\frac{2}{3} = \frac{4}{3}x_e$ of the time in expectation. The feasible region is given in Definition 10. As long as R is nonempty, 1) and 2) are sufficient to give the result since we can induce any distribution on the cycle cut $V \smallsetminus \{r\}$.

Definition 10 (The Feasible Region). *Let*

$$R = \left\{ (p_1, p_2, p_3, p_4) \in \mathbb{R}_+^4 : p_1 + p_2 + p_3 + p_4 = 1, \; p_1 + p_2 = \frac{2}{3}, \; p_2 + p_4 \geq \frac{1}{3} \right\}.$$

See Fig. 5 for an visualization of R in a 2-dimensional space.

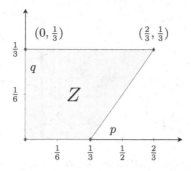

Fig. 5. The feasible region of distributions is $R = \{(p, \frac{2}{3} - p, \frac{1}{3} - q, q) : (p, q) \in Z\}$, where Z is the polytope above.

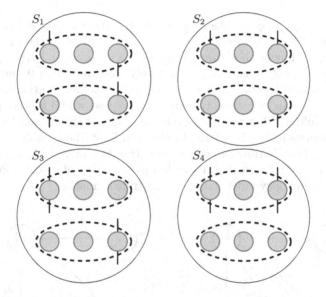

Fig. 6. The patterns and how they map to states of a Markov chain. The states are unchanged regardless of the number of children: they are defined only with respect to which of the edges are in. Note that we ignore doubled edges.

To describe the transitions of the Markov chain, we give (randomized) rules that dictate, for a cycle cut C and a pattern entering it, how to connect up its children. These rules depend on whether C is even or odd. The final form of the Markov chains is illustrated in Fig. 8.[1] The meaning of taking one transition is as follows. Suppose the distribution of patterns entering C is (p_1, p_2, p_3, p_4),

---

[1] In the figure, if there is a variable on an arc, it means that any transition probability in the range of that variable is possible. For example, in P_{even}, we can transition from S_2 to S_1 with probability z for any $z \in [0, 1]$; the transition from S_2 to S_3 then happens with probability $1 - z$.

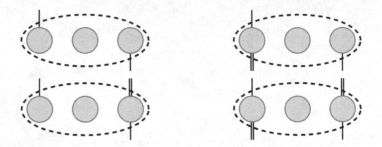

Fig. 7. In our illustrations of the patterns entering a given cycle cut, any edge that is not present may either be unused or doubled. Therefore, all four of the given edge configurations are represented by the upper left most state, S_1.

and suppose (q_1, q_2, q_3, q_4) is the resulting distribution after one transition of a Markov chain. What this means is that for each child of C, the distribution of patterns entering it will be **either** (q_1, q_2, q_3, q_4) or (q_2, q_1, q_3, q_4) depending on if the child is straight or twisted, respectively (see Definition 9 and Fig. 3). In particular, it can be shown that if (q_1, q_2, q_3, q_4) is the distribution induced on a child which is a straight cycle cut, then (q_2, q_1, q_3, q_4) would be the distribution induced on a child which is a twisted cycle cut. Thus, it is sufficient to check that: i) the distributions induced on straight children lie in the feasible region and ii) if (q_1, q_2, q_3, q_4) is a distribution induced on straight children, then (q_2, q_1, q_3, q_4) is also in the feasible region. This corresponds to the set of distributions induced on the children being symmetric under this transformation.[2]

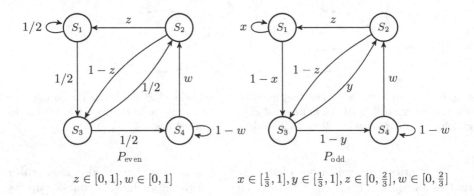

$$z \in [0, 1], w \in [0, 1] \qquad x \in [\tfrac{1}{3}, 1], y \in [\tfrac{1}{3}, 1], z \in [0, \tfrac{2}{3}], w \in [0, \tfrac{2}{3}]$$

Fig. 8. The variables on the arcs indicate that one can feasibly transition according to any probability in the range.

---

[2] Note that the feasible region is not symmetric under this transformation. The distribution induced on the children is thus a symmetric subset of the feasible region.

Proposition 1. *For any cycle cut $C \in \mathcal{H}$ and any distribution of patterns entering C, there is a way to connect its children so that the induced distribution on each child is given by 1) applying the corresponding Markov chain in Fig. 8, and then 2) swapping the first two coordinates if the child is twisted.*

Proof (Sketch). The proof involves going through the 8 cases one by one (depending on the parity of the cut, and which of the 4 states it is in), and showing that in each case, there is a (randomized) rule for connecting the children that achieve the transitions in Fig. 8. To illustrate the main idea, we show the rule in the case that C is even and in state 4.

In this case, the rule for connecting the children of C is illustrated in Fig. 9. Let $w \in [0, 1]$. With probability w, we make all children transition to state 2. To do this, first suppose C has all 4 single edges entering it (the top picture in the left box). In this case, we consider the pairs of edges in $E^{\rightarrow}(C)$ from left to right, and alternate 1) doubling one of the two edges with equal probability (shown by the dotted black edges), and 2) using both edges (shown by the solid black edges). Because C is even, the rightmost pair of edges ends up falling in case 1) of the alternating rule, and so all children transition to state 2. The case where all the edges entering C are used an even number of times (the bottom picture in the left box) is quite similar, except we begin the alternating rule by using both edges.

On the other hand, with probability $1 - w$, we transition back to state 4. This is accomplished by using each pair of edges in the top case of state 4, and by doubling one edge from each pair uniformly at random in the bottom case of state 4. The net transition probabilities are then $(0, w, 0, 1 - w)$, where w can be any number from 0 to 1.   □

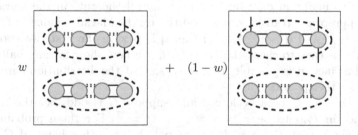

Fig. 9. Transition for state 4 in the even case.

We ensure that in all cases, each edge in $E^{\rightarrow}(C)$ is used $\frac{1}{2}, \frac{1}{2}, 1, 1$ times in expectation if the pattern entering C belongs to state 1, 2, 3, 4, respectively. Therefore, if $\mathbf{p} = (p_1, p_2, p_3, p_4)$ are the probabilities that we are in states 1, 2, 3, 4 respectively, then each edge in $E^{\rightarrow}(C)$ is used exactly $\frac{1}{2}p_1 + \frac{1}{2}p_2 + p_3 + p_4 = 1 - \frac{1}{2}(p_1 + p_2)$ of the time in expectation. Thus to get a $\frac{4}{3}$-approximation, it is necessary that $p_1 + p_2 \geq \frac{2}{3}$. Note that if $\mathbf{p} \in R$, then $p_1 + p_2 = \frac{2}{3}$, so that each edge is used exactly $\frac{2}{3}$ of the time.

To complete the proof, we only need show that if the distribution of patterns entering a cycle cut C belongs to R, then the induced distributions on the children also belong to R. Thus R is sufficient, in sense that if the distribution entering a cycle cut belongs to R, then it is possible to get a $\frac{4}{3}$-approximation all the way down the hierarchy using the Markov chains in Fig. 8. Moreover, we are able to show that R is *necessary*; if the distribution entering a cycle cut does **not** belong to R, then it is impossible to obtain a $\frac{4}{3}$-approximation using our Markov chains. In this sense, R is the *largest* feasible region using our technique.

Theorem 2. *1. (R is **sufficient**) If the distribution of patterns entering a cycle cut belongs to R, then there are feasible Markov chains (among the ones shown in Fig. 8) such that the induced distribution entering each child also belongs to R.*

*2. (R is **necessary**) Suppose the distribution of patterns entering a cycle cut does **not** belong to R. Then it is not possible to obtain a $\frac{4}{3}$-approximation using the Markov chains in Fig. 8.*

Proof (Sketch). For 1), we show that for any $\mathbf{p} \in R$ and for C even or odd, there are feasible values for the transition probabilities of the corresponding Markov chain such that the resulting distribution $\mathbf{q} \in R$ (and also $\mathbf{q}$ with its first two coordinates swapped is in R.) The values of the transition probabilities are derived as a function of $\mathbf{p}$. For 2), we consider an arbitrary distribution $\mathbf{p}$ (not necessarily in R), and let $\mathbf{q}^{(1)}$ and $\mathbf{q}^{(2)}$ be the distributions obtained by applying P_{even} once and twice, respectively. We then argue that $\mathbf{p}$ must belong to R in order for $\mathbf{q}^{(1)}$ and $\mathbf{q}^{(2)}$ to each have their first two coordinates sum to at least $\frac{2}{3}$.   □

Example. To give the reader some more intuition, we give a specific example of how to maintain distributions in R on all the cuts in the hierarchy by choosing appropriate transition probabilities on the Markov chains in Fig. 8. Let $\mathbf{p} = (\frac{4}{9}, \frac{2}{9}, \frac{2}{9}, \frac{1}{9})$ and $\mathbf{q} = (\frac{2}{9}, \frac{4}{9}, \frac{2}{9}, \frac{1}{9})$ (i.e. $\mathbf{q}$ is $\mathbf{p}$ with the first two coordinates swapped). It is easy to check that $\mathbf{p}, \mathbf{q} \in R$. We now show for any half-integral cycle cut instance, it is possible to make it so that the distribution entering any cycle cut is either $\mathbf{p}$ or $\mathbf{q}$.

To see this, let C be a cycle cut and suppose C is odd. Set the transition probabilities in P_{odd} to be $x = y = z = w = \frac{2}{3}$. For these probabilities, it is easy to check that $P_{\text{odd}}\mathbf{p} = P_{\text{odd}}\mathbf{q} = \mathbf{p}$.[3] On the other hand, if C is even, setting $z = w = 1$ in P_{even} gives $P_{\text{even}}\mathbf{p} = \mathbf{p}$, and setting $z = \frac{3}{4}, w = 1$ gives $P_{\text{even}}\mathbf{q} = \mathbf{p}$. Thus, as long as the distribution entering C is $\mathbf{p}$ or $\mathbf{q}$, we can make the distribution on each child of C be either $\mathbf{p}$ or $\mathbf{q}$.

Together with Proposition 1, this already proves a $\frac{4}{3}$-approximation for half-integral cycle cut instances. The additional contribution of Theorem 2 is an exact characterization of the region of distributions that give a $\frac{4}{3}$-approximation using our techniques.

---

[3] In fact, it can be checked that for these probabilities, P_{odd} maps *every* distribution (whose first two coordinates sum to $\frac{2}{3}$), to $\mathbf{p}$.

4   Conclusion and Open Questions

Our result leads to several interesting open questions. One such open question is whether our result extends to the case of cycle cuts for non-half-integral solutions. We believe this to be possible through a more refined understanding of the patterns that result from considering non-half-integral solutions.

Clearly a better understanding of what happens in the case of degree cuts is needed to make substantial progress on the overall half-integral case. We think it is possible to improve incrementally on the 1.4983-approximation of Gupta et al. [6] by using a combination of ideas from this paper with a few other small improvements. Recall that in a degree cut, each vertex has degree four, there are no parallel edges, and every proper cut has at least six edges crossing it. Ideally one would be able to show that any distribution on a parent cut lying in the feasible region of Fig. 5 could be used to induce a distribution on patterns of the children of the degree cut in a subregion of the feasible region with each edge used at most 2/3 of the time; such a result would lead immediately to a 4/3 integrality gap for half-integral instances.

Acknowledgment. The first and third authors would like to thank Anke van Zuylen for early discussions on this problem. The first and third authors were supported in part by NSF grant CCF-2007009. The first author was also supported by NSERC fellowship PGSD3-532673-2019. The second author was supported in part by NSF grants DGE-1762114, CCF-1813135, and CCF-1552097. We would like to thank Martin Drees for his helpful suggestions that allowed us to simplify the proof of the main result.

References

1. Boyd, S., Carr, R.: Finding low cost TSP and 2-matching solutions using certain half-integer subtour vertices. Discret. Optim. **8**, 525–539 (2011)
2. Boyd, S., Sebő, A.: The salesman's improved tours for fundamental classes. Math. Program. **186**, 289–307 (2021)
3. Christofides, N.: Worst case analysis of a new heuristic for the traveling salesman problem. Report 388, Graduate School of Industrial Administration, Carnegie-Mellon University, Pittsburgh, PA (1976)
4. Dantzig, G., Fulkerson, R., Johnson, S.: Solution of a large-scale traveling-salesman problem. J. Oper. Res. Soc. Am. **2**(4), 393–410 (1954). ISSN 00963984. URL https://www.jstor.org/stable/166695
5. Fleiner, T., Frank, A.: A quick proof for the cactus representation of mincuts. Technical Report QP-2009-03, Egerváry Research Group, Budapest (2009). https://www.cs.elte.hu/egres
6. Gupta, A., Lee, E., Li, J., Mucha, M., Newman, H., Sarkar, S.: Matroid-based TSP rounding for half-integral solutions. In: Aardal, K., Sanità, L. (eds.) Integer Programming and Combinatorial Optimization. LNCS, vol. 13265, pp. 305–318 (2022). https://doi.org/10.1007/978-3-031-06901-7_23,See also https://arxiv.org/pdf/2111.09290.pdf
7. Held, M., Karp, R.M.: The traveling-salesman problem and minimum spanning trees. Oper. Res. **18**, 1138–1162 (1971)

8. Karlin, A.R., Klein, N., Gharan, S.O.: A (slightly) improved approximation algorithm for metric tsp. In: STOC. ACM (2021)
9. Karlin, A.R., Klein, N., Gharan, S.O.: An improved approximation algorithm for TSP in the half integral case. In: Makarychev, K., Makarychev, Y., Tulsiani, M., Kamath, G., Chuzhoy, J. (eds.) STOC, pp. 28–39. ACM (2020)
10. Mömke, T., Svensson, O.: Removing and adding edges for the traveling salesman problem. J. ACM, 63 (2016). Article 2
11. Schalekamp, F., Williamson, D.P., van Zuylen, A.: 2-matchings, the traveling salesman problem, and the subtour LP: a proof of the Boyd-Carr conjecture. Math. Oper. Res. **39**(2), 403–417 (2014)
12. Serdyukov, A.: On some extremal walks in graphs. Upravlyaemye Sistemy **17**, 76–79 (1978)
13. Wolsey, L.A.: Heuristic analysis, linear programming and branch and bound. Math. Program. Study **13**, 121–134 (1980)

The Polyhedral Geometry of Truthful Auctions

Michael Joswig[1,2], Max Klimm[1(✉)], and Sylvain Spitz[1]

[1] Technische Universität Berlin, 10623 Berlin, Germany
{joswig,spitz}@math.tu-berlin.de, klimm@tu-berlin.de
[2] Max-Planck Institute for Mathematics in the Sciences, 04103 Leipzig, Germany

Abstract. The difference set of an outcome in an auction is the set of types that the auction mechanism maps to the outcome. We give a complete characterization of the geometry of the difference sets that can appear for a dominant strategy incentive compatible multi-unit auction showing that they correspond to regular subdivisions of the unit cube. This observation is then used to construct mechanisms that are robust in the sense that the set of items allocated to a player does change only slightly when the player's reported type is changed slightly.

1 Introduction

Mechanism design is concerned with the implementation of favorable social outcomes in environments where information is distributed and only released strategically. Specifically, this article is concerned with multi-dimensional mechanism design problems where a set of m items is to be allocated to a set of n players. The attitude of each player for receiving a subset of the items is determined by the so-called *type* of the player and is their private information and not available to the mechanism. In this setting, a mechanism elicits the types from the players, and—based on the reported types—decides on an allocation of the items to the players, and on a price vector that specifies the amount of money that the different players have to pay to the mechanism. In order to incentivize the players to truthfully report their types to the mechanism, one is interested in mechanisms that have the property that no matter what the other players report to the mechanism, no player can benefit from misreporting their type; mechanisms that enjoy this property are called *dominant strategy incentive compatible*, short DSIC. In this paper, we investigate the geometric properties of DSIC mechanisms. Because DSIC mechanisms require truthful reporting of the type no matter of the types declared by the other players, they can be characterized by the one-player mechanisms that arise when the declared valuations of the other players are fixed.

As an example for a mechanism, consider the basic case of a combinatorial auction (see De Vries and Vohra [41] for a survey) where two items are sold to two players with additive valuations. In that case, each player i has a two-parameter type $\theta_i = (\theta_{i,1}, \theta_{i,2})$ where the scalar $\theta_{i,j}$ is the monetary equivalent

© The Author(s), under exclusive license to Springer Nature Switzerland AG 2023
A. Del Pia and V. Kaibel (Eds.): IPCO 2023, LNCS 13904, pp. 231–245, 2023.
https://doi.org/10.1007/978-3-031-32726-1_17

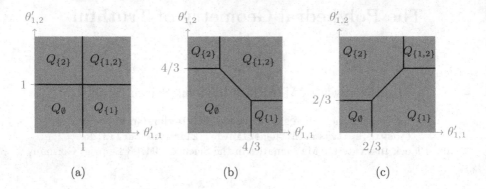

Fig. 1. Difference sets of several mechanisms; cf. [39, Fig. 1].

that player i attaches to receiving item j. For illustration, assume that player 2 reported $\theta'_2 = (1,1)$ and consider the corresponding one-player mechanism for player 1. If each item j is sold independently to the bidder i with the highest reported type $\theta'_{i,j}$ (breaking ties in favor of player 1), we obtain that player 1 receives item j if and only if $\theta'_{1,j} \geq 1$. Geometrically, this one-player mechanism can be represented by its *difference sets* Q_S, $S \in 2^{\{1,2\}}$ where Q_S is equal to the closure of the set of types reported by player 1 so that they get allocated the set of items S. The difference sets were introduced by Vohra [40, p. 41] and reveal valuable information about the properties of the mechanism; see Fig. 1. Figure 1a shows the difference sets of the mechanism selling each item to the highest bidder; Figs. 1b and c show the difference sets of other DSIC mechanisms (not specified here).

Under reasonable assumptions, the difference sets form a polyhedral decomposition of the type space. In this paper, we are interested in characterizing their polyhedral geometry. This is a continuation of work of Vidali [39] who showed that the two combinatorial types shown in Fig. 1b and c are the only cases that can appear for a DSIC mechanism for two items (where the combinatorial type in Fig. 1a is a common degenerate case of both). She then also suggested a similar characterization of the combinatorial types that can appear for three items and asked how these findings can be generalized to more items.

Characterizing the combinatorial types of mechanisms is interesting for a variety of reasons. First, such geometric arguments are often used in order to characterize the set of allocation functions that are implementable by a DSIC mechanism. For instance, the difference sets whose closures have nonempty intersection correspond exactly to two-cycles in an auxiliary network used by Rochet [35] in order to characterize the allocation functions that are implementable by DSIC mechanisms. Second, the combinatorial types can be used to study the sensitivity of mechanisms to deviations in the reported types. As an example consider the mechanism in Fig. 1a. For any $\epsilon > 0$, reporting the type $(1 - \epsilon, 1 - \epsilon)$ yields no item for player 1 while the report of the type $(1 + \epsilon, 1 + \epsilon)$ grants them both items. Put differently, a small change in the reported type may change the outcome from no items being allocated to player 1 to all items being allocated to the same player. This is in contrast with the mechanism shown in Fig. 1c where a small change in the reported type may change the cardinality of the set of allocated items only by 1.[1] Third, the combinatorial types of the mechanism are relevant for the efficiency of the mechanism; see, e.g., [13,39].

Our Results. We give a complete characterization of the combinatorial types of all DSIC combinatorial auctions with m items for any value of m. This answers an open question of Vidali [39] about how to generalize her results for $m = 2$ and $m = 3$ to larger values of m. We employ methods from polyhedral geometry [16] and tropical combinatorics [26]. Our results rest on the observation that a multi-player mechanism is DSIC if and only if its single-player components are DSIC; see [36]. We show that for m items, the combinatorial types of those single-player components are in bijection to equivalence classes of regular subdivisions of the m-dimensional unit cube. We identify the relevant symmetries for exchangeable items and conclude that there are exactly 23 nondegenerate combinatorial types for $m = 3$ and 3,706,261 such types for $m = 4$ (Theorem 9). We then use this characterization to study the optimal sensitivity of mechanisms to slight changes in the reported types. Specifically, we show that for any number of items m, there is a one-player combinatorial auction so that the cardinality of the set of items received by the player changes by at most 1 when the reported type is slightly perturbed (Proposition 11). We also give bounds on a similar measure involving the Hamming distance of the set of received items (Proposition 12). In the full version of this paper, we further show how to apply the same methodology in order to classify the combinatorial types of affine maximizers with n players.

Further Related Work. Rochet's Theorem [35] states that an allocation function is implementable by a DSIC mechanism if and only if the allocation networks of the corresponding one-player mechanisms have no finite cycles of negative lengths. There is a substantial stream of literature exhibiting conditions where it is enough to require conditions on shorter cycles [4,5,8,9,11,18,27]; for instance, it suffices to require the nonnegativity for cycles of length 2 when the preferences are single-peaked [30] or when the type-space is convex [36]. Roberts [34] showed that when the type space of all players is $\mathbb{R}^{\Omega}$, then only affine maximizers are implementable by a DSIC mechanism. Gui et al. [23] and Vohra [40] studied the difference sets Q_A of a mechanism and showed that under reasonable assumptions their closures are polyhedra. Vidali [39] studied the geometry of the polyhedra for the case of two and three items.

In recent years, tropical geometric methods proved to be useful for algorithmic game theory and lead to new results in mechanism design [7,28,37], mean payoff games [1,2], linear optimization [3] and beyond. Beyond the scope of combinatorial auctions, our results are also applicable to the mechanism design problem of scheduling on unrelated machines [12,14,17,21,31].

---

[1] This higher stability of the mechanism in Fig. 1c comes at the expense of a smaller social welfare (i.e., sum of player valuations of the received items) which is maximized for the mechanism in Fig. 1a. Maximizing social welfare, however, is mathematically well-understood since it is achieved by the class of VCG-mechanisms [15,22,38].

2   Preliminaries

In this section, we give a brief overview of basic concepts from mechanism design theory, polyhedral geometry and tropical combinatorics used in this paper. For a more comprehensive treatment we refer to [16, 26, 29, 32].

Mechanism Design. A *multi-dimensional mechanism design problem* consists of a finite set $[m] := \{1, \ldots, m\}$ of *items* and a finite set $[n] := \{1, \ldots, n\}$ of players. Every player i has a *type* $\theta_i = (\theta_{i,1}, \ldots, \theta_{i,m}) \in \mathbb{R}^m$, where the value $\theta_{i,j}$, for $j \in [m]$, is the monetary value player i attaches to receiving item j. A vector $\theta = (\theta_1, \ldots, \theta_n)$ with $\theta_i \in \mathbb{R}^m$ for all $i \in [n]$ is called a *type vector* and $\mathbb{R}^{n \times m}$ is the space of all type vectors. The type θ_i is the private information of player i and unknown to all other players $j \neq i$ and the mechanism designer. Let

$$\Omega = \left\{ A \in \{0,1\}^{n \times m} \;\middle|\; \sum_{i \in [n]} a_{i,j} = 1 \text{ for all } j \in [m] \right\}$$

be the set of allocations of the m items to the n players. The i-th row A_i of an allocation matrix $A \in \Omega$ corresponds to the allocation for the i-th player. A *(direct revelation) auction mechanism* is a tuple $M = (f, p)$ consisting of an *allocation function* $f : \mathbb{R}^{n \times m} \to \Omega$ and a *payment function* $p : \mathbb{R}^{n \times m} \to \mathbb{R}^n$. The mechanism first elicits a claimed type vector $\theta' = (\theta'_1, \ldots, \theta'_n) \in \mathbb{R}^{n \times m}$ where $\theta'_i \in \mathbb{R}^m$ is the type reported by player i. It then chooses an alternative $f(\theta')$ and payments $p(\theta') = (p_1(\theta'), \ldots, p_n(\theta')) \in \mathbb{R}^n$ where $p_i(\theta')$ is the payment from player i to the mechanism. We assume that the players' utilities are *quasi-linear* and the valuations are *additive*, i.e., the utility of player i with type θ_i when the type vector reported to the mechanism is θ' is $u_i(\theta' \mid \theta_i) = f_i(\theta') \cdot \theta_i - p_i(\theta')$, where $f_i(\theta')$ is the i-th row of $f(\theta') = A$. An auction mechanism is called *dominant strategy incentive compatible* (DSIC) or *truthful* if $u_i(\theta \mid \theta_i) \geq u_i((\theta'_i, \theta_{-i}) \mid \theta_i)$, for all $i \in [n]$, $\theta \in \mathbb{R}^{n \times m}$, and $\theta'_i \in \mathbb{R}^m$. Here and throughout, (θ'_i, θ_{-i}) denotes the type vector where player i reports θ'_i and every other player j reports θ_j as in θ. An allocation function is *truthfully implementable*, or just *truthful*, (in weakly dominant strategies) if there is an incentive compatible direct revelation mechanism $M = (f, p)$. For an allocation $A \in \Omega$, let $R_A = \{\theta \in \mathbb{R}^{n \times m} \mid f(\theta) = A\}$ be the preimage of A under f, and let $Q_A = \mathrm{cl}(R_A)$ be the topological closure of R_A. We call Q_A the *difference set* of A.

Polyhedral Geometry and Tropical Combinatorics. We consider the *max-tropical semiring* $(\mathbb{T}, \oplus, \odot)$ with $\mathbb{T} := \mathbb{R} \cup \{-\infty\}$, $a \oplus b := \max\{a, b\}$ and $a \odot b := a + b$. Picking coefficients $\lambda_u \in \mathbb{T}$ for $u \in \mathbb{Z}^m$ such that only finitely many are distinct from $-\infty$ defines an m-variate tropical (Laurent) polynomial, p, whose evaluation at $x \in \mathbb{R}^m$ reads

$$p(x) = \bigoplus_{u \in \mathbb{Z}^m} \lambda_u \odot x^{\odot u} = \max\{\lambda_u + x \cdot u \mid u \in \mathbb{Z}^m\} . \tag{1}$$

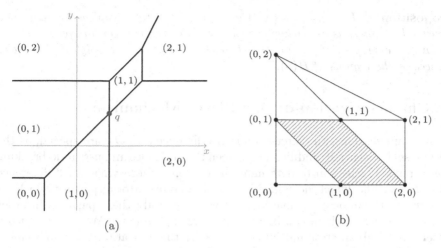

Fig. 2. (a) Tropical hypersurface $H(p)$ and (b) dual regular subdivision of $\mathrm{supp}(p)$, where $p(x,y) = \max\{0, x+1, y+1, 2x, x+y, 2y-1, 2x+y-2\}$. In (a) regions are marked by their support vectors; in (b) the same labels mark vertices of the subdivision. Conversely, e.g., the blue quadrangular cell on the right is dual to the vertex q on the left. (Color figure online)

The *support* of p is the set $\mathrm{supp}(p) = \{u \in \mathbb{Z}^m \mid \lambda_u \neq -\infty\}$. The *tropical hypersurface* $H(p)$ is the set of points $x \in \mathbb{R}^m$ such that the maximum in (1) is attained at least twice. The tropical hypersurface partitions the set $\mathbb{R}^m \setminus H(p)$ into sets in which the maximum of (1) is attained exactly once, for some fixed $u \in \mathbb{Z}^m$; see Fig. 2a. Taking the closure of such a part, we get a *region* of $H(p)$, which is a (possibly unbounded) polyhedron. The *Newton polytope* $\mathcal{N}(p) = \mathrm{conv}(\mathrm{supp}(p))$ of a tropical polynomial p is the convex hull of its support. A finite set $\mathcal{C}$ of polyhedra in $\mathbb{R}^m$ is a *polyhedral complex*, if it is closed with respect to taking faces and if for any two $P, Q \in \mathcal{C}$ the intersection $P \cap Q$ is a face of both P and Q. The polyhedra in $\mathcal{C}$ are the *cells* of $\mathcal{C}$. If every polyhedron in $\mathcal{C}$ is bounded, it is a *polytopal complex*. Further, given a finite set of points $U \subset \mathbb{R}^m$, a polytopal complex $\mathcal{C}$ in $\mathbb{R}^m$ is a *polytopal subdivision* of U if the vertices of all polytopes in $\mathcal{C}$ are points in U and if the union of all polytopes in $\mathcal{C}$ is the convex hull of the points in U. If the cells of a polytopal subdivision are all simplices, it is called a *triangulation*. A subdivision of U is called *regular*, if it can be obtained via a lifting function $\lambda : U \to \mathbb{R}$ on U. Formally, let $P(U, \lambda) := \mathrm{conv}\{(u, \lambda(u)) \in \mathbb{R}^{m+1} \mid u \in U\}$ be the lifted polytope. Its *upper faces* have an outer normal vector with positive last coordinate. Projecting these upper faces by omitting the last coordinate yields a polytopal subdivision of U that is called the regular subdivision of U induced by λ. The following proposition explains the duality between a tropical hypersurface and the regular subdivision of its support. Here we identify a polytopal subdivision with its finite set of cells, partially ordered by inclusion. A proof can be found in [26, Theorem 1.13.]; see Fig. 2 for an example.

Proposition 1. Let $p = \max\{\lambda_u + x \cdot u \mid u \in \mathbb{Z}^m\}$ be a tropical Laurent polynomial. Then there is an inclusion reversing bijection between the regular subdivision of the support of p with respect to $\lambda(u) = \lambda_u$ and the polyhedral complex induced by the regions of $H(p)$.

3   Characterization of One-Player Mechanisms

In many applications, the intersections of difference sets Q_A are special, as the mechanism is essentially indifferent between the outcomes and uses a tie-breaking rule or random selection to determine the outcome. Observing which difference sets intersect and which do not gives rise to a combinatorial pattern which we attribute to the allocation function. We want to study these patterns in order to classify which of them can be attributed to truthfulness. We express such a pattern as an abstract simplicial complex over the allocation space and call it the *indifference complex* of the allocation function.

Formally, an *abstract simplicial complex* over some finite set E is a nonempty set family S of subsets of E, such that for any set $S \in S$ and any subset $T \subseteq S$, we also have $T \in S$. The maximal elements (by inclusion) of an abstract simplicial complex are called *facets*.

Definition 2 (Indifference Complex). The indifference complex $\mathcal{I}(f)$ of an allocation function f is the abstract simplicial complex defined as

$$\mathcal{I}(f) = \left\{ \mathcal{O} \subseteq \Omega \;\middle|\; \bigcap_{A \in \mathcal{O}} Q_A \neq \emptyset \right\}.$$

Note that the indifference complex $\mathcal{I}(f)$ is precisely the nerve complex of the family of difference sets of f; see [10, §10]. Recall that an allocation function f is implementable if there is a DSIC mechanism $M = (f, p)$. Likewise, we call an indifference complex $\mathcal{I}$ implementable, if there is an implementable allocation function f such that $\mathcal{I}(f) = \mathcal{I}$.

We define the *local allocation function* of player i for a given type vector θ_{-i} to be $f_{i,\theta_{-i}}(\theta_i) = A_i$, where A_i is the i-th row of $A = f(\theta_i, \theta_{-i})$. Further, let us fix a payment vector $p \in \mathbb{R}^{2^m}$, which we index by allocations $a \in \{0,1\}^m$. Then, for any type $\theta_i \in \mathbb{R}^m$, we let $u_p(\theta_i) = \max\{\theta_i \cdot a - p_a \mid a \in \{0,1\}^m\}$. The resulting function $u_p : \mathbb{R}^m \to \mathbb{R}$ is a max-tropical polynomial of degree m. We refer to the allocation in $\{0,1\}^m$ which maximizes $u_p(\theta_i)$ as $\arg\max u_p(\theta_i)$. We restate [32, Proposition 9.27], which says that in a truthful setting, the local allocation functions are defined by such tropical polynomials, where the vector p depends only on the types of the other players.

Proposition 3. The allocation function f is truthful, if and only if for all players $i \in [n]$ and all type vectors $\theta \in \mathbb{R}^{n \times m}$, there exists a payment vector $p_i(\theta_{-i}) \in \mathbb{R}^{2^m}$, such that $f_{i,\theta_{-i}}(\theta_i) \in \arg\max u_{p_i(\theta_{-i})}(\theta_i)$.

As an important consequence of Proposition 3, the allocation function f is truthful if and only if all of its local functions $f_{i,\theta_{-i}}$ are truthful. Therefore, for the remainder of this section, we fix a player i and the type vector θ_{-i} of the other players and consider the corresponding one-player mechanism for player i. Note that in this setting, not all items need to be allocated, since the non-allocated items will be distributed among the remaining players. That is, from now on we abuse the notation slightly by considering single-player allocation functions $f : \mathbb{R}^m \to \Omega = \{0,1\}^m$.

Next, we want to discuss the relationship between the indifference complex and the *allocation network*, which is a tool often used to analyze the truthfulness of allocation functions. The latter is the weighted complete directed graph G_f with a node for each allocation $a \in \Omega$ and where the arc lengths are given as $\ell(a, a') = \inf_{\theta \in R_a} \{\theta \cdot a' - \theta \cdot a\}$. The value of $\ell(a, a')$ is the minimal loss of the player's valuation that would occur if the mechanism always chooses allocation a instead of a'. Recall that $R_a = \{\theta \in \mathbb{R}^m \mid f(\theta) = a\}$.

We can link the indifference complex and the allocation network through the following proposition. This generalizes [36, Proposition 5] and also occurs in [39, Lemma 3], without a proof. Here we restate the result in our notation, adding a short proof for the sake of completeness.

Proposition 4. *Let (f, p) be a DSIC mechanism with quasi-linear utilities for one player. Let $C = (a^{(1)}, \ldots, a^{(k)} = a^{(1)})$ be a cycle in the allocation network G_f, such that for each $j \in [k-1]$, we get $Q_{a^{(j)}} \cap Q_{a^{(j+1)}} \neq \emptyset$. Then the length of the cycle C is 0.*

Proof. Let $C = (a^{(1)}, \ldots, a^{(k)})$ be a cycle as in the statement of the proposition. Using [36, Proposition 5], we obtain that for any $\theta \in Q_{a^{(j)}} \cap Q_{a^{(j+1)}}$, the equation $\theta \cdot a^{(j)} - \theta \cdot a^{(j+1)} = \ell(a^{(j)}, a^{(j+1)})$ is satisfied. Since the mechanism (f, p) is truthful and $\theta \in Q_{a^{(j)}} \cap Q_{a^{(j+1)}}$, we obtain $\theta \cdot a^{(j)} - p_{a^{(j)}} = \theta \cdot a^{(j+1)} - p_{a^{(j+1)}}$. Therefore $p_{a^{(j)}} - p_{a^{(j+1)}} = \ell(a^{(j)}, a^{(j+1)})$. Adding up all the lengths of the arcs of C we get 0, which finishes the proof. □

A consequence of Proposition 4 is that all cycles in G_f with the property that all of its edges connect two common nodes of some facet $\mathcal{O} \subseteq \Omega$ of the indifference complex $\mathcal{I}(f)$, have length 0. Especially, each oriented cycle in the one-skeleton of $\mathcal{I}(f)$ is also a zero-cycle in G_f.

For the remainder of this section, our goal is to classify truthful allocation functions for the given type of allocation mechanisms. Recall that Proposition 3 shows that the difference sets of a truthful one-player allocation mechanism are exactly the regions of the tropical utility function of the player. We use this observation to give an alternative proof for the following theorem. It states that there is a bijection between implementable one-player indifference complexes for m items and the regular subdivisions of the m-dimensional cube. A similar result has been shown by Frongillo and Kash [19], who employ power diagrams. The latter are equivalent to regular subdivisions as shown by Aurenhammer [6].[2]

---

[2] Characterizations of implementability in terms of geometric subdivisions of the type space have been used before, e.g., in [24, Proposition 2].

Theorem 5. *An indifference complex $\mathcal{I}$ for m items and one player is implementable if and only if there is a regular subdivision $\mathcal{S}$ of the m-dimensional unit cube, such that the facets of $\mathcal{I}$ are precisely the vertex sets of the maximal cells of $\mathcal{S}$.*

Proof. Let $\mathcal{I}$ be an indifference complex. It is implementable if and only if there exists a truthful allocation function f with $\mathcal{I}(f) = \mathcal{I}$. By Proposition 3, this is equivalent to the fact that there is a payment vector $p \in \mathbb{R}^{2^m}$, such that the difference sets Q_a are exactly the regions of the tropical hypersurface $H(u_p)$. As the Newton polytope of u_p is the unit cube $[0,1]^m$, Proposition 1 provides a duality between the difference sets Q_a and the regular subdivision of $[0,1]^m$ with respect to the lifting $\lambda(a) = -p_a$. Hence, a maximal cell in the regular subdivision with vertices $(a^{(1)}, \ldots, a^{(k)})$ corresponds to a maximal set of allocations such that $\bigcap_{a \in \{a^{(1)}, \ldots, a^{(k)}\}} Q_a \neq \emptyset$. The latter is a facet of $\mathcal{I}(f)$. Conversely, if we start with a regular subdivision $\mathcal{S}$ of the unit cube induced by some lifting λ, setting the prices $p_a = -\lambda(a)$ and defining $f(\theta) \in \arg\max u_p(\theta)$ results in a DSIC mechanism such that the indifference complex $\mathcal{I}(f)$ corresponds to $\mathcal{S}$ in the way described in the statement of the theorem. □

Note that the proof is constructive. Further, Theorem 5 says that the simplicial complex $\mathcal{I}(f)$ is precisely the crosscut complex of the poset of cells of the regular subdivision $\mathcal{S}$ [10, §10]. If $\mathcal{S}$ is a triangulation then its crosscut complex is $\mathcal{S}$ itself, seen as an abstract simplicial complex. The main consequence of Theorem 5 is that for truthful one-player mechanisms with additive and quasi-linear utilities, the partitioning of the type space into difference sets is characterized by the duality to the regular subdivision of the cube, which is captured by the indifference complex.

Example 6. Consider the one-player case with $m = 3$ items and where the player has a type $\theta \in \mathbb{R}^3$. Let f be the local allocation function which we define via $f(\theta) \in \arg\max\{\theta \cdot a - p_a \mid a \in \{0,1\}^3\}$, with $p_{000} = 0$, $p_{100} = p_{010} = p_{001} = 3/7$, $p_{110} = p_{101} = p_{011} = 8/7$, and $p_{111} = 10/7$. Figure 3 shows the subset of the type space for $\theta \in [0,1]^3$. The five facets of $\mathcal{I}(f)$ are $\{0,1,2,4\}$, $\{2,3,4,7\}$, $\{1,4,5,7\}$, $\{1,2,3,7\}$, $\{3,5,6,7\}$; here we use the binary encoding $4a_1 + 2a_2 + a_3$ for the vertex $a \in \{0,1\}^3$. Those cells form a regular triangulation of $[0,1]^3$, which is type F in Fig. 4.

We define two allocation functions $f, g : \mathbb{R}^m \to \Omega$ as *combinatorially equivalent* if their indifference complexes agree; i.e., $\mathcal{I}(f) = \mathcal{I}(g)$. As before we are primarily concerned with the case where $\Omega = \{0,1\}^m$ and the allocation functions are truthful allocations of m items. In this way we can relate the allocation space with regular subdivisions of the cube $[0,1]^m$.

Definition 7. *A truthful allocation function on m items is* nondegenerate *if the associated regular subdivision of the m-cube is a triangulation.*

Triangulations of m-cubes are described in [16, §6.3]. The first two columns of Table 1 summarize the known values of the number of all (regular) triangulations of the m-cube. In particular the second column shows the number of

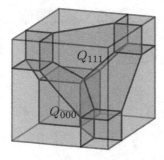

Fig. 3. Subdivision of the subset of the type space, where $\theta \in [0,1]^3$, induced by the allocation function described in Example 6. The region Q_{000} corresponds to the corner in the lower back of the cube and Q_{111} corresponds to the upper front corner. This and other pictures were obtained via `polymake` [20].

Table 1. Triangulations of m-cubes. Orbit sizes refer to regular triangulations

m	All	Regular	Sym(m)-orbits	Γ_m-orbits
2	2	2	2	1
3	74	74	23	6
4	92,487,256	87,959,448	3,706,261	235,277

combinatorial types of nondegenerate truthful allocations. The number of all, not necessarily regular, triangulations of the 4-cube was found by Pournin [33]. The corresponding numbers of triangulations for $m \geq 5$ are unknown.

Remark 8. Any regular subdivision may be refined to a regular triangulation, on the same set of vertices; see [16, Lemma 2.3.15].

Our next goal is to explain the third and fourth columns of Table 1. To this end we need to discuss the symmetries of the cube, which are known. That will be the key to understanding (truthful) allocations of exchangeable items. The automorphism group, Γ_m, of the m-cube $[0,1]^m$ comprises those bijections on the vertex set which map faces to faces. The group Γ_m is known to be a semidirect product of the symmetric group Sym(m) with $\mathbb{Z}_2^m$; its order is $m! \cdot 2^m$. Here the j-th component of $\mathbb{Z}_2^m$ flips the j-th coordinate, and this is a reflection at the affine hyperplane $x_j = \frac{1}{2}$; that map does not have any fixed points among the vertices of $[0,1]^m$. The subgroup $\mathbb{Z}_2^m$ of all coordinate flips acts transitively on the 2^m vertices. The symmetric group Sym(m) naturally acts on the coordinate directions; this is precisely the stabilizer of the origin in Γ_m. Since the cells in each triangulation of $[0,1]^m$ are convex hulls of a subset of the vertices, the group Γ_m also acts on the set of all triangulations of $[0,1]^m$. Moreover, since Γ_m acts via affine maps, it sends regular triangulations to regular triangulations. The stabilizer Sym(m) acts transitively on the $\binom{m}{k}$ vertices of $[0,1]^m$ with exactly k ones. In this way, a Sym(m)-orbit of regular triangulations corresponds to a set of nondegenerate truthful allocation functions for which the

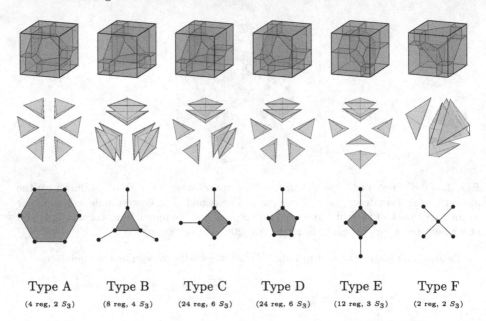

Type A	Type B	Type C	Type D	Type E	Type F
(4 reg, 2 S_3)	(8 reg, 4 S_3)	(24 reg, 6 S_3)	(24 reg, 6 S_3)	(12 reg, 3 S_3)	(2 reg, 2 S_3)

Fig. 4. The combinatorial types of truthful allocation functions corresponding to the six Γ_3-orbits of the 3-cube, together with the corresponding triangulations (exploded) and their tight spans; cf. [16, Fig. 6.35]. The *tight span* of a regular triangulation $\mathcal{S}$ is the subcomplex of bounded cells of the tropical hypersurface dual to $\mathcal{S}$ (seen as an ordinary polyhedral complex); see [26, §10.7]. The numbers below the types show how many regular triangulations or Sym(3)-orbits are of the given type, respectively.

indifference complexes agree, up to permuting the items. We call such allocation functions *combinatorially equivalent for exchangeable items*. The Sym(m)-orbits of regular triangulations have been computed with `mptopcom` [25]; see the third column of Table 1. By Theorem 5 those triangulations bijectively correspond to the implementable indifference complexes. So that computation furnishes a proof of the following result.

Theorem 9. *There are 23 combinatorial types of nondegenerate truthful allocation functions for one player and $m = 3$ exchangeable items. Further, the corresponding count for $m = 4$ yields 3,706,261.*

It makes sense to focus on the combinatorics of triangulations, without paying attention to their interpretations for auctions. This amounts to studying the orbits of the full group Γ_m acting on the set of (regular) triangulations; see the fourth column of Table 1. The six Γ_3-orbits of triangulations of the 3-cube are depicted in Fig. 4. This number expands to 23 if we consider the possible choices of locating the origin. We illustrate the idea for the subdivision of the type space in Fig. 3. Its Γ_3-orbit splits into two Sym(3)-orbits: one from putting the origin in one of the four cubes or in one of the four noncubical cells.

Remark 10. Vidali considered a more restrictive notion of nondegeneracy of allocation functions [39, Definition 8], and in [39, Theorem 1] she arrived at a classification of five types for three items. In our terminology, the number of types is equal to six, which is the count of Γ_3-orbits of regular triangulations of $[0,1]^3$ reported in Table. 1. The missing type in the classification of Vidali is type F (as in Fig. 4), arising from Example 6. Further details will be given in the full version of this paper.

4   Sensitivity of Mechanisms

In this section, we study by how much the allocations for a fixed player may change under a slight modification of the reported type. These changes are measured in the following two ways. For two local allocations $a, b \in \{0,1\}^m$, let the *cardinality distance* be $d_c(a,b) := \big|\,|a|_1 - |b|_1\,\big|$, and let the *Hamming distance* be $d_h(a,b) := |a - b|_1$, where $|\cdot|_1$ is the 1-norm. Note that the cardinality distance is a pseudometric. Let f be an one-player allocation function, we define the *cardinality sensitivity* of f as

$$\mu_c(f) \;=\; \max\big\{ d_c(a,b) \;\big|\; a,b \in F \text{ for some } F \in \mathcal{I}(f)\big\} \;.$$

The *Hamming sensitivity* $\mu_h(f)$ arises in the same way, with d_h instead of d_c. Intuitively, the cardinality sensitivity $\mu_c(f)$ is the maximal amount such that any slight change in the type of the player does not cause her allocated bundle to change its cardinality by more than $\mu_c(f)$. Let Φ_m be the set of truthful allocation functions for one player and m items. We are now interested in computing the values $M_c(m) := \min_{f \in \Phi_m} \mu_c(f)$ and $M_h(m) := \min_{f \in \Phi_m} \mu_h(f)$.

Our strategy to compute these values is as follows. From Theorem 5 we know that the indifference complexes of allocation functions $f \in \Phi_m$ are in bijection with the regular subdivisions of $[0,1]^m$. So we need to identify those subdivisions, for which the maximal distance between any two vertices of one of its cells is minimized. In this way, we can compute $M_c(m)$ exactly, and we give bounds for $M_h(m)$.

Proposition 11. *The minimal cardinality sensitivity of DSIC one-player mechanisms is $M_c(m) = 1$.*

Proof. We first slice the unit cube into the polytopes

$$P_k \;=\; \left\{ x \in [0,1]^m \;\middle|\; k-1 \le \sum_{i=1}^m x_i \le k \right\}, \qquad k = 1, \ldots, m \;.$$

The polytopes $P_1, \ldots, P_m$ form the maximal cells of a polytopal subdivision $\mathcal{S}$ of $[0,1]^m$. That subdivision is regular with height function $\lambda(x) = -\left(\sum_{i=1}^m x_i\right)^2$. This proves the claim, as for each P_k, the difference in the coordinate sums of two of its vertices differ by at most one. $\square$

Note that the height function we used in the proof of the last proposition leads to the mechanism which is defined by the prices $p(a) = \left(\sum_{i=1}^{m} a_i\right)^2$ for the allocations $a \in \{0,1\}^m$.

Proposition 12. *The minimal Hamming sensitivity for DSIC one-player mechanisms on $m \geq 3$ items is bounded by $2 \leq M_h(m) \leq m - 1$.*

Proof. For the lower bound let us consider a triangle with the vertices $a, b, c \in \{0,1\}^m$. If we assume $d_h(a,b) = d_h(a,c) = 1$ then the vertices a and b (resp. a and c) differ by a coordinate flip. Therefore, the vertices b and c differ by either two coordinate flips or none. As $b \neq c$, the former is the case and $d_h(b,c) = 2$. As the maximal cells of a subdivision of $[0,1]^m$ for $m \geq 2$ contain at least three vertices, this proves the lower bound.

For the upper bound, we show that there is a subdivision, $\mathcal{S}$, of $[0,1]^m$ such that no cell of $\mathcal{S}$ contains two antipodal vertices, i.e., two vertices such that their sum equals the all ones vector. We first consider the case where m is odd. For a vertex $x \in \{0,1\}^m$, let $\Delta(x)$ be the *cornered simplex* with apex x. That is, its vertices comprise x and all its neighbors in the vertex-edge graph of the unit cube; cf. [16, Fig. 6.3.1]. Let $\mathcal{S}_m$ be the subdivision of $[0,1]^m$ with the following maximal cells: the *big cell* is the convex hull of all vertices with an even number of ones, and the *small cells* are the cornered simplices $\Delta(x)$, where $x \in \{0,1\}^m$ with $\sum x_i$ odd. The subdivision $\mathcal{S}_3$ is the triangulation of type F in Fig. 4; for $m \geq 5$ the big cell is not a simplex, and so $\mathcal{S}_m$ is not a triangulation in general.

At any rate, the subdivision $\mathcal{S}_m$ is always regular: it is induced by the height function which sends a vertex x to 0, if it has an even number of ones and to -1, if that number is odd. For $m \geq 3$ odd, no antipodal pair of vertices is contained in any cell of $\mathcal{S}_m$, which proves the claim for the uneven case.

If the dimension m is even, we consider the m-dimensional unit cube as a prism over $[0,1]^{m-1}$. Then $m - 1$ is odd, and we can employ the subdivision $\mathcal{S}_{m-1}$ of $[0,1]^{m-1}$ that we discussed before. We obtain a subdivision, $\mathcal{S}_m$, of $[0,1]^m$ whose maximal cells are prisms over the maximal cells of $\mathcal{S}_{m-1}$. The subdivision $\mathcal{S}_m$ is again regular: this can be seen from assigning the vertices $x \times \{0\}$ and $x \times \{1\}$ the same height as the vertex x in $\mathcal{S}_{m-1}$. Now let P be a maximal cell in $\mathcal{S}_{m-1}$, such that $Q = P \times [0,1]$ is a maximal cell of $\mathcal{S}_m$. If Q contained an antipodal pair of vertices, then by removing the last coordinate, we would get an antipodal pair in P, which is absurd. This completes the proof. $\square$

5   Conclusion

We studied DSIC allocation mechanisms where a set of m items is allocated to n players. These mechanisms can be described by the corresponding one-player mechanisms when the types declared by the other players are fixed. For a single player, the allocations correspond to vectors $\{0,1\}^m$, and the combinatorial types of the allocation mechanisms correspond to regular subdivisions of the m-dimensional unit cube. We then used this insight to design mechanisms that are robust in the sense that small changes in the declared type do not lead to a major change in the set of allocated items. In the full version of this paper, we will show how this method can be applied in order to describe affine maximizers with n players.

For multiple copies of items, the deterministic allocations to a single player correspond to a subset of the lattice $\mathbb{N}^m$, and it seems plausible that DSIC mechanisms for such scenarios can also be described by regular subdivisions.

Question 13. How does our approach generalize to allocation mechanisms in a setting with multiple copies of items?

Acknowledgments. We thank Benny Moldovanu for pointing out the work of Frongillo and Kash [19]. Further, we are indebted to three anonymous reviewers for their comments and corrections. This work was supported by Deutsche Forschungsgemeinschaft under Germany's Excellence Strategy, Berlin Mathematics Research Center (Grant EXC-2046/1, project-ID 390685689). M. Joswig has further been supported by "Symbolic Tools in Mathematics and their Application" (TRR 195, project-ID 286237555).

References

1. Akian, M., Gaubert, S., Guterman, A.: Tropical polyhedra are equivalent to mean payoff games. Internat. J. Algebra Comput. **22**(1), 1250001 (2012). https://doi.org/10.1142/S0218196711006674
2. Allamigeon, X., Benchimol, P., Gaubert, S., Joswig, M.: Combinatorial simplex algorithms can solve mean payoff games. SIAM J. Opt. **24**(4), 2096–2117 (2014). https://doi.org/10.1137/140953800
3. Allamigeon, X., Benchimol, P., Gaubert, S., Joswig, M.: What tropical geometry tells us about the complexity of linear programming. SIAM Rev. **63**(1), 123–164 (2021). https://doi.org/10.1137/20M1380211
4. Archera, A., Kleinberg, R.: Truthful germs are contagious: a local-to-global characterization of truthfulness. Games Econ. Behav. **86**, 340–366 (2014). https://doi.org/10.1016/j.geb.2014.01.004
5. Ashlagi, I., Braverman, M., Hassidim, A., Monderer, D.: Monotonicity and implementability. Econometrica **78**(5), 1749–1772 (2010). https://doi.org/10.3982/ECTA8882
6. Aurenhammer, F.: Power diagrams: Properties, algorithms and applications. SIAM J. Comput. **16**(1), 78–96 (1987). https://doi.org/10.1137/0216006
7. Baldwin, E., Klemperer, P.: Understanding preferences: "demand types", and the existence of equilibrium with indivisibilities. Econometrica **87**(3), 867–932 (2019). https://doi.org/10.3982/ECTA13693
8. Berger, A., Müller, R., Naeemi, S.H.: Characterizing implementable allocation rules in multi-dimensional environments. Soc. Choice Welfare **48**(2), 367–383 (2016). https://doi.org/10.1007/s00355-016-1008-6
9. Bikhchandani, S., Chatterji, S., Lavi, R., Mu'alem, A., Nisan, N., Sen, A.: Weak monotonicity characterizes deterministic dominant-strategy implementation. Econometrica **74**(4), 1109–1132 (2006). https://doi.org/10.1111/j.1468-0262.2006.00695.x

10. Björner, A.: Topological methods. In: Handbook of Combinatorics, vol. 1, 2, pp. 1819–1872. Elsevier, Amsterdam (1995)

11. Carbajala, J.C., Müller, R.: Implementability under monotonic transformations in differences. J. Econ. Theory **160**, 114–131 (2015). https://doi.org/10.1016/j.jet.2015.09.001

12. Christodoulou, G., Koutsoupias, E., Kovács, A.: On the Nisan-Ronen conjecture. In: 2021 IEEE 62nd Annual Symposium on Foundations of Computer Science (FOCS), pp. 839–850 (2022). https://doi.org/10.1109/FOCS52979.2021.00086

13. Christodoulou, G., Koutsoupias, E., Vidali, A.: A characterization of 2-player mechanisms for scheduling. In: Proceedings of the 16th Annual European Symposium on Algorithms, (ESA), pp. 297–307 (2008). https://doi.org/10.1007/978-3-540-87744-8_25

14. Christodoulou, G., Koutsoupias, E., Vidali, A.: A lower bound for scheduling mechanisms. Algorithmica **55**(4), 729–740 (2009). https://doi.org/10.1007/s00453-008-9165-3

15. Clarke, E.H.: Multipart pricing of public goods. Public Choice **11**, 17–33 (1971). https://doi.org/10.1007/bf01726210

16. De Loera, J.A., Rambau, J., Santos, F.: Triangulations: Structures for algorithms and applications, Algorithms and Computation in Mathematics, vol. 25. Springer, Berlin (2010). https://doi.org/10.1007/978-3-642-12971-1

17. Dobzinski, S., Shaulker, A.: Improved lower bound for truthful scheduling, abs/2007.04362 (2020)

18. Edelman, P.H., Weymark, J.A.: Dominant strategy implementability and zero length cycles. Econ. Theor. **72**(4), 1091–1120 (2020). https://doi.org/10.1007/s00199-020-01324-7

19. Frongillo, R.M., Kash, I.A.: General truthfulness characterizations via convex analysis. Games Econ. Behav. **130**, 636–662 (2021). https://doi.org/10.1016/j.geb.2021.09.010

20. Gawrilow, E., Joswig, M.: a framework for analyzing convex polytopes. In: Polytopes–combinatorics and computation (Oberwolfach, 1997), DMV Sem., vol. 29, pp. 43–73. Birkhäuser, Basel (2000). https://doi.org/10.1007/978-3-0348-8438-9_2

21. Giannakopoulos, Y., Hammerl, A., Poças, D.: A New Lower Bound for Deterministic Truthful Scheduling. Algorithmica **83**(9), 2895–2913 (2021). https://doi.org/10.1007/s00453-021-00847-2

22. Groves, T.: Incentives in teams. Econometrica **41**, 617–631 (1973). https://doi.org/10.2307/1914085

23. Gui, H., Müller, R., Vohra, R.V.: Dominant strategy mechanisms with multidimensional types. In: Lehmann, D., Müller, R., Sandholm, T. (eds.) Computing and Markets. Dagstuhl Seminar Proceedings, vol. 5011, pp. 1–23 (2005). https://doi.org/10.4230/DagSemProc.05011.8

24. Jehiel, P., Moldovanu, B., Stacchetti, E.: Multidimensional mechanism design for auctions with externalities. J. Econ. Theory **85**(2), 258–293 (1999). https://doi.org/10.1006/jeth.1998.2501

25. Jordan, C., Joswig, M., Kastner, L.: Parallel enumeration of triangulations. Electron. J. Combin. **25**(3), Paper 3.6, 27 (2018). https://doi.org/10.37236/7318

26. Joswig, M.: Essentials of tropical combinatorics. Graduate Studies in Mathematics, American Mathematical Society, Providence, RI (2022)

27. Kushnir, A.I., Lokutsievskiy, L.V.: When is a monotone function cyclically monotone? Theor. Econ. **16**, 853–879 (2021). https://doi.org/10.3982/TE4305

28. Lin, B., Tran, N.M.: Two-player incentive compatible outcome functions are affine maximizers. Linear Algebra Its Appl. **578**, 133–152 (2019). https://doi.org/10.1016/j.laa.2019.04.027

29. Maclagan, D., Sturmfels, B.: Introduction to tropical geometry, Graduate Studies in Mathematics, vol. 161. American Mathematical Society, Providence, RI (2015)

30. Mishra, D., Pramanik, A., Roy, S.: Multidimensional mechanism design in single peaked type spaces. J. Econ. Theory **153**, 103–116 (2014). https://doi.org/10.1016/j.jet.2014.06.002

31. Nisan, N., Ronen, A.: Algorithmic mechanism design. Games Econ. Behav. **35**(1), 166–196 (2001). https://doi.org/10.1006/game.1999.0790

32. Nisan, N., Roughgarden, T., Tardos, E., Vazirani, V.V.: Algorithmic game theory. Cambridge University Press, USA (2007). https://doi.org/10.1017/CBO9780511800481

33. Pournin, L.: The flip-graph of the 4-dimensional cube is connected. Discrete Comput. Geometry **49**(3), 511–530 (2013). https://doi.org/10.1007/s00454-013-9488-y

34. Roberts, K.: The characterization of implementable choice rules. Aggregation and Revelation of Preferences, pp. 321–349 (1979)

35. Rochet, J.C.: A necessary and sufficient condition for rationalizability in a quasi-linear context. J. Math. Econ. **16**(2), 191–200 (1987). https://doi.org/10.1016/0304-4068(87)90007-3

36. Saks, M.E., Yu, L.: Weak monotonicity suffices for truthfulness on convex domains. In: Riedl, J., Kearns, M.J., Reiter, M.K. (eds.) Proceedings of the 6th ACM Conference on Electronic Commerce (EC), pp. 286–293 (2005). https://doi.org/10.1145/1064009.1064040

37. Tran, N.M., Yu, J.: Product-mix auctions and tropical geometry. Math. Oper. Res. **44**(4), 1396–1411 (2019). https://doi.org/10.1287/moor.2018.0975

38. Vickrey, W.: Counterspeculation, auctions, and competitive sealed tenders. J. Finance **16**, 8–37 (1961). https://doi.org/10.1111/j.1540-6261.1961.tb02789.x

39. Vidali, A.: The geometry of truthfulness. In: Leonardi, S. (ed.) Proceedings of the 5th International Workshop on Internet and Network Economics (WINE), pp. 340–350 (2009). https://doi.org/10.1007/978-3-642-10841-9_31

40. Vohra, R.V.: Mechanism design. A Linear Programming Approach, Econometric Society Monographs, vol. 47. Cambridge University Press, Cambridge (2011). https://doi.org/10.1017/CBO9780511835216

41. de Vries, S., Vohra, R.V.: Combinatorial auctions: a survey. INFORMS J. Comput. **15**, 284–309 (2003). https://doi.org/10.1287/ijoc.15.3.284.16077

Competitive Kill-and-Restart and Preemptive Strategies for Non-clairvoyant Scheduling

Sven Jäger[1] [iD], Guillaume Sagnol[2]([✉]) [iD],
Daniel Schmidt genannt Waldschmidt[2] [iD], and Philipp Warode[3] [iD]

[1] RPTU Kaiserslautern-Landau, Paul-Ehrlich-Straße 14, 67663 Kaiserslautern,
Germany
sven.jaeger@rptu.de
[2] TU Berlin, Straße des 17. Juni 136, 10623 Berlin, Germany
{sagnol,dschmidt}@math.tu-berlin.de
[3] HU Berlin, Unter den Linden 6, 10099 Berlin, Germany
philipp.warode@hu-berlin.de

Abstract. We study kill-and-restart and preemptive strategies for the fundamental scheduling problem of minimizing the sum of weighted completion times on a single machine in the non-clairvoyant setting. First, we show a lower bound of 3 for any deterministic non-clairvoyant kill-and-restart strategy. Then, we give for any $b > 1$ a tight analysis for the natural b-scaling kill-and-restart strategy as well as for a randomized variant of it. In particular, we show a competitive ratio of $(1 + 3\sqrt{3}) \approx 6.197$ for the deterministic and of ≈ 3.032 for the randomized strategy by making use of the largest eigenvalue of a Toeplitz matrix. In addition, we show that the preemptive Weighted Shortest Elapsed Time First (WSETF) rule is 2-competitive when jobs are released online, matching the lower bound for the unit weight case with trivial release dates for any non-clairvoyant algorithm. Furthermore, we prove performance guarantees smaller than 10 for adaptions of the b-scaling strategy to online release dates and unweighted jobs on identical parallel machines.

1 Introduction

Minimizing the total weighted completion time on a single processor is one of the most fundamental problems in the field of machine scheduling. The input consists of n jobs with processing times $p_1, \ldots, p_n$ and weights $w_1, \ldots, w_n$, and the task is to sequence them in such a way that the sum of weighted completion times $\sum_{j=1}^{n} w_j C_j$ is minimized. We denote this problem as $1 \,\|\, \sum w_j C_j$.

Full version preprint: http://arxiv.org/abs/2211.02044.

The research of the second, third and fourth authors was supported by the Deutsche Forschungsgemeinschaft (DFG, German Research Foundation) under Germany's Excellence Strategy — The Berlin Mathematics Research Center MATH+ (EXC-2046/1, project ID: 390685689).

A. Del Pia and V. Kaibel (Eds.): IPCO 2023, LNCS 13904, pp. 246–260, 2023.
https://doi.org/10.1007/978-3-031-32726-1_18

Smith [29] showed in the 50's that the optimal schedule is obtained by the Weighted Shortest Processing Time first (WSPT) rule, i.e., jobs are sequenced in non-decreasing order of the ratio of their processing time and their weight.

Reality does not always provide all information beforehand. Around 30 years ago, the non-clairvoyant model, in which the processing time of any job becomes known only upon its completion, was introduced for several scheduling problems [10,25,27]. It is easy to see that no non-preemptive non-clairvoyant algorithm can be constant-competitive for $1 \mid\mid \sum C_j$. In their seminal work, Motwani et al. [25] proved for this problem that allowing preemption breaks the non-constant barrier. Specifically, they showed that the round-robin algorithm is 2-competitive, matching a lower bound for all non-clairvoyant algorithms. This opened up a new research direction, leading to constant-competitive preemptive non-clairvoyant algorithms in more general settings, like weighted jobs [21], multiple machines [6,15,16], precedence constraints [11], and non-trivial release dates. When jobs are released over time, they are assumed to be unknown before their arrivals (online scheduling). No lower bound better than 2 is known for this case, whereas the best known upper bound before this work was 3, see e.g. [23].

But there is a downside of the preemptive paradigm as it uses an unlimited number of interruptions at no cost and has a huge memory requirement to maintain the ability to resume all interrupted jobs. Therefore, we continue by studying the natural class of *kill-and-restart strategies* that—inspired by computer processes—can abort the execution of a job (kill), but when processed again later, the job has to be re-executed from the beginning (restart). It can be considered as an intermediate category of algorithms between preemptive and non-preemptive ones, as on the one hand jobs may be interrupted, and on the other hand all completed jobs have been processed as a whole. Hence, by removing all aborted executions one obtains a non-preemptive schedule. This class of algorithms has already been investigated since the 90's [27], but to the best of our knowledge, the competitive ratio of non-clairvoyant kill-and-restart strategies for the total completion time objective has never been studied.

Our Contribution. We start by strengthening the preemptive lower bound of 2 for the kill-and-restart model.

Theorem 1. *For* $1 \mid\mid \sum C_j$, *no deterministic non-clairvoyant kill-and-restart strategy can achieve a competitive ratio smaller than* $3 - \frac{2}{n+1}$ *on instances with* $n \geq 3$ *jobs, even if every job j has processing time $p_j \geq 1$.*

The main part of this work is devoted to the natural b-scaling strategy $\mathfrak{D}_b$ that repeatedly probes each unfinished job for the time of some integer power of $b > 1$ multiplied by its weight. While the fact that $\mathfrak{D}_2$ is 8-competitive can easily be concluded from the 2-competitiveness of the weighted round-robin algorithm [21], a tight analysis requires more involved techniques.

Theorem 2. *For $b > 1$, $\mathfrak{D}_b$ is $\left(1 + \frac{2b^{3/2}}{b-1}\right)$-competitive for $1 \mid\mid \sum w_j C_j$. Moreover, for all $b > 1$ this bound is tight, even for unit weight instances. In particular, for $b = 3$ the competitive ratio is $1 + 3\sqrt{3} \approx 6.196$.*

Our main technique is to reduce the problem of finding the exact competitive ratio to the computation of the largest eigenvalue of a tridiagonal Toeplitz matrix. Subsequently, we obtain a significantly better exact competitive ratio for a randomized version $\mathfrak{R}_b$ of the b-scaling strategy.

Theorem 3. *For every* $b > 1$, $\mathfrak{R}_b$ *is* $\frac{\sqrt{b}+2b-1}{\sqrt{b}\log(b)}$-*competitive for* $1 \mid\mid \sum w_j C_j$. *Moreover, for all* $b > 1$ *this bound is tight, even for unit weight instances. In particular, for* $b \approx 8.16$ *the competitive ratio is* ≈ 3.032.

The analysis mimics that of the deterministic strategy, but it is necessary to group jobs with similar Smith ratios. This approach leads to the computation of the largest eigenvalues of a sequence of banded symmetric Toeplitz matrices, and the result is obtained by taking its limit.

We then study more general scheduling environments. For the online problem in which jobs are released over time, denoted by $1 \mid r_j$, pmtn $\mid \sum w_j C_j$, we close the gap for the best competitive ratio of preemptive algorithms by analyzing the Weighted Shortest Elapsed Time First rule (WSETF). This algorithm runs at every point in time the job(s) with minimum ratio of elapsed processing time over weight.

Theorem 4. *WSETF is 2-competitive for* $1 \mid r_j$, pmtn $\mid \sum w_j C_j$.

Theorem 4 generalizes the known 2-competitiveness for trivial release dates shown by Kim and Chwa [21]. It also matches the performance guarantee of the best known stochastic online scheduling policy, called F-GIPP [24], for the stochastic variant of the problem, where the probability distributions of the processing times are given at the release dates and the expected objective value is to be minimized. Our improvement upon the analysis of this policy, applied to a single machine, is threefold: First, our strategy does not require any information about the distributions of the processing times, second, we compare to the clairvoyant optimum, while F-GIPP is compared to the optimal non-anticipatory policy, and third, WSETF is more intuitive and easier to implement in applications than the F-GIPP policy.

Using Theorem 4, we then give an upper bound on the competitive ratio of a generalized version of $\mathfrak{D}_b$ for jobs arriving online over time.

Theorem 5. $\mathfrak{D}_b$ *is* $\frac{2b^4}{2b^2-3b+1}$-*competitive for* $1 \mid r_j \mid \sum w_j C_j$. *In particular, for* $b = \frac{9+\sqrt{17}}{8}$, *its performance guarantee is* $\frac{107+51\sqrt{17}}{32} \approx 9.915$.

Finally, we also analyze the unweighted problem $P \mid\mid \sum C_j$ on multiple identical parallel machines.

Theorem 6. $\mathfrak{D}_b$ *is* $\frac{3b^2-b}{b-1}$-*competitive for* $P \mid\mid \sum C_j$. *In particular, for* $b = \frac{3+\sqrt{6}}{3}$, *its performance guarantee is* $5 + 2\sqrt{6} \approx 9.899$.

The proofs of these results are sketched in Sect. 3 to 6 below; full proofs are provided in the preprint [17].

Related Work. The clairvoyant offline variants of all scheduling problems considered in this paper are well understood; either there is a polynomial-time algorithm [8,29], or the problem is strongly NP-hard [22,26] and there is a polynomial-time approximation scheme [1]. In the clairvoyant online model, where the processing times become known at the jobs' release dates, there is a 1.566-competitive deterministic algorithm [28] and a deterministic lower bound of 1.073 [9], when preemption is allowed. For non-preemptive online scheduling the best possible deterministic competitive ratio is exactly 2. [2,14].

In the non-clairvoyant setting no (randomized) non-preemptive algorithm is constant-competitive, so that allowing preemption is crucial. Motwani et al. [25] showed that the simple (non-clairvoyant) round-robin procedure has a competitive ratio of 2 for minimizing the total completion time on identical machines. A weighted variant was presented for a single machine by Kim and Chwa [21] and for identical machines by Beaumont et al. [6]. In the context of non-clairvoyant online scheduling one distinguishes between total (weighted) completion time and total (weighted) flow time. For weighted flow time constant competitiveness is unattainable [25]. Besides work on non-constant competitiveness [7], the problem has been primarily studied in the resource augmentation model [18]. Kim and Chwa and Bansal and Dhamdhere [4] independently showed that WSETF is $(1 + \varepsilon)$-speed $(1 + 1/\varepsilon)$-competitive for weighted flow time on a single machine, entailing a 4-competitive algorithm without speed augmentation for weighted completion time [5]. The proof technique used is, however, not suitable for obtaining better bounds for the weighted completion time objective. For unrelated machines there is a $(1 + \varepsilon)$-speed $O(1/\varepsilon^2)$-competitive algorithm [16].

Shmoys et al. [27] introduced the kill-and-restart model in the context of makespan minimization. We are not aware of any work for the total completion time objective in the non-clairvoyant model. However, for the clairvoyant online model, lower bounds on the competitive ratio of kill-and-restart strategies have been obtained [9,31], and van Stee and La Poutré [30] developed a 3/2-competitive strategy for a single machine, beating the aforementioned best competitive ratio of 2 for non-preemptive algorithms. Motwani et al. also considered preemptive scheduling with a limited number of allowed preemptions, for which they devised algorithms similar to the kill-and-restart strategies presented in this paper.

The kill-and-restart model also shares many similarities with *optimal search problems*, in particular the w-lanes *cow-path* problem. For $w = 2$, deterministic and randomized strategies achieving the best possible competitive ratio are studied in [3,20], respectively. This work has been extended by Kao et al. [19] to the general case $w \in \mathbb{N}$.

2   Preliminaries

We consider the machine scheduling problem of minimizing the weighted sum of completion times on a single machine ($1 \mid\mid \sum w_j C_j$). Formally, an instance $I = (\boldsymbol{p}, \boldsymbol{w})$ consists of a vector of processing times $\boldsymbol{p} = (p_j)_{j=1}^n$ and a vector of weights $\boldsymbol{w} = (w_j)_{j=1}^n$. Sequencing jobs in WSPT order, i.e., ordered non-decreasingly by their Smith ratios p_j/w_j, yields an optimal schedule, denoted by OPT(I). We also denote its objective value by OPT(I).

The focus of our work lies on the analysis of non-clairvoyant strategies. We call a strategy *non-clairvoyant* if it does not use information on the processing time p_j of a job j before j has been completed. A deterministic strategy $\mathfrak{D}$ is said to be *c-competitive* if, for all instances $I = (\boldsymbol{p}, \boldsymbol{w})$, $\mathfrak{D}(I) \le c \cdot \text{OPT}(I)$, where $\mathfrak{D}(I)$ denotes the cost of the strategy $\mathfrak{D}$ for instance I. The *competitive ratio* of $\mathfrak{D}$ is defined as the infimum over all c such that $\mathfrak{D}$ is c-competitive. Similarly, a randomized strategy $\mathfrak{R}$, is said to be *c-competitive* if, for all instances $I = (\boldsymbol{p}, \boldsymbol{w})$, $\mathbb{E}[\mathfrak{R}(I)] \le c \cdot \text{OPT}(I)$, where $\mathbb{E}[\cdot]$ denotes the expected value. The following proposition suggests to consider strategies beyond non-preemptive ones.

Proposition 1. *No randomized non-preemptive non-clairvoyant strategy has a constant competitive ratio for $1 \| \sum C_j$.*

Proof sketch. Consider $n-1$ unit jobs and one job of length n^2, randomize uniformly over all permutations, and use Yao's principle [32]. $\square$

Due to this negative result, we study non-clairvoyant *kill-and-restart strategies* for $1 \| \sum w_j C_j$ that may abort the processing of a job, but when it is processed again later, it has to be executed from the beginning.

Such a strategy performs *probings* (t, j, τ), i.e., it processes at time t job j for a time of $\min\{\tau, p_j\}$. More formally, for a given state consisting of the current time, the set of unfinished jobs and lower bounds on the processing times learned from past probings, a kill-and-restart strategy decides on a family of probings $(t_i, j_i, \tau_i)_{i \in \mathcal{I}}$, such that the intervals $(t_i, t_i + \tau_i)$, $i \in \mathcal{I}$, are disjoint. Whenever a job is completed, i.e., a job is processed completely within one probing, the strategy decides on new probings. We require that probings be chosen independently of the actual processing times of unfinished jobs, ensuring that kill-and-restart strategies are non-clairvoyant. A formal definition is given in the full version.

Observe that such strategies may not be implementable, e.g., on a Turing machine, as the above definition allows for an infinite number of probings in a bounded time range. This subtlety is in fact inherent to all scheduling problems with unknown processing times or search problems with unknown distances. It is not hard to see that no deterministic kill-and-start strategy can be constant-competitive without infinitesimal probing, as there is no lower bound on the processing times at time 0. On the other hand, infinitesimal probing can be avoided if we know a lower bound on the p_j's, thus turning the strategies analyzed in this paper into implementable algorithms.

We denote by $Y_j^{\mathfrak{D}}(I, t)$ the total time for which the machine has been busy processing job j until time t in the schedule constructed by the strategy $\mathfrak{D}$ on the instance I.

3 Lower Bound

In contrast to the lower bound of 2 for preemptive algorithms by Motwani et al. [25], we prove a higher lower bound for kill-and-restart strategies.

Theorem 1. For $1 \,\|\, \sum C_j$, no deterministic non-clairvoyant kill-and-restart strategy can achieve a competitive ratio smaller than $3 - \frac{2}{n+1}$ on instances with $n \geq 3$ jobs, even if every job j has processing time $p_j \geq 1$.

Proof. Let $\varepsilon \in (\frac{2}{n+1}, 1]$ and define $T := \frac{(2-\varepsilon)(n^2+n)}{\varepsilon(n+1)-2}$. Consider an arbitrary deterministic kill-and-restart strategy $\mathfrak{D}$ with the initially chosen family of probings $(t_i, j_i, \tau_i)_{i \in \mathcal{I}}$. Let $Y_j(\theta) := \sum_{i \in \mathcal{I}: t_i < \theta, j_i = j} \min\{\tau_i, \theta - t_i\}$ be the total probing time assigned by $\mathfrak{D}$ to job j up to time θ. We define an instance $I := (\boldsymbol{p}, \boldsymbol{1})$ by distinguishing two cases on the first job j_0 planned to be probed at or after time T. Note that such a job exists, as otherwise $\mathfrak{D}$ does not complete all jobs if processing times are long enough.

If j_0 is probed for a finite amount of time, we denote by $t \geq T$ the end of its probing time. Then, define $p_j := 1 + Y_j(t)$ for all $j \in [n]$. Clearly, no job finishes before t when $\mathfrak{D}$ runs the instance I, hence $\mathfrak{D}(I) \geq nt + \mathrm{OPT}(I)$. On the other hand, it is well known that $\mathrm{OPT}(I) \leq \frac{n+1}{2} \cdot \sum_{j=1}^{n} p_j \leq \frac{n+1}{2}(t+n)$. Therefore, we have $\frac{\mathfrak{D}(I)}{\mathrm{OPT}(I)} \geq 1 + \frac{2nt}{(t+n)(n+1)} \geq 1 + \frac{2nT}{(T+n)(n+1)} = 3 - \varepsilon$.

If j_0 is probed for $\tau = \infty$, i.e., it is processed non-preemptively until its completion, then for each job $j \neq j_0$ we set $p_j := 1 + Y_j(T)$. Denote by OPT' the optimal SPT cost for jobs $[n] \setminus \{j_0\}$, and set $p_{j_0} := 10 \cdot \mathrm{OPT}'$. As j_0 is the first job to complete in I, we clearly have $\mathfrak{D}(I) \geq n \cdot p_{j_0} = 10n \cdot \mathrm{OPT}'$. On the other hand, OPT processes j_0 last, so $\mathrm{OPT}(I) = \mathrm{OPT}' + \sum_{j \neq j_0} p_j + p_{j_0} \leq 12 \cdot \mathrm{OPT}'$. This implies $\frac{\mathfrak{D}(I)}{\mathrm{OPT}(I)} \geq \frac{10n}{12} \geq 3 - \frac{2}{n+1}$, as $n \geq 3$. $\qquad\square$

4   The b-scaling Strategy

The idea of the b-scaling strategy $\mathfrak{D}_b$ for $b > 1$ is simple and quite natural: it proceeds by rounds $q \in \mathbb{Z}$. In round q every non-completed job j is probed (once) for $w_j b^q$, in the order of job indices. To execute $\mathfrak{D}_b$, we can store for each job its *rank* at time t, i.e., the largest q such that it was probed for $w_j b^{q-1}$ until t. At any end of a probing, $\mathfrak{D}_b$ probes the job j with minimum rank q and minimum index for time $w_j b^q$. We also introduce a randomized variant $\mathfrak{R}_b$ of the strategy $\mathfrak{D}_b$. Randomization occurs in two places: First the jobs are reordered according to a permutation Σ drawn uniformly at random from $\mathcal{S}_n$ at the beginning. Second, we replace the probing time $w_j b^q$ of the qth round by $w_j b^{q+\Xi}$ with a random uniform offset $\Xi \sim \mathcal{U}([0, 1])$.

In general, $\mathfrak{D}_b$ starts with infinitesimally small probings at time 0 in rounds $q \to -\infty$. As discussed earlier, this is not implementable. However, if a lower bound of $w_j b^{q_{\min}}$ on every processing time p_j is known, the algorithm can start with round $q = q_{\min}$.

4.1   The Deterministic b-scaling Strategy

We compute tight bounds for the competitive ratio of $\mathfrak{D}_b$ for $1 \,\|\, \sum w_j C_j$. For an instance $I := (\boldsymbol{p}, \boldsymbol{w})$, we denote by $s_j := p_j / w_j$ the *Smith ratio* of job $j \in [n]$ and by $D_{jk} := Y_j^{\mathfrak{D}_b}(I, C_k^{\mathfrak{D}_b}(I))$ the amount of time spent probing job j before the completion of job k. For all $j, k \in [n]$ we define the weighted mutual delay Δ_{jk} by $\Delta_{jk} := w_k D_{jk} + w_j D_{kj}$ if $j \neq k$ and $\Delta_{jj} := w_j D_{jj}$. Thus, it holds

$$\mathfrak{D}_b(I) = \sum_{j=1}^{n} w_j C_j^{\mathfrak{D}_b}(I) = \sum_{j=1}^{n} w_j \sum_{k=1}^{n} D_{kj} = \sum_{1 \leq j \leq k \leq n} \Delta_{jk}.$$

We first provide an overestimator of Δ_{jk} that is piecewise linear in s_j and s_k.

Lemma 1. *Define the function* $F \colon \{(s, s') \in \mathbb{R}_{>0}^2 : s \leq s'\} \to \mathbb{R}$ *by*

$$F(s, s') := \begin{cases} \frac{2b^{\lfloor \log_b s \rfloor + 1}}{b - 1} + s' & \text{if } \lfloor \log_b(s) \rfloor = \lfloor \log_b(s') \rfloor \\ b^{\lfloor \log_b s \rfloor + 1} \cdot (\frac{2}{b-1} + 1) + s & \text{otherwise.} \end{cases}$$

Then for all $j, k \in [n]$ *such that* $s_j \leq s_k$, *it holds* $\Delta_{jk} \leq w_j w_k F(s_j, s_k)$.

Proof sketch. Let $q_j := \lceil \log_b(s_j) \rceil - 1$, so that $b^{q_j} < s_j \leq b^{q_j+1}$. By distinguishing between the case where jobs j, k complete in the same round, i.e., $\lceil \log_b(s_j) \rceil = \lceil \log_b(s_k) \rceil$, and the case where k completes in a later round, we obtain an upper bound of the form $\Delta_{jk} \leq w_j w_k \tilde{F}(s_j, s_k)$, where $\tilde{F}$ is defined as F except that all occurrences of floor operations $\lfloor \log_b(s) \rfloor$, $s \in \{s_j, s_k\}$ are replaced by $\lceil \log_b(s) \rceil - 1$. The result follows by observing that $\tilde{F}$ is non-decreasing with respect to both its arguments and taking its upper semi-continuous envelope. □

Summing the bounds of the previous lemma yields

$$\mathfrak{D}_b(\boldsymbol{p}, \boldsymbol{w}) \leq \sum_{1 \leq j \leq k \leq n} w_j w_k \, F\big(\min(s_j, s_k)\big), \max(s_j, s_k)\big) =: U(\boldsymbol{p}, \boldsymbol{w}). \quad (1)$$

We next prove a lemma showing that for bounding the ratio U/OPT we can restrict to instances in which all Smith ratios are integer powers of b.

Lemma 2. *For any instance* $(\boldsymbol{p}, \boldsymbol{w})$, *there exists another instance* $(\boldsymbol{p}', \boldsymbol{w})$ *with* $p_j' = w_j b^{q_j}$ *for some* $q_j \in \mathbb{Z}$, *for all* $j \in [n]$, *such that*

$$\frac{U(\boldsymbol{p}, \boldsymbol{w})}{\mathrm{OPT}(\boldsymbol{p}, \boldsymbol{w})} \leq \frac{U(\boldsymbol{p}', \boldsymbol{w})}{\mathrm{OPT}(\boldsymbol{p}', \boldsymbol{w})}.$$

Proof sketch. Let $Q := \{\log_b(s_j) : j \in [n]\} \setminus \mathbb{Z}$. We construct the vector $\boldsymbol{p}'$ by sequentially rounding the Smith ratios of a subset of jobs $J_q = \{j \in [n] : s_j = b^q\}$ with $q := \min Q$ to a larger or a smaller value, in such a way that the ratio U/OPT does not decrease, and we repeat this until $Q = \emptyset$. Specifically, for $\delta \in \mathbb{R}$ we define the modified processing times $p_j(\delta) = p_j + w_j \delta \mathbb{1}_{J_q}(j)$. Both $\delta \mapsto U(\boldsymbol{p}(\delta), \boldsymbol{w})$ and $\delta \mapsto \mathrm{OPT}(\boldsymbol{p}(\delta), \boldsymbol{w})$ are linear in a neighborhood of 0 in which the order of the Smith ratios does not change and no job changes the round where it completes. Thus, $\delta \mapsto \frac{U(\boldsymbol{p}(\delta), \boldsymbol{w})}{\mathrm{OPT}(\boldsymbol{p}(\delta), \boldsymbol{w})}$ is a monotone rational function in this neighborhood, so δ can be increased or decreased to a value such that $|Q|$ is decremented by 1, without decreasing our bound on the competitive ratio. □

The next lemma gives a handy upper bound for the competitive ratio of $\mathfrak{D}_b$ relying on the ratio of two quadratic forms.

Lemma 3. *For every $L \in \mathbb{N}$, let $\boldsymbol{A}_L := \left(\frac{1}{2}b^{\min(i,j)-1} \cdot \mathbb{1}_{\{i \neq j\}}\right)_{1 \leq i,j \leq L}$ and $\boldsymbol{B}_L := \left(\frac{1}{2}b^{\min(i,j)-1}\right)_{1 \leq i,j \leq L}$ be symmetric matrices. For any instances $(\boldsymbol{p}, \boldsymbol{w})$ there exists an integer L and a vector $\boldsymbol{x} \in \mathbb{R}_{\geq 0}^L$ such that*

$$\frac{\mathfrak{D}_b(\boldsymbol{p}, \boldsymbol{w})}{\mathrm{OPT}(\boldsymbol{p}, \boldsymbol{w})} \leq \frac{2b}{b-1} + 1 + b \cdot \frac{\boldsymbol{x}^\top \boldsymbol{A}_L \boldsymbol{x}}{\boldsymbol{x}^\top \boldsymbol{B}_L \boldsymbol{x}}.$$

Proof sketch. Let $i_0 \in \mathbb{Z}$ and $L \in \mathbb{N}$ be such that the Smith ratio of each job is of the form $b^{i_0-1+\ell}$ for some $\ell \in [L]$ in the instance $(\boldsymbol{p}', \boldsymbol{w})$ from Lemma 2. Let J_ℓ be the subset of jobs with Smith ratio equal to $b^{i_0-1+\ell}$, for $\ell \in [L]$, and define the vectors $\boldsymbol{x}, \boldsymbol{y} \in \mathbb{R}^L$ such that $x_\ell = \sum_{j \in J_\ell} w_j$ and $y_\ell = \sum_{j \in J_\ell} w_j^2$, respectively. We show that $\mathrm{OPT}(\boldsymbol{p}', \boldsymbol{w}) \geq b^{i_0} \boldsymbol{x}^\top \boldsymbol{B}_L \boldsymbol{x}$ and that $U(\boldsymbol{p}', \boldsymbol{w}) = (\frac{2b}{b-1}+1) \cdot \mathrm{OPT}(\boldsymbol{p}', \boldsymbol{w}) + b^{i_0+1}\boldsymbol{x}^\top \boldsymbol{A}_L \boldsymbol{x}$. Then, the result follows from (1) and Lemma 2. $\qquad\square$

In order to determine an upper bound for the competitive ratio of $\mathfrak{D}_b$, we need to bound the ratio of the two quadratic forms in the bound of Lemma 3. To this end, we bound the maximum eigenvalue of the matrix $\boldsymbol{Z}_L := \boldsymbol{Y}_L^{-\top} \boldsymbol{A}_L \boldsymbol{Y}_L^{-1}$, where $\boldsymbol{Y}_L^\top \boldsymbol{Y}_L = \boldsymbol{B}_L$ is the Cholesky decomposition of the matrix $\boldsymbol{B}_L$. An explicit computation of the matrix $\boldsymbol{Z}_L$ reveals that it is tridiagonal. In particular, the principal submatrix of $\boldsymbol{Z}_L$ obtained by deleting the first row and first column is a tridiagonal Toeplitz matrix, which we refer to as $\boldsymbol{T}_{L-1}$.

Theorem 2. *For $b > 1$, $\mathfrak{D}_b$ is $\left(1 + \frac{2b^{3/2}}{b-1}\right)$-competitive for $1 \,||\, \sum w_j C_j$. Moreover, for all $b > 1$ this bound is tight, even for unit weight instances. In particular, for $b = 3$ the competitive ratio is $1 + 3\sqrt{3} \approx 6.196$.*

Proof sketch. Due to Lemma 3, it only remains to bound $\rho_L := \sup_{\boldsymbol{x} \in \mathbb{R}^L} \frac{\boldsymbol{x}^\top \boldsymbol{A}_L \boldsymbol{x}}{\boldsymbol{x}^\top \boldsymbol{B}_L \boldsymbol{x}}$. As described above, we can express ρ_L as the largest eigenvalue of the matrix $\boldsymbol{Z}_L$, whose principal submatrix $\boldsymbol{T}_{L-1}$ is a tridiagonal Toeplitz matrix. This has $-2/(b-1)$ on the main diagonal and $\sqrt{b}/(b-1)$ on both adjacent diagonals. We show that the largest eigenvalue of $\boldsymbol{Z}_L$ converges to the same value as the largest eigenvalue of $\boldsymbol{T}_{L-1}$, which has the closed form $\frac{-2}{b-1} + \frac{2\sqrt{b}}{b-1} \cdot \cos \frac{\pi}{L} \xrightarrow{L \to \infty} \frac{-2}{b-1} + \frac{2\sqrt{b}}{b-1}$. Therefore, we obtain by Lemma 3 that $\mathfrak{D}_b$ is $\left(1 + \frac{2b^{3/2}}{b-1}\right)$-competitive. This ratio is minimized for $b = 3$, yielding the desired bound.

For the tightness part, we define the vector $\boldsymbol{x}_L \in \mathbb{R}_{\geq 0}^L$ by

$$x_{L,\ell} = 2(Lb^{\ell-1}(b-1))^{-1/2} \cdot \max\left(0, \sqrt{b} \cdot \sin\left(\frac{(\ell-1)\pi}{L}\right) - \sin\left(\frac{\ell\pi}{L}\right)\right), \; \forall \ell \in [L],$$

and show that $\frac{\boldsymbol{x}_L^\top \boldsymbol{A}_L \boldsymbol{x}_L}{\boldsymbol{x}_L^\top \boldsymbol{B}_L \boldsymbol{x}_L}$ converges to $\frac{2(\sqrt{b}-1)}{b-1}$ as $L \to \infty$. The above formula was obtained by transforming the eigenvector belonging to $\lambda_{\max}(\boldsymbol{T}_{L-1})$. Then, for $t > 0$ and $\varepsilon > 0$ we construct an instance with $n_{L,\ell} = \lfloor t \cdot x_{L,\ell} \rfloor$ jobs of unit weight and processing times equal to $b^\ell + \varepsilon$, for $\ell = 1, \ldots, L$, ordered in such a way that in each round, the jobs that finish are executed after all failed probings. A careful analysis of the weighted mutual delays Δ_{jk} shows that $\mathfrak{D}_b/\text{OPT} = \frac{2b}{b-1} + b \cdot \frac{n_L^\top A_L n_L + a_L^\top n_L}{n_L^\top B_L n_L + b_L^\top n_L} + o_{\varepsilon \to 0}(1)$ for some vectors $a_L, b_L \in \mathbb{R}^L$. Finally, the result follows by letting $\varepsilon \to 0$ and $t \to \infty$ and using $n_L = t(x_L + o_{t \to \infty}(1))$. $\square$

4.2   The Randomized b-scaling Strategy

We now consider the randomized variant $\mathfrak{R}_b$ of the strategy $\mathfrak{D}_b$, in which jobs are ordered according to a random permutation Σ and probed for $w_j b^{q+\Xi}$ in round $q \in \mathbb{Z}$, with $\Xi \sim \mathcal{U}([0,1])$. As in the analysis of the deterministic strategy, we start with a lemma giving an overestimator of Δ_{jk} for jobs j and k such that $s_j \le s_k$. This time, our overestimator is not piecewise linear in s_j and s_k anymore, but depends on a concave function applied to the ratio $\frac{s_k}{s_j} \ge 1$. The next lemma follows from standard calculations involving integrals of the form $\int_\alpha^\beta b^\xi \, d\xi = \frac{b^\beta - b^\alpha}{\log b}$ and case distinctions on the rounds in which j and k complete.

Lemma 4. *Let* $f(\alpha) := \frac{1+\alpha}{2} + \frac{2}{\log b} + \frac{\alpha-1}{2\log b} \cdot (1 - \log(\alpha))$ *for* $\alpha \in [1, b]$. *For all* $j \ne k$ *such that* $s_j \le s_k$ *it holds*

$$\mathbb{E}[\Delta_{jj}] = w_j^2 \, s_j \cdot \left(1 + \frac{1}{\log b}\right) \le w_j^2 \, s_j \cdot f(1) \quad \text{and} \quad \mathbb{E}[\Delta_{jk}] = w_j w_k s_j \cdot f\left(\min\left(b, \frac{s_k}{s_j}\right)\right).$$

The difficulty of proving the main result of this subsection resides in the fact that we cannot reduce to a worst-case instance in which all Smith ratios are integer powers of b. Instead, we push the technique used for Theorem 2 to the limit, by partitioning the set of jobs according to the interval of the form $[b^{i/K}, b^{(i+1)/K})$ containing their Smith ratio, and letting $K \to \infty$. This leads to the analysis of a Toeplitz matrix which is not tridiagonal anymore but has a bandwitdth of $2K - 1$. While the maximum eigenvalue of this matrix does not have a closed-form expression for $K > 1$, its limit for $L \to \infty$ can be computed using the Fourier series associated with this matrix.

Theorem 4. *For every* $b > 1$, $\mathfrak{R}_b$ *is* $\frac{\sqrt{b}+2b-1}{\sqrt{b}\log(b)}$-*competitive for* $1 || \sum w_j C_j$. *Moreover, for all* $b > 1$ *this bound is tight, even for unit weight instances. In particular, for* $b \approx 8.16$ *the competitive ratio is* ≈ 3.032.

Proof sketch. For $K \in \mathbb{N}$, let $\beta = b^{1/K}$. We group the set of all jobs into sets $J_k = \{j \in [n] : \beta^k \le s_j < \beta^{k+1}\}$ for all $k \in \mathbb{Z}$. Using Lemma 4 and calculations similar to those used in the proof of Theorem 2, we show that

$$\frac{\mathfrak{R}_b(p, w)}{\text{OPT}(p, w)} \le \beta\big(f(\beta) + \lambda_{\max}(Z)\big),$$

where $Z := \sum_{i=1}^{K-1} (f(\beta^{i+1}) - f(\beta^i)) Z_i$ and Z_i is a sparse symmetric matrix having non-zero entries only on its $\pm i$th and $\pm(i-1)$th superdiagonals for all

$i \in [K-1]$. The principal submatrix of Z obtained by removing its first row and column is a Toeplitz matrix T of size $L \times L$. Then, we show by using a Schur complement that the above bound is smaller than $\frac{\sqrt{b}+2b-1}{\sqrt{b}\log(b)}$ when both L and K grow to ∞, with $L/K \to \infty$. To construct a matching lower bound on the competitive ratio, we have to use an approximate eigenvector $\hat{z} \in \mathbb{R}^L$ of T because no closed form is available if $K > 1$. We set $z_\ell := (\frac{L+1}{2})^{-1/2}\sin\left(\frac{\ell\pi}{L+1}\right)$ for $\ell \in [L]$ and show that $\hat{z}^\top T\hat{z}/\|\hat{z}\|^2 \xrightarrow{L\to\infty} \lambda_{\max}(T)$ by using the Fourier series associated with T. The rest of the proof mimics the steps used in in Theorem 2, where b is replaced by $\beta = b^{1/K}$ and we let $K \to \infty$. $\square$

5 Weighted Shortest Elapsed Time First

In this we consider the online time model, where each job j arrives at its release date r_j and is not known before that time. Thus, an instance for our problem is now given by a triple $I = (p, w, r)$ of processing times, weights, and release dates of all jobs. Intuitively, the classical Weighted Shortest Elapsed Time First (WSETF) rule is the limit for $\varepsilon \to 0$ of the algorithm that divides the time into time slices of length ε and in each time slice processes a job with minimum ratio of elapsed processing time over weight. To formalize this limit process we allow fractional schedules S that, at every point in time t, assign each job j a rate $y_j^S(t) \in [0,1]$ so that $\sum_{j=1}^n y_j^S(t) \le 1$ for all $t \in \mathbb{R}_{\ge 0}$ and $y_j^S(t) = 0$ if $t < r_j$ or $t > C_j^S(I)$, where $C_j^S(I)$ is the smallest t such that $Y_j^S(I,t) := \int_0^t y_j^S(s)\,ds \ge p_j$. At any time t let $J(t)$ be the set of all released and unfinished jobs, and let $A(t)$ be the set of all jobs from $J(t)$ that currently have minimum ratio of elapsed time over weight. Then WSETF sets the rate for all jobs $j \in A(t)$ to $y_j^{\text{WSETF}}(t) := w_j/(\sum_{k\in A(t)} w_k)$ and to 0 for all other jobs. In other words, WSETF always distributes the available processor rate among the jobs in $J(t)$ so as to maximize $\min_{j\in J(t)} Y_j^{\text{WSETF}}(I,t)/w_j$.

The following gives the tight competitive ratio of WSETF for non-clairvoyant online scheduling on a single machine.

Theorem 5. WSETF is 2-competitive for $1\,|\,r_j,\,\text{pmtn}\,|\,\sum w_j C_j$.

Proof sketch. To bound the optimal objective value from below, we consider the *mean busy times* $M_j^S(I) := \int_0^\infty t \cdot y_j^S(t)\,dt$ of all jobs j in the optimal schedule. It is well known [12,13] that the the sum of weighted mean busy times is minimized by the Preemptive WSPT (PWSPT) rule, which always processes an available job with smallest index (i.e. with smallest Smith ratio p_j/w_j). Therefore, it suffices to show that $\sum_{j=1}^n w_j \cdot C_j^{\text{WSETF}}(I) \le 2\cdot\sum_{j=1}^n w_j \cdot M_j^{\text{PWSPT}}(I)$.

For instances I_0 with trivial release dates, the weighted delay of each job in the WSETF schedule compared to the optimal WSPT schedule is exactly its processing time multiplied with the total weight of jobs with larger indices, or in other words, its weighted completion time is

$$w_j \cdot C_j^{\mathrm{WSETF}}(I_0) = w_j \cdot C_j^{\mathrm{WSPT}}(I_0) + \sum_{k=j+1}^{n} w_k \cdot p_j \qquad (2)$$

$$= w_j \cdot M_j^{\mathrm{WSPT}}(I_0) + \underbrace{\left(\frac{w_j}{2} + \sum_{k=j+1}^{n} w_k\right) \cdot p_j}_{(*)}.$$

In order to extend this observation to instances I with release dates, we define for each job j an auxiliary instance $I(j)$ with trivial release dates and relate the completion times of j in the WSETF and PWSPT schedules for I to the completion times in the corresponding schedules for $I(j)$. We then bound the difference $w_j(C_j^{\mathrm{WSETF}}(I) - M_j^{\mathrm{PWSPT}}(I))$ by an expression generalizing $(*)$. To this end, we apply (Eq. 2) to the instance $I(j)$ and use the fact that each deviation of the PWSPT schedule from the WSPT schedule for the instance without release dates increases the total weighted mean busy time. Finally, we show that the sum of the obtained bounds over all jobs is equal to the sum of weighted mean busy times in the PWSPT schedule.   □

6   Upper Bounds for More General Settings

In this section, we give upper bounds on the competitive ratio of the b-scaling strategy for $1 \mid r_j \mid \sum w_j C_j$ and $\mathrm{P} \| \sum C_j$. Let $I = (\boldsymbol{p}, \boldsymbol{w}, \boldsymbol{r}, m)$ denote an instance on m identical parallel machines in which each job j has processing time p_j, weight w_j and release date r_j. The overall idea is to compare the schedule of $\mathfrak{D}_b$ to schedules of WSETF and round-robin (RR) for the release date and the parallel machine case, respectively, since, by Theorem 4 and [25], both strategies are 2-competitive. To this end, we need to consider modified instances with increased processing times and release dates. The following straightforward lemma bounds the increase of the optimal costs under these modifications.

Lemma 5. *Let $I = (\boldsymbol{p}, \boldsymbol{w}, \boldsymbol{r}, m)$ and $I' = (\boldsymbol{p'}, \boldsymbol{w}, \boldsymbol{r'}, m)$ be two instances with $\boldsymbol{p'} \le \alpha \boldsymbol{p}$ and $\boldsymbol{r'} \le \alpha \boldsymbol{r}$. Then, we have $\mathrm{OPT}(I') \le \alpha \cdot \mathrm{OPT}(I)$.*

For $1 \mid r_j \mid \sum w_j C_j$ we extend $\mathfrak{D}_b$ as follows: At the end of a probing, probe the job with minimum rank and index that is released and not completed.

Theorem 5. $\mathfrak{D}_b$ *is* $\frac{2b^4}{2b^2-3b+1}$*-competitive for* $1 \mid r_j \mid \sum w_j C_j$. *In particular, for* $b = \frac{9+\sqrt{17}}{8}$, *its performance guarantee is* $\frac{107+51\sqrt{17}}{32} \approx 9.915$.

We prove a slightly stronger result by bounding the ratio of $\mathfrak{D}_b(I)$ to the cost of an optimal *preemptive* schedule for I.

Proof sketch. Let $I = (\boldsymbol{p}, \boldsymbol{w}, \boldsymbol{r}, 1)$ be an arbitrary instance. As a first step we construct an auxiliary instance $I' = (\boldsymbol{p'}, \boldsymbol{w}, \boldsymbol{r'}, 1)$ as follows: We define processing times $p_j \le p_j' \le bp_j$ such that $p_j'/w_j = b^{q_j}$ with $q_j = \lceil \log_b(p_j/w_j) \rceil$, i.e., all

Smith ratios in the instance I' are integer powers of b. Further, we define new release dates $r'_j \geq r_j$ by either setting r'_j to the end of the probing that runs at r_j in the schedule of I', whenever such a probing exists, or $r'_j = r_j$ otherwise. By construction, we have $r'_j \leq \frac{b^3}{2b-1} r_j$ and $p'_j \leq bp_j \leq \frac{b^3}{2b-1} p_j$ for any job j. Therefore, by Lemma 5, we have $\mathrm{OPT}(I') \leq \frac{b^3}{2b-1}\mathrm{OPT}(I)$. Moreover, we obtain $\mathfrak{D}_b(I) \leq \mathfrak{D}_b(I')$, as the sequence of probings in both schedules is the same and the processing times in I' are longer.

Next, we consider another instance $I'' = (\boldsymbol{p''}, \boldsymbol{w}, \boldsymbol{r'}, 1)$ with processing times $p''_j := Y_j^{\mathfrak{D}_b}(I', C_j^{\mathfrak{D}_b}(I')) = \sum_{i=-\infty}^{q_j} b^i = \frac{b^{q_j+1}}{b-1} = \frac{b}{b-1} p'_j$. We show inductively that, by construction, at any completion time of a job j in the schedule of WSETF for I'', all already released, not completed jobs with minimum rank q_j were probed by $\mathfrak{D}_b$ for an amount of $w_j b^{q_j}$. Therefore, by definition of I' and I'' we have $C_j^{\mathfrak{D}_b}(I') \leq C_j^{\mathrm{WSETF}}(I'')$ and hence, $\mathfrak{D}_b(I') \leq \mathrm{WSETF}(I'')$. Altogether, we obtain

$$\mathfrak{D}_b(I) \leq \mathfrak{D}_b(I') \leq \mathrm{WSETF}(I'') \leq 2\mathrm{OPT}(I'')$$
$$\leq \frac{2b}{b-1}\mathrm{OPT}(I') \leq \frac{2b^4}{2b^2 - 3b + 1}\mathrm{OPT}(I),$$

applying Lemma 5 a second time.   □

For $\mathrm{P} \,||\, \sum C_j$ we extend $\mathfrak{D}_b$ as follows: probe each job for b^q in a list scheduling manner. If at most m jobs remain, process each job on a distinct machine until completion, otherwise increase q by 1 and repeat.

Theorem 6. $\mathfrak{D}_b$ is $\frac{3b^2-b}{b-1}$-competitive for $\mathrm{P} \,||\, \sum C_j$. In particular, for $b = \frac{3+\sqrt{6}}{3}$, its performance guarantee is $5 + 2\sqrt{6} \approx 9.899$.

Proof sketch. Let $I = (\boldsymbol{p}, 1, 0, m)$ be an instance and define a new instance $I' = (\boldsymbol{p'}, 1, 0, m)$ with processing times $p'_j = b^{q_j}$ where $q_j = \lceil \log_b p_j \rceil$. Note that $p'_j \leq bp_j$ and $\mathfrak{D}_b(I) \leq \mathfrak{D}_b(I')$. For the schedule of $\mathfrak{D}_b$ on I', let $T'_i(q)$ denote the point in time, when the last probing of length b^q on machine i ends.

Next, we define another instance $I'' = (\boldsymbol{p''}, 1, 0, m)$, where the processing times p''_j are defined to be the exactly the elapsed time of j in the schedule of $\mathfrak{D}_b$ for I' at its completion time. We consider the schedule of RR on I'' and denote by $T''(q)$ the point in time where the elapsed time of all non-completed jobs is exactly $\frac{b^{q+1}}{b-1}$. By induction, we show that $T''(q) = \frac{1}{m}\sum_{i=1}^{m} T'_i(q)$. This identity allows us to relate $C_j^{\mathfrak{D}_b}(I')$ and $C_j^{\mathrm{RR}}(I'')$. In particular, we obtain $\sum_j C_j^{\mathfrak{D}_b}(I') \leq \sum_j C_j^{\mathrm{RR}}(I'') + p'_j$. The 2-competitiveness of RR and Lemma 5 yield

$$\mathfrak{D}_b(I) \leq \mathfrak{D}_b(I') \leq \mathrm{RR}(I'') + \sum_j p'_j \leq 2\mathrm{OPT}(I'') + \mathrm{OPT}(I')$$
$$\leq \left(\frac{2b}{b-1} + 1\right)\mathrm{OPT}(I') \leq \left(\frac{2b}{b-1} + 1\right) \cdot b \cdot \mathrm{OPT}(I).$$

□

7   Conclusion

We studied kill-and-restart as well as preemptive strategies for the problem of minimizing the sum of weighted completion times and gave a tight analysis of the deterministic and randomized version of the natural b-scaling strategy for $1 \, || \, \sum w_j C_j$ as well as of WSETF for $1 \, | \, r_j, \text{pmtn} \, | \, \sum w_j C_j$.

We hope that this work might lay a basis for obtaining tight bounds on the performance of the b-scaling strategy for more general settings such as non-trivial release dates and parallel machines. Moreover, we think that the class of kill-and-restart strategies combines the best of two worlds. On the one hand, they allow for interruptions leading to small competitive ratios in contrast to non-preemptive algorithms, on the other hand, they reflect the non-preemptive property of only completing a job if it has been processed as a whole.

Acknowledgements. We thank Sungjin Im for helpful comments on an earlier version of this manuscript.

References

1. Afrati, F., et al.: Approximation schemes for minimizing average weighted completion time with release dates. In: 40th Annual IEEE Symposium on Foundations of Computer Science (FOCS), pp. 32–43. IEEE (1999). https://doi.org/10.1109/SFFCS.1999.814574

2. Anderson, E.J., Potts, C.N.: Online scheduling of a single machine to minimize total weighted completion time. Math. Oper. Res. **29**(3), 686–697 (2004). https://doi.org/10.1287/moor.1040.0092

3. Baeza-Yates, R.A., Culberson, J.C., Rawlins, G.J.E.: Searching in the plane. Inf. Comput. **106**(2), 234–252 (1993). https://doi.org/10.1006/inco.1993.1054

4. Bansal, N., Dhamdhere, K.: Minimizing weighted flow time. ACM Trans. Algorithms **3**(4), 39:1–39:14 (2007). https://doi.org/10.1145/1290672.1290676

5. Bansal, N., Pruhs, K.: Server scheduling in the weighted ℓ_p-norm. In: Farach-Colton, Martín (ed.) LATIN 2004. LNCS, vol. 2976, pp. 434–443. Springer, Heidelberg (2004). https://doi.org/10.1007/978-3-540-24698-5_47

6. Beaumont, O., Bonichon, N., Eyraud-Dubois, L., Marchal, L.: Minimizing weighted mean completion time for malleable tasks scheduling. In: 26th International Symposium on Parallel and Distributed Processing (IPDPS), pp. 273–284. IEEE (2012). https://doi.org/10.1109/ipdps.2012.34

7. Becchetti, L., Leonardi, S.: Nonclairvoyant scheduling to minimize the total flow time on single and parallel machines. J. ACM **51**(4), 517–539 (2004). https://doi.org/10.1145/1008731.1008732

8. Conway, R.W., Maxwell, W.L., Miller, L.W.: Theory of Scheduling. Addison-Wesley Publishing Company, Boston (1967)

9. Epstein, L., van Stee, R.: Lower bounds for on-line single-machine scheduling. Theor. Comput. Sci. **299**(1), 439–450 (2003). https://doi.org/10.1016/S0304-3975(02)00488-7

10. Feldmann, A., Sgall, J., Teng, S.H.: Dynamic scheduling on parallel machines. Theor. Comput. Sci. **130**(1), 49–72 (1994). https://doi.org/10.1016/0304-3975(94)90152-X

11. Garg, N., Gupta, A., Kumar, A., Singla, S.: Non-clairvoyant precedence constrained scheduling. In: Baier, C., Chatzigiannakis, I., Flocchini, P., Leonardi, S. (eds.) 46th International Colloquium on Automata, Languages, and Programming (ICALP). LIPIcs, vol. 132, pp. 63:1–63:14 (2019). https://doi.org/10.4230/LIPIcs.ICALP. 2019.63

12. Goemans, M.X.: A supermodular relaxation for scheduling with release dates. In: Cunningham, W.H., McCormick, S.T., Queyranne, M. (eds.) IPCO 1996. LNCS, vol. 1084, pp. 288–300. Springer, Heidelberg (1996). https://doi.org/10.1007/3-540-61310-2_22

13. Goemans, M.X.: Improved approximation algorthims for scheduling with release dates. In: Proceedings of the Eighth Annual ACM-SIAM Symposium Discrete Algorithms (SODA), pp. 591–598. SIAM (1997)

14. Hoogeveen, J.A., Vestjens, A.P.A.: Optimal on-line algorithms for single-machine scheduling. In: Cunningham, W.H., McCormick, S.T., Queyranne, M. (eds.) IPCO 1996. LNCS, vol. 1084, pp. 404–414. Springer, Heidelberg (1996). https://doi.org/10.1007/3-540-61310-2_30

15. Im, S., Kulkarni, J., Munagala, K.: Competitive algorithms from competitive equilibria: non-clairvoyant scheduling under polyhedral constraints. J. ACM 65(1), 1–33 (2017). https://doi.org/10.1145/3136754

16. Im, S., Kulkarni, J., Munagala, K., Pruhs, K.: SelfishMigrate: A scalable algorithm for non-clairvoyantly scheduling heterogeneous processors. In: 55th Annual IEEE Symposium on Foundations of Computer Science (FOCS), pp. 531–540 (2014). https://doi.org/10.1109/FOCS.2014.63

17. Jäger, S., Sagnol, G., Schmidt genannt Waldschmidt, D., Warode, P.: Competitive kill-and-restart and preemptive strategies for non-clairvoyant scheduling (2022). https://doi.org/10.48550/ARXIV.2211.02044

18. Kalyanasundaram, B., Pruhs, K.: Speed is as powerful as clairvoyance. J. ACM 47(4), 617–643 (2000). https://doi.org/10.1145/347476.347479

19. Kao, M.Y., Ma, Y., Sipser, M., Yin, Y.: Optimal constructions of hybrid algorithms. J. Alg. 29(1), 142–164 (1998). https://doi.org/10.1006/jagm.1998.0959

20. Kao, M.Y., Reif, J.H., Tate, S.R.: Searching in an unknown environment: an optimal randomized algorithm for the cow-path problem. Inf. Comput. 131(1), 63–79 (1996). https://doi.org/10.1006/inco.1996.0092

21. Kim, J., Chwa, K.: Non-clairvoyant scheduling for weighted flow time. Inf. Process. Lett. 87(1), 31–37 (2003). https://doi.org/10.1016/S0020-0190(03)00231-X

22. Labetoulle, J., Lawler, E.L., Lenstra, J.K., Rinnooy Kan, A.H.G.: Preemptive scheduling of uniform machines subject to release dates. In: Pulleyblank, W.R. (ed.) Progress in Combinatorial Optimization, pp. 245–261. Academic Press (1984). https://doi.org/10.1016/B978-0-12-566780-7.50020-9

23. Lindermayr, A., Megow, N.: Permutation predictions for non-clairvoyant scheduling. In: Proceedings of the 34th Symposium on Parallelism in Algorithms and Architectures (SPAA), pp. 357–368 (2022). https://doi.org/10.1145/3490148. 3538579

24. Megow, N., Vredeveld, T.: A tight 2-approximation for preemptive stochastic scheduling. Math. Oper. Res. 39(4), 1297–1310 (2014). https://doi.org/10.1287/moor.2014.0653

25. Motwani, R., Phillips, S., Torng, E.: Nonclairvoyant scheduling. Theor. Comput. Sci. 130(1), 17–47 (1994). https://doi.org/10.1016/0304-3975(94)90151-1

26. Rinnooy Kan, A.H.G.: Machine Scheduling Problems: Classification, Complexity and Computations. Martinus Nijhoff (1976). https://doi.org/10.1007/978-1-4613-4383-7

27. Shmoys, D.B., Wein, J., Williamson, D.P.: Scheduling parallel machines on-line. SIAM J. Comput. **24**(6), 1313–1331 (1995). https://doi.org/10.1137/S0097539793248317
28. Sitters, R.: Competitive analysis of preemptive single-machine scheduling. Oper. Res. Lett. **38**(6), 585–588 (2010). https://doi.org/10.1016/j.orl.2010.08.012
29. Smith, W.E.: Various optimizers for single-stage production. Nav. Res. Logist. Q. **3**(1–2), 59–66 (1956). https://doi.org/10.1002/nav.3800030106
30. van Stee, R., La Poutré, H.: Minimizing the total completion time on-line on a single machine, using restarts. J. Alg. **57**(2), 95–129 (2005). https://doi.org/10.1016/j.jalgor.2004.10.001
31. Vestjens, A.P.A.: On-line machine scheduling. Ph.D. thesis, Technische Universiteit Eindhoven (1997). https://doi.org/10.6100/IR500043, https://pure.tue.nl/ws/files/1545064/500043.pdf
32. Yao, A.C.C.: Probabilistic computations: toward a unified measure of complexity. In: 18th Annual IEEE Symposium on Foundations of Computer Science (SFCS), pp. 222–227 (1977). https://doi.org/10.1109/SFCS.1977.24

A Deterministic Better-than-3/2 Approximation Algorithm for Metric TSP

Anna R. Karlin, Nathan Klein[(✉)], and Shayan Oveis Gharan

University of Washington, Seattle, USA
{karlin,nwklein,shayan}@cs.washington.edu

Abstract. We show that the max entropy algorithm can be derandomized (with respect to a particular objective function) to give a deterministic $3/2 - \epsilon$ approximation algorithm for metric TSP for some $\epsilon > 10^{-36}$.

To obtain our result, we apply the method of conditional expectation to an objective function constructed in prior work which was used to certify that the expected cost of the algorithm is at most $3/2 - \epsilon$ times the cost of an optimal solution to the subtour elimination LP. The proof in this work involves showing that the expected value of this objective function can be computed in polynomial time (at all stages of the algorithm's execution).

1 Introduction

One of the most fundamental problems in combinatorial optimization is the traveling salesperson problem (TSP), formalized as early as 1832 (c.f. [App+07, Ch 1]). In an instance of TSP we are given a set of n cities V along with their pairwise symmetric distances, $c : V \times V \to \mathbb{R}_{\geq 0}$. The goal is to find a Hamiltonian cycle of minimum cost. In the metric TSP problem, which we study here, the distances satisfy the triangle inequality. Therefore, the problem is equivalent to finding a closed Eulerian connected walk of minimum cost.

It is NP-hard to approximate TSP within a factor of $\frac{123}{122}$ [KLS15]. An algorithm of Christofides-Serdyukov [Chr76, Ser78] from four decades ago gives a $\frac{3}{2}$-approximation for TSP. Over the years there have been numerous attempts to improve the Christofides-Serdyukov algorithm and exciting progress has been made for various special cases of metric TSP, e.g., [OSS11, MS11, Muc12, SV12, HNR21, KKO20, HN19, Gup+21]. Recently, [KKO21] gave the first improvement for the general case by demonstrating that the so-called "max entropy" algorithm of the third author, Saberi, and Singh [OSS11] gives a randomized $\frac{3}{2} - \epsilon$ approximation for some $\epsilon > 10^{-36}$.

The method introduced in [KKO21] exploits the optimum solution to the following linear programming relaxation of metric TSP studied by [DFJ59, HK70, GB93], also known as the subtour elimination LP:

$$\min \quad \sum_{u,v} x_{\{u,v\}} c(u,v)$$

$$\text{s.t.,} \quad \sum_u x_{\{u,v\}} = 2 \qquad\qquad \forall v \in V,$$

$$\sum_{u \in S, v \notin S} x_{\{u,v\}} \geq 2, \quad \forall S \subsetneq V, S \neq \emptyset$$

$$x_{\{u,v\}} \geq 0 \qquad\qquad \forall u,v \in V. \tag{1}$$

However, [KKO21] had two shortcomings. First, it did not show that the integrality gap of the subtour elimination polytope is bounded below $\frac{3}{2}$. Second, it was randomized, and the analysis in that work was by nature "non-constructive" in the sense that it used the optimal solution; thus it was not clear how to to derandomize it using the method of conditional expectation. Other methods of derandomization seem at the moment out of reach and may require algorithmic breakthroughs. A followup work, [KKO22], remedied the first shortcoming by showing an improved integrality gap. While it did not address the question of derandomization, a byproduct of that work is an analysis of the max entropy algorithm which is in principle polynomially-time computable as it avoids looking at OPT. The purpose of the present work is to show that this analysis can indeed be done in polynomial-time, from which the following can be deduced (remedying the second shortcoming of [KKO21]):

Theorem 1. *Let x be a solution to LP (1) for a TSP instance. For some absolute constant $\epsilon > 10^{-36}$, there is a deterministic algorithm (in particular, a derandomized version of max entropy) which outputs a TSP tour with cost at most $\frac{3}{2} - \epsilon$ times the cost of x.*

Thus, this work in some sense completes the exploratory program concerning whether the max entropy algorithm for TSP beats 3/2 (initiated by [OSS11] in 2011), as now the above two weaknesses of [KKO21] have been addressed. Of course, much work remains in determining the true approximation factor of the algorithm; in this regard we are only at the tip of the iceburg.

Using the recent exciting work of Traub, Vygen, and Zenklusen reducing path TSP to TSP [TVZ20] our theorem also implies that there is a deterministic $\frac{3}{2} - \epsilon$ approximation algorithm for path TSP.

1.1   High Level Proof Overview

The high level strategy for derandomizing the max entropy algorithm is to use the method of conditional expectation on an objective function given by the analysis in [KKO22].

The max entropy algorithm, similar to Christofides' algorithm, first selects a spanning tree and then adds a minimum cost matching on the odd vertices of the tree. While Christofides selects a minimum cost spanning tree, here the spanning tree is sampled from a distribution. In particular, after solving the natural LP relaxation for the problem to obtain a fractional solution x, a tree is sampled from the distribution μ which has maximal entropy subject to the constraint $\mathbb{P}_{T \sim \mu}[e \in T] = x_e$ for all $e \in E$ (with possibly some exponentially small error in these constraints). [KKO21,KKO22] construct a so-called "slack" vector which is used to show the expected cost of the matching (over the randomness of the trees) is at most $\frac{1}{2} - \epsilon$ times the cost of an optimal solution to the LP. Given a solution x to LP (1) these works imply that there is a random vector m as a function of the tree $T \sim \mu$ such that:

(1) The cost of the minimum cost matching on the odd vertices of tree T is at most $c(m)$ (with probability 1), and
(2) $\mathbb{E}_{T \sim \mu}[c(m)] \leq (\frac{1}{2} - \epsilon)c(x)$.

Let $\mathcal{C} = \mathbb{E}_{T \sim \mu}[c(T) + c(m)]$. This will be the objective function to which we will apply the method of conditional expectation. Since the expected cost of the tree T is $c(x)$, as $\mathbb{P}_{T \sim \mu}[e \in T] = x_e$, by (2) $\mathcal{C}$ is at most $(\frac{3}{2} - \epsilon)c(x)$. Since by (1) for a given tree T, $c(T) + c(m)$ is an upper bound on the cost of the output of the algorithm (with probability 1), this shows that the expected cost of the algorithm is bounded strictly below $3/2$.

Ideally, one would like μ to have polynomial sized support. Then one could simply check the cost of the output of the algorithm on every tree in the support, and the above would guarantee that some tree gives a better-than-$3/2$ approximation. However, the max entropy distribution can have exponential sized support, and it's not clear how to find a similarly behaved distribution with polynomial sized support.

Instead, let $\mathcal{T}_{partial}$ be the family of all **partial** settings of the edges of the graph to 0 or 1 where the edges set to 1 are acyclic. For $Set = \{X_{e_1}, \ldots, X_{e_i}\} \in \mathcal{T}_{partial}$, and $1 \leq j \leq i$, we use X_{e_j} to indicate whether e_j is set to 1 or 0.

The method of conditional expectations is then used as follows: Process the edges in an arbitrary order $e_1, \ldots, e_m$ and for each edge e_i:

(1) Assume we inductively have chosen a valid assignment $Set \in \mathcal{T}_{partial}$ to edges $e_1, \ldots, e_{i-1}$.
(2) Let $Set^+ = Set \cup \{X_{e_i} = 1\}$. Compute $\mathcal{C}^+ = \mathbb{E}_{T \sim \mu}[c(T) + c(m) \mid Set^+]$. Similarly, let $Set^- = Set \cup \{X_{e_i} = 0\}$ and compute $\mathcal{C}^- = \mathbb{E}_{T \sim \mu}[c(T) + c(m) \mid Set^-]$.
(3) Let $Set \leftarrow Set^+$ or $Set \leftarrow Set^-$ depending on which quantity is smaller.

After a tree is obtained, add the minimum cost matching on the odd vertices of T. The resulting algorithm is shown in Algorithm 3 (see Algorithm 2 for its instantiation in a simple case).

As $\mathcal{C} \leq (\frac{3}{2} - \epsilon)c(x)$, this algorithm succeeds with probability 1. We only need to show it can be made to run in polynomial time. Since we can compute the expected cost of the tree conditioned on Set using linearity of expectation and the matrix tree theorem (Sect. 2.2), it remains to show that $\mathbb{E}_{T \sim \mu}[c(m)|Set]$ can be computed deterministically and efficiently for any $Set \in \mathcal{T}_{partial}$.

Key Contributions. The key contribution of this paper is to show how to do this computation efficiently, which is based on two observations:

(1) The first is that the vector m (whose cost upper bounds the cost of the minimum cost matching on the odd vertices of the tree) can be written as the (weighted) sum of indicators of events that depend on the sampled tree T, and each of these events happens only when a constant number of (not necessarily disjoint) sets of edges have certain parities or certain sizes.
(2) The second is that the probability of any such event can be deterministically computed in polynomial time by evaluating the generating polynomial of all spanning trees at certain points in $\mathbb{C}^E$, see Lemma 10.

Structure of the Paper. After reviewing some preliminaries, in Sect. 3 we review the matrix tree theorem and show (as a warmup) how to compute the probability two (not necessarily disjoint) sets of edges both have an even number of edges in the sampled tree. In Sect. 4, we then give a complete description and proof of a deterministic algorithm for the special "degree cut" case of TSP. Unlike the subsequent sections of the paper, Sect. 4 is self-contained and thus directed towards readers looking for more high-level intuition or those not familiar with [KKO21,KKO22]. In Sect. 5 we show (2) from above and give the deterministic algorithm in the general case. The remainder of the paper then involves proving (1) for the general definition of m from [KKO21,KKO22].

2 Preliminaries

2.1 Notation

For a set of edges $A \subseteq E$ and (a tree) $T \subseteq E$, we write $A_T = |A \cap T|$. For a tree T, we will say a cut $S \subseteq V$ is *odd in T* if $\delta(S)_T$ is odd and *even in T* otherwise. If the tree is understood we will simply say even or odd. We use $\delta(S) = \{\{u,v\} \in E : |\{u,v\} \cap S| = 1\}$ to denote the set of edges that leave S, and $E(S) = \{\{u,v\} \in E : |\{u,v\} \cap S| = 2\}$ to denote the set of edges inside of S.

For a set $A \subseteq E$ and a vector $x \in \mathbb{R}^{|E|}$ we write $x(A) := \sum_{e \in A} x_e$.

2.2 Randomized Algorithm of [KKO21]

Let x^0 be an optimum solution of LP (1). Without loss of generality we assume x^0 has an edge $e_0 = \{u_0, v_0\}$ with $x^0_{e_0} = 1, c(e_0) = 0$. (To justify this, consider the following process: given x^0, pick an arbitrary node, u, split it into two nodes u_0, v_0 and set $x_{\{u_0,v_0\}} = 1, c(e_0) = 0$ and assign half of every edge incident to u to u_0 and the other half to v_0.)

Let $E_0 = E \cup \{e_0\}$ be the support of x^0 and let x be x^0 restricted to E and $G = (V, E)$. By Lemma 1 x^0 restricted to E is in the spanning tree polytope (2) of G. We write $G = (V, E, x)$ to denote the (undirected) graph G together with special vertices u_0, v_0 and the weight function $x : E \to \mathbb{R}_{\geq 0}$. Similarly, let $G_0 = (V, E_0, x^0)$ and let $G_{/e_0} = G_0/\{e_0\}$, i.e. $G_{/e_0}$ is the graph G_0 with the edge e_0 contracted.

Definition 1. *For a vector $\lambda : E \to \mathbb{R}_{\geq 0}$, a λ-uniform distribution μ_λ over spanning trees of $G = (V, E)$ is a distribution where for every spanning tree $T \subseteq E$, $\mathbb{P}_{\mu_\lambda}[T] = \frac{\prod_{e \in T} \lambda_e}{\sum_{T'} \prod_{e \in T'} \lambda_e}$.*

Theorem 2 ([Asa+10]). *Let z be a point in the spanning tree polytope (see (2)) of a graph $G = (V, E)$. For any $\epsilon > 0$, a vector $\lambda : E \to \mathbb{R}_{\geq 0}$ can be found such that the corresponding λ-uniform spanning tree distribution, μ_λ, satisfies*

$$\sum_{T \in \mathcal{T} : T \ni e} \mathbb{P}_{\mu_\lambda}[T] \leq (1 + \varepsilon) z_e, \quad \forall e \in E,$$

i.e., the marginals are approximately preserved. In the above $\mathcal{T}$ is the set of all spanning trees of (V, E). The algorithm is deterministic and running time is polynomial in $n = |V|$, $-\log \min_{e \in E} z_e$ and $\log(1/\epsilon)$.

[KKO22] showed that the following (randomized) max entropy algorithm has expected cost of the output is at most $(\frac{3}{2} - \epsilon)c(x)$.

Algorithm 1. (Randomized) Max Entropy Algorithm for TSP

Find an optimum solution x^0 of Eq. (1), and let $e_0 = \{u_0, v_0\}$ be an edge with $x_{e_0}^0 = 1, c(e_0) = 0$.
Let $E_0 = E \cup \{e_0\}$ be the support of x^0 and x be x^0 restricted to E and $G = (V, E)$.
Find a vector $\lambda : E \to \mathbb{R}_{\geq 0}$ such that for any $e \in E$, $\mathbb{P}_{T \sim \mu_\lambda}[e \in T] = x_e(1 \pm 2^{-n})$.
Sample a tree $T \sim \mu_\lambda$.
Let M be the minimum cost matching on odd degree vertices of T.
Output $T \cup M$.

2.3 Polyhedral Background

For any graph $G = (V, E)$, Edmonds [Edm70] gave the following description for the convex hull of spanning trees of a graph $G = (V, E)$, known as the *spanning tree polytope.*

$$z(E) = |V| - 1, \quad z(E(S)) \leq |S| - 1 \quad \forall S \subseteq V, \quad z_e \geq 0 \quad \forall e \in E. \quad (2)$$

Edmonds [Edm70] proved that the extreme point solutions of this polytope are the characteristic vectors of the spanning trees of G.

Lemma 1 ([KKO21, Fact 2.1]). *Let x^0 be a feasible solution of (1) such that $x_{e_0}^0 = 1$ with support $E_0 = E \cup \{e_0\}$. Let x be x^0 restricted to E; then x is in the spanning tree polytope of $G = (V, E)$.*

Since $c(e_0) = 0$, the following fact is immediate.

Lemma 2. *Let $G = (V, E, x)$ where x is in the spanning tree polytope. If μ is any distribution of spanning trees with marginals x then $\mathbb{E}_{T \sim \mu}[c(T \cup e_0)] = c(x)$.*

To bound the cost of the min-cost matching on the set $O(T)$ of odd degree vertices of the tree T, we use the following characterization of the $O(T)$-join polyhedron due to Edmonds and Johnson [EJ73].

Proposition 1. *For any graph $G = (V, E)$, cost function $c : E \to \mathbb{R}_+$, and a set $O \subseteq V$ with an even number of vertices, the minimum weight of an O-join equals the optimum value of the following integral linear program.*

$$
\begin{aligned}
\min \quad & c(y) \quad s.t. \\
& y(\delta(S)) \geq 1 \quad S \subseteq V, |S \cap O| \; odd \quad y_e \geq 0 \; \forall e \in E
\end{aligned}
\tag{3}
$$

3   Computing Probabilities

The deterministic algorithm depends on the computation of various probabilities and conditional expectations. In this section (and additionally later in Sect. 5), we show to do these calculations efficiently.

3.1   Notation

Let $\mathcal{B}_E$ be the set of all probability measures on the Boolean algebra $2^{|E|}$. Let $\mu \in \mathcal{B}_E$. The generating polynomial $g_\mu : \mathbb{R}[\{z_e\}_{e \in E}]$ of μ is defined as follows:

$$
g_\mu(z) := \sum_S \mu(S) \prod_{e \in S} z_e.
$$

3.2   Matrix Tree Theorem

Let $G = (V, E)$ with $|V| = n$. For $e = (u, v)$ we let $L_e = (\mathbf{1}_u - \mathbf{1}_v)(\mathbf{1}_u - \mathbf{1}_v)^T$ be the Laplacian of e. Recall Kirchhoff's matrix tree theorem:

Theorem 3 (Matrix tree theorem). *For a graph $G = (V, E)$ let $g_T \in \mathbb{R}[z_{e_1}, \ldots, z_{e_m}] = \sum_{T \in \mathcal{T}} z^T$ be the generating polynomial of the spanning trees of G.*

Then, we have

$$
g_T(\{z_e\}_{e \in E}) = \frac{1}{n} \det(\sum_{e \in E} z_e L_e + \mathbf{1}\mathbf{1}^T / n).
$$

Given a vector $\lambda \in \mathbb{R}^{|E|}$ and a set $S \subseteq E$, let $\lambda^S := \prod_{i \in S} \lambda_i$. Recall that the λ-uniform distribution μ_λ is the probability distribution over spanning trees where the probability of every tree T is λ^T. Then the generating polynomial of μ_λ is

$$
g_{\mu_\lambda}(z) = \sum_{T \in \mathcal{T}} \lambda^T z^T = g_T(\{\lambda_e z_e\}_{e \in E}) = \frac{1}{n} \det\left(\sum_{e \in E} z_e \lambda_e L_e + \mathbf{1}\mathbf{1}^T / n\right)
$$

and can be evaluated at any $z \in \mathbb{C}^E$ efficiently using a determinant computation.

Thus we can compute $\mathbb{P}_{T \sim \mu} [e \in T]$ by computing the sum of the probabilities of trees in the graph $G/\{e\}$, i.e. the graph with e contracted, as follows:

$$\mathbb{P}_{T \sim \mu} [e \in T] = 1 - \mathbb{P}_{T \sim \mu} [e \notin T] = 1 - \sum_{T \in \mathcal{T}:e \notin T} \lambda^T$$

where to compute the sum in the RHS we evaluate g_{μ_λ} at $z_e = 0, z_f = 1$ for all $f \neq e$. Thus,

Lemma 3. *Given a λ-uniform distribution μ_λ over spanning trees, for every edge e, we can compute $\mathbb{P}_{T \sim \mu_\lambda} [e \in T]$ in polynomial time.*

Given some $Set \in \mathcal{T}_{partial}$, we contract each edge e with $X_e = 1$ in Set and delete each edge e with $X_e = 0$ in Set. Let G' be the resulting graph with n' vertices, with corresponding $\lambda'_e \propto \lambda_e$ for all $e \in G'$ normalized such that $\sum_{T' \in G'} \lambda'^T = 1$.

Remark 1. A vector $\lambda \in \mathbb{R}^{|E|}$ is normalized by setting $\lambda'_e = \lambda_e / \left(\sum_T \lambda^T \right)^{1/n-1}$ i.e., $\lambda'_e = \lambda_e / g_T(\{\lambda_e\}_{e \in E})^{1/n-1}$. Thus at the cost of another application of the matrix-tree theorem, we assume without loss of generality that we are always dealing with λ values that are normalized.

Putting the previous facts together, we obtain

Lemma 4. *Given a λ-uniform distribution μ_λ and some $Set \in \mathcal{T}_{partial}$, we can compute a vector λ' such that $\mu_{\lambda'} = \mu_{\lambda | Set}$.*

3.3   Computing Parities in a Simple Case

Lemma 5. *Let $A, B \subseteq E$ and μ_λ be a λ-uniform distribution over spanning trees. Then, we can compute $\mathbb{P}_{T \sim \mu_\lambda} [A_T, B_T \text{ even}]$ in polynomial time.*

Proof. First observe that

$$\mathbb{I}\{A_T, B_T \text{ even}\} = \frac{1}{4}(1 + (-1)^{A_T} + (-1)^{B_T} + (-1)^{((A \smallsetminus B) \cup (B \smallsetminus A))_T})$$

One can easily check that if A_T and B_T are even, this is 1, and otherwise it is 0.

To compute $\mathbb{P}_{T \sim \mu_\lambda} [A \text{ and } B \text{ even in } T]$ it is enough to compute the expected value of this indicator. By linearity of expectation it is therefore enough to compute the expectation of $(-1)^{F_T}$ for any set $F \subseteq E$. We can do this using Theorem 3. Setting $z_e^F = -1$ if $e \in F$ and $z_e^F = +1$ otherwise, we exactly have:

$$g_{\mu_\lambda}(z^F) = \sum_{T \in \mathcal{T}} (-1)^{F_T} \lambda^T = \mathbb{E}_{T \sim \mu_\lambda} \left[(-1)^{F_T} \right].$$

The lemma follows.

Remark 2. We can use the same approach to compute $\mathbb{P}_{T \sim \mu_\lambda} [A_T \text{ odd, } B_T \text{ even}]$ or the probability that both are odd. All we need to do is to multiply $(-1)^{A_T}$ with a -1 if A_T needs to be odd (and similarly for B_T), and $(-1)^{((A \setminus B) \cup (B \setminus A))_T}$ with a -1 if we are looking for different parities in A_T, B_T.

Given some $Set \in \mathcal{T}_{partial}$, by Lemma 4 we can compute $\mu_{\lambda'} = \mu_{\lambda | Set}$. Applying the above lemma to $\mu_{\lambda'}$, it follows (after appropriately updating the parities to account for edges set to 1 in Set):

Corollary 1. *Let $A, B \subseteq E$. We can compute $\mathbb{P}_{T \sim \mu} [A \text{ and } B \text{ even in } T \mid Set]$ in polynomial time.*

4   A Deterministic Algorithm in the Degree Cut Case

As a warmup, in this section we show how to implement the deterministic algorithm for the so-called *"degree cut case,"* i.e., when for every set of vertices S with $2 \leq |S| \leq n - 2$ we have $x(\delta(S)) \geq 2 + \eta$ for some absolute constant $\eta > 0$. See Algorithm 2.

Algorithm 2. A Deterministic Approximation Algorithm for Metric TSP in the Degree Cut Case

---

1: Given a solution x^0 of the LP (1), with an edge e_0 with $x_{e_0} = 1$.
2: Let G be the support graph of x.
3: Find a vector $\lambda : E \to \mathbb{R}_{\geq 0}$ such that for any $e \in E$, $\mathbb{P}_{T \sim \mu_\lambda} [e \in T] = x_e (1 \pm 2^{-n})$ (see Section 2.2).
4: Initialize $Set := \emptyset$
5: **while** there exists $e \neq e_0$ not set in Set **do**
6:     Let $Set^+ := Set \cup \{X_e = 1\}$ and let $Set^- := Set \cup \{X_e = 0\}$;
7:     **if** $\mathbb{E}_{T \sim \mu_\lambda} [c(T) + c(m) \mid Set^+] \leq \mathbb{E}_{T \sim \mu_\lambda} [c(T) + c(m) \mid Set^-]$ (m from Definition 2) **then**
8:         $Set := Set^+$;
9:     **else**
10:         $Set := Set^-$;
11:     **end if**
12: **end while**
13: Return $T = \{e : X_e = 1 \text{ in } Set\}$ together with min cost matching on odd degree vertices of T.

Construction of the Matching Vector. We describe a simple construction for the matching vector $m : \mathcal{T} \to \mathbb{R}^{|E|}$ for the degree cut case. It will ensure that for a tree T, m is in the $O(T)$-Join polyhedron where $O(T)$ is the set of odd vertices of T (we emphasize that m is a function of T). Therefore, $c(m)$ is an upper bound on the cost of the minimum cost matching on the odd vertices of T as desired.

Let $p = 2 \cdot 10^{-10}$ (note that we have not optimized this constant and in the degree cut case it can be greatly improved). We say that an edge $e = (u,v)$ is **good** if
$\mathbb{P}_{T \sim \mu}[u,v \text{ both even in } T] \geq p$, where we say a vertex v is even in a tree T if $\delta(v)_T$ is even. The vector m will consist of the convex combination of two feasible points in the $O(T)$-Join polyhedron, g and b (where g is for "good" edges and b is for "bad" edges).

For a tree T and an edge $e = (u,v)$ we let:

$$g_e = \begin{cases} \frac{1}{2+\eta}x_e & \text{If } u \text{ and } v \text{ are both even in } T \\ \frac{1}{2}x_e & \text{Otherwise} \end{cases}$$

Lemma 6. *g is in the $O(T)$-Join polyhedron.*

Proof. First, consider any cut consisting of a single vertex v (or its complement). If v is odd, we need to ensure that $g(\delta(v)) \geq 1$. If v is odd, then $g_e = x_e/2$ for all $e \in \delta(v)$, so this follows from the fact that $x(\delta(v)) = 2$.

Now consider any cut S with $2 \leq |S| \leq n-2$. We now argue that $g(\delta(S)) \geq 1$ with probability 1. This follows from the fact that:

$$g(\delta(S)) \geq \frac{1}{2+\eta}x(\delta(S)) \geq \frac{1}{2+\eta}(2+\eta) = 1,$$

where we use that every cut S with $2 \leq |S| \leq n-2$ has $x(\delta(S)) \geq 2+\eta$.

We now design our second vector b. For a tree T and an edge $e = (u,v)$ we let:

$$b_e = \begin{cases} \frac{1+\eta}{2+\eta}x_e & \text{If } e \text{ is good} \\ \frac{1}{2+\eta}x_e & \text{If } e \text{ is bad} \end{cases}$$

We will crucially use the following:

Corollary 2 (Corollary of Theorem 5.14 from[KKO21]). *Let v be a vertex. Then, if G_v is the set of good edges adjacent to v, $x(G_v) \geq 1$.*

In [KKO21], it is shown that if x_e is bounded away from $1/2$, then e is a good edge. Furthermore, for any two edges e and f adjacent to v with $x_e \approx x_f \approx 1/2$, at least one is good. So, v can have only one bad edge which has fraction about $1/2$, giving the above corollary (therefore it is even true that $x(G_v) \geq 3/2 - \gamma$ for some small $\gamma > 0$).

Given this, we can show the following:

Lemma 7. *b is in the $O(T)$-Join polyhedron.*

Proof. For any non-vertex cut, similar to above, the $O(T)$-Join constraint is easily satisfied. For a vertex cut v, we use that by the above theorem the x weight of the set of good edges adjacent to v is at least 1. Therefore, $b(v) \geq \frac{1+\eta}{2+\eta} + \frac{1}{2+\eta} = 1$.

Definition 2 (Matching vector m in the degree cut case). *Let* $m = \alpha b + (1 - \alpha)g$, *for some* $0 < \alpha < 1$ *we choose in the next subsection. Since* b *and* g *are both in the* $O(T)$-*Join polyhedron, so is* m.

Lemma 8. *For any good edge* e, $\mathbb{E}\,[g_e] \leq (\frac{1}{2} - \frac{\eta p}{4 + 2\eta})x_e$.

Proof. Let $p_e = \mathbb{P}_{T \sim \mu}\,[u, v \text{ even}]$. We can compute:

$$\mathbb{E}\,[g_e] = \left(\frac{p_e}{2 + \eta} + \frac{1 - p_e}{2}\right)x_e \leq \left(\frac{p}{2 + \eta} + \frac{1 - p}{2}\right)x_e = \left(\frac{1}{2} - \frac{\eta p}{4 + 2\eta}\right)x_e,$$

as desired.

Therefore, for any good edge e,

$$\mathbb{E}\,[m_e] \leq \left(\alpha\left(\frac{1 + \eta}{2 + \eta}\right) + (1 - \alpha)\left(\frac{1}{2} - \frac{\eta p}{4 + 2\eta}\right)\right)x_e$$

For any bad edge e, we have

$$\mathbb{E}\,[m_e] \leq \left(\frac{\alpha}{2 + \eta} + \frac{1 - \alpha}{2}\right)x_e$$

To make the two equal, we set $\alpha = \frac{p}{2+p}$. Therefore,

$$\mathbb{E}\,[m_e] \leq \left(\frac{p/(2 + p)}{2 + \eta} + \frac{1 - p/(2 + p)}{2}\right)x_e < \left(\frac{1}{2} - \frac{p\eta}{9}\right)x_e$$

for all edges e. Since η, p are absolute constants, this is at most $(\frac{1}{2} - \epsilon)x_e$ for some absolute constant $\epsilon > 0$. Therefore the randomized algorithm has expected cost at most $(\frac{3}{2} - \epsilon)c(x)$, which is enough to prove that Algorithm 2 deterministically finds a tree plus a matching whose cost is at most $(\frac{3}{2} - \epsilon)c(x)$. Thus the only remaining question is the computational complexity of Algorithm 2, which we address now.

Computing $\mathbb{E}\,[c(T) + c(m) \mid Set]$. Now that we have explained the construction of m, we observe that there is a simple deterministic algorithm to compute $\mathbb{E}\,[c(T) + c(m) \mid Set]$ in polynomial time.

First, compute $\mathbb{E}\,[c(T) \mid Set]$. By linearity of expectation it is enough to compute $\mathbb{P}\,[e \in T \mid Set]$ for all $e \in E$. To do this, we first apply Lemma 4 to find λ' such that $\mu_{\lambda'} = \mu_{\lambda|Set}$ and then apply Lemma 3.

Now to compute $\mathbb{E}\,[c(m) \mid Set]$, it suffices to compute $\mathbb{E}\,[m_e \mid Set]$ for any $Set \in \mathcal{T}_{partial}$, $\mathbb{P}\,[e \in T \mid Set]$ and any $e = (u, v)$. Given the definition of m, the only event depending on the tree is the event $\mathbb{P}\,[u, v \text{ even} \mid Set]$. This can be computed with Corollary 1.

Algorithm 3. A Deterministic Approximation Algorithm for Metric TSP

1: Given a solution x^0 of the LP (1), with an edge e_0 with $x_{e_0} = 1$.
2: Let G be the support graph of x.
3: Find a vector $\lambda : E \to \mathbb{R}_{\geq 0}$ such that for any $e \in E$, $\mathbb{P}_{T \sim \mu_\lambda}[e \in T] = x_e(1 \pm 2^{-n})$
4: Perform Preprocessing Steps 1, 2, 3, 4, 5, and 6
5: Initialize $Set := \emptyset$.
6: **while** there exists $e \neq e_0$ not set in Set **do**
7:     Let $Set^+ := Set \cup \{X_e = 1\}$ and let $Set^- := Set \cup \{X_e = 0\}$;
8:     Compute $S^+ = \mathsf{EE}T \sim \mu_\lambda c(T) \mid Set^+ + \sum_{e \in E} \mathbb{E}_{c(s^*)}(e, Set^+) + \mathbb{E}_{c(s)}(e, Set^+)$.
9:     Compute $S^- = \mathsf{EE}T \sim \mu_\lambda c(T) \mid Set^- + \sum_{e \in E} \mathbb{E}_{c(s^*)}(e, Set^-) + \mathbb{E}_{c(s)}(e, Set^-)$.
10:     If $S^+ \leq S^-$, let $Set := Set^+$. Otherwise let $Set := Set^-$.
11: **end while**
12: Return $T = \{e : X_e = 1 \text{ in } Set\}$ together with min cost matching on odd degree vertices of T.

---

5   General Case

The matching vector m in the general case, [KKO22, Thm 6.1], can be written as $s + s^* + \frac{1}{2}x$ where s, s^* are functions of the tree $T \sim \mu_\lambda$ and some independent Bernoullis $\mathcal{B}$. Roughly speaking, the (slack) vector $s^* : E \to \mathbb{R}_{\geq 0}$ takes care of matching constraints for near minimum cuts that are crossed and the (slack) vector $s : E \to \mathbb{R}$ takes care of the constraints corresponding to cuts which are not crossed. Most importantly, the guarantee is that for a fixed tree T the expectation of $c(s) + c(s^*) + \frac{1}{2}c(x)$ over the Bernoullis is at least $c(M)$ where M is the minimum cost matching on the odd vertices of T. Furthermore, $\mathbb{E}[c(s) + c(s^*)] \leq -\epsilon c(x)$ which is the necessary bound to begin applying the method of conditional expectation in Algorithm 3.

Remark 3. The definitions of s and s^*, the proof that $\mathbb{E}[c(s) + c(s^*)] \leq -\epsilon c(x)$, and the proof that $x/2 + \mathbb{E}[s + s^* \mid T]$ is in the $O(T)$-join polyhedron come from [KKO21,KKO22]. Here, we will review how to construct the random slack vectors s, s^* for a given spanning tree T and then explain how to efficiently compute $\mathbb{E}[c(s) + c(s^*) \mid Set]$ deterministically for any $Set \in \mathcal{T}_{partial}$.

Unfortunately, a reader who has not read [KKO21,KKO22] may not be able to understand the motivation behind the details of the construction of s, s^*. However, ?? and ?? are self-contained in the sense that a reader should be able to verify that $\mathbb{E}[c(s) + c(s^*) \mid Set]$ can be computed efficiently and deterministically.

Our theorem boils down to showing the following two lemmas:

Lemma 9. *For any $Set \in \mathcal{T}_{partial}$, there is a polynomial time deterministic algorithm that computes:*

(1) $\mathbb{E}_{T \sim \mu_\lambda}[c(s^*) \mid Set]$ (shown in $\mathbb{E}_{c(s^*)}(e, Set)$)
(2) $\mathbb{E}_{T \sim \mu_\lambda}[c(s) \mid Set]$ (shown in $\mathbb{E}_{c(s)}(e, Set)$)

The crux of proving the above lemma is to show that for a given edge e and any Set, each of $\mathbb{E}[s_e^* \mid Set]$ and $\mathbb{E}[s_e \mid Set]$ can be written as the (weighted) sum of indicators of events that depend on the sampled tree T, and each of these events happens only when a constant number of (not necessarily disjoint) sets of edges have certain parities or certain sizes. Technically speaking, these weighted sums are non-trivial for some of the events defined in [KKO21, KKO22]. Given that, the following is enough to prove Lemma 9, as it gives a deterministic algorithm to compute the probability that a collection of (not necessarily disjoint) sets of edges have certain parities or certain sizes.

(1) of Lemma 9 is proved in ??, and (2) in ??. The algorithm for each part requires a series of preprocessing steps and function definitions that we have marked with gray boxes. In each section, the final procedure to calculate the expected cost of the slack vector is given in a yellow box at the end of the corresponding section.

Lemma 10. *Given a probability distribution $\mu : 2^{[n]} \to \mathbb{R}_{\geq 0}$ and an oracle O that can evaluate $g_\mu(z_1, \ldots, z_n)$ at any $z_1, \ldots, z_n \in \mathbb{C}$. Let $E_1, \ldots, E_k$ be a collection of (not necessarily disjoint) subsets of $[n]$ and $(\sigma_1, \ldots, \sigma_k) \in \mathbb{F}_{m_1} \times \cdots \times \mathbb{F}_{m_k}$. Then, we can compute,*

$$\mathbb{P}_{T \sim \mu}\left[(E_i)_T = \sigma_i (mod\ m_i), \forall 1 \leq i \leq k\right].$$

in $N := m_1 \ldots m_k$-many calls to the oracle.[1]

Proof. For each of the sets E_i, define a variable x_i, and substitute $\prod_j x_j^{\mathbb{I}\{e \in E_j\}}$ for z_e into the polynomial g_μ and call the resulting polynomial g. Then

$$g(x_1, \ldots, x_k) = \sum_{S \in supp(\mu)} \mathbb{P}[S] \prod_{i=1}^k x_i^{(E_i)_S}$$

Where recall $(E_i)_S = |E_i \cap S|$. Now, let $\omega_i := e^{\frac{2\pi\sqrt{-1}}{m_i}}$. We claim that

$$\frac{1}{m_1 \cdots m_k} \sum_{(e_1, \ldots, e_k) \in \mathbb{F}_{r_1} \times \cdots \times \mathbb{F}_{r_k}} \prod_{i=1}^k \omega_i^{-e_i \sigma_i} g(\omega_1^{e_1}, \ldots, \omega_k^{e_k})$$

$$= \mathbb{P}_{S \sim \mu}\left[(E_i)_S \equiv \sigma_i \bmod m_i, \forall 1 \leq i \leq k\right]$$

So the algorithm only needs to call the oracle N many times to compute the sum in the LHS.

---

[1] Note that since we are dealing with irrational numbers, we will not be able to compute this probability exactly. However by doing all calculations with $poly(n, N)$ bits of precision we can ensure our estimate has exponentially small error which will suffice to get the bounds we need later.

To see this identity, notice that we can write the LHS as

$$\frac{1}{m_1 \cdots m_k} \sum_{(e_1,\ldots,e_k) \in \mathbb{F}_{m_1} \times \cdots \times \mathbb{F}_{m_k}} \sum_{S \in supp(\mu)} \mathbb{P}[S] \prod_{i=1}^{k} \omega_i^{-e_i\sigma_i + e_i(E_i)_S}$$

$$= \sum_{S \in supp(\mu)} \mathbb{P}[S] \prod_{i=1}^{k} \left(\frac{1}{m_i} \sum_{e_i \in \mathbb{F}_{m_i}} \omega_i^{((E_i)_S - \sigma_i)e_i} \right)$$

$$= \sum_{S \in supp(\mu)} \mathbb{P}[S] \prod_{i=1}^{k} \mathbb{I}\{(E_i)_S - \sigma_i \equiv 0 \bmod \sigma_i\}$$

where the last equality uses that ω_i is the m_i'th root of unity. The RHS is exactly equal to the probability that $(E_i)_S \equiv \sigma_i \bmod m_i$ for all i.

Remark 4. When we apply this lemma in this paper, we will always let k be a constant and $m_i \leq |V|$ for all i. Thus, it will always use a polynomial number of calls to an oracle evaluating the generating polynomial of a spanning tree distribution μ_λ. By Theorem 3, for any $z \in \mathbb{C}^{|E|}$:

$$g_{\mu_\lambda}(\{z_e\}_{e \in E}) = \frac{1}{n} \det(\sum_{e \in E} \lambda_e z_e L_e + 11^T/n),$$

which can be computed in polynomial time.

Corollary 3. *Let μ_λ be a λ-uniform spanning tree distribution and let $Set \in \mathcal{T}_{partial}$. Then, let $E_1, \ldots, E_k$ be a collection of (not necessarily disjoint) subsets of $[n]$ and $(\sigma_1, \ldots, \sigma_k) \in \mathbb{F}_{m_1} \times \cdots \times \mathbb{F}_{m_k}$. Then, we can compute,*

$$\mathbb{P}_{T \sim \mu_\lambda}[(E_i)_T = \sigma_i \quad (\bmod m_i), \forall 1 \leq i \leq k \mid Set].$$

in $N := m_1 \ldots m_k$-many calls to the oracle.

Proof. Construct a new graph G' by contracting all edges with $X_e = 1$ in Set and deleting all edges with $X_e = 0$. We then update all σ_i by subtracting the number of edges that are set to 1 in E_i by Set. Then we apply Lemma 10 to the λ-uniform spanning tree distribution over G' with the updated σ and the same m.

References

[App+07] Applegate, D.L., Bixby, R.E., Chvatal, V., Cook, W.J.: The Traveling Salesman Problem: A Computational Study (Princeton Series in Applied Mathematics). Princeton University Press, Princeton, NJ, USA (2007)

[Asa+10] Asadpour, A., Goemans, M.X., Madry, A., Gharan, S.O., Saberi, A.: An o(log n/ log log n) approximation algorithm for the asymmetric traveling salesman problem. In: SODA, pp. 379–389 (2010)

[Chr76] Nicos Christofides. Worst case analysis of a new heuristic for the traveling salesman problem. Report 388, Graduate School of Industrial Administration, Carnegie-Mellon University, Pittsburgh, PA, 1976

[DFJ59] Dantzig, G.B., Fulkerson, D.R., Johnson, S.: On a linear programming combinatorial approach to the traveling salesman problem. OR 7, 58–66 (1959)

[Edm70] Edmonds, J.: Submodular functions, matroids and certain polyhedra. In: Combinatorial Structures and Their Applications, pp. 69–87, New York, NY, USA (1970). Gordon and Breach

[EJ73]   Edmonds, J., Johnson, E.L.: Matching, Euler tours and the Chinese postman. Math. Program. **5**(1), 88–124 (1973)

[GB93]   Goemans, M., Bertsimas, D.: Survivable network, linear programming relaxations and the parsimonious property. Math. Program. **60**, 06 (1993)

[Gup+21] Gupta, A., Lee, E., Li, J., Mucha, M., Newman, H., Sarkar, S.: Matroid-based TSP rounding for half-integral solutions. CoRR, abs/2111.09290 (2021)

[HK70]   Held, M., Karp, R.M.: The traveling salesman problem and minimum spanning trees. Oper. Res. **18**, 1138–1162 (1970)

[HN19]   Haddadan, A., Newman, A.: Towards improving christofides algorithm for half-integer TSP. In: Bender, M.A., Svensson, O., Herman, G., editors, ESA, vol. 144 of LIPIcs, pp. 56:1–56:12. Schloss Dagstuhl - Leibniz-Zentrum für Informatik (2019)

[HNR21] Haddadan, A., Newman, A., Ravi, R.: Shorter tours and longer detours: uniform covers and a bit beyond. Math. Program. **185**(1–2), 245–273 (2021)

[KKO20] Karlin, A.R., Klein, N., Gharan, S.O.: An improved approximation algorithm for TSP in the half integral case. In: Makarychev, K., Makarychev, Y., Tulsiani, M., Kamath, G., Chuzhoy, J., editors, STOC, pp. 28–39. ACM (2020)

[KKO21] Karlin, A.R., Klein, N., Gharan, S.O.: A (slightly) improved approximation algorithm for metric tsp. In: STOC. ACM (2021)

[KKO22] Karlin, A., Klein, N., Gharan, S.O.: A (slightly) improved bound on the integrality gap of the subtour LP for tsp. In: FOCS, pp. 844–855. IEEE Computer Society (2022)

[KLS15] Karpinski, M., Lampis, M., Schmied, R.: New inapproximability bounds for TSP. J. Comput. Syst. Sci. **81**(8), 1665–1677 (2015)

[MS11]   Moemke, T., Svensson, O.: Approximating graphic tsp by matchings. In: FOCS, pp. 560–569 (2011)

[Muc12] Mucha, M.: $\frac{13}{9}$-approximation for graphic TSP. In: STACS, pp. 30–41 (2012)

[OSS11] Gharan, S.O., Saberi, A., Singh, M.: A randomized rounding approach to the traveling salesman problem. In: FOCS, pp. 550–559. IEEE Computer Society (2011)

[Ser78] Serdyukov, A.I.: O nekotorykh ekstremal'nykh obkhodakh v grafakh. Upravlyaemye sistemy **17**, 76–79 (1978)

[SV12]   Sebö, A., Vygen, J.: Shorter tours by nicer ears: CoRR abs/1201.1870 (2012)

[TVZ20] Traub, V., Vygen, J., Zenklusen, R.: Reducing path TSP to TSP. In: Makarychev, K., Makarychev, Y., Tulsiani, M., Kamath, G., Chuzhoy, J., editors, STOC, pp. 14–27. ACM (2020)

Monoidal Strengthening of Simple $\mathcal{V}$-Polyhedral Disjunctive Cuts

Aleksandr M. Kazachkov[1]($\boxtimes$)(iD) and Egon Balas[2]

[1] University of Florida, Gainesville, FL, USA
akazachkov@ufl.edu
[2] Carnegie Mellon University, Pittsburgh, PA, USA
eb17@andrew.cmu.edu

Abstract. Disjunctive cutting planes can tighten a relaxation of a mixed-integer linear program. Traditionally, such cuts are obtained by solving a higher-dimensional linear program, whose additional variables cause the procedure to be computationally prohibitive. Adopting a $\mathcal{V}$-polyhedral perspective is a practical alternative that enables the separation of disjunctive cuts via a linear program with only as many variables as the original problem. The drawback is that the classical approach of monoidal strengthening cannot be directly employed without the values of the extra variables appearing in the extended formulation. We derive how to compute these values from a solution to the linear program generating $\mathcal{V}$-polyhedral disjunctive cuts. We then present computational experiments with monoidal strengthening of cuts from disjunctions with as many as 64 terms. Some instances are dramatically impacted, with strengthening increasing the gap closed by the cuts from 0 to 100%. However, for larger disjunctions, monoidal strengthening appears to be less effective, for which we identify a potential cause.

1 Introduction

Disjunction-based cutting planes, or *disjunctive cuts*, are a strong class of valid inequalities for mixed-integer programming problems, which can be used as a framework for analyzing or generating general-purpose cuts [8]. Their strength comes at a high computational cost, due to which only very special cases of disjunctive cuts have been deployed in optimization solvers. As a step towards practicality, Balas and Kazachkov [10] introduce a relaxation-based $\mathcal{V}$-polyhedral paradigm for disjunctive cuts, which trades off some theoretical strength for computational efficiency. The approach selects a small number of points and rays whose convex hull forms a relaxation of the disjunction; as a result, some potential cuts are no longer valid, but strong cuts are nevertheless guaranteed to be

---

E. Balas passed away during the preparation of this manuscript, which started when both authors were at Carnegie Mellon University. The core ideas and early results are documented in the PhD dissertation of Kazachkov [37, Chapter 5]. A.M. Kazachkov completed the computational experiments, analysis, and writing independently.

A. Del Pia and V. Kaibel (Eds.): IPCO 2023, LNCS 13904, pp. 275–290, 2023.
https://doi.org/10.1007/978-3-031-32726-1_20

obtainable. Further, cuts from this relaxation, called $\mathcal{V}$-*polyhedral (disjunctive)* *cuts* (VPCs), can be generated via a relatively compact linear program, called the *point-ray linear program* (PRLP), compared to the usual higher-dimensional *cut-generating linear program* (CGLP) for disjunctive cuts [8,14,15]. Hence, with VPCs, it is more computationally efficient to improve the disjunction by adding terms and increase the relaxation quality, thereby accessing disjunctive cuts that differ substantially from the families of cuts typically applied in solvers.

VPCs improve the average *(integrality) gap closed* substantially relative to *Gomory mixed-integer cuts* (GMICs) and other standard cuts in solvers. However, the computational experiments by Balas and Kazachkov [10] reveal a curiosity: there are instances for which GMICs (which can be derived as cuts from a two-term disjunction) remain stronger than VPCs even when using large variable disjunctions. For example, for the instance 10teams, originally part of the 3rd Mixed Integer Programming Library (MIPLIB) [18], GMICs close 100% of the integrality gap, while VPCs from a 64-term disjunction close 0% of the gap.

A potential explanation for this phenomenon is that GMICs benefit from a *strengthening* procedure that cannot be directly applied to VPCs. Specifically, the GMIC two-term disjunction can be obtained via *monoidal strengthening* of a disjunction on a single variable [9,12,38]. Monoidal strengthening of cuts from more general disjunctions is also possible, but the procedure ostensibly requires a *simple* disjunction, where each term only imposes a single new constraint. This is not a theoretical barrier, as any cut from a general disjunction can also be derived from a simple disjunction obtained from the general one by aggregating the constraints defining each disjunctive term. The multipliers for this aggregation are precisely the Farkas certificate for the validity of the cut. The key challenge for VPCs is that this certificate is not readily available, because the PRLP only has variables for the cut coefficients, compared to the CGLP that explicitly includes variables for the Farkas multipliers. Our contributions, summarized next, are to identify a way to efficiently apply monoidal strengthening for the particular version of the VPC framework introduced in Balas and Kazachkov [10], as well as to implement and computationally evaluate this strengthening idea.

Contributions. Given a VPC, one can solve the CGLP with cut coefficients fixed and retrieve the required values of the aggregation multipliers, in order to apply monoidal strengthening. Unfortunately, the computational effort associated to this is likely to be prohibitive. Our first contribution, discussed in Sect. 3, is observing that solving the CGLP is unnecessary: it suffices to use the inverse of an easily-identified nonsingular matrix per disjunctive term. Furthermore, for the type of *simple* VPCs proposed and tested by Balas and Kazachkov [10], this inverse is readily available within the cut generation process.

Next, in Sect. 4, we discuss computational experiments with strengthening simple VPCs on a set of benchmark instances. We compare the strength to unstrengthened VPCs and to GMICs, for disjunctions ranging in size up to 64 terms. We find that strengthening can significantly improve the gap closed for some instances. Furthermore, we see that GMICs and unstrengthened VPCs tend to be complementary in terms of which instances they benefit, but applying monoidal strengthening enables the two families to be simultaneously effective for more instances. The results are most striking for two-term disjunctions, in which strengthened VPCs close 40% more gap than unstrengthened VPCs, on average. For example, returning to the instance 10teams, the VPCs from a single variable disjunction close 0% of the integrality gap, but this value goes to 100% after strengthening the cuts. However, as the size of the disjunction increases, the relative improvement by strengthening becomes smaller. Our final contribution, in Sect. 5, is identifying a theoretical source of this weakness.

Related Work. A focal point in the literature on monoidal strengthening for disjunctive cuts [9] (see also Balas [8, Section 7]) is the special case of *split disjunctions*, which are *parallel* two-term disjunctions that are used for GMICs and related cut families. In this context, the use of the CGLP leads to *lift-and-project cuts* (L&PCs) [14], to which monoidal strengthening can be applied [15, Section 2.4]. The family of strengthened L&PCs is equivalent to GMICs, as shown by Balas and Perregaard [12], and to mixed-integer rounding inequalities [45, 46], as discussed in Cornuéjols and Li [23]. Balas and Perregaard [12] provide an appealing geometric interpretation of this connection via intersection cuts [7]: every undominated L&PC can be derived as an intersection cut from a basis in the original problem space. As a result, L&PCs can be generated without explicitly building the CGLP and without hindering a posteriori strengthening of the cuts. Bonami [19] presents a different method for separating L&PCs in the original space of variables that is also amenable to strengthening. Avoiding formulating the higher-dimensional CGLP is the key advance that has enabled the effective inclusion of L&PCs in several solvers.

Sidestepping the CGLP continues to be crucial to move beyond split disjunctions. However, the aforementioned approaches [12,19] rely on properties of the split set; for example, with general disjunctions, there exist cuts that dominate all intersection cuts [5,11,40], so one cannot hope to merely pivot among bases in the original space. Nonetheless, a stream of work [20,36,40] extends cut generation in the original space to general two-term disjunctions, and monoidal strengthening applies to the resulting cuts [28]. No further extension of this technique to more general disjunctions has been reported in the literature.

This motivates the use of VPCs, due to the PRLP's advantage of having the same number of variables as the original problem. The difficulty is that a description of a polyhedron using points and rays may be exponentially larger than using inequalities, causing exponentially many constraints in the PRLP. This naturally leads to row generation in prior work by Perregaard and Balas [48] and Louveaux et al. [44] when invoking the $\mathcal{V}$-polyhedral perspective. In the experiments by Perregaard and Balas [48], for disjunctions with 16 terms, separating cuts via the PRLP with row generation is an order of magnitude faster than via the CGLP. Nonetheless, row generation is time consuming, as multiple PRLPs must be solved to find one valid inequality.

The remedy by Balas and Kazachkov [10] is to construct a relaxation of each disjunctive term, where the resulting PRLP has few rows and immediately produces valid cuts. This is successful at quickly generating cuts from large disjunctions, but the average gap closed by the cuts alone is less than that from GMICs. It is only when VPCs and GMICs are used together that a marked improvement in gap closed is observed, which shows that VPCs affect a different region of the relaxation than GMICs. However, as mentioned with the 10teams instance in which GMICs close all of the gap, while VPCs close none, the results also suggest that the absence of strengthening for VPCs is a significant deficiency.

As discussed, the vanilla monoidal strengthening presented by Balas and Jeroslow [9] does not directly apply to VPCs due to the lack of the values of the aggregation multipliers. Balas and Qualizza [13, Section 6] show that a cross-polytope disjunction, arising from using multiple rows of the simplex tableau, can be strengthened by *modularizing* the inequalities defining the disjunction, replacing the coefficients of integer-restricted nonbasic variables, and they prove the form of the optimal strengthening for the two-row case.

An alternative to monoidal strengthening is the group-theoretic approach [32, 35], equivalent to monoidal strengthening under some conditions. Specifically, "trivial lifting" has been applied to simple disjunctions [16,24–26,49]. Evaluating the trivial lifting is expensive in general [30], and it does not directly apply to arbitrary disjunctive cuts.

While this paper exclusively approaches disjunctive cut generation via the VPC framework, there exist other methods for producing strong disjunctive cuts without solving the higher-dimensional CGLP. Any such approach could potentially benefit from the efficient computation of a Farkas certificate. For example, a common technique in the literature is to use a disjunction to strengthen cuts via tilting, which has been applied to linear and nonlinear integer optimization problems [37,39,42,47].

2   Notation and Background

Our target is to find strong valid cuts to tighten the natural linear relaxation of the mixed-integer linear program below, given rational data:

$$\min_{x \in \mathbb{R}^n} \quad c^\mathsf{T} x$$

$$
\begin{aligned}
A_i.x \geq b_i \quad &\text{for } i \in [q], \\
x_j \geq 0 \quad &\text{for } j \in [n], \\
x_j \in \mathbb{Z} \quad &\text{for } j \in \mathcal{I}.
\end{aligned}
\qquad \text{(IP)}
$$

Here, $[n] := \{1, \ldots, n\}$ for any integer n, and $\mathcal{I} \subseteq [n]$ is the set of integer-restricted variables. For a given matrix A, we denote the ith row by "$A_i.$" and the jth column by "$A_{.j}$". Let P_I denote the feasible region of (IP), and let $P := \{x \in \mathbb{R}^n_{\geq 0} : Ax \geq b\}$.

One way to strengthen the formulation P (with respect to P_I) is to use logical conditions to formulate a *disjunction*, from which valid inequalities for P_I can then be derived. Suppose $\vee_{t \in \mathcal{T}}(D^t x \geq D_0^t)$ is a valid disjunction, in the sense that $P_I \subseteq \cup_{t \in \mathcal{T}}\{x \in \mathbb{R}^n : D^t x \geq D_0^t\}$. Let $Q^t := \{x \in P : D^t x \geq D_0^t\}$. This is an *$\mathcal{H}$-polyhedral (inequality) description*. We assume $Q^t \neq \emptyset$ for all $t \in \mathcal{T}$.

Let $P^t := \{x \in \mathbb{R}^n : A^t x \geq b^t\}$ denote a relaxation of Q^t, where $A^t x \geq b^t$ is defined by a subset of the constraints defining Q^t. For the VPC procedure, we must ensure that P^t has relatively few extreme points and rays, i.e., it has a compact $\mathcal{V}$-*polyhedral description* $(\mathcal{P}^t, \mathcal{R}^t)$, so that $P^t = \text{conv}(\mathcal{P}^t) + \text{cone}(\mathcal{R}^t)$. Define the *disjunctive hull* $P_D := \text{cl conv}(\cup_{t \in \mathcal{T}} P^t)$, which can be described by the point-ray collection $(\mathcal{P}, \mathcal{R}) := (\cup_{t \in \mathcal{T}} \mathcal{P}^t, \cup_{t \in \mathcal{T}} \mathcal{R}^t)$. For $t \in \mathcal{T}$, let q'_t be the number of rows of A^t. We first summarize some important disjunctive programming concepts and the two cut-generating paradigms that we are relating.

CGLP. One way to generate valid cuts for P_D is through the CGLP, which is an application of *disjunctive programming duality* [8, Section 4]. Specifically, an inequality $\alpha^\intercal x \geq \beta$ is valid for P_D if and only if the inequality is valid for each P^t, $t \in \mathcal{T}$. Consequently, by Farkas's lemma [27], $\alpha^\intercal x \geq \beta$ is valid for P_D if and only if the following system is feasible, in variables $(\alpha, \beta, \{v^t\}_{t \in \mathcal{T}})$, where $v^t \in \mathbb{R}^{1 \times q'_t}$ is a row vector of appropriate length for each $t \in \mathcal{T}$:

$$\left. \begin{array}{c} \alpha^\intercal = v^t A^t \\ \beta \leq v^t b^t \\ v^t \in \mathbb{R}^{q'_t}_{\geq 0} \end{array} \right\} \quad \text{for all } t \in \mathcal{T}. \tag{1}$$

We refer to $\{v^t\}_{t \in \mathcal{T}}$ as the *Farkas certificate* for the validity of $\alpha^\intercal x \geq \beta$ for P_D.

To generate cuts with (1), one typically maximizes the violation with respect to a P_I-infeasible point, after adding a normalization, which can be a crucial choice [29]. For example, the constant of the cut can be fixed to $\bar{\beta} \in \mathbb{R}$:

$$\left\{ (\alpha, \{v^t\}_{t \in \mathcal{T}}) : (\alpha, \bar{\beta}, \{v^t\}_{t \in \mathcal{T}}) \text{ is feasible to (1)} \right\}. \tag{CGLP($\bar{\beta}$)}$$

PRLP. An alternative way to generate disjunctive cuts is through the *reverse polar* of P_D [8, Section 5], which is defined with respect to a given $\bar{\beta} \in \mathbb{R}$ as

$$\left\{ \alpha \in \mathbb{R}^n : \alpha^\intercal x \geq \bar{\beta} \text{ for all } x \in P_D \right\}.$$

Clearly this captures all of the valid inequalities for P_D whose constant is equal to $\bar{\beta}$. Since $x \in P_D$ if and only if $x \in \text{conv}(\mathcal{P}) + \text{cone}(\mathcal{R})$, it holds that $\alpha^\intercal x \geq \bar{\beta}$ is valid for P_D if and only if it is satisfied by all of the points and rays in $(\mathcal{P}, \mathcal{R})$. This yields the system (PRLP($\bar{\beta}$)), in variables $\alpha \in \mathbb{R}^n$, for a fixed $\bar{\beta}$:

$$\begin{array}{ll} \alpha^\intercal p \geq \bar{\beta} & \text{for all } p \in \mathcal{P} \\ \alpha^\intercal r \geq 0 & \text{for all } r \in \mathcal{R}. \end{array} \quad \text{(PRLP($\bar{\beta}$))}$$

The feasible solutions to (PRLP($\bar{\beta}$)) are what we refer to as VPCs.

As discussed, the advantage of (PRLP($\bar{\beta}$)) over (CGLP($\bar{\beta}$)) is the absence of the Farkas multipliers as variables, so VPCs are generated without requiring a lifted space. As we see next, the disadvantage to (PRLP($\bar{\beta}$)) is that these missing variables are used in strengthening the cuts after they are generated.

Monoidal Strengthening. Balas and Jeroslow [9] strengthen cuts with a *monoid*:

$$\mathrm{M} := \left\{ m \in \mathbb{Z}^{|\mathcal{T}|} : \sum_{t \in \mathcal{T}} m_t \geq 0 \right\}. \tag{M}$$

It is also assumed that, for each $t \in \mathcal{T}$, there exists a finite lower bound vector ℓ^t such that $D^t x \geq \ell^t$ for all $x \in P_I$. Let $\Delta^t := D_0^t - \ell^t$.

To strengthen the cut, we improve the underlying disjunction. Specifically, given a valid disjunction $\vee_{t \in \mathcal{T}} (D^t x \geq D_0^t)$, for any $m \in \mathrm{M}$ and $k \in \mathcal{I}$, the disjunction $\vee_{t \in \mathcal{T}} (\tilde{D}^t x \geq \tilde{D}_0^t)$ is also valid, where $\tilde{D}_{\cdot k}^t := D_{\cdot k}^t + \Delta^t m_t$, and $\tilde{D}_{\cdot j}^t = D_{\cdot j}^t$ for all $j \neq k$. The strengthened cut is obtained by applying the Farkas certificate of the unstrengthened cut to the strengthened disjunction.

Let q_t denote the number of constraints in $D^t x \geq D_0^t$ for term $t \in \mathcal{T}$. Given row vectors $(u^t, u_0^t) \in \mathbb{R}_{\geq 0}^{1 \times q} \times \mathbb{R}_{\geq 0}^{1 \times q_t}$, define

$$\alpha_k^t := u^t A_{\cdot k} + u_0^t D_{\cdot k}^t. \tag{α_k^t}$$

Then (using an appropriate CGLP) the cut $\alpha^\top x \geq \beta$ is valid for P_D, where

$$\alpha_k := \max_{t \in \mathcal{T}} \{\alpha_k^t\} \qquad \text{and} \qquad \beta := \min_{t \in \mathcal{T}} \{u^t b + u_0^t D_0^t\}.$$

(The above applies to cuts valid for $\vee_{t \in \mathcal{T}} Q^t$; for P_D, assume a value of zero for the multipliers on constraints of Q^t that are not present in P^t.) Define $\hat{u}_k^t := \alpha_k - \alpha_k^t$. We now apply monoidal strengthening to the cut $\alpha^\top x \geq \beta$.

Theorem 1 ([9, **Theorem 3**]). *Given $(u^t, u_0^t) \in \mathbb{R}_{\geq 0}^{1 \times q} \times \mathbb{R}_{\geq 0}^{1 \times q_t}$ for $t \in \mathcal{T}$, the inequality $\tilde{\alpha}^\top x \geq \beta$ is valid for P_I, where $\tilde{\alpha}_k := \alpha_k$ for $k \notin \mathcal{I}$, and, for $k \in \mathcal{I}$,*

$$\tilde{\alpha}_k := \inf_{m \in \mathrm{M}} \max_{t \in \mathcal{T}} \{\alpha_k^t + u_0^t \Delta^t m_t\} = \alpha_k + \inf_{m \in \mathrm{M}} \max_{t \in \mathcal{T}} \{-\hat{u}_k^t + u_0^t \Delta^t m_t\}.$$

Thus, the Farkas certificate $\{(u^t, u_0^t)\}_{t \in \mathcal{T}}$ is used for monoidal strengthening. Computing these values without solving the CGLP is our next target.

3   Correspondence Between PRLP and CGLP Solutions

Let $\bar{\alpha}^\top x \geq \bar{\beta}$ be a valid inequality for P_D, corresponding to a feasible solution to $(\mathrm{PRLP}(\bar{\beta}))$. Our goal is to compute Farkas multipliers certifying the cut's validity without explicitly solving the CGLP. While one can solve for values v^t that satisfy $\bar{\alpha}^\top = v^t A^t$, $\bar{\beta} = v^t b^t$, $v^t \geq 0$, we provide an improvement via basic linear programming concepts. We first present a special case in Sects. 3.1 and 3.2, when the disjunctive terms P^t are not primal degenerate, a condition that is satisfied by the VPC procedure implemented for our experiments. Then, Sect. 3.3 discusses a challenge posed by the general case.

We assume that $\bar{\alpha}^\top x \geq \bar{\beta}$ is supporting for all terms in $\mathcal{T}$. This is for ease of notation, as otherwise we would need to add an index t to the constant side. Concretely, the assumption is without loss of generality because, for any term

$t \in \mathcal{T}$, we can increase the constant side of the cut until we obtain an inequality $\bar{\alpha}^\mathsf{T} x \geq \bar{\beta}_t$ that is supporting for term t, though perhaps invalid for other terms. The value of $\bar{\beta}_t$ can be quickly calculated by taking the dot product of $\bar{\alpha}$ with every point in $\mathcal{P}^t$. We can then find a certificate v^t of the validity of $\bar{\alpha}^\mathsf{T} x \geq \bar{\beta}_t$ for P^t, which also serves as a certificate for the weaker inequality $\bar{\alpha}^\mathsf{T} x \geq \bar{\beta}$. We state, without proof, a slightly more general version of this in Lemma 2.

Lemma 2. *For $t \in \mathcal{T}$, let $C^t \supseteq P^t$ and $\bar{\beta}_t \geq \bar{\beta}$ such that $\bar{\alpha}^\mathsf{T} x \geq \bar{\beta}_t$ is valid for C^t. Then, given any Farkas certificate for the validity of the inequality $\bar{\alpha}^\mathsf{T} x \geq \bar{\beta}_t$ for C^t, the same multipliers certify that $\bar{\alpha}^\mathsf{T} x \geq \bar{\beta}$ is valid for P^t.* □

For convenience, we introduce extra notation to refer to the feasible region of Q^t as $\hat{A}^t x \geq \hat{b}^t$, and we define the number of these constraints as $\hat{q}_t := q + q_t + n$. For $N \subseteq [\hat{q}_t]$, define $\hat{A}^t_N x \geq \hat{b}^t_N$ as the constraints of Q^t indexed by N.

3.1 Simple VPCs

Our experimental setup in Sect. 4 follows that of Balas and Kazachkov [10], who focus on a variant of the VPC framework called *simple* VPCs. Let p^t be a vertex of Q^t, for $t \in \mathcal{T}$. There exists a *cobasis* for p^t, a set of n linearly independent constraints among those defining Q^t that are tight at p^t. Let $N^t \subseteq [\hat{q}_t]$ denote the indices of these n constraints, and define the *basis cone* $C^t := \{x \in \mathbb{R}^n : \hat{A}^t_{N^t} x \geq \hat{b}^t_{N^t}\}$. The inequality $\bar{\alpha}^\mathsf{T} x \geq \bar{\beta}$ is a simple VPC if P^t is a basis cone for each term. The (translated) cone C^t has a particularly easy $\mathcal{V}$-polyhedral representation: there is a single extreme point p^t, and there are n extreme rays $\{r^i\}_{i \in [n]}$. The ith extreme ray of C^t corresponds to increasing the "slack" on the ith constraint defining C^t [21, Chapter 6]. Lemma 3 states that, for simple VPCs, the values of the variables $\{v^t\}_{t \in \mathcal{T}}$ to (CGLP($\bar{\beta}$)) can be computed via the dot product of the cut coefficients with the rays of C^t.

Lemma 3. *Let C^t be a basis cone defined by N^t, the indices of n linearly independent constraints of Q^t. If $\bar{\alpha}^\mathsf{T} x \geq \bar{\beta}$ is valid for C^t, then the multiplier on constraint $i \in [n]$ of C^t has value $v^t_i = \bar{\alpha}^\mathsf{T} r^i$, where r^i is column i of $(\hat{A}^t_{N^t})^{-1}$.*

Proof. Add nonnegative slack variables $s^t_{N^t}$ for each row indexed by N^t, so that $\hat{A}^t_{N^t} x - s^t_{N^t} = b^t_{N^t}$. Then observe that, being a cobasis, $\hat{A}^t_{N^t}$ is invertible, so $x = (\hat{A}^t_{N^t})^{-1} b^t_{N^t} + (\hat{A}^t_{N^t})^{-1} s^t_{N^t} = p^t + \sum_{i \in N^t} r^i s^t_i$. The last equality follows from the derivation of the rays of C^t; see, for example, Conforti et al. [21, Chapter 6]. □

Therefore, for simple VPCs, the Farkas certificate can be computed with no extra effort when given the point-ray representation of P_D. Moreover, Balas and Kazachkov [10] obtain simple VPCs from the leaf nodes of a partial branch-and-bound tree and use p^t as the optimal solution to the linear relaxation at each leaf; implemented carefully, this can further reduce the computational load for generating then strengthening VPCs, as the values of the rays can be read from the optimal tableau, which is typically readily available from a solver.

3.2   Relaxations Without Primal Degeneracy

Suppose the relaxation $P^t \supseteq Q^t$ is a *simple polyhedron*, in which every extreme point and ray is defined by a *unique* basis [50]. The basis cone C^t used for simple VPCs is one example. While the basis cone setting may seem quite narrow, it turns out to encompass more general situations. Specifically, there always exists a basis cone $C^t \supseteq P^t$ such that $\bar{\alpha}^\top x \geq \bar{\beta}$ is valid and supporting for C^t.

Lemma 4. *Let P^t be a simple polyhedron, and suppose the point-ray collection $(\mathcal{P}^t, \mathcal{R}^t)$ satisfies $P^t = \mathrm{conv}(\mathcal{P}^t) + \mathrm{cone}(\mathcal{R}^t)$. Let $\bar{\alpha}^\top x \geq \bar{\beta}$ be a valid inequality for P^t. Then there exists a vertex $p^t \in \mathcal{P}^t$ such that $\bar{\alpha}^\top x \geq \bar{\beta}$ is valid for the basis cone C^t associated to p^t, defined with respect to the constraints of P^t.*

Proof. Let p^t be an optimal solution to $\min_x\{\bar{\alpha}^\top x : x \in P^t\} = \min_p\{\bar{\alpha}^\top p : p \in \mathcal{P}^t\}$. Define $\bar{\beta}_t := \bar{\alpha}^\top p$. Note that the rays in $\mathcal{R}^t$ need not be considered, as the optimization problem must be bounded since $\bar{\alpha}^\top x \geq \bar{\beta}_t$ is valid for all $x \in P^t$. The point p^t has a unique basis, so the basis cone C^t is defined by the (precisely) n constraints of P^t that are tight at p^t. Optimality of p^t implies all reduced costs are nonnegative. It follows that $\bar{\alpha}^\top r \geq 0$ every ray $r \in C^t$. Since $\bar{\alpha}^\top p^t = \bar{\beta}_t \geq \bar{\beta}$, the inequality $\bar{\alpha}^\top x \geq \bar{\beta}$ is valid for C^t.   □

Therefore, we can invoke Lemmas 2 and 3 to find the Farkas certificate for this case. Note that, when the given point-ray collection only contains extreme points and rays, the rays of C^t for any basis cone of the simple polyhedron P^t can be computed as the rays $\mathcal{R}^t$, along with the directions $p - p^t$ for every point $p \in \mathcal{P}^t$ that is adjacent (one pivot away) from p^t.

3.3   Relaxations with Primal Degeneracy

Up to now, we have made the convenient assumption that the relaxation P^t is a simple polyhedron. More generally, there always exists a basis cone C^t, such that a cut valid for P^t is valid for C^t. With Example 5, we illustrate the complication if $\bar{\alpha}^\top x \geq \bar{\beta}$ is supporting at a primal degenerate point of P^t: a basis for that point needs to be chosen carefully, as the inequality may not be valid for some basis cones. It can be computationally involved to find a valid basis in these situations, which prevents a direct application of our approach relying on simple polyhedra. The purpose of this example is to highlight a crucial obstacle to a complete correspondence between PRLP and CGLP solutions, but we do not further investigate the nondegenerate case in this paper.

Example 5. Figure 1 shows a polyhedron P, defined as the feasible solutions to

$$-(13/8)x_1 - (1/4)x_2 - x_3 \geq -15/8 \tag{c1}$$
$$(1/2)x_1 + x_2 \geq 1/2 \tag{c2}$$
$$(1/2)x_1 - x_3 \geq -3/4 \tag{c3}$$
$$(1/2)x_1 - x_2 \geq -1/2 \tag{c4}$$
$$x_2 \geq 0. \tag{c5}$$

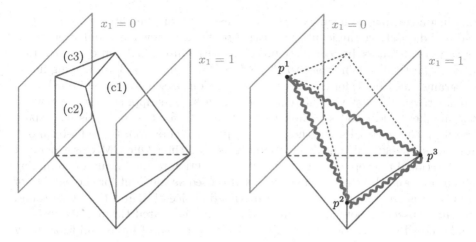

Fig. 1. Example 5: Disjunctive terms with primal degeneracy, despite a nondegenerate initial polyhedron. The VPC is the red wavy line in the second panel. (Color figure online)

A valid cut from the disjunction $(-x_1 \geq 0) \vee (x_1 \geq 1)$ has coefficients $\bar{\alpha}^\mathsf{T} = (-5/8, -1/4, -1)$ and constant $\bar{\beta} = -7/8$. The cut, depicted in the right panel, is incident to point $p^1 = (0, 1/2, 3/4)$ on $P^1 := \{x \in P : -x_1 \geq 0\}$. This point is tight for four inequalities: three defining P (constraints (c2)-(c4)), and the disjunction-defining inequality $-x_1 \geq 0$. Note that P is simple, but P^1 is not.

To construct the cobasis N^1, such that the inequality is valid for the associated basis cone C^1, we must select three linearly independent constraints among those that are tight at p^1. One of the inequalities must be $-x_1 \geq 0$, as otherwise we have not imposed the disjunction at all (but we also know the cut is not valid for P). It can be verified that the only valid choice for this example is N^1 containing the indices for (c3), (c4), and the disjunctive inequality $-x_1 \geq 0$. ∎

4   Computational Experiments

We implement monoidal strengthening for simple VPCs, building on the code used by Balas and Kazachkov [10] from https://github.com/akazachk/vpc. Our goal for the computational study is to measure the effect of monoidal strengthening on the *percent integrality gap closed* by VPCs, compared to unstrengthened VPCs and GMICs, and evaluated across different disjunction sizes.

The code is run on HiPerGator, a shared cluster through Research Computing at the University of Florida. The computational setup is nearly identical to the one described in Balas and Kazachkov [10, Section 5 and Appendix C]. We select instances from the union of the MIPLIB [4,17,18,31,41], CORAL [22], and NEOS sets, restricted to those with at most 5,000 rows and columns and based on other criteria given in [10, Appendix C]. This yields 332 instances suitable for gap closed comparisons. However, we only report on 274 of these

332 instances, due to memory resource constraints on the cluster. Despite this reduced dataset, we can identify recurring patterns in how monoidal strengthening affects instances. Instances are presolved with Gurobi [34], but cut generation is done via the C++ interface to COIN-OR [43], using Clp [3] for solving linear programs and Cbc [1] for constructing disjunctions based on partial branch-and-bound trees. We test six different disjunction sizes, stopping branching when the number of leaf nodes (disjunctive terms) is 2^ℓ for $\ell \in [6]$. Thus, we report results with monoidal strengthening of disjunctive cuts from up to 64-term disjunctions, though only one disjunction is used at a time. One GMIC is generated per fractional integer variable at an optimal solution to the linear programming relaxation, and the number of GMICs is also used as the limit for the number of VPCs we generate for that instance per fixed choice of disjunction. One round of cuts is used for both procedures. GMICs are generated through CglGMI [2], while the VPC generation procedure is identical to that of Balas and Kazachkov [10], with strengthening applied afterwards.

While Lemma 3 enables us to calculate the values of the Farkas multipliers via the rays of each relaxation P^t, and these values are readily available based on how we built the PRLP, we do not avail of this connection. Instead, we calculate $v^t = \bar{\alpha}^\mathsf{T}(A^t)^{-1}$. This approach is still more direct than solving a feasibility version of $(\mathrm{CGLP}(\bar{\beta}))$ with $\bar{\alpha}$ fixed. We opt for numerical safety for this exploratory investigation, so we use the Eigen library [33] to recompute the inverse of A^t rather than reading from the Cbc / Clp internal basis inverse for each term.

We report the average percent integrality gap closed by VPCs and GMICs in Table 1. The first six data rows contain the results for each fixed disjunction size. The penultimate data row, labeled "Best", uses the highest gap closed per instance across all disjunctions. The last data row, labeled "Wins", reports the number of instances for which the "Best" gap closed is at least 10^{-3} higher than the gap closed by GMICs. In the columns, we refer to GMICs by "G", unstrengthened VPCs by "V", strengthened VPCs by "V$^+$". The columns "G+V" and "G+V$^+$" refer to GMICs applied together with VPCs. There are two sets of instances: "All" reports on all 274 instances, while "$\geq$10%" reports on the 97 instances for which unstrengthened VPCs alone close at least 10% of the integrality gap for the "Best" values.

In terms of overall gap closed, despite the monoidal strengthening procedure, as reported by Balas and Kazachkov [10], VPCs alone do not outperform GMICs for the "All" set, but using VPCs and GMICs together provides around 40% improvement in gap closed relative to GMICs alone. While adding VPCs with GMICs might double the number of cuts, one round of VPCs continues to close substantial more gap even after multiple rounds of solver-default cuts [10]. Hence, VPCs tighten the relaxation in different regions relative to GMICs. This is also highlighted by the "$\geq$10%" set, which are instances for which VPCs have strong performance; for this set, GMICs are relatively weaker, with the best VPCs per instance (used alone) providing a 75% improvement in average percent gap closed over GMICs alone. We also see this in the "Wins" row: for the "$\geq$10%" set, VPCs alone outperform GMICs for 73 of the 97 instances in the set.

Table 1. Average percent gap closed by VPCs and GMICs according to the number of leaf nodes used to construct the partial branch-and-bound tree. "Best" refers to the maximum gap closed per instance across all partial tree sizes.

	All					$\geq 10\%$				
	G	V	V^+	G+V	$G+V^+$	G	V	V^+	G+V	$G+V^+$
2 leaves	17.21	2.28	3.25	17.95	18.18	16.29	5.34	6.47	18.13	18.59
4 leaves	17.21	3.35	3.72	18.37	18.54	16.29	7.81	8.35	19.14	19.48
8 leaves	17.21	4.51	4.76	18.98	19.15	16.29	10.84	11.16	20.66	20.91
16 leaves	17.21	6.41	6.57	20.54	20.67	16.29	15.81	16.05	24.86	25.04
32 leaves	17.21	8.78	8.97	22.31	22.48	16.29	21.82	22.28	29.59	29.97
64 leaves	17.21	10.46	10.57	23.72	23.83	16.29	25.59	25.85	32.90	33.14
Best	17.21	11.93	12.57	24.67	24.89	16.29	29.26	29.53	35.27	35.59
Wins		103	104	185	190		73	73	94	94

Next, we summarize observations about the effect of monoidal strengthening. We start with the first data row, in which VPCs are derived from one split disjunction per instance. For the set "All", monoidal strengthening affects the gap closed by VPCs for 87 instances and increases the average gap closed by VPCs by ~1% from 2.28% to 3.25%, a 40% relative improvement. For the set "$\geq 10\%$", the corresponding relative improvement is 20%.

Although the two-term case is encouraging, and a similar relative improvement in gap closed would be substantial for larger disjunctions, this unfortunately does not materialize. From Table 1, we see that as the disjunction size increases, the contribution of monoidal strengthening tends to further diminish, with an absolute improvement in gap closed of only 0.1% for VPCs from a 64-term disjunction. We will discuss a potential cause for this in the next section.

We now compare the columns "$G+V^+$" to "G+V". On the set "All", even for split disjunctions, the effect of strengthening is minimal when VPCs are combined with GMICs, with strengthening only yielding an additional 0.23% in percent gap closed, preserving around 23% of the improvement between "V^+" and "V". For larger disjunctions, while the absolute increase in gap closed by strengthened VPCs is small, over 80% of that improvement is preserved when adding GMICs together with VPCs.

A closer examination of the results supports the hypothesis that monoidal strengthening is a key factor enabling GMICs to close more gap than VPCs. We sort the instances by the increase in gap closed by strengthened VPCs compared to unstrengthened ones, using the best gap closed across all disjunction sizes, per column. Table 2 shows the top ten instances, sorted by the last column, which calculates the difference between "V^+" and "V". The table includes the instance 10teams discussed earlier, as well as six other instances for which unstrengthened VPCs close at most 5% of the gap. We see that monoidal strengthening of VPCs bridges a large portion of the difference with GMICs for these instances. For **neos-1281048**, the situation is reversed: 121 GMICs close no gap while 29

Table 2. Percent gap closed for instances where strengthening VPCs works best.

Instance	G	V	V$^+$	G+V	G+V$^+$	V$^+$−V
10teams	100.00	0.00	100.00	100.00	100.00	100.00
neos-1281048	0.00	17.09	29.36	17.09	29.36	12.27
neos-1599274	34.65	0.00	11.19	34.65	34.65	11.19
f2gap401600	62.97	2.53	11.34	63.31	71.77	8.80
prod2	2.31	27.60	35.90	27.63	35.91	8.29
neos-942830	6.25	0.00	6.25	6.25	6.25	6.25
p0548	48.62	3.28	9.03	49.03	55.11	5.75
mkc	6.08	2.60	6.56	6.35	9.61	3.96
f2gap201600	60.27	8.58	12.13	60.27	60.27	3.56
neos-4333596-skien	20.84	7.05	9.83	20.84	20.85	2.78

unstrengthened VPCs close 17% of the gap, which is further improved to 29% after strengthening. From this table, we also observe the phenomenon that the value in column "G+V" is typically either entirely due to GMICs or to VPCs, but which cuts are more important varies by instance. The situation remains similar for the column "G+V$^+$", though now we find several cases (f2gap401600, p0548, mkc) in which the two cut families add to each other.

While running time is not our focus, and the shared computing environment makes wall clock times unreliable, Table 3 provides the average number of seconds for a single run of each instance, including generating then strengthening VPCs. On average, cut generation takes, in total, from less than a second for two-term disjunctions to 50 s for 16-term disjunctions, 150 s for 32-term disjunctions, and nearly 9 min for 64-term disjunctions. The time per cut, on average, is less than 0.1 s for two-term disjunctions, ranging up to 9 s for 32 terms and over 30 s for 64 terms.

5   Choosing a Relaxation Amenable to Strengthening

In this section, we examine a potential cause of the diminishing effect of monoidal strengthening with larger disjunctions. From Theorem 1, given an initial cut $\alpha^\top x \geq \beta$, we can strengthen coefficient α_k, $k \in \mathcal{I}$, to

$$\widetilde{\alpha}_k = \alpha_k + \inf_{m \in \mathbb{M}} \max_{t \in \mathcal{T}} \left\{ -\hat{u}_k^t + u_0^t \Delta^t m_t \right\},$$

Table 3. Average time (seconds) to generate the cuts in column V$^+$ of Table 1.

Statistic	Set	2 leaves	4 leaves	8 leaves	16 leaves	32 leaves	64 leaves
Cut time (s)	All	0.76	6.39	15.33	49.90	149.84	525.78
	≥10%	0.92	9.31	21.06	130.45	273.51	521.99
Time/cut (s)	All	0.08	0.39	0.97	2.65	9.00	30.54
	≥10%	0.07	0.35	0.79	2.46	7.75	20.19

where $\hat{u}_k^t = \alpha_k - (u^t A_{.k} + u_0^t D_{.k}^t)$ is the slack on the CGLP constraint $\alpha_k \geq u^t A_{.k} + u_0^t D_{.k}^t$. Equivalently, $\hat{u}_k^t$ is the Farkas multiplier for the nonnegativity constraint $x_k \geq 0$. The next lemma restates the (known) reason that a *nonbasic* integral variable k is required for monoidal strengthening.

Lemma 6. *If $\hat{u}_k^t = 0$, then $\widetilde{\alpha}_k = \alpha_k$.*

Proof. In this case, $\widetilde{\alpha}_k = \alpha_k + \inf_{m \in \mathbb{M}} \max_{t \in \mathcal{T}} \{u_0^t \Delta^t m_t\}$. Since $\sum_{t \in \mathcal{T}} m_t \geq 0$ for every $m \in \mathbb{M}$, and $u_0^t \Delta^t \geq 0$, the optimal solution is $m = 0$. $\square$

In the correspondence in Sect. 3, we ultimately find a point $p^t \in P^t$ such that $\bar{\alpha}^\top p^t = \bar{\beta}_t$, where $\bar{\beta}_t = \min_p\{\bar{\alpha}^\top p : p \in \mathcal{P}^t\}$. We then compute a basis cone at p^t for which the cut is valid and use this (translated) cone to compute the values of the Farkas certificate. However, by complementary slackness, if $p_k^t > 0$, then necessarily $\hat{u}_k^t = 0$.

Although at first this appears simultaneously unfortunate and unavoidable, there are two potential remedies. First, there may be dual degeneracy in the choice of p^t: each such point can lead to a different Farkas certificate and therefore a different strengthening. Second, as observed by Balas and Qualizza [6], "sometimes weaking a disjunction helps the strengthening". Though in that context, the weakening involves adding terms to the disjunction, the sentiment applies to our setting as well: if $\bar{\beta}_t > \bar{\beta}$, then one can seek a different, potentially infeasible, basis of Q^t in which more integer variables are nonbasic and $\bar{\alpha}^\top x \geq \bar{\beta}$ is still valid for the associated basis cone.

The computational results support the above intuition. When VPCs are generated from a split disjunction, on average, around 95% of the generated cuts per instance have any coefficient strengthened with the monoidal technique. This decreases to 85% for 64-term disjunctions. Furthermore, on average among VPCs to which strengthening has been applied, 20% of the cut coefficients are strengthened for split disjunctions, while this value steadily decreases as disjunction size increases, so among the analogous VPCs from 64-term disjunctions, only 10% of the coefficients are strengthened.

6   Conclusion

We show that strengthening cuts from general disjunctions is possible without explicitly solving a higher-dimensional CGLP, and that this strengthening can have a high impact for certain instances. However, several challenges are also highlighted for future work. First, the strengthening does not work well on average for larger disjunctions. While we propose a viable explanation and remedy, it is computationally demanding and requires development. Second, the optimal monoidal strengthening involves solving an integer program per cut; this is a relatively small and easy problem, but it nonetheless can be slow for larger disjunctions, as suggested by Table 3, which includes strengthening time. One can reduce this load by selectively strengthening only the most promising cuts, identified by theoretical properties or good heuristics, or to forego optimality in the strengthened cut coefficients. Our computational results indicate that VPCs and GMICs seem to have complementary affects; understanding this better is an opportunity to more widely adopt disjunctive cuts.

References

1. COIN-OR Branch and Cut. https://github.com/coin-or/Cbc
2. COIN-OR Cut Generation Library. https://github.com/coin-or/Cgl
3. COIN-OR Linear Programming. https://github.com/coin-or/Clp
4. Achterberg, T., Koch, T., Martin, A.: MIPLIB 2003. Oper. Res. Lett. **34**(4), 361–372 (2006)
5. Andersen, K., Cornuéjols, G., Li, Y.: Split closure and intersection cuts. Math. Program., 102(3, Ser. A), 457–493 (2005)
6. Balas, E., Qualizza, A.: Monoidal cut strengthening revisited. Discrete Optim. **9**(1), 40–49 (2012)
7. Balas, E.: Intersection cuts–a new type of cutting planes for integer programming. Oper. Res. **19**(1), 19–39 (1971)
8. Balas, E.: Disjunctive programming. Ann. Discrete Math. **5**, 3–51 (1979)
9. Balas, E., Jeroslow, R.G.: Strengthening cuts for mixed integer programs. Eur. J. Oper. Res. **4**(4), 224–234 (1980)
10. Balas, E., Kazachkov, A.M.: $\mathcal{V}$-polyhedral disjunctive cuts (2022). https://arxiv.org/abs/2207.13619
11. Balas, E., Kis, T.: On the relationship between standard intersection cuts, lift-and-project cuts and generalized intersection cuts. Math. Program., 1–30 (2016)
12. Balas, E., Perregaard, M.: A precise correspondence between lift-and-project cuts, simple disjunctive cuts, and mixed integer Gomory cuts for 0-1 programming. Math. Program. **94**(2–3, Ser. B), 221–245 (2003). The Aussois 2000 Workshop in Combinatorial Optimization
13. Balas, E., Qualizza, A.: Intersection cuts from multiple rows: a disjunctive programming approach. EURO J. Computat. Optim. **1**(1), 3–49 (2013)
14. Balas, E., Ceria, S., Cornuéjols, G.: A lift-and-project cutting plane algorithm for mixed 0-1 programs. Math. Program. **58**(3, Ser. A), 295–324 (1993)
15. Balas, E., Ceria, S., Cornuéjols, G.: Mixed 0-1 programming by lift-and-project in a branch-and-cut framework. Man. Sci. **42**(9), 1229–1246 (1996)
16. Basu, A., Bonami, P., Cornuéjols, G., Margot, F.: Experiments with two-row cuts from degenerate tableaux. INFORMS J. Comput. **23**(4), 578–590 (2011)
17. Bixby, R.E., Boyd, E.A., Indovina, R.R.: MIPLIB: a test set of mixed integer programming problems. SIAM News **25**, 16 (1992)
18. Bixby, R.E., Ceria, S., McZeal, C.M., Savelsbergh, M.W.P.: An updated mixed integer programming library: MIPLIB 3.0. Optima, 58, 12–15, 6 (1998)
19. Bonami, P.: On optimizing over lift-and-project closures. Math. Program. Comput. **4**(2), 151–179 (2012)
20. Bonami, P., Conforti, M., Cornuéjols, G., Molinaro, M., Zambelli, G.: Cutting planes from two-term disjunctions. Oper. Res. Lett. **41**(5), 442–444 (2013)
21. Conforti, M., Cornuéjols, G., Zambelli, G.: Integer Programming, vol. 271 of Graduate Texts in Mathematics. Springer, Cham (2014). https://doi.org/10.1007/978-3-319-11008-0
22. CORAL. Computational Optimization Research at Lehigh. MIP instances. https://coral.ise.lehigh.edu/data-sets/mixed-integer-instances/ (2020). Accessed Sept 2020

23. Cornuéjols, G., Li, Y.: Elementary closures for integer programs. Oper. Res. Lett. **28**(1), 1–8 (2001)
24. Dey, S.S., Wolsey, L.A.: Two row mixed-integer cuts via lifting. Math. Program. **124**(1–2, Ser. B), 143–174 (2010)
25. Dey, S.S., Lodi, A., Tramontani, A., Wolsey, L.A.: On the practical strength of two-row tableau cuts. INFORMS J. Comput. **26**(2), 222–237 (2014)
26. Espinoza, D.G.: Computing with multi-row Gomory cuts. Oper. Res. Lett. **38**(2), 115–120 (2010)
27. Farkas, J.: Theorie der einfachen Ungleichungen. J. Reine Angew. Math. **124**, 1–27 (1902)
28. Fischer, T., Pfetsch, M.E.: Monoidal cut strengthening and generalized mixed-integer rounding for disjunctions and complementarity constraints. Oper. Res. Lett. **45**(6), 556–560 (2017)
29. Fischetti, M., Lodi, A., Tramontani, A.: On the separation of disjunctive cuts. Math. Program. **128**(1–2, Ser. A), 205–230 (2011)
30. Fukasawa, R., Poirrier, L., Xavier, Á.S.: The (not so) trivial lifting in two dimensions. Math. Program. Comp. **11**(2), 211–235 (2019)
31. Gleixner, A., et al.: MIPLIB 2017: Data-Driven compilation of the 6th mixed-integer programming library. Math. Prog. Comp., (2021)
32. Gomory, R.E., Johnson, E.L.: Some continuous functions related to corner polyhedra. Math. Program. **3**(1), 23–85 (1972)
33. Guennebaud, G., et al.: Eigen v3. http://eigen.tuxfamily.org (2010)
34. Gurobi Optimization, LLC. Gurobi Optimizer Reference Manual (2022)
35. Johnson, E.L.: On the group problem for mixed integer programming. Math. Program. Stud. **2**, 137–179 (1974)
36. Júdice, J.J., Sherali, H.D., Ribeiro, I.M., Faustino, A.M.: A complementarity-based partitioning and disjunctive cut algorithm for mathematical programming problems with equilibrium constraints. J. Global Optim. **36**(1), 89–114 (2006)
37. Kazachkov, A.M.: Non-Recursive Cut Generation. PhD thesis, Carnegie Mellon University (2018)
38. Kazachkov, A.M., Serrano, F.: Monoidal cut strengthening. In: Prokopyev, O., Pardalos, P.M., editors, Encyclopedia of Optimization. Springer, US, Boston, MA. Under review
39. Kılınç, M., Linderoth, J., Luedtke, J., Miller, A.: Strong-branching inequalities for convex mixed integer nonlinear programs. Comput. Optim. Appl. **59**(3), 639–665 (2014). https://doi.org/10.1007/s10589-014-9690-8
40. Kis, T.: Lift-and-project for general two-term disjunctions. Discrete Optim. **12**, 98–114 (2014)
41. Koch, T., Achterberg, T., Andersen, E., Bastert, O., Berthold, T., Bixby, R.E., et al.: MIPLIB 2010: mixed integer programming library version 5. Math. Program. Comput. **3**(2), 103–163 (2011)
42. Kronqvist, J., Misener, R.: A disjunctive cut strengthening technique for convex MINLP. Optim. Eng. **22**(3), 1315–1345 (2021)
43. Lougee-Heimer, R.: The Common Optimization INterface for Operations Research: promoting open-source software in the operations research community. IBM J. Res. Dev. 47 (2003)
44. Louveaux, Q., Poirrier, L., Salvagnin, D.: The strength of multi-row models. Math. Program. Comput. **7**(2), 113–148 (2015)
45. Nemhauser, G.L., Wolsey, L.A.: Integer and combinatorial optimization. Wiley-Interscience Series in Discrete Mathematics and Optimization. John Wiley & Sons Inc, New York (1988)

46. Nemhauser, G.L., Wolsey, L.A.: A recursive procedure to generate all cuts for 0-1 mixed integer programs. Math. Program. **46**(1), 379–390 (1990)
47. Perregaard, M.: Generating Disjunctive Cuts for Mixed Integer Programs. PhD thesis, Carnegie Mellon University, 9 (2003)
48. Perregaard, M., Balas, E.: Generating cuts from multiple-term disjunctions. In: Aardal, K., Gerards, B. (eds.) IPCO 2001. LNCS, vol. 2081, pp. 348–360. Springer, Heidelberg (2001). https://doi.org/10.1007/3-540-45535-3_27
49. Xavier, Á.S., Fukasawa, R., Poirrier, L.: Multirow intersection cuts based on the infinity norm. INFORMS J. Comput. **33**(4), 1624–1643 (2021)
50. Ziegler, G.M.: Lectures on Polytopes, vol. 152 of Graduate Texts in Mathematics. Springer-Verlag, New York (1995). https://doi.org/10.1007/978-1-4613-8431-1

Optimal General Factor Problem
and Jump System Intersection

Yusuke Kobayashi$^{(\boxtimes)}$ (ID)

Kyoto University, Kyoto, Japan
yusuke@kurims.kyoto-u.ac.jp

Abstract. In the optimal general factor problem, given a graph $G = (V, E)$ and a set $B(v) \subseteq \mathbb{Z}$ of integers for each $v \in V$, we seek for an edge subset F of maximum cardinality subject to $d_F(v) \in B(v)$ for $v \in V$, where $d_F(v)$ denotes the number of edges in F incident to v. A recent crucial work by Dudycz and Paluch shows that this problem can be solved in polynomial time if each $B(v)$ has no gap of length more than one. While their algorithm is very simple, its correctness proof is quite complicated. In this paper, we formulate the optimal general factor problem as the jump system intersection, and reveal when the algorithm by Dudycz and Paluch can be applied to this abstract form of the problem. By using this abstraction, we give another correctness proof of the algorithm, which is simpler than the original one. We also extend our result to the valuated case.

1 Introduction

1.1 General Factor Problem

Matching in graphs is one of the most well-studied topics in combinatorial optimization. Since a maximum matching algorithm was proposed by Edmonds [6] in 1960s, a lot of generalizations of the matching problem have been proposed and studied in the literature. Among them, we focus on the *general factor problem*, which contains several important problems as special cases. In the general factor problem (or also called *B-factor problem*), we are given a graph $G = (V, E)$ and a set $B(v) \subseteq \mathbb{Z}$ of integers for each $v \in V$. The objective is to find an edge subset $F \subseteq E$ such that $d_F(v) \in B(v)$ for any $v \in V$ if it exists, where $d_F(v)$ denotes the number of edges in F incident to v. Such an edge set is called a *B-factor*.

Since the general factor problem is NP-hard in general (e.g. it contains the 3-edge-coloring problem [13]), polynomially solvable special cases have attracted attention. A B-factor amounts to a perfect matching if $B(v) = \{1\}$ for each $v \in V$, and it is called a *b-factor* if $B(v) = \{b(v)\}$ for each $v \in V$, where $b: V \to \mathbb{Z}$. For $a, b: V \to \mathbb{Z}$, if $B(v) = \{a(v), a(v)+1, a(v)+2, \ldots, b(v)-1, b(v)\}$ (resp, $B(v) = \{a(v), a(v)+2, a(v)+4, \ldots, b(v)-2, b(v)\}$) for $v \in V$, then a B-factor is called an *(a, b)-factor* (resp. an *(a, b)-parity factor*). It is well-known

The full version is available at arXiv [10].

that, in the above cases, we can find a B-factor in polynomial time by using a maximum matching algorithm; see [13] and [23, Section 35]. Note that the parity constraint can be dealt with by adding $\frac{1}{2}(b(v) - a(v))$ self-loops to each $v \in V$ and modifying $B(v)$. Another special case is the *antifactor problem*, in which $B(v) = \{0, 1, 2, \ldots, d_E(v)\} \setminus \{\alpha_v\}$ for some $\alpha_v \in \{0, 1, 2, \ldots, d_E(v)\}$, that is, exactly one value is forbidden for each $v \in V$. Graphs with an antifactor were characterized by Lovász [14]. The *edge-and-triangle partitioning problem* is to cover all the vertices in a graph by edges and triangles that are mutually disjoint, which can be easily reduced to the general factor problem with $B(v) = \{1\}, \{0, 2\}$, or $\{0, 2, 3\}$. The edge-and-triangle partitioning problem is known to be solvable in polynomial time [4].

All the above polynomially solvable cases have a property that each $B(v)$ has no *gap* of length more than one. Here, $B(v) \subseteq \mathbb{Z}$ is said to have *a gap of length p* if there exists $\alpha \in B(v)$ such that $\alpha+1, \alpha+2, \ldots, \alpha+p \notin B(v)$ and $\alpha+p+1 \in B(v)$. It turns out that this is a key property to design a polynomial-time algorithm. Indeed, Cornuéjols [3] gave a polynomial-time algorithm for the general factor problem with this property and Sebő [24] gave a good characterization.

An optimization variant of the general factor problem has also attracted attention, which we call the *optimal general factor problem* (or the *optimal general matching problem*). In the problem, given a graph $G = (V, E)$ and a set $B(v) \subseteq \mathbb{Z}$ of integers for each $v \in V$, we seek for a B-factor of maximum cardinality. It is the maximum matching problem if $B(v) = \{0, 1\}$, and is the maximum b-matching problem if $B(v) = \{0, 1, \ldots, b(v)\}$, both of which can be solved in polynomial time. In the same way as the search problem described above, we can find a maximum (a, b)-factor (or (a, b)-parity factor) in polynomial time; see [23, Section 35]. The optimization variant of the edge-and-triangle partitioning problem was studied with the name of the *simplex matching problem*, and a polynomial-time algorithm was designed for this problem [1]; see also [22].

Recently, Dudycz and Paluch [5] showed that the optimal general factor problem can be solved in polynomial time if each $B(v)$ has no gap of length more than one. This is definitely a crucial result in this area, because it is a generalization of all the above results. While their algorithm is very simple, its correctness proof is quite complicated.

1.2 Jump System Intersection

In this paper, we introduce an abstract form of the optimal general factor problem by using the concept of *jump systems* introduced by Bouchet and Cunningham [2] (see also [9,17]). Let V be a finite set. For $x, y \in \mathbb{Z}^V$, we say that $s \in \mathbb{Z}^V$ is an (x, y)-step if $\|s\|_1 = 1$ and $\|(x + s) - y\|_1 = \|x - y\|_1 - 1$. A non-empty subset $J \subseteq \mathbb{Z}^V$ is called a *jump system* if it satisfies the following property:

(JUMP) For any $x, y \in J$ and for any (x, y)-step s, either $x + s \in J$ or there exists an $(x + s, y)$-step t such that $x + s + t \in J$.

Typical examples of jump systems include matroids, delta-matroids, integral polymatroids (or submodular systems [7]), and degree sequences of subgraphs.

When $J \subseteq \mathbb{Z}$ is one-dimensional, one can see that J is a jump system if and only if it has no gap of length more than one. One can also see that the direct product of one-dimensional jump systems is also a jump system. We consider the optimization problem over the intersection of two jump systems, where one is the direct product of one-dimensional jump systems.

JUMP SYSTEM INTERSECTION
Input. A jump system $J \subseteq \mathbb{Z}^V$, a finite one-dimensional jump system $B(v) \subseteq \mathbb{Z}$
  for each $v \in V$, and a vector $c \in \mathbb{Z}^V$.
Problem. Find a vector $x \in J \cap B$ maximizing $c^\top x$, where $B \subseteq \mathbb{Z}^V$ is the direct
  product of $B(v)$'s.

If J consists of degree sequences of subgraphs, i.e., $J = \{d_F \in \mathbb{Z}^V \mid F \subseteq E\}$, and $c(v) = 1$ for $v \in V$, then the problem amounts to the optimal general factor problem, which can be solved in polynomial time [5]. On the other hand, if J is a 2-polymatroid and $B(v) = \{0, 2\}$ for each $v \in V$, then the problem amounts to the *matroid matching problem* [15] or the *matroid parity problem* [12]. This implies that the problem cannot be solved in polynomial time if J is given as a membership oracle [8,16]; see also [18].

A similar problem is to determine whether the intersection of two jump systems J_1 and J_2 is empty or not, which is also hard in general. This problem was studied in [17] as a membership problem of $J_1 - J_2 := \{x - y \mid x \in J_1, y \in J_2\}$, because $J_1 \cap J_2 \neq \emptyset$ if and only if $\mathbf{0} \in J_1 - J_2$.

1.3   Our Contribution: Jump System with SBO Property

A natural question is why the optimal general factor problem can be solved efficiently, while the general setting of JUMP SYSTEM INTERSECTION is hard. In this paper, we answer this question by revealing the properties of J that are essential in the argument in [5].

For a positive integer ℓ, we denote $\{1, 2, \ldots, \ell\}$ by $[\ell]$. For $x, y \in \mathbb{Z}^V$, we say that a multiset $\{p_1, \ldots, p_\ell\}$ of vectors is a 2-*step decomposition of* $y - x$ if $p_i \in \mathbb{Z}^V$ and $\|p_i\|_1 = 2$ for each $i \in [\ell]$, $\|y - x\|_1 = 2\ell$, and $y - x = \sum_{i \in [\ell]} p_i$. A non-empty subset $J \subseteq \mathbb{Z}^V$ is called a *jump system with SBO property*[1] if it satisfies the following property:

(SBO-JUMP) For any $x, y \in J$, there exists a 2-step decomposition $\{p_1, \ldots, p_\ell\}$
  of $y - x$ such that $x + \sum_{i \in I} p_i \in J$ for any $I \subseteq [\ell]$.

We can see that (SBO-JUMP) implies (JUMP). To see this, for given $x, y \in J$, suppose that there exist vectors $p_1, \ldots, p_\ell \in \mathbb{Z}^V$ satisfying the conditions in (SBO-JUMP). Then, for any (x, y)-step s, there exists an $(x + s, y)$-step t such that $s + t = p_i$ for some $i \in [\ell]$, and hence $x + s + t = x + p_i \in J$. Therefore, if J is a jump system with SBO property, then it is a jump system such that $\sum_{v \in V} x(v)$ has the same parity for any $x \in J$, which is called a *constant parity jump system*. See [21] for a characterization of constant parity jump systems.

We now give a few examples of jump systems with SBO property.

---

[1] SBO stands for *strongly base orderable* (see Example 1).

Example 1. A matroid $M = (S, \mathcal{B})$ with a ground set S and a base family $\mathcal{B}$ is called *strongly base orderable* if, for any bases $B_1, B_2 \in \mathcal{B}$, there exists a bijection $f \colon B_1 \backslash B_2 \to B_2 \backslash B_1$ such that $(B_1 \backslash X) \cup \{f(x) \mid x \in X\} \in \mathcal{B}$ for any $X \subseteq B_1 \backslash B_2$ (see e.g., [23, Section 42.6c]). By definition, the characteristic vectors of the bases of a strongly base orderable matroid satisfy (SBO-JUMP).

Note that the characteristic vectors of the bases do not satisfy (SBO-JUMP) if the matroid is not strongly base orderable, which implies that the class of jump systems with SBO property is strictly smaller than that of constant parity jump systems. By merging some elements in Example 1, we obtain the following example, which was studied for linear matroids in a problem similar to JUMP SYSTEM INTERSECTION [25].

Example 2. Let $M = (S, \mathcal{B})$ be a strongly base orderable matroid and let $(S_1, S_2, \ldots, S_n)$ be a partition of S. Then, $J = \{x \in \mathbb{Z}^n \mid B \in \mathcal{B}, x(i) = |B \cap S_i|$ for $i \in [n]\}$ satisfies (SBO-JUMP).

Another example is the set of the degree sequences of subgraphs.

Example 3. Let $G = (V, E)$ be a graph and let J be the set of the degree sequences of subgraphs, i.e., $J = \{d_F \mid F \subseteq E\}$. Then, J satisfies (SBO-JUMP). To see this, for $x, y \in J$, let $M, N \subseteq E$ be edge sets with $d_M = x$ and $d_N = y$. Then, the symmetric difference of M and N can be decomposed into alternating paths $P_1, \ldots, P_\ell$ and alternating cycles such that $\{d_{N \cap P_i} - d_{M \cap P_i} \mid i \in [\ell]\}$ is a 2-step decomposition of $y - x$. Note that each P_i is regarded as an edge subset. Let $p_i := d_{N \cap P_i} - d_{M \cap P_i}$ for $i \in [\ell]$. For any $I \subseteq [\ell]$, $x + \sum_{i \in I} p_i$ is the degree sequence of the symmetric difference of M and $\bigcup_{i \in [I]} P_i$, and hence it is in J.

Our contribution is to introduce the jump system with SBO property and show that (SBO-JUMP) is crucial when we apply the algorithm in [5] for JUMP SYSTEM INTERSECTION. For $\alpha, \beta \in \mathbb{Z}$ with $\alpha \leq \beta$ that have the same parity, a set $\{\alpha, \alpha + 2, \ldots, \beta - 2, \beta\}$ is called a *parity interval*. The main result in this paper is stated as follows.

Theorem 1. *There is an algorithm for* JUMP SYSTEM INTERSECTION *whose running time is polynomial in* $\sum_{v \in V} \sum_{\alpha \in B(v)} \log(|\alpha| + 1) + \sum_{v \in V} \log(|c(v)| + 1)$ *if the following properties hold:*

(C1) *a feasible solution* $x_0 \in J \cap B$ *is given,*
(C2) *J satisfies (SBO-JUMP), and*
(C3) *for any direct product* $B' \subseteq \mathbb{Z}^V$ *of parity intervals, there is an oracle for finding a vector* $x \in J \cap B'$ *maximizing* $c^\top x$.

Note that no explicit representation of J is required in this theorem. We only need the oracle in Condition (C3). Note also that Condition (C3) implies the existence of the membership oracle of J.

When J is the set of the degree sequences of subgraphs, we see that J satisfies (C1)–(C3) as follows. It was shown by Cornuéjols [3] that a feasible solution $x_0 \in J \cap B$ in (C1) can be found in polynomial time, and (C2) holds by Example 3.

The subproblem in (C3) is to find a maximum (a, b)-parity factor, which can be solved in polynomial time.

Our proof for Theorem 1 is based on the argument of Dudycz and Paluch [5]. While their algorithm is very simple, the correctness proof is quite complicated. In particular, an involved case analysis is required to prove a key lemma [5, Lemma 2]. Our technical contribution in this paper is to give a new simpler proof of this lemma in a slightly different form (Lemma 1). In our proof, we use several properties that are peculiar to our problem formulation (see Sect. 4.1), which is an advantage of introducing the abstract form of the optimal general factor problem. We also show that a scaling technique used in [5] is not required in the algorithm, which is another contribution of this paper.

We also introduce a quantitative extension of (SBO-JUMP), and extend Theorem 1 to a valuated variant of JUMP SYSTEM INTERSECTION; see Theorem 2.

1.4 Organization

The rest of this paper is organized as follows. Some preliminaries are given in Sect. 2. In Sect. 3, we describe our algorithm and prove its correctness by using a key technical lemma (Lemma 1). A proof of Lemma 1 is given in Sect. 4, where properties shown in Sect. 4.1 play important roles to simplify the argument. In Sect. 5, we extend our results to the valuated case and show that a polynomial-time algorithm for the weighted general factor problem is derived from our results. Proofs of theorems/lemmas marked with ($\star$) are omitted due to the page limitation and given in the full version [10].

2 Preliminaries

Let V be a finite set. For $v \in V$, let $\chi_v \in \mathbb{Z}^V$ denote the characteristic vector of v, that is, $\chi_v(v) = 1$ and $\chi_v(u) = 0$ for $u \in V \setminus \{v\}$. For each $v \in V$, we are given a non-empty finite set $B(v) \subseteq \mathbb{Z}$ that has no gap of length more than one, i.e., $B(v)$ is a one-dimensional jump system. Throughout this paper, let $B \subseteq \mathbb{Z}^V$ be the direct product of $B(v)$'s, i.e., $B := \{x \in \mathbb{Z}^V \mid x(v) \in B(v) \text{ for any } v \in V\}$. For $x \in \mathbb{Z}^V$, we denote $\min B \le x \le \max B$ if $\min B(v) \le x(v) \le \max B(v)$ for every $v \in V$. For $x \in \mathbb{Z}^V$, we define $q(x) = |\{v \in V \mid x(v) \notin B(v)\}|$. Note that, if $\min B \le x \le \max B$, then $q(x) := \min_{y \in B} \|x - y\|_1$, because each $B(v)$ has no gap of length greater than one. Recall that a parity interval is a subset of $\mathbb{Z}$ that is of the form $\{\alpha, \alpha + 2, \ldots, \beta - 2, \beta\}$. For $v \in V$, we see that $B(v)$ is uniquely partitioned into inclusionwise maximal parity intervals (see Fig. 1), which we call *maximal parity intervals* of $B(v)$. For $\alpha, \beta \in \mathbb{Z}$ with $\min B(v) \le \alpha \le \beta \le \max B(v)$, we define $\text{dist}_{B(v)}(\alpha, \beta)$ as the number of maximal parity intervals of $B(v)$ intersecting $[\alpha, \beta]$ minus one. In other words, $\text{dist}_{B(v)}(\alpha, \beta)$ is the number of pairs of consecutive integers in $B(v) \cap [\alpha, \beta]$. We also define

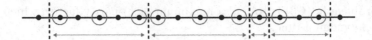

Fig. 1. Blue circles are elements in $B(v)$ and red arrows are maximal parity intervals. (Color figure online)

$\mathrm{dist}_{B(v)}(\beta, \alpha) := \mathrm{dist}_{B(v)}(\alpha, \beta)$. For $x, y \in \mathbb{Z}^V$ with $\min B \leq x, y \leq \max B$, we define $\mathrm{dist}_B(x, y) := \sum_{v \in V} \mathrm{dist}_{B(v)}(x(v), y(v))$; see Fig. 2. Note that dist_B satisfies the triangle inequality.

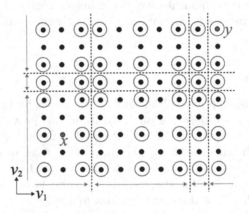

Fig. 2. In this two-dimensional example, $\mathrm{dist}_{B(v_1)}(x(v_1), y(v_1)) = 3$, $\mathrm{dist}_{B(v_2)}(x(v_2), y(v_2)) = 2$, $\mathrm{dist}_B(x, y) = 5$, $\|x - y\|_1 = 14$, $q(x) = 1$, and $q(y) = 0$.

3   Algorithm and Correctness

Our algorithm for JUMP SYSTEM INTERSECTION is basically the same as [5]. We first initialize the vector $x := x_0$, where x_0 is as in Condition (C1) in Theorem 1. In each iteration, we compute a vector $x' \in J \cap B$ maximizing $c^\top x'$ subject to $\mathrm{dist}_B(x, x') \leq 2$. If $c^\top x' = c^\top x$, then the algorithm terminates by returning x. Otherwise, we replace x with x' and repeat the procedure. See Algorithm 1 for a pseudocode of the algorithm.

In the correctness proof, we use the following key lemma, whose proof is given in Sect. 4. Note again that giving a simpler proof for this lemma is a technical contribution of this paper.

Lemma 1. *Let* $x, y \in B$ *be vectors with* $\mathrm{dist}_B(x, y) = 4$, *let* $\{p_1, \ldots, p_\ell\}$ *be a 2-step decomposition of* $y - x$, *and let* $w_i \in \mathbb{R}$ *for* $i \in [\ell]$. *Then, there exists a set* $I \subseteq [\ell]$ *such that* $z := x + \sum_{i \in I} p_i$ *is contained in* B, $\mathrm{dist}_B(x, z) = 2$, *and* $\sum_{i \in I} w_i \geq \min\{0, \sum_{i \in [\ell]} w_i\}$.

Algorithm 1: Algorithm for JUMP SYSTEM INTERSECTION

Input: $J, B, c,$ and x_0.
Output: $x \in J \cap B$ maximizing $c^\top x$.

1   $x \leftarrow x_0$;
2   **while** true **do**
3   Find a vector $x' \in J \cap B$ maximizing $c^\top x'$ subject to $\text{dist}_B(x, x') \le 2$;
4   **if** $c^\top x' = c^\top x$ **then**
5   **return** x
6      $x \leftarrow x'$;

Remark 1. In Lemma 1, the roles of x and y are symmetric by changing the signs of p_i and w_i, because $\bar{I} := [\ell] \setminus I$ satisfies the following:

- $x + \sum_{i \in I} p_i = y + \sum_{i \in \bar{I}} (-p_i)$,
- $\text{dist}_B(x, z) = 2 \iff \text{dist}_B(y, z) = 2$, and
- $\sum_{i \in I} w_i \ge \min\{0, \sum_{i \in [\ell]} w_i\} \iff \sum_{i \in \bar{I}} (-w_i) \ge \min\{0, \sum_{i \in [\ell]} (-w_i)\}$.

Let $w \in \mathbb{R}^\ell$ be the vector consisting of w_i's, and denote $w(I) := \sum_{i \in I} w_i$ for $I \subseteq [\ell]$. We next show the following lemma. Note that almost the same result is shown for degree sequences in [5, Lemma 1].

Lemma 2. *Let k be a positive integer. Let $x, y \in B$ be vectors with $\text{dist}_B(x, y) = 2k$ and let $\{p_1, \ldots, p_\ell\}$ be a 2-step decomposition of $y - x$. Then, there exist index sets $\emptyset = I_0 \subsetneq I_1 \subsetneq I_2 \subsetneq \cdots \subsetneq I_k = [\ell]$ such that $z_j := x + \sum_{i \in I_j} p_i$ is contained in B and $\text{dist}_B(z_{j-1}, z_j) = 2$ for $j \in [k]$.*

Proof. It suffices to construct $I_1 \subseteq [\ell]$ satisfying the conditions, because $I_2,$ $I_3, \ldots, I_{k-1}$ can be constructed in this order in the same way.

By changing the direction of axes if necessary, we may assume that $x(v) \le y(v)$ for every $v \in V$. Then, each p_i is equal to $\chi_a + \chi_b$ for some $a, b \in V$ (possibly $a = b$). For $z \in \mathbb{Z}^V$, we denote $\phi(z) := (\text{dist}_B(x, z), q(z)) \in \mathbb{Z}_{\ge 0}^2$. In order to construct I_1, we start with $I := I_0 = \emptyset$ and add an element one by one to I. During the procedure, we keep $\phi(z) \in \{(0,0), (0,2), (1,1), (2,0)\}$, where $z := x + \sum_{i \in I} p_i$. Note that $\phi(z) = (0,0)$ when I is initialized to I_0.

If $\phi(z) = (2,0)$, then $I_1 := I$ clearly satisfies the conditions. Otherwise, it holds that $\phi(z) \in \{(0,0), (0,2), (1,1)\}$. In this case, we show that there exists an index $i \in [\ell] \setminus I$ such that $\phi(z + p_i) \in \{(0,0), (0,2), (1,1), (2,0)\}$ by the following case analysis.

- Suppose that $\phi(z) = (0,0)$. Let i be an arbitrary index in $[\ell] \setminus I$. Then, $p_i = \chi_a + \chi_b$ for some $a, b \in V$ (possibly $a = b$). We see that $\phi(z + \chi_a) \in \{(0,1), (1,0)\}$, and hence $\phi(z + p_i) = \phi(z + \chi_a + \chi_b) \in \{(0,0), (0,2), (1,1), (2,0)\}$.
- Suppose that $\phi(z) = (0,2)$. Then, $z + \chi_a + \chi_b \in B$ for some distinct $a, b \in V$ such that $z(a) < y(a)$ and $z(b) < y(b)$. Let i be an index in $[\ell] \setminus I$ such that $p_i = \chi_a + \chi_c$ for some $c \in V$ (possibly $c = a$ or $c = b$). Then, we see that $\phi(z + \chi_a) = (0,1)$, and hence $\phi(z + p_i) = \phi(z + \chi_a + \chi_c) \in \{(0,0), (0,2), (1,1)\}$.

– Suppose that $\phi(z) = (1,1)$. Then, $z + \chi_a \in B$ for some $a \in V$ with $z(a) < y(a)$. Let i be an index in $[\ell] \setminus I$ such that $p_i = \chi_a + \chi_b$ for some $b \in V$ (possibly $b = a$). Then, we see that $\phi(z + \chi_a) = (1,0)$, and hence $\phi(z + p_i) = \phi(z + \chi_a + \chi_b) \in \{(1,1),(2,0)\}$.

If $\phi(z + p_i) = (2,0)$, then $I_1 := I \cup \{i\}$ satisfies the conditions. Otherwise, we replace I with $I \cup \{i\}$ and repeat the procedure. Since $[\ell]$ is finite, this process terminates by finding a desired index set I_1, which completes the proof. □

By using Lemmas 1 and 2, we can evaluate the improvement of the objective value in each iteration of Algorithm 1 as follows.

Lemma 3. *Let J be a jump system with SBO property, let $x^* \in J \cap B$ be an optimal solution of* JUMP SYSTEM INTERSECTION, *and let $x \in J \cap B$ be a vector with $x \neq x^*$. Let $x' \in J \cap B$ be a vector maximizing $c^\top x'$ subject to* $\mathrm{dist}_B(x, x') \leq 2$. *Then, $c^\top x' - c^\top x \geq \frac{2}{\|x^* - x\|_1}(c^\top x^* - c^\top x)$.*

Proof. If $\mathrm{dist}_B(x, x^*) \leq 2$, then the inequality is obvious. Since $\mathrm{dist}_B(x, x^*)$ is even, suppose that $\mathrm{dist}_B(x, x^*) \geq 4$. Since $x, x^* \in J$, there exists a 2-step decomposition $\{p_1, \ldots, p_\ell\}$ of $x^* - x$ that satisfies the conditions in (SBO-JUMP). For $i \in [\ell]$, we define $w_i = c^\top p_i - \frac{c^\top x^* - c^\top x}{\ell} + \varepsilon$, where ε is a sufficiently small positive number (e.g. $\varepsilon = \frac{1}{(\ell+1)^2}$) that is used to break ties. Observe that, for $I, I' \subseteq [\ell]$ with $|I| \neq |I'|$, $w(I) \neq w(I')$ holds because of ε. By Lemma 2, there exist index sets $\emptyset = I_0 \subsetneq I_1 \subsetneq I_2 \subsetneq \cdots \subsetneq I_k = [\ell]$ such that $z_j := x + \sum_{i \in I_j} p_i$ is contained in B and $\mathrm{dist}_B(z_{j-1}, z_j) = 2$ for $j \in [k]$. We choose $I_1, I_2, \ldots, I_{k-1}$ so that $(w(I_1), w(I_2), \ldots, w(I_{k-1}))$ is lexicographically maximum. Note that $z_j \in J$ for $j \in [k]$ by (SBO-JUMP).

Let $j \in [k]$ be the minimum index such that $w(I_{j-1}) < w(I_j)$. Note that such j must exist, because $w(I_0) = 0 < \varepsilon\ell = w(I_k)$. Assume that $j \neq 1$. Then, the minimality of j shows that $w(I_{j-2}) > w(I_{j-1}) < w(I_j)$, where we note that $w(I_{j-2}) \neq w(I_{j-1})$ as $|I_{j-2}| \neq |I_{j-1}|$. By applying Lemma 1 to a 2-step decomposition $\{p_i \mid i \in I_j \setminus I_{j-2}\}$ of $z_j - z_{j-2}$, we obtain an index set $I \subseteq I_j \setminus I_{j-2}$ such that $z'_{j-1} := z_{j-2} + \sum_{i \in I} p_i$ is contained in B, $\mathrm{dist}_B(z_{j-2}, z'_{j-1}) = 2$, and $w(I) \geq \min\{0, w(I_j \setminus I_{j-2})\}$. Let $I'_{j-1} := I_{j-2} \cup I$. By $z'_{j-1} = x + \sum_{i \in I'_{j-1}} p_i$ and (SBO-JUMP), we see that $z'_{j-1} \in J$. Furthermore, we obtain

$$w(I'_{j-1}) = w(I_{j-2}) + w(I) \geq \min\{w(I_{j-2}), w(I_j)\} > w(I_{j-1}),$$

which contradicts the choice of I_{j-1}.

Therefore, we obtain $j = 1$, that is, $0 = w(I_0) < w(I_1)$. Since

$$0 < w(I_1) = \sum_{i \in I_1}\left(c^\top p_i - \frac{c^\top x^* - c^\top x}{\ell} + \varepsilon\right)$$

$$= c^\top z_1 - c^\top x - \left(\frac{c^\top x^* - c^\top x}{\ell} - \varepsilon\right)|I_1|$$

and ε is sufficiently small, we obtain

$$c^\top z_1 - c^\top x \geq \frac{(c^\top x^* - c^\top x)|I_1|}{\ell}.$$

We also see that $c^\top x' \geq c^\top z_1$, because $z_1 \in J \cap B$ and $\operatorname{dist}_B(x, z_1) \leq 2$. By combining these inequalities with $|I_1| \geq 1$ and $\ell = \frac{\|x^* - x\|_1}{2}$, we obtain $c^\top x' - c^\top x \geq \frac{2}{\|x^* - x\|_1}(c^\top x^* - c^\top x)$. □

This implies that the global optimality is guaranteed by the local optimality.

Corollary 1. *In an instance of* JUMP SYSTEM INTERSECTION *with (C2), a feasible solution* $x \in J \cap B$ *maximizes* $c^\top x$ *if and only if* $c^\top x \geq c^\top x'$ *for any* $x' \in J \cap B$ *with* $\operatorname{dist}_B(x, x') \leq 2$.

We are now ready to prove the correctness of Algorithm 1.

Proof (Proof of Theorem 1). We first show that each iteration of Algorithm 1 runs in polynomial time. For $x, x' \in B$ with $\operatorname{dist}_B(x, x') \leq 2$, we see that $x(v)$ and $x'(v)$ are contained in the same maximal parity interval of $B(v)$ for any $v \in V$ except at most two elements. Thus, for $x \in B$, $\{x' \in B \mid \operatorname{dist}_B(x, x') \leq 2\}$ can be partitioned into $O(n^2)$ sets, each of which is a direct product of parity intervals. Therefore, we can find a vector $x' \in J \cap B$ maximizing $c^\top x'$ subject to $\operatorname{dist}_B(x, x') \leq 2$ by using the oracle in Condition (C3), $O(n^2)$ times.

We next evaluate the number of iterations in the algorithm. Let OPT be the optimal value of the problem and let $B_{\text{size}} := \sum_{v \in V} |B(v)|$. Since J is a jump system with SBO property by Condition (C2), we can apply Lemma 3. By this lemma, if x is replaced with x' in line 6 of Algorithm 1, then

$$\text{OPT} - c^\top x' \leq \left(1 - \frac{2}{\|x^* - x\|_1}\right)(\text{OPT} - c^\top x) \leq \left(1 - \frac{1}{B_{\text{size}}}\right)(\text{OPT} - c^\top x),$$

that is, the gap to the optimal value decreases by a factor of at most $1 - \frac{1}{B_{\text{size}}}$. Therefore, by repeating this procedure $O(B_{\text{size}} \log(\text{OPT} - c^\top x_0))$ times, the algorithm terminates and returns an optimal solution.

This shows that Algorithm 1 solves JUMP SYSTEM INTERSECTION in polynomial time. □

4   Outline of the Proof of Lemma 1

4.1   Minimal Counterexample

This section gives an outline of the proof of Lemma 1. A tuple $(x, y, (p_i)_{i \in [\ell]}, w)$ is called an *instance* and a set I satisfying the conditions is called a *solution*. To derive a contradiction, assume that Lemma 1 does not hold. Suppose that $(x, y, (p_i)_{i \in [\ell]}, w)$ is a counterexample that minimizes $\|y - x\|_1$. Among such counterexamples, we choose one that minimizes $|\{(p_i, w_i) \mid i \in [\ell]\}|$, that is, we minimize the number of different (p_i, w_i) pairs. Such $(x, y, (p_i)_{i\in[\ell]}, w)$ is called a *minimal counterexample*. Define $U \subseteq V$ as $U := \{v \in V \mid \text{dist}_{B(v)}(x(v), y(v)) \geq 1\}$. By changing the direction of axes if necessary, we may assume that $x(v) \leq y(v)$ for every $v \in V$. Then, each p_i is equal to $\chi_a + \chi_b$ for some $a, b \in V$ (possibly $a = b$). We show some properties of the minimal counterexample. Our argument becomes simpler with the aid of these properties.

Lemma 4. *For any $i \in [\ell]$, $p_i = \chi_a + \chi_b$ for some $a, b \in U$ (possibly $a = b$). Consequently, $x(v) = y(v)$ for all $v \in V \setminus U$.*

Proof. Assume to the contrary that there exists $i \in [\ell]$ such that $p_i = \chi_a + \chi_c$ for some $a \in V$ and for some $c \in V \setminus U$.

Suppose that $a = c$, i.e., $p_i = 2\chi_c$. We consider a new instance by removing p_i and replacing y with $y - 2\chi_c \in B$. By the minimality of the counterexample, the obtained instance has a solution $I \subseteq [\ell] \setminus \{i\}$, which implies that $w(I) \geq 0$ or $w(I) \geq w([\ell] \setminus \{i\})$. Then, $I' := I$ is a solution of the original instance in the former case and $I' := I \cup \{i\}$ is a solution of the original instance in the latter case, which is a contradiction.

Suppose next that $a \neq c$. Since $\text{dist}_{B(c)}(x(c), y(c)) = 0$ and $x(c), y(c) \in B(c)$, we see that $x(c)$ and $y(c)$ have the same parity. Thus, there exists $i' \in [\ell] \setminus \{i\}$ such that $p_{i'} = \chi_b + \chi_c$ for some $b \in V \setminus \{c\}$. We merge p_i and $p_{i'}$ as follows: replace p_i and $p_{i'}$ with a new vector $p_{i''} := \chi_a + \chi_b$ whose weight is $w_i + w_{i'}$, and replace y with $y - 2\chi_c \in B$. By the minimality of the counterexample, the obtained instance has a solution $I \subseteq ([\ell] \setminus \{i, i'\}) \cup \{i''\}$. Then, we see that the set

$$I' := \begin{cases} (I \setminus \{i''\}) \cup \{i, i'\} & \text{if } i'' \in I, \\ I & \text{otherwise} \end{cases}$$

is a solution of the original instance, which is a contradiction.   □

Lemma 5. (⋆) *For any $i \in [\ell]$, $p_i \neq 2\chi_a$ for $a \in U$ with $\text{dist}_{B(a)}(x(a), y(a)) = 1$.*

Lemma 6. *For any $i, j \in [\ell]$ with $p_i = p_j$, it holds that $w_i = w_j$.*

Proof. Let $(x, y, (p_i)_{i\in[\ell]}, w)$ be a minimal counterexample of Lemma 1, and assume that $p_i = p_j$ does not imply $w_i = w_j$. Let $I^* \subseteq [\ell]$ be a maximal index set such that $p_i = p_j$ for any $i, j \in I^*$ and $w_i \neq w_j$ for some $i, j \in I^*$. We denote $I^* = \{i_1, i_2, \ldots, i_t\}$, where $w_{i_1} \geq w_{i_2} \geq \cdots \geq w_{i_t}$. Let $w^* := \frac{1}{t} w(I^*)$. Define $w'_i := w^*$ for $i \in I^*$ and $w'_i := w_i$ for $i \in [\ell] \setminus I^*$. We note that $w'([\ell]) = w([\ell])$. If there exists a solution $I' \subseteq [\ell]$ for a new instance $(x, y, (p_i)_{i\in[\ell]}, w')$, then $I := (I' \setminus I^*) \cup \{i_1, i_2, \ldots, i_{|I'\cap I^*|}\}$ is a solution for the original instance $(x, y, (p_i)_{i\in[\ell]}, w)$, because $w_{i_1} + w_{i_2} + \cdots + w_{i_{|I'\cap I^*|}} \geq |I' \cap I^*| \cdot w^* = w'(I' \cap I^*)$ implies that $w(I) \geq w(I')$. This shows that instance $(x, y, (p_i)_{i\in[\ell]}, w')$ has no solution, and hence it is a counterexample. Since $|\{(p_i, w'_i) \mid i \in [\ell]\}| < |\{(p_i, w_i) \mid i \in [\ell]\}|$, this contradics the minimality of $(x, y, (p_i)_{i\in[\ell]}, w)$.   □

Let $I^+ := \{i \in [\ell] \mid w_i > 0\}$ and $z^+ := x + \sum_{i \in I^+} p_i$. By Lemma 6, we observe the following.

Observation 1. *For any $i \in I^+$ and for any $j \in [\ell] \setminus I^+$, it holds that $p_i \neq p_j$.*

Since $x(v) = y(v) = z^+(v)$ for $v \in V \setminus U$ by Lemma 4, it holds that $q(z^+) \leq |U| \leq \operatorname{dist}_B(x, y) = 4$. We derive a contradiction for the cases when $|U| = 4$, $|U| = 3$, and $|U| \leq 2$, separately. In this extended abstract we only consider the case when $|U| = 3$ as a demonstration. The other cases are dealt with in the full version [10].

In the case analysis, we use the following lemma, which is obtained by the same argument as Lemma 2. Here, we denote $\phi(z) := (\operatorname{dist}_B(x, z), q(z)) \in \mathbb{Z}_{\geq 0}^2$ for $z \in \mathbb{Z}^V$.

Lemma 7. *Let $I_0 \subseteq [\ell]$ be an index set such that $z_0 := x + \sum_{i \in I_0} p_i$ satisfies $\phi(z_0) \in \{(0,0), (0,2), (1,1), (2,0)\}$. Then, there exists an index set $I \subseteq [\ell]$ with $I_0 \subseteq I$ such that $z := x + \sum_{i \in I} p_i$ is contained in B and $\operatorname{dist}_B(x, z) = 2$, i.e., $\phi(z) = (2,0)$.*

4.2   Part of Case Analysis: $|U| = 3$

In this extended abstract, we only consider the case when $|U| = 3$. Let $U = \{v_1, v_2, v_3\}$ such that $\operatorname{dist}_{B(v_1)}(x(v_1), y(v_1)) = \operatorname{dist}_{B(v_2)}(x(v_2), y(v_2)) = 1$ and $\operatorname{dist}_{B(v_3)}(x(v_3), y(v_3)) = 2$. By Lemmas 4 and 5, for any $i \in [\ell]$, either $p_i = \chi_a + \chi_b$ for some distinct $a, b \in U$ or $p_i = 2\chi_{v_3}$.

Since $\operatorname{dist}_B(x, z^+) + \operatorname{dist}_B(y, z^+) = 4$, by changing the roles of x and y if necessary (see Remark 1), we may assume that $\operatorname{dist}_B(x, z^+) \leq 2$.[2] Furthermore, since $\|x - z^+\|_1$ is even, we see that $\operatorname{dist}_B(x, z^+) + q(z^+)$ is even. Therefore, the pair $\phi(z^+) := (\operatorname{dist}_B(x, z^+), q(z^+))$ is one of the following: $(0,0), (0,2), (1,1), (1,3)$, $(2,0)$, and $(2,2)$, where we note that $q(z^+) \leq |U| = 3$. We derive a contradiction by considering each case separately.

Case 1: $\phi(z^+) = (0,0), (0,2), (1,1),$ or $(2,0)$.

By Lemma 7, there exists an index set $I \subseteq [\ell]$ with $I^+ \subseteq I$ such that $z := x + \sum_{i \in I} p_i$ is contained in B and $\operatorname{dist}_B(x, z) = 2$. Since $w_i \leq 0$ for each $i \in [\ell] \setminus I$, we obtain $w(I) \geq w([\ell])$, and hence I is a solution of Lemma 1. This is a contradiction.

Case 2: $\phi(z^+) = (1,3)$.

In this case, $z^+(v) \notin B(v)$ for $v \in U$. Since $z^+(v_1) \neq y(v_1)$, there exists $i \in [\ell] \setminus I^+$ such that $p_i = \chi_{v_1} + \chi_u$ for some $u \in \{v_2, v_3\}$. Since $\phi(z^+ + p_i) = (1,1)$, by Lemma 7, there exists an index set $I \subseteq [\ell]$ with $I^+ \cup \{i\} \subseteq I$ such that $z := x + \sum_{j \in I} p_j$ is contained in B and $\operatorname{dist}_B(x, z) = 2$. We see that such I is a solution of Lemma 1 in the same way as Case 1, which is a contradiction.

---

[2] If we change the roles of x and y, then $I^- := \{i \in [\ell] \mid w_i < 0\}$ and $z^- := y - \sum_{i \in I^-} p_i$ play the roles of I^+ and z^+, respectively. We see that if $\operatorname{dist}_B(x, z^+) \geq 3$, then $\operatorname{dist}_B(y, z^-) \leq \operatorname{dist}_B(y, z^+) = 4 - \operatorname{dist}_B(x, z^+) \leq 1$.

Case 3: $\phi(z^+) = (2,2)$.

Since $q(z^+) = 2$ and $|U| = 3$, at least one of $z^+(v_1) \notin B(v_1)$ and $z^+(v_2) \notin B(v_2)$ holds. By changing the roles of v_1 and v_2 if necessary, we may assume that $z^+(v_1) \notin B(v_1)$. Let $v^* \in \{v_2, v_3\}$ be the other element such that $z^+(v^*) \notin B(v^*)$. Since $z^+(v_1) \neq x(v_1)$, there exists $i_1 \in I^+$ such that $p_{i_1} = \chi_{v_1} + \chi_u$ for some $u \in \{v_2, v_3\}$. Similarly, since $z^+(v_1) \neq y(v_1)$, there exists $i_2 \in [\ell] \setminus I^+$ such that $p_{i_2} = \chi_{v_1} + \chi_u$ for some $u \in \{v_2, v_3\}$. By Observation 1, either $p_{i_1} = \chi_{v_1} + \chi_{v^*}$ or $p_{i_2} = \chi_{v_1} + \chi_{v^*}$ holds (Fig. 3). If $p_{i_1} = \chi_{v_1} + \chi_{v^*}$, then $I := I^+ \setminus \{i_1\}$ is a solution, because $w(I) \geq 0$, which is a contradiction; see Fig. 3 (left two). If $p_{i_2} = \chi_{v_1} + \chi_{v^*}$, then $I := I^+ \cup \{i_2\}$ is a solution, because $w(I) \geq w([\ell])$, which is a contradiction; see Fig. 3 (right two).

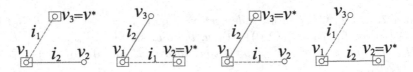

Fig. 3. Possible situations in Case 3. A blue edge (u,v) corresponds to an element $i \in [\ell] \setminus I^+$ with $p_i = \chi_u + \chi_v$, a red dashed edge (u,v) corresponds to an element $i \in I^+$ with $p_i = \chi_u + \chi_v$, and a vertex $v \in V$ in a rectangle satisfies that $z^+(v) \notin B(v)$. (Color figure online)

5   Extension to Valuated Problem

In this section, we consider a valuated version of JUMP SYSTEM INTERSECTION.

VALUATED JUMP SYSTEM INTERSECTION

Input. A function $f : J \to \mathbb{Z}$ on a jump system $J \subseteq \mathbb{Z}^V$ and a finite one-dimensional jump system $B(v) \subseteq \mathbb{Z}$ for each $v \in V$.

Problem. Find a vector $x \in J \cap B$ maximizing $f(x)$, where $B \subseteq \mathbb{Z}^V$ is the direct product of $B(v)$'s.

Note that f and J may be given in an implicit way, e.g., by an oracle. To simplify the notation, we extend the domain of f to $\mathbb{Z}^V$ by setting $f(x) = -\infty$ for $x \in \mathbb{Z}^V \setminus J$. The following property is a quantitative extension of (SBO-JUMP).

(SBO-M-JUMP) For any $x, y \in J$, there exist real values $g_1, \ldots, g_\ell$ and a 2-step decomposition $\{p_1, \ldots, p_\ell\}$ of $y - x$ such that $f(x + \sum_{i \in I} p_i) \geq f(x) + \sum_{i \in I} g_i$ for any $I \subseteq [\ell]$ and $f(y) = f(x) + \sum_{i \in [\ell]} g_i$.

Note that we use "M" in the name of the exchange axiom, because it defines a subclass of *M-concave functions on constant parity jump systems* [21]; see Remark 2 below. We can see that if f satisfies (SBO-M-JUMP), then its effective domain $J := \{x \in \mathbb{Z}^V \mid f(x) > -\infty\}$ satisfies (SBO-JUMP). By using (SBO-M-JUMP), we generalize Theorem 1 as follows.

Theorem 2. ($\star$) *There is an algorithm for* VALUATED JUMP SYSTEM INTER-SECTION *whose running time is polynomial in* $\sum_{v \in V} \sum_{\alpha \in B(v)} \log(|\alpha| + 1) +$ $\max_{x \in J} \log(|f(x)| + 1)$ *if the following properties hold:*

(C1') *a vector* $x_0 \in J \cap B$ *is given,*
(C2') *f satisfies (SBO-M-JUMP), and*
(C3') *for any direct product* $B' \subseteq \mathbb{Z}^V$ *of parity intervals, there is an oracle for finding a vector* $x \in J \cap B'$ *maximizing* $f(x)$.

Remark 2. Functions with (SBO-M-JUMP) form a subclass of M-concave functions on constant parity jump systems studied in the context of discrete convex analysis [11,19–21]. For $J \subseteq \mathbb{Z}^V$, a function $f \colon J \to \mathbb{Z}$ is called an *M-concave function on a constant parity jump system* [21] if it satisfies the following exchange axiom.

(M-JUMP) For any $x, y \in J$ and for any (x, y)-step s, there exists an $(x+s, y)$-step t such that $f(x + s + t) + f(y - s - t) \geq f(x) + f(y)$.

We can see that (SBO-M-JUMP) implies (M-JUMP) as follows. For $x, y \in J$, suppose that there exist a 2-step decomposition $\{p_1, \ldots, p_\ell\}$ of $y - x$ and $g_i \in \mathbb{R}$ for $i \in [\ell]$ satisfying the conditions in (SBO-M-JUMP). For any (x, y)-step s, there exists an $(x + s, y)$-step t such that $s + t = p_i$ for some $i \in [\ell]$. Such t satisfies the conditions in (M-JUMP), because

$$f(x + s + t) + f(y - s - t) = f(x + p_i) + f\left(x + \sum_{j \in [\ell] \setminus \{i\}} p_j\right)$$

$$\geq (f(x) + g_i) + \left(f(x) + \sum_{j \in [\ell] \setminus \{i\}} g_j\right) = f(x) + f(y).$$

6   Weighted Optimal General Factor Problem

It was shown by Dudycz and Paluch [5] that the edge-weighted variant of the optimal general factor problem can also be solved in polynomial time if each $B(v)$ has no gap of length more than one. Formally, in the *weighted optimal general factor problem*, given a graph $G = (V, E)$, an edge weight $w(e) \in \mathbb{Z}$ for $e \in E$, and a set $B(v) \subseteq \mathbb{Z}$ of integers for each $v \in V$, we seek for a B-factor $F \subseteq E$ that maximizes its total weight $\sum_{e \in F} w(e)$, where we denote $w(F) := \sum_{e \in F} w(e)$. Their algorithm consists of local improvement steps used in Algorithm 1 and a scaling technique.

In what follows in this section, we show that the polynomial solvability of the weighted optimal general factor problem is derived from Theorem 2.

Theorem 3 (Dudycz and Paluch [5]). *The weighted optimal general factor problem can be solved in polynomial time if each* $B(v)$ *has no gap of length more than one.*

Proof. Let $G = (V, E)$, w, and B be an instance of the weighted optimal general factor problem such that each $B(v)$ has no gap of length more than one. Let $J := \{d_F \mid F \subseteq E\}$, and define $f \colon J \to \mathbb{Z}$ by $f(x) := \max\{w(F) \mid d_F = x, F \subseteq E\}$ for $x \in J$.

We now show (C1'), (C2'), and (C3') in Theorem 2. Since an edge set $F_0 \subseteq E$ with $d_{F_0} \in B$ can be found in polynomial time by the algorithm of Cornuéjols [3] (if it exists), we obtain $x_0 := d_{F_0}$ satisfying the condition in (C1'). The subproblem in (C3') is to find an (a, b)-factor with parity constraints that maximizes the total edge weight, which can be solved in polynomial time; see [23, Section 35]. To see (C2'), for $x, y \in J$, let $M, N \subseteq E$ be edge sets such that $d_M = x$, $d_N = y$, $w(M) = f(x)$, and $w(N) = f(y)$. As in Example 3, the symmetric difference of M and N can be decomposed into alternating paths $P_1, \ldots, P_\ell$ and alternating cycles such that $\{d_{N \cap P_i} - d_{M \cap P_i} \mid i \in [\ell]\}$ is a 2-step decomposition of $y - x$. For $i \in [\ell]$, let $p_i := d_{N \cap P_i} - d_{M \cap P_i}$ and $g_i := w(N \cap P_i) - w(M \cap P_i)$. For $I \subseteq [\ell]$, let $F_I \subseteq E$ be the symmetric difference of M and $\bigcup_{i \in [I]} P_i$. Then, since $d_{F_I} = x + \sum_{i \in I} p_i$ and $w(F_I) = f(x) + \sum_{i \in I} g_i$, we obtain $f(x + \sum_{i \in I} p_i) \geq f(x) + \sum_{i \in I} g_i$. This shows (C2').

By Theorem 2, we can find $x^* \in J \cap B$ maximizing $f(x^*)$ in polynomial time. Furthermore, an edge set $F^* \subseteq E$ satisfying $w(F^*) = f(x^*)$ and $d_{F^*} = x^*$ can also be found in polynomial time by a weighted b-factor algorithm. By definition, such F^* is an optimal solution of the weighted optimal general factor problem. □

7   Concluding Remarks

In this paper, we have revealed that (SBO-JUMP) is a key property to obtain a polynomial time-algorithm for JUMP SYSTEM INTERSECTION, which is an abstract form of the optimal general factor problem. By using this abstraction, we have obtained a simpler correctness proof for the polynomial solvability of the optimal general factor problem. We have also extended the results to the valuated case.

There are some possible directions for future research. It is nice if we obtain more examples of jump systems satisfying (SBO-JUMP) other than Examples 1–3. It is open whether JUMP SYSTEM INTERSECTION can be solved in polynomial time if each $B(v)$ is given as a union of parity intervals. It is also a natural open problem whether we can obtain a strongly polynomial-time algorithm for the weighted general factor problem. Finally, it is interesting to find a new property of J other than (SBO-JUMP) that enables us to design a different polynomial-time algorithm.

Acknowledgements. The author thanks Kenjiro Takazawa for his helpful comments. This work is supported by JSPS KAKENHI grant numbers 20K11692 and 20H05795, Japan.

References

1. Anshelevich, E., Karagiozova, A.: Terminal backup, 3D matching, and covering cubic graphs. SIAM J. Comput. **40**, 678–708 (2011)
2. Bouchet, A., Cunningham, W.H.: Delta-matroids, jump systems, and bisubmodular polyhedra. SIAM J. Discret. Math. **8**, 17–32 (1995)
3. Cornuéjols, G.: General factors of graphs. J. Comb. Theory Ser. B **45**(2), 185–198 (1988)
4. Cornuéjols, G., Hartvigsen, D., Pulleyblank, W.: Packing subgraphs in a graph. Oper. Res. Lett. **1**(4), 139–143 (1982)
5. Dudycz, S., Paluch, K.E.: Optimal general matchings. In: WG 2018. LNCS, vol. 11159, pp. 176–189. Springer, Cham (2018). arXiv version is available at http://arxiv.org/abs/1706.07418. https://doi.org/10.1007/978-3-030-00256-5_15
6. Edmonds, J.: Paths, trees, and flowers. Can. J. Math. **17**, 449–467 (1965)
7. Fujishige, S.: Submodular Functions and Optimization, 2nd edn, vol. 58. Annals of Discrete Mathematics. Elsevier, Amsterdam (2005)
8. Jensen, P.M., Korte, B.: Complexity of matroid property algorithms. SIAM J. Comput. **11**, 184–190 (1982)
9. Kabadi, S.N., Sridhar, R.: Δ-matroid and jump system. J. Appl. Math. Decis. Sci. **2005**(2), 95–106 (2005)
10. Kobayashi, Y.: Optimal general factor problem and jump system intersection. arXiv:2209.00779 (2022)
11. Kobayashi, Y., Murota, K., Tanaka, K.: Operations on M-convex functions on jump systems. SIAM J. Discret. Math. **21**, 107–129 (2007)
12. Lawler, E.L.: Combinatorial Optimization – Networks and Matroids. Holt, Rinehalt, and Winston, New York (1976)
13. Lovász, L.: The factorization of graphs. II. Acta Mathematica Academiae Scientiarum Hungarica **23**, 223–246 (1972)
14. Lovász, L.: Antifactors of graphs. Period. Math. Hung. **4**, 121–123 (1973)
15. Lovász, L.: The matroid matching problem. Algebr. Methods Graph Theory Colloq. Math. Soc. János Bolyai **25**, 495–517 (1978)
16. Lovász, L.: Matroid matching and some applications. J. Comb. Theory Ser. B **28**, 208–236 (1980)
17. Lovász, L.: The membership problem in jump systems. J. Comb. Theory Ser. B **70**, 45–66 (1997)
18. Lovász, L., Plummer, M.D.: Matching Theory. North-Holland, Amsterdam (1986)
19. Minamikawa, N., Shioura, A.: Time bounds of basic steepest descent algorithms for M-convex function minimization and related problems. J. Oper. Res. Soc. Jpn. **64**(2), 45–60 (2021)
20. Murota, K.: Discrete Convex Analysis. SIAM, Philadelphia (2003)
21. Murota, K.: M-convex functions on jump systems: a general framework for min-square graph factor. SIAM J. Discret. Math. **20**, 213–226 (2006)
22. Pap, G.: A TDI description of restricted 2-matching polytopes. In: Bienstock, D., Nemhauser, G. (eds.) IPCO 2004. LNCS, vol. 3064, pp. 139–151. Springer, Heidelberg (2004). https://doi.org/10.1007/978-3-540-25960-2_11
23. Schrijver, A.: Combinatorial Optimization. Polyhedra and Efficiency. Algorithms and Combinatorics, vol. 24. Springer, Heidelberg (2003)
24. Sebő, A.: General antifactors of graphs. J. Comb. Theory Ser.B **58**(2), 174–184 (1993)
25. Szabó, J.: Matroid parity and jump systems: A solution to a conjecture of Recski. SIAM J. Discret. Math. **22**(3), 854–860 (2008)

Decomposition of Probability Marginals for Security Games in Abstract Networks

Jannik Matuschke[✉]

KU Leuven, Leuven, Belgium
jannik.matuschke@kuleuven.be

Abstract. Given a set system $(E, \mathcal{P})$, let $\pi \in [0,1]^{\mathcal{P}}$ be a vector of requirement values on the sets and let $\rho \in [0,1]^E$ be a vector of probability
marginals with $\sum_{e \in P} \rho_e \geq \pi_P$ for all $P \in \mathcal{P}$. We study the question under
which conditions the marginals ρ can be decomposed into a probability distribution on the subsets of E such that the resulting random set intersects
each $P \in \mathcal{P}$ with probability at least π_P.

Extending a result by Dahan, Amin, and Jaillet [3] motivated by a network security game in directed acyclic graphs, we show that such a distribution exists if $\mathcal{P}$ is an *abstract network* and the requirements are of
the form $\pi_P = 1 - \sum_{e \in P} \mu_e$ for some $\mu \in [0,1]^E$. Our proof yields an
explicit description of a feasible distribution that can be computed efficiently. As a consequence, equilibria for the security game studied in [3] can
be efficiently computed even when the underlying digraph contains cycles.
As a subroutine of our algorithm, we provide a combinatorial algorithm
for computing shortest paths in abstract networks, partially answering an
open question by McCormick [14]. We further show that a conservation
law proposed in [3] for requirement functions in partially ordered sets can
be reduced to the setting of affine requirements described above.

1 Introduction

Consider a set system $(E, \mathcal{P})$, where E is a finite ground set and $\mathcal{P} \subseteq 2^E$ is a collection of subsets of E. Given *probability marginals* $\rho \in [0,1]^E$ and *requirements*
$\pi \in [0,1]^{\mathcal{P}}$, we are interested in finding a probability distribution on the power
set 2^E of E that is consistent with these marginals and that ensures that each
set in $P \in \mathcal{P}$ is hit with probability at least π_P. In other words, we are looking
for a solution x to the system

$$\sum_{S \subseteq E : e \in S} x_S = \rho_e \quad \forall e \in E, \tag{1}$$

$$\sum_{S \subseteq E : S \cap P \neq \emptyset} x_S \geq \pi_P \quad \forall P \in \mathcal{P}, \tag{2}$$

$$\sum_{S \subseteq E} x_S = 1, \tag{3}$$

$$x_S \geq 0 \quad \forall S \subseteq E. \tag{4}$$

Throughout this paper, we will call a distribution x fulfilling (1) to (4) a
feasible decomposition of ρ for $(E, \mathcal{P})$ and π, and we will say that the marginals
ρ are *feasible for $(E, \mathcal{P})$ and π* if such a feasible decomposition exists.

© The Author(s), under exclusive license to Springer Nature Switzerland AG 2023
A. Del Pia and V. Kaibel (Eds.): IPCO 2023, LNCS 13904, pp. 306–318, 2023.
https://doi.org/10.1007/978-3-031-32726-1_22

A necessary condition for the existence of a feasible decomposition is that the marginals suffice to cover each set of the system individually, i.e.,

$$\sum_{e \in P} \rho_e \geq \pi_P \qquad \forall P \in \mathcal{P}. \tag{$\star$}$$

We are particularly interested in identifying classes of systems and requirement functions for which $(\star)$ is not only a necessary but also a sufficient condition. For such systems, the set of distributions on 2^E fulfilling (2) can be described by the corresponding polytope of feasible marginals defined by $(\star)$, which is of exponentially lower dimension.

1.1 Motivation

A natural application for feasible decompositions in the setting described above lies in network security games; see, e.g., [1–3, 8, 16, 17] for various examples and applications of network security games. In fact, such a game was also the motivation of Dahan, Amin, and Jaillet [3], who originally introduced the decomposition setting described above. We will discuss their game in detail in Sect. 5. Here, we describe a simpler yet relevant problem as an illustrative example.

Consider the following game played on a set system $(E, \mathcal{P})$, where each element $e \in E$ is equipped with a usage cost $c_e \geq 0$ and an inspection cost $d_e \geq 0$. A *defender* D determines a random subset S of elements from E to inspect at cost $\sum_{e \in S} d_e$ (e.g., a set of links of a network at which passing traffic is monitored). She anticipates that an *attacker* A is planning to carry out an illegal action, where A chooses a set in $P \in \mathcal{P}$ (e.g., a route in the network along which he smuggles contraband), for which he will receive utility $U_1 - \sum_{e \in P} c_e$ for some constant $U_1 > 0$. However, if P intersects with the random set S of elements inspected by D, then A is discovered while carrying out his illegal action, reducing his utility by a penalty $U_2 \geq U_1$. The attacker also has the option to not carry out any attack, resulting in utility 0. Thus, A will refrain from using $P \in \mathcal{P}$ if the probability that $S \cap P \neq \emptyset$ exceeds $\pi_P := (U_1 - \sum_{e \in P} c_e)/U_2$.

A natural goal for D is to discourage A from attempting any attack at all, while keeping the incurred inspection cost as small as possible. Note that the randomized strategies that achieve this goal correspond exactly to vectors x that minimize $\sum_{S \subseteq E} \sum_{e \in S} d_e x_S$ subject to constraints (2) to (4). Unfortunately, the corresponding LP has both an exponential number of variables and an exponential number of constraints in the size of the ground set E.

However, assume that we can establish the following three properties for our set system: (i) condition $(\star)$ is sufficient for the feasibility of marginals, (ii) we can efficiently compute the corresponding feasible decompositions, and (iii) given $\gamma \in \mathbb{R}_+^E$, we can efficiently solve $\min_{P \in \mathcal{P}} \sum_{e \in P} \gamma_e$. Then (i) allows us to formulate D's problem in terms of the marginals, i.e., $\min_{\rho \in [0,1]^E} \sum_{e \in E} d_e \rho_e$ subject to constraints $(\star)$, (iii) allows us to separate the linear constraints $(\star)$ and obtain optimal marginals ρ, and (ii) allows us to turn these marginals into a distribution corresponding to an optimal inspection strategy for the defender D. In this paper, we will establish all three conditions for a generic type of set systems called *abstract networks*.

1.2 Abstract Networks

An *abstract network* consists of a set system $(E, \mathcal{P})$ where each set $P \in \mathcal{P}$ (also referred to as an *(abstract) path*) is equipped with an internal linear order $\preceq_P$ of its elements, such that for all $P, Q \in \mathcal{P}$ and all $e \in P \cap Q$ there is a abstract path $R \in \mathcal{P}$ with $R \subseteq \{p \in P : p \preceq_P e\} \cup \{q \in Q : e \preceq_Q q\}$. Given $P, Q \in \mathcal{P}$ and $e \in P \cap Q$, we use the notation $P \times_e Q$ to denote an arbitrary but fixed feasible choice for such an $R \in \mathcal{P}$.

Intuitively, this definition is an abstraction of the property of digraphs that one can construct a new path by concatenating a prefix and a suffix of two intersecting paths. Interesting special cases of abstract networks include $\mathcal{P}$ being the set of maximal chains in a partially ordered set $(E, \preceq)$ (here, $\preceq_P$ is simply the restriction of $\preceq$ to P) and $\mathcal{P}$ being the set of simple s-t-paths in a digraph $D = (V, A)$ (here, $E = V \cup A$ and each path is identified with the sequence of its nodes and arcs). We remark that in both cases, the order $\preceq_{P \times_e Q}$ is consistent with $\preceq_P$ and $\preceq_Q$, which is not a general requirement for abstract networks; see, e.g., [10] for examples of abstract networks where this is not the case.

Abstract networks were introduced by Hofmann [7] to illustrate the generality of Ford and Fulkerson's [6] max-flow/min-cut theorem.[1] McCormick [14] provided a combinatorial algorithm for computing maximum flows in abstract networks using a *membership oracle* that, given $F \subseteq E$, returns $P \in \mathcal{P}$ with $P \subseteq F$ together with the corresponding order $\preceq_P$, or certifies that no such P exists. Martens and McCormick [12] later extended this result by giving a combinatorial algorithm for a weighted version of the problem, using a stronger oracle. Applications of abstract networks include, e.g., line planning for public transit systems [11] and route assignment in evacuation planning [9, 15].

1.3 Previous Results

Dahan et al. [3] studied the case where $\mathcal{P}$ is the set of maximal chains of a partially ordered set (poset), or, equivalently, the set of s-t-paths in a directed acyclic graph (DAG). They showed that $(\star)$ is sufficient for the existence of a feasible distribution when the requirements fulfill the following conservation law:

$$\pi_P + \pi_Q = \pi_{P \times_e Q} + \pi_{Q \times_e P} \quad \forall P, Q \in \mathcal{P}, e \in P \cap Q. \tag{C}$$

Although their result is algorithmic, the corresponding algorithm requires explicitly enumerating all maximal chains and hence does not run in polynomial time in the size of E. However, Dahan et al. [3] provide a polynomial-time algorithm for the case of *affine* requirements, in which there exists a vector $\mu \in [0, 1]^E$ such that the requirements are of the form

$$\pi_P = 1 - \sum_{e \in P} \mu(e) \quad \forall P \in \mathcal{P}. \tag{A}$$

---

[1] Given an abstract network $(E, \mathcal{P})$ with capacities $u \in \mathbb{R}_+^E$, a flow is a vector $f \in \mathbb{R}^+$ fulfilling capacity constraints $\sum_{P \in \mathcal{P}: e \in P} f_P \leq u_e$ for all $e \in E$. The maximum abstract flow problem asks for a flow of maximum value $\sum_{P \in \mathcal{P}} f_P$. Hoffman [7] showed that the corresponding dual linear program is totally dual integral (even in a more general weighted setting), thus generalizing the max-flow/min-cut theorem.

As a consequence of this latter result, the authors were able to characterize Nash equilibria for their network security game (which is a flow-interdiction game played on s-t-paths in a digraph) by means of a compact arc-flow LP formulation and compute such equilibria in polynomial time, under the condition that the underlying digraph is acyclic. Indeed, this positive result is particularly surprising, as similar—and seemingly simpler—flow-interdiction games had previously been shown to be NP-hard, even on DAGs [5].

1.4 Our Results

We extend the results of Dahan et al. [3] for posets/DAGs in multiple directions:

1. For the affine requirements case (A), we show that $(\star)$ is a sufficient condition for the feasibility of marginals when $(E, \mathcal{P})$ is an abstract network, by providing an explicit description of a feasible decomposition for this case, based on a natural generalization of shortest-path distances to abstract networks (see Sect. 2). The described solutions have the property that the sets in their support can be represented by an interval matrix. A special case of this result is the case where $\mathcal{P}$ is the set of s-t-paths in a digraph (which is not necessarily acyclic). In this case, a feasible decomposition can be computed efficiently by a single run of a standard shortest-path algorithm.
2. We also provide an algorithm for efficiently computing the corresponding feasible decompositions for the general case of an arbitrary abstract network given by a membership oracle (see Sect. 3). This algorithm makes use of the following result as a subroutine.
3. We provide a combinatorial strongly polynomial algorithm for computing shortest paths in abstract networks when $\mathcal{P}$ is given by a membership oracle (see Sect. 4). Beyond its relevance for the present work, this result also gives a partial answer to an open question by McCormick [14], who conjectured that such an algorithm might enable a strongly polynomial algorithm for computing maximum flows in abstract networks.
4. As a consequence of our results, Nash equilibria for the network security game studied by Dahan et al. [3] can be described by a compact polyhedron and computed efficiently even when the game is played on an abstract network, including the case of a digraph with cycles (see Sect. 5).
5. We further show that the conservation law (C) proposed in [3] for maximal chains in posets can be reduced to the affine requirements case (A) (see Sect. 6). We provide a polynomial-time algorithm for computing the corresponding weights μ when the requirements π are given by a value oracle. As a consequence, the corresponding feasible decompositions can be computed efficiently in this case as well.
6. Finally, we discuss other types of set systems (see Sect. 7). We observe that $(\star)$ is not sufficient for the feasibility of the marginals when $\mathcal{P}$ consists of the bases of a matroid, perfect matchings of a bipartite graph, or paths in a multicommodity network. We further show that deciding whether a given set of marginals is feasible is NP-hard in general, even when $\mathcal{P}$ is given by an explicit list of small sets and the requirements are all equal to 1.

1.5   Notation

Before we discuss our results in detail, we introduce some useful notation concerning abstract networks. Let $(E, \mathcal{P})$ be an abstract network. For $P \in \mathcal{P}$ and $e \in P$, we use the following notation to denote prefixes of P ending at e and suffixes of P starting at e, respectively:

$$[P, e] := \{p \in P : p \preceq_P e\} \qquad [e, P] := \{p \in P : e \preceq_P p\}$$
$$(P, e) := \{p \in P : p \prec_P e\} \qquad (e, P) := \{p \in P : e \prec_P p\}$$

For any path $P \in \mathcal{P}$, we further denote the minimal and maximal element of P with respect to $\preceq_P$ by s_P and t_P, respectively.

Throughout the paper, proofs of results marked with ($\blacklozenge$) can be found in the full version [13].

2   Feasible Decompositions in Abstract Networks

In this section we prove the following theorem, providing an explicit description of feasible decompositions of marginals in abstract networks assuming that requirements are of the form (A) and fulfill the necessary condition ($\star$). The construction, described in the following theorem, is based on a natural generalization of shortest-path distances in abstract networks.

Theorem 1. *Let $(E, \mathcal{P})$ be an abstract network and let $\rho, \mu \in [0, 1]^E$ fulfilling condition ($\star$), i.e., $\sum_{e \in P} \rho_e \geq \pi_P := 1 - \sum_{e \in P} \mu_e$ for all $P \in \mathcal{P}$. Define*

$$\alpha_e' := \min \left\{ \textstyle\sum_{f \in (Q,e)} \mu_f + \rho_f : Q \in \mathcal{P}, e \in Q \right\} \quad \text{and} \quad \alpha_e := \min \{\alpha_e', 1 - \rho_e\}$$

for $e \in E$. For $\tau \sim U[0, 1]$ drawn uniformly at random from $[0, 1]$, let

$$S_\tau := \{e \in E : \alpha_e \leq \tau < \alpha_e + \rho_e\}.$$

Then x defined by $x_S := \mathbf{Pr}[S_\tau = S]$ for $S \subseteq E$ is a feasible decomposition of ρ for $(E, \mathcal{P})$ and π.

Intuitively, the values α_e' for $e \in E$ in the construction above correspond to the "shortest-path distance" to element e in the abstract network $(E, \mathcal{P})$, with the truncation of α_e at $1 - \rho_e$ ensuring that $[\alpha_e, \alpha_e + \rho_e] \subseteq [0, 1]$. Before we prove Theorem 1, let us first discuss some of its implications.

Interval Structure and Explicit Computation of x. Given the vector α, the non-zero entries of x can be easily determined in polynomial time. Indeed, note that the set $\Lambda := \{\alpha_e, \alpha_e + \rho_e : e \in E\}$ induces a partition of $[0, 1]$ into at most $2|E| + 1$ intervals (each with two consecutive values from $\Lambda \cup \{0, 1\}$ as its endpoints), such that $S_{\tau'} = S_{\tau''}$ whenever τ' and τ'' are in the same interval. Thus, there are at most $2|E| + 1$ non-zero entries in x, whose values can be determined by sorting Λ, determining all corresponding intervals, computing S_τ for one τ in each of these intervals, and then, for each occurring set S, setting x_S to the total length of all intervals in which this set is attained.

Special Case: Directed Graphs. Consider the case where $\mathcal{P}$ is the set of simple s-t-paths in a digraph $D = (V, A)$ and $E = V \cup A$. For $v \in V$, let $\mathcal{P}_{sv}$ denote the set of simple s-v-paths in D. If we are given explicit access to D (rather than accessing $\mathcal{P}$ via a membership oracle), we can compute feasible decompositions as follows. Without loss of generality, we can assume that for any $v \in V$ and $Q \in \mathcal{P}_{sv}$ there is $Q' \in \mathcal{P}_{st}$ with $Q \subseteq Q'$.[2] Then $\alpha'_v = \min_{Q \in \mathcal{P}_{sv}} \sum_{f \in Q \setminus \{v\}} \mu_f + \rho_f$ for $v \in V$ and $\alpha'_a = \min_{Q \in \mathcal{P}_{sv}} \sum_{f \in Q} \mu_f + \rho_f$ for $a = (v, w) \in A$. Hence, the vector α' corresponds to shortest-path distances in D with respect to $\rho + \mu$ (with costs on both arcs and nodes). Both α' and the corresponding feasible decomposition of ρ can be computed by a single run of Dijkstra's [4] algorithm.

Computing feasible decompositions in the general case of arbitrary abstract networks is more involved. We show how this can be achieved in Sect. 3.

Proof of Theorem 1. We show that x as constructed in Theorem 1 is a feasible decomposition. Note that x fulfills (3) and (4) by construction. Note further that x fulfills (1) because $\sum_{S \subseteq E : e \in S} x_S = \mathbf{Pr}\left[e \in S_\tau \right] = \mathbf{Pr}\left[\alpha_e \leq \tau < \alpha_e + \rho_e \right] = \rho_e$ for all $e \in E$, where the second identity follows from $0 \leq \alpha_e \leq 1 - \rho_e$. It remains to prove that x fulfills (2). The following lemma will be helpful in this endeavour.

Lemma 2. *Given $(E, \mathcal{P})$, ρ, μ, and α as described in Theorem 1, the following two conditions are fulfilled for every $P \in \mathcal{P}$:*

1. $\alpha_{t_P} + \mu_{t_P} + \rho_{t_P} \geq 1$ and
2. for every $e \in P \setminus \{t_P\}$ there is $e' \in (e, P)$ with $\alpha_{e'} \leq \alpha_e + \mu_e + \rho_e$.

Proof. We first show statement 1. By contradiction assume $\alpha_{t_P} + \mu_{t_P} + \rho_{t_P} < 1$. Let $Q \in \mathcal{P}$ with $t_P \in Q$ and $\sum_{f \in (Q, t_P)} \mu_f + \rho_f = \alpha_{t_P}$ and let $R := Q \times_{t_P} P$. Note that $R \subseteq [Q, t_P]$ and hence $\sum_{e \in R} \mu_e + \rho_e \leq \alpha_{t_P} + \mu_{t_P} + \rho_{t_P} < 1$, implying $\sum_{e \in R} \rho_e < 1 - \sum_{e \in R} \mu_e$, a contradiction to $(\star)$.

We now turn to statement 2. If $\alpha_e \geq 1 - \mu_e - \rho_e$, then the statement follows with $e' = t_P$ because $\alpha_{t_P} \leq 1 \leq \alpha_e + \mu_e + \rho_e$. Thus assume $\alpha_e < 1 - \mu_e - \rho_e$ and let $Q \in \mathcal{P}$ with $\alpha_e = \sum_{f \in (Q, e)} \mu_f + \rho_f$. Let $R := Q \times_e P$. By $(\star)$ we observe that $\sum_{f \in R} \mu_f + \rho_f \geq 1 > \alpha_e + \mu_e + \rho_e$, which implies $R \setminus [Q, e] \neq \emptyset$ because $\mu, \rho \geq 0$. Thus, let $e' \in R \setminus [Q, e]$ be minimal with respect to $\prec_R$. Observe that $R \setminus [Q, e] \subseteq (e, P)$ and hence $e' \in (e, P)$. The statement then follows from

$$\alpha_{e'} \leq \sum_{f \in (R, e')} \mu_f + \rho_f \leq \sum_{f \in [Q, e]} \mu_f + \rho_f = \alpha_e + \mu_e + \rho_e,$$

where the second inequality is due to the fact that $(R, e') \subseteq [Q, e]$ by choice of e' and the fact that $\mu, \rho \geq 0$. ∎

With the help of Lemma 2, we can prove that x fulfills (2) as follows. Let $P \in \mathcal{P}$. For $e \in P$ define

$$\phi(e) := \mathbf{Pr}\left[S_\tau \cap [P, e] \neq \emptyset \wedge \tau \leq \alpha_e + \rho_e \right] + \sum_{f \in [P, e]} \mu_f.$$

Let $F := \{e \in P : \phi(e) \geq \alpha_e + \mu_e + \rho_e\}$. We will show that $t_P \in F$. Note that this suffices to prove (2), because the definition of F together with statement 1 of Lemma 2 imply $\phi(t_P) \geq \alpha_{t_P} + \mu_{t_P} + \rho_{t_P} \geq 1$, which in turn yields

$$\sum_{S \subseteq E: S \cap P \neq \emptyset} x_S = \mathbf{Pr}\left[S_\tau \cap P \neq \emptyset\right] \geq \phi(t_P) - \sum_{f \in P} \mu_f \geq 1 - \sum_{f \in P} \mu_f = \pi_P.$$

We proceed to show $t_P \in F$. By contradiction assume this is not the case. Note that $F \neq \emptyset$ because $\alpha_{s_P} = 0$ and $\phi(s_P) = \mathbf{Pr}\left[s_P \in S_\tau\right] + \mu_{s_P} = \rho_{s_P} + \mu_{s_P}$. Thus let $e \in F$ be maximal with respect to $\prec_P$. Because $e \neq t_P$, we can invoke statement 2 of Lemma 2 and obtain $e' \in (e, P)$ with

$$\alpha_{e'} \leq \alpha_e + \mu_e + \rho_e. \tag{5}$$

We will show that $e' \in F$, contradicting our choice of e. Note that the definition of ϕ and the fact that $e' \succ_P e$ imply

$$\begin{aligned}\phi(e') &\geq \phi(e) + \mathbf{Pr}\left[e' \in S_\tau \wedge \tau > \alpha_e + \rho_e\right] + \mu_{e'} \\ &\geq \alpha_e + \mu_e + \rho_e + \mathbf{Pr}\left[e' \in S_\tau \wedge \tau > \alpha_e + \rho_e\right] + \mu_{e'}, \end{aligned} \tag{6}$$

where the second inequality follows from $e \in F$. Moreover, observe that $e' \in S_\tau$ if and only if $\alpha_{e'} \leq \tau < \alpha_{e'} + \rho_{e'}$ and hence

$$\begin{aligned}\mathbf{Pr}\left[e' \in S_\tau \wedge \tau > \alpha_e + \rho_e\right] &= \alpha_{e'} + \rho_{e'} - \max\{\alpha_{e'}, \alpha_e + \rho_e\} \\ &\geq \alpha_{e'} + \rho_{e'} - (\alpha_e + \mu_e + \rho_e), \end{aligned}$$

where the inequality follows from (5). Combining this bound with (6) yields $\phi(e') \geq \alpha_{e'} + \mu_{e'} + \rho_{e'}$ and hence $e' \in F$, contradicting our choice of e and completing the proof of Theorem 1. □

3   Computing Feasible Decompositions

Complementing our existence result from the previous section, we now discuss how to compute corresponding feasible decompositions. We will assume that the ground set E is given explicitly, while the set of abstract paths $\mathcal{P}$ is given by a membership oracle that, given $F \subseteq E$, either returns $P \in \mathcal{P}$ with $P \subseteq F$ and the corresponding order $\preceq_P$, or confirms that no $P \in \mathcal{P}$ with $P \subseteq F$ exists.

By our arguments in Sect. 2, it suffices to compute the values of α_e for all $e \in E$. Unfortunately, a complication arises in that even finding a path containing a certain element $e \in E$ is NP-hard.[3] However, as we show below, it is possible to identify a subset $U \subseteq E$ for which we can compute the values of α, while the elements in $E \setminus U$ turn out to be redundant w.r.t. the feasibility of the marginals. From this, we obtain the following theorem.

---

[3] Note that even for the special case where $\mathcal{P}$ corresponds to the set of simple s-t-paths in a digraph, finding $P \in \mathcal{P}$ containing a certain arc e is equivalent to the 2-disjoint path problem (for P to be simple, its prefix up to the tail of e and its suffix starting from the head of e must be disjoint). Simply side-stepping this issue by introducing additional elements as done in the second remark after Theorem 1 is not possible here, because we are restricted to accessing $\mathcal{P}$ only via the membership oracle.

Theorem 3. *There is an algorithm that, given an abstract network $(E, \mathcal{P})$ via a membership oracle and $\rho, \mu \in [0, 1]^E$ such that $\sum_{e \in P} \rho_e \geq \pi_P := 1 - \sum_{e \in P} \mu_e$ for all $P \in \mathcal{P}$, computes a feasible decomposition of ρ for $(E, \mathcal{P})$ and π in time $\mathcal{O}(|E|^3 \cdot T_{\mathcal{P}})$, where $T_{\mathcal{P}}$ denotes the time for a call to the membership oracle of $\mathcal{P}$.*

The Algorithm. Theorem 3 is established via Algorithm 1, which computes values $\bar{\alpha}_e$ for elements e in a subset $U \subseteq E$ as follows. Starting from $U = \emptyset$, the algorithm iteratively computes a path P minimizing $\sum_{f \in P \cap U} \mu_f + \rho_f$ and adds the first element e of $P \setminus U$ to U, determining $\bar{\alpha}_e$ based on the length of (P, e).

Algorithm 1: Computing a feasible decomposition

Initialize $U := \emptyset$.

while $\min_{P \in \mathcal{P}} \sum_{f \in P \cap U} \mu_f + \rho_f < 1$ **do**

> Let $P \in \operatorname{argmin}_{P \in \mathcal{P}} \sum_{f \in P \cap U} \mu_f + \rho_f$.
> Let $e := \min_{\preceq_P} P \setminus U$.
> Set $U := U \cup \{e\}$ and $\bar{\alpha}_e := \min \left\{ \sum_{f \in (P, e)} \mu_f + \rho_f,\ 1 - \rho_e \right\}$.

return $\bar{\alpha}, U$

Analysis. First note that in every iteration of the while loop, the set $P \setminus U$ is nonempty because $\sum_{f \in P} \mu_f + \rho_f \geq 1$ by the assumption on the input in Theorem 3. Hence the algorithm is well-defined and terminates after at most $|E|$ iterations. We further remark that finding $P \in \mathcal{P}$ minimizing $\sum_{e \in P \cap U} \rho_f + \mu_f$ can be done in time $\mathcal{O}(|E|^2 T_{\mathcal{P}})$ using the Algorithm 2 described in Sect. 4. The following lemma then suffices to complete the proof of Theorem 3.

Lemma 4 (♦). *Let $\bar{\alpha}, U$ be the output of Algorithm 1 and define $\bar{\rho}_e := \rho_e$ and $\bar{\mu}_e := \mu_e$ for $e \in U$ and $\bar{\rho}_e := 0$ and $\bar{\mu}_e := 0$ for $e \in E \setminus U$. Then*

1. $\sum_{e \in P} \bar{\rho}_e \geq \bar{\pi}_P := 1 - \sum_{e \in P} \bar{\mu}_e$ *for all* $P \in \mathcal{P}$ *and*

2. $\bar{\alpha}_e = \min \left\{ \sum_{f \in (Q, e)} \bar{\mu}_f + \bar{\rho}_f : Q \in \mathcal{P}, e \in Q \right\} \cup \{1 - \bar{\rho}_e\}$ *for all* $e \in U$.

Indeed, observe that Lemma 4 together with Theorem 1 implies that $\bar{\alpha}$ induces a feasible decomposition $\bar{x}$ of $\bar{\rho}$ for $(E, \mathcal{P})$ and $\bar{\pi}$. Because $\bar{\rho}_e \leq \rho_e$ for all $e \in E$ and $\bar{\pi}_P \geq \pi_P$ for all $P \in \mathcal{P}$, this decomposition can be extended to a feasible decomposition of ρ for $(E, \mathcal{P})$ and π by arbitrarily incorporating the elements from $E \setminus U$. This completes the proof of Theorem 3.

4   Computing Shortest Paths in Abstract Networks

In this section, we consider the following natural generalization of the classic shortest s-t-path problem in digraphs: Given an abstract network $(E, \mathcal{P})$ and a cost vector $\gamma \in \mathbb{R}_+^E$, find a path $P \in \mathcal{P}$ minimizing $\sum_{e \in P} \gamma_e$. We provide a combinatorial, strongly polynomial algorithm for this problem, accessing $\mathcal{P}$ only via a membership oracle. In fact, the question for such an algorithm was already raised by McCormick [14], who conjectured that it can be used to turn (an adaptation of) his combinatorial, but only weakly polynomial algorithm for the maximum abstract flow problem into a strongly polynomial one. Our results show that such a shortest-path algorithm indeed exists, but leave it open how to use it to improve the running time of the maximum abstract flow algorithm.

Theorem 5. *There is an algorithm that, given an abstract network $(E, \mathcal{P})$ via a membership oracle and $\gamma \in \mathbb{R}_+^E$ computes $P \in \mathcal{P}$ minimizing $\sum_{e \in P} \gamma_e$ in time $\mathcal{O}(|E|^2 \cdot T_{\mathcal{P}})$, where $T_{\mathcal{P}}$ denotes the time for a call to the membership oracle of $\mathcal{P}$.*

The Algorithm. For notational convenience, we assume that there is $s, t \in E$ with $s_P = s$ and $t_P = t$ for all $P \in \mathcal{P}$. Note that this assumption is without loss of generality, as it can be ensured by adding dummy elements s and t to E and including them at the start and end of each path, respectively.

The algorithm is formally described as Algorithm 2. It can be seen as a natural extension of Dijkstra's [4] algorithm in that it maintains for each element $e \in E$ a (possibly infinite) label ψ_e indicating the length of the shortest segment $[Q_e, e]$ for some $Q_e \in \mathcal{P}$ with $e \in Q_e$ found so far, and in that its outer loop iteratively chooses an element with currently smallest label for processing. However, updating these labels is more involved, as an abstract network does not provide local concepts such as "the set of arcs leaving a node". In its inner loop, the algorithm therefore carefully tries to extend the segment Q_e for the currently processed element e to find new shortest segments $Q_{e'}$ for other elements e'.

Algorithm 2: Computing a shortest path in an abstract network

---

Initialize $T := \emptyset$, $\psi_s := \gamma_s$, and $\psi_e := \infty$ for all $e \in E \setminus \{s\}$.
Let $Q_s \in \mathcal{P}$.
while $\psi_t > \min_{f \in E \setminus T} \psi_f$ **do**
  Let $e \in \operatorname{argmin}_{f \in E \setminus T} \psi_f$.
  Let $F := (E \setminus T) \cup [Q_e, e]$.
  while *there is $P \in \mathcal{P}$ with $P \subseteq F$* **do**
    Let $e' := \min_{\preceq_P} P \setminus [Q_e, e]$.
    Set $F := F \setminus \{e'\}$.
    if $\sum_{f \in [P, e']} \gamma_f < \psi_{e'}$ **then**
      Set $\psi_{e'} := \sum_{f \in [P, e']} \gamma_f$ and $Q_{e'} := P$.
  Set $T := T \cup \{e\}$.
return Q_t

Analysis. The proof of the correctness of Algorithm 2 crucially relies on the following lemma, which essentially certifies that the algorithm does not overlook any shorter path segments when processing an element.

Lemma 6 (♦). *Algorithm 2 maintains the following invariant: For all $P \in \mathcal{P}$, there is $e \in P$ with $[e, P] \cap T = \emptyset$ and $\psi_e \leq \sum_{f \in [P, e]} \gamma_f$.*

Proof of Theorem 5. When Algorithm 2 terminates, $\psi_t \leq \psi_f$ for all $f \in E \setminus T$ by the termination criterion of the outer while loop. Let $P \in \mathcal{P}$. By Lemma 6 there is an element $e \in P \setminus T$ with $\psi_e \leq \sum_{f \in [P,e]} \gamma_f$. Note that this implies $\sum_{f \in Q_t} \gamma_f = \psi_t \leq \psi_e = \sum_{f \in [P,e]} \gamma_f \leq \sum_{f \in P} \gamma_f$, where the last inequality uses the fact that $\gamma_f \geq 0$ for all $f \in E$. We conclude that the path Q_t returned by the algorithm is indeed a shortest path.

To see that the algorithm terminates in polynomial time, observe that the outer while loop stops after at most $|E| - 1$ iterations, as in each iteration an element from $E \setminus \{t\}$ is added to T and the termination criterion is fulfilled if $T = E \setminus \{t\}$. Furthermore, each iteration of the inner while loop removes an element from F and hence after at most $|F| \leq |E|$ iterations no path $P \subseteq F$ exists anymore, implying that the inner while loop terminates. $\square$

5   Dahan et al.'s Network Security Game

Dahan et al. [3] studied the following network security game. The input is a set system $(E, \mathcal{P})$ with capacities $u \in \mathbb{R}_+^E$, transportation cost $c \in \mathbb{R}_+^E$ and interdiction costs $d \in \mathbb{R}_+^E$. There are two players: the *routing entity* R, whose strategy space is the set of *flows* $F := \{f \in \mathbb{R}_+^{\mathcal{P}} : \sum_{P \in \mathcal{P}: e \in P} f_P \leq u_e \; \forall e \in E\}$, and the *interdictor* I, who selects a subset of elements $S \subseteq E$ to interdict, with the intuition that all flow on interdicted elements is disrupted. Given strategies $f \in F$ and $S \subseteq E$, the payoffs for R and I, respectively, are given by

$$P_R(f, S) := \sum_{P \in \mathcal{P}: P \cap S = \emptyset} f_P - \sum_{P \in \mathcal{P}} \sum_{e \in P} c_e f_P \text{ and}$$
$$P_I(f, S) := \sum_{P \in \mathcal{P}: P \cap S \neq \emptyset} f_P - \sum_{e \in S} d_e,$$

respectively. That is, R's payoff is the total amount of non-disrupted flow, reduced by the cost for sending flow f, while I's payoff is the total amount of flow that is disrupted, reduced by the interdiction cost for the set S.

We are interested in finding (mixed) Nash equilibria (NE) for this game, i.e., random distributions σ_R and σ_I over the strategy spaces of I and R, respectively, such that no player can improve their expected payoff by unilateral deviation. However, the efficient computation of such equilibria is hampered by the fact that the strategy spaces of both players are of exponential size/dimension in the size of the ground set E. To overcome this issue, Dahan et al. [3] proposed to consider the following pair of primal and dual linear programs:

$$[\text{LP}_R] \; \max \sum_{P \in \mathcal{P}} \pi_P^c f_P \qquad\qquad [\text{LP}_I] \; \min \sum_{e \in E} u_e \mu_e + d_e \rho_e$$

$$\text{s.t.} \sum_{P \in \mathcal{P}: e \in P} f_P \leq u_e \; \forall e \in E \qquad \text{s.t.} \sum_{e \in P} \mu_e + \rho_e \geq \pi_P^c \; \forall P \in \mathcal{P}$$

$$\sum_{P \in \mathcal{P}: e \in P} f_P \leq d_e \; \forall e \in E \qquad\qquad \mu \geq 0$$

$$f \geq 0 \qquad\qquad\qquad\qquad\quad \rho \geq 0,$$

where $\pi_P^c := 1 - \sum_{a \in P} c_a$. Dahan et al. [3] showed the following result.

Theorem 7 (Dahan et al. [3]). *Let f^* and (μ^*, ρ^*) be optimal solutions to $[LP_R]$ and $[LP_I]$, respectively. Let σ_I be a feasible decomposition of ρ^* for $(E, \mathcal{P})$ and $\pi_P := \pi_P^c - \sum_{e \in P} \mu_e^*$ and let σ_R be a distribution over F with $\sum_{f \in F} \sigma_{R,f} f_P = f_P^*$. Then (σ_R, σ_I) is a Nash equilibrium.*

In particular, note that any feasible solution to $[LP_I]$ defines marginals ρ that fulfil $(\star)$ for $\pi_P := \pi_P^c - \sum_{e \in P} \mu_e$. Hence, if condition $(\star)$ is sufficient for feasibility of marginals in the set system $\mathcal{P}$ under affine requirements, any pair of optimal solutions to the LPs induces a Nash equilibrium. If we can moreover efficiently compute optimal solutions to the LPs and the corresponding feasible decompositions, we can efficiently find a Nash equilibrium.

Dahan et al. [3] showed that this is possible when $\mathcal{P}$ is the set of s-t-paths in a DAG. Hence NE for the game can be found efficiently in that setting. This positive result is particularly interesting because NE are hard to compute for the variant of the game in which the interdictor is limited by a budget, even when interdiction costs are uniform, transportation costs are zero, and the game is played on a DAG [5]. Our results in Sects. 2 to 4 imply that all three conditions for the computability of NE are also met when $(E, \mathcal{P})$ is an abstract network (note that we can use Algorithm 2 to separate the constraints of $[LP_I]$). Hence we can compute Nash equilibria for the above game when $(E, \mathcal{P})$ is an abstract network given by a membership oracle, in time polynomial in $|E|$, including the case where the game is played on a digraph with cycles.

We remark that Dahan et al. [3] also showed that, if there is at least one dual solution with a decomposition that assigns positive probability to the empty set, then all NE of the game are of the form described in Theorem 7. They showed that this condition is always fulfilled in the DAG case when all transportation costs are positive. Via a small adjustment to our construction in Sect. 2, the same result can be proven for the case of abstract networks ($\blacklozenge$).

6   The Conservation Law for Partially Ordered Sets

As discussed in Sect. 1, Dahan et al. [3] established the sufficiency of $(\star)$ in partially ordered sets not only for the case of affine requirements (A) but also for the case where requirements fulfill the conservation law (C). However, they left it open whether it is possible to efficiently compute the corresponding decompositions in the latter case. In this section, we resolve this question by showing that the conservation law (C) for maximal chains in a poset can be reduced to the case of affine requirements (A) in the corresponding Hasse diagram,[4] for which a feasible decomposition then can be computed efficiently.

Theorem 8 ($\blacklozenge$). *Let $D = (V, A)$ be a directed acyclic graph, let $s, t \in V$, and let $\mathcal{P} \subseteq 2^{V \cup A}$ be the set of s-t-paths in D. Let $\pi \in [0, 1]^{\mathcal{P}}$ such that (C) is fulfilled. Then there exists $\mu \in [0, 1]^{V \cup A}$ such that $\pi_P = 1 - \sum_{e \in P} \mu_e$. Furthermore, μ can be computed in strongly polynomial time in $|V|$ and $|A|$ when π is given by an oracle that, given $P \in \mathcal{P}$, returns π_P.*

---

[4] See [13] for details on why the transformation to the Hasse diagram is necessary.

Proof (sketch). By Farkas' lemma, the existence of μ is equivalent to showing that $\sum_{P \in \mathcal{P}} (1 - \pi_P) y_P \geq 0$ for every $y \in \mathbb{R}^{\mathcal{P}}$ with $\sum_{P \in \mathcal{P}: e \in P} y_P \geq 0$ for all $e \in V \cup A$. This property can be established by iteratively applying (C) to transform y into a nonnegative vector without changing $\sum_{P \in \mathcal{P}} (1 - \pi_P) y_P$. □

7 Other Set Systems

The results in this paper lead to the question whether sufficiency of $(\star)$ and computability of feasible decompositions can be established for other set systems, beyond abstract networks. We give negative answers for several natural candidates of such systems and point out interesting questions for future research.

Sufficiency of $(\star)$ ($\blacklozenge$). There are simple counterexamples for the sufficiency of $(\star)$ in the following cases, even when assuming that $\pi \equiv 1$: when $\mathcal{P}$ is the set of bases of a matroid; when $\mathcal{P}$ is the set of perfect matchings in a bipartite graph; when $\mathcal{P}$ is the set of s_i-t_i-paths in a digraph with multiple terminal pairs (s_i, t_i). An interesting question in this context is whether we can describe the systems for which $(\star)$ is sufficient by means of forbidden substructures.

Approximately Feasible Decompositions ($\blacklozenge$). Given the non-existence result mentioned above, one may be interested in finding decompositions that satisfy the requirements at least approximately. We say a decomposition x of marginals ρ is *β-approximately feasible*, for $\beta \in [0,1]$, if it fulfills (1), (3), (4) and $\sum_{S \subseteq E: S \cap P \neq \emptyset} x_S \geq \beta \cdot \pi_P$ for all $P \in \mathcal{P}$. Indeed, if marginals ρ fulfill $(\star)$ for requirements π, a $(1 - 1/e)$-approximately feasible decomposition always exists: Simply include each element $e \in E$ in the random set independently with probability ρ_e. An interesting question for future research is whether better guarantees may be achieved for some classes of systems.

Computing Feasible Decompositions and Optimization ($\blacklozenge$). For a given instance, we may also be interested in finding a decomposition of the given marginals that is β-approximately feasible for the largest possible value of β. Note that this also includes the case of finding a feasible decomposition if it exists (resulting in $\beta = 1$). Unfortunately, this latter problem is NP-complete, even in quite restricted cases, as evidenced by the theorem below. However, note that this hardness result still leaves room for approximating the best possible β.

Theorem 9 ($\blacklozenge$). *The following decision problem is NP-complete: Given a set system $(E, \mathcal{P})$ with $|P| = 3$ for all $P \in \mathcal{P}$ and marginals $\rho \in [0,1]^E$, is there a feasible decomposition of ρ for $(E, \mathcal{P})$ and requirement vector $\pi \equiv 1$?*

Acknowledgements. The author thanks three anonymous reviewers for numerous helpful suggestions that improved the manuscript. This work has been supported by the special research fund of KU Leuven (project C14/22/026).

References

1. Bertsimas, D., Nasrabadi, E., Orlin, J.B.: On the power of randomization in network interdiction. Oper. Res. Lett. **44**, 114–120 (2016)
2. Correa, J., Harks, T., Kreuzen, V.J., Matuschke, J.: Fare evasion in transit networks. Oper. Res. **65**, 165–183 (2017)
3. Dahan, M., Amin, S., Jaillet, P.: Probability distributions on partially ordered sets and network interdiction games. Math. Oper. Res. **47**, 458–484 (2022)
4. Dijkstra, E.W.: A note on two problems in connexion with graphs. Numer. Math. **269**, 271 (1959)
5. Disser, Y., Matuschke, J.: The complexity of computing a robust flow. Oper. Res. Lett. **48**, 18–23 (2020)
6. Ford, L.R., Fulkerson, D.R.: Maximal flow through a network. Can. J. Math. **8**, 399–404 (1956)
7. Hoffman, A.J.: A generalization of max flow–min cut. Math. Program. **6**, 352–359 (1974)
8. Holzmann, T., Smith, J.C.: The shortest path interdiction problem with randomized interdiction strategies: Complexity and algorithms. Oper. Res. **69**, 82–99 (2021)
9. Kappmeier, J.P.W.: Generalizations of flows over time with applications in evacuation optimization, Ph. D. thesis, TU Berlin (2015)
10. Kappmeier, J.P.W., Matuschke, J., Peis, B.: Abstract flows over time: a first step towards solving dynamic packing problems. Theoret. Comput. Sci. **544**, 74–83 (2014)
11. Karbstein, M.: Line planning and connectivity, Ph. D. thesis, TU Berlin (2013)
12. Martens, M., McCormick, S.T.: A polynomial algorithm for weighted abstract flow. In: Lodi, A., Panconesi, A., Rinaldi, G. (eds.) IPCO 2008. LNCS, vol. 5035, pp. 97–111. Springer, Heidelberg (2008). https://doi.org/10.1007/978-3-540-68891-4_7
13. Matuschke, J.: Decomposition of probability marginals for security games in abstract networks. Tech. rep., arXiv:2211.04922 (2022)
14. McCormick, S.T.: A polynomial algorithm for abstract maximum flow. In: Proceedings of the 7th annual ACM-SIAM Symposium on Discrete Algorithms, pp. 490–497 (1996)
15. Pyakurel, U., Khanal, D.P., Dhamala, T.N.: Abstract network flow with intermediate storage for evacuation planning. Eur. J. Oper. Res. **305**, 1178–1193 (2022)
16. Szeszlér, D.: Security games on matroids. Math. Program. **161**, 347–364 (2017)
17. Tambe, M.: Security and game theory: Algorithms, deployed systems, lessons learned. Cambridge University Press (2011)

Set Selection Under Explorable Stochastic Uncertainty via Covering Techniques

Nicole Megow[ID] and Jens Schlöter[(✉)][ID]

Faculty of Mathematics and Computer Science, University of Bremen,
Bremen, Germany
{nmegow,jschloet}@uni-bremen.de

Abstract. Given subsets of uncertain values, we study the problem of identifying the subset of minimum total value (sum of the uncertain values) by querying as few values as possible. This *set selection problem* falls into the field of *explorable uncertainty* and is of intrinsic importance therein as it implies strong adversarial lower bounds for a wide range of interesting combinatorial problems such as knapsack and matchings. We consider a stochastic problem variant and give algorithms that, in expectation, improve upon these adversarial lower bounds. The key to our results is to prove a strong structural connection to a seemingly unrelated covering problem with uncertainty in the constraints via a linear programming formulation. We exploit this connection to derive an algorithmic framework that can be used to solve both problems under uncertainty, obtaining nearly tight bounds on the competitive ratio. This is the first non-trivial stochastic result concerning the sum of unknown values without further structure known for the set. With our novel methods, we lay the foundations for solving more general problems in the area of explorable uncertainty.

Keywords: explorable uncertainty · queries · set selection · set cover

1 Introduction

In the setting of *explorable uncertainty*, we consider optimization problems with uncertainty in numeric input parameters. Instead of having access to the precise numeric values, we are given *uncertainty intervals* that contain the precise values. Each uncertainty interval can be *queried*, which reveals the corresponding precise value. The goal is to adaptively query intervals until we have sufficient information to optimally (or approximately) solve the underlying optimization problem, while minimizing the number of queries.

We mainly consider the *set selection problem* (MINSET) under explorable uncertainty. In this problem, we are given a set of n uncertain values represented by uncertainty intervals $\mathcal{I} = \{I_1, \ldots, I_n\}$ and a family of m sets $\mathcal{S} = \{S_1, \ldots, S_m\}$ with $S \subseteq \mathcal{I}$ for all $S \in \mathcal{S}$. A value w_i lies in its uncertainty interval I_i, is initially unknown, and can be revealed via a query. The value of an $S \in \mathcal{S}$ is $w(S) = \sum_{I_i \in S} w_i$. The goal is to determine a subset of minimum value as well as its value by using a minimal number of queries. It can be seen as an integer linear program (ILP) with uncertainty in the coefficients of the objective function:

$$\min \sum_{j=1}^{m} x_j \sum_{I_i \in S_j} w_i$$
$$\text{s.t.} \sum_{j=1}^{m} x_j = 1 \qquad\qquad\qquad (1)$$
$$x_j \in \{0,1\} \quad \forall j \in \{1,\ldots,m\}.$$

Since the w_i's are uncertain, we might have to execute queries to determine an optimal solution to (1). We refer to this problem as MINSET under uncertainty.

In this paper, we consider the stochastic problem variant, where all values w_i are drawn independently at random from their intervals I_i according to unknown distributions d_i. As there are instances that cannot be solved without querying the entire input, we analyze an algorithm ALG in terms of its *competitive ratio*: for the set of problem instances $\mathcal{J}$, it is defined as $\max_{J \in \mathcal{J}} \mathbb{E}[\text{ALG}(J)]/\mathbb{E}[\text{OPT}(J)]$, where ALG$(J)$ is the number of queries needed by ALG to solve instance J, and OPT(J) is the minimum number of queries necessary to solve the instance.

MINSET is a fundamental problem and of intrinsic importance within the field of explorable uncertainty. The majority of existing works considers the *adversarial setting*, where query outcomes are not stochastic but returned in a worst-case manner. Selection type problems have been studied in the adversarial setting and constant (matching) upper and lower bounds are known, e.g., for selecting the minimum [19], the k-th smallest element [13,19], a minimum spanning tree [10,12,18,23], sorting [17] and geometric problems [5]. However, these problems essentially boil down to comparing *single* uncertainty intervals and identifying the minimum of two unknown values. Once we have to compare two (even disjoint) sets and the corresponding *sums* of unknown values, no deterministic algorithm can have a better adversarial competitive ratio than n, the number of uncertainty intervals. This has been shown by Erlebach et al. [11] for MINSET, and it implies strong adversarial lower bounds for classical combinatorial problems, such as, knapsack and matchings [25], as well as solving ILPs with uncertainty in the cost coefficients as in (1) [25]. As a main result, we provide substantially better algorithms for MINSET under stochastic uncertainty. This is a key step for breaching adversarial lower bounds for a wide range of problems.

For the *stochastic setting*, the only related results we are aware of concern sorting [6] and the problem of finding the minimum in each set of a given collection of sets [2]. Asking for the *sum* of unknown values is substantially different.

The Covering Point of View. Our key observation is that we can view MIN-SET as a covering problem with uncertainty in the *constraints*. To see this, we focus on the structure of the uncertainty intervals and how a query affects it. We assume that each interval $I_i \in \mathcal{I}$ is either open (*non-trivial*) or *trivial*, i.e., $I_i = (L_i, U_i)$ or $I_i = \{w_i\}$; a standard technical assumption in explorable uncertainty. In the latter case, $L_i = U_i = w_i$. We call L_i and U_i *lower and upper limit*. For a set $S \in \mathcal{S}$, we define the *initial lower limit* $L_S = \sum_{I_i \in S} L_i$ and *initial upper limit* $U_S = \sum_{I_i \in S} U_i$. Clearly, $w(S) \in (L_S, U_S)$.

As the intervals (L_S, U_S) of the sets $S \in \mathcal{S}$ can overlap, we might have to execute queries to determine the set of minimum value. A query to an interval

I_i reveals the precise value w_i and, thus, replaces both, L_i and U_i, with w_i. In a sense, a query to an $I_i \in S$ reduces the range (L_S, U_S) in which $w(S)$ might lie by increasing L_S by $w_i - L_i$ and decreasing U_S by $U_i - w_i$. Let $L_S(Q)$ and $U_S(Q)$ denote the limits of set S *after* querying a set of intervals $Q \subseteq \mathcal{I}$.

For a MINSET instance $(\mathcal{I}, \mathcal{S})$, let $w^* = \min_{S \in \mathcal{S}} w(S)$ be the initially uncertain minimum set value. To solve the problem, we have to adaptively query a set of intervals Q until $U_{S^*}(Q) = L_{S^*}(Q) = w^*$ holds for some $S^* \in \mathcal{S}$ and $L_S(Q) \geq w^*$ holds for all $S \in \mathcal{S}$. Only then, we know for sure that w^* is indeed the minimum set value and that S^* achieves this value. The following ILP with $a_i = w_i - L_i$ for all $I_i \in \mathcal{I}$ and $b_S = w^* - L_S$ for all $S \in \mathcal{S}$ formulates this problem:

$$
\begin{aligned}
\min \ & \textstyle\sum_{I_i \in \mathcal{I}} x_i \\
\text{s.t.} \ & \textstyle\sum_{I_i \in S} x_i \cdot a_i \geq b_S \ \forall S \in \mathcal{S} \\
& x_i \in \{0,1\} \qquad \forall I_i \in \mathcal{I}
\end{aligned}
\qquad \text{(MINSETIP)}
$$

Observe that this ILP is a special case of the *multiset multicover problem* (see, e.g., [26]). If $a_i = w_i - L_i = 1$ for all $I_i \in \mathcal{I}$ and $b_S = w^* - L_S = 1$ for all $S \in \mathcal{S}$, then the problem is exactly the classical SETCOVER problem with $\mathcal{I}$ corresponding to the SETCOVER *sets* and $\mathcal{S}$ corresponding to the SETCOVER *elements*.

The optimal solution to (MINSETIP) is the optimal query set for the corresponding MINSET instance. Under uncertainty however, the coefficients $a_i = w_i - L_i$ and right-hand sides $b_S = w^* - L_S$ are unknown. We only know that $a_i \in (L_i - L_i, U_i - L_i) = (0, U_i - L_i)$ as $a_i = (w_i - L_i)$ and $w_i \in (L_i, U_i)$. In a sense, to solve MINSET under uncertainty, we have to solve (MINSETIP) with uncertainty in the coefficients and irrevocable decisions. For the rest of the paper, we interpret MINSET under uncertainty in exactly that way: We have to solve (MINSETIP) without knowing the coefficients in the constraints. Whenever we *irrevocably* add an interval I_i to our solution (i.e., set x_i to 1), the information on the coefficients (in form of w_i) is revealed. Our goal is to add elements to our solution until it becomes feasible for (MINSETIP), and to minimize the number of added elements. In this interpretation, the terms "querying an element" and "adding it to the solution" are interchangeable, and we use them as such.

Our main contribution is an algorithmic framework that exploits techniques for classical covering problems and adapts them to handle uncertainty in the coefficients a_i and the right-hand sides b_S. This framework allows us to obtain improved results for MINSET under uncertainty and other covering problems.

Our Results. We design a polynomial-time algorithm for MINSET under stochastic uncertainty with competitive ratio $\mathcal{O}(\frac{1}{\tau} \cdot \log^2 m)$, where m is the number of sets (number of constraints in (MINSETIP)) and parameter τ characterizes how "balanced" the distributions of values within the given intervals are. More precisely, $\tau = \min_{I_i \in \mathcal{I}} \tau_i$ and τ_i is the probability that w_i is larger than the center of I_i (e.g., for uniform distributions $\tau = \frac{1}{2}$). This is the first stochastic result in explorable uncertainty concerning the sum of unknown values and it builds on new methods that shall be useful for solving more general problems in this field. The ratio is independent of the number of elements, n. In particular for a small number of sets, m, this is a significant improvement upon the adversarial lower bound of n [11]. Dependencies on parameters such as τ are quite standard and necessary [3, 4, 15, 22, 29]. For example, in [22] the upper bounds depend on the probability to draw the largest value of the uncertainty interval, which is an even stricter assumption that does not translate to open intervals. Our results translate also to the maximization variant of MinSet; see full version [24].

We remark that the hidden constants in the performance bounds depend on the upper limits of the given intervals. Assuming those to be constant is also a common assumption; see, e.g., [22]. Even greedy algorithms for covering problems similar to (MinSetIP) *without* uncertainty have such dependencies [9, 26, 28].

As MinSet contains the classical SetCover, an approximation factor better than $\mathcal{O}(\log m)$ is unlikely, unless P=NP [8]. We show that this holds also in the stochastic setting, even for uniform distributions. We also show a lower bound of $\frac{1}{\tau}$ for MinSet under stochastic explorable uncertainty, even for pairwise disjoint sets. Thus, the dependency on $\log m$ and $\frac{1}{\tau}$ in our results is necessary.

In the special case that all given sets are disjoint, we provide a simpler algorithm with competitive ratio $\frac{2}{\tau}$. This is a gigantic improvement compared to the adversarial setting, where the lower bound of n holds even for disjoint sets [11].

Algorithmically, we exploit the covering point of view to introduce a class of greedy algorithms that use the same basic strategy as the classical SetCover greedy algorithm [7]. However, we do not have sufficient information to compute and query an exact greedy choice under uncertainty as this choice depends on uncertain parameters. Instead, we show that it is sufficient to query a small number of elements that together achieve a similar greedy value to the exact greedy choice. If we do this repeatedly and the number of queries per iteration is small in expectation, then we achieve guarantees comparable to the approximation factor of a greedy algorithm with full information. It is worth noting that this way of comparing an algorithm to the optimal solution is a novelty in explorable uncertainty as all previous algorithms for adversarial explorable uncertainty (MinSet and other problems) exploit *witness sets*. A witness set is a set of queries Q such that each feasible solution has to query at least one element of Q, which allows to compare an algorithm with an optimal solution.

Our results translate to other covering problems under uncertainty. In particular, for (MinSetIP) under uncertainty with *deterministic right-hand sides*, we give a simplified algorithm with improved competitive ratio $\mathcal{O}(\frac{1}{\tau} \cdot \log m)$. For a slightly different balancing parameter, this holds even for the more general variant, where a variable can have different coefficients for different constraints, each with an individual uncertainty interval and distribution; see full version [24].

All missing proofs are provided in the full version [24].

Further Previous Work. For adversarial MinSet under uncertainty, Erlebach et al. [11] show a (best possible) competitive ratio of $2d$, where d is the cardinality of the largest set. In the lower bound instances, $d \in \Omega(n)$. The algorithm repeatedly queries disjoint witness sets of size at most $2d$. This result was stated for the setting, in which it is not necessary to determine the value of the minimal set; if the value has to be determined, the bounds change to d.

Further related work on MINSET includes the result by Yamaguchi and Maehara [22], who consider packing ILPs with stochastic uncertainty in the cost coefficients, which can be queried. They present a framework for solving several problems and bound the absolute number of iterations that it requires to solve them, instead of the competitive ratio. However, we show in the full version [24] that their algorithm has competitive ratio $\Omega(n)$ for MINSET, even for uniform distributions. Thus, it does not improve upon the adversarial lower bound.

Wang et al. [30] also consider selection-type problems in a somewhat related model. They consider different constraints on the set of queries that, in a way, imply a budget on the number of queries. They solve optimization problems with respect to this budget, which has a very different flavor than our setting.

While we are not aware of previous work on covering problems with value-queries and uncertainty in the constraints, there is related work on queries that reveal the *existence* of edges in a graph instead of numeric values [3,4,15,29]. Furthermore, there is related work on covering problems in different stochastic settings (see, e.g., [1,14,16,27]).

2   Algorithmic Framework

In this section, we present our algorithmic framework. To illustrate the main ideas, we first consider the offline variant of MINSET and give hardness results.

2.1   Offline Problems and Hardness of Approximation

We refer to the problem of solving (MINSETIP) with full knowledge of the precise values w_i (and w^*) as *offline*. For MINSET under uncertainty, we say a solution is optimal, if it is an optimal solution for the corresponding offline problem. We use OPT to refer to an optimal solution and its objective value.

Offline MINSET contains SETCOVER and, thus, is as hard to approximate. This result transfers to the stochastic setting, even with uniform distributions. Thus, an approximation factor better than $\mathcal{O}(\log m)$ is unlikely, unless P=NP [8].

On the positive side, we can approximate offline MINSET by adapting covering results [7,9,20,21,26]. In particular, we want to use greedy algorithms that iteratively and irrevocably add elements to the solution that are selected by a certain greedy criterion. Recall that "adding an element to the solution" corresponds to both, setting the variable x_i of an interval $I_i \in \mathcal{I}$ in (MINSETIP) to one and querying I_i. As the greedy criterion for adding an element depends on previously added elements, we define a version of the ILP parametrized by the set Q of elements that have already been added to the solution and adjust the right-hand sides to the remaining covering requirements after adding Q. To that end, let $b_S(Q) = \max\{b_S - \sum_{I_i \in Q \cap S} a_i, 0\}$ for $a_i = w_i - L_i$ and $b_S = w^* - L_S$.

$$
\begin{aligned}
\min \ & \sum_{I_i \in \mathcal{I} \setminus Q} x_i \\
\text{s.t.} \ & \sum_{I_i \in S \setminus Q} x_i \cdot a_i \geq b_S(Q) \ \forall S \in \mathcal{S} \qquad \text{(MINSETIP-Q)}\\
& x_i \qquad\qquad\quad \in \{0,1\} \ \forall I_i \in \mathcal{I} \setminus Q
\end{aligned}
$$

Based on the ILP and the sum of right-hand sides $b(Q) = \sum_{S \in \mathcal{S}} b_S(Q)$, we adjust an algorithm for multiset multicover by Dobson [9] to our setting.

The Offline Algorithm scales the coefficients to a' and b' such that all non-zero left-hand side coefficients are at least 1 (we refer to such instances as *scaled*). Then it greedily adds the element to the solution that reduces the right-hand sides the most, i.e., the interval $I_i \in \mathcal{I} \setminus Q$ that maximizes the *greedy value* $g_c(Q, I_i) = b'(Q) - b'(Q \cup \{I_i\})$. For a subset $G \subseteq \mathcal{I}$, we define $g_c(Q, G) = b'(Q) - b'(Q \cup G)$.

After $b'_S(Q) < 1$ for all $S \in \mathcal{S}$, we can exploit that all non-zero coefficients a'_i are at least one. This means that adding an element $I_i \in \mathcal{I} \setminus Q$ satisfies all remaining constraints of sets S with $I_i \in S$. Thus, the remaining problem reduces to a SETCOVER instance, which can be solved by using the classical greedy algorithm by Chvatal [7]. This algorithm greedily adds the element I_i that maximizes greedy value $g_s(Q, I_i) = A(Q) - A(Q \cup \{I_i\})$ with $A(Q) = |\{S \in \mathcal{S} \mid b'_S(Q) > 0\}|$, i.e., the element that satisfies the largest number of constraints that are not already satisfied by Q. For subsets $G \subseteq \mathcal{I}$, we define $g_s(Q, G) = A(Q) - A(Q \cup G)$.

Theorem 1 (Follows from *[9]*). *The Offline Algorithm is a polynomial-time $\mathcal{O}(\log m)$-approximation for offline MINSET. The precise approximation factor is $\rho(\gamma) = \lceil \ln(\gamma \cdot m \cdot \max_S(w^* - L_S)) \rceil + \lceil \ln(m) \rceil$ with $s_{\min} = \min_{I_i \in \mathcal{I}: (w_i - L_i) > 0}(w_i - L_i)$, $\gamma = 1/s_{\min}$ and $m = |\mathcal{S}|$.*

We will state the competitive ratios of our algorithms in terms of ρ. To that end, define $\bar{\rho}(\gamma) = \lceil \ln(\gamma \cdot m \cdot \max_{S,S'}(U_S - L_{S'})) \rceil + \lceil \ln(m) \rceil$, which is an upper bound on $\rho(\gamma)$. Under uncertainty, we compare against $\bar{\rho}$ to avoid the random variable w^*. For constant U_i's, $\bar{\rho}$ and ρ are asymptotically the same.

2.2 Algorithmic Framework

To solve MINSET under uncertainty, we ideally would like to apply the Offline Algorithm. However, since the coefficients $a_i = w_i - L_i$ and $b_S = w^* - L_S$ are unknown, we cannot do so as we cannot compute the greedy values g_c or g_s.

While we cannot precisely compute the greedy choice, our strategy is to *approximate* it and to show that approximating it is sufficient to obtain the desired guarantees. To make this more precise, consider an *iterative* algorithm for (MINSETIP) that iteratively adds pairwise disjoint subsets $G_1, \ldots, G_h$ of $\mathcal{I}$ to the solution. For each j, let $Q_j = \bigcup_{1 \le j' \le j-1} G_{j'}$, i.e., Q_j contains the elements that have been added to the solution before G_j. If the combined greedy value of G_j is within a factor of α to the best greedy value for the problem instance *after* adding Q_j, then we say that G_j α-approximates the greedy choice. The following definition makes this more precise while taking into account that there are two different greedy values g_c and g_s (cf. the Offline Algorithm).

Definition 1. *For a scaled instance of (MINSETIP), some $Q \subseteq \mathcal{I}$, and an $\alpha \ge 1$, a subset $G \subseteq \mathcal{I} \setminus Q$ α-approximates the current greedy choice (as characterized by Q) if one of the following conditions holds:*

1. $b'_S(Q) < 1$ *for all* $S \in \mathcal{S}$ *and* $g_s(Q, G) \geq \frac{1}{\alpha} \cdot \max_{I_i \in \mathcal{I} \setminus Q} g_s(Q, I_i)$.
2. $b'(Q) \geq 1$ *and* $g_c(Q, G) \geq \frac{1}{\alpha} \cdot \max_{I_i \in \mathcal{I} \setminus Q} g_c(Q, I_i)$.

We bound the number of iterations j in which G_j α-approximates the current greedy choice via an adjusted SETCOVER greedy analysis.

Lemma 1. *Consider an arbitrary algorithm for (*MINSETIP*) that scales the coefficients by factor* γ *and iteratively adds disjoint subsets* $G_1, \ldots, G_h$ *of* $\mathcal{I}$ *to the solution until the instance is solved. The number of groups* G_j *that* α-*approximate the current greedy choice (after adding* Q_j*) is at most* $\alpha \cdot \rho(\gamma) \cdot \mathrm{OPT}$.

The lemma states that the number of such groups G_j is within a factor of α of the performance guarantee $\rho(\gamma)$ of the offline greedy algorithm. If each G_j α-approximates its greedy choice, the iterative algorithm achieves an approximation factor of $\max_j |G_j| \cdot \alpha \cdot \rho(\gamma)$. Thus, approximating the greedy choices by a constant factor using a constant group size is sufficient to only lose a constant factor compared to the offline greedy algorithm.

This insight gives us a framework to solve MINSET under uncertainty. Recall that the w_i's (and by extension the a_i's and b_S's) are uncertain and only revealed once we irrevocably add an $I_i \in \mathcal{I}$ to the solution. We refer to a revealed w_i as a *query result*, and to a fixed set of revealed w_i's for all $I_i \in \mathcal{I}$ as a *realization of query results*. Consider an iterative algorithm. The sets G_j can be computed and queried adaptively and are allowed to depend on (random) query results from previous iterations. Hence, $X_j = |G_j|$ is a random variable. Let Y_j be an indicator variable denoting whether the algorithm executes iteration j ($Y_j = 1$) or terminates beforehand ($Y_j = 0$). We define the following class of iterative algorithms and show that algorithms from this class achieve certain guarantees.

Definition 2. *An iterative algorithm is* (α, β, γ)-GREEDY *if it satisfies:*

1. *For every realization of query results; each* G_j α-*approximates the greedy choice as characterized by* Q_j *on the instance with coefficients scaled by* γ.
2. $\mathbb{E}[X_j \mid Y_j = 1] \leq \beta$ *holds for all iterations* j.

Theorem 2. *Each* (α, β, γ)-GREEDY *algorithm for* MINSET *under uncertainty achieves a competitive ratio of* $\alpha \cdot \beta \cdot \bar{\rho}(\gamma) \in \mathcal{O}(\alpha \cdot \beta \cdot \log m)$.

Proof. Consider an (α, β, γ)-GREEDY algorithm ALG for MINSET. Its expected cost is $\mathbb{E}[\mathrm{ALG}] = \sum_j \mathbb{E}[X_j] = \sum_j \mathbb{P}[Y_j = 1]\,\mathbb{E}[X_j \mid Y_j = 1] + \mathbb{P}[Y_j = 0]\,\mathbb{E}[X_j \mid Y_j = 0]$. As $\mathbb{E}[X_j \mid Y_j = 0] = 0$ (if the algorithm terminates before iteration j, it adds no more elements and, thus, $X_j = 0$), the equality reduces to $\mathbb{E}[\mathrm{ALG}] = \sum_j \mathbb{P}[Y_j = 1]\,\mathbb{E}[X_j \mid Y_j = 1]$. By Definition 2, this implies $\mathbb{E}[\mathrm{ALG}] \leq \beta \sum_j \mathbb{P}[Y_j = 1]$.

It remains to bound $\sum_j \mathbb{P}[Y_j = 1]$, which is the expected number of iterations of ALG. Consider a fixed realization of query results. By the first property of Definition 2, each G_j α-approximates its greedy choice for the (MINSETIP) instance of the realization scaled by factor γ. Thus, Lemma 1 implies that the number of iterations is at most $\alpha\rho(\gamma)\mathrm{OPT}$, which is upper bounded by $\alpha\bar{\rho}(\gamma)\mathrm{OPT}$. As

---

Algorithm 1: MINSET with deterministic right-hand sides.

Input: Instance of MINSET with deterministic right-hand sides.

1 $Q = \emptyset$; Scale a and b by $\frac{2}{s_{\min}}$ to a' and b' for $s_{\min} = \min_{I_i \in \mathcal{I}: U_i - L_i > 0} U_i - L_i$;

2 **while** *the problem is not solved* **do**

3   **if** $b'(Q) \geq 1$ **then** $g = \bar{g}_c$ **else** $g = \bar{g}_s$;

4   **repeat**

5 | $I_i \leftarrow \arg\max_{I_j \in \mathcal{I} \setminus Q} g(Q, I_j)$; Query I_i; $Q \leftarrow Q \cup \{I_i\}$;

6   **until** *the problem is solved or* $w_i - L_i \geq \frac{1}{2} \cdot (U_i - L_i)$;

---

this upper bound on the number of iterations holds for every realization and OPT is the only random variable of that term (as we substituted ρ by $\bar{\rho}$), we get $\sum_j \mathbb{P}[Y_j = 1] \leq \alpha \bar{\rho}(\gamma) \mathbb{E}[\text{OPT}]$, which implies $\mathbb{E}[\text{ALG}] \leq \alpha \beta \bar{\rho}(\gamma) \mathbb{E}[\text{OPT}]$. □

3 MINSET with Deterministic Right-Hand Sides

We consider a variant of MINSET under uncertainty, where the right-hand sides b_S of the ILP representation (MINSETIP) are deterministic and explicit part of the input. Thus, only the coefficients $a_i = (w_i - L_i)$ remain uncertain within the interval $(0, U_i - L_i)$. For this variant, the instance might have no feasible solution. In that case, we require every algorithm (incl. OPT) to reduce the covering requirements as much as possible. Recall that in the stochastic setting the balancing parameter is $\tau = \min_{I_i \in \mathcal{I}} \tau_i$ for $\tau_i = \mathbb{P}[w_i \geq \frac{U_i + L_i}{2}]$.

Theorem 3. *There is a polynomial-time algorithm for* MINSET *under uncertainty with deterministic right-hand sides and a competitive ratio of* $\frac{2}{\tau} \cdot \rho(\gamma) \in \mathcal{O}(\frac{1}{\tau} \cdot \log m)$ *with* $\gamma = 2/s_{\min}$ *for* $s_{\min} = \min_{I_i \in \mathcal{I}: U_i - L_i > 0} U_i - L_i$.

The algorithm loses only a factor $\frac{2}{\tau}$ compared to the greedy approximation factor $\rho(\gamma)$ on the offline problem. We show the theorem by proving that Algorithm 1 is an (α, β, γ)-GREEDY algorithm for $\alpha = 2$, $\beta = \frac{1}{\tau}$ and $\gamma = \frac{2}{s_{\min}}$ with $s_{\min} = \min_{I_i \in \mathcal{I}: U_i - L_i > 0} U_i - L_i$. Using Theorem 2, this implies the theorem.

The algorithm scales the instance by factor γ; a' and b' refer to the scaled coefficients. The idea is to execute the Offline Algorithm under the assumption that $a_i = U_i - L_i$ (and $a'_i = \gamma a_i$) for all $I_i \in \mathcal{I}$ that were not yet added to the solution. As $a_i = (w_i - L_i) \in (0, U_i - L_i)$, this means that we assume a_i to have the largest possible value. Consequently, $s_{\min}$ is the smallest (non-zero) coefficient a_i under this assumption. The algorithm computes the greedy choice based on the *optimistic greedy values*

$$\bar{g}_c(Q, I_i) = \sum_{S \in \mathcal{S}: I_i \in S} b'_S(Q) - \max\{0, b'_S(Q) - \gamma(U_i - L_i)\}$$

(if $b'(Q) \geq 1$) and

$$\bar{g}_s(Q, I_i) = |\{S \in \mathcal{S}: I_i \in S \mid b'_S(Q) > 0 \wedge b'_S(Q) - \gamma(U_i - L_i) \leq 0\}|$$

(otherwise). These are the greedy values under the assumption $a_i = U_i - L_i$. We call them optimistic as they might overestimate but never underestimate the actual greedy values. For subsets $G \subseteq \mathcal{I}$, we define $\bar{g}_s(Q, G)$ and $\bar{g}_c(Q, G)$ analogously.

In contrast to g_s and g_c, Algorithm 1 has sufficient information to compute $\bar{g}_s$ and $\bar{g}_c$, and the best greedy choice based on the optimistic greedy values. The algorithm is designed to find, in each iteration, an element I_i with $w_i - L_i \geq \frac{U_i - L_i}{2}$. We show that (i) this ensures that each iteration 2-approximates the greedy choice and (ii) that finding such an element takes only $\frac{1}{\tau}$ tries in expectation. This suffices to apply Theorem 2.

To show (ii), we can observe that $w_i \geq \frac{(U_i + L_i)}{2}$ implies $w_i - L_i \geq \frac{1}{2}(U_i - L_i)$ and that $\mathbb{P}[w_i \geq \frac{(U_i + L_i)}{2}] \geq \tau$ holds by assumption. Thus, we find an interval I_i satisfying $w_i - L_i \geq \frac{1}{2}(U_i - L_i)$ with probability at least τ. This implies that, given an iteration of the while-loop is started, it in expectation takes $\frac{1}{\tau}$ tries to find such an interval I_i. This is exactly the second property of Definition 2.

To prove (i), we use the next lemma, which shows that the optimistic greedy value of an I_i with $w_i - L_i \geq \frac{1}{2}(U_i - L_i)$ is close to the actual greedy value.

Lemma 2. *Consider an instance of (*MINSETIP*) scaled by* $\gamma = \frac{2}{s_{\min}}$ *and some* $Q \subseteq \mathcal{I}$*. If* $w_i - L_i \geq (U_i - L_i)/2$ *for an* $I_i \in \mathcal{I} \setminus Q$*, then* $g_c(Q, I_i) \geq \bar{g}_c(Q, I_i)/2$*. If additionally* $b'(Q) < 1$*, then* $g_s(Q, I_i) = \bar{g}_s(Q, I_i)$

Proof. The statement regarding g_c and $\bar{g}_c$ holds directly by definition. To show the second statement, we use the assumption $b'(Q) < 1$ and the choice of γ.

From $b'(Q) < 1$ follows $b'_S(Q) < 1$ for all $S \in \mathcal{S}$. As $w_i - L_i \geq \frac{U_i - L_i}{2}$, we have $a_i = w_i - L_i \geq \frac{U_i - L_i}{2} \geq \frac{s_{\min}}{2}$ and, therefore, $a'_i = \gamma a_i = \frac{2}{s_{\min}} a_i \geq 1$. This means that adding I_i to the solution satisfies *all* constraints for sets S with $I_i \in S$ that are not yet satisfied by Q. Thus, the optimistic greedy value $\bar{g}_s(Q, I_i)$ and the real greedy value $g_s(Q, I_i)$ are the same, i.e., $\bar{g}_s(Q_j, I_i) = g_s(Q, I_i)$, as adding I_i cannot satisfy more constraints even if the coefficient a_i was $U_i - L_i$. $\square$

As the algorithm always queries the interval I_i with the best optimistic greedy value, Lemma 2 shows that the last query of the iteration 2-approximates the greedy choice after querying the set Q of all previous queries. This implies (i) and Property 1 of Definition 2. By Theorem 2 this suffices to prove Theorem 3.

4   MINSET Under Uncertainty

We consider MINSET under uncertainty and prove the following main result.

Theorem 4. *There is a polynomial-time algorithm for* MINSET *under uncertainty with a competitive ratio of* $\mathcal{O}(\frac{1}{\tau} \log m \cdot \bar{\rho}(\gamma)) \subseteq \mathcal{O}(\frac{1}{\tau} \cdot \log^2 m)$ *with* $\gamma = 2/s_{\min}$ *for* $s_{\min} = \min_{I_i \in \mathcal{I}: U_i - L_i > 0} U_i - L_i$*.*

Exploiting Theorem 2, we prove the statement by providing Algorithm 2 and showing that it is an (α, β, γ)-GREEDY algorithm for $\alpha = 2$, $\gamma = 2/s_{\min}$ and

$\beta = \frac{1}{\tau}(\lceil \log_{1.5}(m \cdot 2(\max_{I_i \in \mathcal{I}}(U_i - L_i))/s_{\min}) \rceil + \lceil \log_2(m) \rceil)$. Note that α and γ are defined as in the previous section and will be used analogously. For β on the other hand, we require a larger value to adjust for the additional uncertainty in the right-hand sides $b_S = w^* - L_S$ as the minimum set value w^* is unknown. Notice that we do not have sufficient information to execute Algorithm 1, since we need the right-hand side values to compute even the optimistic greedy values.

To handle this additional uncertainty, we want to ensure that each iteration of our algorithm α-approximates the greedy choice for *each* possible value of w^*. To do so, we compute and query the best optimistic greedy choice for several carefully selected possible values w^*. To state our algorithm, we define a parametrized variant of (MINSETIP) that states the problem under the assumptions that $w^* = w$ for some w and that the set $Q \subseteq \mathcal{I}$ has already been queried. The coefficients are scaled to $a'_i = (2/s_{\min})(w_i - L_i)$ and $b'_S(Q, w) = \max\{(2/s_{\min})(w - L_S) - \sum_{I_i \in Q \cap S} a'_i, 0\}$. As before, let $b'(Q, w) = \sum_{S \in \mathcal{S}} b'_S(Q, w)$.

$$
\begin{aligned}
\min \quad & \textstyle\sum_{I_i \in \mathcal{I} \setminus Q} x_i \\
\text{s.t.} \quad & \textstyle\sum_{I_i \in S \setminus Q} x_i \cdot a'_i \geq b'_S(Q, w) \; \forall S \in \mathcal{S} \qquad\qquad \text{(MINSETIP-Qw)} \\
& x_i \in \{0, 1\} \quad \forall I_i \in \mathcal{I}
\end{aligned}
$$

As the right-hand sides are unknown, we define the greedy values for every possible value w for w^*. To that end, let $g_c(Q, I_i, w) = b'(Q, w) - b'(Q \cup \{I_i\}, w)$ and $g_s(Q, I_i, w) = A(Q, w) - A(Q \cup \{I_i\}, w)$, where $A(Q, w) = |\{S \in \mathcal{S} \mid b'_S(Q, w) > 0\}|$. As before, $g_c(Q, I_i, w)$ and $g_s(Q, I_i, w)$ describe how much adding I_i to the solution reduces the sum of right-hand sides and the number of non-satisfied constraints, respectively; now under the assumption that $w^* = w$. The optimistic greedy values $\bar{g}_c(Q, I_i, w)$ and $\bar{g}_s(Q, I_i, w)$ for an $I_i \in \mathcal{I}$ are defined analogously but again assume that $a_i = U_i - L_i$.

Similar to Algorithm 1, we would like to repeatedly compute and query the best optimistic greedy choice until the queried I_i satisfies $w_i - L_i \geq \frac{U_i - L_i}{2}$ (cf. the repeat-statement). However, we cannot decide which greedy value, $\bar{g}_c$ or $\bar{g}_s$, to use as deciding whether $b'_S(Q, w^*) < 1$ depends on the unknown w^*. Instead, we compute and query the best optimistic greedy choice for both greedy values (cf. the for-loop). Even then, the best greedy choice still depends on the unknown right-hand sides. Thus, we compute and query the best optimistic greedy choice for several carefully selected values w (cf. the inner while-loop) to make sure that the queries of the iteration approximate the greedy choice for every possible w^*. Additionally, we want to ensure that we use at most β queries in expectation.

Consider an iteration of the outer while-loop with $g = \bar{g}_c$, and let Q' denote the set of queries that were executed before the start of the iteration. Since we only care about the greedy value g_c if there exists some $S \in \mathcal{S}$ with $b'_S(Q) > 1$ (otherwise we use $\bar{g}_s$ and g_s instead), we assume that this is the case. If not, we use a separate analysis for the for-loop iteration with $g = \bar{g}_s$.

Our goal for the iteration is to query a set of intervals $\bar{Q}$ that 2-approximates the best greedy choice I^* after querying Q, i.e., it has a greedy value $g_c(\bar{Q}, Q, w^*) \geq \frac{1}{2} g_c(I^*, Q, w^*)$. To achieve this for the unknown w^*, the algorithm uses the parameter d, which is initialized with 1 (cf. Line 5), the minimum

---

Algorithm 2: Algorithm for MINSET under uncertainty.

Input: Instance of MINSET under uncertainty.

1 Scale all coefficients with $\gamma = 2/s_{\min}$ for $s_{\min} = \min_{I_i \in \mathcal{I}:\ (U_i - L_i) > 0}(U_i - L_i)$;

2 $Q \leftarrow \emptyset$, $w_{\min} \leftarrow$ minimum possible value w^* (keep up-to-date);

3 **while** *the problem is not solved* **do**

4   **foreach** g *from the ordered list* $\bar{g}_c, \bar{g}_s$ **do**

5           $d \leftarrow 1$; $Q' \leftarrow Q$;

6   **if** $g = \bar{g}_c$ **then** $w_{\max} \leftarrow$ max possible value w^*;

7   **else** $w_{\max} \leftarrow$ max w s.t. $b'_S(Q, w) < 1$ for all $S \in \mathcal{S}$;

8   **while** $\exists w_{\min} \leq w \leq w_{\max}$ *such that* $\max_{I_h \in \mathcal{I} \setminus Q} g(Q, I_h, w) \geq d$ **do**

9   **repeat**

10                   $w \leftarrow \min w_{\min} \leq w \leq w_{\max}$ s.t. $\max_{I_h \in \mathcal{I} \setminus Q} g(Q, w, I_h) \geq d$;

11                   $I_i \leftarrow \arg\max_{I_h \in \mathcal{I} \setminus Q} g(Q, I_h, w)$; Query I_i; $Q \leftarrow Q \cup \{I_i\}$;

12                   $Q_{1/2} \leftarrow \{I_j \in Q \setminus Q' \mid w_j - L_j \geq \frac{U_j - L_j}{2}\}$;

13   **if** $g = \bar{g}_c$ **then** $d \leftarrow g_c(Q', Q_{1/2}, w)$ **else** $d \leftarrow g_s(Q', Q_{1/2}, w)$;

14   **until** $w_i - L_i \geq \frac{U_i - L_i}{2}$ *or* $\nexists w \leq w_{\max}$: $\max_{I_h \in \mathcal{I} \setminus Q} g(Q, w, I_h) \geq d$;

---

possible value for $\bar{g}_c(I^*, Q, w^*)$ under the assumption that there exists some $S \in \mathcal{S}$ with $b'_S(Q) > 1$. In an iteration of the inner while-loop, the algorithm repeatedly picks the minimal value w such that the best current optimistic greedy choice has optimistic greedy value of at least d (cf. Line 10). If no such value exists, then the loop terminates (cf. Lines 8, 14). Afterwards, it queries the corresponding best optimistic greedy choice I_i for the selected value w (cf. Line 11). Similar to the algorithms of the previous section, this is done repeatedly until $w_i - L_i \geq (U_i - L_i)/2$.

The key idea to achieve the 2-approximation with an expected number of queries that does not exceed β, is to always reset the value d to $g_c(Q', Q_{1/2}, w)$, where $Q_{1/2}$ is the subset of all intervals I_j that have already been queried in the current iteration of the outer while-loop and satisfy $w_j - L_j \geq (U_j - L_j)/2$ (cf. Lines 12, 13). This can be seen as an implicit doubling strategy to search for an unknown value. It leads to an exponential increase of d over the iterations of the inner while-loop, which will allow us to bound their number. With the following lemma, we prove that this choice of d also ensures that the queries of the iteration indeed 2-approximate the best greedy choice for w^* if there exists a $S \in \mathcal{S}$ with $b'_S(Q', w^*) \geq 1$. If there is no such set, we can use a similar proof. For an iteration j of the outer while-loop, let G_j be the set of queries during the iteration and let $Q_j = \bigcup_{j' < j} G_{j'}$ denote the queries before the iteration (cf. Q' in the algorithm).

Lemma 3. *If there is an $S \in \mathcal{S}$ with $b'_S(Q_j, w^*) \geq 1$, then G_j 2-approximates the greedy choice for the scaled instance with $w = w^*$ after querying Q_j.*

Proof. Consider the subset $\bar{G}_j \subseteq G_j$ of queries that were executed with $g = \bar{g}_c$ before the increasing value w (cf. Line 10) surpasses w^*. That is, $\bar{G}_j$ only contains intervals that were queried for a current value $w \leq w^*$. Let $\bar{I}_i$ be the element of $\bar{G}_j$ that is queried last. Finally, let $\bar{d}_j$ denote the value d computed by the algorithm in Line 13 directly after querying $\bar{I}_i$. We continue to show that G_j 2-approximates the greedy choice of (MINSETIP-QW) for $Q = Q_j$ and $w = w^*$.

Observe that $g_c(Q_j, \bar{G}_j, w^*) \geq \bar{d}_j$. To see this, recall that $\bar{d}_j$ was computed in Line 13 after $\bar{I}_i$ was queried. Thus, $\bar{d}_j = g_c(Q', Q_{1/2}, w)$ for $Q' = Q_j$, $Q_{1/2} = \{I_j \in \bar{G}_j \mid w_j - L_j \geq \frac{U_j - L_j}{2}\}$ and some value w with $w \leq w^*$ by assumption. Since $w^* \geq w$ and $Q_{1/2} \subseteq \bar{G}_j$, the greedy value $g_c(Q_j, \bar{G}_j, w^*)$ can never be smaller than $\bar{d}_j = g_c(Q', Q_{1/2}, w)$, which implies $g_c(Q_j, \bar{G}_j, w^*) \geq \bar{d}_j$.

We continue by showing that $d^* \leq 2 \cdot g_c(Q_j, \bar{G}_j, w^*)$ holds for the best greedy value d^* at the start of the iteration, i.e., $d^* = \max_{I_i \in \mathcal{I} \setminus Q_j} g_c(Q_j, I_i, w^*)$. As $\bar{G}_j \subseteq G_j$, this implies $d^* \leq 2 \cdot g_c(Q_j, G_j, w^*)$ and G_j satisfies Definition 1.

To upper bound d^*, first observe that the best optimistic greedy value d' after querying $\bar{G}_j \cup Q_j$ is smaller than $\bar{d}_j$, i.e., $d' = \max_{I_i \in \mathcal{I} \setminus (Q_j \cup \bar{G}_j)} \bar{g}_c(Q_j \cup \bar{G}_j, I_i, w^*) < \bar{d}_j$. This follows directly from Line 10 as $\bar{I}_i$ is the last query for a value $w \leq w^*$ by assumption. As $g_c(Q_j, \bar{G}_j, w^*) \geq \bar{d}_j$, we get $g_c(Q_j, \bar{G}_j, w^*) \geq d'$.

By definition of g_c, the best greedy value after querying Q_j can never be larger than the sum of the greedy value of $\bar{G}_j$ after querying Q_j and the best greedy value after querying $\bar{G}_j \cup Q_j$. Thus, we have $d^* \leq g_c(Q_j, \bar{G}_j, w^*) + d' \leq 2 \cdot g_c(Q_j, \bar{G}_j, w^*)$, which concludes the proof. $\square$

Using a similar proof, we show an analogous lemma for the case where $b'_S(Q', w^*) < 1$ for all $S \in \mathcal{S}$, which then implies Property 1 of Definition 2. To prove Theorem 4, it remains to show Property 2 of Definition 2. The proof idea is to show that parameter d increases by a factor of at least 1.5 in each iteration of the inner while-loop. As $\bar{g}_c(Q, I_i, w)$ is upper bounded by $m(2/s_{\min}) \max_{I_i \in \mathcal{I}}(U_i - L_i)$, this means the inner loop executes at most $\lceil \log_{1.5}(m(2/s_{\min}) \max_{I_i \in \mathcal{I}}(U_i - L_i)) \rceil$ iterations for $g = \bar{g}_c$. For $g = \bar{g}_s$, we can argue in a similar way that at most $\lceil \log_2(m) \rceil$ iterations are executed. Similar to the previous section, we can also show that each iteration of the inner while-loop executes at most $\frac{1}{\tau}$ queries in expectation. Combining these insights, we can bound the expected number of queries during an execution of the outer while-loop by β. Formally proving the increase of d by a factor of 1.5 requires to take care of several technical challenges. The basic idea is to exploit that the interval I_i queried in Line 11 has an optimistic greedy value of at least d. If it satisfies $w_i - L_i \geq (U_i - L_i)$, Lemma 2 implies that the real greedy value is at least $d/2$. When d is recomputed in Line 13, then I_i is a new member of the set $Q_{1/2}$ and leads to the increase of $d/2$.

5   Disjoint MINSET

In *disjoint* MINSET, all given sets are pairwise disjoint, i.e., $S \cap S' = \emptyset$ for all $S, S' \in \mathcal{S}$ with $S \neq S'$. Disjoint MINSET is of particular interest as it gives lower bounds for several problems under adversarial explorable uncertainty, cf. [11, 25].

Theorem 5. *There is a polynomial-time algorithm for disjoint MINSET under uncertainty with competitive ratio $\frac{2}{\tau}$ and a nearly matching lower bound of $\frac{1}{\tau}$.*

In disjoint MINSET, each I_i occurs in exactly one constraint for one set S in the corresponding (MINSETIP). Thus, each set S defines a disjoint subproblem and the optimal solution OPT of the instance is the union of optimal solutions for the subproblems. The optimal solution for a subproblem S is to query the elements of $I_i \in S$ in order of non-decreasing $(w_i - L_i)$ until the sum of those coefficients is at least $(w^* - L_S)$. Under uncertainty, we adapt this strategy and query in order of non-decreasing $(U_i - L_i)$. Since we do not know w^*, we do not know when to stop querying in a subproblem. We handle this by only querying in the set S of minimum current lower limit as the subproblem for this set is clearly not yet solved. Combining these insights with the ideas of the previous sections, one can prove that this algorithm achieves the upper bound.

Final remarks

We provide the first results for MINSET under stochastic explorable uncertainty and break, in expectation, adversarial lower bound instances for a number of problems (matching, knapsack, solving ILPs [25]). We handle for the first time uncertainty in the constraints, develop new techniques beyond witness sets and lay the foundation for solving more general problems. We leave open whether the second log factor in our main result, Theorem 4, is necessary. Furthermore, the best competitive ratio achievable in exponential running time remains open.

References

1. Agarwal, A., Assadi, S., Khanna, S.: Stochastic submodular cover with limited adaptivity. In: SODA, pp. 323–342. SIAM (2019). https://doi.org/10.1137/1.9781611975482.21
2. Bampis, E., Dürr, C., Erlebach, T., de Lima, M.S., Megow, N., Schlöter, J.: Orienting (hyper)graphs under explorable stochastic uncertainty. In: ESA. LIPIcs, vol. 204, pp. 1–18 (2021). https://doi.org/10.4230/LIPIcs.ESA.2021.10
3. Behnezhad, S., Blum, A., Derakhshan, M.: Stochastic vertex cover with few queries. In: SODA, pp. 1808–1846. SIAM (2022)
4. Blum, A., Dickerson, J.P., Haghtalab, N., Procaccia, A.D., Sandholm, T., Sharma, A.: Ignorance is almost bliss: Near-optimal stochastic matching with few queries. Oper. Res. **68**(1), 16–34 (2020)
5. Bruce, R., Hoffmann, M., Krizanc, D., Raman, R.: Efficient update strategies for geometric computing with uncertainty. Theory Comput. Syst. **38**(4), 411–423 (2005). https://doi.org/10.1007/s00224-004-1180-4
6. Chaplick, S., Halldórsson, M.M., de Lima, M.S., Tonoyan, T.: Query minimization under stochastic uncertainty. Theor. Comput. Sci. **895**, 75–95 (2021)
7. Chvatal, V.: A greedy heuristic for the set-covering problem. Math. Oper. Res. **4**(3), 233–235 (1979)
8. Dinur, I., Steurer, D.: Analytical approach to parallel repetition. In: STOC, pp. 624–633. ACM (2014)
9. Dobson, G.: Worst-case analysis of greedy heuristics for integer programming with nonnegative data. Math. Oper. Res. **7**(4), 515–531 (1982)

10. Erlebach, T., Hoffmann, M.: Minimum spanning tree verification under uncertainty. In: Kratsch, D., Todinca, I. (eds.) WG 2014. LNCS, vol. 8747, pp. 164–175. Springer, Cham (2014). https://doi.org/10.1007/978-3-319-12340-0_14

11. Erlebach, T., Hoffmann, M., Kammer, F.: Query-competitive algorithms for cheapest set problems under uncertainty. Theoret. Comput. Sci. **613**, 51–64 (2016). https://doi.org/10.1016/j.tcs.2015.11.025

12. Erlebach, T., de Lima, M.S., Megow, N., Schlöter, J.: Learning-augmented query policies for minimum spanning tree with uncertainty. In: ESA. LIPIcs, vol. 244, pp. 1–18. Schloss Dagstuhl - Leibniz-Zentrum für Informatik (2022)

13. Feder, T., Motwani, R., Panigrahy, R., Olston, C., Widom, J.: Computing the median with uncertainty. SIAM J. Comput. **32**(2), 538–547 (2003). https://doi.org/10.1137/S0097539701395668

14. Ghuge, R., Gupta, A., Nagarajan, V.: The power of adaptivity for stochastic submodular cover. In: ICML. Proceedings of Machine Learning Research, vol. 139, pp. 3702–3712. PMLR (2021)

15. Goemans, M.X., Vondrák, J.: Covering minimum spanning trees of random subgraphs. Random Struct. Algorithms **29**(3), 257–276 (2006)

16. Goemans, M., Vondrák, J.: Stochastic covering and adaptivity. In: Correa, J.R., Hevia, A., Kiwi, M. (eds.) LATIN 2006. LNCS, vol. 3887, pp. 532–543. Springer, Heidelberg (2006). https://doi.org/10.1007/11682462_50

17. Halldórsson, M.M., de Lima, M.S.: Query-competitive sorting with uncertainty. In: MFCS. LIPIcs, vol. 138, pp. 1–15 (2019). https://doi.org/10.4230/LIPIcs.MFCS.2019.7

18. Hoffmann, M., Erlebach, T., Krizanc, D., Mihalák, M., Raman, R.: Computing minimum spanning trees with uncertainty. In: STACS. LIPIcs, vol. 1, pp. 277–288. Schloss Dagstuhl - Leibniz-Zentrum für Informatik, Germany (2008). https://arxiv.org/abs/0802.2855

19. Kahan, S.: A model for data in motion. In: STOC, pp. 267–277. ACM (1991). https://doi.org/10.1145/103418.103449

20. Kolliopoulos, S.G., Young, N.E.: Tight approximation results for general covering integer programs. In: FOCS, pp. 522–528. IEEE Computer Society (2001). https://doi.org/10.1109/SFCS.2001.959928

21. Kolliopoulos, S.G., Young, N.E.: Approximation algorithms for covering/packing integer programs. J. Comput. Syst. Sci. **71**(4), 495–505 (2005). https://doi.org/10.1016/j.jcss.2005.05.002

22. Maehara, T., Yamaguchi, Y.: Stochastic packing integer programs with few queries. Math. Programm. **182**, 141–174 (2019). https://doi.org/10.1007/s10107-019-01388-x

23. Megow, N., Meißner, J., Skutella, M.: Randomization helps computing a minimum spanning tree under uncertainty. SIAM J. Comput. **46**(4), 1217–1240 (2017). https://doi.org/10.1137/16M1088375

24. Megow, N., Schlöter, J.: Set selection under explorable stochastic uncertainty via covering techniques (2022). https://doi.org/10.48550/ARXIV.2211.01097

25. Meißner, J.: Uncertainty exploration: algorithms, competitive analysis, and computational experiments, Ph. D. thesis, Technischen Universität Berlin (2018). https://doi.org/10.14279/depositonce-7327

26. Rajagopalan, S., Vazirani, V.V.: Primal-dual RNC approximation algorithms for set cover and covering integer programs. SIAM J. Comput. **28**(2), 525–540 (1998)

27. Shmoys, D.B., Swamy, C.: Stochastic optimization is (almost) as easy as deterministic optimization. In: FOCS, pp. 228–237. IEEE Computer Society (2004). https://doi.org/10.1109/FOCS.2004.62

28. Vazirani, V.V.: Approximation algorithms, vol. 1. Springer, Heidelberg (2001). https://doi.org/10.1007/978-3-662-04565-7
29. Vondrák, J.: Shortest-path metric approximation for random subgraphs. Random Struct. Algorithms **30**(1–2), 95–104 (2007)
30. Wang, W., Gupta, A., Williams, J.: Probing to minimize. In: ITCS. LIPIcs, vol. 215, pp. 1–123. Schloss Dagstuhl - Leibniz-Zentrum für Informatik (2022)

Towards a Characterization of Maximal Quadratic-Free Sets

Gonzalo Muñoz[1], Joseph Paat[2(✉)], and Felipe Serrano[3]

[1] Engineering Sciences Institute, Universidad de O'Higgins, Rancagua, Chile
`gonzalo.munoz@uoh.cl`
[2] Sauder School of Business, University of British Columbia, Vancouver, BC, Canada
`joseph.paat@sauder.ubc.ca`
[3] I2DAMO GmbH, Berlin, Germany
`serrano@i2damo.de`

Abstract. In 1971, Balas introduced intersection cuts as a method for generating cutting planes in integer optimization. These cuts are derived from convex S-free sets, and inclusion-wise maximal S-free sets yield the strongest intersection cuts. When S is a lattice, maximal S-free sets are well-studied. In this work, we provide a new characterization of maximal S-free sets, for arbitrary S, based on sequences that 'expose' inequalities defining the S-free set; these exposing sequences generalize the notion of blocking points when S is a lattice. We then apply our characterization to partially characterize maximal S-free polyhedra when S is defined by a homogeneous quadratic inequality. Our results generate new families of maximal quadratic-free sets and considerably generalize some of the constructions by Muñoz and Serrano (IPCO 2020), who first introduced maximal quadratic-free sets.

1 Introduction

Given a closed set $S \subseteq \mathbb{R}^d$, we say that a closed convex set $C \subseteq \mathbb{R}^d$ is S-*free* if its interior $\mathrm{intr}(C)$ contains no points in S. The family of S-free sets form the foundation of *intersection cuts* for mathematical programs of the form

$$\min\{\mathbf{c}^\top \mathbf{x} : \mathbf{x} \in S \cap P\}, \qquad (1)$$

where $\mathbf{c} \in \mathbb{R}^d$ and $P \subseteq \mathbb{R}^d$ is closed and convex. Intersection cuts were introduced by Balas [7] when S is a lattice and by Tuy [27] when S is a reverse convex set. Since then, intersection cuts have been well-studied; see, e.g., [1,2,4,13,14,16,22]. For a general reference on intersection cuts, we point to [18, Chapter 6]. One important feature of intersection cuts is the following: if one solves an LP relaxation of (1) and obtains a vertex $\mathbf{x} \notin S$, the construction of an S-free set that contains $\mathbf{x}$ in its interior ensures the separation of $\mathbf{x}$.

The family of inclusion-wise *maximal* S-free sets play an important role as they generate the strongest intersection cuts, and they also serve as optimality certificates for mixed integer programs; see [6,9,25]. For the case when S is a

A. Del Pia and V. Kaibel (Eds.): IPCO 2023, LNCS 13904, pp. 334–347, 2023.
https://doi.org/10.1007/978-3-031-32726-1_24

lattice, Lovász [21] demonstrates that full-dimensional maximal S-free sets are polyhedra with integer points in the relative interior of each facet; see also [3]. For extensions of Lovász's result to lattice points in linear subspaces and rational polyhedra see [11,12]. Connections between maximal S-free sets and the Helly number of S have been established in [5,9,19]. There are some characterizations of maximality beyond the lattice setting, e.g., Bienstock et al. [13,14] characterize various maximal S-free sets when S is the set of rank 1 real-valued symmetric matrices.

Muñoz and Serrano [23,24] are the first to study the setting when S is defined by an arbitrary quadratic inequality. They develop methods for proving maximality when S is defined by either a single homogeneous or a single non-homogeneous quadratic inequality. A computational implementation of the resulting intersection cuts was developed in [15], with favorable results. In this paper, we focus on the homogeneous setting, i.e., $S = \left\{ \mathbf{s} \in \mathbb{R}^p : \mathbf{s}^\top \mathbf{A} \mathbf{s} \leq 0 \right\}$ for $\mathbf{A} \in \mathbb{R}^{p \times p}$. Maximal S-free sets derived in this setting extend to (not necessarily maximal) sets in the non-homogeneous setting. Indeed, an $\widehat{S}$-free set for the non-homogeneous setting $\widehat{S} := \left\{ \mathbf{s} \in \mathbb{R}^p : \mathbf{s}^\top \mathbf{A} \mathbf{s} + \mathbf{g}^\top \mathbf{s} + h \leq 0 \right\}$ can be constructed by taking an S'-free set for the homogeneous set $S' := \left\{ (\mathbf{s}, z) \in \mathbb{R}^p \times \mathbb{R} : \mathbf{s}^\top \mathbf{A} \mathbf{s} + (\mathbf{g}^\top \mathbf{s}) z + h z^2 \leq 0 \right\}$ and intersecting the S'-free set with $z = 1$. To simplify our presentation in the homogeneous setting, we follow reductions in [23,24] to assume that the homogeneous setting has the form

$$Q := \{ (\mathbf{x}, \mathbf{y}) \in \mathbb{R}^n \times \mathbb{R}^m : \|\mathbf{x}\| \leq \|\mathbf{y}\| \},$$

where $\| \cdot \|$ is the ℓ_2-norm. We replace S (resp. S-free) with Q (resp. Q-free) to highlight that we are looking at *quadratic-free sets*. We use this definition of Q for the rest of the paper, and refer to any set C that is Q-free as *homogeneous quadratic-free*.

Among their results, Muñoz and Serrano prove that a particular homogeneous quadratic-free set is maximal [24, Theorem 2.1]. One of the motivations in this paper is to provide more general characterizations of maximal Q-free sets that can be used to generate alternative families of them and, consequently, new families of cutting planes for quadratically-constrained problems. In order to derive our characterizations, we also derive a new characterization of maximality for S-free sets when S is an arbitrary closed set.

Notation. We use $\mathrm{conv}(\mathcal{X})$ and $\mathrm{cone}(\mathcal{X})$ to denote the closed convex hull and closed convex conic hull of a set $\mathcal{X} \subseteq \mathbb{R}^d$. For background on convexity including common definitions, we point to [8,26]. For each $d \in \mathbb{N}$, we set $D^d := \{ \mathbf{x} \in \mathbb{R}^d : \|\mathbf{x}\| = 1 \}$.

1.1 Contributions

We characterize Q-free sets by representing them as an intersection of half-spaces of the form $\Gamma(\boldsymbol{\beta})^\top \mathbf{x} - \boldsymbol{\beta}^\top \mathbf{y} \geq 0$, where $\Gamma : D^m \to D^n$. Given such a Γ function, we define

$$C_\Gamma := \{ (\mathbf{x}, \mathbf{y}) \in \mathbb{R}^n \times \mathbb{R}^m : \Gamma(\boldsymbol{\beta})^\top \mathbf{x} - \boldsymbol{\beta}^\top \mathbf{y} \geq 0 \ \ \forall \, \boldsymbol{\beta} \in D^m \}. \tag{2}$$

Note that C_Γ is always convex and Q-free: by the Cauchy-Schwarz inequality, any $(\mathbf{x}, \mathbf{y})$ in the interior of C_Γ satisfies $\boldsymbol{\beta}^\top \mathbf{y} < \|\mathbf{x}\|$ for all $\boldsymbol{\beta} \in D^m$ which implies that $\|\mathbf{y}\| < \|\mathbf{x}\|$.

Our first main result shows that this 'standard form' C_Γ is necessary for a Q-free set to be maximal.

Theorem 1 (A necessary condition for maximality). *Let C be a full-dimensional closed convex maximal Q-free set. There exists a function $\Gamma : D^m \to D^n$ such that $C = C_\Gamma$.*

Our next results provide partial converses of Theorem 1. For each result, we assume something about the 'expansivity' of Γ.

Definition 1 (Non-expansive, isometric, strictly non-expansive). *Let $\Gamma : D^m \to D^n$. We say that $\boldsymbol{\beta}, \boldsymbol{\beta}' \in D^m$ are **isometric** if $\|\Gamma(\boldsymbol{\beta}) - \Gamma(\boldsymbol{\beta}')\| = \|\boldsymbol{\beta} - \boldsymbol{\beta}'\|$. Additionally, we say that Γ is*

- **non-expansive** *if $\|\Gamma(\boldsymbol{\beta}) - \Gamma(\boldsymbol{\beta}')\| \leq \|\boldsymbol{\beta} - \boldsymbol{\beta}'\|$ for all $\boldsymbol{\beta}, \boldsymbol{\beta}' \in D^m$.*
- **isometric** *if $\|\Gamma(\boldsymbol{\beta}) - \Gamma(\boldsymbol{\beta}')\| = \|\boldsymbol{\beta} - \boldsymbol{\beta}'\|$ for all $\boldsymbol{\beta}, \boldsymbol{\beta}' \in D^m$.*
- **strictly non-expansive** *if $\|\Gamma(\boldsymbol{\beta}) - \Gamma(\boldsymbol{\beta}')\| < \|\boldsymbol{\beta} - \boldsymbol{\beta}'\|$ for all $\boldsymbol{\beta}, \boldsymbol{\beta}' \in D^m$, $\boldsymbol{\beta} \neq \boldsymbol{\beta}'$.*

Observe that $\Gamma : D^m \to D^n$ is non-expansive (respectively, isometric or strictly non-expansive) if and only if $\boldsymbol{\beta}^\top \boldsymbol{\beta}' \leq \Gamma(\boldsymbol{\beta})^\top \Gamma(\boldsymbol{\beta}')$ for all $\boldsymbol{\beta}, \boldsymbol{\beta}' \in D^m$ (respectively, $\boldsymbol{\beta}^\top \boldsymbol{\beta}' = \Gamma(\boldsymbol{\beta})^\top \Gamma(\boldsymbol{\beta}')$ or $\boldsymbol{\beta}^\top \boldsymbol{\beta}' < \Gamma(\boldsymbol{\beta})^\top \Gamma(\boldsymbol{\beta}')$) because $\boldsymbol{\beta}, \boldsymbol{\beta}' \in D^m$ and $\Gamma(\boldsymbol{\beta}), \Gamma(\boldsymbol{\beta}') \in D^n$.

Theorem 2 (First sufficient condition for maximality). *Let $\Gamma : D^m \to D^n$ and define C_Γ as in (2). If Γ is strictly non-expansive, then C_Γ is a full-dimensional maximal Q-free set.*

Theorem 2 generalizes [24, Theorem 2.1], where Γ is a constant function. Example 1 in Section 2 illustrates the construction of a maximal Q-free set using a non-constant Γ function that is strictly non-expansive.

We conjecture that requiring Γ non-expansive is a sufficient condition for a full-dimensional C_Γ to be maximal in general. We take a step in this direction with our third result.

Theorem 3 (Second sufficient condition for maximality). *Let $\Gamma : D^m \to D^n$ and define C_Γ as in (2). If Γ is non-expansive and C_Γ is a full-dimensional polyhedron, then C_Γ is a maximal Q-free set.*

We complement Theorem 3 by characterizing when C_Γ is a polyhedron.

Theorem 4 (A characterization of polyhedrality). *Let $\Gamma : D^m \to D^n$ be non-expansive and define C_Γ as in (2). C_Γ is a polyhedron if and only if there is a finite set $I \subseteq D^m$ such that for every $\overline{\boldsymbol{\beta}} \in D^m$ there exists a set $J \subseteq I$ of pairwise isometric points satisfying $\overline{\boldsymbol{\beta}} \in \text{cone}(J)$. Moreover, $C_\Gamma = \{(\mathbf{x}, \mathbf{y}) \in \mathbb{R}^n \times \mathbb{R}^m : \Gamma(\boldsymbol{\beta})^\top \mathbf{x} - \boldsymbol{\beta}^\top \mathbf{y} \geq 0 \ \forall \boldsymbol{\beta} \in I\}$.*

Examples 2 and 3 in Section 2 show the construction of polyhedral maximal Q-free sets using a non-expansive Γ.

Note that the aforementioned conjecture—that Γ non-expansive is a sufficient condition for maximality of C_Γ—would subsume Theorems 2 and 3. We comment more on this conjecture in Remark 2 (Section 7), specifically, we discuss the main issues of a potential direct generalization of our current proofs.

We point out one difference between maximal Q-free polyhedra and maximal S-free polyhedra when S is a lattice. In Lovász's characterization, the number of facets of a maximal S-free polyhedra is upper bounded by a function of the dimension; this is not the case for maximal Q-free polyhedra, which can have an arbitrary number of facets (see Example 3 in Section 2). We also note that Theorem 4 can be used to construct, starting from a set $I \subseteq D^m$, a function Γ that yields a maximal Q-free polyhedron (see Example 3 for an illustration of this).

Underlying the proofs of Theorem 2 and 3 is our final main result, which is a general characterization of maximality of S-free sets. We turn once again to the case when S is a lattice to motivate this result. Lovász proved that if C is maximal S-free, then C is a polyhedron and every facet F of C contains a lattice point $\mathbf{z}^F$ in its relative interior. The point $\mathbf{z}^F$ is similar to a 'blocking point' used to generate maximal S-free sets when S is a mixed integer set through lifting [10,17,20]. In order for C to be S-free in the lattice setting, each $\mathbf{z}^F$ must be separated from C by a facet defining inequality; the inequality defining F is the unique facet separating $\mathbf{z}^F$ from C. Thus, in a way $\mathbf{z}^F$ 'exposes' the facet. The notion of exposing points is considered by Muñoz and Serrano [24], where they argue that if every inequality defining a Q-free set C has an exposing point, then C is maximal. However, there are maximal Q-free sets defined by inequalities that do not have exposing points; see Example 2 in Section 2. A generalization of this is the notion of an exposing sequence.

Definition 2 (Exposing sequence). *Let $C \subseteq \mathbb{R}^d$ be a convex set and $\boldsymbol{\alpha}^\top \mathbf{x} \leq \alpha_0$, with $\boldsymbol{\alpha} \neq \mathbf{0}$, a valid inequality for C. A sequence $(\mathbf{x}^t)_{t=1}^\infty$ in $\mathbb{R}^d$ is an **exposing sequence** for $\boldsymbol{\alpha}^\top \mathbf{x} \leq \alpha_0$ if $\lim_{t\to\infty}(\boldsymbol{\delta}^t, \delta_0^t) = (\boldsymbol{\alpha}, \alpha_0)$ for every sequence $((\boldsymbol{\delta}^t, \delta_0^t))_{t=1}^\infty$ in $\mathbb{R}^d \times \mathbb{R}$ such that $\|\boldsymbol{\delta}^t\| = \|\boldsymbol{\alpha}\|$, $\boldsymbol{\delta}^{t\top}\mathbf{x} \leq \delta_0^t$ is a valid inequality for C, and $\boldsymbol{\delta}^{t\top}\mathbf{x}^t \geq \delta_0^t$ for each t.*

Remark 1. Muñoz and Serrano [24] define a notion of 'exposing sequence at infinity'. This is more restrictive than Definition 2, and moreover, it can be shown that an exposing sequence at infinity reduces to an exposing point for Q, because Q is a cone.

Theorem 5. *Let $S \subseteq \mathbb{R}^d$ be closed, and let $C \subseteq \mathbb{R}^d$ be a closed convex full-dimensional S-free set. C is maximal S-free if and only if there exists a set $I \subseteq \mathbb{R}^d \times \mathbb{R}$ such that*

$$C = \{\mathbf{x} \in \mathbb{R}^d : \boldsymbol{\alpha}^\top \mathbf{x} \leq \alpha_0 \ \forall \ (\boldsymbol{\alpha}, \alpha_0) \in I\}$$

and each $(\boldsymbol{\alpha}, \alpha_0) \in I$ has an exposing sequence $(\mathbf{x}^t)_{t=1}^\infty$ in S.

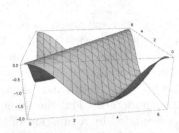
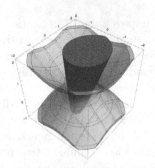

(a)   Plot   of $\beta(\theta_1)^\top\beta(\theta_2)$ –
$\Gamma(\beta(\theta_1))^\top\Gamma(\beta(\theta_2))$, which is non-
positive and takes the value 0 only
when $\theta_1 = \theta_2$ (mod 2π). This shows Γ
is strictly non-expansive.

(b) 3-dimensional slices of the 4-
dimensional sets Q (boundary in orange) and C_Γ (red).

Fig. 1. Construction of the maximal Q-free set in Example 1. (Color figure online)

2   Examples of Maximal Homogeneous Quadratic-Free Sets

Example 1. Our first example illustrates a non-polyhedral Q-free set using Theorem 2. Consider $n = m = 2$. We construct a function Γ using polar coordi-
nates: for $\theta \in [0, 2\pi]$, we define $\gamma(\theta) = -\theta(\theta - 2\pi)/(4\pi)$ and define Γ such that
$\beta(\theta) := (\cos(\theta), \sin(\theta)) \mapsto \Gamma(\beta(\theta)) := (\cos(\gamma(\theta)), \sin(\gamma(\theta)))$. It can be shown
that Γ is strictly non-expansive: in Figure 1a we illustrate this. Therefore, The-
orem 2 implies that C_Γ is maximal Q-free. Figure 1b shows a 3-dimensional slice
of this 4-dimensional C_Γ. Note how the non-differentiability of γ at 0 translates
into a non-smooth region of the set.

Example 2. Suppose $n = m$ and define $\Gamma(\beta) = |\beta|$, where the absolute value is
taken component-wise. The reverse triangle inequality $||a| - |b|| \le |a - b|$ implies
Γ is non-expansive. Set $I = \{e^1, \ldots, e^m, -e^1, \ldots, -e^m\} \subseteq D^m$ where $\mathbf{e}^i \in \mathbb{Z}^m$ is
the ith standard unit vector. Each $\beta \in D^m$ is in cone(J) generated by a linearly
independent set $J \subseteq I$. By definition of I and because J is linearly independent,
each $\beta, \beta' \in J$ are isometric. Theorem 4 ensures that C_Γ is polyhedral. Moreover,
from Theorem 4 we see that

$$C_\Gamma = \{(\mathbf{x}, \mathbf{y}) \in \mathbb{R}^n \times \mathbb{R}^m : \Gamma(\beta)^\top\mathbf{x} - \beta^\top\mathbf{y} \ge 0 \ \forall \beta \in I\}$$
$$= \{(\mathbf{x}, \mathbf{y}) \in \mathbb{R}^n \times \mathbb{R}^m : x_i \ge |y_i| \ \forall i \in \{1, \ldots, m\}\}.$$

C_Γ is full-dimensional, and therefore Theorem 3 ensures maximality of C_Γ.
Figure 2 illustrates a 3-dimensional slice of the 4-dimensional sets Q and C_Γ
obtained for $n = m = 2$. The maximality of this example could not have been
proved with the results of [24]: it can be seen that $C_\Gamma \cap Q = \{(\mathbf{x}, \mathbf{y}) \in \mathbb{R}^n \times \mathbb{R}^m :
x_i = |y_i| \ \forall i \in \{1, \ldots, m\}\}$, therefore every facet of C_Γ intersects Q and, more

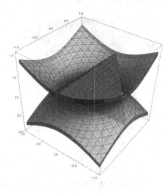

Fig. 2. 3-dimensional slices of the 4-dimensional sets Q (boundary in orange) and C_Γ (red) obtained when $n = m = 2$ in Example 2. (Color figure online)

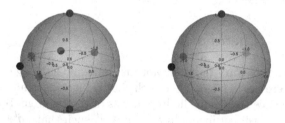

Fig. 3. Representation of Γ in the 6-dimensional set of Example 3. A point on the left plot represents a β that gets mapped by Γ to the point on the right plot of the same color. (Color figure online)

importantly, any $(\mathbf{x}, \mathbf{y}) \in C_\Gamma \cap Q$ is contained in m different facets of C_Γ. This means that there is no exposing point in $C_\Gamma \cap Q$ for any of the facets of C_Γ.

Example 3. In Figure 3 (left) consider $\{\pm \mathbf{e}^3\}$ (in blue), $\{\pm \mathbf{e}^1, -\mathbf{e}^2\}$ (in red), $\{\mathbf{e}^2\}$ (in green), and $\{-1/\sqrt{2} \cdot \mathbf{e}^1 \pm 1/\sqrt{2} \cdot \mathbf{e}^2\}$ (in black) in D^3; let I be the set of these 8 points. We define Γ to map $\{\pm \mathbf{e}^3\}$ to $\mathbf{e}^3$, $\{\pm \mathbf{e}^1, -\mathbf{e}^2\}$ to $-\mathbf{e}^2$, $\{\mathbf{e}^2\}$ to $-\mathbf{e}^1$, and $\{-1/\sqrt{2} \cdot \mathbf{e}^1 \pm 1/\sqrt{2} \cdot \mathbf{e}^2\}$ to $-1/\sqrt{2} \cdot \mathbf{e}^1 + 1/\sqrt{2} \cdot \mathbf{e}^2$; see Figure 3 (right). Each $\beta \in D^3$ (left) is a conic combination of at most three points in I that are pairwise isometric. One can extend Γ from I to a non-expansive function on D^3 through a conic interpolation; see Lemma 2. Theorem 3 then implies that

$$C_\Gamma = \{(\mathbf{x}, \mathbf{y}) \in \mathbb{R}^3 \times \mathbb{R}^3 : \Gamma(\beta)^\top \mathbf{x} - \beta^\top \mathbf{y} \geq 0 \ \forall \beta \in I\}.$$

is maximal Q-free. Figure 4 shows a 3-dimensional slice of the 6-dimensional C_Γ.

The construction in this example can be generalized to an arbitrarily large set I as long as their conic combinations generate D^m. This produces a maximal Q-free polyhedra with arbitrarily many facets.

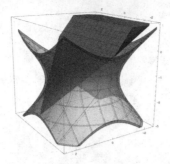

Fig. 4. 3-dimensional slices of the 6-dimensional sets Q (boundary in orange) and C_Γ (red) obtained using the Γ function depicted in Figure 3. We note that maximality may not be evident in this picture since maximality is not preserved when taking slices. (Color figure online)

3   A Proof of Theorem 5

($\Leftarrow$) Assume to the contrary that C is not maximal and let $K \supsetneq C$ be a convex S-free set. Then there exists $(\overline{\alpha}, \overline{\alpha}_0) \in I$ such that $\overline{\alpha}^\top \mathbf{x} \le \overline{\alpha}_0$ is not valid for K. Let $(\mathbf{x}^t)_{t=0}^\infty$ be a sequence in S as in the hypothesis of the theorem. Since $\mathbf{x}^t \in S$ and K is S-free, there exists an inequality $\delta^{t\top} \mathbf{x} \le \delta_0^t$ such that $\|\delta^t\| = \|\overline{\alpha}\|$, is valid for K, and $\delta^{t\top} \mathbf{x}^t \ge \delta_0^t$. The inequality $\delta^{t\top} \mathbf{x} \le \delta_0^t$ is valid for C because $C \subsetneq K$. By the definition of exposing sequence, we have $\lim_{t\to\infty}(\delta^t, \delta_0^t) = (\overline{\alpha}, \overline{\alpha}_0)$, so $\overline{\alpha}^\top \mathbf{x} \le \overline{\alpha}_0$ is valid for K. This is a contradiction.

($\Rightarrow$) Let $C^\circ \subseteq \mathbb{R}^d \times \mathbb{R}$ be the polar of C, i.e., the set of coefficients corresponding to valid inequalities for C:

$$C = \{\mathbf{x} \in \mathbb{R}^d : \alpha^\top \mathbf{x} \le \alpha_0 \ \forall \ (\alpha, \alpha_0) \in C^\circ\}.$$

The set C° is a closed convex cone, and C° is pointed because C is full-dimensional. Since C° is a closed pointed cone, there is a set $I \subseteq C^\circ$ generating the extreme rays of C°; see, e.g., [8, Page 67]. We have $C^\circ = \text{cone}(I)$, so

$$C = \{\mathbf{x} \in \mathbb{R}^d : \alpha^\top \mathbf{x} \le \alpha_0 \ \forall \ (\alpha, \alpha_0) \in I\}.$$

Let $(\overline{\alpha}, \overline{\alpha}_0) \in I$ and define

$$K_t := \{(\alpha, \alpha_0) \in \mathbb{R}^d \times \mathbb{R} : \|(\overline{\alpha}, \overline{\alpha}_0) - (\alpha, \alpha_0)\| < {}^1\!/t\},$$
$$C_t := \{\mathbf{x} \in \mathbb{R}^d : \alpha^\top \mathbf{x} \le \alpha_0 \ \forall \ (\alpha, \alpha_0) \in I \setminus K_t\}.$$

We proceed to show that $C \subsetneq C_t$. Assume to the contrary that $C = C_t$. This implies that $I \subseteq \text{cone}(I \setminus K_t)$. However, since K_t is a ball of positive radius (and thus full-dimensional), this means that $(\overline{\alpha}, \overline{\alpha}_0)$ does not generate an extreme ray of C°. Hence, $C \subsetneq C_t$.

Due to the maximality of C, there is a vector $\mathbf{x}^t \in \text{intr}(C_t) \cap S$. If $\delta^{t\top} \mathbf{x} \le \delta_0^t$ is valid for C and $\delta^{t\top} \mathbf{x}^t \ge \delta_0^t$, then $(\delta^t, \delta_0^t) \in K_t$. Thus, $\lim_{t\to\infty}(\delta^t, \delta^t) = (\overline{\alpha}, \overline{\alpha}_0)$.

4   A Proof of Theorem 1

To prove Theorem 1, we write $Q = \bigcup_{\beta \in D^m} Q_\beta$, where

$$Q_\beta := \{(\mathbf{x}, \mathbf{y}) \in \mathbb{R}^n \times \mathbb{R}^m : \|\mathbf{x}\| - \beta^\top \mathbf{y} \le 0\}$$
$$= \{(\mathbf{x}, \mathbf{y}) \in \mathbb{R}^n \times \mathbb{R}^m : \gamma^\top \mathbf{x} - \beta^\top \mathbf{y} \le 0 \ \forall \, \gamma \in D^n\}. \tag{3}$$

Note that each Q_β is convex. The following lemma is used to prove Theorem 1. The proof follows from Theorem 17.3 [26].

Lemma 1. *Let $\beta \in D^m$. Every tight valid inequality for Q_β has the form (possibly after scaling by a positive number) $\gamma^\top \mathbf{x} - \beta^\top \mathbf{y} \le 0$ for some $\|\gamma\| \le 1$.*

We remark that this lemma, while not hard to prove, does not follow immediately from (3); Lemma 1 refers to *every* tight valid inequality, which, in principle, can include inequalities not explicitly considered in the description (3).

Proof (of Theorem 1). For each $\beta \in D^m$, there is a hyperplane separating C and Q_β because both sets are convex and C is Q-free. For each $\beta \in D^m$, by Lemma 1 we can take the corresponding inequality to be $\gamma^\top \mathbf{x} - \beta^\top \mathbf{y} \ge 0$ for $\|\gamma\| \le 1$. From this discussion, it follows that we can define a function $\gamma : D^m \to \{\mathbf{x} \in \mathbb{R}^n : \|\mathbf{x}\| \le 1\}$ such that for each β, we have that $\gamma(\beta)^\top \mathbf{x} - \beta^\top \mathbf{y} \ge 0$ is valid for C and separates Q_β. Thus,

$$C \subseteq \{(\mathbf{x}, \mathbf{y}) \in \mathbb{R}^n \times \mathbb{R}^m : \gamma(\beta)^\top \mathbf{x} - \beta^\top \mathbf{y} \ge 0 \ \forall \, \beta \in D^m\}. \tag{4}$$

Each Q_β is separated from the set on the right-hand side of (4), implying that it is Q-free. By the maximality of C, we have that (4) is an equality.

We now show that, since C is a maximal Q-free set, we can further restrict $\gamma(\beta)$ to have unit norm[1]. For this, consider the pair of valid inequalities for C:

$$\gamma(\beta)^\top \mathbf{x} - \beta^\top \mathbf{y} \ge 0 \qquad \text{and} \qquad \gamma(-\beta)^\top \mathbf{x} - (-\beta)^\top \mathbf{y} \ge 0$$

for each $\beta \in D^m$. Multiplying the first inequality by $\lambda + 1$, the second one by λ with $\lambda \ge 0$, and adding them we obtain the following valid inequality for C:

$$(\lambda(\gamma(\beta) + \gamma(-\beta)) + \gamma(\beta))^\top \mathbf{x} - \beta^\top \mathbf{y} \ge 0.$$

Notice that $\gamma(\beta) + \gamma(-\beta) \ne \mathbf{0}$ for each β as otherwise $-(\gamma(-\beta)^\top \mathbf{x} - (-\beta)^\top \mathbf{y}) = \gamma(\beta)^\top \mathbf{x} - \beta^\top \mathbf{y} \ge 0$ is valid for C implying that C satisfies an equation and is not full-dimensional; this is a contradiction. So, there exists $\lambda(\beta) \ge 0$ such that $\|\lambda(\beta)(\gamma(\beta) + \gamma(-\beta)) + \gamma(\beta)\| = 1$. Define $\Gamma : D^m \to D^n$ by

$$\Gamma(\beta) := \lambda(\beta)(\gamma(\beta) + \gamma(-\beta)) + \gamma(\beta).$$

---

[1] Note that Lemma 1 does not directly imply that $\gamma(\beta)$ can be assumed to have unit norm. Moreover, one could produce a (not necessarily maximal) Q-free set with γ that satisfies $\|\gamma(\beta)\| < 1$ for some β.

Therefore,

$$C \subseteq \{(\mathbf{x}, \mathbf{y}) \in \mathbb{R}^n \times \mathbb{R}^m : \Gamma(\boldsymbol{\beta})^\top \mathbf{x} - \boldsymbol{\beta}^\top \mathbf{y} \geq 0 \ \forall \ \boldsymbol{\beta} \in D^m\} = C_\Gamma. \qquad (5)$$

Again using that the right-hand side in (5) is Q-free and C is maximal, we conclude that (5) is an equality. $\qquad \square$

5   A Proof of Theorem 2

We know the set C_Γ is Q-free. To prove maximality, we use Theorem 3.1 from Muñoz and Serrano [24], which states that it is enough to show that every inequality has an exposing point, that is, that for every $\boldsymbol{\beta}_0 \in D^m$ there exist a vector $(\mathbf{x}_0, \mathbf{y}_0) \in Q$ such that $\Gamma(\boldsymbol{\beta}_0)^\top \mathbf{x}_0 - \boldsymbol{\beta}_0^\top \mathbf{y}_0 = 0$ and $\Gamma(\boldsymbol{\beta})^\top \mathbf{x}_0 - \boldsymbol{\beta}^\top \mathbf{y}_0 > 0$ for all $\boldsymbol{\beta} \in D^m \backslash \{\boldsymbol{\beta}_0\}$. The point $(\mathbf{x}_0, \mathbf{y}_0)$ is called an *exposing point* of $\Gamma(\boldsymbol{\beta}_0)^\top \mathbf{x} - \boldsymbol{\beta}^\top \mathbf{y} \geq 0$. In terms of Theorem 5, an exposing point is equivalent to an exposing sequence for $\Gamma(\boldsymbol{\beta}_0)^\top \mathbf{x} - \boldsymbol{\beta}_0^\top \mathbf{y} \geq 0$ defined by the constant sequence $((\mathbf{x}_0, \mathbf{y}_0))_{t=1}^\infty$.

Let $\boldsymbol{\beta}_0 \in D^m$ and set $(\mathbf{x}_0, \mathbf{y}_0) := (\Gamma(\boldsymbol{\beta}_0), \boldsymbol{\beta}_0)$. It is easy to verify that $(\mathbf{x}_0, \mathbf{y}_0) \in Q$ and $\Gamma(\boldsymbol{\beta}_0)^\top \mathbf{x}_0 - \boldsymbol{\beta}_0^\top \mathbf{y}_0 = 0$. Furthermore, for each $\boldsymbol{\beta} \in D^m \setminus \{\boldsymbol{\beta}_0\}$, we have $\Gamma(\boldsymbol{\beta})^\top \mathbf{x}_0 - \boldsymbol{\beta}^\top \mathbf{y}_0 > 0$ because Γ is strictly non-expansive. Thus, Theorem 3.1 in [24] implies that C_Γ is a maximal Q-free set. The existence of an exposing point immediately implies that C_Γ is full-dimensional. For example, any strict convex combination of two exposing points is in the interior of C_Γ.

6   Preliminary Results on Non-expansive Functions

In this section we collect a variety of lemmata to prove the main theorems. Throughout this section, assume $\Gamma : D^m \to D^n$ is non-expansive.

Lemma 2. *Let $I \subseteq D^m$ be a finite set of pairwise isometric points. The following properties hold true:*

1. *If $\sum_{\boldsymbol{\beta} \in I} \epsilon_\beta \boldsymbol{\beta} \in D^m$, where $\epsilon_\beta \geq 0$ for each $\boldsymbol{\beta} \in I$, then $\sum_{\boldsymbol{\beta} \in I} \epsilon_\beta \Gamma(\boldsymbol{\beta}) \in D^n$ and $\Gamma(\sum_{\boldsymbol{\beta} \in I} \epsilon_\beta \boldsymbol{\beta}) = \sum_{\boldsymbol{\beta} \in I} \epsilon_\beta \Gamma(\boldsymbol{\beta})$.*
2. *If $\sum_{\boldsymbol{\beta} \in I} \epsilon_\beta \Gamma(\boldsymbol{\beta}) \in D^n$, where $\epsilon_\beta \geq 0$ for each $\boldsymbol{\beta} \in I$, then $\sum_{\boldsymbol{\beta} \in I} \epsilon_\beta \boldsymbol{\beta} \in D^m$ and $\Gamma(\sum_{\boldsymbol{\beta} \in I} \epsilon_\beta \boldsymbol{\beta}) = \sum_{\boldsymbol{\beta} \in I} \epsilon_\beta \Gamma(\boldsymbol{\beta})$.*

Proof. Set $\widehat{\boldsymbol{\beta}} := \sum_{\boldsymbol{\beta} \in I} \epsilon_\beta \boldsymbol{\beta}$. Using the isometry of points in I, we have

$$\| \sum_{\boldsymbol{\beta} \in I} \epsilon_\beta \Gamma(\boldsymbol{\beta}) \|^2 = \sum_{\boldsymbol{\beta}, \boldsymbol{\beta}' \in I} \epsilon_\beta \epsilon_{\beta'} \Gamma(\boldsymbol{\beta})^\top \Gamma(\boldsymbol{\beta}') = \sum_{\boldsymbol{\beta}, \boldsymbol{\beta}' \in I} \epsilon_\beta \epsilon_{\beta'} \boldsymbol{\beta}^\top \boldsymbol{\beta}' = \|\widehat{\boldsymbol{\beta}}\|^2.$$

In the case of Property 1, we assume $\widehat{\boldsymbol{\beta}} \in D^m$, so the previous equation proves that $\| \sum_{\boldsymbol{\beta} \in I} \epsilon_\beta \Gamma(\boldsymbol{\beta}) \| = 1$. In the case of Property 2, we assume $\| \sum_{\boldsymbol{\beta} \in I} \epsilon_\beta \Gamma(\boldsymbol{\beta}) \|^2 = 1$, so the previous equation proves that $\widehat{\boldsymbol{\beta}} \in D^m$. Therefore, it remains to show that $\Gamma(\widehat{\boldsymbol{\beta}}) = \sum_{\boldsymbol{\beta} \in I} \epsilon_\beta \Gamma(\boldsymbol{\beta})$ in both cases; we prove these simultaneously.

Using the non-expansive property of Γ and the nonnegativity of ϵ_β, we have

$$1 = \widehat{\beta}^\top\widehat{\beta} = \sum_{\beta\in I}\epsilon_\beta\widehat{\beta}^\top\beta \le \sum_{\beta\in I}\epsilon_\beta\Gamma(\widehat{\beta})^\top\Gamma(\beta) = \Gamma(\widehat{\beta})^\top\left(\sum_{\beta\in I}\epsilon_\beta\Gamma(\beta)\right).$$

The Cauchy-Schwarz inequality implies that $1 = \Gamma(\widehat{\beta})^\top(\sum_{\beta\in I}\epsilon_\beta\Gamma(\beta))$. Since both vectors have unit norm, we conclude that $\Gamma(\widehat{\beta}) = \sum_{\beta\in I}\epsilon_\beta\Gamma(\beta)$. □

The following lemma is helpful when analyzing full-dimensional Q-free sets.

Lemma 3. *Define C_Γ as in (2). Assume C_Γ is full-dimensional and let $I \subseteq D^m$ be a finite set. If $\sum_{\beta\in I}\lambda_\beta\beta = \mathbf{0}$, where $\lambda_\beta > 0$ for each $\beta \in I$, then $\sum_{\beta\in I}\lambda_\beta\Gamma(\beta) \ne \mathbf{0}$.*

Proof. Assume to the contrary that $\sum_{\beta\in I}\lambda_\beta\Gamma(\beta) = \mathbf{0}$. Fix $\beta' \in I$. We have $\beta' = -\sum_{\beta\in I\setminus\{\beta'\}}(\lambda_\beta/\lambda_{\beta'})\beta$ and $\Gamma(\beta') = -\sum_{\beta\in I\setminus\{\beta'\}}(\lambda_\beta/\lambda_{\beta'})\Gamma(\beta)$. The following inequalities are both valid for C_Γ:

$$(\sum_{\beta\in I\setminus\{\beta'\}}\tfrac{\lambda_\beta}{\lambda_{\beta'}}\Gamma(\beta))^\top x - (\sum_{\beta\in I\setminus\{\beta'\}}\tfrac{\lambda_\beta}{\lambda_{\beta'}}\beta)^\top y \ge 0$$
$$\Gamma(\beta')^\top x - \beta'^\top y = -(\sum_{\beta\in I\setminus\{\beta'\}}\tfrac{\lambda_\beta}{\lambda_{\beta'}}\Gamma(\beta))^\top x + (\sum_{\beta\in I\setminus\{\beta'\}}\tfrac{\lambda_\beta}{\lambda_{\beta'}}\beta)^\top y \ge 0.$$

Thus, C_Γ satisfies an equation contradicting that it is full-dimensional. □

The next lemma, which is known from convexity theory, will allow us to simplify the description of C_Γ in the proofs of Theorems 3 and 4. The proof follows from Theorem 17.3 [26].

Lemma 4. *Define C_Γ as in (2). If C_Γ is full-dimensional and $\overline{\gamma}^\top x - \overline{\beta}^\top y \ge 0$ is valid for C_Γ, then $(\overline{\gamma},\overline{\beta}) \in \mathrm{cone}(\{(\Gamma(\beta),\beta): \beta \in D^n\})$.*

Our final lemma states that if an inequality $\Gamma(\overline{\beta})^\top x - \overline{\beta}^\top y \ge 0$ is implied by other inequalities of the same form indexed by $I \subseteq D^m$, then $\overline{\beta}$ must be isometric with $\beta \in I$. This will be used in the proof of Theorem 4 to help establish that we have a covering of D^m by isometric points.

Lemma 5. *Let $\overline{\beta} \in D^m$, $I \subseteq D^m$ be a finite set, and $\lambda_\beta > 0$ for each $\beta \in I$ be such that $(\Gamma(\overline{\beta}),\overline{\beta}) = \sum_{\beta\in I}\lambda_\beta(\Gamma(\beta),\beta)$. Then $\overline{\beta}$ and β are isometric for each $\beta \in I$.*

Proof. Notice that

$$0 = \Gamma(\overline{\beta})^\top\Gamma(\overline{\beta}) - \overline{\beta}^\top\overline{\beta} = (\sum_{\beta\in I}\lambda_\beta\Gamma(\beta))^\top\Gamma(\overline{\beta}) - (\sum_{\beta\in I}\lambda_\beta\beta)^\top\overline{\beta}$$
$$= \sum_{\beta\in I}\lambda_\beta(\Gamma(\beta)^\top\Gamma(\overline{\beta}) - \beta^\top\overline{\beta}),$$

where the first equality follow from $\overline{\beta} \in D^m$ and $\Gamma(\overline{\beta}) \in D^n$. Due to the non-expansiveness of Γ, every summand is non-negative. Since the sum is 0, every summand must be 0. As $\lambda_\beta > 0$, we conclude that $\Gamma(\beta)^\top\Gamma(\overline{\beta}) = \beta^\top\overline{\beta}$. □

7   A Proof of Theorem 3

From our discussion in Section 1.1, we know the set C_Γ is always Q-free. Since we assume C_Γ is a polyhedron, it admits a finite description using facet inequalities. By applying Lemma 4 to each facet inequality, we can assume that there is a finite set $I \subseteq D^m$ such that

$$C_\Gamma = \{(\mathbf{x}, \mathbf{y}) \in \mathbb{R}^n \times \mathbb{R}^m : \Gamma(\beta)^\top \mathbf{x} - \beta^\top \mathbf{y} \geq 0 \ \forall \, \beta \in I\}.$$

We assume that $\Gamma(\beta)^\top \mathbf{x} - \beta^\top \mathbf{y} \geq 0$ defines a facet of C_Γ for each $\beta \in I$. We will prove that this representation of C_Γ suffices to prove maximality in Theorem 3 using Theorem 5. To this end, let $\overline{\beta} \in I$. According to Theorem 5, it suffices to show that $\Gamma(\overline{\beta})^\top \mathbf{x} - \overline{\beta}^\top \mathbf{y} \geq 0$ has an exposing sequence. For $t \in \mathbb{N}$, define

$$\mathbf{x}^t := \Gamma(\overline{\beta}) + \tfrac{\sqrt{2t+1}}{t}\,\overline{\Gamma} \quad \text{and} \quad \mathbf{y}^t := (1 + \tfrac{1}{t})\,\overline{\beta}, \tag{6}$$

where $\overline{\Gamma} \in D^n$ will be chosen in Claim 1 so that $\Gamma(\overline{\beta})^\top \overline{\Gamma} = 0$. Using this property, the inclusion $\overline{\beta} \in D^m$ and $\overline{\Gamma}, \Gamma(\overline{\beta}) \in D^n$, we see that $(\mathbf{x}^t, \mathbf{y}^t) \in Q$.

Consider a bounded sequence of inequalities $\gamma^{t\top} \mathbf{x} - \alpha^{t\top} \mathbf{y} \geq 0$, where $t \in \mathbb{N}$, that are satisfied by points in C_Γ and such that $0 \geq \gamma^{t\top} \mathbf{x}^t - \alpha^{t\top} \mathbf{y}^t$. By the Farkas Lemma, there exist numbers $\lambda_{\beta,t} \geq 0$ for each $\beta \in I$ such that

$$(\gamma^t, \alpha^t) = \sum_{\beta \in I} \lambda_{\beta,t}(\Gamma(\beta), \beta). \tag{7}$$

After normalizing (γ^t, α^t), we may assume $\|(\gamma^t, \alpha^t)\| = \|(\Gamma(\overline{\beta}), \overline{\beta})\|$ for all t. Furthermore, according to Carathéodory's theorem we may assume that for each t the set $\{(\Gamma(\beta), \beta) : \lambda_{\beta,t} > 0\}$ is linearly independent. Consequently, there exists $\tau > 0$ such that $\lambda_{\beta,t} \leq \tau$ for each $\beta \in I$ and $t \in \mathbb{N}$.

In order to demonstrate that $((\mathbf{x}^t, \mathbf{y}^t))_{t=1}^\infty$ is an exposing sequence, we prove

$$\lim_{t\to\infty} \sum_{\beta \in I} \lambda_{\beta,t}(\Gamma(\beta), \beta) = (\Gamma(\overline{\beta}), \overline{\beta}).$$

To this end, it suffices to prove $\lim_{t\to\infty} \lambda_{\beta,t} = 0$ for each $\beta \in I \setminus \{\overline{\beta}\}$. We will choose $\overline{\Gamma}$ so that this condition is met. Note that

$$0 \geq \gamma^{t\top} \mathbf{x}^t - \alpha^{t\top} \mathbf{y}^t = \sum_{\beta \in I} \lambda_{\beta,t}\big((\Gamma(\beta)^\top \Gamma(\overline{\beta}) - \beta^\top \overline{\beta}) + \tfrac{\sqrt{2t+1}}{t}\,\overline{\Gamma}^\top \Gamma(\beta) - \tfrac{1}{t}\beta^\top \overline{\beta}\big).$$

Multiplying through by t, we have

$$0 \geq \sum_{\beta \in I} \lambda_{\beta,t}\big(t(\Gamma(\beta)^\top \Gamma(\overline{\beta}) - \beta^\top \overline{\beta}) + \sqrt{2t+1}\,\overline{\Gamma}^\top \Gamma(\beta) - \beta^\top \overline{\beta}\big). \tag{8}$$

Claim 1. $\overline{\Gamma}$ can be chosen such that $\overline{\Gamma}^\top \Gamma(\overline{\beta}) = 0$ and for each $\beta \in I \setminus \{\overline{\beta}\}$

$$\lim_{t\to\infty} t(\Gamma(\beta)^\top \Gamma(\overline{\beta}) - \beta^\top \overline{\beta}) + \sqrt{2t+1}\,\overline{\Gamma}^\top \Gamma(\beta) - \beta^\top \overline{\beta} = \infty.$$

Proof of Claim. Regardless of $\overline{\Gamma}$, if β is such that $\Gamma(\beta)^\top \Gamma(\overline{\beta}) - \beta^\top \overline{\beta} > 0$, then the limit goes to ∞ because the term t dominates $\sqrt{2t+1}$.

In what remains, we need to choose $\overline{\Gamma}$ so that $\Gamma(\beta)^\top \overline{\Gamma} > 0$ for all $\beta \in I \setminus \{\overline{\beta}\}$ satisfying $\Gamma(\beta)^\top \Gamma(\overline{\beta}) - \beta^\top \overline{\beta} = 0$. If we establish this, then the limit tends to infinity because of the term $\sqrt{2t+1}$. Define

$$J := \{\beta \in I \setminus \{\overline{\beta}\} : \ \Gamma(\beta)^\top \Gamma(\overline{\beta}) - \beta^\top \overline{\beta} = 0\}.$$

We consider two cases: whether $\Gamma(\overline{\beta}) \in \text{cone}(\Gamma(J))$ or not.

Case 1. Assume $\Gamma(\overline{\beta}) \notin \text{cone}\{\Gamma(\beta) : \beta \in J\}$. Define $K := \text{cone}(\Gamma(J \cup \{\overline{\beta}\}))$. If K is not pointed, then $\mathbf{0} \in \mathbb{R}^n$ is a non-trivial conic combination of the generators of K. By Lemma 2, $\mathbf{0} \in \mathbb{R}^m$ can be obtained using the same conic multipliers applied to vectors in $J \cup \{\overline{\beta}\}$. However, Lemma 3 implies that C_Γ is not full-dimensional, which is a contradiction. Hence, K is pointed.

The fact that J is finite together with $\Gamma(\overline{\beta}) \notin \text{cone}\{\Gamma(\beta) : \beta \in J\}$ implies that $\Gamma(\overline{\beta})$ generates an extreme ray of K and there is no $\beta \in J$ that generates the same extreme ray. Thus, by the separating hyperplane theorem there exists some $\overline{\Gamma} \in D^n$ such that $\overline{\Gamma}^\top \Gamma(\overline{\beta}) = 0 < \overline{\Gamma}^\top \Gamma(\beta)$ for all $\beta \in J$, as desired.

Case 2. Assume $\Gamma(\overline{\beta}) \in \text{cone}\{\Gamma(\beta) : \beta \in J\}$. There exists a set $H \subseteq J$ and numbers $\epsilon_\beta > 0$ for each $\beta \in H$ such that $\Gamma(\overline{\beta}) = \sum_{\beta \in H} \epsilon_\beta \Gamma(\beta)$. Define $\widehat{\beta} := \sum_{\beta \in H} \epsilon_\beta \beta$. Using the non-expansive property of Γ, we have

$$\|\widehat{\beta}\|^2 = \sum_{\beta, \beta' \in H} \epsilon_\beta \epsilon_{\beta'} \beta^\top \beta' \leq \sum_{\beta, \beta' \in H} \epsilon_\beta \epsilon_{\beta'} \Gamma(\beta)^\top \Gamma(\beta') = \|\Gamma(\overline{\beta})\|^2 = 1.$$

Using the isometry with $\overline{\beta}$ and each $\beta \in H$, we then have

$$\widehat{\beta}^\top \overline{\beta} = \sum_{\beta \in H} \epsilon_\beta \beta^\top \overline{\beta} = \sum_{\beta \in H} \epsilon_\beta \Gamma(\beta)^\top \Gamma(\overline{\beta}) = \Gamma(\overline{\beta})^\top \Gamma(\overline{\beta}) = 1.$$

Thus, we have equality in the Cauchy-Schwarz inequality $|\widehat{\beta}^\top \overline{\beta}| \leq \|\widehat{\beta}\| \|\overline{\beta}\| \leq 1$, so $\widehat{\beta} = \beta^1$. Thus, $(\Gamma(\overline{\beta}), \overline{\beta}) = \sum_{\beta \in H} \epsilon_\beta (\Gamma(\beta), \beta)$ contradicting that $\Gamma(\overline{\beta})^\top \mathbf{x} - \overline{\beta}^\top \mathbf{y} \geq 0$ defines a facet of C_Γ.

Choose $\overline{\Gamma} \in D^n$ according to Claim 1. For each $t \in \mathbb{N}$, we have

$$\lambda_{\beta,t} \big(t(\Gamma(\overline{\beta})^\top \Gamma(\overline{\beta}) - \overline{\beta}^\top \overline{\beta}) + \sqrt{2t+1}\, \overline{\Gamma}^\top \Gamma(\overline{\beta}) - \overline{\beta}^\top \overline{\beta} \big) = -\lambda_{\beta,t} \geq -\tau.$$

Together with (8), this implies

$$0 \geq -\tau + \sum_{\beta \in I \setminus \{\overline{\beta}\}} \lambda_{\beta,t} \big(t(\Gamma(\beta)^\top \Gamma(\overline{\beta}) - \beta^\top \overline{\beta}) + \sqrt{2t+1}\, \overline{\Gamma}^\top \Gamma(\beta) - \beta^\top \overline{\beta} \big). \quad (9)$$

For $\beta \in I \setminus \{\overline{\beta}\}$, if $\lambda_{\beta,t}$ does not go to 0 as t tends to ∞, then Claim 1 implies that the righthand side of (9) will go to ∞, which is a contradiction. Hence, (7) tends to $(\Gamma(\beta), \beta)$ as t tends to ∞. $\qquad\square$

Remark 2. As we mentioned in Section 1.1, we conjecture that Theorem 3 is generalizable to a set C_Γ that is not necessarily a polyhedron. With this in mind, a natural question is how reliant on polyhedrality the proof of this section is. Various points of the proof can be adapted to handle a non-polyhedral case: for example, a similar expression to (7) can be obtained for an infinite I. However, one the key steps that heavily uses finiteness is the construction of $\overline{\Gamma}$ using a *strict* separating hyperplane in Case 1 of Claim 1. It is not clear if such $\overline{\Gamma}$ exists in a general case, and the proof may need a different approach.

346   G. Muñoz et al.

8   A Proof of Theorem 4

($\Leftarrow$) We show that if $\overline{\beta} \in D^m \setminus I$, then $\Gamma(\overline{\beta})^\top \mathbf{x} - \overline{\beta}^\top \mathbf{y} \geq 0$ is implied by the inequalities indexed by I. Let $\overline{\beta} \in D^m \setminus I$. By assumption, there exists a set $J \subseteq I$ of pairwise isometric points satisfying $\overline{\beta} \in \mathrm{cone}(J)$. Hence, there exist $\lambda_\beta \geq 0$ for each $\beta \in J$ such that $\overline{\beta} = \sum_{\beta \in J} \lambda_\beta \beta$. We have $\Gamma(\overline{\beta}) = \sum_{\beta \in J} \lambda_\beta \Gamma(\beta)$ by Lemma 2. This shows that $(\Gamma(\overline{\beta}), \overline{\beta}) \in \mathrm{cone}(\{(\Gamma(\beta), \beta) : \beta \in J\})$. Hence, $\Gamma(\overline{\beta})^\top \mathbf{x} - \overline{\beta}^\top \mathbf{y} \geq 0$ is implied by the inequalities indexed by I.

($\Rightarrow$) C_Γ is a polyhedron, so by Lemma 4 there is a finite representation

$$C_\Gamma = \left\{ (\mathbf{x}, \mathbf{y}) \in \mathbb{R}^n \times \mathbb{R}^n : \Gamma(\beta)^\top \mathbf{x} - \beta^\top \mathbf{y} \geq 0 \ \forall \beta \in I \right\}.$$

Let $\overline{\beta} \in D^m \setminus I$. The inequality $\Gamma(\overline{\beta})^\top \mathbf{x} - \overline{\beta}^\top \mathbf{y} \geq 0$ is valid for C_Γ, so there exists a set $J \subseteq I$ and positive coefficients λ_β for each $\beta \in J$ such that

$$(\Gamma(\overline{\beta}), \overline{\beta}) = \sum_{\beta \in J} \lambda_\beta (\Gamma(\beta), \beta).$$

Lemma 5 states that $\overline{\beta}^\top \beta = \Gamma(\overline{\beta})^\top \Gamma(\beta)$ for all $\beta \in J$. For each $\beta' \in J$, we have

$$\Gamma(\beta')^\top \Gamma(\overline{\beta}) - \beta'^\top \overline{\beta} = \sum_{\beta \in J} \lambda_\beta (\Gamma(\beta')^\top \Gamma(\beta) - \beta'^\top \beta).$$

The left-hand side is 0 because $\overline{\beta}$ and β' are isometric, and every summand on the right-hand side is nonnegative because $\lambda_\beta > 0$ and $\Gamma(\beta')^\top \Gamma(\beta) - \beta'^\top \beta \geq 0$ by the non-expansive property of Γ. Hence, $\Gamma(\beta')^\top \Gamma(\beta) = \beta'^\top \beta$ for all $\beta \in J$. As β' was arbitrarily chosen in J, we see that all elements of J are pairwise isometric and $\overline{\beta} \in \mathrm{cone}(J)$.

Acknowledgements. The second author was supported by a Natural Sciences and Engineering Research Council of Canada (NSERC) Discovery Grant [RGPIN-2021-02475]. The authors would like to thank the three anonymous reviewers for their valuable feedback.

References

1. Andersen, K., Jensen, A.N.: Intersection cuts for mixed integer conic quadratic sets. In: Goemans, M., Correa, J. (eds.) IPCO 2013. LNCS, vol. 7801, pp. 37–48. Springer, Heidelberg (2013). https://doi.org/10.1007/978-3-642-36694-9_4
2. Andersen, K., Louveaux, Q., Weismantel, R., Wolsey, L.: Cutting planes from two rows of the simplex tableau. In: Proceedings of Integer Programming and Combinatorial Optimization (IPCO), pp. 1–15 (2007)
3. Averkov, G.: A proof of Lovász's theorem on maximal lattice-free sets. Contrib. Algebra Geom. (2013)
4. Averkov, G., Basu, A., Paat, J.: Approximation of corner polyhedra with families of intersection cuts. SIAM J. Optim. **28**(1), 904–929 (2018)
5. Averkov, G.: On maximal s-free sets and the Helly number for the family of s-convex sets. SIAM J. Discret. Math. **27**(3), 1610–1624 (2013)

6. Baes, M., Oertel, T., Weismantel, R.: Duality for mixed-integer convex minimization. Math. Program. **158**, 547–564 (2016)
7. Balas, E.: Intersection cuts - a new type of cutting planes for integer programming. Oper. Res. (1971)
8. Barvinok, A.: A course in convexity. Am. Math. Soc. (2002)
9. Basu, A., Conforti, M., Cornuéjols, G., Weismantel, R., Weltge, S.: Optimality certificates for convex minimization and Helly numbers. Oper. Res. Lett. **45**(6), 671–674 (2017)
10. Basu, A., Dey, S., Paat, J.: Nonunique lifting of integer variables in minimal inequalities. SIAM J. Discret. Math. (2019)
11. Basu, A., Conforti, M., Cornuéjols, G., Zambelli, G.: Maximal lattice-free convex sets in linear subspaces. Math. Oper. Res. **35**(3), 704–720 (2010)
12. Basu, A., Conforti, M., Cornuéjols, G., Zambelli, G.: Minimal inequalities for an infinite relaxation of integer programs. SIAM J. Discret. Math. **24**(1), 158–168 (2010)
13. Bienstock, D., Chen, C., Muñoz, G.: Intersection cuts for polynomial optimization. In: Lodi, A., Nagarajan, V. (eds.) IPCO 2019. LNCS, vol. 11480, pp. 72–87. Springer, Cham (2019). https://doi.org/10.1007/978-3-030-17953-3_6
14. Bienstock, D., Chen, C., Muñoz, G.: Outer-product-free sets for polynomial optimization and oracle-based cuts. Math. Program. **183**, 105–148 (2020)
15. Chmiela, A., Muñoz, G., Serrano, F.: On the implementation and strengthening of intersection cuts for QCQPs. Math. Program. 1–38 (2022)
16. Conforti, M., Cornuéjols, G., Daniilidis, A., Lemaréchal, C., Malick, J.: Cut-generating functions and S-free sets. Math. Oper. Res. (2014)
17. Conforti, M., Cornuéjols, G., Zambelli, G.: A geometric perspective on lifting. Oper. Res. **59**(3), 569–577 (2011)
18. Conforti, M., Cornuéjols, G., Zambelli, G.: Integer Programming. Springer, Cham (2014). https://doi.org/10.1007/978-3-319-11008-0
19. Conforti, M., Summa, M.D.: Maximal s-free convex sets and the Helly number. SIAM J. Discret. Math. **30**(4), 2206–2216 (2016)
20. Dey, S., Wolsey, L.: Two row mixed-integer cuts via lifting. Math. Program. **124**, 143–174 (2010)
21. Lovász, L.: Geometry of numbers and integer programming. In: Iri, M., Tanabe, K. (eds.) Mathematical Programming: Recent Developments and Applications, pp. 177–201. Kluwer Academic Publishers, Amsterdam (1989)
22. Modaresi, S., Kılınç, M., Vielma, J.: Intersection cuts for nonlinear integer programming convexification techniques for structured sets. Math. Program. (2016)
23. Muñoz, G., Serrano, F.: Maximal quadratic-free sets. In: Proceedings of the International Conference on Integer Programming and Combinatorial Optimization, pp. 307–321 (2020)
24. Muñoz, G., Serrano, F.: Maximal quadratic-free sets. Math. Program. **192**, 229–270 (2022)
25. Paat, J., Schlöter, M., Speakman, E.: Constructing lattice-free gradient polyhedra in dimension two. Math. Program. **192**(1), 293–317 (2022)
26. Rockafellar, R.T.: Convex Analysis. Princeton University Press, Princeton (1970)
27. Tuy, H.: Concave minimization under linear constraints with special structure. Dokl. Akad. Nauk SSSR **159**, 32–35 (1964)

Compressing Branch-and-Bound Trees

Gonzalo Muñoz[1], Joseph Paat[2(✉)], and Álinson S. Xavier[3]

[1] Institute of Engineering Sciences, Universidad de O'Higgins, Rancagua, Chile
`gonzalo.munoz@uoh.cl`
[2] Sauder School of Business, University of British Columbia, Vancouver, BC, Canada
`joseph.paat@sauder.ubc.ca`
[3] Energy Systems and Infrastructure Analysis Division, Argonne National Laboratory, Lemont, IL, USA
`axavier@anl.gov`

Abstract. A branch-and-bound (BB) tree certifies a dual bound on the value of an integer program. In this work, we introduce the tree compression problem (TCP): *Given a BB tree T that certifies a dual bound, can we obtain a smaller tree with the same (or stronger) bound by either (1) applying a different disjunction at some node in T or (2) removing leaves from T?* We believe such post-hoc analysis of BB trees may assist in identifying helpful general disjunctions in BB algorithms. We initiate our study by considering computational complexity and limitations of TCP. We then conduct experiments to evaluate the compressibility of realistic branch-and-bound trees generated by commonly-used branching strategies, using both an exact and a heuristic compression algorithm.

1   Introduction

Consider an integer linear programming (IP) problem

$$\min\{\mathbf{c}^\top \mathbf{x} : \ \mathbf{x} \in \mathcal{P} \cap \mathbb{Z}^n\}, \tag{1}$$

where $\mathbf{c} \in \mathbb{Q}^n$ and $\mathcal{P} := \{\mathbf{x} \in \mathbb{R}^n : \ \mathbf{A}\mathbf{x} \leq \mathbf{b}\}$ for $\mathbf{A} \in \mathbb{Q}^{m \times n}$ and $\mathbf{b} \in \mathbb{Q}^m$. Primal bounds on (1) can be certified by integer feasible solutions $\mathbf{z} \in \mathcal{P} \cap \mathbb{Z}^n$. Dual bounds on (1), on the other hand, are typically certified using **branch-and-bound (BB) trees**. A BB tree is a graph-theoretical tree T where each node v corresponds to a polyhedron $\mathcal{Q}(v)$, with the root corresponding to $\mathcal{P}$. Moreover, v is either a leaf, or it has exactly two children corresponding to the polyhedra defined by applying a disjunction $(\boldsymbol{\pi}^\top \mathbf{x} \leq \pi_0) \vee (\boldsymbol{\pi}^\top \mathbf{x} \geq \pi_0 + 1)$ to $\mathcal{Q}(v)$, where we call $\boldsymbol{\pi} \in \mathbb{Z}^n$ the **branching direction** and $\pi_0 \in \mathbb{Z}$. If we solve the corresponding linear programs over all leaves of T, then the smallest value obtained over all leaves yields a dual bound for (1). See Sect. 2 for a formal definition of BB trees and the dual bound.

In order to generate a BB tree, one must identify a strategy for selecting a leaf of the tree and a strategy for selecting a disjunction to apply. See [22] for a survey on different strategies. In practical implementations of the BB method, the only allowed directions are typically $\{e^1, \ldots, e^n\}$, in which case we say the algorithm uses *variable disjunctions*. However, many results explore the benefit of additional directions: various subsets of $\{-1, 0, 1\}^n$ are explored in [25, 27, 30]; directions derived from mixed integer Gomory cuts are explored in [9, 19]; directions derived using basis reduction techniques are explored in [1, 26]; Mahajan and Ralphs [23] solve a subproblem to find a disjunction that closes the duality gap by a certain amount. The largest set of directions is the set $\mathbb{Z}^n$, in which case the algorithm uses *general disjunctions*.

Although a larger set of allowable directions provides more flexibility, it has been repeatedly verified that searching through this set during the execution of the algorithm can be computationally expensive [15, 23]. The work in this paper is motivated by a different approach to identify meaningful directions. Given a tree T produced using some set of allowable directions $\mathcal{D} \subseteq \mathbb{Z}^n$, we ask if T can be "compressed" into a smaller tree with the same (or stronger) dual bound by using a potentially larger set of directions $\mathcal{D}' \supseteq \mathcal{D}$, and a limited set of transformations. This post-hoc compression analysis is more restricted and allows one to use a global view of the tree to identify potentially meaningful branching directions, as opposed to the dynamic approach. We believe this compression question may help produce small trees to be used as better certificates [7] or as training data for learn-to-branch strategies.

Related Work. To the best of our knowledge, this is the first piece of work to study the tree compression problem. A related question is the minimum size of a BB tree certifying optimality or infeasibility of (1); we use some of these results in our own work. Chvátal [8] and Jeroslow [18] gives examples of IPs that require a BB tree whose size is exponential in the number of variables n when only variable directions $\mathcal{D} = \{e^1, \ldots, e^n\}$ are used to generate disjunctions. There are examples where an exponential lower bound in n cannot be avoided even with general disjunctions [10, 11]. Basu et al. [3] consider the set $\mathcal{D}_s$ of directions whose support is at most s; they prove that if $s \in O(1)$, then a BB tree proving infeasibility of Jeroslow's instance has exponential in n many nodes [3]. For an interesting perspective on provable upper bounds, Dey et al. [12] relate the size of BB trees generated using full strong branching and variable disjunctions to the additive integrality gap for certain classes of instances like vertex cover.

For complexity results, Pfetsch et al. [16] show that is it NP-hard to find the smallest BB tree generated using only variable disjunctions. Mahajan and Ralphs [24] show that it is NP-complete to decide whether there is a general disjunction proving infeasibility at the root node. They also provide a MIP that can be solved at a node in a BB tree to yield a disjunction maximizing the dual bound improvement.

The tree compression problem is a post-hoc analysis of a BB tree. A similar kind of analysis is done in backdoor branching, where one explores a tree T to find small paths from the root to the optimal solution with the ultimate aim to identify good branching decisions to make next time the algorithm is run on a similar IP [14, 20]. The major difference between backdoor branching and the compression question is that the former only considers finding a path existing in the tree while the latter considers how to modify a tree to create short paths. Another form of post-hoc analysis is tree balancing, where the goal is to transform a tree T proving integer infeasibility into a new tree with the same dual bound whose size is polynomial in $|T|$ and whose depth is polylogarithmic in $|T|$; see, e.g., [4] for a discussion on balancing and stabbing planes. A major difference between the balancing question and the compression question is that the former is allowed to grow the tree along branches while the latter is not.

Our Contributions. We introduce the tree compression problem in Sect. 2. In Theorem 1, we show that the problem is NP-Complete when $\mathcal{D} = \mathbb{Z}^n$ and $\mathbf{c} = \mathbf{0}$. We then demonstrate in Theorem 2 that tree compression does not always give the smallest BB tree meeting a certain dual bound. In fact, we give an example of a BB tree T of size $|T| \geq 2^{n+1} - 1$ that cannot be compressed to a BB tree with fewer than $(2^n - 1)/n$ nodes, but there is a different BB tree with the same root and dual bound with only 7 nodes. These results appear in Sect. 3.

We next provide extensive computational results on the compression problem. We look at BB trees from MIPLIB 3.0 [6] instances generated using *full strong branching*, the state-of-the-art variable branching strategy with respect to tree size, and *reliability branching with plunging*, often considered the state-of-the-art branching strategy with respect to running time. We first compress these trees using a computationally-expensive exact algorithm based on a MIP formulation by Mahajan and Ralphs [23,24]. We then evaluate how much of this compression is achievable in a short amount of time, by applying a heuristic algorithm based on the iterative procedure introduced by Owen and Mehrota [27]. Overall, we see that many MIPLIB 3.0 trees can be significantly compressed. Moreover, we find that the heuristic procedure achieves good compression. These algorithms and results are described in Sects. 4 and 5, respectively.

2 The Tree Compression Problem (TCP)

We define a **branch-and-bound (BB) tree** as a graph-theoretical rooted tree where each node v corresponds to a polyhedron $\mathcal{Q}(v)$, and the root node r corresponds to $\mathcal{Q}(r) = \mathcal{P}$. Furthermore, each node v is either a leaf, or it has exactly two children corresponding to the polyhedra

$$\mathcal{Q}(v) \cap \{\mathbf{x} \in \mathbb{R}^n : \boldsymbol{\pi}^\top \mathbf{x} \leq \pi_0\} \quad \text{and} \quad \mathcal{Q}(v) \cap \{\mathbf{x} \in \mathbb{R}^n : \boldsymbol{\pi}^\top \mathbf{x} \geq \pi_0 + 1\}, \quad (2)$$

where $\boldsymbol{\pi} \in \mathbb{Z}^n$ is called the **branching direction** and $\pi_0 \in \mathbb{Z}$. The **dual bound** relative to $\mathbf{c} \in \mathbb{Q}^n$ provided by a BB tree T is

$$d(T, \mathbf{c}) := \min_{v \in L(T)} \min\{\mathbf{c}^\top \mathbf{x} : \mathbf{x} \in \mathcal{Q}(v)\},$$

where $L(T)$ is the set of leaves of T. Define $d(T, \mathbf{c}) = \infty$ if $\mathcal{Q}(v) = \emptyset$ for each $v \in L(T)$, and $d(T, \mathbf{c}) = -\infty$ if $\mathbf{x} \mapsto \mathbf{c}^\top \mathbf{x}$ is unbounded from below over $\mathcal{Q}(v)$ for some $v \in L(T)$. For simplicity, our definition allows BB trees that have multiple nodes corresponding to the same polyhedron, although such trees would typically not be generated by well-designed BB algorithms. We also do not require the tree to certify infeasibility or optimality of (1), to allow trees generated by partial (e.g. time- or node-limited) runs of the BB method.

Let T be a BB tree and $v \in T$ be a non-leaf node. Our notion of compression is based on two operations on T. For $(\boldsymbol{\pi}, \pi_0) \in \mathbb{Z}^n \times \mathbb{Z}$, let

$$\text{replace}(T, v, \boldsymbol{\pi}, \pi_0)$$

denote the BB tree obtained from T by replacing all descendants of v with the two new children defined by applying the disjunction $(\boldsymbol{\pi}^\top \mathbf{x} \leq \pi_0) \vee (\boldsymbol{\pi}^\top \mathbf{x} \geq \pi_0 + 1)$ to $\mathcal{Q}(v)$, i.e., the two new children are the polyhedra in (2). We use

$$\text{drop}(T, v)$$

to denote the BB tree obtained from T by removing all descendants of v.

We refer to the number of nodes in T as the **size** of T and denote it by $|T|$. We say that a BB tree T' is a **compression** of T if there exists a sequence of BB trees $T_1 = T, T_2, \ldots, T_k = T'$ such that for each $i \in \{2, \ldots, k\}$ we have that

1. Either $T_i = \text{drop}(T_{i-1}, v)$ for some $v \in T_{i-1}$, or $T_i = \text{replace}(T_{i-1}, v, \boldsymbol{\pi}, \pi_0)$ for some $v \in T_{i-1}$ and $(\boldsymbol{\pi}, \pi_0) \in \mathbb{Z}^n \times \mathbb{Z}$.
2. $|T_i| < |T_{i-1}|$ and $d(T_i, \mathbf{c}) \geq d(T_{i-1}, \mathbf{c})$.

Note that the definition of compression depends on the dual bound of T. Also, observe that the replacement operation only acts on non-leaf nodes and thus only produces children of non-leaf nodes. Consequently, leaf nodes of a BB tree will either remain leaf nodes or disappear from the tree during the compression process. Given that the replacement operation creates two new nodes that are leaves themselves, the previous discussion implies that any new (potentially dense) disjunctions introduced in the compression process appear near the bottom of the tree. See [5,15,28] for comments on potential drawbacks of dense inequalities.

As an example of these definitions, consider $\mathcal{P} := [0, 1/5]^2$ and the following BB tree T (disjunctions are indicated on edges and polyhedra in the nodes):

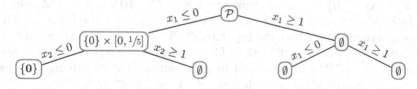

Let $\mathbf{c} = (-1, -1)$; we have $d(T, \mathbf{c}) = 0$. We can compress T with the drop operation at the right child v_2 of the root r; see figure (a). We can compress T with the replace operation at the root with $\boldsymbol{\pi} = -\mathbf{c}$ and $\pi_0 = 0$; see figure (b). It can be checked that $d(\text{drop}(T, v_2), \mathbf{c}) = d(\text{replace}(T, r, \boldsymbol{\pi}, 0), \mathbf{c}) = 0$.

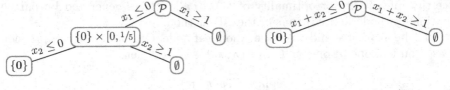

(a) The BB tree drop(T, v_2) (b) The BB tree replace$(T, r, \boldsymbol{\pi}, 0)$

The example illustrates that strict dual improvement is not necessary in the compression process. It is possible for the dual bound to improve during the compression process; e.g., use the same example except replace $\mathcal{P}$ by the triangle with vertices $(-1/2, -1/2)$, $(-1/2, 1)$, $(1, -1/2)$. For an example of an invalid compression operation, one can replace $1/5$ in the original example by $1/2$; here replace$(T, r, \boldsymbol{\pi}, 0)$ would no longer be a compression because we deteriorate the lower bound.

The **tree compression problem (TCP)** with respect to a set of allowable directions $\mathcal{D}$ is defined as follows: *Given a BB tree T and an objective vector $\mathbf{c} \in \mathbb{Q}^n$, is there a compression of T where the replacement operation only uses branching directions in $\mathcal{D}$?* There is also an optimization version of this question in which we try to compress T as much as possible. Section 3 considers this decision problem (showing this is NP-Complete) and the optimization problem (showing limitations of compression). Our computational results in Sects. 4 and 5 consider the optimization problem. As seen in the previous example, the choice of $\mathcal{D}$ influences the compression question; the BB tree in figure (a) is the best compression if $\mathcal{D}$ only contains unit vectors while the BB tree in figure (b) is the best compression if $\mathcal{D}$ contains the all-ones vector.

3   Complexity Results and Lower Bounds

We show (TCP) is NP-Complete when $\mathcal{D} = \mathbb{Z}^n$ and $\mathbf{c} = \mathbf{0}$. Our proof uses a reduction from the NP-Complete problem of disjunctive infeasibility **(DI)** [24, Proposition 3.2]: *Given $\mathbf{A} \in \mathbb{Q}^{m \times n}$ and $\mathbf{b} \in \mathbb{Q}^n$ defining a polyhedron $\mathcal{S} = \{\mathbf{x} \in \mathbb{R}^n : \mathbf{A}\mathbf{x} \leq \mathbf{b}\}$, decide if there exists $\boldsymbol{\pi} \in \mathbb{Z}^n \setminus \{\mathbf{0}\}$ and $\pi_0 \in \mathbb{Z}$ such that $\mathcal{S} \subseteq \{\mathbf{x} \in \mathbb{R}^n : \pi_0 < \boldsymbol{\pi}^\top \mathbf{x} < \pi_0 + 1\}$.* Keep in mind that the input to (DI) is a single polyhedron whereas the input to (TCP) is an entire BB tree. For this reason, if $\mathcal{D}$ is a smaller set in the (TCP) definition, e.g., vectors of bounded support, then (TCP) can be solved in polynomial time by solving a series of fixed dimension MIPs, one at each node of T; see [24, §2.1].

Theorem 1. (TCP) *is NP-Complete when $\mathcal{D} = \mathbb{Z}^n$ and $\mathbf{c} = \mathbf{0}$.*

Proof. We briefly argue (TCP) is in NP when $\mathcal{D} = \mathbb{Z}^n$ and $\mathbf{c} = \mathbf{0}$. Let T be a BB tree that can be compressed. Either $d(T, \mathbf{0}) = 0$, which happens if $\mathcal{Q}(v) \neq \emptyset$ for some $v \in L(T)$, or $d(T, \mathbf{0}) = \infty$, which happens if $\mathcal{Q}(v) = \emptyset$ for all $v \in L(T)$. If a non-leaf node v of T satisfies $\mathcal{Q}(v) = \emptyset$, then $T' = \text{drop}(T, v)$ is a compression of T whose size is polynomial in the size of T. Suppose $d(T, \mathbf{0}) = \infty$. Since T

can be compressed, there exists $v \in T$ and $(\pi, \pi_0) \in \mathbb{Z}^n \times \mathbb{Z}$ such that applying the disjunction $(\pi^\top \mathbf{x} \le \pi_0) \vee (\pi^\top \mathbf{x} \ge \pi_0 + 1)$ to $\mathcal{Q}(v)$ will yield two empty polyhedra. Mahajan and Ralphs demonstrate that finding such a disjunction is in NP [24, §3]. In particular, there is a compression $T' = \mathrm{replace}(T, v, \pi, \pi_0)$ of T whose size is polynomial in the size of T. This shows that (TCP) is in NP.

Consider an instance $(\mathbf{A}, \mathbf{b})$ of (DI). Let $\mathbf{x}^* \in \mathcal{S} \setminus \mathbb{Z}^n$; this can be found in polynomial time unless $\mathcal{S}$ is empty (in which case the answer to (DI) is 'yes') or a single integer vector (in which case the answer is 'no'). Without loss of generality, $x_1^* \notin \mathbb{Z}$. We lift $\mathcal{S}$ into $\mathbb{R}^{n+1}$ to create an instance of (TCP). We write a point in $\mathbb{R}^{n+1}$ as $(\mathbf{x}, y) \in \mathbb{R}^n \times \mathbb{R}$. Define

$$\mathcal{P} := \mathrm{conv}\left(\{(\mathbf{x}^*, 0), (\mathbf{x}^*, 1)\} \cup \{(\mathbf{x}, {}^1\!/\!{}_2) : \mathbf{x} \in \mathcal{S}\}\right)$$

We build a BB tree T with root node r and $\mathcal{Q}(r) = \mathcal{P}$. Branch on the disjunction $(y \le 0) \vee (y \ge 1)$ at r to obtain v_1 and v_2:

$$\mathcal{Q}(v_1) := \{(\mathbf{x}, y) \in \mathcal{P} : y \le 0\} = \{(\mathbf{x}^*, 0)\}$$
$$\mathcal{Q}(v_2) := \{(\mathbf{x}, y) \in \mathcal{P} : y \ge 1\} = \{(\mathbf{x}^*, 1)\}.$$

Branch on v_1 and v_2 using $(x_1 \le \lfloor x_1^* \rfloor) \vee (x_1 \ge \lceil x_1^* \rceil)$ to obtain v_3, v_4, v_5, v_6:

$$\mathcal{Q}(v_3) := \{(\mathbf{x}, y) \in \mathcal{P} : y \le 0 \text{ and } x_1 \le \lfloor x_1^* \rfloor\} = \emptyset$$
$$\mathcal{Q}(v_4) := \{(\mathbf{x}, y) \in \mathcal{P} : y \le 0 \text{ and } x_1 \ge \lceil x_1^* \rceil\} = \emptyset$$
$$\mathcal{Q}(v_5) := \{(\mathbf{x}, y) \in \mathcal{P} : y \ge 1 \text{ and } x_1 \le \lfloor x_1^* \rfloor\} = \emptyset$$
$$\mathcal{Q}(v_6) := \{(\mathbf{x}, y) \in \mathcal{P} : y \ge 1 \text{ and } x_1 \ge \lceil x_1^* \rceil\} = \emptyset.$$

T has 7 nodes, and the four leaves v_3, v_4, v_5, v_6 have corresponding polyhedra that are empty. The encoding size of T is polynomial in the encoding size of $\mathcal{S}$.

If (DI) has a 'yes' answer with certificate $\pi \in \mathbb{Z}^n \setminus \{\mathbf{0}\}$ and $\pi_0 \in \mathbb{Z}$, then

$$\mathcal{P} \subseteq \mathcal{S} \times \mathbb{R} \subseteq \{(\mathbf{x}, y) \in \mathbb{R}^n \times \mathbb{R} : \pi_0 < \pi^\top \mathbf{x} < \pi_0 + 1\}.$$

Hence, the answer to (TCP) is 'yes' because $\mathrm{replace}(T, r, (\pi, 0), \pi_0)$ is a compression of T. Assume (TCP) has a 'yes' answer. The drop operation can only be applied to r, v_1 or v_2, and doing so to any of these does not compress the tree because the dual bound decreases. So, the 'yes' answer must come from the replace operation. In order to decrease the size of the tree, which is required for compression, the replace operation must be applied at r. Therefore, there is $(\pi, \pi_{n+1}) \in \mathbb{Z}^n \times \mathbb{Z}$ and $\pi_0 \in \mathbb{Z}$ such that $\pi_0 < \pi^\top \mathbf{x} + \pi_{n+1} y < \pi_0 + 1$ for all $(\mathbf{x}, y) \in \mathcal{P}$. Note that $\pi \ne \mathbf{0}$ and $\pi_{n+1} = 0$ as otherwise $(\mathbf{x}^*, 0)$ or $(\mathbf{x}^*, 1)$ violates one of these inequalities. The tuple (π, π_0) provides a 'yes' answer to (DI). $\square$

Note that (TCP) can be answered in polynomial time if the set $\mathcal{D}$ of directions allowed in the replacement operation is polynomial in the size of T, e.g., $\mathcal{D} = \{\mathbf{e}^1, \ldots, \mathbf{e}^n\}$. Indeed, one can try the drop operation at each node and the replace operation for each node-direction pair $(v, \mathbf{d})$; this requires polynomial time because the size of $\mathcal{D}$ is polynomial in the size of T.

The next theorem shows that tree compression does not always yield the smallest tree for a given dual bound.

Theorem 2. *For $n \geq 2$, there exists a polytope $\mathcal{P} \subseteq \mathbb{R}^{n+1}$ and a BB tree T with root polyhedron $\mathcal{P}$ such that*

1. *$|T| \geq 2^{n+1} - 1$ and $d(T, \mathbf{0}) = \infty$.*
2. *T cannot be compressed to a tree with fewer than $(2^n - 1)/n$ nodes.*
3. *There exists a tree T' with root $\mathcal{P}$, $|T'| = 7$ and $d(T, \mathbf{0}) = d(T', \mathbf{0})$.*

Proof. Let $\overline{\mathcal{P}} \subseteq [0,1]^n$ be a polytope satisfying $\overline{\mathcal{P}} \cap \mathbb{Z}^n = \emptyset$ and if a tree $\overline{T}$ with root $\overline{\mathcal{P}}$ satisfies $d(\overline{T}, \mathbf{0}) = \infty$, then $|\overline{T}| \geq 2^{n+1} - 1$. One such $\overline{\mathcal{P}}$ comes from [11, Proposition 3]. Let $\overline{T}$ be a BB tree of minimal size with root $\overline{\mathcal{P}}$ and $d(\overline{T}, \mathbf{0}) = \infty$. Minimality implies that only the leaves correspond to empty polyhedra. There exist $(|\overline{T}| - 1)/2 \geq 2^n - 1$ non-empty non-leaf nodes in $\overline{T}$. For each non-empty node $\overline{v} \in \overline{T}$, we have $\mathcal{Q}(\overline{v}) \setminus \mathbb{Z}^n = \mathcal{Q}(\overline{v}) \neq \emptyset$. So, there exists $i^* \in \{1, \ldots, n\}$ such that at least $(2^n - 1)/n$ nodes in $\overline{T}$ whose corresponding polyhedron contains a point with i^*th component in $(0,1)$. We denote the set of these nodes as

$$\overline{N} := \{\overline{v} \in \overline{T} : \exists\, \mathbf{x} \in \mathcal{Q}(\overline{v}) \text{ with } x_{i^*} \in (0,1)\}.$$

For each $\overline{v} \in \overline{N}$, arbitrarily choose a point in $\mathcal{Q}(\overline{v})$ whose i^*th component is in $(0,1)$ and call this point $\mathbf{x}(\overline{v})$. Define

$$\mathcal{P} := \text{conv}(\{(\mathbf{x}(\overline{v}), t) : \overline{v} \in \overline{N} \text{ and } t \in \{0,1\}\} \cup (\overline{\mathcal{P}} \times \{1/2\})).$$

Note that $\mathcal{P} \cap \mathbb{Z}^{n+1} = \emptyset$. Create a BB tree T' with root polyhedron $\mathcal{P}$ and $d(T', \mathbf{0}) = \infty$ by first branching on $(x_{n+1} \leq 0) \vee (x_{n+1} \geq 1)$; the polyhedra of the resulting children are $\text{conv}\{(\mathbf{x}(\overline{v}), j) : \overline{v} \in \overline{N}\}$ for $j \in \{0,1\}$. Given that $x(\overline{v})_{i^*} \in (0,1)$ for each $\overline{v} \in \overline{N}$, we can branch on each $\text{conv}\{(\mathbf{x}(\overline{v}), j) : \overline{v} \in \overline{N}\}$ using $(x_{i^*} \leq 0) \vee (x_{i^*} \geq 1)$ to obtain all empty children nodes. This proves 3.

We define T in the theorem by lifting $\overline{T}$. More precisely, extend every disjunction $(\overline{\boldsymbol{\pi}}^\top \mathbf{x} \leq \pi_0) \vee (\overline{\boldsymbol{\pi}}^\top \mathbf{x} \geq \pi_0 + 1)$ in $\overline{T}$ to a disjunction $(\boldsymbol{\pi}^\top \mathbf{x} \leq \pi_0) \vee (\boldsymbol{\pi}^\top \mathbf{x} \geq \pi_0 + 1)$, where $\boldsymbol{\pi} := (\overline{\boldsymbol{\pi}}, 0)$. Thus, $|T| = |\overline{T}| \geq 2^{n+1} - 1$. Furthermore, $\mathcal{P} \subseteq \overline{\mathcal{P}} \times \mathbb{R}$, so $d(T, \mathbf{0}) = \infty$ because $d(\overline{T}, \mathbf{0}) = \infty$. Thus, T satisfies 1.

It remains to prove 2. Every point in $\mathcal{P}$ is of the form $(\mathbf{x}', \alpha)$, where $\mathbf{x}'$ in $\overline{\mathcal{P}}$ and $\alpha \in [0,1]$. Assume to the contrary that T can be compressed via the drop operation. The corresponding node in $\overline{T}$ can also be dropped. However, this contradicts the minimality of $\overline{T}$. We claim that if $v \in T$ corresponds to a node in $\overline{v} \in \overline{N}$, then T cannot be compressed at v using the replace operation. Suppose there exists $v \in T$ corresponding to a node $\overline{v} \in \overline{N}$ and a disjunction $(\boldsymbol{\pi}^\top \mathbf{x} + \pi_{n+1} x_{n+1} \leq \pi_0) \vee (\boldsymbol{\pi}^\top \mathbf{x} + \pi_{n+1} x_{n+1} \geq \pi_0 + 1)$ that we can use to compress T at v via the replace operation, i.e., $\mathcal{Q}(v) \subseteq \{(\mathbf{x}, \alpha) \in \mathbb{R}^n \times \mathbb{R} : \pi_0 < \boldsymbol{\pi}^\top \mathbf{x} + \pi_{n+1}\alpha < \pi_0 + 1\}$. If $\pi_{n+1} = 0$, then this disjunction can be projected to $\overline{T}$ to compress it, contradicting the minimality of $\overline{T}$. So, $\pi_{n+1} \neq 0$. For each $\alpha \in [0,1]$ the point $(\mathbf{x}(\overline{v}), \alpha)$ satisfies $\pi_0 - \boldsymbol{\pi}^\top \mathbf{x}(\overline{v}) < \pi_{n+1}\alpha < \pi_0 - \boldsymbol{\pi}^\top \mathbf{x}(\overline{v}) + 1$. But this cannot be satisfied if we plug in $\alpha = 0$ and $\alpha = 1$ because $\pi_{n+1} \in \mathbb{Z}$. Thus, the replace operation can only be applied to nodes in T that do not correspond to nodes in $\overline{N}$. We have $|\overline{N}| \geq (2^n - 1)/n$, so T cannot be compressed to fewer than $(2^n - 1)/n$ nodes, which proves 2.   □

4   Compression Algorithms

In this section we introduce two compression algorithms and later evaluate their performance. Let T be a BB tree and $c \in \mathbb{Q}^n$. For both algorithms, the general approach we follow is: (1) Traverse T starting from the root. We may skip leaves, since these are not compressible; (2) If the minimum of $\mathbf{x} \mapsto \mathbf{c}^\top \mathbf{x}$ over $\mathcal{Q}(v)$ is greater than or equal to $d(T, \mathbf{c})$ then we apply drop(T, v); (3) Otherwise, we search for $(\boldsymbol{\pi}, \pi_0) \in \mathbb{Z}^n \times \mathbb{Z}$ such that $T' = \text{replace}(T, v, \boldsymbol{\pi}, \pi_0)$ satisfies $d(T, \mathbf{c}) \geq d(T', \mathbf{c})$. In the following, we provide two methods for Step (3), which is the bottleneck of the procedure.

4.1   An Exact Method

A BB tree replace$(T, v, \boldsymbol{\pi}, \pi_0)$ is a compression of T if and only if $\min\{\mathbf{c}^\top \mathbf{x} : \mathbf{x} \in \mathcal{Q}(v), \boldsymbol{\pi}^\top \boldsymbol{x} \leq \pi_0\} \geq d(T, \mathbf{c})$ and $\min\{\mathbf{c}^\top \mathbf{x} : \mathbf{x} \in \mathcal{Q}(v), \boldsymbol{\pi}^\top \boldsymbol{x} \geq \pi_0 + 1\} \geq d(T, \mathbf{c})$. Mahajan and Ralphs [23] proposed MIP formulation that can be used to find such $(\boldsymbol{\pi}, \pi_0)$; the only difference is that they used it for finding a general disjunction that could provide the best possible dual improvement when branching, but we can easily adapt it to our compression task. The resulting model we use is

$$\max_{\substack{\delta, \mathbf{p}, \mathbf{q}, \boldsymbol{\pi}, \\ \pi_0, s_L, s_R}} \left\{ \delta : \begin{array}{c} \mathbf{A}^\top \mathbf{p} - s_L \mathbf{c} - \boldsymbol{\pi} = 0, \ \mathbf{p}^\top \mathbf{b} - d(T, \mathbf{c}) s_L - \pi_0 \geq \delta \\ \mathbf{A}^\top \mathbf{q} - s_R \mathbf{c} + \boldsymbol{\pi} = 0, \ \mathbf{q}^\top \mathbf{b} - d(T, \mathbf{c}) s_R - \pi_0 \geq \delta - 1 \\ \mathbf{p}, \mathbf{q} \geq 0, \ s_L, s_R \geq 0, \ \boldsymbol{\pi} \in \mathbb{Z}^n, \ \pi_0 \in \mathbb{Z} \end{array} \right\} \quad (3)$$

Any feasible solution with $\delta > 0$ produces a tuple $(\boldsymbol{\pi}, \pi_0)$ that we can use in the replace operation. Conversely, if no such δ exists, neither does a suitable disjunction; see [23]. Model (3) can be costly to solve in practice. However, if given enough time, one can be certain that it will yield an algorithm capable of compressing T as much as possible.

4.2   A Heuristic Method

Many heuristic methods for finding good branching directions have been proposed in the literature (e.g. [9,16,19,27]) and can be readily used for tree compression. Here, we adapt a procedure in Owen and Mehrota [27] that iteratively improves variable directions by changing one coefficient at a time.

To outline the method, assume we have solved the LP relaxation at a node v. The first step is to find the best variable direction $\boldsymbol{\pi} \in \{\mathbf{e}^1, \ldots, \mathbf{e}^n\}$. Suppose $\boldsymbol{\pi}^\top \mathbf{x} \leq \pi_0$ is the side of the disjunction with the smallest optimal value. We add this constraint to the node LP and re-solve it to obtain a fractional solution $\bar{\mathbf{x}}$. For each fractional component $\bar{x}_i$, we then evaluate the branching directions $\boldsymbol{\pi} + \mathbf{e}^i$ and $\boldsymbol{\pi} - \mathbf{e}^i$. If one of these directions yields a better dual bound than $\boldsymbol{\pi}$, then we replace $\boldsymbol{\pi}$ by it and repeat the procedure until $\boldsymbol{\pi}$ can no longer be improved. At the end, if the bound provided by $\boldsymbol{\pi}$ is better than the tree bound, we apply replace$(T, v, \boldsymbol{\pi}, \pi_0)$.

Unlike the previous exact method, this iterative method provides no guarantees that a suitable disjunction will be found, even if it exists, and therefore may not achieve the best compression. However, it is typically much faster.

5   Computational Experiments

In this section, we attempt to compress MIPLIB 3.0 trees using the methods described in the previous section. Our main goal is to evaluate, without taking running time into consideration, how compressible are realistic BB trees generated by two commonly-used branching strategies — *full strong branching (FSB)* and *reliability branching with plunging (RB)*. Our secondary goal is to estimate how much of this compression can be achieved in shorter and more practical running times.

5.1   Methodology

For each branching strategy and for each MIPLIB 3.0 instance, we started by generating a BB tree using a custom textbook implementation of the BB method. We chose MIPLIB 3.0, instead of larger benchmark sets, so that we could compute large FSB trees for all instances and could obtain accurate results for the exact compression method. We used a custom implementation of the BB method, instead of exporting the tree generated by a commercial MIP solver, so that we could easily understand how exactly the tree is generated and control every aspect of the algorithm. The implementation is written in Julia 1.8 and has been made publicly available as part of the open-source MIPLearn software package [29]. It relies on an external LP solver, accessed through JuMP [13] and Math-OptInterface [21], to solve the LP relaxation of each BB node and to evaluate strong branching decisions. In our experiments, we used Gurobi 9.5 [17] with default settings as the LP solver. When generating the trees, we provided the optimal value to the BB method and imposed a 10,000-node limit. No time limit was imposed, and no presolve or cutting planes were applied.

After the trees were generated, they were then compressed by the exact and the heuristic methods described in Sect. 4. Both methods were implemented in Python 3.10 and `gurobipy`. The nodes were traversed using depth-first search. For the exact method, we imposed a 24-h limit on the entire procedure and a 20-min limit on each individual MIP. For the heuristic method, we imposed a 15-min limit on the entire procedure and no time limits on individual nodes. All MIPs and LPs were solved with Gurobi 9.5 with default settings. The experiments were run on a dedicated desktop computer (AMD Ryzen 9 7950X, 4.5/5.7 GHz, 16 cores, 32 threads, 128 GB DDR5), and 32 trees were compressed in parallel at a time; each compression was single-threaded.

5.2   Full Strong Branching Results

Full strong branching (FSB) is a strategy which solves, at each node of the BB tree, two LPs for each fractional variable, then picks as the branching variable the one that presents the best overall improvement to dual bound [2]. FSB is often paired, as we do in our experiments, with *best-bound node selection* rule, which always picks, as the BB node to process next, an unexplored leaf node that has minimal optimal value. Although computationally expensive, FSB is typically

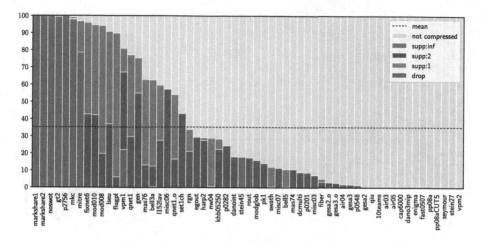

Fig. 1. Compressibility of FSB trees (exact method, 24-h limit).

considered the state-of-the-art branching strategy in terms of node count, so one would naturally expect such trees to be hard to compress.

Figure 1 shows the exact compressibility of FSB trees under different restrictions on the support size of the disjunction. Specifically, supp:inf corresponds to the exact method based on Model (3), whereas supp:1 and supp:2 use the same model, but impose the additional constraint that at most 1 or 2 coefficients of π, respectively, can be non-zero. Method drop is the method in which we are only allowed to drop nodes, not replace them. In the chart, the compressibility of different methods is superimposed, with the weaker methods in the foreground and the stronger methods in the background. The y-axis indicates how small is the resulting tree, with larger values indicating higher compression. For example, on instance vmp1, methods drop, supp:2 and supp:inf were able to reduce the tree by 22.2%, 67.1% and 80.9%, respectively. Method supp:1 does not appear in the chart because it was not able to improve upon drop. The line shows the average compression obtained by the strongest method across all instances.

Our first insight from Fig. 1 is that many FSB trees can be significantly compressed, despite the notorious tree-size efficiency of this branching rule. On average, supp:inf was able to reduce tree size by 35.2%, with the ratio exceeding 50% for 20 (out of 59) instances. We also note, from the figure, that a large support size is required for obtaining the best results, although a restricted support size still provides significant compression. On average, supp:2 compressed the trees by 24.0%, which is still considerable, although being well below supp:inf. Method supp:1, on the other hand, never outperformed drop. This was expected, as it can be easily shown that trees generated by FSB (with best-bound) on a particular set of candidate branching directions can never be compressed (beyond dropping nodes) based on the same set of directions. Also as a direct consequence of using the best-bound node selection rule, we observed that, for the vast majority of instances, few nodes could be dropped. On average, drop was only able

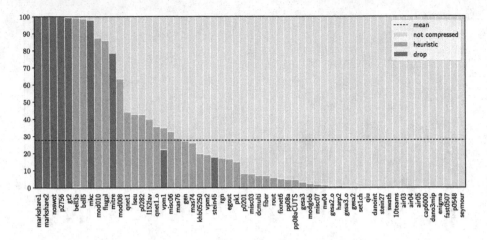

Fig. 2. Compressibility of FSB trees (heuristic method, 15-min limit).

to compress the trees by 12.1% on average, with the compression being near zero for 50 instances. Finally, despite the positive average compression results for `supp:inf`, we do note that a large number of trees could not be meaningfully compressed. Specifically, `supp:inf` presented a compression ratio below 5% for 19 instances, which may indicate that trees for certain classes of problems are hard to compress. Furthermore, `supp:inf` took an exceedingly long average time of 47,153 s, with 25 instances hitting the 24-h limit.

We now focus on more practical tree compression algorithms. Figure 2 shows the performance of the heuristic method, outlined in Subsect. 4.2, on the same BB trees, with a 15-min limit. We see that the heuristic method is able to obtain compression ratios comparable to `supp:inf` in a reasonable amount of time. On average, `heuristic` took 493 s to run (95x faster than the exact method), and reduced tree size by 27.7% (7.5% points lower). We conclude that FSB trees are compressible not only in a theoretical sense, but also in practice. We also note that `heuristic` outperformed `supp:inf` for 12 instances, sometimes by a significant margin. Notable examples include instances `bell5`, `bell3a`, `vpm2`, `p0282` and `mas74`, where the margin exceeded 15% points. This is possible due to the time limits imposed on `supp:inf`.

5.3   Reliability Branching with Plunging

Reliability branching (RB) is a strategy that attempts to accelerate FSB by skipping strong branching computations for variables that already have reliable pseudocosts [2]. In our experiments, the pseudocost of a variable is considered reliable if it is based on 10 or more strong branching evaluations. RB has been shown to perform well on a variety of real-world instances and it is often considered the state-of-the-art branching rule in terms of running time. *Plunging* is a modification to node selection which attempts to exploit the fact that sequentially solving two LPs that are similar can done much faster than solving two LPs

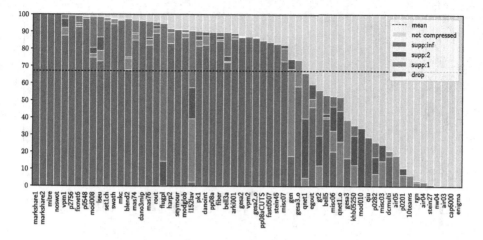

Fig. 3. Compressibility of RB trees (exact method, 24-h limit).

that are significantly different. When plunging is enabled, the BB method picks, as the node to explore next, one of the children of the most-recently explored node, falling back to *best-bound node selection* when both children are pruned. Our motivation for analyzing RB trees with plunging is that we expect such trees to resemble the ones generated by typical state-of-the-art MIP solvers.

Figure 3 shows the exact compressibility of RB trees, under different support size restrictions. The first notable fact is that RB trees are, as expected, much more compressible than FSB trees. On average, drop, supp:1, supp:2 and supp:inf were able to reduce tree size by 51.9%, 57.3%, 61.5% and 66.3%, respectively. Method supp:inf presented compression ratio above 50% for 42 (out of 59) instances, and above 80% for 34 instances. The strong performance of drop can be directly attributed to plunging. While the technique may be helpful when solving MIPs, we observed that it leads to the exploration of areas in the tree that do not contribute to its overall dual bound, and which can be dropped in a post-hoc analysis. As with previous experiments, the best compression results were obtained with larger support sizes, although, in this case, the benefits of unbounded support were not as large as before, in relative terms. Method supp:1, unlike in previous experiments, provided significant compression in a number of instances (e.g. gen, 1152lav, qnet1_o), and a modest average improvement over drop. We attribute this to suboptimal variable branching decisions made by RB, which is also expected. As in the previous case, we do note that supp:inf failed to meaningfully compress a few instances, and it was overall prohibitively slow, requiring 45,256 s on average.

Finally, Fig. 4 shows the performance of the heuristic method on RB trees. Similarly to the results in the previous section, the heuristic method presented very strong performance, obtaining compression ratios that approached or even exceed those of the exact method, in much smaller running times. Method heuristic took an average of 335 s (134x faster) and obtained an average com-

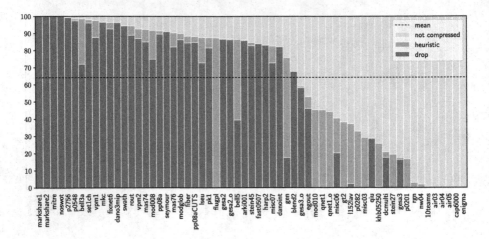

Fig. 4. Compressibility of RB trees (heuristic method, 15-min limit).

pression ratio of 63.7% (2.5% points lower). We conclude that BB trees generated by node and variable selection rules that focus on MIP solution time tend to be highly compressible, in both a theoretical and a practical sense.

6   Future Work

We have formally introduced the tree compression problem, and we demonstrated through experiments how much trees can be compressed. There are many open questions that we believe warrant future research. First, is there a family of problems for which BB trees generated, say using strong branching, can be provably compressed? Second, for a tree generated using branching directions in a set $\mathcal{D}$, how compressible is the tree using directions in the Minkowski Sum $\mathcal{D} + \mathcal{D}$? In particular when $\mathcal{D}$ is the set of variable disjunctions, a positive result may indicate sparse disjunctions that are useful in a BB tree. This would complement our current computational results on disjunctions of support size 2. Third, given that the compression algorithm is based (partially) on general disjunctions which can be seen as splits, is there a relationship between the strength of split cuts at the root and the compressibility of a BB tree? Finally, could the general disjunctions found by the compression algorithm be useful in solving similar MIP instances?

Acknowledgements. J. Paat was supported by a Natural Sciences and Engineering Research Council of Canada Discovery Grant [RGPIN-2021-02475]. Á.S. Xavier was partially supported by the U.S. Department of Energy Office of Electricity. The authors want to thank the referees, whose comments improved the overall presentation of the paper, led to better bounds in Theorem 2, and identified directions of future work.

References

1. Aardal, K., Lenstra, A.: Hard equality constrained integer knapsacks. Math. Oper. Res. **29**, 724–738 (2004)
2. Achterberg, T., Koch, T., Martin, A.: Branching rules revisited. Oper. Res. Lett. **33**(1), 42–54 (2005)
3. Basu, A., Conforti, M., Di Summa, M., Jiang, H.: Complexity of branch-and-bound and cutting planes in mixed-integer optimization - II. In: Singh, M., Williamson, D.P. (eds.) IPCO 2021. LNCS, vol. 12707, pp. 383–398. Springer, Cham (2021). https://doi.org/10.1007/978-3-030-73879-2_27
4. Beame, P., et al.: Stabbing planes. In: Karlin, A.R. (ed.) 9th Innovations in Theoretical Computer Science Conference (ITCS 2018). Leibniz International Proceedings in Informatics (LIPIcs), vol. 94, pp. 10:1–10:20. Schloss Dagstuhl-Leibniz-Zentrum fuer Informatik, Dagstuhl (2018). https://doi.org/10.4230/LIPIcs.ITCS.2018.10, http://drops.dagstuhl.de/opus/volltexte/2018/8341
5. Bixby, R.: Solving real-world linear programs: a decade and more of progress. Oper. Res. **50**(1), 3–15 (2002)
6. Bixby, R., Boyd, E., Indovina, R.: MIPLIB: a test set of mixed integer programming problems. SIAM News (1992)
7. Cheung, K.K.H., Gleixner, A., Steffy, D.E.: Verifying integer programming results. In: Eisenbrand, F., Koenemann, J. (eds.) IPCO 2017. LNCS, vol. 10328, pp. 148–160. Springer, Cham (2017). https://doi.org/10.1007/978-3-319-59250-3_13
8. Chvátal, V.: Hard knapsack problems. Oper. Res. **28**, 1402–1411 (1980)
9. Cornuéjols, G., Liberti, L., Nannicini, G.: Improved strategies for branching on general disjunctions. Math. Program. **130**, 225–247 (2011)
10. Dadush, D., Tiwari, S.: On the complexity of branching proofs. In: Saraf, S. (ed.) 35th Computational Complexity Conference (CCC 2020). Leibniz International Proceedings in Informatics (LIPIcs), vol. 169, pp. 34:1–34:35. Schloss Dagstuhl-Leibniz-Zentrum für Informatik, Dagstuhl (2020)
11. Dey, S., Dubey, Y., Molinaro, M.: Lower bounds on the size of general branch-and-bound trees. Math. Program. (2022)
12. Dey, S., Dubey, Y., Molinaro, M., Shah, P.: A theoretical and computational analysis of full strong-branching. arXiv:2110.10754 (2021)
13. Dunning, I., Huchette, J., Lubin, M.: Jump: a modeling language for mathematical optimization. SIAM Rev. **59**(2), 295–320 (2017). https://doi.org/10.1137/15M1020575
14. Fischetti, M., Monaci, M.: Backdoor branching. INFORMS J. Comput. **25**(4), 693–700 (2018)
15. Gamrath, G., Melchiori, A., Berthold, T., Gleixner, A.M., Salvagnin, D.: Branching on multi-aggregated variables. In: Michel, L. (ed.) CPAIOR 2015. LNCS, vol. 9075, pp. 141–156. Springer, Cham (2015). https://doi.org/10.1007/978-3-319-18008-3_10
16. Gläser, M., Pfetsch, M.: On the complexity of finding shortest variable disjunction branch-and-bound proofs. In: Aardal, K., Sanità, L. (eds.) IPCO 2022. LNCS, vol. 13265, pp. 291–304. Springer, Cham (2022). https://doi.org/10.1007/978-3-031-06901-7_22
17. Gurobi Optimization: Gurobi Optimizer (Version 9.5). https://www.gurobi.com/products/gurobi-optimizer/. Accessed 4 Nov 2022
18. Jeroslow, R.: Trivial integer programs unsolvble by branch-and-bound. Math. Program. **6**, 105–109 (1974)

19. Karamanov, M., Cornuéjols, G.: Branching on general disjunctions. Math. Program. **128**, 403–436 (2011)
20. Khalil, E., Vaezipoor, P., Dilkina, B.: Finding backdoors to integer programs: a Monte Carlo tree search framework. In: Proceedings of AAAI (2022)
21. Legat, B., Dowson, O., Dias Garcia, J., Lubin, M.: MathOptInterface: a data structure for mathematical optimization problems. INFORMS J. Comput. **34**(2), 672–689 (2021). https://doi.org/10.1287/ijoc.2021.1067
22. Linderoth, J., Savelsbergh, M.: A computational study of search strategies for mixed integer programming. INFORMS J. Comput. **11**(2), 173–187 (1999)
23. Mahajan, A., Ralphs, T.: Experiments with branching using general disjunctions. In: Proceedings of Operations Research and Cyber-Infrastructure, pp. 101–118 (2009)
24. Mahajan, A., Ralphs, T.: On the complexity of selecting disjunctions in integer programming. SIAM J. Optim. **20**(5), 2181–2198 (2010)
25. Mahmoud, H., Chinneck, J.: Achieving MILP feasibility quickly using general disjunctions. Comput. Oper. Res. **40**, 2094–2102 (2013)
26. Mehrotra, S., Li, Z.: Branching on hyperplane methods for mixed integer linear and convex programming using adjoint lattices. J. Glob. Optim. (2010)
27. Owen, J., Mehrotra, S.: Experimental results on using general disjunctions in branch-and-bound for general-integer linear programs. Comput. Optim. Appl. **20**, 159–170 (2001)
28. Walter, M.: Sparsity of lift-and-project cutting planes. In: Helber, S., et al. (eds.) Operations Research Proceedings 2012, pp. 9–14. Springer, Cham (2014). https://doi.org/10.1007/978-3-319-00795-3_2
29. Xavier, A.S., Qiu, F.: MIPLearn: a framework for learning-enhanced mixed-integer optimization (Julia interface) (2022). https://github.com/ANL-CEEESA/MIPLearn.jl
30. Yang, Y., Boland, N., Savelsbergh, M.: Multivariable branching: a 0–1 knapsack problem case study. INFORMS J. Comput. **33**(4), 1354–1367 (2021)

Exploiting the Polyhedral Geometry
of Stochastic Linear Bilevel Programming

Gonzalo Muñoz$^{(\boxtimes)}$, David Salas, and Anton Svensson

Instituto de Ciencias de la Ingeniería, Universidad de O'Higgins, Rancagua, Chile
{gonzalo.munoz,david.salas,anton.svensson}@uoh.cl

Abstract. We study linear bilevel programming problems whose lower-level objective is given by a random cost vector with known distribution. We consider the case where this distribution is nonatomic, allowing to pose the problem of the leader using vertex-supported beliefs in the sense of [29]. We prove that, under suitable assumptions, this formulation turns out to be piecewise affine over the so-called chamber complex of the feasible set of the high point relaxation. We propose two algorithmic approaches to solve general problems enjoying this last property. The first one is based on enumerating the vertices of the chamber complex. The second one is a Monte-Carlo approximation scheme based on the fact that randomly drawn points of the domain lie, with probability 1, in the interior of full-dimensional chambers, where the problem (restricted to this chamber) can be reduced to a linear program.

Keywords: Bilevel Programming · Bayesian Approach · Chamber complex · Enumeration algorithm · Monte-Carlo algorithm

1 Introduction

Stackelberg games, also referred to as bilevel programming problems, were first introduced by H. von Stackelberg in [31]. In this seminal work, an economic equilibrium problem between two firms was studied, under the particularity that one of them, the leader, is able to anticipate the decisions of the other one, the follower. Bilevel programming is an active field of research, and we refer the reader to the monographs [10,11] for comprehensive introductions, and to [12] for recent developments. In the last decade, researchers have started to consider uncertainty in Stackelberg games. A recent survey by Beck, Ljubić and Schmidt [3] provides an overview of new questions and recent contributions on this topic.

One model that considers uncertainty in Stackelberg games is the Bayesian approach [26,29]. The starting point is that for any given leader's decision x, the leader only knows that the reaction y of the follower is selected from a set $Y(x)$,

The first author was supported by FONDECYT Iniciación 11190515 (ANID-Chile).
The second author was supported by the Center of Mathematical Modeling (CMM)
FB210005 BASAL funds for centers of excellence (ANID-Chile), and the grant FONDE-CYT Iniciación 11220586 (ANID-Chile). The third author was supported by the grant FONDECYT postdoctorado 3210735 (ANID-Chile).

A. Del Pia and V. Kaibel (Eds.): IPCO 2023, LNCS 13904, pp. 363–377, 2023.
https://doi.org/10.1007/978-3-031-32726-1_26

hence being y a decision-dependent uncertainty parameter. The leader endows the set $Y(x)$ with a probability distribution β_x which models how the leader believes that the possible responses of the follower are distributed. Note that the classical optimistic and pessimistic approaches of bilevel programming are included in this setting, under quite mild assumptions (see [29]).

Uncertainty in the data of the lower-level has been considered by Claus for linear bilevel programming from a variational perspective considering risk measures (see the survey [6] by Burtscheidt, and the references therein, and the recent works [7,8]). In [19], Ivanov considered the cost function of the follower as a bilinear form $\langle Ax + \xi(\omega), y \rangle$. Recently, in [5], Buchheim, Henke and Irmai considered a bilevel version of the continuous knapsack problem with uncertainty on the follower's objective.

In this work, we consider a linear bilevel programming problem where the lower-level objective is uncertain for the leader but follows a prior known distribution (as the particular case studied in [5]). We study the problem from a Bayesian approach perspective [29], and by means of the so-called *chamber complex* of a polytope (see Sect. 4), which subdivides the space of the leader's decisions in a meaningful way. The idea of using the chamber complex to understand geometrical properties of optimization problems under uncertainty is not new, but it is recent. To the best of our knowledge, the first work that does this is [14] (see also [13]), on which multistage stochastic linear optimization is studied. However, the techniques there cannot be extended to Stackelberg games. Due to space constraints, we do not provide all details in this extended abstract. We refer the reader to our full-length preprint [27].

1.1 Problem Formulation and Contributions

Our study focuses on the setting of linear bilevel programming, i.e., the objective functions and constraints of the problem are all linear. More precisely, we aim to study the problem where the leader decides a vector $x \in \mathbb{R}^{n_x}$ that solves

$$(\mathcal{P}) := \begin{cases} \min_{x} \ \langle d_1, x \rangle + \mathbb{E}[\langle d_2, y(x, \omega) \rangle] \\[2mm] s.t. \ x \in X \\[2mm] y(x, \omega) \text{ solves } \begin{cases} \min_{y} \ \langle c(\omega), y \rangle \\ s.t. \ \begin{array}{l} y \in \mathbb{R}^{n_y}, \\ Ax + By \leq b, \end{array} \end{cases} \quad \omega \in \Omega \text{ a.s.} \end{cases} \tag{1}$$

where $A \in \mathbb{R}^{m \times n_x}$, $B \in \mathbb{R}^{m \times n_y}$, $b \in \mathbb{R}^m$, $d_1 \in \mathbb{R}^{n_x}$, $d_2 \in \mathbb{R}^{n_y}$, $c : \Omega \to \mathbb{S}_{n_y}$ is a random vector over a probability space $(\Omega, \Sigma, \mathbb{P})$ with values in the unit sphere of $\mathbb{R}^{n_y}$, and $X \subset \mathbb{R}^{n_x}$ is a nonempty polytope.

The notation carries the usual ambiguity of bilevel problems, which appears whenever the lower-level optimal response $y(x, \omega)$ is not uniquely determined for some $x \in X$. However, we focus our attention here on costs whose distributions are nonatomic (in a sense we will specify later on) which implies that, with probability 1, $y(x, \omega)$ is unique for all $x \in X$.

Our main contributions regarding this problem are: (a) To rewrite (1) using a Bayesian approach formulation and a sample average approximation for it; (b) to show the structure of the leader's objective function and its relation to the chamber complex of the feasible set of the high-point relaxation; and (c) to exploit these structures in a mixed-integer-programming-based algorithm and in a Monte-Carlo algorithm that can tackle (1).

2 Preliminaries

For an integer $n \in \mathbb{N}$, we write $[n] := \{1, \ldots, n\}$. Throughout this work, we will consider Euclidean spaces $\mathbb{R}^n$ endowed with their usual norm $\|\cdot\|$ and their inner product $\langle \cdot, \cdot \rangle$. We denote by $\mathbb{S}_n$ the unit sphere of $\mathbb{R}^n$. For a set $O \subset \mathbb{R}^n$, we denote the affine dimension of O as $\dim(O)$, which corresponds to the dimension of the affine envelope of O, denoted by $\mathrm{aff}(O)$. The relative interior of O is denoted by $\mathrm{ri}(O)$. We write $\mathbb{1}_O$ to denote the indicator function of a set $O \subset \mathbb{R}^n$, having value 1 on O and 0 elsewhere.

In general, we follow the standard notation of mathematical programming. For a polyhedron P, we denote by $\mathscr{F}(P)$ the collection of all faces of P, and by $\mathrm{ext}(P)$ the vertices (extreme points) of P. For any $k \in \{0, \ldots, n\}$ we will write

$$\mathscr{F}_{\leq k}(P) := \{F \in \mathscr{F}(P) \ : \ \dim(F) \leq k\}. \tag{2}$$

If there is no ambiguity, we might simply write $\mathscr{F}$ and $\mathscr{F}_{\leq k}$. For a convex set $C \subset \mathbb{R}^n$ and a point $x \in C$, we write $N_C(x)$ to denote the normal cone of C at x, i.e.,

$$N_C(x) = \{d \in \mathbb{R}^n \ : \ \langle d, y - x \rangle \leq 0 \quad \forall y \in C\}. \tag{3}$$

Motivated by the structure of problem (1), we define the polyhedron D as the feasible region of the high-point relaxation of the problem (see, e.g., [21]), i.e.,

$$D = \{(x, y) \in \mathbb{R}^{n_x} \times \mathbb{R}^{n_y} \ : \ Ax + By \leq b\}. \tag{4}$$

It will be assumed throughout the paper that D is full dimensional. We do not lose generality since it is always possible to embed D into $\mathbb{R}^{\dim(D)}$. We will also assume that D is compact, i.e. it is a polytope. Finally, by moving the leader's constraints to the follower's problem, we assume without losing any generality that

$$X = \{x \in \mathbb{R}^{n_x} \ : \ \exists y \in \mathbb{R}^{n_y} \text{ such that } Ax + By \leq b\}. \tag{5}$$

In the latter assumption about X we are simply stating that the lower-level problem is feasible for any feasible choice of x and restricting 'unilaterally' the x coordinate in D which does not change the lower-level problem. We define the ambient space Y for the follower's decision vector as

$$Y = \{y \in \mathbb{R}^{n_y} \ : \ \exists x \in \mathbb{R}^{n_x} \text{ such that } Ax + By \leq b\}. \tag{6}$$

Note that, since D is full-dimensional and compact in $\mathbb{R}^{n_x} \times \mathbb{R}^{n_y}$, both X and Y are full-dimensional (in $\mathbb{R}^{n_x}$ and $\mathbb{R}^{n_y}$, respectively) and compact as well. We write $\pi : \mathbb{R}^{n_x} \times \mathbb{R}^{n_y} \to \mathbb{R}^{n_x}$ to denote the parallel projection given by $\pi(x, y) = x$. In particular, equation (5) can be written as $X = \pi(D)$.

Given nonempty sets $U \subset \mathbb{R}^{n_x}$ and $V \subset \mathbb{R}^{n_y}$ we write $M : U \rightrightarrows V$ to denote a set-valued map, i.e., a function assigning to each $x \in U$ a (possibly empty) set $M(x)$ in V. In this work, we consider the set-valued map $S : X \rightrightarrows \mathbb{R}^{n_y}$ defined as

$$S(x) = \{y \in \mathbb{R}^{n_y} : (x, y) \in D\}. \tag{7}$$

We call S the *fiber map* of D (through the projection π). Clearly, S has nonempty convex and compact values, and $S(X) = Y$, as given by equation (6).

3   Vertex-Supported Beliefs and Bayesian Formulation

In what follows, we write $\mathcal{B}(Y)$ to denote the Borel σ-algebra of Y and $\mathscr{P}(Y)$ to denote the family of all (Borel) probability measures on Y. We endow $\mathscr{P}(Y)$ with the topology of weak convergence (see, e.g., [22, Chapter 13]). Accordingly, we will say that a measure-valued map $h : X \to \mathscr{P}(Y)$ is weak continuous if it is so for this topology (which coincides with the weak* topology when looking the space of measures as the dual space of the space of continuous functions, see [22, Remark 13.14]).

Recall from [29] that for a set-valued map $M : X \rightrightarrows Y$ with closed and nonempty values, a map $\beta : X \to \mathscr{P}(Y)$ is said to be *a belief over M* if for every $x \in X$, the measure $\beta(x) = \beta_x$ concentrates on $M(x)$, i.e., $\beta_x(M(x)) = 1$. By identifying $\mathscr{P}(M(x))$ with its natural injection into $\mathscr{P}(Y)$, a belief over M can be understood as a selection of $\{\mathscr{P}(M(x)) : x \in X\}$.

Let $(\Omega, \Sigma, \mathbb{P})$ be a probability space. To model the cost function of the lower-level problem in (1), we will consider only random vectors $c : \Omega \to \mathbb{S}_{n_y}$ with nonatomic distributions, in the sense that

$$\forall O \in \mathcal{B}(\mathbb{S}_{n_y}) : \quad \mathcal{H}^{n_y - 1}(O) = 0 \implies \mathbb{P}[c(\omega) \in O] = 0, \tag{8}$$

where $\mathcal{B}(\mathbb{S}_{n_y})$ stands for the Borel σ-algebra of $\mathbb{S}_{n_y}$, and $\mathcal{H}^{n_y - 1}$ denotes the $(n_y - 1)$-Hausdorff measure over $(\mathbb{S}_{n_y}, \mathcal{B}(\mathbb{S}_{n_y}))$. In other words, the probability measure $\mathbb{P} \circ c^{-1}$ is absolutely continuous with respect to $\mathcal{H}^{n_y - 1}$. Note that with this definition, any random vector $c : \Omega \to \mathbb{R}^{n_y}$, which has an absolutely continuous distribution with respect to the Lebesgue measure $\mathcal{L}^{n_y}$, induces an equivalent random vector $\bar{c} : \Omega \to \mathbb{S}_{n_y}$ given by $\bar{c}(\omega) = \frac{c(\omega)}{\|c(\omega)\|}$. This new random vector is well-defined almost surely in Ω, except for the negligible set $N = c^{-1}(0)$, and using $c(\cdot)$ or $\bar{c}(\cdot)$ in problem (1) is equivalent.

Now, to understand the distribution of the optimal response $y(x, \omega)$ induced by the random vector $c : \Omega \to \mathbb{S}_{n_y}$, we consider a belief $\beta : X \to \mathscr{P}(Y)$ over the fiber map $S : X \rightrightarrows Y$ given by

$$d\beta_x(y) = p_c(x, y) := \mathbb{P}[-c(\omega) \in N_{S(x)}(y)]. \tag{9}$$

Note that $\mathbb{P}[-c(\omega) \in N_{S(x)}(y)] = \mathbb{P}[-c(\omega) \in \operatorname{int}(N_{S(x)}(y)) \cap \mathbb{S}_{n_y}]$ for any point $(x, y) \in D$, and that $\operatorname{int}(N_{S(x)}(y))$ is nonempty only if y is an extreme point of $S(x)$. By putting together both observations, one can easily deduce that for each $x \in X$, $p(x, \cdot)$ is a discrete density function whose support is contained in $\text{ext}(S(x))$. Therefore, the belief β is given by

$$\beta_x(O) = \sum_{y \in \text{ext}(S(x))} p_c(x, y) \mathbb{1}_O(y), \quad \forall O \in \mathcal{B}(Y). \tag{10}$$

We call β the *vertex-supported belief induced by* c. With this construction, we can rewrite problem (1) as

$$(\mathcal{P}) := \min_{x \in X} \langle d_1, x \rangle + \mathbb{E}_{\beta_x}[\langle d_2, y \rangle] \tag{11}$$

where $\mathbb{E}_{\beta_x}[\langle d_2, y \rangle] = \sum_{y \in \text{ext}(S(x))} \langle d_2, y \rangle p_c(x, y)$. Our goal in this work is to study problem (1) by profiting from the Bayesian formulation (11), in the sense of [29].

3.1 Sample Average Formulation

Problem (11) has an intrinsic difficulty which consists in how to evaluate the objective function $\theta(x) = \langle d_1, x \rangle + \mathbb{E}_{\beta_x}[\langle d_2, y \rangle]$. To make an exact evaluation of θ at a point $x \in X$ one would require to compute the set of all vertices $y_1, \ldots, y_k$ of $S(x)$ (having a positive probability of being optimal for $c(\omega)$) and to compute the corresponding probabilities $p_c(x, y_1), \ldots, p_c(x, y_k)$, defined as the "sizes" of the respective normal cones at each vertex y_i. This is not always possible.

To deal with this issue, we consider the well-known sample average approximation (SAA) method for stochastic optimization (see, e.g., [18,30]). That is, we take an i.i.d sample $\{\hat{c}_1, \ldots, \hat{c}_{N_0}\}$ of the random lower-level cost $c(\cdot)$, where each sample unit is drawn following the (known) distribution of $c(\cdot)$, and try to solve the (now deterministic) problem

$$(\hat{\mathcal{P}}) := \begin{cases} \min_x \langle d_1, x \rangle + \frac{1}{N_0} \sum_{i=1}^{N_0} \langle d_2, y_i(x) \rangle \\ \text{s.t.} \ \ x \in X \\ \forall i \in [N_0], \ y_i(x) \text{ solves } \begin{cases} \min_y \langle \hat{c}_i, y \rangle \\ \text{s.t.} \ \ Ax + By \leq b. \end{cases} \end{cases} \tag{12}$$

Proposition 1 ([27]). *Assume that* $c(\cdot)$ *has a nonatomic distribution over* $\mathbb{S}_{n_y}$, *in the sense of (8). Then, with probability 1, we have that for each* $i \in [N_0]$ *and each* $x \in X$ *the set* $\text{argmin}_y \{\langle \hat{c}_i, y \rangle \ : \ Ax + By \leq b\}$ *is a singleton.*

Based on the above proposition, for the sample $\{\hat{c}_1, \ldots, \hat{c}_{N_0}\}$ we can, for each $i \in [N_0]$, define the mapping $x \mapsto y_i(x)$ where $y_i(x)$ is the unique element $\text{argmin}_y \{\langle \hat{c}_i, y \rangle \ : \ Ax + By \leq b\}$. Thus, almost surely, problem (12) is well-defined. We close this section by stating that problem (12) is a consistent approximation of the original problem (11).

Proposition 2 ([27]). *Let $x_{N_0}^*$ be an optimal solution of (12) for a sample of size N_0, and let $\nu_{N_0}^*$ be the associated optimal value. Let S be the set of solutions of (1) and let $\bar{\nu}$ be its optimal value. Then, $d(x_{N_0}^*, S) := \inf_{x \in S} \|x_{N_0}^* - x\| \xrightarrow{N_0 \to \infty} 0$, and $\nu_{N_0}^* \xrightarrow{N_0 \to \infty} \bar{\nu}$, both with probability 1.*

4   Geometrical Structure of Vertex-Supported Beliefs

Here, we recall the definition of a chamber complex, frequently used in some fields of mathematics, like computational geometry (see, e.g., [9]).

Definition 1 (Chamber complex). *Let $D \subset \mathbb{R}^{n_x} \times \mathbb{R}^{n_y}$ be a polyhedron as described in (4). For each $x \in X = \pi(D)$, we define the chamber of x as*

$$\sigma(x) = \bigcap \{\pi(F) : F \in \mathscr{F}(D), x \in \pi(F)\}. \tag{13}$$

*The **chamber complex**, is then given by the (finite) collection of chambers, i.e., $\mathscr{C}(D) = \{\sigma(x) : x \in X\}$.*

For a more comprehensive exposition of the chamber complex $\mathscr{C}(D)$ and their many interesting properties, we refer the reader to [9,13,14,27] and references therein. The next proposition shows that to compute a chamber it is enough to consider faces of D with dimensions up to n_x instead of the collection of all faces.

Proposition 3 ([27]). *For any $x \in X$, one has that*

$$\sigma(x) = \bigcap \{\pi(F) : F \in \mathscr{F}, x \in \pi(F), \dim(F) \le n_x\}.$$

While the previous result narrows the class of faces that are needed to compute a chamber, we may still need faces of drastically different dimensions. The next example illustrates this phenomenon.

Example 1. Consider the polytope $D := \{(x, y) \in \mathbb{R}^2 \times \mathbb{R} : |x_1| \le y \le 1 - |x_2|\}$, whose vertices are $(0, \pm 1, 0)$ and $(\pm 1, 0, 1)$. Clearly, $(0, 0)$ and $(1, 0)$ are minimal chambers, however, $(0, 0)$ is not a projection of a vertex of D, while $(1, 0)$ cannot be obtained using only projections of facets.

It is well-known (see, e.g., [14]) that the family $\{ri(K) : K \in \mathcal{C}(D)\}$ is a partition of X. With this in mind, we introduce the following definition.

Definition 2. *A function $f : X \to \mathbb{R}$ is said to be piecewise linear over the chamber complex $\mathscr{C}(D)$ if there exists a sequence of pairs $\{(d_K, a_K) : K \in \mathscr{C}(D)\} \subset \mathbb{R}^{n_x} \times \mathbb{R}$ such that*

$$f(x) = \sum_{K \in \mathscr{C}(D)} (\langle d_K, x \rangle + a_K) \mathbb{1}_{ri(K)}(x), \qquad \forall x \in X. \tag{14}$$

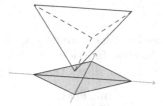

Fig. 1. The polytope D of Example 1 and its chamber complex whose vertex $(0,0)$ is not a projection of a vertex of D.

In what follows, it will be useful to distinguish minimal chambers and maximal chambers, with respect to the inclusion ordering. The former are characterized by having nonempty interior, while the latter are the vertices in the chamber complex.

Definition 3. *Let* $D \subset \mathbb{R}^{n_x} \times \mathbb{R}^{n_y}$ *be a polyhedron as described in* (4). *We define*

$$\mathcal{K}(D) := \{K \in \mathcal{C}(D) \; : \; \text{int}(K) \neq \emptyset\}, \tag{15}$$

$$\mathcal{V}(D) := \{v \in X \; : \; \{v\} \in \mathcal{C}(D)\}. \tag{16}$$

We call $\mathcal{K}(D)$ *the family of full-dimensional chambers of* D, *and* $\mathcal{V}(D)$, *the vertices of the chamber complex* $\mathcal{C}(D)$ *(which correspond to the family of zero-dimensional chambers of* D).

It is worth mentioning that $\mathcal{K}(D)$ is a covering of X, i.e., $X \subset \bigcup \mathcal{K}(D)$. This follows from the fact that each chamber is contained in a full-dimensional chamber. A very straightforward proposition (see [27]) is that

$$\mathcal{V}(D) = \bigcup_{K \in \mathcal{K}(D)} \text{ext}(K). \tag{17}$$

A direct implication of this observation is the following corollary, which is one of the keystones of the enumeration algorithm that we propose.

Corollary 1. *If a function* $f : X \to \mathbb{R}$ *is continuous and piecewise linear over the chamber complex* $\mathcal{C}(D)$, *then it has at least one minimizer in* $\mathcal{V}(D)$.

We finish this section with our main theorem.

Theorem 1 (See [27]). *Consider a random cost* c *with nonatomic distribution and* $\beta : X \to \mathscr{P}(Y)$ *the vertex-supported belief induced by* c *over* S *as defined in* (10). *Then,*

1. β *is weakly continuous, and thus for any lower semicontinuous function* $f : X \times Y \to \mathbb{R}$, *the problem* $\min_{x \in X} \mathbb{E}_{\beta_x}[f(x, \cdot)]$ *has a solution.*
2. *The function* $x \in X \mapsto \langle d_1, x \rangle + \mathbb{E}_{\beta_x}[\langle d_2, y \rangle]$ *is continuous and piecewise linear over* $\mathcal{C}(D)$.
3. *For almost every sample* $\{\hat{c}_1, \ldots, \hat{c}_{N_0}\}$, *the function* $x \in X \mapsto \langle d_1, x \rangle + \frac{1}{N_0} \sum_{i=1}^{N_0} \langle d_2, y_i(x) \rangle$ *is well-defined, continuous and piecewise linear over* $\mathcal{C}(D)$.

In particular, problem (1) *has at least one solution over* $\mathcal{V}(D)$.

5   Algorithms

The rest of the work is focused on algorithms to solve problem (11) in the cases when we can evaluate the objective function $x \mapsto \langle d_1, x \rangle + \mathbb{E}_{\beta_x}[\langle d_2, y \rangle]$, or solve problem (12), otherwise. Theorem 1 shows that both problems have the form

$$\min_{x \in X} \theta(x), \tag{18}$$

where $\theta : X \to \mathbb{R}$ is a continuous function and it is piecewise linear over the chamber complex $\mathscr{C}(D)$ (with probability 1, in the case of problem (12)). Thus, we provide algorithms to solve this generic problem.

5.1   Enumeration Algorithm

Corollary 1 gives us a natural strategy to solve problem (18): It is enough to compute the chamber vertices $\mathscr{V}(D)$ and evaluate the corresponding objective function θ at each one of them. In this section we provide an enumeration algorithm to compute $\mathscr{V}(D)$ by sequentially solving mixed-integer programming problems which are formulated using $\mathscr{F}_{\leq n_x} := \{F \in \mathscr{F} : \dim(F) \leq n_x\}$, as shown in Proposition 3. We remind the reader that, due to the discussion in Example 1, we may need faces of different dimensions to compute $\mathscr{V}(D)$. This is why we rely on the full set $\mathscr{F}_{\leq n_x}$.

Remark 1. Computing $\mathscr{V}(D)$ is at least as hard as computing all vertices of a polytope. Indeed, given an arbitrary (full-dimensional) polytope $P \subseteq \mathbb{R}^n$, one may consider $D = P \times [0,1]$. The vertices of P correspond exactly to $\mathscr{V}(D)$. To the best of our knowledge, the complexity of finding all vertices of a polytope P is currently unknown; however, for a polyhedron P (not necessarily bounded), it is known that it is NP-complete to decide, given a subset of vertices of P, if there is a new vertex of P to add to the collection [20]. Therefore, we can expect that computing $\mathscr{V}(D)$ will be computationally expensive.

For $x \in X$, let us define the *label* of x as the set $\ell(x) := \{F \in \mathscr{F}_{\leq n_x} : x \in \pi(F)\}$. Endowing the set of all (finitely many) labels with the order of the inclusion, one can show (see [27]) that

$$x \in \mathscr{V}(D) \iff \ell(x) \text{ is a maximal label.} \tag{19}$$

Intuitively, this states that a vertex of a chamber is obtained by intersecting as many projections of faces as possible (of dimension ranging from 0 to n_x). Using this result, we can generate an element of $\mathscr{V}(D)$ through a MILP formulation that finds an x such that $\ell(x)$ is a maximal label. The following formulation achieves this.

$$\max_{z,x,y} \quad \sum_{F \in \mathscr{F}_{\leq n_x}} z_F \tag{20a}$$

$$\text{s.t.} \quad Ax + By_F \leq b \qquad\qquad \forall F \in \mathscr{F}_{\leq n_x} \tag{20b}$$

$$\qquad A_F x + B_F y_F \geq b_F - M(1 - z_F) \qquad \forall F \in \mathscr{F}_{\leq n_x} \tag{20c}$$

$$\qquad z_F \in \{0,1\} \qquad\qquad\qquad \forall F \in \mathscr{F}_{\leq n_x} \tag{20d}$$

Here, y and z stand for the vectors $(y_F : F \in \mathscr{F}_{\leq n_x})$ and $(z_F : F \in \mathscr{F}_{\leq n_x})$, respectively. For each $F \in \mathscr{F}_{\leq n_x}$, A_F, B_F and b_F are submatrices of A, B and b such that $F = \{(x,y) \in D : A_F x + B_F y = b\}$. Finally, M is a vector of m positive values such that $A_i x + B_i y - b_i \geq -M_i$, for all $(x,y) \in D$. This vector M is well defined when D is a polytope and can be easily computed using m linear programs. Formulation (20) is straightforward. It tries to "activate" as many faces as possible such that the intersection of their projection is non-empty.

Remark 2. We note that, while conceptually simple, (20) depends on the enumeration of all faces $\mathscr{F}_{\leq n_x}$ which can be a highly challenging task. Other approaches below also rely on this. We have not yet devised a mechanism around this potentially expensive enumeration; moreover, in many of our computational experiments (Sect. 6), our approaches are actively making use of a significant portion of $\mathscr{F}_{\leq n_x}$. For this reason, we can only handle small-size problems in this work. We discuss potential future avenues for solving these limitations in Sect. 6. Note that these difficulties are in line with the discussion of Remark 1.

Let (z^*, x^*, y^*) be an optimal solution of (20). It is not hard to see that x^* is an element of $\mathscr{V}(D)$, thus, we can collect it and focus on generating a new element of $\mathscr{V}(D)$. Noting that $\ell(x^*) = \{F \in \mathscr{F}_{\leq n_x} : z_F^* = 1\}$, we see that such new element can be obtained by adding the following inequality to (20)

$$\sum_{F \in \mathscr{F}_{\leq n_x} : z_F^* = 0} z_F \geq 1. \tag{21}$$

Since $\ell(x^*)$ is a maximal label, we can easily see that constraint (21) is removing only the single element x^* of $\mathscr{V}(D)$ from (20). This is a so-called "no-good cut". This procedure can be iterated until the optimization problem becomes infeasible. In our computational experiments, however, we noted that detecting infeasibility was particularly challenging for the optimization solver, and thus devised an alternative strategy: we add a new binary variable s that can relax (21) when needed, and whose value will define the stopping criterion.

Under these considerations, we next present the precise model we use. Suppose we have partially generated a set $V \subseteq \mathscr{V}(D)$, we generate an element of $\mathscr{V}(D) \setminus V$ or determine $V = \mathscr{V}(D)$ using the following optimization problem

$$\max_{z,s,x,y} \quad \sum_{F \in \mathscr{F}_{\leq n_x}} z_F \tag{22a}$$

$$\text{s.t.} \quad Ax + B y_F \leq b \qquad\qquad \forall F \in \mathscr{F}_{\leq n_x} \tag{22b}$$

$$A_F x + B_F y_F \geq b_F - M(1 - z_F) \qquad \forall F \in \mathscr{F}_{\leq n_x} \tag{22c}$$

$$\sum_{F \notin \ell(v)} z_F + s \geq 1 \qquad\qquad \forall v \in V \tag{22d}$$

$$\sum_{F \in \mathscr{F}_{\leq n_x}} z_F \leq |\mathscr{F}_{\leq n_x}|(1 - s) \tag{22e}$$

$$z_F \in \{0,1\} \qquad\qquad \forall F \in \mathscr{F}_{\leq n_x} \tag{22f}$$

$$s \in \{0,1\} \tag{22g}$$

Algorithm 1: Chamber vertex enumeration algorithm

1 **Input:** A, B, b defining a polytope $D = \{(x, y) \in \mathbb{R}^{n_x} \times \mathbb{R}^{n_y} : Ax + By \leq b\}$;
2 Set $V = \emptyset$, $s^* = 0$;
3 Compute $\mathscr{F}_{\leq n_x}(D)$;
4 **while** *true* **do**
5 $\quad$ Solve problem (22) and obtain an optimal solution (z^*, s^*, x^*, y^*);
6 $\quad$ **if** $s^* = 0$ **then**
7 $\quad\quad$ | $\quad$ break;
8 $\quad$ **end**
9 $\quad$ $V \leftarrow V \cup \{x^*\}$;
10 $\quad$ Store $\ell(x^*) = \{F \in \mathscr{F}_{\leq n_x}(D) : z_F^* = 1\}$;
11 **end**
$\quad$ **Result:** The set $V = \mathscr{V}(D)$

Lemma 1 (see [27]). *Problem (22) is always feasible provided that $D \neq \emptyset$. Moreover, in an optimal solution (z^*, s^*, x^*, y^*), then one (and only one) of the following situations hold (a) $s^* = 0$ and $x^* \in \mathscr{V}(D) \setminus V$, or (b) $s^* = 1$ and $\mathscr{V}(D) = V$.*

In Algorithm 1 we formalize our enumeration procedure. To solve problem (18), it is enough to run Algorithm 1, and evaluate θ over the set $V = \mathscr{V}(D)$.

5.2 Monte-Carlo Approximation Scheme

The previous enumeration algorithm of Sect. 5.1 has several drawbacks. First, it requires (in practice) computing the whole collection of faces of D, which might depend exponentially on the whole dimension $n_x + n_y$. And even with the collection of faces at hand, computing all chamber vertices in $\mathscr{V}(D)$ can be hard. Moreover, $\mathscr{V}(D)$ might be exponentially large.

Another approach, that we explore in this section, is to try to compute the collection of full-dimensional chambers $\mathscr{K}(D)$. Despite the fact that $\mathscr{K}(D)$ might still be exponentially large, in some cases it is considerably smaller than $\mathscr{C}(D)$. Moreover drawing points x uniformly over X yield that $\sigma(x) \in \mathscr{K}(D)$ almost surely. Indeed, this follows from Proposition 2.6 in [27] and the facts that there are finitely many chambers in $\mathscr{K}(D)$, and only those in $\mathscr{K}(D)$ are not negligible.

To simplify the exposition, we will assume that $X \subset [0, 1]^{n_x}$ and we will write $X^c = [0, 1]^{n_x} \setminus X$. To be able to consider samples in $[0, 1]^{n_x}$, we identify $\ell(x) = \emptyset$ for all $x \in X^c$. We base our algorithm in the following lemma.

Lemma 2 (see [27]). *Let $\ell = \ell(\bar{x})$ for some $\bar{x} \in X$ and assume that $K_\ell := \sigma(\bar{x})$ is a full-dimensional chamber. Then, $\ell(x) \subset \mathscr{F}_{n_x} := \{F \in \mathscr{F} : \dim(F) = n_x\}$. Moreover, for each $j \in [n_x]$, the following linear problem*

$$\begin{cases} \max_{t, (y_F)_{F \in \ell}} t \\ \text{s.t.} \quad (\bar{x} + te_j, y_F) \in F, \forall F \in \ell, \end{cases} \tag{23}$$

Algorithm 2: Monte-Carlo algorithm

---

1 **Input:** A, B, b defining a polytope $D = \{(x, y) \in \mathbb{R}^{n_x} \times \mathbb{R}^{n_y} : Ax + By \leq b\}$,
 $\theta : X \to \mathbb{R}$ continuous and piecewise linear over $\mathscr{C}(D)$;
2 Generate a (uniformly iid) training sample S of size N over $[0, 1]^{n_x}$;
3 Set **List** $= \emptyset$, $\hat{x} = NaN$, $\hat{\theta} = \infty$;
4 Compute $\mathscr{F}_{n_x}$;
5 **foreach** $\xi \in S$ **do**
6 $\quad$ Compute $\ell = \{F \in \mathscr{F}_{n_x} : \xi \in \pi(F)\}$;
7 $\quad$ **if** $\ell \in List$ *or* $\ell = \emptyset$ **then**
8 $\quad\quad$ | continue;
9 $\quad$ **end**
10 $\quad$ **List** $\leftarrow$ **List** $\cup \{\ell\}$; Compute d_ℓ as in Lemma 2;
11 $\quad$ Solve the linear problem

$$\begin{cases} \min\limits_{x, (y_F)_{F \in \ell}} & \langle d_\ell, x \rangle \\ s.t. & (x, y_F) \in F, \forall F \in \ell \end{cases}$$

$\quad$ finding a solution $\hat{x}_\ell$ and set the value $\hat{\theta}_\ell = \theta(\hat{x}_\ell)$;
12 $\quad$ **if** $\hat{\theta}_\ell < \hat{\theta}$ **then**
13 $\quad\quad$ | $\hat{x} \leftarrow \hat{x}_\ell$, $\hat{\theta} \leftarrow \hat{\theta}_\ell$;
14 $\quad$ **end**
15 **end**

Result: The pair solution-value $(\hat{x}, \hat{\theta})$ for problem (18).

---

has a solution $t_j^* > 0$. Finally, for every function $\theta : X \to \mathbb{R}$ continuous and piecewise linear over the chamber complex $\mathscr{C}(D)$, the vector $d_\ell := d_{K_\ell} \in \mathbb{R}^{n_x}$ such that $\theta\big|_{K_\ell} = \langle d_\ell, \cdot \rangle + a_{K_\ell}$ (for some $a_{K_\ell} \in \mathbb{R}$) can be computed as

$$d_\ell = \left(\frac{\theta(\bar{x} + t_1^* e_1) - \theta(\bar{x})}{t_1^*}, \dots, \frac{\theta(\bar{x} + t_{n_x}^* e_{n_x}) - \theta(\bar{x})}{t_{n_x}^*} \right)^\top .$$

With this result, we can establish a Monte-Carlo algorithm (see Algorithm 2) to approximate the solution of problem (18): we randomly draw points $\bar{x}$ from $[0, 1]^{n_x}$, compute $\ell(\bar{x})$, and use Lemma 2 to optimize θ over $\sigma(\bar{x})$ via an LP formulation. Note that, to compute labels, we only need (with probability 1) access to $\mathscr{F}_{n_x}$, which might depend exponentially only on the lower-level dimension n_y and not $n_x + n_y$ (see [27]). The main drawback of Algorithm 2 is that we cannot ensure in general that the result $(\hat{x}, \hat{\theta})$ is an optimal solution of problem (18) or not. A measurement in terms of unseen chambers is proposed in [27] to quantify how good the solution $(\hat{x}, \hat{\theta})$ might be.

6 Numerical Experiments

We implemented both Algorithms 1 and 2 in Julia 1.8.2 [4], using Polymake [15] to compute the faces of a polytope and Gurobi 9.5.2 [16] to solve (22) and any auxiliary LP. Our code is publicly available in https://github.com/g-munoz/bilevelbayesian. All experiments were run single-threaded on a Linux machine with an Intel Xeon Silver 4210 2.2G CPU and 128 GB RAM. The main objectives behind these experiments are (1) to determine how Algorithm 1 scales and (2) how well the Monte-Carlo algorithm performs in comparison to the exact method. A global time limit of 15 min was set for Algorithm 1; in case this time limit is met, only the chamber vertices that were found are used.

We focus our attention in sample average formulations, as in (12), where the lower-level cost is assumed to have a uniform distribution over the unit sphere. We use instances from two publicly available libraries: BOLib [32] and the bilevel instances in [1], which we call CoralLib. Since our approach relies on computing a (possibly exponentially) large number of faces, we can only consider low-dimensional instances at the moment: we restrict to $n_x + n_y \leq 10$.

Additionally, we consider randomly generated instances of the stochastic bilevel continuous knapsack problem [5]. These instances have the form:

$$
\begin{aligned}
&\max_{x} -\delta x + d^\top y \\
&s.t. \ \ x \in [L, U] \\
&y(x, \omega) \text{ solves } \begin{cases} \min_{y} \ \langle c(\omega), y \rangle \\ s.t. \ \ \begin{aligned} a^\top y &\leq x, \qquad \omega \in \Omega \text{ a.s.} \\ y &\in [0, 1]^{n_y} \end{aligned} \end{cases}
\end{aligned}
\tag{24}
$$

In our experiments, we consider a to be a random non-negative vector, $\delta = 1/4$ and d the vector of ones. We call *Knapsack_i* an instance generated for $n_y = i$. While these instances have a more efficient algorithm for them than the one presented here (see [5]), they are helpful in showing how well our general-purpose Monte-Carlo algorithm performs.

In all experiments, we used a sample of size 100 for the follower's cost vector. The same sample is used in both algorithms to better compare their performance. Additionally, in Algorithm 2 we used samples of size 200 for the domain X. In Table 1, we compare the performance of both methods.

The gap measures how far the value of the Monte-Carlo algorithm is from the exact method, i.e., if val_i is the value obtained by Algorithm i, then the gap is Gap $= |val_1|^{-1}(val_2 - val_1)$. Since we ran Algorithm 1 with a time limit, it may be that Gap < 0, which indicates the Monte-Carlo algorithm performing better than the exact method.

The results in Table 1 clearly shows an advantage of the Monte-Carlo approach over the exact method. The Monte-Carlo approach was able to meet or surpass the value of the exact method in almost all cases. In the largest examples, the Monte-Carlo had a much better performance, in some cases providing much better solutions than the exact method in shorter running times.

The main (and clear) challenge for this work is scalability: while these results show short running times, these are all instances of small dimensions. The main bottleneck currently is the enumeration of the faces of a polytope. In the case of Algorithm 1, there does not seem to be much hope in improving this substan-

Table 1. Summary of results for Algorithms 1 and 2 for selected BOLib instances [32], CoralLib instances [1] and Knapsack instances [5]. The "Size" of the instance is $(n_x + n_y, m)$. The "Obj gap" column shows the gap between the values found for both algorithms; a negative gap indicates the stochastic method performed better. The "Error" column shows the upper estimation of the *volume of unseen chambers* during the sampling process (see [27] for the details). The columns labeled "Computation Times" contain the running times (in seconds) for the computation of all the faces, the execution of Algorithm 1 and of Algorithm 2. The columns labeled "Used faces" contain the number of faces that were explicitly used during the execution of each algorithm.

Instance	Size	$|\mathscr{F}_{\leq n_x}|$	$|\mathscr{F}_{n_x}|$	Obj gap	Error	Computation Times Faces	Alg. 1	Alg. 2	Used Faces Alg. 1	Alg 2.
BOLib/AnandalinghamWhite1990	(2,7)	12	6	0%	0%	1.4	3.8	7.2	12	5
BOLib/Bard1984a	(2,6)	10	5	0%	0%	3.1	6.8	7.9	10	5
BOLib/Bard1984b	(2,6)	10	5	0%	0%	1.4	4.1	7.5	10	5
BOLib/Bard1991Ex2	(3,6)	14	9	0%	0%	1.4	4.2	8.2	14	6
BOLib/BardFalk1982Ex2	(4,7)	45	17	0%	45%	1.7	4.3	7.6	45	5
BOLib/BenAyedBlair1990a	(3,6)	20	12	0%	0%	1.4	4.2	8.5	20	4
BOLib/BenAyedBlair1990b	(2,5)	6	3	0%	0%	1.5	4.3	8.0	6	3
BOLib/BialasKarwan1984a	(3,8)	20	12	0%	0%	1.6	4.4	8.4	20	10
BOLib/BialasKarwan1984b	(2,7)	12	6	0%	0%	1.5	4.3	8.0	12	5
BOLib/CandlerTownsley1982	(5,8)	111	48	1%	37%	1.8	7.9	18.5	111	16
BOLib/ClarkWesterberg1988	(2,3)	6	3	0%	0%	1.5	4.3	8.0	6	3
BOLib/ClarkWesterberg1990b	(3,7)	15	9	0%	0%	1.5	4.4	8.3	15	9
BOLib/GlackinEtal2009	(3,6)	20	5	0%	48%	1.6	4.8	8.0	20	3
BOLib/HaurieSavardWhite1990	(2,4)	8	4	0%	0%	1.6	4.5	8.3	8	4
BOLib/HuHuangZhang2009	(3,6)	20	12	0%	0%	1.6	4.4	8.6	20	7
BOLib/LanWenShihLee2007	(2,8)	14	7	0%	1%	1.6	4.5	8.2	14	6
BOLib/LiuHart1994	(2,5)	10	5	0%	0%	1.6	4.4	8.6	10	4
BOLib/MershaDempe2006Ex1	(2,6)	8	4	0%	0%	1.7	4.8	8.6	8	4
BOLib/MershaDempe2006Ex2	(2,7)	10	5	0%	64%	1.5	4.3	6.4	10	3
BOLib/TuyEtal1993	(4,7)	45	17	0%	54%	1.7	4.5	7.6	45	5
BOLib/TuyEtal1994	(4,8)	72	24	0%	46%	1.6	4.5	7.5	72	6
BOLib/VisweswaranEtal1996	(2,6)	8	4	0%	0%	1.4	4.3	7.9	8	4
BOLib/WangJiaoLi2005	(3,7)	23	14	0%	0%	1.5	4.4	8.6	23	5
CoralLib/linderoth	(6,15)	545	51	0%	74%	1.4	148.0	7.4	545	7
CoralLib/moore90_2	(2,7)	12	6	0%	0%	1.4	4.0	7.6	12	5
CoralLib/moore90	(2,8)	8	4	0%	0%	1.6	4.4	8.1	8	4
Knapsack_6	(7,15)	574	447	0%	0%	3.2	117.1	19.2	574	255
Knapsack_7	(8,17)	1278	1023	0%	0%	2.5	626.9	38.5	1278	575
Knapsack_8	(9,19)	2814	2303	-14%	0%	4.9	914.7	82.4	1990	1279
Knapsack_9	(10,21)	6142	5119	-380%	0%	15.4	984.2	181.4	3346	2815

tially: note in Table 1 that in all but the bottom two entries[1], Algorithm 1 used all available faces. This is because the algorithm heavily relies on *maximal labels*, which is important in our procedure to not repeat chambers when enumerating.

---

[1] The last two entries of Table 1 correspond to cases where $\mathscr{F}_{\leq n_x}$ was not fully computed due to the time limit.

Nonetheless, we still believe Algorithm 1 can be useful as a baseline that has optimality guarantees.

Algorithm 2, however, could potentially be improved significantly. First of all, recall that this approach only uses the faces of dimension n_x (i.e. $\mathscr{F}_{n_x}$), which can be considerably smaller than $\mathscr{F}_{\leq n_x}$ (see columns 3 and 4 of Table 1). Therefore, a more intricate enumeration that exploits this could be devised. Additionally, and perhaps more importantly, note that in the instances where $|\mathscr{F}_{n_x}|$ is not too small (say, more than 40) Algorithm 2 only uses a fraction of $\mathscr{F}_{n_x}$ in its execution. This indicates that one could heavily restrict the faces to consider initially and generate more on-the-fly, much like in a column generation approach. Another potential improvement path is exploiting more structure of a particular family of instances, which may indicate which are the faces that one would truly need.

References

1. Coral bilevel optimization problem library. https://coral.ise.lehigh.edu/data-sets/bilevel-instances/ Accessed 3 Nov 2022
2. Bazaraa, M.S., Sherali, H.D., Shetty, C. M.: Nonlinear programming: theory and algorithms. John Wiley Sons (2013)
3. Beck, Y., Ljubić, I., Schmidt, M.: A survey on bilevel optimization under uncertainty. European J. Oper. Res., (2023) (In Press)
4. Bezanson, J., Edelman, A., Karpinski, S., Shah, V.B.: Julia: A fresh approach to numerical computing. SIAM review **59**(1), 65–98 (2017)
5. Buchheim, C., Henke, D., Irmai, J.: The stochastic bilevel continuous knapsack problem with uncertain follower's objective. J. Optim. Theory. Appl. **194**, 521–542 (2022)
6. Burtscheidt, J., Claus, M.: Bilevel Linear Optimization Under Uncertainty. In: Dempe, S., Zemkoho, A. (eds.) Bilevel Optimization. SOIA, vol. 161, pp. 485–511. Springer, Cham (2020). https://doi.org/10.1007/978-3-030-52119-6_17
7. Claus, M.: On continuity in risk-averse bilevel stochastic linear programming with random lower level objective function. Oper. Res. Lett. **49**(3), 412–417 (2021)
8. Claus, M.: Existence of solutions for a class of bilevel stochastic linear programs. European J. Oper. Res. **299**(2), 542–549 (2022)
9. De Loera, J., Rambau, J., Santos, F.: Triangulations: structures for algorithms and applications, volume 25. Springer Science Business Media (2010) https://doi.org/10.1007/978-3-642-12971-1
10. Dempe, S.: Foundations of bilevel programming. Springer Science Business Media (2002) https://doi.org/10.1007/b101970
11. Dempe, S., Kalashnikov, V., Pérez-Valdés, G.A., Kalashnykova, N.: Bilevel programming problems. Energy Systems. Springer, Heidelberg, 2015. Theory, algorithms and applications to energy networks https://doi.org/10.1007/978-3-662-45827-3
12. Dempe, S., Zemkoho, A. (eds.): SOIA, vol. 161. Springer, Cham (2020). https://doi.org/10.1007/978-3-030-52119-6
13. Forcier, M.: Multistage stochastic optimization and polyhedral geometry. PhD. Thesis, École de Ponts - ParisTech (2022)

14. Forcier, M., Gaubert, S., Leclère, V.: Exact quantization of multistage stochastic linear problems (2021) (preprint - arXiv:2107.09566)
15. Gawrilow, E., Joswig, M., polymake: a framework for analyzing convex polytopes. In Polytopes–combinatorics and computation of DMV Sem (Oberwolfach, 1997), 29, pp. 43–73. Birkhäuser, Basel, (2000)
16. Gurobi Optimization, LLC. Gurobi Optimizer Reference Manual (2022)
17. Hiriart-Urruty, J.-B., Lemaréchal, J.-B.: Convex analysis and minimization algorithms I: Fundamentals, volume 305. Springer science business media (2013) https://doi.org/10.1007/978-3-662-02796-7
18. Homem-de Mello, T., Bayraksan, G.: Monte Carlo sampling-based methods for stochastic optimization. Surv. Oper. Res. Manag. Sci., 19(1), 56–85 (2014)
19. Ivanov, S.V.: A bilevel programming problem with random parameters in the follower's objective function. Diskretn. Anal. Issled. Oper. **25**(4), 27–45 (2018)
20. Khachiyan, L., Boros, E., Borys, K., Gurvich, V., Elbassioni, K.:Generating all vertices of a polyhedron is hard. In 20th Anniversary Volume, 1–17. Springer (2009) https://doi.org/10.1007/s00454-008-9050-5
21. Kleinert, T., Labbé, M., Ljubić, I., Schmidt, M.: A survey on mixed-integer programming techniques in bilevel optimization. EURO J. Comput. Optim. **9**, 100007 (2021)
22. Klenke, A: Probability Theory: a Comprehensive Course. Springer (2014) https://doi.org/10.1007/978-1-4471-5361-0
23. Leobacher, G., Pillichshammer, F.: Introduction to quasi-Monte Carlo integration and applications. Compact Textbooks in Mathematics. Birkhäuser/Springer, Cham (2014)
24. Lu, S., Robinson, S.M.: Normal fans of polyhedral convex sets: structures and connections. Set-Valued Anal. **16**(2–3), 281–305 (2008)
25. Mak, W.-K., Morton, D.P., Wood, R.K.: Monte Carlo bounding techniques for determining solution quality in stochastic programs. Oper. Res. Lett. **24**(1–2), 47–56 (1999)
26. Mallozzi, L., Morgan, J.: Hierarchical Systems with Weighted Reaction Set, pp. 271–282. Springer, US, Boston, MA, (1996) https://doi.org/10.1007/978-1-4899-0289-4_19
27. Muñoz, G., Salas, D., Svensson, A.: Exploiting the polyhedral geometry of stochastic linear bilevel programming (2023). (preprint - arXiv:2211.02268. Former title: Linear bilevel programming with uncertain lower-level costs)
28. Rambau, J., Ziegler, G.M.: Projections of polytopes and the generalized baues conjecture. Discrete Comput. Geom. **16**(3), 215–237 (1996)
29. Salas, D., Svensson, A.: Existence of solutions for deterministic bilevel games under a general bayesian approach (2020) (preprint - arXiv:2010.05368)
30. Shapiro, A., Dentcheva, D., Ruszczyński, A.: Lectures on stochastic programming–modeling and theory, volume 28 of MOS-SIAM Series on Optimization. 3rd eds Society for Industrial and Applied Mathematics (SIAM), Philadelphia, PA; Mathematical Optimization Society, Philadelphia, PA, (2021)
31. Stackelberg, V.H.: Marktform und Gleichgewitch. Springer (1934) https://doi.org/10.1007/978-3-642-12586-7
32. Zhou, S., Zemkoho, A.B., Tin, A.: BOLIB: Bilevel Optimization LIBrary of Test Problems. In: Dempe, S., Zemkoho, A. (eds.) Bilevel Optimization. SOIA, vol. 161, pp. 563–580. Springer, Cham (2020). https://doi.org/10.1007/978-3-030-52119-6_19

Towards an Optimal Contention Resolution Scheme for Matchings

Pranav Nuti[✉] and Jan Vondrák

Stanford University, Stanford, CA 94305, USA
{pranavn,jvondrak}@stanford.edu

Abstract. In this paper, we study contention resolution schemes for matchings. Given a fractional matching x and a random set $R(x)$ where each edge e appears independently with probability x_e, we want to select a matching $M \subseteq R(x)$ such that $\Pr[e \in M \mid e \in R(x)] \geq c$, for c as large as possible. We call such a selection method a c-balanced contention resolution scheme.

Our main results are (i) an asymptotically (in the limit as $\|x\|_\infty$ goes to 0) optimal $\simeq 0.544$-balanced contention resolution scheme for general matchings, and (ii) a 0.509-balanced contention resolution scheme for bipartite matchings. To the best of our knowledge, this result establishes for the first time, in any natural relaxation of a combinatorial optimization problem, a separation between (i) offline and random order online contention resolution schemes, and (ii) monotone and non-monotone contention resolution schemes.

Keywords: Contention resolution · Matching · Random graphs

1 Introduction

Suppose that there are n employees looking for jobs. Each employee likes a random set of jobs which, on average, has cardinality one. n jobs are available in total, and no job is especially popular amongst the employees, though some employees might have a strong preference for some particular jobs. We would like to match the employees to jobs.

We are immediately faced with many natural questions: On average, what fraction of employees can we match to a job they like? Can we match employees to jobs in a fair way, without partially favoring any particular employee? What if no employee has a strong preference for any particular job? Is it easier to match employees if we learn about their preferences all at once, rather than if we learn

---

The full version of this paper can be found at https://arxiv.org/abs/2211.03599. It contains omitted proofs, discussion of the relationship of this paper with van der Waerden's conjecture, and an application of our contention resolution scheme to a combinatorial allocation problem.

J. Vondrák—Supported by NSF Award 2127781.

A. Del Pia and V. Kaibel (Eds.): IPCO 2023, LNCS 13904, pp. 378–392, 2023.
https://doi.org/10.1007/978-3-031-32726-1_27

about them in an online fashion? Our paper provides answers to these questions, through the lens of contention resolution schemes.

Contention resolution schemes aim to solve the following problem: Given a family of feasible sets $\mathcal{F} \subset 2^E$ and a random set R sampled from a distribution on 2^E, how can we choose a feasible subset $I \subseteq R, I \in \mathcal{F}$, so that each element from R is picked with some guaranteed conditional probability: $\Pr[e \in I \mid e \in R] \geq c$ for some fixed $c > 0$ and all $e \in E$? We call such a scheme c-balanced. This condition is a kind of fairness constraint, ensuring every element e has a reasonable chance of making it into I.

In this paper, we think about E as the set of edges in a graph, and $\mathcal{F}$ as the set of matchings of the graph. The constant c is the conditional probability with which we can ensure an edge ends up in the matching I we pick, given it appears in R.

A natural assumption on the random set R is that it comes from a product distribution with marginal probabilities x_e such that x is in a polytope corresponding to the family $\mathcal{F}$ (either the exact convex hull, or a suitable relaxation, depending on the application), i.e., roughly speaking, R on average, is in $\mathcal{F}$. For matchings on graphs, this corresponds to an assumption that each edge e appears in R independently with probability x_e, and the vector $(x_e)_{e \in E}$ belongs to the matching polytope, i.e, is a *fractional matching*.

The formal notion of contention resolution was first investigated as a tool for randomized rounding. In this setting, we have an optimization problem subject to a constraint, and x represents a fractional solution to a relaxation of the problem. Contention resolution is one of the phases of a randomized rounding approach to converting this fractional solution into an integral solution: First we generate a random set R, by sampling each element e independently with probability x_e, and then we select a subset of R which satisfies the desired constraint. The flexibility of the approach enables its wide applicability in combinatorial optimization.

This approach was introduced by Feige [2], who developed a contention resolution scheme (CRS) for matchings on the restricted class of star graphs, in the context of an application to combinatorial auctions. CRSs were then investigated more systematically in [4] in the context of submodular optimization. In particular, an optimal $(1 - 1/e)$-balanced CRS was identified in [4] for the case where $\mathcal{F}$ forms a matroid. The $1 - 1/e$ factor is optimal even for $\mathcal{F} = \{I : |I| \leq 1\}$.

For applications in submodular optimization, it turns out that an additional property of *monotonicity* is often useful: A CRS is called monotone, if for every element e, the probability that e is selected from a set R is non-increasing as a function on the sets R containing e. This property is generally needed for the analysis of randomized rounding with a submodular objective function [4]. However, for some applications it is not necessary that a CRS is monotone; in particular it was not needed in Feige's original application in [2], and it is also unnecessary for a related application that we present in the full version of this paper.

Contention resolution has also been studied in online settings (where it has seen applications to prophet inequalities and sequential pricing problems, for example) with either adversarial or random ordering of elements [6,8–10,13]. For example, for matroids there is a 1/2-balanced adversarial order online CRS [9]. We do not investigate online contention resolution here, but we should mention that in prior results, random order online contention resolution schemes (RCRS) are able to match the best known offline results: For matroids, there is a $(1 - 1/e)$-balanced RCRS, due to an elegant LP duality connection with prophet inequalities [9].

The situation is much more complicated when $\mathcal{F}$ encodes constraints such as matchings and the optimal factors are generally unknown. The cases of bipartite and general matchings have attracted attention due to their fundamental nature and their frequent appearance in applications. We can think of matching constraints as an intersection of two matroid constraints, and for an intersection of k matroid constraints, there is a $\frac{1}{k+1}$-balanced RCRS [8]; in particular, this gives a 1/3-balanced RCRS (and hence also an offline CRS with the same factor) for bipartite matchings.

Recent work in both offline and online settings has significantly improved the factor of 1/3. In the offline setting, [7] gives a $(1 - e^{-2})/2 \simeq 0.432$-balanced scheme for general matchings, which can be improved slightly further [11]. Very interestingly, [11] identifies the optimal monotone scheme for bipartite matchings, which achieves a balancedness of $\simeq 0.476$. Nevertheless, the optimal CRSs for bipartite and general matchings are still unknown. The primary reason that it seems to be harder to obtain the optimal non-monotone scheme is that decisions on whether an edge should be included in the matching need not be local (i.e., a function of the edge's immediate neighborhood), and it is harder to analyze the behavior of an algorithm that makes non-local decisions. In terms of impossibility results, an upper bound of $\simeq 0.544$ follows from a classical paper of Karp and Sipser [1], as discussed in [7].

In the online setting, the best known CRSs are due to recent results in [12]: In the random order case, they provide a 0.474-balanced scheme for general matchings and a 0.476-balanced scheme for bipartite matchings, and in the bipartite case, they also establish an upper bound of 0.5. Notably, the 0.474-balanced scheme is in fact the best known CRS for general matchings, whether offline or online. In the adversarial order case, they provide a 0.344-balanced scheme for general matchings and a 0.349-balanced scheme for bipartite matchings.

1.1   Our Results

To explain our results, we start by formally setting up some notation. Given a graph $G = (V, E)$, a fractional matching is a point $x \in [0,1]^E$ in the matching polytope, i.e., a point in the convex hull of vectors $\mathbf{1}_M$ for all matchings M in G. For a fractional matching x, let x_{uv} be the component of x corresponding to the edge (u, v).

The problem we are interested in studying is:

Contention Resolution for Matchings. We are given a fractional matching x, and a random set $R(x)$ where edges appear independently with probabilities x_{uv}. Our goal is to choose a matching $M \subseteq R(x)$ such that for every edge (u,v),

$$\Pr[(u,v) \in M \mid (u,v) \in R(x)] \geq c.$$

Such a scheme is called c-balanced, and we want to find a scheme with c as large as possible. The main questions we ask are:

(i) Is there a contention resolution scheme for matchings achieving the upper bound $\simeq 0.544$ of Karp and Sipser?
(ii) Is there a separation between the optimal c for online and offline contention resolution schemes?
(iii) Is there a separation between the optimal c for monotone and non-monotone contention resolution schemes?

In this paper, we prove the following results. The first result, which applies to both bipartite and non-bipartite matchings, is an attempt to answer (i).

Theorem 1. *Assuming that x is a fractional matching such that $\|x\|_\infty \leq \epsilon$, there is $(\gamma - f(\epsilon))$-balanced contention resolution scheme, where $\gamma \simeq 0.544$ is the impossibility bound of Karp and Sipser and $\lim_{\epsilon \to 0} f(\epsilon) = 0$.*

For fractional matchings without any assumption on their ℓ_∞ norm[1], we present an improved CRS in the bipartite case.

Theorem 2. *There is 0.509-balanced contention resolution scheme for bipartite matchings.*

This theorem answers questions (ii) and (iii), since the optimal RCRS for bipartite matchings is at most 0.5-balanced, and the optimal monotone CRS for bipartite matchings is $\simeq 0.476$-balanced. Our theorem thus establishes separations that, to our knowledge, have not been demonstrated in any other natural relaxations of combinatorial optimization problems before. (Note that for matroids, the known optimal $(1 - 1/e)$-balanced schemes are monotone.)

Returning to the context we started this paper with, our results establish that we can match more than half of all the employees to jobs they like without partially favoring any particular employee, and in case no employee has a strong preference for any particular job, we can do better, and match 54% of employees to jobs. This is a significant improvement over what we can do if we learn the employees preferences in an online fashion.

We should also mention here the important concept of a *correlation gap*. Informally, the correlation gap measures how much we might lose while optimizing a function, in the worst case, by assuming that the distributions that define

---

[1] It might appear from the work of Bruggmann and Zenklusen (see lemma 7 in [11]) that the assumption of $\|x\|_\infty \leq \epsilon$ should be easy to drop from Theorem 1. This would be the case if our theorem applied to graphs with parallel edges, which unfortunately, it does not.

the function are independent rather than correlated. In the context of bipartite matchings, the correlation gap is defined as the minimum possible ratio between $\mathbb{E}[\max\{\sum_{e \in M} w_e : M \subseteq R(x), M \text{ is a matching}\}]$ and $\sum_{e \in E} w_e x_e$, where x is a fractional bipartite matching and w is any vector of weights. By LP duality (see [4]), Theorem 2 also provides (the best known) lower bound of 0.509 on the correlation gap for bipartite matchings.

In light of Theorem 1, we believe that the correlation gap for bipartite (and perhaps even non-bipartite) matchings is indeed the Karp-Sipser bound of $\gamma \simeq 0.544$, and the optimal CRS is γ-balanced.

1.2   Our Techniques

Our Theorem 1 follows from an improved and simplified analysis of Karp and Sipser's algorithm [1] for constructing matchings by adding random edges adjacent to leaves. While we utilize many of the ideas from Karp and Sipser's paper, our analysis of the algorithm is an improvement in several ways:

- We obtain a contention resolution scheme, while Karp and Sipser only compute the expected size of the maximum matching. This yields the somewhat surprising conclusion that Karp and Sipser's algorithm works just as well for weighted matchings as it does for unweighted matchings.
- We avoid Karp and Sipser's (technically complicated) use of the so-called differential equation method. We also avoid the use of generating functions, another method used recently to calculate the expected size of the maximum matching in random graphs [5].
- We obtain results for any random graph $R(x)$ constructed from a fractional matching satisfying $\|x\|_\infty \leq \epsilon$, unlike Karp and Sipser who only consider the Erdos-Renyi random graph $G_{n,c/n}$.

Many previous results require that there be some kind of symmetry in the random graph to obtain bounds on the size of the matching. We stress that we do not need to make any such assumption on $R(x)$.

We do need to assume that $\|x\|_\infty \leq \epsilon$. This assumption is useful because it ensures that the neighbourhood of any particular edge looks like a random tree. A closely related assumption ("local weak convergence") has been considered previously in the literature. This assumption, together with recursive distributional equations, is used to formalize various statistical mechanical heuristics regarding matchings in random graphs. Most related to our work is the work of Bordenave, Lelarge, and Salez [3]. Once again, the advantage of our method is that we obtain a CRS (as opposed to computing the expected size of the maximum matching) and we avoid the use of technically complicated tools.

These improvements come at a cost–we assume that the average degree of each vertex is less than or equal to 1. The theoretical and practical significance of this case, and the importance of contention resolution schemes, make this trade-off a good choice.

Our Theorem 2 requires several new techniques, although the basic idea can be traced back to Karp and Sipser as well: When deciding which edge incident to a vertex we should add to a matching, it is beneficial to pick an edge which is adjacent to a leaf, since it doesn't block us from adding other edges into the matching. It turns out that in general, it is actually better not to follow this rule absolutely (at least in our analysis) but we still pick degree-1 edges with significant priority over other edges.

We present two different schemes using these ideas; the first one is simpler and achieves a factor $\simeq 0.480$ (already establishing the separation between monotone and non-monotone schemes). An interesting feature of this scheme is that it can be implemented as a parallel algorithm with each vertex independently making decisions about whether to include an edge adjacent to it in the matching by looking only at its immediate neighborhood. The best schemes known previously did not have this useful property. Our more complicated scheme achieves a factor $\simeq 0.509$ (thus demonstrating a separation between offline CRSs and RCRSs).

Both schemes rely on an extended version of contention resolution for choosing 1 element from a possibly *correlated distribution*, which we present in Sect. 3, and the 0.509-balanced scheme uses the FKG inequality to handle correlations between edges in the final stage.

Throughout this paper, even though we state our theorems for fractional matchings x, we will actually only need to assume that x satisfies the vertex constraints $\sum_v x_{uv} \leq 1$. Furthermore, we can always assume that x satisfies $\sum_v x_{uv} = 1$ for every u. We can achieve this by adding vertices and edges with probabilities such that the edge probabilities at each vertex add up to 1; this only makes the task of designing a CRS more difficult.

2   An Optimal CRS When $\|x\|_\infty \to 0$

2.1   The Karp-Sipser Algorithm

The Karp-Sipser algorithm is a method to select a matching in a graph. Given a graph G, the algorithm deletes all the degree 0 vertices, selects a random degree 1 vertex (if one exists), and adds the edge adjacent to it to the matching. Then, it deletes all the edges adjacent to the edge just added to the matching, and recurses on the newly obtained graph G'. Note that unlike in the paper of Karp and Sipser, we do *not* use a two stage process to generate the matching.

An attractive feature of the Karp-Sipser algorithm is that it doesn't "make any mistakes". This is because for any vertex v of degree 1 in a graph G, G has a maximum matching in which v is matched.

If an edge is deleted by the algorithm at some stage, we will say that *it disappears*. We also say that a *vertex is added to the matching* if an edge adjacent to it is added to the matching. Before we discuss the analysis of the algorithm, we take a brief detour.

2.2   Random Trees

Consider the following method to generate a random tree in steps. Fix two special vertices, u and v, and draw an edge between them. In step i, for each vertex at the depth $i - 1$, independently sample a Poisson random variable with mean 1, and add as many children to the vertex as the obtained sample. Stop at step j if there are no vertices at depth $j - 1$. Let us call the random tree generated by this process T.

Since the two subtrees of u and v are independent copies of a Galton-Watson process with 1 expected child at each node, it is straightforward to prove that this process terminates with probability 1. So it is almost always true that this process produces a finite tree.

The following lemma explains why we care about the process T: Up to small errors, it describes the distribution of the connected component containing a given edge (u, v) in $R(x)$. We omit the proof which involves a coupling argument and an application of Le Cam's theorem.

Lemma 1. *Let x be a fractional matching with $\sum_w x_{vw} = 1$ for every vertex v, and $\|x\|_\infty \le \epsilon$. Let $R(x)$ be the corresponding random graph. Let us condition on $(u, v) \in R(x)$ and define $N((u, v))$ to be the connected component in $R(x)$ containing (u, v). Let T be a random tree produced by the process described above and T_0 be any finite realization of the process. Then*

$$|\Pr[N((u, v)) = T_0 \mid (u, v) \in R(x)] - \Pr[T = T_0]| = O(\epsilon|T_0|^2).$$

We stress that the lemma is only true for graphs without parallel edges. For graphs with parallel edges, the lemma fails, even for the simple case of a graph with only two vertices. The use of this lemma makes it impossible to apply the work of Bruggmann and Zenklusen (lemma 7 in [11]).

2.3   The Karp-Sipser Algorithm on Trees

It is easy to prove by induction (using the fact that trees always have degree 1 vertices) that in an execution of the Karp-Sipser algorithm on a forest, an edge must eventually either disappear, or else, is added to the matching. Together with the fact that the Karp-Sipser algorithm does not make mistakes, this shows that the Karp-Sipser algorithm finds a maximum matching in a tree.

Given a tree, we would like to be able to analyze which vertices and edges end up in the matching the algorithm selects, independent of the random choices the algorithm makes. To that end, consider the following algorithm to label the vertices of a tree (this is similar to, but not exactly same as the scheme in [1]):

Root the tree at an arbitrary vertex. Starting at the maximum possible depth, look at all the vertices at a fixed depth. If a vertex has no L children (this can perhaps be true vacuously), label it L. Else, label it W. Iteratively label vertices higher in the tree, until the root of the tree receives a label.

The following claims are true (regardless of the chosen root, and regardless of the random choices the algorithm makes):

1. If an edge between a W parent and an L child disappears, it must be because the W vertex was added to the matching.
2. Every W vertex is added to the matching.

3. Every edge between two W vertices disappears.

Proof of claim 1. Suppose by way of contradiction that an edge between a W parent and L child disappears because the L vertex was added to the matching. Certainly, this does not happen in the first step of the execution of the algorithm. Consider the very first time it happens.

The L vertex must have been added to the matching through a W labelled child it has. This W vertex must have degree 1, and so the edge connecting it to an L child must have disappeared. This contradicts our assumption of the original edge being the first edge between a W parent and an L child that has disappeared because the L vertex was added to the matching.   □

Proof of claim 2. Every W vertex has an edge connecting it to an L child; either that edge disappears, and the claim follows by claim 1, or that edge is added to the matching and the claim still follows.   □

Proof of claim 3. Suppose by way of contradiction that an edge between two W vertices is added to the matching. Consider the state of the graph just before this edge is added. One of the vertices must have degree 1, so an edge connecting to its L child must have disappeared. But the only way such an edge can disappear is by the W vertex being added to the matching, contradiction! □

2.4 Putting It Together

We can now calculate the probability with which the Karp-Sipser algorithm, when executed on the random tree T, adds the special edge between u and v to the matching.

To this end, first label the trees rooted at u and v using the procedure described in the previous section (imagining the special edge connecting u and v does not exist, and we are just labelling two different rooted trees).

Let us first calculate the probability λ that u is labelled L:

$$\lambda = \Pr[u \text{ is labelled L}]$$
$$= \Pr[u \text{ has no children labelled L}]$$
$$= \sum_{k=0}^{\infty} \Pr[u \text{ has } k \text{ children and none of them are labelled L}]$$
$$= \sum_{k=0}^{\infty} \frac{e^{-1}}{k!}(1-\lambda)^k = e^{-1} \cdot e^{1-\lambda} = e^{-\lambda}$$

λ is thus the unique real number which solves the equation $x = e^{-x}$.

Second, let us calculate the probability that the edge between u and v is added to the matching, and v is labelled L. Imagine now rooting the random tree T at u. This does not change the label of any of the vertices except possibly u which is now labelled W.

This means that u must end up in the matching. None of the edges connecting u with any of its W children end up in the matching. All the edges connecting u

with any of its L children, and the special edge between u and v are completely symmetric from the standpoint of the execution of the Karp-Sipser algorithm. Therefore,

Pr[(u, v) is added to the matching, v is labelled L]

$$= \sum_{k=0}^{\infty} \Pr[(u, v) \text{ is added to the matching, } v \text{ is labelled L, } u \text{ has } k \text{ L children}]$$

$$= \sum_{k=0}^{\infty} \frac{\lambda}{k+1} \sum_{r=k}^{\infty} \binom{r}{k} \lambda^k (1-\lambda)^{r-k} \frac{e^{-1}}{r!} = \sum_{k=0}^{\infty} \frac{\lambda^{k+1} e^{-\lambda}}{(k+1)!} = (e^\lambda - 1)e^{-\lambda} = 1 - \lambda$$

Third, note that if we initially labelled both u and v L, then (u, v) must end up in the matching. This is because if we imagine rooting the tree at u, u is labelled W, so ends up in the matching, but the only way this can happen is if (u, v) ends up in the matching since all of its other children are labelled W.

Fourth, note that if we initially labelled both u and v W, then (u, v) must disappear. This is because if we imagine rooting the tree at u, the labelling remains the same, and every edge between W vertices disappears.

Finally, we can compute the probability that the special edge (u, v) ends up in the matching selected by Karp-Sipser as the sum of the probabilities of the edge ending up in the matching when u and v are labelled (respectively) L and L, L and W, W and L, and W and W. This is equal to $2(1 - \lambda) - \lambda^2$.

Theorem 1 now follows from a careful application of Lemma 1. Lemma 1 involves establishing a close correspondence between the neighborhood of an edge in the Galton-Watson process and the neighborhood of an edge in $R(x)$. This correspondence can be exploited to obtain a correspondence in the behavior of the Karp-Sipser algorithm in the two settings.

We omit the details, but we show here how Lemma 1 can be applied to prove the weaker statement that for fractional matchings x such that $\|x\|_\infty \to 0$, the neighborhood of any edge (u, v) does not contain a cycle with high probability. This weaker statement contains the main idea of the proof of Theorem 1.

Lemma 1 implies that for any set of finite trees $F = \{T_1, T_2, \ldots, T_m\}$ that are realizations of the random process T,

$$|\Pr[N((u, v)) \in F \mid (u, v) \in R(x)] - \Pr[T \in F]| = O\left(\epsilon \sum |T_i|^2\right)$$

and hence it follows that

$$|\Pr[N((u, v)) \notin F \mid (u, v) \in R(x)] - (1 - \Pr[T \in F])| = O\left(\epsilon \sum |T_i|^2\right)$$

and so, for any F we have

$$\lim_{\epsilon \to 0} \Pr[N((u, v)) \text{ contains a cycle} \mid (u, v) \in R(x)]$$

$$\leq \lim_{\epsilon \to 0} \Pr[N((u, v)) \notin F \mid (u, v) \in R(x)]$$

$$= 1 - \Pr[T \in F]$$

Since we know that T produces a finite tree with probability 1, we can take F larger and larger to prove that

$$\lim_{\epsilon \to 0} \Pr[N((u,v)) \text{ contains a cycle} \mid (u,v) \in R(x)] = 0$$

3   Improved CRSs for Bipartite Matchings

Now we turn to contention resolution for bipartite matchings, without any assumption on the ℓ_∞ norm of the fractional matching.

A basic building block of our CRSs is the following theorem which establishes the existence of a scheme for choosing 1 out of n elements (historically the first CRS [2]).

Theorem 3. *Suppose that $\mathcal{D}$ is a distribution on 2^E such that for every set $S \subseteq E$,*

$$\Pr_{R \sim \mathcal{D}}[S \cap R \neq \emptyset] \geq \sum_{i \in S} \beta_i.$$

Then there is a monotone contention resolution scheme for choosing one element $e(R)$ from $R \sim \mathcal{D}$ such that $\Pr[e(R) = i] \geq \beta_i$ for every $i \in E$.

This theorem can be proved using a max-flow/min-cut argument, as briefly discussed in [2]. We omit the proof.

3.1   A 0.480-Balanced Scheme for Bipartite Matchings

Before going to the proof of Theorem 2, we will show the existence of a simple $2(1 - e^{-1/e}) - e^{-2}$-balanced contention resolution scheme for bipartite matchings[2]. We remark that $2(1 - e^{-1/e}) - e^{-2} \geq 0.480$ and hence this already beats the *optimal monotone scheme* for bipartite matchings. By necessity, our scheme is non-monotone and this result establishes a strict separation between monotone and non-monotone schemes for bipartite matchings.

The Simple Scheme:

1. For each edge (u,v) (with probability x_{uv} of appearing), independently declare it active with probability $\frac{1 - e^{-x_{uv}}}{x_{uv}}$ given it appears.
2. For each vertex u, call an active edge (u,v) "available at vertex u" if v has no other edges adjacent to it which are active.
3. Using a contention resolution scheme for 1 element, select one of the available edges at each vertex. Ensure that an edge (u,v) gets selected at vertex u with probability at least $x_{uv}\left((1 - e^{-1/e}) - \frac{1}{2e^2}\right) + \frac{e^{2x_{uv}} - e^{x_{uv}}}{2e^2}$.
4. The set of edges selected at all the vertices form the matching M.

---

[2] Note the similarity in the expression to $2(1 - \lambda) - \lambda^2 = 2(1 - e^{-\lambda}) - \lambda^2$, the constant from our previous analysis. This similarity is not a coincidence, and we can think of the scheme we describe as first-order approximation to Karp-Sipser.

The first step of the scheme is a kind of pre-processing which ensures that edges with high probabilities of appearing don't destroy the chances of neighbouring edges getting picked by the scheme. This strategy has also been used in the literature previously [11]. The second step is a first attempt at using the idea that it is always useful to add vertices with degree 1 into the matching. This does not seem to have been explicitly exploited by previous CRSs.

To analyze the algorithm, note firstly that it is easy to see that the selected edges really do form a matching. Secondly, let us note that the probability that an edge (u, v) is available at a vertex u is

$$F(x_{uv}) = \left(1 - e^{-x_{uv}}\right) \prod_{u' \in \delta(v), u' \neq u} e^{-x_{u'v}} = \left(1 - e^{-x_{uv}}\right) e^{-(1-x_{uv})} = \frac{e^{x_{uv}} - 1}{e}$$

and similarly, the probability that an edge (u, v) is isolated amongst active edges is

$$\left(1 - e^{-x_{uv}}\right) \left(e^{-(1-x_{uv})}\right)^2 = \frac{e^{2x_{uv}} - e^{x_{uv}}}{e^2}$$

Hence, if a CRS of the sort used in step 3 exists, the desired result follows since

$$\Pr[(u, v) \in M]$$
$$= 2 \Pr[(u, v) \text{ is selected at } u] - \Pr[(u, v) \text{ is selected at both } u, v]$$
$$= 2 \Pr[(u, v) \text{ is selected at } u] - \Pr[(u, v) \text{ is isolated amongst active edges}]$$
$$= 2x_{uv} \left((1 - e^{-1/e}) - \frac{1}{2e^2}\right) + \frac{e^{2x_{uv}} - e^{x_{uv}}}{e^2} - \frac{e^{2x_{uv}} - e^{x_{uv}}}{e^2}$$
$$= 2x_{uv} \left((1 - e^{-1/e}) - \frac{1}{2e^2}\right)$$

Furthermore whether (u, v_i) is available at u is independent of whether (u, v_j) is available at u (this is where we use the fact that the graph is bipartite; note that we actually only need to assume that the graph is triangle-free; it is unclear how to drop this assumption). Therefore, it follows from Theorem 3 that if the probability of (u, v_i) appearing is x_i (short for x_{uv_i}), the existence of the required CRS depends only on whether

$$\Pr[\text{at least one of } (u, v_i) \text{ is available}] = 1 - \prod_{i=1}^{n}(1 - F(x_i))$$
$$\geq \sum_{i=1}^{n} x_i \left((1 - e^{-1/e}) - \frac{1}{2e^2}\right) + \frac{e^{2x_i} - e^{x_i}}{2e^2}$$

Therefore, the desired result follows from the following lemma, whose proof we omit.

Lemma 2. *If $\sum x_i \leq 1, x_i \geq 0$, and $F(x) = \frac{e^x - 1}{e}$, then*

$$1 - \prod_{i=1}^{n}(1 - F(x_i)) \geq \sum_{i=1}^{n} x_i \left((1 - e^{-1/e}) - \frac{1}{2e^2}\right) + \frac{e^{2x_i} - e^{x_i}}{2e^2}$$

3.2   A 0.509-Balanced Scheme for Bipartite Matchings

In this section we present our best CRS for bipartite matchings which beats the best possible RCRS. We call the set of vertices on the left V_1, and on the right, V_2. There are two main ideas. One idea is to select edges in two stages, with the first stage devoted to running contention resolution on the edges at each vertex in V_1, and the second stage is devoted to running contention resolution on the edges picked in stage 1 at each vertex in V_2.

The other main idea involves noticing that edges (u, v) such that v has degree 1 in R have no competition in the second stage and hence should be preferentially selected in the first stage, since if selected they will certainly survive in our matching.

The Red/Blue/Gray Scheme:

1. Recall that $\Pr[(u, v) \in R] = x_{uv}$. Decide for each edge $(u, v) \in R$ independently at random whether to mark it **gray**, so that $\Pr[(u, v) \text{ is gray}] = x_{uv} - (1 - e^{-x_{uv}})$. We call the edges $(u, v) \in R$ which are not gray *active*.
2. For each (u, v) such that (u, v) is the only active edge incident to v, decide independently at random whether to mark (u, v) **red**, so that $\Pr[(u, v) \text{ is red}] = 1 - e^{-x_{uv}/e}$. Mark all other active edges **blue**. We have $\Pr[(u, v) \text{ is blue}] = e^{-x_{uv}/e} - e^{-x_{uv}}$.
3. For each $u \in V_1$, if there are any red edges incident to u, perform contention resolution to include one of them in R_1, so that $\Pr[(u, v) \in R_1] = (1 - e^{-1/e})x_{uv}$.
4. For each $u \in V_1$, if there are no red edges incident to u, and there are some blue edges incident to u, perform contention resolution to include one of them in R_2, so that $\Pr[(u, v) \in R_2] = (e^{-1/e} - e^{-1})x_{uv}$.
5. For each $u \in V_1$, if there are no active (red or blue) edges incident to u, and there are some gray edges incident to u, perform contention resolution to include one of them in R_3, so that $\Pr[(u, v) \in R_3] \geq \frac{1}{2e}x_{uv}^2$.
6. Finally, for each $v \in V_2$, perform contention resolution among all edges in $R_1 \cup R_2 \cup R_3$ incident to v, to include one of them in M, so that

$$\Pr[(u, v) \in M] \geq 0.509 \, x_{uv}.$$

Implicit in step 2 is the claim that the definition of red edges is valid. Implicit in each of steps 3, 4, 5, and 6 above is a claim that there exists a certain contention resolution scheme for choosing 1 out of n elements. The existence of such schemes can be proved by applying Theorem 3, if we can calculate the probability that at least one of a subset of edges "is available for consideration at that stage" (i.e., is red, is blue, etc.).

For steps 3, 4, and 5, this quantity is fairly simple to calculate, because all the choices (to designate edges red or blue or gray) are made independently. The following lemma, whose proof we omit, summarizes the claimed existences in these cases.

Lemma 3. *The definition of red edges is valid. There are CRSs among red, blue, and gray edges such that* $\Pr[(u, v) \in R_1] = (1 - e^{-1/e})x_{uv}, \Pr[(u, v) \in R_2] = (e^{-1/e} - e^{-1})x_{uv},$ *and* $\Pr[(u, v) \in R_3] \geq \frac{1}{2e}x_{uv}^2$ *for every edge* (u, v).

To finish, we need to analyze Step 6 of the algorithm, which is contention resolution among all the surviving edges on the right-hand side. Here, there can be at most one red edge incident to a vertex $v \in V_2$, possibly multiple gray edges which appear independently, and possibly multiple blue edges whose survival up to this stage of the scheme is correlated. This correlation causes the main trouble in our analysis of this final step, because it makes it harder to calculate the probability that at least one of a subset of edges incident at a vertex v is in $R_1 \cup R_2 \cup R_3$. Ideally, we would like to prove that the appearance of blue edges satisfies some form of negative correlation. At the moment, we are able to prove only *pairwise* negative correlation which is sufficient to achieve the factor of 0.509. A stronger correlation result (for example negative cylinder dependence) would lead to an improved factor. We include a full proof of the following lemma due to its conceptual importance in the overall proof.

Lemma 4. *For any two incident edges* (u, v) *and* (u', v),

$$\Pr[(u, v) \in R_2 \ \& \ (u', v) \in R_2 \mid (u, v), (u', v) \text{ are blue}]$$

$$\leq \Pr[(u, v) \in R_2 \mid (u, v), (u', v) \text{ are blue}] \cdot \Pr[(u', v) \in R_2 \mid (u, v), (u', v) \text{ are blue}].$$

Proof. Define $\Gamma(u) = \{v' : (u, v') \text{ active}\}$ and $\Gamma(u') = \{v' : (u', v') \text{ active}\}$. Note that conditioning on $(u, v), (u', v)$ being blue edges is the same as conditioning on $v \in \Gamma(u) \cap \Gamma(u')$, because edges $(u, v), (u', v)$ being active also means that they must be blue.

We claim that conditioned on $\Gamma(u), \Gamma(u')$ such that $v \in \Gamma(u) \cap \Gamma(u')$, the probability that $(u, v) \in R_2$ is decreasing in $\Gamma(u)$ and increasing in $\Gamma(u')$, while conversely the probability that $(u', v) \in R_2$ is increasing in $\Gamma(u)$ and decreasing in $\Gamma(u')$.

We prove this by considering a fixed choice of the active edges incident to $V_1 \setminus \{u, u'\}$, and at the end averaging over these choices. Consider $\Gamma(u), \Gamma(u')$ where $v \in \Gamma(u) \cap \Gamma(u')$. For (u, v) to be selected in R_2, there cannot be any red edge incident to u. The only candidates for such red edges are $(u, \tilde{v})$ where $\tilde{v} \in \Gamma(u) \setminus \Gamma(u')$, because edges incident to $\Gamma(u) \cap \Gamma(u')$ are blue by definition. For each $\tilde{v} \in \Gamma(u) \setminus \Gamma(u')$, $(u, \tilde{v})$ is red if $\tilde{v}$ does not have any other incident active edges (and an additional independent coin flip succeeds, as defined in Step 2). Clearly, the event of no red edge incident to u is monotonically decreasing in $\Gamma(u) \setminus \Gamma(u')$.

In case there is no red edge incident to u, we perform contention resolution among the blue edges incident to u, which are all the edges $(u, v'), v' \in \Gamma(u)$. Since this scheme is monotone, the probability of survival is monotonically decreasing in $\Gamma(u)$. This monotonicity property also remains preserved when we average over the choices of active edges incident to $V_1 \setminus \{u, u'\}$. Overall, the probability of (u, v) surviving in R_2 is monotonically decreasing in $\Gamma(u)$ and increasing in $\Gamma(u')$. Symmetrically, the probability of (u', v) surviving in R_2 is monotonically decreasing in $\Gamma(u')$ and increasing in $\Gamma(u)$.

Given this monotonicity property, we use the FKG inequality to prove our result. The appearances of vertices in $\Gamma(u)$ and $\Gamma(u')$ are independent (since these are determined by the edges incident to u and u' respectively). Let us define $\gamma \in \{0, 1\}^{2n}$ where $\gamma_i = 0$ if $i \in \Gamma(u)$ and $\gamma_{n+i} = 1$ if $i \in \Gamma(u')$. As we argued, $\Pr[(u, v) \in R_2 \mid \gamma]$ is increasing in γ and $\Pr[(u', v) \in R_2 \mid \gamma]$ is decreasing in γ, for all γ consistent with $v \in \Gamma(u) \cap \Gamma(u')$. Therefore, by the FKG inequality applied to this subspace (conditioned on $v \in \Gamma(u) \cap \Gamma(u')$),

$$\Pr[(u, v) \in R_2 \ \& \ (u', v) \in R_2 \mid v \in \Gamma(u) \cap \Gamma(u')]$$

$$\leq \Pr[(u, v) \in R_2 \mid v \in \Gamma(u) \cap \Gamma(u')] \cdot \Pr[(u', v) \in R_2 \mid v \in \Gamma(u) \cap \Gamma(u')].$$

as desired. □

The main takeaway from this lemma is that if we let β be a lower bound on the probability that an edge survives in R_2 given that it is blue, then, conditioned on having at least two active (and hence blue) edges at a vertex v in V_2 in step 6, the probability that one of them survives in R_2 is at least $2\beta - \beta^2$. In the final analysis of step 6, if there are more than 2 blue edges at a vertex v, we only use two of them. This allows us to establish our desired conclusion.

Theorem 4. *There is a CRS in Step 6 which achieves a factor of* 0.509.

To prove the theorem, we consider a vertex $v \in V_2$ and all the edges incident to v which are in $R_1 \cup R_2 \cup R_3$ (i.e. survived contention resolution on the left-hand side). To show that the required kind of CRS exists, we consider a subset of edges S incident to v, and compute the probability that at least one of them survives as follows:

$$\begin{aligned} \Pr[S \cap (R_1 \cup R_2 \cup R_3) \neq \emptyset] = \ &\Pr[S \cap R_1 \neq \emptyset] \\ &+ \Pr[S \cap R_1 = \emptyset \ \& \ S \cap R_2 \neq \emptyset] \\ &+ \Pr[S \cap (R_1 \cup R_2) = \emptyset \ \& \ S \cap R_3 \neq \emptyset]. \end{aligned}$$

We finish the proof by establishing lower bounds on each of the three terms and applying Theorem 3. Lemma 4 is useful in the analysis of the second term. We omit the proof here. We refer the reader to the full version for details.

Acknowledgements. We would like to thank Chandra Chekuri for stimulating discussions.

References

1. Karp, R.M., and Sipser, M.: Maximum matchings in sparse random graphs. In: 22nd Annual Symposium on Foundations of Computer Science, Nashville, Tennessee, USA, 28–30 October 1981, pp. 364–375. IEEE Computer Society (1981)

2. Feige, U.: On maximizing welfare when utility functions are subadditive. SIAM J. Comput. **39**(1), 122–142 (2009)
3. Bordenave, C., Lelarge, M., Salez, J.: Matchings on innite graphs. Probab. Theory Relat. Fields **157**(1), 183–208 (2013)
4. Chekuri, C., Vondrák, J., Zenklusen, R.: Submodular function maximization via the multilinear relaxation and contention resolution schemes. SIAM J. Comput. **43**(6), 1831–1879 (2014)
5. Balister, P., and Gerke, S.: Controllability and matchings in random bipartite graphs. In: Czumaj, A., Georgakopoulos, A., Král, D., Lozin, V., Pikhurko, O. (eds.) Surveys in Combinatorics 2015. London Mathematical Society Lecture Note Series. Cambridge University Press, pp. 119–146 (2015)
6. Feldman, M., Svensson, O., Zenklusen, R.: Online contention resolution schemes. In: Krauthgamer, R. (ed.) Proceedings of the Twenty-Seventh Annual ACM-SIAM Symposium on Discrete Algorithms, SODA 2016, Arlington, VA, USA, 10–12 January 2016, pp. 1014–1033. SIAM (2016)
7. Guruganesh, G., Lee, E.: Understanding the correlation gap for matchings. In: Lokam, S.V., and Ramanujam, R. (eds.) 37th IARCS Annual Conference on Foundations of Software Technology and Theoretical Computer Science, FSTTCS 2017, 11–15 December 2017, Kanpur, India. LIPIcs, 1–15. Schloss Dagstuhl - Leibniz-Zentrum für Informatik (2017)
8. Adamczyk, M., Wlodarczyk, M.: Random order contention resolution schemes. In: Thorup, M. (ed.) 59th IEEE Annual Symposium on Foundations of Computer Science, FOCS 2018, Paris, France, 7–9 October 2018, pp. 790–801. IEEE Computer Society (2018)
9. Lee, E., Singla, S.: Optimal online contention resolution schemes via ex-ante prophet inequalities. In: Azar, Y., Bast, H., and Herman, G. (eds.) 26th Annual European Symposium on Algorithms, ESA 2018, 20–22 August 2018, Helsinki, Finland. LIPIcs, 1–14. Schloss Dagstuhl - Leibniz-Zentrum für Informatik (2018)
10. Fu, H., Tang, Z.G., Wu, H., Wu, J., Zhang, Q.: Random order vertex arrival contention resolution schemes for matching, with applications. In: Bansal, N., Merelli, E., Worrell, J. (eds.) 48th International Colloquium on Automata, Languages, and Programming, ICALP 2021, 12–16 July 2021, Glasgow, Scotland (Virtual Conference). LIPIcs, 1–20. Schloss Dagstuhl - Leibniz-Zentrum für Informatik (2021)
11. Bruggmann, S., Zenklusen, R.: An optimal monotone contention resolution scheme for bipartite matchings via a polyhedral viewpoint. Math. Program. **191**(2), 795–845 (2022)
12. MacRury, C., Ma, W., Grammel, N.: On (random-order) online contention resolution schemes for the matching polytope of (bipartite) graphs (2022). https://arxiv.org/abs/2209.07520
13. Pollner, T., Roghani, M., Saberi, A., Wajc, D.: Improved online contention resolution for matchings and applications to the gig economy. In: Proceedings of the 23rd ACM Conference on Economics and Computation. EC 2022, pp. 321–322. Association for Computing Machinery, Boulder, CO, USA (2022)

Advances on Strictly Δ-Modular IPs

Martin Nägele[1]([✉]) [iD], Christian Nöbel[2] [iD], Richard Santiago[2] [iD],
and Rico Zenklusen[2] [iD]

[1] Research Institute for Discrete Mathematics and Hausdorff Center for
Mathematics, University of Bonn, Bonn, Germany
mnaegele@uni-bonn.de
[2] Department of Mathematics, ETH Zurich, Zurich, Switzerland
{cnoebel,rtorres,ricoz}@ethz.ch

Abstract. There has been significant work recently on integer programs (IPs) $\min\{c^\top x \colon Ax \leq b, \, x \in \mathbb{Z}^n\}$ with a constraint marix A with bounded subdeterminants. This is motivated by a well-known conjecture claiming that, for any constant $\Delta \in \mathbb{Z}_{>0}$, Δ-modular IPs are efficiently solvable, which are IPs where the constraint matrix $A \in \mathbb{Z}^{m \times n}$ has full column rank and all $n \times n$ minors of A are within $\{-\Delta, \ldots, \Delta\}$. Previous progress on this question, in particular for $\Delta = 2$, relies on algorithms that solve an important special case, namely *strictly Δ-modular IPs*, which further restrict the $n \times n$ minors of A to be within $\{-\Delta, 0, \Delta\}$. Even for $\Delta = 2$, such problems include well-known combinatorial optimization problems like the minimum odd/even cut problem. The conjecture remains open even for strictly Δ-modular IPs. Prior advances were restricted to prime Δ, which allows for employing strong number-theoretic results.

In this work, we make first progress beyond the prime case by presenting techniques not relying on such strong number-theoretic prime results. In particular, our approach implies that there is a randomized algorithm to check feasibility of strictly Δ-modular IPs in strongly polynomial time if $\Delta \leq 4$.

Keywords: Bounded subdeterminants · Congruency constraints

1 Introduction

Integer Programs (IPs) $\min\{c^\top x \colon Ax \leq b, \, x \in \mathbb{Z}^n\}$ are a central NP-hard problem class in Combinatorial Optimization. There is substantial prior work

Funded through the Swiss National Science Foundation grants 200021_184622 and P500PT_206742, the European Research Council (ERC) under the European Union's Horizon 2020 research and innovation programme (grant agreement No 817750), and the Deutsche Forschungsgemeinschaft (DFG, German Research Foundation) under Germany's Excellence Strategy – EXZ-2047/1 – 390685813.

A. Del Pia and V. Kaibel (Eds.): IPCO 2023, LNCS 13904, pp. 393–407, 2023.
https://doi.org/10.1007/978-3-031-32726-1_28

and interest in identifying special classes of polynomial-time solvable IPs while remaining as general as possible. One of the best-known such classes are IPs with a constraint matrix that is *totally unimodular* (TU), i.e., the determinant of any of its square submatrices is within $\{-1, 0, 1\}$. A long-standing open conjecture in the field is whether this result can be generalized to Δ-modular constraint matrices for constant Δ. Here, we say that a matrix $A \in \mathbb{Z}^{k \times n}$ is Δ-*modular* if it has full column rank and all $n \times n$ submatrices have determinants in $\{-\Delta, \ldots, \Delta\}$.[1] For brevity, we call an IP with Δ-modular constraint matrix a Δ-*modular IP*. We recap the above-mentioned conjecture below. Unfortunately, we do not know its precise origin; it may be considered folklore in the field.

Conjecture 1. For constant $\Delta \in \mathbb{Z}_{\geq 0}$, Δ-modular IPs can be solved in polynomial time.

First progress on Conjecture 1 was made by Artmann, Weismantel, and Zenklusen [3], who showed that it holds for $\Delta = 2$ (the bimodular case). Fiorini, Joret, Weltge, and Yuditsky [11] show that the conjecture is true for an arbitrary constant Δ under the extra condition that the constraint matrix has at most two non-zero entries per row or column. Through a non-trivial extension of the techniques in [3], it was shown by Nägele, Santiago, and Zenklusen [24] that there is a randomized algorithm to check feasibility of an IP with a strictly 3-modular constraint matrix in polynomial time. Here, a matrix $A \in \mathbb{Z}^{k \times n}$ is called *strictly* Δ-*modular* if it has full column rank and all its $n \times n$ submatrices have determinants in $\{-\Delta, 0, \Delta\}$.

As a key ingredient, all these prior approaches solve certain combinatorial optimization problems with congruency constraints. This is not surprising, as even strictly Δ-modular IPs include the following class of *MCCTU problems*:[2]

Multi-Congruency-Constrained TU Problem (MCCTU): Let $T \in \mathbb{Z}^{k \times n}$ be TU, $b \in \mathbb{Z}^k$, $c \in \mathbb{R}^k$, $m \in \mathbb{Z}_{>0}^q$, $\gamma_i \in \mathbb{Z}^n$ for $i \in [q]$, $r \in \mathbb{Z}^q$. Solve
$$\min\{c^\top x : Tx \leq b, \ \gamma_i^\top x \equiv r_i \pmod{*}m_i \ \forall i \in [q], \ x \in \mathbb{Z}^n\} \ .$$

Unless mentioned otherwise, we assume that in the context of MCCTU problems, q and m_i are constant. Even MCCTU with just a single congruency constraint, i.e., $q = 1$, already contains the classical and well-studied odd and even cut problems, and, more generally, the problem of finding a minimum cut whose number of vertices is $r \pmod m$. (See [5, 14, 19, 25, 26, 29] for related work.) It can also

---

[1] A weaker variant of the conjecture claims efficient solvability of IPs with *totally* Δ-*modular* constraint matrices, where all subdeterminants are bounded by Δ in absolute value. The conjecture involving Δ-modular matrices implies the weaker variant. Indeed, an IP $\min\{c^\top x : Ax \leq b, x \in \mathbb{Z}^n\}$ with a totally Δ-modular constraint matrix can be reformulated as $\min\{c^\top(x^+ - x^-) : A(x^+ - x^-) \leq b, x^+, x^- \in \mathbb{Z}_{\geq 0}^n\}$. It is not hard to see that the constraint matrix of the new LP remains totally Δ-modular; moreover, it has full column rank because of the non-negativity constraints.

[2] To capture an MCCTU problem as a strictly Δ-modular IP, replace each congruency constraint $\gamma_i^\top x \equiv r_i \pmod{m_i}$ by an equality constraint $\gamma_i^\top x + m_i y_i = r$ with $y_i \in \mathbb{Z}$. The corresponding constraint matrix is strictly Δ-modular for $\Delta = \prod_{i=1}^q m_i$.

capture the minimum T-join problem, congruency-constrained flow problems, and many other problems linked to TU matrices.

Combinatorial optimization problems with congruency constraints are highly non-trivial and many open questions remain. As they are already captured by strictly Δ-modular IPs, this motivates the following weakening of Conjecture 1.

Conjecture 2. Strictly Δ-modular IPs can be solved in polynomial time for constant $\Delta \in \mathbb{Z}_{\geq 0}$.

Even resolving this weaker conjecture would settle several open problems, including congruency-constrained min cuts (in both directed and undirected graphs), or the problem of efficiently and deterministically finding a perfect matching in a red/blue edge-colored bipartite graph such that the number of red matching edges is r (mod m). (This is a simplified version of the famous red-blue matching problem, where the task is to find a perfect matching with a specified number of red edges; for both versions, randomized algorithms are known.) Interestingly, for the bimodular case ($\Delta = 2$), a result by Veselov and Chirkov [33] implies that Conjecture 1 and Conjecture 2 are equivalent (see [3]).

Our goal is to shed further light on Conjecture 2 and overcome some important hurdles of prior approaches. In a first step, we note that a positive resolution of Conjecture 2 does not only imply efficient solvability of MCCTU problems, but also vice versa, and this reduction works in strongly polynomial time.

Lemma 1. *Let $\Delta > 0$. Every strictly Δ-modular IP can, in strongly polynomial time, be reduced to an MCCTU problem with moduli m_i such that $\Delta = \prod_{i=1}^{q} m_i$.*

Without the strongly polynomial time condition, this also follows from very recent work of Gribanov, Shumilov, Malyshev, and Pardalos [15, Lemma 4].

Further, we are interested in making progress regarding the feasibility version of Conjecture 2, i.e., efficiently deciding whether a strictly Δ-modular IP is feasible. Prior approaches settle this question for $\Delta = 2$ [3] and—using a randomized algorithm—for $\Delta = 3$ [24]. A main hurdle to extend these is that they crucially rely on Δ being prime, for example through the use of the Cauchy-Davenport Theorem. Our main contribution here is to address this. In particular, we can check feasibility for $\Delta = 4$ with a randomized algorithm, which is the first result in this context for non-prime Δ. More importantly, our techniques will hopefully prove useful for future advances on this challenging question.

Theorem 1. *There exists a strongly polynomial-time randomized algorithm to find a feasible solution of a strictly 4-modular IP, or detect that it is infeasible.*

We remark that the randomization appearing in the above theorem comes from the fact that one building block of our result is a reduction to a problem class that includes the aforementioned congruency-constrained red/blue-perfect matching problem, for which only randomized approaches are known.

1.1 Group-Constrained Problems and Proof Strategy for Theorem 1

To show Theorem 1, we exploit its close connection to MCCTU. Capturing the congruency constraints of an MCCTU problem through an abelian group constraint, we attain the following *group-constrained TU feasibility problem.*

> **Group-Constrained TU Feasibility (GCTUF):** Let $T \in \mathbb{R}^{k \times n}$ be a TU matrix, let $b \in \mathbb{Z}^k$, let $(G, +)$ be a finite abelian group, and let $\gamma \in G^n$ and $r \in G$. The task is to show infeasibility or find a solution of the system
> $$Tx \le b, \ \gamma^\top x = r, \ x \in \mathbb{Z}^n \ .$$

Here, the scalar product $\gamma^\top x$ denotes the linear combination of the group elements $\gamma_1, \ldots, \gamma_n$ with multiplicities $x_1, \ldots, x_n$ in G. Group constraints generalize congruency constraints, which are obtained in the special case where G is cyclic. More generally, by the fundamental theorem of finite abelian groups, a finite abelian group G is, up to isomorphism, a direct product of cyclic groups. Hence, a group constraint can be interpreted as a set of congruency constraints and vice versa. Thus, GCTUF and MCCTU feasibility are two views on the same problem. We stick to GCTUF mostly for convenience of notation. Moreover, the GCTUF setting also allows for an elegant use of group-related results later on. One may assume that the group is given through its multiplication table (the *Cayley table*). In fact, the precise group representation is not of great importance to us. Concretely, for constant Δ, strictly Δ-modular IP feasibility problems reduce to GCTUF problems with a constant size group. Many of our polynomial-time algorithmic results can even be extended to settings where the group size is not part of the input, and access to group operations is provided through an oracle.

By Lemma 1 and the aforementioned equivalent viewpoint of multiple congruency constraints and a group constraint, in order to prove Theorem 1, it is enough for us to show the equivalent statement below.

Theorem 2. *There exists a strongly polynomial time randomized algorithm for GCTUF problems with a group of cardinality at most 4.*

On a high level, we follow a well-known strategy for TU-related problems by employing Seymour's decomposition [31] to decompose the problem into problems on simpler, more structured TU matrices. (See, e.g., [1,3,9,24].) Roughly speaking, a TU matrix is either very structured—in which case we call it a *base block*—or can be decomposed into smaller TU matrices through a small set of well-defined operations. (See the discussion following Theorem 7.) The use of Seymour's decomposition typically comes with two main challenges, namely (i) solving the base block cases, and (ii) propagating solutions of the base block cases back through the decomposition efficiently to solve the original problem. First, we show that this propagation can be done efficiently for our problem.

Theorem 3. *Let G be an abelian group of size at most 4. Given an oracle for solving base block GCTUF problems with group G, we can solve GCTUF problems with group G in strongly polynomial time with strongly polynomially many calls to the oracle.*

In fact, our approach underlying Theorems 2 and 3 operates in a hierarchy of GCTUF problems with increasingly relaxed group constraints of the form $\gamma^\top x \in R$ for subsets $R \subseteq G$ of increasing size, and allows for proving the above results for such relaxed GCTUF problems for arbitrary constant-size groups G as long as $|G| - |R| \leq 3$. (See Sect. 3 for more details.) In principle, this is along the lines of the approach to congruency-constrained TU problems in [24], but incorporates the new viewpoint of group constraints, and additionally improves over earlier results in two ways: First, our approach applies to arbitrary finite abelian groups, while previous setups heavily relied on the group cardinality being a prime. Secondly, in the setting with relaxed group constraints, we extend the admissible range of $|G| - |R|$ by one, thus proceeding further in the hierarchy of GCTUF problems, and newly covering GCTUF problems with groups of cardinality 4.

Besides being a key part of our approach, Theorem 3 underlines that base block GCTUF problems are not merely special cases, but play a key role in progress on general GCTUF problems. There are only two non-trivial types of such base block GCTUF problems, namely when the constraint matrix is a so-called *network matrix* or a transpose thereof. Both cases cover combinatorial problems that are interesting on their own, and their complexity status remains open to date. If the constraint matrix is a network matrix, GCTUF can be cast as a circulation problem with a group constraint. By reducing to and exploiting results of Camerini, Galbiati, and Maffioli [7] on exact perfect matching problems, a randomized algorithm for the congruency-constrained case has been presented in [24]. We observe that these results extend to the group-constrained setting. The other base block case, where the constraint matrix is the transpose of a network matrix, can be cast as a group-constrained directed minimum cut problem by leveraging a result in [24]. Prior work combined this reduction with results on congruency-constrained submodular minimization [25] to solve the optimization version of the problem for congruency-constraints of prime power modulus. We show that the feasibility question on this base block can be solved efficiently on any finite abelian group of constant order, thus circumventing the prime power restriction that is intrinsic in prior approaches.

Theorem 4. *Let G be a finite abelian group. There is a strongly polynomial time algorithm for solving GCTUF problems with group G where the constraint matrix is the transpose of a network matrix.*

1.2 Further Related Work

The parameter Δ has been studied from various viewpoints. While efficient recognition of (totally) Δ-modular matrices is open for any $\Delta \geq 2$, approaches to approximate the largest subdeterminant in absolute value were studied [8,27]. Also, focusing on more restricted subdeterminant patterns proved useful [2,12,33]. Aiming at generalizing a bound of Heller [21] for $\Delta = 1$, bounds on the maximum number of rows of a Δ-modular matrix were obtained [4,13,23]. Also, the influence of the parameter Δ on structure and properties of IPs and polyhedra is multi-faceted (see, e.g., [6,10,16–18,22,28,32] and references therein).

1.3   Structure of the Paper

In Sect. 2, we prove Theorem 4. Section 3 illustrates our approach and new contributions towards Theorem 3 on a more technical level, and explains the main new ingredients of our proof. Due to space constraints, some proofs are deferred to a long version of this paper, including the proof of Lemma 1.

2   GCTUF with Transposed Network Constraint Matrices

In the setting with a congruency constraint instead of a group constraint, [24] shows that every base block problem with a constraint matrix that is a transposed network matrix can be reduced to a node-weighted minimization problem over a lattice with a congruency constraint,[3] i.e., a problem of the form

$$\min\{w(S)\colon S \in \mathcal{L},\ \gamma(S) \equiv r(\mathrm{mod}\ m)\}\ ,\tag{1}$$

where $\mathcal{L} \subseteq 2^N$ is a lattice on some finite ground set N, $\gamma\colon N \to \mathbb{Z}$, $r \in \mathbb{Z}$, $m \in \mathbb{Z}_{>0}$, $w\colon N \to \mathbb{R}$, and we use $\gamma(S) \coloneqq \sum_{v \in S} \gamma(v)$ as well as $w(S) \coloneqq \sum_{v \in S} w(v)$.[4] Being a special case of congruency-constrained submodular minimization, it is known that such problems, and thus the corresponding congruency-constrained TU problems with a transposed network constraint matrix, can be solved in strongly polynomial time for constant prime power moduli m, while the case of general constant composite moduli remains open [25]. The progress on GCTUF, particularly the reduction to base block feasibility problems through Theorem 3 and its generalization (Theorem 8 in Sect. 3), motivates studying these reductions and results in the feasibility setting and with a group constraint instead of a congruency constraint, giving rise to the following problem.

> **Group-Constrained Lattice Feasibility (GCLF):** Let N be a finite set, $\mathcal{L} \subseteq 2^N$ a lattice, $(G, +)$ a finite abelian group, $\gamma\colon N \to G$, $r \in G$. The task is to find $X \in \mathcal{L}$ with $\gamma(X) = r$, or decide infeasibility.

We observe that the reduction in [24] from congruency-constrained TU problems with transposed network constraint matrices to problems of the form given in (1) extends to the group-constrained case. In particular, we obtain the following result in the feasibility setting.

Proposition 1. *Let G be a finite abelian group. Any GCTUF problem with group G and a constraint matrix that is a transposed network matrix can in strongly polynomial time be reduced to a GCLF problem with group G.*

---

[3] In fact, the proof in [24] claims a reduction to a submodular minimization problem, but shows the stronger one presented here.

[4] We recall that a *lattice* $\mathcal{L} \subseteq 2^N$ is a set family such that for any $A, B \in \mathcal{L}$, we have $A \cap B, A \cup B \in \mathcal{L}$. We assume such a lattice to be given by a compact encoding in a directed acyclic graph H on the vertex set N such that $X \subseteq N$ is an element of the lattice if and only if $\delta_H^-(X) = \emptyset$ (cf. [20, Section 10.3]). Here, as usual, in a digraph $G = (V, A)$ and for $X \subseteq V$, we denote by $\delta^+(X)$ and $\delta^-(X)$ the arcs in A leaving and entering X, respectively. Moreover, we write $\delta^\pm(v) \coloneqq \delta^\pm(\{v\})$ for $v \in V$.

Thus, it remains to study GCLF problems. Interestingly, for the pure feasibility question, we can circumvent the barriers present in the optimization setting, and obtain the following result through a concise argument.

Theorem 5. *Let G be a finite abelian group. GCLF problems with group G can be solved in strongly polynomial time.*

Clearly, Proposition 1 and Theorem 5 together imply Theorem 4. The main observation towards a proof of Theorem 5 is the following elementary lemma.

Lemma 2. *Let G be a finite abelian group, and let $\gamma_1, \ldots, \gamma_\ell \in G$. If $\ell \geq |G|$, then there is a non-empty subset $I \subseteq [\ell]$ such that $\sum_{i \in I} \gamma_i = 0$.*

Proof. Either $s_i := \sum_{j \leq i} \gamma_j = 0$ for some $i \in [\ell]$, or there exist $i < j$ with $s_i = s_j$; hence $I = [i]$ or $I = \{i+1, \ldots, j\}$, respectively, has the desired properties. $\square$

To prove Theorem 5, we work with a representation of the lattice $\mathcal{L}$ through an acyclic digraph H (see Footnote 4). We exploit that every $X \in \mathcal{L}$ is uniquely defined by the subset $C_X := \{x \in X : \delta^+(x) \subseteq \delta^+(X)\}$.

Proof of Theorem 5. We claim that if the given GCLF problem is feasible, there is a feasible X with $|C_X| < |G|$. If so, we obtain an efficient procedure for GCLF with group G through enumerating all such C_X and checking if $\gamma(X) = r$. To prove the claim, assume for contradiction that it is wrong, and let $X \in \mathcal{L}$ be minimal with $\gamma(X) = r$. Then $|C_X| \geq |G|$, and applying Lemma 2 to C_X gives a non-empty subset $Y \subseteq C_X$ with $\gamma(Y) = 0$. Thus, $X \setminus Y$ is a strictly smaller lattice element with $\gamma(X \setminus Y) = \gamma(X) - \gamma(Y) = \gamma(X) = r$, a contradiction. $\square$

3 Overview of Our Techniques Leading to Theorem 3

In order to tackle GCTUF problems, following ideas from [24], we introduce a hierarchy of slightly relaxed GCTUF problems by weakening the group constraint.

> **R-Group-Constrained TU Feasibility (R-GCTUF):** Let $T \in \{-1, 0, 1\}^{k \times n}$ be TU, $b \in \mathbb{Z}^k$, let $(G, +)$ be a finite abelian group, $\gamma \in G^n$ and $R \subseteq G$. The task is to show infeasibility or find a solution of
> $$Tx \leq b, \ \gamma^\top x \in R, \ x \in \mathbb{Z}^n .$$

Here, we typically call R the set of target elements. The above setup allows us to measure progress between GCTUF (the case of $|R| = 1$) and an unconstrained IP with TU constraint matrix (captured by setting $R = G$). In particular, the difficulty of an R-GCTUF problem increases as the size of R, i.e., the number of target elements, decreases. The main parameter capturing this hardness is the depth $d := |G| - |R|$ of the problem. We show the following generalization of Theorem 2.

Theorem 6. *Let G be a finite abelian group. There is a strongly polynomial randomized algorithm solving R-GCTUF problems with group G and $|G| - |R| \leq 3$.*

Our argument uses Seymour's decomposition theorem. To this end, for matrices $A \in \mathbb{Z}^{k_A \times n_A}$ and $B \in \mathbb{Z}^{k_B \times n_B}$ as well as vectors e, f, g, and h of appropriate size, we recall that the 3-*sum* of $\begin{pmatrix} A & e & e \\ h^\top & 0 & 1 \end{pmatrix}$ and $\begin{pmatrix} 0 & 1 & f^\top \\ g & g & B \end{pmatrix}$ is $\begin{pmatrix} A & ef^\top \\ gh^\top & B \end{pmatrix}$.[5]

Theorem 7 (Seymour's Decomposition). *Let $T \in \mathbb{Z}^{k \times n}$ be TU. Then either (i) T is a base block matrix, or (ii) T can, possibly after row and column permutations and pivoting once, be decomposed into a 3-sum of TU matrices with $n_A, n_B \geq 2$. Additionally, we can in time $\mathrm{poly}(n)$ decide which of the cases holds and determine the involved matrices.*

Item (i) covers three types of matrices: network matrices, transposes thereof, and matrices obtainable through basic operations from one of two specific 5×5 TU matrices. (For more details on Seymour's decomposition, see, e.g., [30], and for a version tailored to our setting, see [24, Theorem 2.2].) By combining results for base blocks from [24] with our results from Sect. 2, it follows that GCTUF problems can be solved in strongly polynomial time if the constraint matrix is a base block matrix; hence dealing with Item (i) above. In Item (ii), the potential pivoting step can be handled by extending a result from [24] to the group setting. Hence, it remains to discuss how to deal with constraint matrices that are 3-sums. We devote the rest of this section to discuss the main ingredients needed to cover this case. Altogether, we proof the following generalization of Theorem 3.

Theorem 8. *Let G be a finite abelian group and $\ell \in \mathbb{Z}_{\geq 1}$ with $\ell \geq |G| - 3$. Given an oracle for solving base block R-GCTUF problems with group G and any $R \subseteq G$ with $|R| \geq \ell$, we can solve R-GCTUF problems with group G and $R \subseteq G$ with $|R| \geq \ell$ in strongly polynomial time with strongly polynomially many calls to the oracle.*

3.1 Reducing to a Simpler Problem When the Target Elements Form a Union of Cosets

If R, the set of target elements, is a union of cosets of the same non-trivial proper subgroup H of G (i.e., it is of the form $R = \bigcup_{i=1}^{k} (g_i + H)$ for some $g_1, \ldots, g_k \in G$, or equivalently, $R = R + H$), we can directly reduce to a simpler problem. Indeed, assume $R = R + H$ for a non-trivial proper subgroup H of G. Then, we can equivalently rewrite the R-GCTUF problem with a group constraint in the quotient group G/H and new target set $R' = R/H$. The depth of the new problem in the corresponding hierarchy is $d' = |G/H| - |R/H| = \frac{|G| - |R|}{|H|} < |G| - |R|$, so we end up with an easier problem. Since existence of such a subgroup H can be checked efficiently (given that G has constant size), we can always determine upfront whether the R-GCTUF problem at hand is reducible, and if so, reduce it to a simpler R-GCTUF problem. Thus, for the rest of this section we assume R is not a union of cosets. This assumption allows us to apply a special case of the Cauchy-Davenport theorem that holds despite the fact that the group order is not assumed to be prime. We refer to Lemma 3 for details.

---

[5] For simplicity, we use a notion of a 3-sum that allows one or both of $ef^\top$ and $gh^\top$ to be zero matrices. Typically, those cases would be called 2- and 1-sums, respectively.

3.2   Decomposing the Problem

We now focus on an R-GCTUF problem with a constraint matrix T that can be decomposed into a 3-sum of the form $T = \begin{pmatrix} A & ef^\top \\ gh^\top & B \end{pmatrix}$. The decomposition allows for splitting x, b, and γ into two parts accordingly, giving the equivalent formulation

$$\begin{pmatrix} A & ef^\top \\ gh^\top & B \end{pmatrix} \cdot \begin{pmatrix} x_A \\ x_B \end{pmatrix} \leq \begin{pmatrix} b_A \\ b_B \end{pmatrix} \quad , \quad \gamma_A^\top x_A + \gamma_B^\top x_B \in R \ , \quad \begin{matrix} x_A \in \mathbb{Z}^{n_A} \\ x_B \in \mathbb{Z}^{n_B} \end{matrix} \quad . \quad (2)$$

In the inequality system, the variables x_A and x_B interact only through the rank-one blocks $ef^\top$ and $gh^\top$. Fixing values of $\alpha := f^\top x_B$ and $\beta := h^\top x_A$ allows for rephrasing (2) through the following two almost independent problems

$$\begin{matrix} Ax_A \leq b_A - \alpha e & h^\top x_A = \beta \\ x_A \in \mathbb{Z}^{n_A} \end{matrix} \quad , \quad \text{and} \quad \begin{matrix} Bx_B \leq b_B - \beta g & f^\top x_B = \alpha \\ x_B \in \mathbb{Z}^{n_B} \end{matrix} \quad , \quad (3)$$

where we seek to find solutions x_A and x_B such that their corresponding group elements $r_A := \gamma_A^\top x_A$ and $r_B := \gamma_B^\top x_B$, respectively, satisfy $r_A + r_B \in R$. Hence, this desired relation between the target elements r_A and r_B is the only dependence between the two problems once α and β are fixed. We assume without loss of generality that A has no fewer columns than B, and refer to the problem on the left as the *A-problem*, and the problem on the right as the *B-problem*. We denote by Π the set of all $(\alpha, \beta) \in \mathbb{Z}^2$ such that both the A- and B-problem are feasible. (Note that both problems are described through a TU constraint matrix; hence, feasibility can be checked efficiently.) Also, for $(\alpha, \beta) \in \Pi$, let $\pi_A(\alpha, \beta) \subseteq G$ be all group elements $r_A \in G$ for which there is a solution x_A to the A-problem with $\gamma^\top x_A = r_A$, and define π_B analogously. We refer to π_A and π_B as *patterns*. Hence, (2) is feasible if and only if there is a pair $(\alpha, \beta) \in \Pi$ such that, for some $r_A \in \pi_A(\alpha, \beta)$ and $r_B \in \pi_B(\alpha, \beta)$, we have $r_A + r_B \in R$. Thus, patterns contain all information needed to decide feasibility.

Using techniques from [24], we can restrict our search for feasible solutions to a constant-size subset $\widehat{\Pi} \subseteq \Pi$. More precisely, for $i \in \{0, 1, 2\}$, we can in strongly polynomial time find $\ell_i, u_i \in \mathbb{Z}$ with $u_i - \ell_i \leq d$ such that if (2) is feasible, then there is a pair (α, β) in

$$\widehat{\Pi} := \{(\alpha, \beta) \in \mathbb{Z}^2 : \ell_0 \leq \alpha + \beta \leq u_0, \ell_1 \leq \alpha \leq u_1, \ell_2 \leq \beta \leq u_2\} \quad (4)$$

for which there is a solution x_A to the A-problem and a solution x_B to the B-problem with $\gamma^\top x_A + \gamma^\top x_B \in R$. Therefore, the challenges lie less in the size of Π, but rather in how to obtain information on the sets $\pi_A(\alpha, \beta)$ and $\pi_B(\alpha, \beta)$ for pairs $(\alpha, \beta) \in \Pi$. Opposed to previous techniques, which almost solely focused on π_B, we investigate both π_A and π_B and their interplay—see Sect. 3.3.

As B has at most half the columns of the constraint matrix T of the original R-GCTUF problem (2), we can afford (runtime-wise) to recursively call our algorithm multiple times on the B-problem for different targets R_B of the same depth $d = |G| - |R|$ as the original problem, i.e., with $|R_B| = |R|$. (We refrain from using larger depths, as GCTUF become harder with increasing depth.) This allows us to compute a set $\bar{\pi}_B(\alpha, \beta) \subseteq \pi_B(\alpha, \beta)$ of size $|\bar{\pi}_B(\alpha, \beta))| = \min\{d + 1, \pi_B(\alpha, \beta)\}$. Indeed, we can start with $\bar{\pi}_B(\alpha, \beta) = \emptyset$ and, as long as $|\bar{\pi}_B(\alpha, \beta)| < \min\{d + 1, \pi_B(\alpha, \beta)\}$, we solve an R_B-GCTUF B-problem (i.e., we look for a B-problem solution x_B with $\gamma^\top x_B \in R_B$) with $R_B = G \backslash \bar{\pi}_B(\alpha, \beta)$ being a set of size at least $|G| - d$. If $R_B \cap \pi_B(\alpha, \beta) \neq \emptyset$, then we find an element in $R_B \cap \pi_B(\alpha, \beta)$ that can be added to $\bar{\pi}_B(\alpha, \beta)$ and we repeat; otherwise, $R_B \cap \pi_B(\alpha, \beta) = \emptyset$ and we know that we computed $\bar{\pi}_B(\alpha, \beta) = \pi_B(\alpha, \beta)$.

To the contrary, note that the A-problem may be almost as big as the original GCTUF problem (possibly with just two fewer columns). Hence, here we cannot afford (runtime-wise) a similar computation as for the B-problem. However, we can afford to solve multiple R_A-GCTUF A-problems of smaller depth, i.e., $|R_A| > |R|$, because the runtime decreases significantly with decreasing depth. By using the same approach as in the B-problem, but with sets R_A of size $|R_A| \geq |R|+1$, we obtain a set $\bar{\pi}_A(\alpha, \beta) \subseteq \pi_A(\alpha, \beta)$ of size $|\bar{\pi}_A(\alpha, \beta)| = \min\{d, \pi_A(\alpha, \beta)\}$.

Let us next take a closer look at patterns. Fix some $(\alpha, \beta) \in \Pi$ and let $\pi_A(\alpha, \beta) = \{r_A^1, \dots, r_A^{\ell_A}\}$ for some $\ell_A \geq 1$ and pairwise different $r_A^i \in G$, and let $x_A^1, \dots, x_A^{\ell_A}$ be corresponding solutions of the A-problem with $\gamma_A^\top x_A^i = r_A^i$. Define ℓ_B, r_B^i, and x_B^i analogously. Observe that if $\ell_A \leq d$ and $\ell_B \leq d + 1$, we have $\bar{\pi}_X(\alpha, \beta) = \pi_X(\alpha, \beta)$ for both $X \in \{A, B\}$. Hence, we can compute all feasible group elements and check explicitly whether $r_A^i + r_B^j \in R$ for some $i \in [\ell_A]$ and $j \in [\ell_B]$, i.e., whether a solution exists. If $\ell_B \geq d + 1$, we can (independently of ℓ_A) even show that there always exists a feasible solution, and we can also find one: Indeed, we can compute $d + 1$ solutions $x^i := (x_A^i, x_B^i)$ with pairwise different sums $r_A^i + r_B^i \in G$, at least one of which must satisfy $r_A^1 + r_B^i \in R$. If $\ell_A \geq d$ and $\ell_B \geq 2$, we can argue similarly: We show that among any d elements of $\bar{\pi}_A(\alpha, \beta)$, and any two elements of $\bar{\pi}_B(\alpha, \beta)$ (which we can compute), there is a pair r_A^i, r_B^j with $r_A^i + r_B^j \in R$. Note that while for groups of prime order this can be shown via the Cauchy-Davenport theorem, the above result does not hold in general. We show, however, that as long as R is not a union of cosets in G, we can recover the implication (cf. Section 3.1 for why this assumption is legit).

Lemma 3. *Let G be a finite abelian group, and let $R \subseteq G$ be such that $R \neq R + H$ for any non-trivial subgroup H of G. Then, for any subsets $X, Y \subseteq G$ with $|X| = |G| - |R|$ and $|Y| \geq 2$, we have $(X + Y) \cap R \neq \emptyset$.*

Proof. Let $b_1, b_2 \in Y$ with $b_1 \neq b_2$, and set $h = b_1 - b_2$. Assume $(X + Y) \cap R = \emptyset$. Then $|X| = |G| - |R|$ implies $|X + Y| = |X|$. Thus, $X + b_1 = X + b_2$ and hence $X = X + h$. Iterating gives $X = X + \langle h \rangle$, where $\langle h \rangle$ denotes the subgroup generated by h. As $R = G \setminus (X + b_1)$, we get $R = R + \langle h \rangle$, a contradiction. $\square$

The following observation summarizes the above discussion.

Observation 1 *Let $(\alpha, \beta) \in \widehat{\Pi}$. If $|\bar{\pi}_A(\alpha, \beta)| \leq d - 1$ or $|\bar{\pi}_B(\alpha, \beta)| \geq 2$, we can immediately determine whether a feasible solution to the original R-GCTUF problem exists for such (α, β), and if so, obtain one by combining solutions computed for the A- and B-subproblem when determining $\bar{\pi}_A$ and $\bar{\pi}_B$.*

Thus, the only case in which we cannot immediately check whether a feasible solution exists for some (α, β), is when $\ell_B = 1$ and $\ell_A \geq d + 1$ (which imply $|\bar{\pi}_A(\alpha, \beta)| = d$ and $|\bar{\pi}_B(\alpha, \beta)| = 1$). This is the only case where we may have $(\pi_A(\alpha, \beta) + \pi_B(\alpha, \beta)) \cap R \neq \emptyset$ but $(\bar{\pi}_A(\alpha, \beta) + \bar{\pi}_B(\alpha, \beta)) \cap R = \emptyset$, in which case we say that (α, β) contains a *hidden solution*.

3.3   Handling Patterns

In the rest of this section, we describe how our new techniques allow for overcoming barriers restricting previous approaches to depth $d = 2$. Recall that we focus on a constant size subset $\widehat{\Pi}$ as defined in (4). We call sets of this form, for any choice of ℓ_i and u_i, *pattern shapes*, and denote by

$$\mathcal{D} := \left\{ \pm\left(\begin{smallmatrix}1\\0\end{smallmatrix}\right), \pm\left(\begin{smallmatrix}0\\1\end{smallmatrix}\right), \pm\left(\begin{smallmatrix}1\\-1\end{smallmatrix}\right) \right\} \tag{5}$$

the possible edge directions of $\mathrm{conv}(\widehat{\Pi})$. Focusing on $\widehat{\Pi}$ allows for efficiently computing $\bar{\pi}_X(\alpha, \beta)$ for $X \in \{A, B\}$ and all $(\alpha, \beta) \in \widehat{\Pi}$ to the extent discussed earlier. In order to proceed, we use a structural result from [24], called averaging, that allows us to relate solutions—and thus elements of π_X—across different (α, β). Despite being true in more generality, the exposition here requires the following special case only.

Proposition 2 ([24, **special case of Lemma 5.3**]). *Consider an R-GCTUF problem as described in* (2). *Let* $X \in \{A, B\}$, $v \in \mathcal{D}$, *and* $(\alpha, \beta) \in \widehat{\Pi}$ *with* $(\alpha, \beta) + 2v \in \widehat{\Pi}$. *Given a solution* x_1 *of the X-problem for* (α, β) *and, similarly,* x_2 *for* $(\alpha, \beta) + 2v$, *there are solutions* x_3, x_4 *for the X-problem for* $(\alpha, \beta) + v$ *such that* $x_1 + x_2 = x_3 + x_4$.

We remark that the proof of the above result for congruency-constrained problems given in [24] only exploits that congruency-constraints are linear constraints; therefore, the result carries over to group-constraints seamlessly.

In previous approaches for depth $d = 2$, it was enough to only compute a single element from π_A (e.g., by solving the A-problem after dropping the group constraint). Concretely, consider patterns of the shape as given in Fig. 1. For $d = 2$, Proposition 2 can be used to show that, if there is a hidden feasible solution for $(\alpha, \beta) = (0, 0)$ or $(\alpha, \beta) = (2, 0)$, then there must also be a feasible solution for $(\alpha, \beta) = (1, 0)$. The example in Fig. 1 shows that this is no longer true if the depth d exceeds 2, as only $(\alpha, \beta) = (0, 0)$ admits a feasible solution.

This problem can be circumvented by analyzing the A-pattern $\bar{\pi}_A$. As argued in Sect. 3.2, if a pair (α, β) has a hidden solution, then $|\pi_A(\alpha, \beta)| \geq d + 1$ (and hence $|\bar{\pi}_A(\alpha, \beta)| = d$), hence we assume that there exists at least one such pair. The following result uses averaging (i.e., Proposition 2) to show that pairs (α', β') adjacent to such a pair (α, β) containing a hidden solution also have large $\bar{\pi}_A(\alpha', \beta')$.

Lemma 4. *Let* $d \in \{1, 2, 3\}$, $v \in \mathcal{D}$, *and* $(\alpha, \beta) \in \widehat{\Pi}$ *such that* $|\pi_A(\alpha, \beta)| \geq d + 1$ *and* $(\alpha, \beta) + 2v \in \widehat{\Pi}$. *Then* $|\bar{\pi}_A((\alpha, \beta) + v)| = d$.

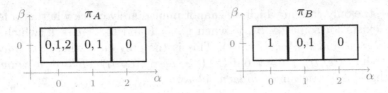

Fig. 1. Possible patterns π_A and π_B for a problem with group $G = {}^{\mathbb{Z}}\!/_{4\mathbb{Z}}$. Every square corresponds to a pair $(\alpha, \beta) \in \widehat{\Pi}$, and the numbers in the box indicate elements of $\pi_A(\alpha, \beta)$ and $\pi_B(\alpha, \beta)$, respectively. For $R = \{3\}$, there is a feasible solution with $(\alpha, \beta) = (0,0)$, but this cannot be detected without studying π_A.

Proof. It is enough to show that $|\pi_A((\alpha, \beta) + v)| \geq d$. To this end, for each of the at least $d + 1$ elements $r \in \pi_A(\alpha, \beta)$, let x_1^r be a corresponding solution of the A-problem, and let x_2 denote any fixed solution for the A-problem on the pair $(\alpha, \beta) + 2v$. Proposition 2 applied to x_1^r and x_2 gives solutions x_3^r and x_4^r corresponding to elements $\gamma_A^\top x_3^r, \gamma_A^\top x_4^r \in \pi_A((\alpha, \beta) + v)$ with $\gamma_A^\top x_3^r + \gamma_A^\top x_4^r$ taking at least $d + 1$ different values. Assume for the sake of deriving a contradiction that $|\pi_A((\alpha, \beta) + v)| \leq d - 1$. Then, since the number of different sums of pairs of elements in $\pi_A((\alpha, \beta) + v)$ is bounded by $\binom{d-1}{2} + d - 1 = (d-1)d/2 < d + 1$ for $d \in \{1, 2, 3\}$, this contradicts the above construction.   □

Remark 1. For depth $d = 4$, one can find GCTUF problems with $G = {}^{\mathbb{Z}}\!/_{5\mathbb{Z}}$ and patterns that fail to satisfy Lemma 4; we present one such example in Fig. 2. Moreover, we remark that Lemma 4 is the only place in our proofs where we use the assumption that $d = |G| - |R| \leq 3$.

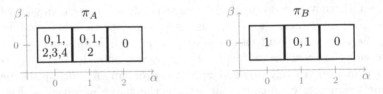

Fig. 2. Possible patterns π_A and π_B for a problem with group $G = {}^{\mathbb{Z}}\!/_{5\mathbb{Z}}$. Every square corresponds to a pair $(\alpha, \beta) \in \widehat{\Pi}$, and the numbers in the box indicate the elements of $\pi_A(\alpha, \beta)$ and $\pi_B(\alpha, \beta)$, respectively. For $d = 4$, Lemma 4 fails to hold for $(\alpha, \beta) = (0,0)$ and $v = (1,0)$.

The main application of Lemma 4 is the following: If, on top of the assumption in Lemma 4, $|\pi_B((\alpha, \beta) + v)| \geq 2$ holds, then Lemma 3 guarantees $(\bar{\pi}_A((\alpha, \beta) + v) + \bar{\pi}_B((\alpha, \beta) + v)) \cap R \neq \emptyset$. From now on, we analyze both the A- and B-patterns in detail, in particular through averaging, to guarantee the aforementioned non-empty intersection and thus find a solution, or identify additional properties that lead to progress. To distinguish cases of different pattern structure, we need the following definition (see Fig. 3 for an illustration).

Definition 1. *Let $\mathcal{D}$ be as in* (5). *We call* $(\alpha, \beta) \in \widehat{\Pi}$ *an interior pair if* $(\alpha, \beta) + v \in \widehat{\Pi}$ *for all* $v \in \mathcal{D}$, *a border pair if* $(\alpha, \beta) \pm v \in \widehat{\Pi}$ *for exactly two* $v \in \mathcal{D}$, *and a vertex pair if it is not an interior or border pair.*

Note that for a border pair (α, β), due to symmetry, the two directions $v \in \mathcal{D}$ with $(\alpha, \beta) \pm v \in \widehat{\Pi}$ will always be antiparallel, i.e., v and $-v$ for some $v \in \mathcal{D}$. To continue, the four types of patterns we distinguish are the following: (I) $|\pi_B(\alpha, \beta)| = 1$ for all $(\alpha, \beta) \in \widehat{\Pi}$, or this is not the case, and (II) Π has an interior pair, or (III) Π has no interior but border pairs, or (IV) Π has only vertex pairs. We sketch how to proceed for each of the types and present the detailed discussion in the long version.

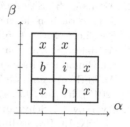

Fig. 3. A pattern shape with an interior, border, and vertex pairs (marked i, b, and w, respectively).

Patterns of type I. In a type I pattern, techniques of [24] enable reducing the problem to a new GCTUF problem with same G and $|R|$, and at least one variable less, thus allowing to make progress.

Patterns of type II. If $\widehat{\Pi}$ contains an interior pair and a hidden solution for some pair (α, β), we can use techniques from [24] to find $v \in \mathcal{D}$ such that $(\alpha, \beta) + 2v \in \widehat{\Pi}$ and $|\pi_B((\alpha, \beta) + v)| \geq 2$. Using Lemma 4 gives $|\pi_A((\alpha, \beta) + v)| \geq d$. So by Lemma 3, $(\bar{\pi}_A((\alpha, \beta) + v) + \bar{\pi}_B(\alpha, \beta) + v) \cap R \neq \emptyset$, hence we can find a solution.

Patterns of type III. In this case, we show that if we fail to find a solution in $\bar{\pi}_A + \bar{\pi}_B$, then we can reduce to a smaller pattern shape Π', allowing to induct.

We first deal with the case where there is no border pair (α, β) satisfying $|\pi_B(\alpha, \beta)| \geq 2$. One can show that this implies that all pairs with $|\pi_B(\alpha, \beta)| = 1$ lie on a single tight constraint of $\widehat{\Pi}$. Recall that these are the only pairs that might contain a hidden solution, so we can use as Π' all pairs on this tight constraint.

In the other case, there is a border pair (α, β) with $|\pi_B(\alpha, \beta)| \geq 2$. Assume additionally that there is a hidden solution for a pair sharing a tight constraint of $\widehat{\Pi}$ with (α, β) (note that the latter is unique). Using Lemma 4 and Proposition 2, we get that there must exist a pair (α', β') with $|\pi_A(\alpha', \beta')| \geq d$ and $|\pi_B(\alpha', \beta')| \geq 2$. By Lemma 3, this implies existence of a solution in $\bar{\pi}_A + \bar{\pi}_B$, contradicting the assumption. Thus, there cannot be a hidden solution anywhere on the tight constraint at (α, β), so taking Π and strengthening that constraint by one unit leads to the desired Π'.

Patterns of type IV. For type IV pattern structure, we first observe that, by Observation 1, if there are any solutions for pairs $(\alpha, \beta) \in \widehat{\Pi}$ with $|\pi_B(\alpha, \beta)| \geq 2$, we can also find one efficiently by combining solutions computed for the A- and B-subproblem when determining $\bar{\pi}_A$ and $\bar{\pi}_B$. In the other case, i.e., when no solutions exist for such (α, β), it turns out that a type IV pattern is structured enough to allow a reduction to a type I pattern, analogous to an argument of [24].

References

1. Aprile, M., Fiorini, S.: Regular matroids have polynomial extension complexity. Math. Oper. Res. **47**(1), 540–559 (2021). https://doi.org/10.1287/moor.2021.1137

2. Artmann, S., Eisenbrand, F., Glanzer, C., Oertel, T., Vempala, S., Weismantel, R.: A note on non-degenerate integer programs with small sub-determinants. Oper. Res. Lett. **44**(5), 635–639 (2016). https://doi.org/10.1016/j.orl.2016.07.004

3. Artmann, S., Weismantel, R., and Zenklusen, R.: A Strongly Polynomial Algorithm for Bimodular Integer Linear Programming. In: Proceedings of the 49th Annual ACM Symposium on Theory of Computing (STOC '17), pp. 1206–1219, Montreal (2017). https://doi.org/10.1145/3055399.3055473

4. Averkov, G., Schymura, M.: On the Maximal Number of Columns of a Δ - modular Matrix. In: Proceedings of the 23rd International Conference on Integer Programming and Combinatorial Optimization (IPCO '22), pp. 29–42, Eidhoven (2022). https://doi.org/10.1007/978-3-031-06901-7_3

5. Barahona, F., Conforti, M.: A construction for binary matroids. Discret. Math. **66**(3), 213–218 (1987). https://doi.org/10.1016/0012-365X(87)90097-5

6. Bonifas, N., Di Summa, M., Eisenbrand, F., Hähnle, N., Niemeier, M.: On Sub-determinants and the Diameter of Polyhedra. Discrete Comput. Geometry **52**(1), 102–115 (2014). https://doi.org/10.1007/s00454-014-9601-x

7. Camerini, P.M., Galbiati, G., Maffioli, F.: Random pseudo-polynomial algorithms for exact matroid problems. J. Algorithms **13**, 258–273 (1992). https://doi.org/10.1016/0196-6774(92)90018-8

8. Di Summa, M., Eisenbrand, F., Faenza, Y., Moldenhauer, C.: On Largest Volume Simplices and Sub-determinants. In: Proceedings of the 26th Annual ACMSIAM Symposium on Discrete Algorithms (SODA '15), pp. 315–323, San Diego (2015). https://doi.org/10.1137/1.9781611973730.23

9. Dinitz, M., Kortsarz, G.: Matroid secretary for regular and decomposable matroids. SIAM J. Comput. **43**(5), 1807–1830 (2014). https://doi.org/10.1137/13094030X

10. Eisenbrand, F., Vempala, S.: Geometric random edge. Math. Program. **1**, 325–339 (2016). https://doi.org/10.1007/s10107-016-1089-0

11. Fiorini, S., Joret, G., Weltge, S., and Yuditsky, Y.: Integer programs with bounded subdeterminants and two nonzeros per row. In: Proceedings of the 62nd Annual Symposium on Foundations of Computer Science (FOCS '22), pp. 13–24 (2022). https://doi.org/10.1109/FOCS52979.2021.00011

12. Glanzer, C., Stallknecht, I., and Weismantel, R.: On the recognition of a, b, c-modular matrices. In: Proceedings of the 22nd International Conference on Integer Programming and Combinatorial Optimization (IPCO '21), pp. 238–251, Atlanta (2021). https://doi.org/10.1007/978-3-030-73879-2_17

13. Glanzer, C., Weismantel, R., Zenklusen, R.: On the number of distinct rows of a matrix with bounded subdeterminants. SIAM J. Discret. Math. **32**(3), 1706–1720 (2018). https://doi.org/10.1137/17M1125728

14. Goemans, M.X., Ramakrishnan, V.S.: Minimizing submodular functions over families of sets. Combinatorica **15**(4), 499–513 (1995). https://doi.org/10.1007/BF01192523

15. Gribanov, D., Shumilov, I., Malyshev, D., Pardalos, P.: On Δ-modular integer linear problems in the canonical form and equivalent problems. J. Global Optim. (2022). https://doi.org/10.1007/s10898-022-01165-9

16. Gribanov, D.V.: An FPTAS for the Δ-modular multidimensional knapsack problem. In: Proceedings of the International Conference on Mathematical Optimization Theory and Operations Research (MOTOR), pp. 79–95 (2021). https://doi.org/10.1007/978-3-030-77876-7_6

17. Gribanov, D.V., Zolotykh, N.Y.: On lattice point counting in Δ-modular polyhedra. Optim. Lett. (1), 1–28 (2021). https://doi.org/10.1007/s11590-021-01744-x

18. Gribanov, D.V., Veselov, S.I.: On integer programming with bounded determinants. Optim. Lett. **10**(6), 1169–1177 (2015). https://doi.org/10.1007/s11590-015-0943-y

19. Grötschel, M., Lovász, L., Schrijver, A.: Corrigendum to our paper 'The ellipsoid method and its consequences in combinatorial optimization'. Combinatorica **4**(4), 291–295 (1984). https://doi.org/10.1007/BF02579139

20. Grötschel, M., Lovász, L., Schrijver, A.: Geometric Algorithms and Combinatorial Optimization. Springer, Cham (1993). https://doi.org/10.1007/978-3-642-78240-4

21. Heller, I.: On linear systems with integral valued solutions. Pac. J. Math. **7**(3), 1351–1364 (1957). https://doi.org/10.2140/pjm.1957.7.1351

22. Lee, J., Paat, J., Stallknecht, I., Xu, L.: Improving proximity bounds using sparsity. In: Baïou, M., Gendron, B., Günlük, O., Mahjoub, A.R. (eds.) ISCO 2020. LNCS, vol. 12176, pp. 115–127. Springer, Cham (2020). https://doi.org/10.1007/978-3-030-53262-8_10

23. Lee, J., Paat, J., Stallknecht, I., Xu, L.: Polynomial upper bounds on the number of differing columns of Δ-modular integer programs. Math. Oper. Res. (2022). https://doi.org/10.1287/moor.2022.1339

24. Nägele, M., Santiago, R., Zenklusen, R.: Congruency-constrained TU problems beyond the bimodular case. In: Proceedings of the 33rd Annual ACM-SIAM Symposium on Discrete Algorithms (SODA 2022), pp. 2743–2790 (2022). https://doi.org/10.1137/1.9781611977073.108

25. Nägele, M., Sudakov, B., Zenklusen, R.: Submodular minimization under congruency constraints. Combinatorica **39**(6), 1351–1386 (2019). https://doi.org/10.1007/s00493-019-3900-1

26. Nägele, M., Zenklusen, R.: A new contraction technique with applications to congruency-constrained cuts. Math. Program. (6), 455–481 (2020). https://doi.org/10.1007/s10107-020-01498-x

27. Nikolov, A.: Randomized rounding for the largest simplex problem. In: Proceedings of the 47th Annual ACM Symposium on Theory of Computing (STOC 2015), pp. 861–870, Portland (2015). https://doi.org/10.1145/2746539.2746628

28. Paat, J., Schlöter, M., Weismantel, R.: The integrality number of an integer program. Math. Program. (6), 1–21 (2021). https://doi.org/10.1007/s10107-021-01651-0

29. Padberg, M.W., Rao, M.R.: Odd minimum cut-sets and b-matchings. Math. Oper. Res. **7**(1), 67–80 (1982). https://doi.org/10.1287/moor.7.1.67

30. Schrijver, A.: Theory of Linear and Integer Programming. Wiley, New York (1998)

31. Seymour, P.D.: Decomposition of regular matroids. J. Comb. Theory, Ser. B **28**(3), 305–359 (1980). https://doi.org/10.1016/0095-8956(80)90075-1

32. Tardos, É.: A strongly polynomial algorithm to solve combinatorial linear programs. Oper. Res. **34**(2), 250–256 (1986). https://doi.org/10.1287/opre.34.2.25

33. Veselov, S.I., Chirkov, A.J.: Integer program with bimodular matrix. Discret. Optim. **6**(2), 220–222 (2009). https://doi.org/10.1016/j.disopt.2008.12.002

Cut-Sufficient Directed 2-Commodity Multiflow Topologies

Joseph Poremba$^{(\boxtimes)}$ and F. Bruce Shepherd

Computer Science, University of British Columbia, Vancouver, BC, Canada
{jporemba,fbrucesh}@cs.ubc.ca

Abstract. In multicommodity network flows, a supply-demand graph pair (G, H) (called a *multiflow topology*) is *cut-sufficient* if, for all capacity weights u and demand weights d, the cut condition is enough to guarantee the existence of a feasible multiflow. We characterize the cut-sufficient topologies for two classes of directed 2-commodity flows: *roundtrip* demands, where H is a 2-cycle, and *2-path* demands, where H is a directed path of length two. To do so, we introduce a theory of *relevant minors*. Unlike the undirected setting, for directed graphs the cut-sufficient topologies are not closed under taking minors. They are however closed under taking relevant minors. Respectively, the cut-sufficient topologies for roundtrip and two-path demands are characterized by one and two forbidden relevant minors. As an application of our results, we show that recognizing cut-sufficiency for directed multiflow topologies is NP-hard, even for roundtrip demands. This is in contrast to undirected 2-commodity flows, for which topologies are always cut-sufficient.

Keywords: Network Flows · Multiflows · Cuts · Flow-cut Gap

1 Introduction

Network flows are one of the fundamental areas of combinatorial optimization and we study its feasibility question. Given an edge-capacitated supply network (directed or undirected) $G = (V, E, u)$ and an edge-weighted demand network $H = (V, F, d)$, can we route (fractionally) in G the demands from H without violating G's edge capacities? Perhaps the most natural requirement for this flow to exist is the *cut condition*: for each non-empty $S \subsetneq V$, the total demand on edges in H in the cut induced by S, is at most the total capacity on G's edges in this cut. For undirected graphs this is:

$$d(\delta_H(S)) \leq u(\delta_G(S)) \quad \text{for all } S \subsetneq V, S \neq \emptyset. \tag{1}$$

For directed graphs, we replace δ with δ^+.

The classical Max-Flow Min-Cut (aka Menger's) Theorem asserts that this condition is sufficient in the case where $|E(H)| = 1$. This is not always the case.

The authors are grateful for an NSERC Discovery Grant which supported this work.

For instance, in Fig. 1 the cut condition holds for unit capacities and demands. However, the shortest path between s, t for any demand edge $st \in E(H)$ is 2. Hence any flow that satisfies this demand would require 8 units of capacity, whereas G only has 6. This is a proof that we must scale up some capacity to $\frac{4}{3}$ in order to route H. It is easy to check in fact that we may route H as long as every edge has capacity $\frac{4}{3}$.

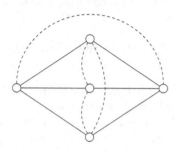

Fig. 1. Flow-cut gap $\frac{4}{3}$. The edges of G are solid and the edges of H are dashed.

This idea of finding the minimum scaling up of capacities so that H is routable is the basis for the notion of flow-cut gap. A *multiflow topology* consists of a pair of graphs (G, H) on the same node set V. We say the topology has *flow-cut gap* $\alpha \geq 1$ if the following holds for every multiflow instance on (G, H), i.e. for every endowment of capacities u and demands d: if (G, H, u, d) satisfies the cut condition, then a multiflow exists if we scale the capacities up to αu. We emphasize that the flow-cut gap is a property of a topology (G, H), while the cut condition and feasibility are properties of a weighted instance (G, H, u, d). The foundational works of Leighton-Rao [21], Auman-Rabani [3], and Linial-London-Rabinovich [22] show that for arbitrary undirected topologies the flow-cut gap is $O(\log n)$; moreover this is tight due to an expander example in [21]. There is also a conjecture which asserts that the so-called *integral flow-cut gap* is quantitatively linked to this standard flow-cut gap [7]. There have also been investigations on establishing constant-factor flow-cut gaps for restricted families [5–7,10,19,20].

Our focus is on topologies that are *cut-sufficient*, that is, multiflow topologies with flow-cut gap 1. Several important classes of undirected topologies have been shown to have this property. The most renowned results are the Max-Flow Min-Cut Theorem ($|E(H)| = 1$), Hu's 2-commodity Theorem ($|E(H)| = 2$) [12], and the Okamura-Seymour Theorem (planar G where all demand edges have their endpoints on a single face) [26]. There is no over-arching theorem, however, that captures all cut-sufficient undirected topologies. On a related note, the problem of recognizing cut-sufficiency has not been studied as far as we know.

The lack of a recognition algorithm for cut-sufficiency is curious since it is a minor-closed property. Namely, if an undirected pair (G, H) has a flow-cut gap $\leq \alpha$, then so does any pair which arises by contracting or deleting edges in G. This is verified by noting that deleting an edge corresponds to setting its capacity to 0 and contracting corresponds to setting its capacity to $+\infty$. Hence it is tempting to claim that there is an algorithm to detect $\leq \alpha$ flow-cut gap topologies, since this minor-closed property should confirm a finite list of "forbidden multiflow topologies". This is not the case however, due to the subtlety that the demand graph is also part of the topology. Specifically, [4] gives a characterization of when the flow-cut gap is 1 for pairs (G, H) where G is series-parallel. They prove there are infinitely many minimal "bad" topologies called the odd spindles. These generalize the graph of Fig. 1 by replacing the "demand triangle" by a "demand odd cycle". We are not aware of an algorithm for recognizing this property for series-parallel topologies.

Of course, when $|E(H)| = 2$, Hu's Theorem gives us a very simple polynomial time recognition algorithm for undirected cut-sufficiency. Given such a topology (G, H), it always outputs YES! We consider the directed analogue of this question and prove the following contrasting result.

Theorem 1. *It is NP-hard to determine whether a directed multiflow topology (G, H) is cut-sufficient, even if H is a 2-cycle.*

In general, there is much less work for directed cut-sufficiency. One beautiful result is a theorem of Nagamochi and Ibaraki [24] which shows that any cut-sufficient directed topology (G, H) is also "integrally cut-sufficient". In other words, if all the capacities and demands are integral, then the cut condition is sufficient to guarantee an integral routing.

Theorem 1 relies on an understanding of the forbidden minors in directed 2-commodity topologies, much along the lines of the undirected series-parallel characterization of Chakrabarti et al. [4]. There are a number of difficulties we face with this approach. Most significantly, directed cut-sufficiency does not enjoy the same minor-closed property as the undirected setting. The key issue is that contracting an edge e in the directed setting may not correspond to defining its capacity to $+\infty$. This is because it may create entirely new paths for flows to use. This requires us to develop a theory of "relevant minors". Namely, a minor (G', H') is relevant if the cut condition in the minor is directly connected to the cut condition in (G, H) with an appropriate setting of edge weights in (G, H). The technical results we need are developed in Sect. 3.

We partition the class of 2-commodity topologies into one of three types. H is said to have *roundtrip* demands if $E(H) = \{(s, t), (t, s)\}$ for distinct nodes $s, t \in V$, it has *2-path demands* if $E(H) = \{(s, t), (t, r)\}$ where s, t, r are distinct nodes, and it has *2-matching demands* if $E(H) = \{(s_1, t_1), (s_2, t_2)\}$ for distinct nodes s_1, s_2, t_1, t_2. There is also the case where H has a single common head or common tail, but these are always cut-sufficient by a simple reduction from the Max-Flow Min-Cut Theorem. Using the notion of relevant minors, we find that the two topologies from Fig. 2a-2b, are the only minimal forbidden relevant minors for roundtrip and 2-path topologies. More precisely we have the following.

Theorem 2. *A directed multiflow topology with roundtrip demands is cut-sufficient if and only if it does not contain the bad dual triangles (Fig. 2a) as a relevant minor.*

Theorem 3. *A directed multiflow topology with 2-path demands is cut-sufficient if and only if it does not contain the bad triangle (Fig. 2b) or the bad dual triangles (Fig. 2a) as a relevant minor.*

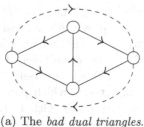

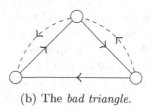

(a) The *bad dual triangles*.

(b) The *bad triangle*.

Fig. 2. The minimal bad roundtrip and 2-path demand topologies. With unit capacities and demands, each satisfies the cut condition but is not feasible.

These results imply that, in the case of roundtrip and 2-path demands, the minimal non-cut-sufficient directed topologies are certified as such by multiflow instances with 0,1 data (i.e., demands and capacities are 0, 1-valued).

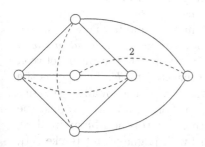

Fig. 3. An undirected topology where the cut condition is sufficient for unit demands, but not in general.

Ultimately, this implies that any non-cut-sufficient topology in these classes is certified by 0,1 demand values and 0,1,+∞ capacity values. This property need not always hold; Fig. 3 is an undirected example where the cut condition is sufficient for unit demand weights but not when one of the demand edges has weight 2. We conjecture - see Sect. 6 - that this 0,1 property is true for all directed 2-commodity topologies. If true, this would give a complete characterization of 2-commodity cut-sufficiency using a result in [27], which establishes that the non-cut-sufficient topologies certified by 0, 1 demands are characterized via the minors in Fig. 2b and 2a.

We achieve Theorems 2 and 3 by showing that, for roundtrip and 2-path demands, if G contains two paths for its different commodities that share an edge, then we find either the bad dual triangles or the bad triangle as a relevant minor. This argument also yields the following characterization of cut-sufficiency, which is in a convenient form to prove Theorem 1.

Theorem 4. *Suppose (G, H) is a directed multiflow topology with roundtrip or 2-path demands, say (s, t) and (t, r) where r may equal s. The topology is cut-sufficient if and only if every st-path is arc-disjoint from every tr-path in G.*

1.1 Other Related Work

There are several other known cut-sufficient undirected classes. Lomonosov and Seymour ([23,30], cf. Corollary 72.2a [29]) yield a characterization of the demand graphs H such that every supply graph G "works" for H, i.e., (G, H) is cut-sufficient for *any* graph G with $V(H) \subseteq V(G)$. They prove that any such H is (a subgraph of) either K_4, C_5 or the union of two stars. Another question asks for which (supply) graphs G is it the case that (G, H) is cut-sufficient for every H which is a subgraph of G; Seymour [31] shows that this is precisely the class of K_5 minor-free graphs. A related result is to characterize which pairs (G, Z) with $Z \subseteq V(G)$ are cut-sufficient for every demand graph H with $V(H) = Z$. In [25] it is shown that this occurs if and only if G is (reducible to) planar and Z is a subset of one face, i.e., essentially Okamura-Seymour instances [26]. We refer the reader to [7] for additional discussion.

Similar to our relevant minors, other works have defined restricted types of directed graph minors to achieve interesting characterizations in other applications. Some examples include butterfly minors (introduced in [13], used in [1,14,17]), strongly-connected contractions [16], d-minors [8], and shallow minors [18]. There is no unified theory of directed minors that is useful in every context. Our context is different than the aforementioned in that we need to consider both G and H to achieve the desired properties (though in Corollary 1, connections to butterfly and strongly-connected contractions can be seen).

A related but different gap result for directed multicommodity flows has been studied that is defined in terms of the *directed non-bipartite sparsest cut problem*. Here, a "cut" is any set $F \subseteq E(G)$, and the sparsity of F is the ratio between $u(F)$ and the total demand of all commodities separated by F. The gap is the smallest ratio, over all u and d, of the maximum flow to the smallest sparsity of a cut. This is the "non-bipartite" flow-cut gap. For undirected graphs, the two gaps coincide, but not so in the directed setting. For this non-bipartite flow-cut gap, a general bound of $O(\sqrt{n})$ was proven by Hajiaghayi and Räcke [11], later improved to $\tilde{O}(n^{11/23})$ by Agarwal et al. [2]. Recently, several results have given improved bounds for certain classes of supply graphs, such as directed series-parallel and graphs of bounded pathwidth [28], and planar [15].

2   Preliminaries

By paths we always mean simple paths, and for directed graphs we mean these to be simple directed paths. We use tail(e) and head(e) to represent the tail and head of an edge e, respectively. For an edge set F, we define tail(F) = {tail(e) : $e \in F$} and head(F) = {head(e) : $e \in F$}.

For multicommodity flows, we always assume integer capacities u and demand weights d. This assumption does not affect the value of the flow-cut gap, or whether a topology is cut-sufficient. We let G_u be the multigraph obtained by splitting each $e \in E(G)$ into $u(e)$ parallel copies (and define H_d similarly).

2.1   Cut-Deceptive Weights and Minors of Multiflow Topologies

For discussion of contraction, we allow parallel edges and loops. We write deletion and contraction of an edge set F as $G - F$ and G/F respectively. Edges have their own identity beyond their incident nodes. Contractions may change the ends of non-contracted edges, but those edges themselves still exist. For notational convenience we still write edges in terms of incident nodes, such as $e = (w, v)$.

A *minor* of an undirected or directed multiflow topology (G, H) is obtained by a sequence of edge deletions from G and H, and contractions of edges in G. Contracting $e \in E(G)$ identifies its ends in *both* G and H, and we denote this topology $(G, H)/e$.

For a multiflow topology (G, H), we say weights (u, d) are *cut-deceptive* if (G, H, u, d) satisfies the cut condition but is not feasible. Hence, a topology is cut-sufficient if and only if it does not have any cut-deceptive weights.

For undirected multiflow topologies, the family of cut-sufficient topologies is closed under taking minors. In particular, if a minor (G', H') has cut-deceptive weights (u, d), then those weights can be extended to cut-deceptive weights (u_{ext}, d_{ext}) for (G, H) as follows.

Definition 1 (Extension of Weights). *Let (G', H') be a minor of an undirected or directed multiflow topology (G, H). Let (u, d) be weights for (G', H'). We define the* extension[1] *of (u, d) to be the weights (u_{ext}, d_{ext}) of (G, H) where:*

- *$u_{ext}(e) = 0$ if $e \in E(G)$ was deleted,*
- *$u_{ext}(e) = +\infty$ if $e \in E(G)$ was contracted,*
- *$u_{ext}(e) = u(e)$ for all other edges $e \in E(G')$,*
- *$d_{ext}(e) = 0$ if $e \in E(H)$ was deleted,*
- *$d_{ext}(e) = d(e)$ for all other edges $e \in E(H')$.*

3 Relevant Minors and Entry-Exit Connected Edge Sets

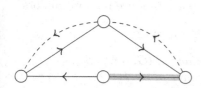

Fig. 4. A cut-sufficient topology, with a non-cut-sufficient minor.

In contrast to the undirected setting, the cut-sufficient directed multiflow topologies are not closed under taking minors. For example, consider the topology in Fig. 4. It is trivially cut-sufficient, since it is impossible to satisfy the cut condition with non-zero demands. However, contracting the highlighted edge yields the bad triangle (Fig. 2b), which is not cut-sufficient. Without restricting the kinds of minors allowed, it does not make sense to speak of forbidden minors.

---

[1] Recall that deletions and contractions commute. The extension is invariant to the specific ordering of these operations. However, there may be operation sequences that are different beyond simple re-ordering but still produce the same minor. For example, a single vertex is a minor of a triangle, but any two edges can be contracted and the third deleted to obtain it. The specific choice affects the extension. It is tedious to instrument this nuance throughout the text, but we associate a minor with a set of deletions and contractions that produce it.

3.1   Relevant Minors

We introduce a theory of *relevant minors* for directed multiflow topologies that gives closure for cut-sufficiency. In particular, if (G', H') is a relevant minor of (G, H), and (G', H') has cut-deceptive weights (u, d), then $(u_{\text{ext}}, d_{\text{ext}})$ are cut-deceptive weights of (G, H).

There are two potential reasons for $(u_{\text{ext}}, d_{\text{ext}})$ to not be cut-deceptive: either $(G, H, u_{\text{ext}}, d_{\text{ext}})$ is feasible, or the cut condition is not satisfied. The feasibility aspect is not a problem: it is easy to observe that if $(G, H, u_{\text{ext}}, d_{\text{ext}})$ is feasible, then (G', H', u, d) is also feasible using the same flow paths (after contracting). Thus the following definition gives us the desired properties.

Definition 2 (Relevant Minor). *A minor (G', H') of directed multiflow topology (G, H) is relevant if, for all weights (u, d) of (G', H') that satisfy the cut condition, $(G, H, u_{\text{ext}}, d_{\text{ext}})$ also satisfies the cut condition.*

Proposition 1. *Let (G', H') be a relevant minor of directed multiflow topology (G, H). If (u, d) are cut-deceptive weights for (G', H'), then (u_{ext}, d_{ext}) are cut-deceptive weights for (G, H). Hence, a directed multiflow topology (G, H) is cut-sufficient if and only if every relevant minor of (G, H) is cut-sufficient.*

As a consequence, we conclude one direction of Theorems 2 and 3: if a directed multiflow topology contains either the bad triangle or the bad dual triangles as a relevant minor, then it is not cut-sufficient.

3.2   Contractions of Entry-Exit Connected Sets

The definition of a relevant minor is abstract and does not immediately suggest how to show that a minor is relevant. We develop some useful tools for this purpose. First, it is easily shown that deletions always produce relevant minors.

Proposition 2. *Let (G, H) be a directed multiflow topology.*

- *For any $e \in E(G)$, $(G - e, H)$ is a relevant minor of (G, H).*
- *For any $e \in E(H)$, $(G, H - e)$ is a relevant minor of (G, H).*

Proof. In either case, let (G', H') be the minor in question. Consider weighted instances (G', H', u, d) and $(G, H, u_{\text{ext}}, d_{\text{ext}})$. Cuts in the two instances that are induced by the same node subset have the same weights. Hence, the cut condition is satisfied in both or neither.

So the only potential issues are contractions. We say an edge or edge set is *safe* if its contraction produces a relevant minor. What might make an edge unsafe? Consider Fig. 4. Contracting the highlighted edge produces a minor that is not relevant. The issue is that new connectivity is created by the contraction, so it is easier to satisfy the cut condition than in the original topology. This notion of safety seems to be intricate and does not satisfy simple properties such as: safe edge sets being closed under unions. While we are progressing towards a full characterization of safety, in this paper, we rely on a sufficient condition for a set of edges to be safe. Specifically, we show that if contracting an edge set does not create new terminal connectivity, then it is safe. We formalize this notion as follows.

Definition 3 (Entry/Exit Points, Entry-Exit Connected). *Let (G, H) be a directed multiflow topology. Let $F \subseteq E(G)$ be weakly connected.*

- *We denote by $\text{Entry}(F)$ the set of nodes $x \in V(F)$ such that there exists an sx-path in $G - F$ for some $s \in \text{tails}(H)$. We call these entry points of F.*
- *We denote by $\text{Exit}(F)$ the set of nodes $y \in V(F)$ such that there exists a yt-path in $G - F$ for some $t \in \text{heads}(H)$. We call these exit points of F.*

We say F is entry-exit connected if $G[F]$ contains an xy-path for every $x \in \text{Entry}(F)$ and $y \in \text{Exit}(F)$.

To prove such sets are safe, we recast the cut condition from discussing density of cuts to discussing paths connecting terminals. We call this new form the *path cut condition*. In essence, the cut condition in (G, H, u, d) is equivalent to the existence of, for every $F \subseteq E(H)$, $d(F)$ many arc-disjoint $\text{tails}(F) - \text{heads}(F)$ paths (i.e., paths between $\text{tails}(F)$ and $\text{heads}(F)$) in G_u.

Definition 4 (Weakly Feasible Routing). *Let $F \subseteq E(H)$ be a demand subset of directed multiflow instance (G, H, u, d). A weakly feasible routing of F is a set $\mathcal{P}_F$ of $d(F)$ arc-disjoint $\text{tails}(F) - \text{heads}(F)$ paths in G_u. Furthermore, the weakly feasible routing is fair or marginal-satisfying if both:*

- *exactly $d\left(\delta_F^+(s)\right)$ paths in $\mathcal{P}_F$ start at s for every $s \in \text{tails}(F)$, and*
- *exactly $d\left(\delta_F^-(t)\right)$ paths in $\mathcal{P}_F$ end at t for every $t \in \text{heads}(F)$.*

On a technical note, this definition is problematic if $\text{tails}(F) \cap \text{heads}(F) \neq \emptyset$. Fortunately, in such a case no cut contains the entirety of F. It is a degenerate case that ultimately can be excluded from the path cut condition[2].

Now, we define the path cut condition. In fact, we define two versions. Each is useful in different circumstances, and both are equivalent to the cut condition.

Definition 5 (Path Cut Condition). *Let (G, H, u, d) be a directed multiflow instance. The (fair) path cut condition is the property that, for every $F \subseteq E(H)$ where $\text{tails}(F) \cap \text{heads}(F) = \emptyset$, there exists a (fair) weakly feasible routing of F.*

Theorem 5. *Let (G, H, u, d) be a directed multiflow instance. The cut condition, the path cut condition, and the fair path cut condition are equivalent.*

The path cut condition gives a convenient way to prove that it is safe to contract entry-exit connected edge sets.

---

[2] An equally valid alternative would be to allow a weakly feasible routing of F to pack infinitely many copies of the length-zero path on a node in $\text{tails}(F) \cap \text{heads}(F)$.

Theorem 6. *Let (G, H) be a directed multiflow topology. If a weakly connected edge set $F \subseteq E(G)$ is entry-exit connected, then F is safe.*

Proof. Let $(G', H') = (G, H)/F$. Let v_F be the identified node for F in (G', H'). Let (u, d) be weights for (G', H') that satisfy the path cut condition.

We prove that $(G, H, u_{\text{ext}}, d_{\text{ext}})$ satisfies the path cut condition by showing that for each $J \subseteq E(H)$ with $\text{tails}_H(J) \cap \text{heads}_H(J) = \emptyset$, there is a weakly feasible routing of J in $(G, H, u_{\text{ext}}, d_{\text{ext}})$. Let $k = d(J) = d_{\text{ext}}(J)$.

For each $x \in \text{Entry}(F)$ and $y \in \text{Exit}(F)$, let $F_{x,y}$ be an xy path in $G[F]$ (which exist, since F is entry-exit connected). Note that in $G_{u_{\text{ext}}}$, there are infinitely many copies of each $F_{x,y}$, since $u_{\text{ext}}(e) = +\infty$ for all $e \in F$. For each x and y, select k of them, say $F_{x,y,1}, \ldots, F_{x,y,k}$.

Recall that, per our convention on minors, the edges of J still exist in H'. However, the ends of edges may change. We split into two cases.

In the first case, suppose there exists $v \in \text{tails}_{H'}(J) \cap \text{heads}_{H'}(J)$. Since $\text{tails}_H(J) \cap \text{heads}_H(J) = \emptyset$, but the intersection is non-empty after contracting F into v_F, it must be that $v = v_F$. Then there exist $s \in \text{tails}_H(J)$ and $t \in \text{heads}_H(J)$ such that $s, t \in V(G[F])$. Note that $s \in \text{Entry}(F)$ and $t \in \text{Exit}(F)$. Then taking $F_{s,t,i}$ for $i = 1, \ldots, k$ gives a weakly feasible routing of F.

In the second case, suppose that $\text{tails}_{H'}(J) \cap \text{heads}_{H'}(J) = \emptyset$. Then, the path cut condition of (G', H', u, d) implies that there is a set $\mathcal{P}'$ of k arc-disjoint $\text{tails}_{H'}(J) - \text{heads}_{H'}(J)$ paths in G'_u.

We map each of the paths $P_1, \ldots, P_k$ in $\mathcal{P}'$ to a $\text{tails}_H(J) - \text{heads}_H(J)$ path in $G_{u_{\text{ext}}}$. Each $P_i \in \mathcal{P}'$ is a path from some $s \in \text{tails}_{H'}(J)$ to some $t \in \text{heads}_{H'}(J)$. If P_i avoids v_F, then $s \in \text{tails}_H(J)$ and $t \in \text{heads}_H(J)$, and we map P_i to itself. If P_i uses v_F, then in $G_{u_{\text{ext}}}$ its edges form two paths: a path X_i from some $\hat{s} \in \text{tails}_H(J)$ to some $x \in \text{Entry}(F)$, and a path Y_i from some $y \in \text{Exit}(F)$ to some $\hat{t} \in \text{heads}_H(J)$ (either path may have length zero). Joining X_i and Y_i with $F_{x,y,i}$ yields a $\text{tails}_H(J) - \text{heads}_H(J)$ path. We map P_i to this path. In this way, each $P_i \in \mathcal{P}'$ maps to a $\text{tails}_H(J) - \text{heads}_H(J)$ path in $G_{u_{\text{ext}}}$ that uses only the edges of P_i and edges in some $F_{x,y,i}$. Then the image of $\mathcal{P}'$ under this mapping is a set of k arc-disjoint $\text{tails}_H(J) - \text{heads}_H(J)$ paths in $G_{u_{\text{ext}}}$, as desired.

There are several special cases of entry-exit connected sets that lead to quick ways to justify safety.

Definition 6. *A subdivision of directed multiflow topology (G, H) is obtained by a sequence of the following operation: select an edge $e \in E(G)$, and replace it with a path of length at least one.*

Corollary 1. *Let (G, H) be a directed multiflow topology.*

1. *If $F \subseteq E(G)$ is strongly connected, then F is safe.*
2. *If $e = (a, b) \in E(G)$ with $\deg_G^+(a) = 1$ and $\deg_G^-(a) = 0$, then e is safe.*
3. *If $e = (a, b) \in E(G)$ with $\deg_G^-(b) = 1$ and $\deg_H^+(b) = 0$, then e is safe.*
4. *If (G, H) is a subdivision of (G', H'), then (G', H') is a relevant minor of (G, H).*

Items 1 and 4 are used to prove Theorem 2, and Item 3 is additionally used to prove Theorem 3, though the proof for the latter is omitted in this paper since it is very similar to the former. We note that these theorems do not require the full generality of this entry-exit connected machinery, and Corollary 1 can be proven in an ad-hoc fashion without appealing to Theorem 6. However it shows that the cases of Corollary 1 are different expressions of the same general connectivity property. More importantly, the entry-exit machinery is general in the sense of not being restricted to a two demand setting. Moreover, the general form of entry-exit connected sets is used in [27] for studying safe contractions when there are 2-matching demands. We expect it to be useful in investigating characterizations of cut-sufficiency for more general H.

4   Characterizations of Cut-Sufficiency

We now apply our theory of relevant minors to prove Theorems 2, 3, and 4, which characterize the cut-sufficient directed multiflow topologies for roundtrip and 2-path demands. What remains to prove is that if there are cut-deceptive weights for a topology, then one of the desired relevant minors exists.

4.1   Opposingly Ordered Paths

The cut condition implies the existence of certain paths in G_u. From there, infeasibility implies particular interactions of these paths, which we use to prove the existence of the desired relevant minor. To that end, we use the following terminology to describe interactions between paths.

Definition 7 (Overlap Segments, Bridges). *Let P and Q be paths in a directed graph that share at least one node.*

- *An* overlap segment *is a maximal common subpath of P and Q. It is* trivial *if it is one node. It is* terminal *if it contains the start or end of either path.*
- *A P-bridge of Q is a maximal subpath B of P that has at least one edge and shares no edges or internal nodes of B with Q.*

In the above, note that P can be written as an alternating sequence of P-bridges of Q and the overlap segments of P and Q. When we say that we are listing objects, such as overlap segments, nodes, or edges, in P-*order*, we mean we list them by their order of occurrence when following the (directed) path P.

Paths can intersect numerous times in varied configurations, so it may not be obvious how to proceed looking for a particular minor. To manage this complexity, we reduce to the case where the overlap segments follow a special pattern.

Definition 8 (Opposingly Ordered). *Let P and Q be paths in a directed graph that share at least one node. We say that P and Q are* opposingly ordered *if the P-order of their overlap segments is the reverse of the Q-order.*

Lemma 1. *Let P be an s_1t_1-path and let Q be an s_2t_2-path in a directed graph, where $s_1 \neq s_2$ and $t_1 \neq t_2$, such that P and Q share an edge. In $P \cup Q$, there exists an s_2t_2-path Q^* such that:*

- *P and Q^* are opposingly ordered, and*
- *P and Q^* share an edge.*

4.2   Characterization for Roundtrip and Two-Path Demands

We now have the tools to prove our characterizations. We begin with roundtrip demands.

Lemma 2. *Let (G, H) be a directed multiflow topology with roundtrip demands between s and t. Suppose G contains an st-path P and a ts-path Q that are not arc-disjoint. Then (G, H) contains the bad dual triangles (Fig. 2a) as a relevant minor.*

Proof. Consider a counterexample that minimizes $|E(P)|+|E(Q)|$. By Lemma 1, we may assume P and Q are opposingly ordered, otherwise we replace Q with Q^*. Note that s and t themselves are trivial terminal overlap segments.

Since P and Q share an edge, there is a non-trivial overlap segment J^*. Note that J^* is not terminal.

We claim J^* is the only non-terminal overlap segment. Suppose there are at least two, for the sake of contradiction. Let J_P be the first non-terminal overlap segment in P-order (last in Q-order), and let J_Q be the first in Q-order (last in P-order). At least one of these two is not J^*. Without loss of generality, suppose $J_P \neq J^*$. Say J_P starts at w and ends at v. Let C be the cycle obtained by starting at s, following the first (in P-order) P-bridge of Q to w, following J_P from w to v, then following the last (in Q-order) Q-bridge of P from v to s.

Consider the topology $(G', H') = (G, H)/C$, which has roundtrip demands. Since directed cycles are strongly connected, this contraction is safe by Corollary 1. Additionally, $P' = P/(P \cap C)$ and $Q' = Q/(Q \cap C)$ are paths for the two commodities. Moreover, they have a non-trivial overlap segment, namely J^*. By minimality, we obtain the desired relevant minor in (G', H'), and hence in (G, H), a contradiction.

So, there is exactly one non-terminal overlap segment J^*, and it is non-trivial. Let w^* and v^* be the first and last nodes of J^*, respectively (in either P-order or Q-order, both are the same for nodes of an overlap segment). The only other overlap segments are the terminal ones, s and t themselves. There are thus exactly two P-bridges of Q, connecting s to w^* and v^* to t, respectively. Similarly there are exactly two Q-bridges of P, connecting t to w^* and v^* to s, respectively. Then, the topology $(P \cup Q, H)$ is exactly a subdivision of the bad dual triangles. By Corollary 1, the bad dual triangles is a relevant minor of $(P \cup Q, H)$ and hence also of (G, H).

Lemma 2 implies that if (G, H) has roundtrip demands and does not contain the bad dual triangles as a relevant minor, then every st-path in G is arc-disjoint from every ts-path. Such topologies are easily verified to be cut-sufficient, since for any weights, any pair of weakly feasible routings for the two commodities together form a feasible routing for the whole instance. Hence Theorem 2 follows, as well as the part of Theorem 4 pertaining to roundtrip demands.

For 2-path demands, we proceed along similar lines as for roundtrip demands. Theorem 3, and the remainder of Theorem 4, is straightforwardly implied by the following lemma whose proof we defer to the full version.

Lemma 3. *Let* (G, H) *be a directed multiflow topology with 2-path demands* (s, t) *and* (t, r). *If* G *has an* st-*path and a* tr-*path that are not arc-disjoint, then* (G, H) *contains either the bad dual triangles or the bad triangle as a relevant minor.*

5   NP-Hardness of Recognizing Cut-Sufficiency

The proofs of Theorems 2 and 3 can be adapted to give a polynomial time algorithm. Given (G, H) with roundtrip or 2-path demands, and weights (u, d) that satisfy the cut condition, the algorithm outputs either a feasible integer routing for (G, H, u, d) or one of the two forbidden relevant minors.

The algorithm does not however determine whether (G, H) is cut-sufficient, since particular (u, d) may be feasible despite (G, H) not being cut-sufficient in general. We show that determining if a directed multiflow topology (G, H) is cut-sufficient (the CutSufficient decision problem) is NP-hard, even if we restrict to roundtrip demands (the CutSufficientRT decision problem).

We reduce from an NP-hard decision problem we call UsefulEdge. Given a directed graph G, distinct nodes $s, t \in V(G)$, and an edge $e \in E(G)$, it asks whether there exists a (simple directed) st-path in G that uses the edge e. The directed 2-node-disjoint path problem, proved by Fortune et al. [9] to be NP-hard, can be reduced to UsefulEdge. Theorem 1 is implied by the following.

Theorem 7. *There is a polynomial time reduction from the* UsefulEdge *problem to the* CutSufficientRT *problem.*

Proof. Given an input $(G, s, t, e = (w, v))$ for UsefulEdge, we construct a multiflow topology (G', H') with roundtrip demands. We obtain G' from G as follows:

1. Subdivide e into $e_1 = (w, w'), e_2 = (w', v'), e_3 = (v', v)$, where w', v' are new nodes and w, v maintain their other incident edges.
2. Add two new nodes s', t', and edges $(s', s), (t, t'), (t', w')$, and (v', s').

Finally, define $H' = (V(G'), \{(s', t'), (t', s')\})$. We claim that (G', H') is not cut-sufficient if and only if there is an st-path in G that uses e, which proves the result. Define Q' to be the path $(t', w'), (w', v'), (v', s')$.

For the "only if" direction, suppose that (G', H') is not cut-sufficient. Then by Theorem 4, G' has some $s't'$-path P' that shares an edge with some $t's'$-path. The only $t's'$-path is Q', and the only possible shared edge is $e_2 = (w', v')$, so P' uses e_1, e_2, e_3. It also uses (s', s) and (t, t'). Swapping e_1, e_2, e_3 for e and removing (s', s) and (t, t') from P', we find an st-path of G that uses e.

For the "if" direction, suppose there is an st-path P in G that uses e. By swapping e for e_1, e_2, e_3 and adding (s', s) and (t, t'), we obtain an $s't'$-path P' in G' that uses e_2. Now, P' shares $e_2 = (w', v')$ with Q'. By Theorem 4, the topology is not cut-sufficient.

6   Towards a Complete 2-Commodity Characterization

For two commodities, the only remaining case is 2-matching demands. We conjecture that the bad dual triangles and the bad triangle are the only forbidden relevant minors for this case.

Conjecture 1. A directed multiflow topology with 2-matching demands (and hence, two demands in general) is cut-sufficient if and only if it does not contain the bad triangle or the bad dual triangles as a relevant minor.

In [27], the following is proved.

Proposition 3. *If directed multiflow topology (G, H) has 2-matching demands and (integer) cut-deceptive weights (u, d) where $d(e) = 1$ for all $e \in E(H)$, then it contains either the bad triangle or the bad dual triangles as a relevant minor.*

The argument considers more intricate interactions of paths arising from the fair path cut condition. It also requires general entry-exit connected contractions, rather than the specialized cases of Corollary 1.

For roundtrip and 2-path demands, there is a clear reduction to the case of unit weight demands: if there is a cut-deceptive weighting (u, d), then by taking any paths P and Q for the two commodities that share an edge, Lemmas 2 and 3 show there is a cut-deceptive weighting for $(P \cup Q, H)$ where the demands are unit. This observation is encapsulated in Theorem 4. However, we do not have so strong a result for 2-matching demands. It is not enough to just take paths for the two commodities that share an edge, as this may not even satisfy the cut condition.

References

1. Adler, I.: Directed tree-width examples. J. Comb. Theory Ser. B **97**(5), 718–725 (2007). https://doi.org/10.1016/j.jctb.2006.12.006, https://linkinghub.elsevier.com/retrieve/pii/S0095895606001444

2. Agarwal, A., Alon, N., Charikar, M.S.: Improved approximation for directed cut problems. In: Proceedings of the Thirty-Ninth Annual ACM Symposium on Theory of Computing - STOC '07, p. 671. ACM Press, San Diego, California, USA (2007). https://doi.org/10.1145/1250790.1250888, http://portal.acm.org/citation.cfm?doid=1250790.1250888

3. Aumann, Y., Rabani, Y.: An o (log k) approximate min-cut max-flow theorem and approximation algorithm. SIAM J. Comput. **27**(1), 291–301 (1998)

4. Chakrabarti, A., Fleischer, L., Weibel, C.: When the cut condition is enough: a complete characterization for multiflow problems in series-parallel networks. In: Proceedings of the Forty-Fourth Annual ACM Symposium on Theory of Computing, pp. 19–26. ACM (2012)

5. Chakrabarti, A., Jaffe, A., Lee, J.R., Vincent, J.: Embeddings of topological graphs: lossy invariants, linearization, and 2-sums. In: 2008 49th Annual IEEE Symposium on Foundations of Computer Science, pp. 761–770. IEEE, Philadelphia, PA, USA, October 2008. https://doi.org/10.1109/FOCS.2008.79, http://ieeexplore.ieee.org/document/4691008/

6. Chekuri, C., Gupta, A., Newman, I., Rabinovich, Y., Sinclair, A.: Embedding k-outerplanar graphs into l1. SIAM J. Discret. Math. **20**(1), 119–136 (2006). https://doi.org/10.1137/S0895480102417379, http://epubs.siam.org/doi/10.1137/S0895480102417379

7. Chekuri, C., Shepherd, F.B., Weibel, C.: Flow-cut gaps for integer and fractional multiflows. J. Comb. Theory Ser. B **103**(2), 248–273 (2013)

8. Deligkas, A., Meir, R.: Directed graph minors and serial-parallel width. In: 43rd International Symposium on Mathematical Foundations of Computer Science, vol. 21, p. 38 (2018)

9. Fortune, S., Hopcroft, J., Wyllie, J.: The directed subgraph homeomorphism problem. Theor. Comput. Sci. **10**(2), 111–121 (1980). https://doi.org/10.1016/0304-3975(80)90009-2, http://www.sciencedirect.com/science/article/pii/0304397580900092

10. Gupta, A., Newman, I., Rabinovich, Y., Sinclair, A.: Cuts, trees and ℓ_1-embeddings of graphs. Combinatorica **24**(2), 233–269 (2004)

11. Hajiaghayi, M.T., Räcke, H.: An $O(\sqrt{n})$-approximation algorithm for directed sparsest cut. Inf. Process. Lett. **97**(4), 156–160 (2006). https://doi.org/10.1016/j.ipl.2005.10.005, https://linkinghub.elsevier.com/retrieve/pii/S0020019005002929

12. Hu, T.C.: Multi-commodity network flows. Oper. Res. **11**(3), 344–360 (1963). https://doi.org/10.1287/opre.11.3.344, http://pubsonline.informs.org/doi/10.1287/opre.11.3.344

13. Johnson, T., Robertson, N., Seymour, P., Thomas, R.: Directed tree-width. J. Comb. Theory Ser. B **82**(1), 138–154 (2001). https://doi.org/10.1006/jctb.2000.2031, https://linkinghub.elsevier.com/retrieve/pii/S0095895600920318

14. Kawarabayashi, K.i., Kreutzer, S.: The directed grid theorem. In: Proceedings of the Forty-Seventh Annual ACM Symposium on Theory of Computing, pp. 655–664. ACM, Portland Oregon USA, June 2015. https://doi.org/10.1145/2746539.2746586, https://dl.acm.org/doi/10.1145/2746539.2746586

15. Kawarabayashi, K.I., Sidiropoulos, A.: Embeddings of planar quasimetrics into directed ℓ_1 and polylogarithmic approximation for directed sparsest-cut. In: 2021 IEEE 62nd Annual Symposium on Foundations of Computer Science (FOCS), pp. 480–491. IEEE, Denver, CO, USA, February 2022. https://doi.org/10.1109/FOCS52979.2021.00055, https://ieeexplore.ieee.org/document/9719783/

16. Kim, I., Seymour, P.: Tournament minors. J. Comb. Theory Ser. B **112**, 138–153 (2015). https://doi.org/10.1016/j.jctb.2014.12.005, https://linkinghub.elsevier.com/retrieve/pii/S0095895614001403

17. Kintali, S., Zhang, Q.: Forbidden directed minors and Kelly-width. Theor. Comput. Sci. **662**, 40–47 (2017). https://doi.org/10.1016/j.tcs.2016.12.008, https://linkinghub.elsevier.com/retrieve/pii/S0304397516307149

18. Kreutzer, S., Tazari, S.: Directed nowhere dense classes of graphs. In: Proceedings of the Twenty-Third Annual ACM-SIAM Symposium on Discrete Algorithms, pp. 1552–1562. Society for Industrial and Applied Mathematics, January 2012. https://doi.org/10.1137/1.9781611973099.123, https://epubs.siam.org/doi/10.1137/1.9781611973099.123

19. Lee, J.R., Raghavendra, P.: Coarse differentiation and multi-flows in planar graphs. Discret. Comput. Geom. **43**(2), 346–362 (2010). https://doi.org/10.1007/s00454-009-9172-4, http://link.springer.com/10.1007/s00454-009-9172-4

20. Lee, J.R., Sidiropoulos, A.: On the geometry of graphs with a forbidden minor. In: Proceedings of the 41st Annual ACM Symposium on Theory of Computing - STOC '09, p. 245. ACM Press, Bethesda, MD, USA (2009). https://doi.org/10.1145/1536414.1536450, http://portal.acm.org/citation.cfm?doid=1536414.1536450

21. Leighton, T., Rao, S.: Multicommodity max-flow min-cut theorems and their use in designing approximation algorithms. J. ACM **46**(6), 787–832 (1999)

22. Linial, N., London, E., Rabinovich, Y.: The geometry of graphs and some of its algorithmic applications. Combinatorica **15**(2), 215–245 (1995)

23. Lomonosov, M.V.: Combinatorial approaches to multiflow problems. North-Holland (1985)

24. Nagamochi, H., Ibaraki, T.: On max-flow min-cut and integral flow properties for multicommodity flows in directed graphs. Inf. Process. Lett. **31**, 279–285 (1989)

25. Naves, G., Shepherd, B.: When do Gomory-Hu subtrees exist? SIAM J. Discret. Math. **36**(3), 1567–1585 (2022)

26. Okamura, H., Seymour, P.D.: Multicommodity flows in planar graphs. J. Comb. Theory Ser. B **31**(1), 75–81 (1981). http://www.sciencedirect.com/science/article/B6WHT-4KBW025-8/2/9b4489ece0a97e9d8340d69948600501

27. Poremba, J.C.: Directed multicommodity flows: cut-sufficiency and forbidden relevant minors. Master's thesis, University of British Columbia (2022)

28. Salmasi, A., Sidiropoulos, A., Sridhar, V.: On constant multi-commodity flow-cut gaps for families of directed minor-free graphs. In: Proceedings of the Thirtieth Annual ACM-SIAM Symposium on Discrete Algorithms, pp. 535–553. SIAM (2019)

29. Schrijver, A.: Combinatorial Optimization: Polyhedra and Efficiency, vol. 24. Springer, Heidelberg (2003)

30. Seymour, P.D.: Four-terminus flows. Networks **10**(1), 79–86 (1980)

31. Seymour, P.D.: Matroids and multicommodity flows. Eur. J. Comb. **2**(3), 257–290 (1981)

Constant-Competitiveness for Random Assignment Matroid Secretary Without Knowing the Matroid

Richard Santiago$^{(\boxtimes)}$, Ivan Sergeev, and Rico Zenklusen

Department of Mathematics, ETH Zurich, Zurich, Switzerland
{rtorres,isergeev,ricoz}@ethz.ch

Abstract. The Matroid Secretary Conjecture is a notorious open problem in online optimization. It claims the existence of an $O(1)$-competitive algorithm for the Matroid Secretary Problem (MSP). Here, the elements of a weighted matroid appear one-by-one, revealing their weight at appearance, and the task is to select elements online with the goal to get an independent set of largest possible weight. $O(1)$-competitive MSP algorithms have so far only been obtained for restricted matroid classes and for MSP variations, including *Random-Assignment* MSP (RA-MSP), where an adversary fixes a number of weights equal to the ground set size of the matroid, which then get assigned randomly to the elements of the ground set. Unfortunately, these approaches heavily rely on knowing the full matroid upfront. This is an arguably undesirable requirement, and there are good reasons to believe that an approach towards resolving the MSP Conjecture should not rely on it. Thus, both Soto [SIAM Journal on Computing 2013] and Oveis Gharan & Vondrak [Algorithmica 2013] raised as an open question whether RA-MSP admits an $O(1)$-competitive algorithm even without knowing the matroid upfront.

In this work, we answer this question affirmatively. Our result makes RA-MSP the first well-known MSP variant with an $O(1)$-competitive algorithm that does not need to know the underlying matroid upfront and without any restriction on the underlying matroid. Our approach is based on first approximately learning the rank-density curve of the matroid, which we then exploit algorithmically.

1 Introduction

The Matroid Secretary Problem (MSP), introduced by Babaioff, Immorlica, and Kleinberg [1], is a natural and well-known generalization of the classical Secretary Problem [6], motivated by strong connections and applications in mechanism design. Formally, MSP is an online selection problem where we are given

This project received funding from Swiss National Science Foundation grant 200021_184622 and the European Research Council (ERC) under the European Union's Horizon 2020 research and innovation programme (grant agreement No 817750).

A. Del Pia and V. Kaibel (Eds.): IPCO 2023, LNCS 13904, pp. 423–437, 2023.
https://doi.org/10.1007/978-3-031-32726-1_30

a matroid $\mathcal{M} = (N, \mathcal{I})$,[1] with elements of unknown weights $w \colon N \to \mathbb{R}_{\geq 0}$ that appear one-by-one in uniformly random order. Whenever an element appears, one has to immediately and irrevocably decide whether to select it, and the goal is to select a set of elements $I \subseteq N$ that (i) is independent, i.e., $I \in \mathcal{I}$, and (ii) has weight $w(I) = \sum_{e \in I} w(e)$ as large as possible. The key challenge in the area is to settle the notorious Matroid Secretary Problem (MSP) Conjecture:

Conjecture 1 *([1]).* There is an $O(1)$-competitive algorithm for MSP.

The best-known procedures for MSP are $O(\log \log(\mathrm{rank}(\mathcal{M})))$-competitive [7,13], where $\mathrm{rank}(\mathcal{M})$ is the rank of the matroid $\mathcal{M}$, i.e., the cardinality of a largest independent set.

Whereas the MSP Conjecture remains open, extensive work in the field has led to constant-competitive algorithms for variants of the problem and restricted settings. This includes constant-competitive algorithms for specific classes of matroids [2,4,5,8,9,11,12,14,16]. Moreover, in terms of natural variations of the problem, Soto [16] showed that constant-competitiveness is achievable in the so-called *Random-Assignment MSP*, RA-MSP for short. Here, an adversary chooses $|N|$ weights, which are then assigned uniformly at random to ground set elements N of the matroid. (Soto's result was later extended by Oveis Gharan and Vondrák [15] to the setting where the arrival order of the elements is adversarial instead of uniformly random.) Constant-competitive algorithms also exist for the *Free Order Model*, where the algorithm can choose the order in which elements appear [9].

Intriguingly, a key aspect of prior advances on constant-competitive algorithms for special cases and variants of MSP is that they heavily rely on knowing the full matroid $\mathcal{M}$ upfront. This is also crucially exploited in Soto's work on RA-MSP. In fact, if the matroid is not known upfront in full, there is no natural variant of MSP for which a constant-competitive algorithm is known.

A high reliance on knowing the matroid $\mathcal{M} = (N, \mathcal{I})$ upfront (except for its size $|N|$) is undesirable when trying to approach the MSP Conjecture, because it is easy to obstruct an MSP instance by adding zero-weight elements. Not surprisingly, all prior advances on the general MSP conjecture, like the above-mentioned $O(\log \log(\mathrm{rank}(\mathcal{M})))$-competitive algorithms [7,13] and also earlier procedures [1,3], only need to know $|N|$ upfront and make calls to an independence oracle on elements revealed so far. Thus, for RA-MSP, it was raised as an open question both in [16] and [15], whether a constant-competitive algorithm exists without knowing the matroid upfront. The key contribution of this work is to affirmatively answer this question, making the random assignment setting the first MSP variant for which a constant-competitive algorithm is known without knowing the matroid and without any restriction on the underlying matroid.

---

[1] A matroid $\mathcal{M}$ is a pair $\mathcal{M} = (N, \mathcal{I})$ where N is a finite set and $\mathcal{I} \subseteq 2^N$ is a non-empty family satisfying: 1) if $A \subseteq B$ and $B \in \mathcal{I}$ then $A \in \mathcal{I}$, and 2) if $A, B \in \mathcal{I}$ and $|B| > |A|$ then $\exists e \in B \setminus A$ such that $A \cup \{e\} \in \mathcal{I}$.

Theorem 1. *There is a constant-competitive algorithm for RA-MSP with only the cardinality of the matroid known upfront.*

Moreover, our result holds in the more general *adversarial order with a sample* setting, where we are allowed to sample a random constant fraction of the elements and all remaining (non-sampled) elements arrive in adversarial order.

As mentioned, when the matroid is fully known upfront, an $O(1)$-competitive algorithm was known for RA-MSP even when the arrival order of all elements is adversarial [15]. Interestingly, for this setting it is known that, without knowing the matroid upfront, no constant-competitive algorithm exists. More precisely, a lower bound on the competitiveness of $\Omega(|N|/\log\log|N|)$ was shown in [15].

Organization of the Paper. We start in Sect. 2 with a brief discussion on the role of (matroid) densities in the context of random assignment models, as our algorithm heavily relies on densities. Decomposing the matroid into parts of different densities has been central in prior advances on RA-MSP. However, this crucially relies on knowing the matroid upfront. We work with a rank-density curve, introduced in Sect. 3.1, which is also unknown upfront; however, we show that it can be learned approximately (in a well-defined sense) by observing a random constant fraction of the elements. Section 3 provides an outline of our approach based on rank-density curves and presents the main ingredients which allow us to derive Theorem 1. Sect. 4 showcases the main technical tool that allows us to approximate the rank-density curve from a sample set. Finally, Sect. 5 discusses our main algorithmic contribution and a sketch of its analysis.

We emphasize that we predominantly focus on providing a simple algorithm and analysis, refraining from optimizing the competitive ratio of our procedure at the cost of complicating the presentation. Moreover, due to space constraints, some proofs are deferred to the long version of this paper.

We assume that all matroids are loopless, i.e., every element is independent by itself. This is without loss of generality, as loops can simply be ignored in matroid secretary problems.

2   Random-Assignment MSP and Densities

A main challenge in the design and analysis of MSP algorithms is how to protect heavier elements (or elements of an offline optimum) from being spanned by lighter ones that are selected earlier during the execution of the algorithm. In the random assignment setting, however, weights are assigned to elements uniformly at random, which allows for shifting the focus from protecting elements based on their weights to protecting elements based on their role in the matroid structure. Intuitively speaking, an element arriving in the future is at a higher risk of being spanned by the algorithm's prior selection if it belongs to an area of the matroid with larger cardinality and smaller rank ("denser" area) than an area with smaller cardinality and larger rank ("sparser" area).

This is formally captured by the notion of density: the *density* of a set $U \subseteq N$ in a matroid $\mathcal{M} = (N, \mathcal{I})$ is $|U|/r(U)$, where $r: 2^N \to \mathbb{Z}_{\geq 0}$ is the rank function of $\mathcal{M}$.[2]

Densities play a crucial role in RA-MSP [15,16]. Indeed, prior approaches decomposed $\mathcal{M}$ into its *principal sequence*, which is the chain $\emptyset \subsetneq S_1 \subsetneq \cdots \subsetneq S_k = N$ of sets of decreasing densities obtained as follows. $S_1 \subseteq N$ is the densest set of $\mathcal{M}$ (in case of ties it is the unique maximal densest set), S_2 is the union of S_1 and the densest set in the matroid obtained from $\mathcal{M}$ after contracting S_1, and so on until a set S_k is obtained with $S_k = N$. Figure 1a shows an example of the principal sequence of a graphic matroid.

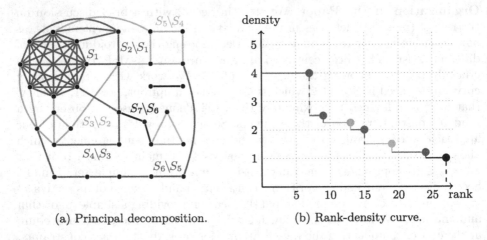

(a) Principal decomposition.   (b) Rank-density curve.

Fig. 1. Figure 1a shows a graph representing a graphic matroid together with its principal sequence $\emptyset \subsetneq S_1 \subsetneq \cdots \subsetneq S_7 = N$, where N are all edges of the graph. Figure 1b shows its rank-density curve. Each step in the rank-density curve (highlighted by a circle) corresponds to one S_i and has y-coordinate equal to the density of $\mathcal{M}_i = (\mathcal{M}/S_{i-1})|_{S_i \setminus S_{i-1}}$ and x-coordinate equal to $r(S_i)$.

Previous approaches then considered, independently for each $i \in [k] := \{1, \ldots, k\}$, the matroid $\mathcal{M}_i := (\mathcal{M}/S_{i-1})|_{S_i \setminus S_{i-1}}$, i.e., the matroid obtained from $\mathcal{M}$ by contracting S_{i-1} and then restricting to $S_i \setminus S_{i-1}$. (By convention, we set $S_0 := \emptyset$.) These matroids are also known as the *principal minors* of $\mathcal{M}$. Given an independent set in each principal minor, their union is guaranteed to be independent in the original matroid $\mathcal{M}$. Prior approaches (see, in particular, [16] for details) then exploited the following two key properties of the principal minors $\mathcal{M}_i$:

---

[2] The rank function $r: 2^N \to \mathbb{Z}_{\geq 0}$ assigns to any set $U \subseteq N$ the cardinality of a maximum cardinality independent set in U, i.e., $r(U) := \max\{|I| : I \subseteq U, I \in \mathcal{I}\}$.

(i) $\sum_{i=1}^{k} \mathbb{E}[w(\mathrm{OPT}(\mathcal{M}_i))] = \Omega(\mathbb{E}[w(\mathrm{OPT}(\mathcal{M}))])$, where $\mathrm{OPT}(\mathcal{M})$ (and analogously $\mathrm{OPT}(\mathcal{M}_i)$) is an (offline) maximum weight independent set in $\mathcal{M}$ and the expectation is over all random weight assignments.

(ii) Each matroid $\mathcal{M}_i$ is *uniformly dense*, which means that the (unique maximal) densest set in $\mathcal{M}_i$ is the whole ground set of $\mathcal{M}_i$.

Property (i) guarantees that, to obtain an $O(1)$-competitive procedure, it suffices to compare against the (offline) optima of the matroids $\mathcal{M}_i$. Combining this with property (ii) implies that it suffices to design a constant-competitive algorithm for uniformly dense matroids. Since uniformly dense matroids behave in many ways very similarly to uniform matroids, which are a special case of uniformly dense matroids, it turns out that the latter admit a simple yet elegant $O(1)$-competitive algorithm. (See [16] for details.)

3 Outline of Our Approach

As discussed, prior approaches [15,16] for RA-MSP heavily rely on knowing the matroid upfront, as they need to construct its principal sequence upfront. A natural approach would be to observe a sample set $S \subseteq N$ containing a constant fraction of all elements and then try to mimic the existing approaches using the principal sequence of $\mathcal{M}|_S$, the matroid $\mathcal{M}$ restricted to the elements in S. A main hurdle lies in how to analyze such a procedure as the principal sequence of $\mathcal{M}|_S$ can differ significantly from the one of $\mathcal{M}$. In particular, one can construct matroids where the density of some parts is likely to be underestimated by a super-constant factor. Moreover, generally $\mathcal{M}|_S$ may have many different densities not present in $\mathcal{M}$ (e.g., when $\mathcal{M}$ is uniformly dense).

We overcome these issues by not dealing with principal sequences directly, but rather using what we call the *rank-density curve* of a matroid, which captures certain key parameters of the principal sequence. As we show, rank-density curves have three useful properties:

(i) They provide a natural way to derive a quantity that both relates to the offline optimum and can be easily compared against to bound the competitiveness of our procedure.

(ii) They can be learned approximately by observing an $O(1)$-fraction of N.

(iii) Approximate rank-density curves can be used algorithmically to protect denser areas from sparser ones without having to know the matroid upfront.

Section 3.1 introduces rank-density curves and shows how they conveniently allow for deriving a quantity that compares against the offline optimum. Section 3.2 then discusses our results on approximately learning rank-density curves and how this can be exploited algorithmically.

3.1 Rank-Density Curves

Given a matroid $\mathcal{M} = (N, \mathcal{I})$, one natural way to define its rank-density curve $\rho_{\mathcal{M}} : \mathbb{R}_{>0} \to \mathbb{R}_{\geq 0}$, is through its principal minors $\mathcal{M}_1, \ldots, \mathcal{M}_k$, which are defined through the principal sequence $\emptyset \subsetneq S_1 \subsetneq \cdots \subsetneq S_k = N$ as explained in Sect. 2. For a value $t \in (0, \text{rank}(\mathcal{M})]$, let $i_t \in [k]$ be the smallest index such that $r(S_{i_t}) > t$. The value $\rho_{\mathcal{M}}(t)$ is then given by the density of $\mathcal{M}_{i_t}$. (See Fig. 1b for an example.) In addition, we set $\rho_{\mathcal{M}}(t) = 0$ for any $t > \text{rank}(\mathcal{M})$.

A formally equivalent way to define $\rho_{\mathcal{M}}$, which is more convenient for what we do later, is as follows. For any $S \subseteq N$ and $\lambda \in \mathbb{R}_{\geq 0}$, we define

$$D_{\mathcal{M}}(S, \lambda) \in \underset{U \subseteq S}{\text{argmax}} \{|U| - \lambda r(U)\} \tag{1}$$

to be the unique maximal maximizer of $\max_{U \subseteq S}\{|U| - \lambda r(U)\}$. It is well-known that each set in the principal sequence $S_1, \ldots, S_k$ is nonempty and of the form $D_{\mathcal{M}}(N, \lambda)$ for $\lambda \in \mathbb{R}_{\geq 0}$. This leads to the following way to define the rank-density curve, which is the one we use in what follows.

Definition 1 (rank-density curve). Let $\mathcal{M} = (N, \mathcal{I})$ be a matroid. Its *rank-density* curve $\rho_{\mathcal{M}} \colon \mathbb{R}_{>0} \to \mathbb{R}_{\geq 0}$ is defined by

$$\rho_{\mathcal{M}}(t) := \begin{cases} \max\{\lambda \in \mathbb{R}_{\geq 0} : r(D_{\mathcal{M}}(N, \lambda)) \geq t\} & \forall t \in (0, \text{rank}(\mathcal{M})] \\ 0 & \forall t > \text{rank}(\mathcal{M}). \end{cases}$$

When the matroid $\mathcal{M}$ is clear from context, we usually simply write ρ instead of $\rho_{\mathcal{M}}$ for its rank-density curve and $D(N, \lambda)$ instead of $D_{\mathcal{M}}(N, \lambda)$. Note that ρ is piecewise constant, left-continuous, and non-increasing. (See Fig. 1b for an example.) If $\mathcal{M}$ is a uniformly dense matroid with density λ, we have $\rho(t) = \lambda$ for $t \in (0, \text{rank}(\mathcal{M})]$.

We now expand on how $\rho_{\mathcal{M}}$ is related to the expected offline optimum value $\mathbb{E}[\text{OPT}(\mathcal{M})]$ of an RA-MSP instance. To this end, we use the function $\eta \colon [0, |N|] \to \mathbb{R}_{\geq 0}$ defined by

$$\eta(a) := \mathbb{E}_{R \sim \text{Unif}(N, \lfloor a \rfloor)} \left[\max_{e \in R} w(e) \right], \tag{2}$$

where $\text{Unif}(N, \lfloor a \rfloor)$ is a uniformly random set of $\lfloor a \rfloor$ many elements out of N; and we set $\eta(a) = 0$ for $a \in [0, 1)$ (i.e., when the set R above is empty) by convention. In words, $\eta(a)$ is the expected maximum weight out of $\lfloor a \rfloor$ weights chosen uniformly at random from all the weights $\{w_e\}_{e \in N}$. Based on this notion, we assign the following value $F(\rho)$ to a rank-density curve ρ:

$$F(\rho) := \int_0^\infty \eta(\rho(t)) dt. \tag{3}$$

Note that as the graph of ρ is a staircase, the above integral is just a finite sum. One can then show that the values of the offline optimum and $F(\rho)$ differ by at most a constant factor — proof deferred.

Lemma 1. *Let $(\mathcal{M}, w)$ be a random-assignment MSP instance. Let $\mathcal{M}'$ be any matroid minor of $\mathcal{M}$ and let F be as defined above. Then $\mathbb{E}[w(\text{OPT}(\mathcal{M}'))] \leq \frac{3e}{e-1} \cdot F(\rho_{\mathcal{M}'})$.*

Thus, to be constant-competitive, it suffices to provide an algorithm returning an independent set of expected weight $\Omega(F(\rho))$.

RA-MSP Subinstances. We will often work with minors of the matroid that is originally given in our RA-MSP instance, and apply certain results to such minors instead of the original matroid. To avoid confusion, we fix throughout the paper one RA-MSP instance with matroid $\mathcal{M}_{\mathrm{orig}} = (N_{\mathrm{orig}}, \mathcal{I}_{\mathrm{orig}})$ and unknown but (adversarially) fixed weights $w \colon N_{\mathrm{orig}} \to \mathbb{R}_{\geq 0}$, and our goal is to design an $O(1)$-competitive algorithm for this instance. The weights w of the original instance are the only weights we consider, even when working with RA-MSP subinstances on minors of $\mathcal{M}_{\mathrm{orig}}$, as their elements also obtain their weights uniformly at random from w. In particular, the function F as defined in (3) is always defined with respect to the original weights w. Many of our statements hold not just for minors of $\mathcal{M}$ but any matroid with weights uniformly drawn from w. For simplicity, we typically also state these results for minors of $\mathcal{M}$.

For a matroid $\mathcal{M} = (N, \mathcal{I})$ with $|N| \leq |N_{\mathrm{orig}}|$, we denote by $(\mathcal{M}, w)$ the RA-MSP instance on the matroid $\mathcal{M}$ obtained by assigning a uniformly random subset of $|N|$ weights among the weights w uniformly at random to the elements in N. Our subinstances will be of this type (with $\mathcal{M}$ being a minor of $\mathcal{M}_{\mathrm{orig}}$). Even though there may be more weights than elements, such instances $(\mathcal{M}, w)$ can indeed be interpreted as RA-MSP instances, as they correspond to the adversary first choosing uniformly at random a subset of $|N|$ weights among the weights in w, which then get assigned uniformly at random to the elements.

3.2 Proof Plan for Theorem 1 via Rank-Density Curves

We now expand on how one can learn an approximation $\tilde{\rho}$ of the rank-density curve $\rho_{\mathcal{M}_{\mathrm{orig}}}$ and how this can be exploited algorithmically to return an independent set of expected weight $\Omega(F(\rho_{\mathcal{M}_{\mathrm{orig}}}))$, which by Lemma 1 implies $O(1)$-competitiveness of the procedure. To this end, we start by formalizing the notion of an *approximate rank-density* curve, which relies on the notion of *downshift*.

Definition 2. Let $\rho \colon \mathbb{R}_{>0} \to \mathbb{R}_{\geq 0}$ be a non-increasing function and let $\alpha, \beta \in \mathbb{R}_{\geq 1}$. The (α, β)-*downshift* $\rho' \colon \mathbb{R}_{>0} \to \mathbb{R}_{\geq 0}$ of ρ is defined via an auxiliary function $\phi \colon \mathbb{R}_{>0} \to \mathbb{R}_{\geq 0}$ as follows:

$$\phi(t) := \begin{cases} \frac{\rho(\alpha)}{\beta} & \forall t \in (0, 1], \\ \frac{\rho(\alpha \cdot t)}{\beta} & \forall t > 1; \end{cases} \qquad \rho'(t) := \begin{cases} 1 & \text{if } \phi(t) \in (0, 1), \\ \phi(t) & \text{otherwise}. \end{cases}$$

Moreover, a function $\tilde{\rho} \colon \mathbb{R}_{>0} \to \mathbb{R}_{\geq 0}$ is called an (α, β)-*approximation* of ρ if it is non-increasing and $\rho' \leq \tilde{\rho} \leq \rho$, where ρ' is the (α, β)-downshift of ρ.

The reason we round up values in $(0, 1)$ in the above definition of downshift, is that while we define the latter for a more general type of curves, throughout the paper we mainly use it with rank-density curves, and density values are always at least one.

One issue when working with an $(O(1), O(1))$-approximation $\tilde{\rho}$ of ρ is that $F(\tilde{\rho})$ may be more than a constant factor smaller than $F(\rho)$ and we thus cannot compare against $F(\tilde{\rho})$ to obtain an $O(1)$-competitive procedure. However, as the following lemma shows, also in this case we can obtain a simple lower bound for the value $F(\tilde{\rho})$ in terms of $F(\rho)$ and the largest weight $w_{\max}$ in w.

Lemma 2. *Let $\mathcal{M}$ be a matroid minor of $\mathcal{M}_{\text{orig}}$, let $\alpha, \beta \in \mathbb{R}_{\geq 1}$, and let $\tilde{\rho}$ be an (α, β)-approximation of $\rho_{\mathcal{M}}$. Then $F(\rho_{\mathcal{M}}) \leq 2\alpha\beta F(\tilde{\rho}) + \alpha w_{\max}$.*

Proof. By the definition of an (α, β)-approximate curve we have

$$F(\tilde{\rho}) = \int_0^\infty \eta(\tilde{\rho}(t))dt \geq \frac{1}{2\alpha\beta} \int_\alpha^\infty \eta(\rho(t))dt = \frac{1}{2\alpha\beta}\left[F(\rho) - \int_0^\alpha \eta(\rho(t))dt\right]$$
$$\geq \frac{1}{2\alpha\beta}\left[F(\rho) - \alpha w_{\max}\right],$$

where the first inequality follows since $\tilde{\rho}$ is an (α, β)-approximation of $\rho_{\mathcal{M}}$ and from properties of η (proof deferred), and the last inequality holds by definition of η. □

A key implication of Lemma 2 is that it suffices to obtain an algorithm that returns an independent set of expected weight $\Omega(F(\tilde{\rho}))$ for some $(O(1), O(1))$-approximation $\tilde{\rho}$ of $\rho_{\mathcal{M}_{\text{orig}}}$. Indeed, Lemma 2 then implies $F(\tilde{\rho}) = \Omega(F(\rho_{\mathcal{M}_{\text{orig}}})) - O(w_{\max})$. By running this algorithm with some probability (say 0.5) and otherwise Dynkin's [6] classical secretary algorithm, which picks the heaviest element with constant probability, an overall algorithm is obtained that returns an independent set of expected weight $\Omega(F(\rho_{\mathcal{M}_{\text{orig}}}))$. Hence, Lemma 2 helps to provide bounds on the competitiveness of algorithms that are competitive with the F-value of an approximate rank-density curve. This technique is also used in the following key statement, which shows that an algorithm with strong guarantees can be obtained if we are given an $(O(1), O(1))$-approximation of the rank-density curve of the matroid on which we work — see Sect. 5.

Theorem 2. *Let $\mathcal{M}$ be a matroid minor of $\mathcal{M}_{\text{orig}}$, and let $\rho_{\mathcal{M}}$ denote the rank-density curve of $\mathcal{M}$. Assume we are given an (α, β)-approximation $\tilde{\rho}$ of $\rho_{\mathcal{M}}$ for integers $\alpha \geq 24$ and $\beta \geq 3$. Then there is an efficient procedure $\text{ALG}(\tilde{\rho}, \alpha, \beta)$ that, when run on the RA-MSP subinstance given by $\mathcal{M}$, returns an independent set I of $\mathcal{M}$ of expected weight at least $\left(\frac{1}{1440e\alpha^2\beta^2}\right)\left(F(\rho_{\mathcal{M}}) - \alpha^2 w_{\max}\right)$.*

The last main ingredient of our approach is to show that such an *accurate* proxy $\tilde{\rho}$ can be computed with constant probability. More precisely, we show that, after observing a sample set S containing every element of N_{orig} independently with probability $1/2$, the rank-density curve of (the observed) $\mathcal{M}_{\text{orig}}|_S$

- is close to the rank-density curve of $\mathcal{M}_{\text{orig}}|_{N_{\text{orig}}\setminus S}$, allowing us to use $\rho_{\mathcal{M}_{\text{orig}}|_S}$ as desired proxy for the RA-MSP subinstance given by $\mathcal{M}_{\text{orig}}|_{N_{\text{orig}}\setminus S}$, and
- is close to the rank-density curve of $\mathcal{M}_{\text{orig}}$, which allows for relating the offline optimum of the RA-MSP subinstance given by $\mathcal{M}_{\text{orig}}|_{N_{\text{orig}}\setminus S}$ to the one of $\mathcal{M}_{\text{orig}}$.

We highlight that the next result is purely structural and hence independent of weights or the MSP setting. See Sect. 4 for details.

Theorem 3. *Let $\mathcal{M} = (N, \mathcal{I})$ be a matroid and $S \subseteq N$ be a random set containing every element of N independently with probability $1/2$. Then, with probability at least $1/100$, $\rho_{\mathcal{M}|_S}$ and $\rho_{\mathcal{M}|_{N \setminus S}}$ are both $(288, 9)$-approximations of $\rho_{\mathcal{M}}$.*

Combining the above results, we get the desired $O(1)$-competitive algorithm.

Proof of Theorem 1. For brevity, let $\mathcal{M} := \mathcal{M}_{\text{orig}}$ and $N := N_{\text{orig}}$ throughout this proof. Recall that by Lemma 1, it suffices to provide an algorithm returning an independent set of expected weight $\Omega(F(\rho_{\mathcal{M}}))$. Consider the following procedure: First observe (without picking any element) a set $S \subseteq N$ containing every element of N independently with probability $1/2$ and let $\tilde{\rho}$ denote the $(288, 9)$-downshift of $\rho_{\mathcal{M}|_S}$. Then run the algorithm described in Theorem 2 on $\mathcal{M}|_{N \setminus S}$ with $\tilde{\rho}$ as the approximate rank-density curve. Let I denote the output of the above procedure and let $\mathcal{A}$ be the event defined in Theorem 3, that is,

$$\mathcal{A} = \{S \subseteq N \colon \rho_{\mathcal{M}|_S} \text{ and } \rho_{\mathcal{M}|_{N \setminus S}} \text{ are } (288, 9)\text{-approximations of } \rho_{\mathcal{M}}\}.$$

Then, observe that for any fixed $S \in \mathcal{A}$, we have

$$\mathbb{E}[w(I) \mid S] \geq \left(\tfrac{1}{1440e \cdot 288^4 \cdot 9^4}\right)\left(F\left(\rho_{\mathcal{M}|_{N \setminus S}}\right) - 288^4 w_{\max}\right)$$
$$\geq \left(\tfrac{1}{2880e \cdot 288^5 \cdot 9^5}\right) F\left(\rho_{\mathcal{M}}\right) - \tfrac{w_{\max}}{720e \cdot 9^3},$$

where the first inequality follows from Theorem 2 and that, for every $S \in \mathcal{A}$, the curve $\tilde{\rho}$ is a $(288^2, 9^2)$-approximation of $\rho_{\mathcal{M}|_{N \setminus S}}$,[3] while the second inequality follows from Lemma 2 and the fact that for every $S \in \mathcal{A}$ the curve $\rho_{\mathcal{M}|_{N \setminus S}}$ is a $(288, 9)$-approximation of $\rho_{\mathcal{M}}$. Moreover, the first inequality uses that conditioning on any fixed $S \in \mathcal{A}$ does not have any impact on the uniform assignment of the weights w to the elements. This holds because the event $\mathcal{A}$ only depends on the sampled elements S but not the weights of its elements. Hence, the RA-MSP subinstance given by $\mathcal{M}|_{N \setminus S}$ on which we use the algorithm described in Theorem 2, indeed assigns weights of w uniformly at random to elements, as required. It then follows that the output of the above procedure satisfies

$$\mathbb{E}[w(I)] \geq \sum_{S \in \mathcal{A}} \mathbb{E}[w(I) \mid S] \Pr[S] \geq \tfrac{1}{100}\left(\left(\tfrac{1}{2880e \cdot 288^5 \cdot 9^5}\right) F\left(\rho_{\mathcal{M}}\right) - \tfrac{w_{\max}}{720e \cdot 9^3}\right),$$

where the last inequality uses that $\Pr[\mathcal{A}] \geq 1/100$ by Theorem 3.

Since running the classical secretary algorithm on $\mathcal{M}_{\text{orig}}$ returns an independent set of expected weight at least $w_{\max}/e$, the desired result now follows by running the procedure described above with probability $1/2$, and running the classical secretary algorithm otherwise.   □

---

[3] $\tilde{\rho}$ is a $(288^2, 9^2)$-approximation of $\rho_{\mathcal{M}|_{N \setminus S}}$ because $\tilde{\rho}$ is the $(288, 9)$-downshift of $\rho_{\mathcal{M}|_S}$, and, both $\rho_{\mathcal{M}|_{N \setminus S}}$ and $\rho_{\mathcal{M}|_S}$ are $(288, 9)$-approximations of $\rho_{\mathcal{M}}$. First, this implies that $\rho_{\mathcal{M}|_{N \setminus S}}$ lies above $\tilde{\rho}$. Second, the approximation parameter $(288^2, 9^2)$ can be derived by observing that the (α_2, β_2)-downshift of the (α_1, β_1)-downshift of some rank-density function is an $(\alpha_1 \alpha_2, \beta_1 \beta_2)$-approximation of that rank-density function — proof deferred.

4   Learning Rank-Density Curves from a Sample

One of the main challenges when designing and analyzing algorithms for MSP is understanding what kind of (and how much) information can be learned about the underlying instance after observing a random sample of it.

In this section, we discuss the main ingredient to show that, with constant probability, after observing a sample set S one can learn a good approximation of the rank-density curve of both $\mathcal{M}$ and $\mathcal{M}|_{N\setminus S}$ — see Theorem 3. However, even if one knew the exact (instead of an approximate) rank-density curve of $\mathcal{M}|_{N\setminus S}$, given that the matroid is not known upfront (and hence neither which elements are associated to each of the different density areas of the curve), it is a priori not clear how to proceed. A second main contribution of this section is to show that the set of elements in $N \setminus S$ that are spanned by a subset of S of a given density is well-structured. In particular, this will allow us to build a (chain) decomposition $\bigoplus_{i=1}^{k} \mathcal{M}_i$ of $\mathcal{M}|_{N\setminus S}$ where all the $\mathcal{M}_i$'s satisfy some desired properties with constant probability — see Sect. 5.1 for details.

The main technical contribution in this section is the following result.

Theorem 4. *Let $\mathcal{M} = (N, \mathcal{I})$ be a matroid containing $3h$ disjoint bases for some $h \in \mathbb{Z}_{\geq 1}$. Let $S \sim B(N, 1/2)$. Then*

$$\Pr\left[|\mathrm{span}(D(S,h)) \setminus S| \leq \tfrac{|N|}{12}\right] \leq \exp\left(-\tfrac{|N|}{144}\right), \tag{4}$$

$$\Pr\left[r(D(S,h)) \leq \tfrac{r(N)}{8}\right] \leq \exp\left(-\tfrac{r(N)}{48}\right). \tag{5}$$

Proof. We prove (4) and defer the proof of (5) to the full version. Let $\mathcal{M}_h = (N, \mathcal{I}_h)$ denote the h-fold union of $\mathcal{M}$ and let r_h denote its rank function. Consider the procedure described in Algorithm 1, which is loosely inspired by [10].

Algorithm 1: Algorithm for lower bounding $\mathbb{E}_S\left[|\mathrm{span}(D(S,h)) \setminus S|\right]$

Set $W \leftarrow \emptyset$, $G \leftarrow \emptyset$, and $C \leftarrow \emptyset$
for *every $e \in N$ considered in an arbitrary order* **do**
    if $W \cup \{e\} \in \mathcal{I}_h$ **then**
        if $e \in S$ **then** Update $W \leftarrow W \cup \{e\}$
    else
        if $e \in S$ **then** Update $C \leftarrow C \cup \{e\}$ **else** Update $G \leftarrow G \cup \{e\}$

Note that the following three properties hold at all times: W, G, and C are pairwise disjoint; $W \subseteq S$ and $C \subseteq S$, while $G \cap S = \emptyset$; and $W \in \mathcal{I}_h$. In addition, by construction, at the end of the procedure we have:

(i) $S = C \uplus W$. Moreover, the random sets G and C have identical distributions, because each element belongs to S with probability $1/2$ independently.

(ii) $G \subseteq \text{span}(D(S, h)) \setminus S$. Because $G \cap S = \emptyset$, it is enough to show $G \subseteq \text{span}(D(S, h))$. Given an arbitrary $e \in G$, by construction we have $W \cup \{e\} \notin \mathcal{I}_h$, i.e., $r_h(W \cup \{e\}) = r_h(W)$. As $W \subseteq S$, this yields $r_h(S \cup \{e\}) = r_h(S)$, which then implies (proof deferred) $e \in \text{span}(D(S, h))$.

As $G \subseteq \text{span}(D(S, h)) \setminus S$, and G and C have the same distribution, we get

$$\Pr\left[|\text{span}(D(S, h) \setminus S)| \leq \tfrac{|N|}{12}\right] \leq \Pr\left[|G| \leq \tfrac{|N|}{12}\right] = \Pr\left[|C| \leq \tfrac{|N|}{12}\right]. \qquad (6)$$

Moreover,

$$|C| = |S| - |W| \geq |S| - hr(N) \geq |S| - |N|/3, \qquad (7)$$

where the equality follows from $S = C \uplus W$, the first inequality from $W \in \mathcal{I}_h$ (which implies $|W| = r_h(W) \leq hr(N)$), and the last one from the fact that $\mathcal{M}$ contains $3h$ many disjoint bases (and hence $|N| \geq 3hr(N)$).

Combining (6) and (7) we obtain

$$\Pr\left[|\text{span}(D(S, h) \setminus S)| \leq \tfrac{|N|}{12}\right] \leq \Pr\left[|S| - \tfrac{|N|}{3} \leq \tfrac{|N|}{12}\right] \leq \Pr\left[|S| \leq \tfrac{5}{6}\mathbb{E}[|S|]\right], \qquad (8)$$

where the second inequality follows from $\mathbb{E}[|S|] = |N|/2$. Relation (4) now follows by applying a Chernoff bound $\Pr[X < (1-\delta)\mathbb{E}[X]] < \exp[-\delta^2\mathbb{E}[X]/2]$ for $X = |S|$ to the right-hand side expression in (8) and using $\mathbb{E}[|S|] = |N|/2$. $\qquad \square$

The proof of Theorem 3 is based on the concentration result (5). In summary, rather than directly showing that $\rho_{\mathcal{M}|_S}$ approximates $\rho_{\mathcal{M}}$ well everywhere, we consider a discrete set of points on $\rho_{\mathcal{M}}$ associated to minors of $\mathcal{M}$ of geometrically increasing ranks. We then apply (5) to these minors and employ a union bound to show that we get a good approximation for these grid points. The union bound works out because the ranks are geometrically increasing and appear in the exponent of the right-hand side of (5). (Complete proof is deferred.)

5   The Main Algorithm and Its Analysis

In this section we describe the procedure from Theorem 2 and discuss the two main ingredients of its analysis. The first one is to show that if the approximate curve $\tilde{\rho}$ is well-structured (in some well-defined sense), then there is an algorithm retrieving a constant factor of $F(\tilde{\rho})$ on expectation — see Theorem 5. The second one is then to show that given any initial approximate curve $\tilde{\rho}$, one can find well-structured curves whose F function value is close to $F(\tilde{\rho})$ — see Theorem 6.

The next result, whose proof is sketched in Sect. 5.1, formalizes the first step above.

Theorem 5. *Let* $\mathcal{M} = (N, \mathcal{I})$ *be a matroid minor of* $\mathcal{M}_{\text{orig}}$, *and let* r *and* $\rho_{\mathcal{M}}$ *denote the rank function and rank-density curve of* $\mathcal{M}$, *respectively. Let* $\overline{\rho} \leq \rho_{\mathcal{M}}$ *be a rank-density curve with densities* $\{\overline{\lambda}_i\}_{i \in [m]}$ *such that the* $\overline{\lambda}_i$ *are powers of some integer* $\beta \geq 3$ *and* $\overline{\lambda}_1 > \cdots > \overline{\lambda}_m \geq 1$. *Assume* $r(D(N, \overline{\lambda}_{i+1})) \geq 24r(D(N, \overline{\lambda}_i/\beta))$ *for* $i \in [m-1]$. *Then there is an efficient procedure* $\text{ALG}(\tilde{\rho}, \beta)$ *that, when run on the RA-MSP subinstance given by* $\mathcal{M}$, *returns an independent set* I *of* $\mathcal{M}$ *of expected weight at least* $(1/180e)F(\overline{\rho})$.

The second main ingredient in the proof of Theorem 2 is the following result. Due to space constraints, we defer its proof to the long version.

Theorem 6. *Let $\mathcal{M} = (N, \mathcal{I})$ be a matroid minor of $\mathcal{M}_{\text{orig}}$, and let r and $\rho_{\mathcal{M}}$ denote the rank function and rank-density curve of $\mathcal{M}$, respectively. Given an (α, β)-approximate curve $\tilde{\rho}$ of $\rho_{\mathcal{M}}$ with $\alpha \in \mathbb{R}_{\geq 24}$ and $\beta \in \mathbb{Z}_{\geq 3}$, there is a procedure $\text{ALG}(\tilde{\rho}, \alpha, \beta)$ returning rank-density curves $\overline{\rho}, \overline{\rho}_1, \overline{\rho}_2, \overline{\rho}_3, \overline{\rho}_4$ such that:*

(i) $\overline{\rho}$ is an (α^2, β^2)-approximation of $\rho_{\mathcal{M}}$.
(ii) $\sum_{i \in [4]} F(\overline{\rho}_i) \geq F(\overline{\rho})$.
(iii) For each $i \in [4]$, $\overline{\rho}_i$ satisfies the following properties: Let $\{\mu_j\}_{j \in [\ell]}$ be the densities of $\overline{\rho}_i$, then all the μ_j are powers of $\beta \geq 3$, and $r(D(N, \mu_{j+1})) \geq \alpha r(D(N, \mu_j/\beta)) \geq 24 r(D(N, \mu_j/\beta))$ for $j \in [\ell - 1]$. Moreover, $\overline{\rho}_i \leq \rho_{\mathcal{M}}$.

We now show how Theorem 5 and Theorem 6 combined imply Theorem 2.

Proof of Theorem 2. Given an (α, β)-approximation $\tilde{\rho}$ of $\rho_{\mathcal{M}}$, first run the procedure from Theorem 6 to get curves $\overline{\rho}, \overline{\rho}_1, \overline{\rho}_2, \overline{\rho}_3, \overline{\rho}_4$. Then choose an index $i \in [4]$ uniformly at random and run the procedure from Theorem 5 on $\overline{\rho}_i$ to get an independent set with expected weight at least

$$\frac{1}{180e}\left(\frac{1}{4}\sum_{i=1}^{4} F(\overline{\rho}_i)\right) \geq \frac{1}{720e}F(\overline{\rho}) \geq \frac{1}{1440e\alpha^2\beta^2}\left[F(\rho_{\mathcal{M}}) - \alpha^2 w_{\max}\right],$$

where the last inequality uses Lemma 2 and the fact that $\overline{\rho}$ is an (α^2, β^2)-approximation of $\rho_{\mathcal{M}}$. $\qquad\square$

Thus, to show Theorem 2, it remains to prove Theorem 5.

5.1 Proof (Sketch) of Theorem 5

Throughout this section we use the notation and assumptions from Theorem 5.

We prove the theorem in two steps. First, we argue that after observing a sample set S, we can build a chain $\bigoplus_{i=1}^{k} \mathcal{M}_i$ of $\mathcal{M}|_{N \setminus S}$ satisfying certain properties with at least constant probability. Then we argue that, given such a chain, there is a procedure returning an independent set I of $\mathcal{M}$ with $\mathbb{E}[w(I)] = \Omega(F(\overline{\rho}))$, leading to the desired result. We start by discussing the former claim.

Given a sample set $S \subseteq N$, we build a chain of matroids as follows. For $i \in [m]$ let

$$N_i := \text{span}(D(S, \overline{\lambda}_i/\beta)) \setminus (S \cup \text{span}(D(S, \overline{\lambda}_{i-1}/\beta))), \text{ and}$$
$$\mathcal{M}_i := (\mathcal{M}/\text{span}(D(S, \overline{\lambda}_{i-1}/\beta)))|_{N_i}, \tag{9}$$

where $D(S, \overline{\lambda}_0/\beta) = \emptyset$ by convention.

In addition, for every $i \in [m]$ let $\overline{N}_i := D(N, \overline{\lambda}_i)$, and define $\overline{\Lambda} := \{i \in [m]: r(\overline{N}_i) \geq 24, \ \overline{\lambda}_i \geq \beta\}$. Note that $\overline{\Lambda}$ and the $\overline{N}_i$'s do not depend on the sample set S. Moreover, from the assumptions of Theorem 5 it follows that $\overline{\Lambda} \supseteq [m] \setminus \{1, m\}$. The next result shows that with constant probability, the sample set S is such that for each $i \in \overline{\Lambda}$, the set N_i contains a subset U_i of large rank and density; more precisely, $r(U_i) \geq \Omega(r(\overline{N}_i))$ and $|U_i|/r(U_i) \geq \Omega(\overline{\lambda}_i)$.

Lemma 3. *Let* $S \sim B(N, 1/2)$, *and let* N_i, $\overline{N}_i$, *and* $\overline{\Lambda}$ *be as defined above. Then, with probability at least* $1/3$, *every* N_i *with* $i \in \overline{\Lambda}$ *contains* $\overline{\lambda}_i$ *disjoint independent sets* $I_1, \ldots, I_{\overline{\lambda}_i}$ *such that* $\sum_{j \in [\overline{\lambda}_i]} |I_j| \geq (1/24)\overline{\lambda}_i r(\overline{N}_i)$.

The second main ingredient in the proof is to show that the above result can be exploited algorithmically. More precisely, we prove the following.

Lemma 4. *Let* $\mathcal{M} = (N, \mathcal{I})$ *be a matroid minor of* $\mathcal{M}_{\mathrm{orig}}$ *containing* h *disjoint independent sets* $I_1, \ldots, I_h$ *such that* $s := (1/h)\sum_{j=1}^{h} |I_j| \geq 1$. *Then there is a procedure that, when run on the RA-MSP subinstance given by* $\mathcal{M}$, *and with only* h *given upfront, returns an independent set of* $\mathcal{M}$ *with expected weight at least* $(s/2e)\eta(h)$. *This is still the case even if the elements of* $\mathcal{M}$ *are revealed in adversarial (rather than uniformly random) order.*

We can now combine Lemmas 3 and 4 to prove Theorem 5 as follows.

Proof of Theorem 5. Let $\mathrm{OSP}(\mathcal{M}, h)$ denote the online selection procedure described in Lemma 4. Additionally, for $i \in [m]$, let r_i denote the coefficient of $\eta(\overline{\lambda}_i)$ in $F(\overline{p})$. Hence, $F(\overline{p}) = \sum_{i=1}^{m} r_i \eta(\overline{\lambda}_i)$. Consider the following algorithm: choose and execute one of the three branches presented below with probability $12/15$, $2/15$, and $1/15$, respectively.

(i) Observe $S \sim B(N, 1/2)$, construct the chain $\bigoplus_{i=1}^{k} \mathcal{M}_i$ as defined in (9), and run $\mathrm{OSP}(\mathcal{M}_i, \overline{\lambda}_i)$ for every $i \in [m]$ (independently in parallel), returning all the picked elements.
(ii) Run the classical secretary algorithm on $\mathcal{M}$ without observing anything and return the picked element (if any).
(iii) Run $\mathrm{OSP}(\mathcal{M}, 1)$ without observing anything and return all picked elements.

Suppose we execute branch (i). By Lemma 3, with probability at least $1/3$, every $\mathcal{M}_i$ with $i \in \overline{\Lambda}$ satisfies the conditions of Lemma 4 with parameters $h = \overline{\lambda}_i$ and $s = (1/24)r(\overline{N}_i)$. Note that $s \geq 1$ holds given that $r(\overline{N}_i) \geq 24$ for all $i \in \overline{\Lambda}$. As additionally all matroids in the chain form a direct sum, executing the first branch of the algorithm returns an independent set with expected weight at least

$$\frac{1}{3} \sum_{i \in \overline{\Lambda}} \frac{1}{2e} \cdot \frac{r(\overline{N}_i)}{24} \eta(\overline{\lambda}_i) = \frac{1}{144e} \sum_{i \in \overline{\Lambda}} r(\overline{N}_i)\eta(\overline{\lambda}_i) \geq \frac{1}{144e} \sum_{i \in \overline{\Lambda}} r_i \eta(\overline{\lambda}_i),$$

where the inequality follows from $\overline{p} \leq \rho_{\mathcal{M}}$ and $\overline{N}_i = D(N, \overline{\lambda}_i)$ for every $i \in [m]$.

Therefore, if $i \in \overline{\Lambda}$, then the corresponding term $r_i \eta(\overline{\lambda}_i)$ in $F(\overline{p})$ is accounted for by branch (i). Thus it only remains to consider $i \in [m] \setminus \overline{\Lambda} \subseteq \{1, m\}$.

Assume first that $1 \notin \overline{\Lambda}$. In this case, we must have $r(\overline{N}_1) < 24$. Since the expected weight yielded by running the classical secretary algorithm is at least

$\eta(|N|)/e$, and $\eta(|N|) \geq \eta(\overline{\lambda}_1)$, then by running branch (ii) the expected weight of the output set is at least

$$\frac{\eta(|N|)}{e} \geq \frac{1}{e} \cdot \frac{r(\overline{N}_1)\eta(\overline{\lambda}_1)}{r(\overline{N}_1)} \geq \frac{1}{23e}r_1\eta(\overline{\lambda}_1),$$

where the last inequality follows from $r_1 \leq r(\overline{N}_1) \leq 23$.

Finally, assume that $m \notin \overline{\Lambda}$. Then $\overline{\lambda}_m = 1$, in which case running branch (iii) yields

$$\mathbb{E}[w(\mathrm{OSP}(\mathcal{M}, 1))] \geq \frac{1}{2e}r(N)\eta(1) \geq \frac{1}{2e}r(\overline{N}_m)\eta(\overline{\lambda}_m),$$

where the first inequality holds by Lemma 4 with $h = 1$ and $s = r(N) \geq 1$, as any basis of $\mathcal{M}$ is an independent set of rank $r(N)$, and the second inequality holds because $r(N) \geq r(\overline{N}_m)$ and $\overline{\lambda}_m = 1$.

The desired lower bound on the expected weight of the set returned by the algorithm now follows by combining the above results with the respective probabilities that each branch is executed.   □

To sum up, we discuss that our main result (i.e., Theorem 1) still holds in the more general adversarial order with a sample setting, where we are allowed to sample a set $S \subseteq N$ containing every element of N independently with probability $1/2$, and the remaining (non-sampled) elements arrive in adversarial order.

In order to see this, first note that the only place in the proof of Theorem 5 where we use that the non-sampled elements (i.e., $N \setminus S$) arrive in random order, is to argue that when running the classical secretary algorithm in branch (ii) we obtain an expected weight of at least $w_{\max}/e$. Indeed, branches (i) and (iii) rely on running the procedure from Lemma 4, whose guarantees hold in the case where the elements arrive in adversarial order. However, note that running the classical secretary procedure in the above adversarial order with a sample setting outputs an element with expected weight of at least $w_{\max}/4$. Indeed, the probability of selecting $w_{\max}$ in the latter setting is at least the probability of the event that $w_{\max}$ is not sampled and the second largest weight is; which occurs with probability $1/4$. Thus, Theorem 5 holds (up to possibly a slightly worse constant) in the adversarial order with a sample setting.

Next, observe that this implies that Theorem 2 also holds in the above setting (again, up to possibly a slightly worse constant). This follows because its proof relies on combining the procedures from Theorems 5 and 6, and the latter is completely oblivious to the arrival order of the elements.

Finally, note that the proof of Theorem 1 uses the procedure from Theorem 5 and the classical secretary algorithm. Because (as discussed above) both of these algorithms have very similar guarantees in the adversarial order with a sample setting to the ones shown in this paper for random order, the claim follows.

References

1. Babaioff, M., Immorlica, N., Kleinberg, R.: Matroids, secretary problems, and online mechanisms. In: Symposium on Discrete Algorithms (SODA 2007), pp. 434–443 (2007)
2. Babaioff, M., et al.: Matroid secretary problems. J. ACM **65**(6), 1–26 (2018)
3. Chakraborty, S., Lachish, O.: Improved competitive ratio for the matroid secretary problem. In: Proceedings of the 23rd Annual ACM-SIAM Symposium on Discrete Algorithms (SODA), pp. 1702–1712 (2012)
4. Dimitrov, N.B., Plaxton, C.G.: Competitive weighted matching in transversal matroids. Algorithmica **62**(1), 333–348 (2012)
5. Dinitz, M., Kortsarz, G.: Matroid secretary for regular and decomposable matroids. In: Proceedings of the 24th Annual ACM-SIAM Symposium on Discrete Algorithms (SODA), pp. 108–117 (2013)
6. Evgenii Borisovich Dynkin: The optimum choice of the instant for stopping a Markov process. Soviet Math. **4**, 627–629 (1963)
7. Feldman, M., Svensson, O., Zenklusen, R.: A simple O(log log(rank))-competitive algorithm for the matroid secretary problem. Math. Oper. Res. **43**(2), 638–650 (2018)
8. Im, S., Wang, Y.: Secretary problems: laminar matroid and interval scheduling. In: Proceedings of the 22nd Annual ACM-SIAM Symposium on Discrete Algorithms (SODA), pp. 1265–1274 (2011)
9. Jaillet, P., Soto, J.A., Zenklusen, R.: Advances on matroid secretary problems: free order model and laminar case. In: Goemans, M., Correa, J. (eds.) IPCO 2013. LNCS, vol. 7801, pp. 254–265. Springer, Heidelberg (2013). https://doi.org/10.1007/978-3-642-36694-9_22
10. Karger, D.: Random sampling and greedy sparsification for matroid optimization problems. Math. Program. 82 (1998). https://doi.org/10.1007/BF01585865
11. Kesselheim, T., et al.: An optimal online algorithm forweighted bipartite matching and extensions to combinatorial auctions. In: Proceedings of the 21st Annual European Symposium on Algorithms (ESA), pp. 589–600 (2013)
12. Korula, N., Pál, M.: Algorithms for secretary problems on graphs and hypergraphs. In: Proceedings of the 36th International Colloquium on Automata, Languages and Programming (ICALP), pp. 508–520 (2009)
13. Lachish, O.: O(log log(rank)) competitive ratio for the matroid secretary problem. In: IEEE 55th Annual Symposium on Foundations of Computer Science. IEEE 2014, pp. 326–335 (2014)
14. Ma, T., Tang, B., Wang, Y.: The simulated greedy algorithm for several submodular matroid secretary problems. In: Proceedings of the 30th International Symposium on Theoretical Aspects of Computer Science (STACS), pp. 478–489 (2013)
15. Gharan, S.O., Vondrák, J.: On variants of the matroid secretary problem. Algorithmica **67**(4), 472–497 (2013)
16. Soto, J.A.: Matroid secretary problem in the random-assignment model. SIAM J. Comput. **42**(1), 178–211 (2013)

A Fast Combinatorial Algorithm for the Bilevel Knapsack Problem with Interdiction Constraints

Noah Weninger[✉][iD] and Ricardo Fukasawa[iD]

University of Waterloo, Waterloo, ON, Canada
{nweninger,rfukasawa}@uwaterloo.ca

Abstract. We consider the bilevel knapsack problem with interdiction constraints, a fundamental bilevel integer programming problem which generalizes the 0-1 knapsack problem. In this problem, there are two knapsacks and n items. The objective is to select some items to pack into the first knapsack such that the maximum profit attainable from packing some of the remaining items into the second knapsack is minimized. We present a combinatorial branch-and-bound algorithm which outperforms the current state-of-the-art solution method in computational experiments by 4.5 times on average for all instances reported in the literature. On many of the harder instances, our algorithm is hundreds of times faster, and we solved 53 of the 72 previously unsolved instances. Our result relies fundamentally on a new dynamic programming algorithm which computes very strong lower bounds. This dynamic program solves a relaxation of the problem from bilevel to $2n$-level where the items are processed in an online fashion. The relaxation is easier to solve but approximates the original problem surprisingly well in practice. We believe that this same technique may be useful for other interdiction problems.

Keywords: Bilevel programming · Interdiction · Knapsack problem · Combinatorial algorithm · Dynamic programming · Branch and bound

1 Introduction

Bilevel integer programming (BIP), a generalization of integer programming (IP) to two-round two-player games, has been increasingly studied due to its wide real-world applicability [5,12,17]. In the BIP model, there are two IPs, called the *upper level* (or *leader*) and *lower level* (or *follower*), which share some variables between them. The objective is to optimize the upper level IP but with the constraint that the shared variables must be optimal for the lower level IP. The term *interdiction* is used to describe bilevel problems in which the upper level IP has the capability to block access to some resources used by the lower level IP. The upper level is typically interested in blocking resources in a way that produces the worst possible outcome for the lower level IP. For instance, the resources may be nodes or edges in a graph, or items to be packed into a knapsack. These problems often arise in military defense settings (e.g., see [17]).

© The Author(s), under exclusive license to Springer Nature Switzerland AG 2023
A. Del Pia and V. Kaibel (Eds.): IPCO 2023, LNCS 13904, pp. 438–452, 2023.
https://doi.org/10.1007/978-3-031-32726-1_31

In this paper we study the bilevel knapsack problem with interdiction constraints (BKP), which was introduced by DeNegre in 2011 [6]. This problem is a natural extension of the 0-1 knapsack problem (KP) to the bilevel setting. Formally, we are given n items. Each item $i \in \{1, \ldots, n\}$ has an associated profit $p_i \in \mathbb{Z}_{>0}$, upper-level weight $w_i^U \in \mathbb{Z}_{>0}$ and lower-level weight $w_i^L \in \mathbb{Z}_{\geq 0}$. The upper-level knapsack has capacity $C^U \in \mathbb{Z}_{\geq 0}$ and the lower-level knapsack has capacity $C^L \in \mathbb{Z}_{\geq 0}$. We use the standard notation: for a vector x and set S we let $x(S) := \sum_{i \in S} x_i$. The problem BKP can then be stated as follows:

$$\min_{X \in \mathcal{U}} \max_{Y \in \mathcal{L}(X)} p(Y) \qquad \text{(objective)}$$

$$\text{where } \mathcal{U} = \left\{ X \subseteq \{1, \ldots, n\} : w^U(X) \leq C^U \right\}, \qquad \text{(upper level)}$$

$$\text{and } \mathcal{L}(X) = \left\{ Y \subseteq \{1, \ldots, n\} \setminus X : w^L(Y) \leq C^L \right\}. \qquad \text{(lower level)}$$

We call a solution (X, Y) *feasible* if $X \in \mathcal{U}$ and $Y \in \text{argmax}\{p(\hat{Y}) : \hat{Y} \in \mathcal{L}(X)\}$. A solution (X, Y) is *optimal* if it minimizes $p(Y)$ over all feasible solutions. Note that determining whether (X, Y) is feasible is weakly NP-Hard.

Given that "the knapsack problem is believed to be one of the 'easier' NP-hard problems," [16] one may propose that BKP may also be one of the 'easier' Σ_2^p-hard problems. While this may indeed be the case, unlike KP, which admits a pseudopolynomial time algorithm, BKP remains NP-complete when the input is described in unary and thus has no pseudopolynomial time algorithm unless P = NP [1]. In addition, BKP is a Σ_2^p-complete problem, which means it cannot even be modelled as an IP with polynomially many variables and constraints, unless the polynomial hierarchy collapses. A recent positive theoretical result for BKP is a polynomial-time approximation scheme [3].

This theoretical hardness seemed to have been confirmed by the struggle of computational approaches to solve small instances. Until recently, proposed algorithms – either generic BIP algorithms [6,8,19] or more specific algorithms for BKP (or slight generalizations of it) [2,9,13] – were only able to solve instances with at most 55 items. A breakthrough result came in a paper by Della Croce and Scatamacchia [4], that proposed a BKP-specific algorithm (henceforth referred to as *DCS*) which was able to solve instances containing up to 500 items.

It is worth noting that all papers prior to DCS only consider instances which were generated in an *uncorrelated* fashion, meaning that weights and profits were chosen uniformly at random with no correlation between the values. The DCS algorithm is able to solve uncorrelated instances with 500 items in less than a minute, but its performance drops significantly even for weakly correlated instances, and most strongly correlated instances remain unsolved after an hour of computing time. These results seem to mimic what is known for KP: uncorrelated KP instances are some of the easiest types of instances to solve [16] and early KP algorithms such as expknap [15] could quickly solve uncorrelated instances but struggled with strongly correlated ones.

A common aspect among all methods in the literature is that they rely fundamentally on MIP solvers. In this paper, we present a simple combinatorial branch-and-bound algorithm for solving BKP. Our algorithm improves on the performance of the DCS algorithm for 94% of instances, even achieving a speedup of orders of magnitude in many cases. Furthermore, our algorithm appears to be largely impervious to correlation: it solves strongly correlated instances with ease, only significantly slowing down when the lower-level weights equal the profits (i.e., the subset sum case). In Sect. 2, we describe our algorithm. Our algorithm relies fundamentally on a new strong lower bound computed by dynamic programming which we present in Sect. 3. Section 4 details our computational experiments. We conclude in Sect. 5 with some directions for future research. We note that some proofs were omitted for brevity.

2 A Combinatorial Algorithm for BKP

In this section we describe our exact solution method for BKP. At a high level, the algorithm is essentially just standard depth-first branch-and-bound. Our strong lower bound, defined later in Sect. 3, is essential for reducing the search space. To begin formalizing this, we first define the notion of a subproblem.

Definition 1. *A subproblem* (X, i) *consists of some* $i \in \{1, \ldots, n + 1\}$ *and set of items* $X \subseteq \{1, \ldots i - 1\}$ *such that* $X \in \mathcal{U}$.

Note that this definition depends on the ordering of the items, which throughout the paper we assume to be such that $\frac{p_1}{w_1^L} \geq \frac{p_2}{w_2^L} \geq \cdots \geq \frac{p_n}{w_n^L}$ with ties broken by placing items with larger p_i first. These subproblems will form the nodes of the branch-and-bound tree; $(\emptyset, 1)$ is the root node, and for every $X \in \mathcal{U}$, $(X, n+1)$ is a leaf. Every non-leaf subproblem (X, i) has the child $(X, i+1)$, which represents omitting item i from the upper-level solution. Non-leaf subproblems (X, i) with $X \cup \{i\} \in \mathcal{U}$ have an additional child $(X \cup \{i\}, i + 1)$ which represents including item i in the upper-level solution.

The algorithm simply starts at the root and traverses the subproblems in a depth-first manner, preferring the child $(X \cup \{i\}, i + 1)$ if it exists because it is more likely to lead to a good solution. Every time the search reaches a leaf $(X, n + 1)$, we solve the knapsack problem $\max\{p(Y) : Y \in \mathcal{L}(X)\}$ to get a feasible solution, and updating the incumbent if appropriate.

2.1 The Bound Test

At each node (X, i) of the branch-and-bound, we find a lower bound on the optimal value of that subproblem by the use of a *bound test* algorithm, which tests lower bounds against a known incumbent solution value z^*.

The lower bound used to prune a subproblem is computed in three steps: (1) we solve a knapsack problem on items $\{1, \ldots, i-1\} \setminus X$, (2) we compute a lower bound for BKP restricted to items $\{i, \ldots, n\}$, and (3) we combine (1) and (2) into a lower bound for the descendants of (X, i).

For step (1), we define a function $K(\bar{X}, c)$, which, for a given $\bar{X} \subseteq \{1, \ldots, n\}$ and $c \geq 0$, returns the optimal value of the knapsack problem with weights w^L, profits p, and capacity c, under the restriction that items in $\bar{X}$ cannot be used:

$$K(\bar{X},c) = \max\left\{p(Y) : Y \subseteq \{1,\ldots,n\} \setminus \bar{X} \text{ and } w^L(Y) \leq c\right\}.$$

For step (2), we need a function $\omega(i, c^U, c^L)$ which is a lower bound on BKP but with upper-level capacity c^U, lower-level capacity c^L, and restricted to items $\{i,\ldots,n\}$. So, formally, ω must satisfy

$$\omega(i,c^U,c^L) \leq \min\{K(X' \cup \{1,\ldots,i-1\}, c^L) : X' \subseteq \{i,\ldots,n\}, w^U(X') \leq c^U\}.$$

We will define precisely what ω is in Sect. 3; for now, we only need to know that it has this property. We now prove the following lemma, which describes how to achieve step (3).

Lemma 1. *Let (X,i) be a subproblem. For all $c \in \{0,\ldots,C^L\}$,*

$$K\left(X \cup \{i,\ldots,n\}, c\right) + \omega\left(i, C^U - w^U(X), C^L - c\right)$$
$$\leq \min\left\{p(\bar{Y}) : (\bar{X},\bar{Y}) \text{ is feasible for BKP and } \bar{X} \cap \{1,\ldots,i-1\} = X\right\}.$$

Proof. First, note that for any $X' \subseteq \{i,\ldots,n\}$,

$$K(X \cup \{i,\ldots,n\}, c) + K(X' \cup \{1,\ldots,i-1\}, C^L - c) \leq K(X \cup X', C^L).$$

Thus, if we let $\chi' := \{X' \subseteq \{i,\ldots,n\} : w^U(X') \leq C^U - w^U(X)\}$ and take the minimum with respect to χ' we get

$$K(X\cup\{i,\ldots,n\},c)+\omega(i, C^U - w^U(X), C^L - c) \leq \min\{K(X\cup X', C^L) : X' \in \chi'\}.$$

and now just note that this last term is equal to

$$\min\left\{p(\bar{Y}) : (\bar{X},\bar{Y}) \text{ is feasible for BKP and } \bar{X} \cap \{1,\ldots,i-1\} = X\right\}. \quad \square$$

From this, it follows that for any $c \in \{0,\ldots,C^L\}$, if we have

$$K\left(X \cup \{i,\ldots,n\}, c\right) + \omega\left(i, C^U - w^U(X), C^L - c\right) \geq z^*$$

then we can prune subproblem (X,i). We also note that the lower bound still remains valid if we replace $K\left(X \cup \{i,\ldots,n\}, c\right)$ by a feasible solution to that problem. This is what is done in the function `BoundTest` in Algorithm 1 whose correctness is established by the following lemma.

Lemma 2. *If `BoundTest(X, i)` returns true, then subproblem (X,i) can be pruned.*

We end with an important consideration regarding the efficient implementation of Algorithm 1. The greedy part of the bound test (Lines 2 to 4) appears to require time $O(n)$. However, considering how we choose to branch, the values w_g and p_g can be computed in time $O(1)$ given their values for the parent subproblem. To determine $K(X \cup \{i,\ldots,n\}, c)$, we use the standard dynamic program (DP) for knapsack. However, for each bound test, it is only necessary to compute a single row of a DP table (i.e., fill in all C^L capacity values for the row associated with item i) from the row computed in the parent subproblem. By doing this, the entire `BoundTest` function will run in time $O(C^L)$. Furthermore, when the branch-and-bound reaches a leaf, the knapsack solution needed to update the upper bound will already have been found by the bound test.

Algorithm 1: Returns true if the subproblem (X, i) can be pruned.

Precondition: z^* is the value of the best incumbent solution

1 **function** BoundTest(X, i)

2    $\quad w_g, p_g \leftarrow 0;$

3    $\quad$ **for** $j = 1, \ldots, i-1$ **do**

4    $\quad\quad \lfloor$ **if** $j \notin X$ *and* $w_g + w_j^L \leq C^L$ **then** $w_g \leftarrow w_g + w_j^L$, $p_g \leftarrow p_g + p_j;$

5    $\quad$ **if** $p_g + \omega(i, C^U - w^U(X), C^L - w_g) \geq z^*$ **then return true;**

6    $\quad$ **for** $c = 0, \ldots, C^L$ **do**

7    $\quad\quad \lfloor$ **if** $K(X \cup \{i, \ldots, n\}, c) + \omega(i, C^U - w^U(X), C^L - c) \geq z^*$ **then return true;**

8    $\quad$ **return false;**

2.2   Computing Initial Bounds

In our algorithm, a strong initial upper bound z^* can help decrease the size of the search tree. For this we use a simple heuristic we call GreedyHeuristic. GreedyHeuristic works in two steps. First, an upper level set $\bar{X}$ is picked by solving $\max\{p(X) : X \in \mathcal{U}\}$. Then the lower level solution $\bar{Y}$ is picked by solving $\max\{p(Y) : Y \in \mathcal{L}(\bar{X})\}$. We say GreedyHeuristic returns $(\bar{X}, \bar{Y}, \bar{z})$, where $\bar{z}$ is the objective value of the solution $(\bar{X}, \bar{Y})$. We now establish a case in which GreedyHeuristic actually returns an optimal solution.

Lemma 3. GreedyHeuristic() *returns an optimal solution if there exists an optimal solution* (X, Y) *for BKP where* $Y = \{1, \ldots, n\} \setminus X$.

The proof is skipped for brevity, but we remark that previous work has noted that BKP is very easily solved for such instances [2,4]. Following [4], we use an LP to detect some cases where the GreedyHeuristic is optimal. The below formulation $LB(i)$ is a simplified version of the LP in [4].

$$LB(i) = \min \quad \textstyle\sum_{j=1}^{i-1} p_j(1 - x_j)$$

$$\text{such that} \quad \textstyle\sum_{j=1}^{i-1} w_j^U x_j \leq C^U$$

$$C^L - w_i^L + 1 \leq \textstyle\sum_{j=1}^{i-1} w_j^L(1 - x_j) \leq C^L$$

$$0 \leq x \leq 1, x \in \mathbb{R}^{i-1}$$

We define $LB(i) = \infty$ if the LP is infeasible. This LP is used by the following lemma. We skip the proof for brevity.

Lemma 4. *Suppose* GreedyHeuristic() *returns* $(\bar{X}, \bar{Y}, \bar{z})$. *If* $\bar{z} \leq \min\{LB(c) : 1 \leq c \leq n\}$ *then* $(\bar{X}, \bar{Y})$ *is optimal for BKP.*

Before starting our branch-and-bound algorithm, we run GreedyHeuristic and check if it is optimal using Lemma 4. This enables us to quickly solve trivial instances without needing to run our main algorithm.

3   Lower Bound

In this section we define the lower bound $\omega(i, c^U, c^L)$ that we use in our algorithm. Recall that $\omega(i, c^U, c^L)$ must lower bound the restriction of BKP where we can only use items $\{i, \ldots, n\}$, have upper-level capacity c^U, and lower-level capacity c^L. Our lower bound is based on dynamic programming (DP), which computes $\omega(i, c^U, c^L)$ for all parameter values with time and space complexity $O(nC^U C^L)$.

The main idea for the lower bound is to obtain good feasible solutions for the lower-level problem. Perhaps the most obvious way is to assume that the lower-level problem finds a greedy solution. It is not hard to see why this is a lower bound: a greedy lower-level solution will always achieve profit at most that of an optimal lower-level solution. We can compute this lower bound with the following recursively-defined DP algorithm:

$$
\omega_g(i, c^U, c^L) =
\begin{cases}
\infty & \text{if } c^U < 0, \\
-\infty & \text{if } c^L < 0, \\
0 & \text{if } c^U \geq 0, c^L \geq 0 \text{ and } i > n, \\
\omega_g(i+1, c^U, c^L) & \text{if } c^U \geq 0, w_i^L > c^L \text{ and } i \leq n. \\
\min \left\{ \begin{array}{l} \omega_g(i+1, c^U - w_i^U, c^L), \\ \omega_g(i+1, c^U, c^L - w_i^L) + p_i \end{array} \right\} & \text{if } c^U \geq 0, w_i^L \leq c^L \text{ and } i \leq n.
\end{cases}
$$

The first three expressions are to take care of trivial cases. The fourth case skips any item which cannot fit in the lower-level knapsack, as it would be pointless for the upper level to take such an item. The fifth case picks the worse (i.e., better for the upper level) out of the two possible greedy solutions from the two children nodes of subproblem (X, i).

This lower bound already has very good performance in practice. However, we can do better by making a deceptively simple modification: giving the lower level the option to ignore an item. This modification produces our strong DP lower bound, ω, which is equal to ω_g in the first three cases, but instead of the last two cases we get:

$$
\omega(i, c^U, c^L) = \min \left\{ \begin{array}{l} \omega(i+1, c^U - w_i^U, c^L), \\ \max \left\{ \begin{array}{l} \omega(i+1, c^U, c^L - w_i^L) + p_i, \\ \omega(i+1, c^U, c^L) \end{array} \right\} \end{array} \right\} \quad \begin{array}{l} \text{if } c^U \geq 0, c^L \geq 0 \\ \text{and } i \leq n. \end{array}
$$

It is not a hard exercise to show that $\omega_g(i, c^U, c^L) \leq \omega(i, c^U, c^L)$ for all $1 \leq i \leq n$, $0 \leq c^U \leq C^U$ and $0 \leq c^L \leq C^L$. Extrapolating our intuition about ω_g, formulation ω appears to actually find optimal lower-level solutions, so one might guess that $\omega(1, C^U, C^L)$ is actually optimal for BKP, if it weren't that this is impossible unless P = NP [1]. The subtlety is that by giving the lower level a choice of whether to take an item, we have also given the upper level the power to react to that choice. Specifically, the upper level choice of whether to take item i can depend on how much capacity the lower level has used on items $\{1, \ldots, i-1\}$. Evidently, this is not permitted by the definition of BKP, which dictates that the upper level solution is completely decided prior to choosing the lower level solution. However, our experiments show that this actually gives the upper level an extremely small amount of additional power in practice.

The lower bound ω may also be interpreted as a relaxation from a 2-round game to a $2n$-round game. This may seem to be making the problem more difficult, but each round is greatly simplified, so the problem becomes easier to solve. This $2n$-round game is as follows. In round $2i-1$, the leader (the upper level player) decides whether to include the item i. In round $2i$, the follower (the lower level player) responds to the leader's decision: if item i is still available, then the follower decides whether to include item i. The score of the game is simply the total profit of all items chosen by the follower. It is straightforward to see that the minimax value of this game (i.e., the score given that both players follow an optimal strategy) is equal to $\omega(1, C^U, C^L)$.

We now show formally that $\omega(1, C^U, C^L)$ lower bounds the optimal objective value of BKP. To this end we define ω_X, a modified version of ω where instead of the minimization in the case where $c^U \geq 0$, $c^L \geq 0$ and $i \leq n$, the choice is made depending on whether $i \in X$ for some given set X. $\omega_X(i, c^U, c^L)$ is equal to ω_g in the first three cases, but replaces the last two cases with:

$$
\omega_X(i, c^U, c^L) =
\begin{cases}
\omega_X(i+1, c^U - w_i^U, c^L) & \text{if } c^U \geq 0, c^L \geq 0, i \leq n \text{ and } i \in X, \\
\max \left\{ \begin{aligned} &\omega_X(i+1, c^U, c^L - w_i^L) + p_i, \\ &\omega_X(i+1, c^U, c^L) \end{aligned} \right\} & \text{if } c^U \geq 0, c^L \geq 0, i \leq n \text{ and } i \notin X.
\end{cases}
$$

With this simple modification, we claim that $\omega_X(1, C^U, C^L) = \max\{p(Y) : Y \in \mathcal{L}(X)\}$ (and similarly for other i, c^U, and c^L). To formalize this, we show that ω_X and K (as defined in Sect. 2.1) are equivalent in the following sense.

Lemma 5. *For all* $1 \leq i \leq n$, $X \subseteq \{i, \ldots, n\}$, $c^U \geq w^U(X)$ *and* $c^L \geq 0$,

$$
\omega_X(i, c^U, c^L) = K(X \cup \{1, \ldots, i-1\}, c^L).
$$

Proof. Given that $c^U \geq w^U(X)$, the case $c^U < 0$ can not occur in the expansion of $\omega_X(i, c^U, c^L)$, so $\omega_X(i, c^U, c^L) = \omega_X(i, \infty, c^L)$. Consider the 0-1 knapsack problem with profits p' and weights w' formed by taking $p' = p$ and $w' = w^L$ except with $p_j' = w_j' = 0$ for items $j \in X$. We can then simplify the definition of $\omega_X(i, \infty, c^L)$ by using p' and w' to effectively skip items in X:

$$
\omega_X(i, \infty, c^L) =
\begin{cases}
-\infty & \text{if } c^L < 0, \\
0 & \text{if } c^L \geq 0 \text{ and } i > n, \\
\max \left\{ \begin{aligned} &\omega_X(i+1, \infty, c^L - w_i') + p_i', \\ &\omega_X(i+1, \infty, c^L) \end{aligned} \right\} & \text{if } c^L \geq 0 \text{ and } i \leq n.
\end{cases}
$$

The recursive definition of $\omega_X(i, \infty, c^L)$ above describes the standard DP algorithm for 0-1 knapsack with capacity c^L, profits p' and weights w' but restricted to items $\{i, \ldots, n\}$; this is the same problem which is solved by $K(X \cup \{1, \ldots, i-1\}, c^L)$.   □

We now establish that $\omega(i, c^U, c^L)$ is a lower bound as desired.

Theorem 1. *For all* $1 \leq i \leq n$, $c^U \geq 0$ *and* $c^L \geq 0$,

$$\omega(i, c^U, c^L) \leq \min_{X \subseteq \{i,\ldots,n\} \,:\, w^U(X) \leq c^U} K(X \cup \{1, \ldots, i-1\}, c^L).$$

Proof. By definition, $\omega_X(i, c^U, c^L) = \infty$ if $w^U(X) > c^U$, so

$$\min_{X \subseteq \{i,\ldots,n\}} \omega_X(i, c^U, c^L) = \min_{X \subseteq \{i,\ldots,n\} \,:\, w^U(X) \leq c^U} \omega_X(i, c^U, c^L)$$

$$= \min_{X \subseteq \{i,\ldots,n\} \,:\, w^U(X) \leq c^U} K(X \cup \{1, \ldots, i-1\}, c^L).$$

where the last equality follows from Lemma 5. Therefore, it suffices to show that $\omega(i, c^U, c^L) \leq \min_{X \subseteq \{i,\ldots,n\}} \omega_X(i, c^U, c^L)$. The proof is by induction on i from $n+1$ to 1. Let $c^U \geq 0$ and $c^L \geq 0$ be arbitrary. Our inductive hypothesis is that $\omega(i, c^U, c^L) \leq \min_{X \subseteq \{i,\ldots,n\}} \omega_X(i, c^U, c^L)$. For the base case, where $i = n+1$, by definition we have $\omega(i, c^U, c^L) = \omega_X(i, c^U, c^L) = 0$ for any $X \subseteq \{i, \ldots, n\} = \emptyset$. Now we prove the inductive case. Let $1 \leq i \leq n$ be arbitrary and assume that the inductive hypothesis holds for $i+1$, with every $c^U \geq 0$ and $c^L \geq 0$. We present only the case where $w_i^U \leq c^U$ and $w_i^L \leq c^L$. The remaining cases (where $w_i^U > c^U$ or $w_i^L > c^L$) are just simpler versions of this.

$$\omega(i, c^U, c^L) = \min \left\{ \begin{array}{l} \omega(i+1, c^U - w_i^U, c^L), \\ \max\left\{ \omega(i+1, c^U, c^L - w_i^L) + p_i, \omega(i+1, c^U, c^L) \right\} \end{array} \right\}$$

$$\leq \min \left\{ \begin{array}{l} \displaystyle\min_{X \subseteq \{i+1,\ldots,n\}} \omega_X(i+1, c^U - w_i^U, c^L), \\[1ex] \max \left\{ \begin{array}{l} \displaystyle\min_{X \subseteq \{i+1,\ldots,n\}} \omega_X(i+1, c^U, c^L - w_i^L) + p_i, \\[1ex] \displaystyle\min_{X \subseteq \{i+1,\ldots,n\}} \omega_X(i+1, c^U, c^L) \end{array} \right\} \end{array} \right\}$$

$$\leq \min \left\{ \begin{array}{l} \displaystyle\min_{X \subseteq \{i+1,\ldots,n\}} \omega_X(i+1, c^U - w_i^U, c^L), \\[1ex] \displaystyle\min_{X \subseteq \{i+1,\ldots,n\}} \max \left\{ \begin{array}{l} \omega_X(i+1, c^U, c^L - w_i^L) + p_i, \\ \omega_X(i+1, c^U, c^L) \end{array} \right\} \end{array} \right\}$$

$$= \min_{X \subseteq \{i+1,\ldots,n\}} \min \left\{ \begin{array}{l} \omega_X(i+1, c^U - w_i^U, c^L), \\[1ex] \max \left\{ \begin{array}{l} \omega_X(i+1, c^U, c^L - w_i^L) + p_i, \\ \omega_X(i+1, c^U, c^L) \end{array} \right\} \end{array} \right\}$$

$$= \min_{X \subseteq \{i,\ldots,n\}} \omega_X(i, c^U, c^L).$$ □

Note that in particular, this implies that $\omega(1, C^U, C^L) \leq \min_{X \in \mathcal{U}} K(X, C^L) = \min_{X \in \mathcal{U}} \max_{Y \in \mathcal{L}(X)} p(Y)$, i.e., $\omega(1, C^U, C^L)$ is a lower bound for BKP.

We end this section with a simple observation. The approach we derived for our problem was based on obtaining good feasible solutions to the lower problem. Now, if the lower problem is already NP-hard, one may ask how useful can an approximate solution to the lower level be. For this, we consider a very generic problem:

$$z^* := \min_{x \in \mathcal{U}} \max_{y \in \mathcal{L}(x)} c(x, y) \tag{1}$$

For each $x \in \mathcal{U}$, assume there exists $y \in \mathcal{L}(x)$ that maximizes the inner problem. Let $y^*(x)$ be such a maximizer of $c(x, y)$ for $y \in \mathcal{L}(x)$. The following lemma then shows that if we can solve the problem with an approximate lower level, instead of an exact one, we get an approximate solution to (1).

Lemma 6. *Suppose we have a function $f(x)$ such that for all $x \in \mathcal{U}$:*

- *$f(x) \in \mathcal{L}(x)$, and*
- *$c(x, f(x)) \leq c(x, y^*(x)) \leq \alpha c(x, f(x))$, for some $\alpha \geq 1$.*

Let $\tilde{x} \in \arg\min_{x \in \mathcal{U}} c(x, f(x))$. Then $c(\tilde{x}, y^*(\tilde{x})) \leq \alpha z^*$.

Proof. Let $(x^*, y^*(x^*))$ be the optimal solution to (1). Then

$$\frac{1}{\alpha} c(\tilde{x}, y^*(\tilde{x})) \leq c(\tilde{x}, f(\tilde{x})) \leq c(x^*, f(x^*)) \leq c(x^*, y^*(x^*)) = z^*. \qquad \square$$

While this does not immediately give an approximation algorithm for the problem, we believe it may be useful to simplify some Σ_2^p-hard bilevel interdiction problems and, for that reason, we include this lemma in this work. Note that an analogous result can be also derived for a $\max - \min$ problem.

4   Computational Results

In this section, we perform computational experiments to compare our algorithm (*Comb*) with the method from [4] (*DCS*). Given that the superiority of the DCS algorithm over other approaches has been well demonstrated we do not compare our algorithm directly to the prior works [2,6,8,9,18,19].

4.1   Implementation

We were unable to obtain either source code or a binary from the authors of [4], so we reimplemented their algorithm. We use Gurobi 9.5 instead of CPLEX 12.9, and obviously run it on a different machine, so an exact replication of their results is nearly impossible. Nonetheless, we found our reimplementation produces results very similar to what is reported in [4], and even solves three additional instances which were not solved in [4]. Therefore, we believe that any comparison with our version of the DCS algorithm is reasonably fair.

Both Comb and DCS were run using 16 threads. However, not all parts of the algorithms were parallelized. Specifically, in the DCS implementation, the only part which is parallelized is the MIP solver. In the implementation of Comb, we only parallelized two parts: the computation of the lower bound ω and the computation of the initial lower bound $\min\{LB(c) : 1 \leq c \leq n\}$.

Our code is implemented in C++ and relies on OpenMP 4.5 for parallelism, Gurobi 9.5 for solving MIPs, and the implementation of the combo knapsack algorithm [14] from [11]. The code was executed on a Linux machine with four 16-core Intel Xeon Gold 6142 CPUs @ 2.60 GHz and 256 GB of RAM. All code and instances are available at https://github.com/nwoeanhinnogaehr/bkpsolver.

4.2 Instances

Our test set contains all instances described in the literature [2,4,6,10,19] and 1660 new instances which we generated. The first 1500 were generated as follows. For each $n \in \{10, 25, 50, 10^2, 10^3, 10^4\}$, INS $\in \{0.5, 1, 1.5, 2, 2.5, 3, 3.5, 4, 4.5, 5\}$ and $R \in \{10, 25, 50, 100, 1000\}$, we generated five instances according to five different methods, which we call classes 1-5. All weights and profits were selected uniformly at random in the range $[1, R]$, but for some of the five classes, we equated w^L, w^U or p with each other:

1. w^L, w^U and p are independent (uncorrelated)
2. $w^L = p$ but w^U is independent (lower subset-sum)
3. $w^U = p$ but w^L is independent (upper subset-sum)
4. $w^L = w^U = p$ (both subset-sum)
5. $w^L = w^U$ but p is independent (equal weights)

The capacities are chosen as follows. Let $C^L = \lceil \text{INS}/11 \cdot \sum_i w_i^L \rceil$ and choose C^U uniformly at random in the range $[C^L - 10, C^L + 10]$. If there is any item with $w_i^L < C^L$ or $w_i^U < C^U$, then we increase the appropriate capacity so that this is not the case. This is the same way that the capacities are selected in [2,4,10], except that we exclude instances that would almost certainly be solved by the initial bound test and we include half integral values of INS. Note that the easiest and hardest instances reported in the literature were uncorrelated and lower subset-sum, respectively [4]. Hence, we expect these new instances to capture both best-case and worst-case behavior from the solvers.

The remaining 160 instances were intended to test the case where the capacity is very large but the number of items is small. These instances were generated following the same scheme as above except that we only generated uncorrelated instances, and we chose $n \in \{5, 10, 20, 30\}$ and $R \in \{10^3, 10^4, 10^5, 10^6\}$. To the best of our knowledge, instances with such large capacity have not been evaluated previously.

4.3 Results

Our results on instances from the literature are summarized in Table 1. To best match the test environment used for the original DCS implementation, we ran

Table 1. Results for instances from the literature, grouped by instance type.

Group	Num	DCS				Comb			
		Opt	Best	Avg	Max	Opt	Best	Avg	Max
uncorrelated	940	940	66	2.32	15.73	940	874	0.31	6.48
weak correlated	50	50	0	13.49	72.64	50	50	0.26	3.54
strong correlated	50	41	0	689.58	3,600	50	50	0.34	3.98
inverse strong corr.	50	38	0	919.91	3,600	50	50	1.07	34.69
almost strong corr.	50	40	0	815.4	3,600	50	50	0.24	3.16
subset-sum	50	35	0	1,087.18	3,600	42	42	586.29	3,600
even-odd subset-sum	50	36	0	1,033.98	3,600	42	42	581.42	3,600
even-odd strong corr.	50	41	0	747.12	3,600	50	50	0.61	17.21
similar weight uncorr.	50	50	0	22.89	79.85	50	50	0.05	0.08

the tests with a time limit of 1 h, and used the same parameters for the DCS algorithm as reported by the authors [4]. For each instance group and each solver, we reported the number of instances solved to optimality (column Opt), the number of instances on which the solver took strictly less time than the other solver (column Best), the average wall-clock running time in seconds (column Avg) and the maximum wall-clock running time in seconds (column Max). Note that measuring wall-clock time as opposed to CPU time only disadvantages our algorithm, if anything, because the DCS implementation utilizes all 16 threads for a large proportion of the time due to parallelization within Gurobi, whereas the same is not true for our algorithm. Overall, our solver had better performance on 1258 of the 1340 instances (about 94%), achieving about 4.5 times better performance on average, and solving 53 of the 69 instances which our DCS implementation did not (the original DCS implementation did not solve 72 instances). These results demonstrate the remarkable advantage that Comb has on hard instances. DCS struggles with all instances involving strong or subset-sum correlation, but Comb only significantly slows down for subset-sum instances.

In Fig. 1, we plot a performance profile for instances from the literature comparing the DCS algorithm to some variants of our algorithm. This type of graph plots, for each instance, the ratio of each algorithm's performance to the performance of the best algorithm for that instance. The instances are sorted by difficulty. Note that instances which timed out are counted as 3600 s seconds. For a comprehensive introduction to performance profiles, see [7]. The two variants of Comb included are Comb-weak, which uses the lower bound ω_g instead of ω, and Comb-greedy, which uses a greedy lower bound test, i.e., where Lines 6 to 7 are omitted from Algorithm 1. The graph indicates that while Comb does better with more threads and the main variant is best, the single-threaded version and the variants still outperform 16-thread DCS. Although it is not depicted in the performance profile, we also tested variants with different item orderings, and found the one described in Sect. 2 to be the best. This is somewhat expected as this is the same ordering that gives rise to the greedy algorithm for 0-1 knapsack.

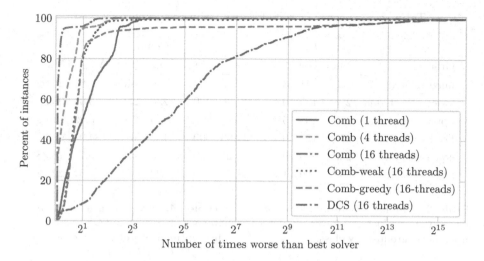

Fig. 1. Performance profile for all instances from the literature.

We now turn our attention to the new instances. Due to the large number of new instances and high difficulty, we used a reduced time limit of 15 min (900 s) in order to complete the testing in a timely fashion. For these tests we use the same DCS parameters used by the DCS authors for testing their own instances [4].

The results for the new instances are summarized in Tables 2 and 3. Note that there are 300 instances of each group, but Table 2 only instances for which our test machine had enough memory to store the DP table used in Comb. The performance of DCS is reported on the remaining instances in Table 3. We can see that Comb offers a significant speed improvement for all classes, and although both solvers are roughly equally capable of solving instances in the uncorrelated, upper subset-sum, equal weights, and large capacity classes, Comb solved 122 more of the instances in the lower subset-sum and both subset-sum classes than DCS. Extrapolating from the results in Table 2, we suspect that given sufficient memory, our solver would be able to solve many more of these instances with better performance than DCS.

In Tables 5 and 6, we report some statistics collected during the tests of our algorithm. In these tables, RootTime is the average number of (wall clock) seconds required to perform the initial bound test and compute the DP table, OptTime is the average number of seconds that branch-and-bound takes to find an optimal solution (excluding RootTime), ProofTime is the average number of seconds needed to prove optimality after an optimal solution is found, Nodes is the average number of nodes searched by branch-and-bound, and RootGap% is $100 \cdot (\bar{z} - \omega(1, C^U, C^L))/\bar{z}$ where $\bar{z}$ is the value of the solution returned by `GreedyHeuristic`. These tables only consider instances which fit in memory and did not time out, as some of the columns are undefined otherwise. Considering all

Table 2. Summary of results for new instances, grouped by class.

Group	Num	DCS				Comb			
		Opt	Best	Avg	Max	Opt	Best	Avg	Max
uncorrelated	243	241	3	12.42	900	243	240	0.96	24.02
lower subset-sum	256	173	0	320.36	900	236	236	73.62	900
upper subset-sum	235	235	5	2.7	89.25	235	230	0.79	21.08
both subset-sum	235	130	0	423.83	900	189	189	184.9	900
equal weights	236	236	9	2.2	120.55	236	227	0.7	18.38
large capacity	92	90	1	38.78	900	92	91	2.31	22.72

Table 3. Summary of results for DCS on new instances which our solver could not fit in memory, grouped by class.

Group	Num	Opt	Avg	Max
uncorrelated	57	41	487.85	900
lower subset-sum	44	1	886.75	900
upper subset-sum	65	59	283.34	900
both subset-sum	65	0	900	900
equal weights	64	62	175.92	900
large capacity	68	5	864.33	900

Table 4. An instance with n items, $C^U = n-1$ and $C^L = n$ that has optimal objective value $n-1$ but $\omega(1, C^U, C^L) = 1$.

item no.	p	w^U	w^L
1	1	1	1
$\vdots$	$\vdots$	$\vdots$	$\vdots$
n-1	1	1	1
n	n-1	n-1	n

instances which fit in memory, the average root gap is 3.22%, and the maximum is 57.89%.

Evidently, the root gap is typically very small and in fact $\omega(1, C^U, C^L) \geq 0.5\,\mathrm{OPT}$ in all tested instances. However, the worst case approximation factor of ω is actually unbounded. Table 4 describes a family of instances with n items for which $\mathrm{OPT} = (n-1)\omega(1, C^U, C^L)$. Despite this, branch-and-bound is able to solve these instances using only $O(n)$ nodes. On the other hand, it is interesting to note that the subset-sum instances typically have very small root gaps, but solving them to optimality is evidently very hard.

Evidently, the main disadvantage of our algorithm is its high memory usage. In our solver, we use a few simple tricks to reduce memory usage slightly: when possible, we store the DP table entries as 16-bit integers, and we avoid computing table entries for capacity values which cannot be seen in any feasible solution. Other optimizations to reduce memory usage are certainly possible as well, such as a DP-with-lists type approach, but we have not implemented this.

Table 5. Statistics from our solver, for instances from the literature.

Group	RootTime	OptTime	ProofTime	NumNodes	RootGap%
uncorrelated	0.3	0.03	0.02	12,053.23	5.4
weak correlated	0.25	0.04	0.03	7,130.98	2.04
strong correlated	0.26	0.03	0.11	571,473.6	2.98
inverse strong corr.	0.37	0.04	0.73	4,751,888.08	0.39
almost strong corr.	0.24	0.03	0.03	735.76	2.99
subset-sum	0.08	0.05	12.22	104,067,718.79	0.02
even-odd subset-sum	0.07	0.04	6.42	61,246,957.64	0.02
even-odd strong corr.	0.26	0.03	0.37	2,551,213.42	2.95
similar weight uncorr.	0.05	0.05	0.05	0	0

Table 6. Statistics from our solver, for new instances.

Group	RootTime	OptTime	ProofTime	Nodes	RootGap%
uncorrelated	0.86	0.27	0.17	73,616.37	9.86
lower subset-sum	0.95	0.8	3.43	11,293,085.12	1.27
upper subset-sum	0.79	0	0	182.99	5.75
both subset-sum	0.16	3.02	7.68	117,664,085.82	3.44
equal weights	0.7	0.02	0.02	14.65	1.64
large capacity	2.31	0	0	29.82	10.15

5   Conclusion

We presented a new combinatorial algorithm for solving BKP that is on average 4.5 times better, and achieves up to 3 orders of magnitude improvement in runtime over the performance of the previous state-of-the-art algorithm, DCS. The only disadvantage of our algorithm that we identified in computational testing is the high memory usage. Because of this, if memory is limited and time is not a concern, it may be a better idea to use DCS. However, if there is any correlation between the lower-level weights and profits, DCS is unlikely to solve the instance in any reasonable amount of time, so it is preferable to use our algorithm on a machine with a large amount of memory, and/or to use additional implementation tricks to reduce the memory usage.

For future work, it would be of interest to investigate whether our lower bound can be strengthened further (say, to an $O(1)$-approximation). We expect that it would be straightforward to generalize this work to the multidimensional variant of BKP (i.e., where there are multiple knapsack constrains at each level), although the issues with high memory usage would likely become worse in this setting. It may also be straightforward to apply this technique to covering interdiction problems. Beyond this, we suspect that a similar lower bound and search algorithm can be used to efficiently solve a variety of interdiction problems.

References

1. Caprara, A., Carvalho, M., Lodi, A., Woeginger, G.J.: A study on the computational complexity of the bilevel knapsack problem. SIAM J. Optim. **24**(2), 823–838 (2014)
2. Caprara, A., Carvalho, M., Lodi, A., Woeginger, G.J.: Bilevel knapsack with interdiction constraints. INFORMS J. Comput. **28**(2), 319–333 (2016)
3. Chen, L., Wu, X., Zhang, G.: Approximation algorithms for interdiction problem with packing constraints. arXiv preprint arXiv:2204.11106 (2022)
4. Della Croce, F., Scatamacchia, R.: An exact approach for the bilevel knapsack problem with interdiction constraints and extensions. Math. Program. **183**(1), 249–281 (2020)
5. Dempe, S.: Bilevel optimization: theory, algorithms, applications and a bibliography. In: Dempe, S., Zemkoho, A. (eds.) Bilevel Optimization. SOIA, vol. 161, pp. 581–672. Springer, Cham (2020). https://doi.org/10.1007/978-3-030-52119-6_20
6. DeNegre, S.: Interdiction and discrete bilevel linear programming, Ph. D. thesis, Lehigh University (2011)
7. Dolan, E.D., Moré, J.J.: Benchmarking optimization software with performance profiles. Math. Program. **91**(2), 201–213 (2002)
8. Fischetti, M., Ljubić, I., Monaci, M., Sinnl, M.: A new general-purpose algorithm for mixed-integer bilevel linear programs. Oper. Res. **65**(6), 1615–1637 (2017)
9. Fischetti, M., Ljubic, I., Monaci, M., Sinnl, M.: Interdiction games and monotonicity, with application to knapsack problems. INFORMS J. Comput. **31**, 390–410 (2019)
10. Fischetti, M., Monaci, M., Sinnl, M.: A dynamic reformulation heuristic for generalized interdiction problems. Eur. J. Oper. Res. **267**, 40–51 (2018)
11. Fontan, F.: Knapsack solver (Github repository). https://github.com/fontanf/knapsacksolver (2017)
12. Kleinert, T., Labbé, M., Ljubić, I., Schmidt, M.: A survey on mixed-integer programming techniques in bilevel optimization. EURO J. Comput. Optimiz. **9**, 100007 (2021)
13. Lozano, L., Bergman, D., Cire, A.A.: Constrained shortest-path reformulations for discrete bilevel and robust optimization. arXiv preprint arXiv:2206.12962 (2022)
14. Martello, S., Pisinger, D., Toth, P.: Dynamic programming and strong bounds for the 0–1 knapsack problem. Manage. Sci. **45**(3), 414–424 (1999)
15. Pisinger, D.: An expanding-core algorithm for the exact 0–1 knapsack problem. Eur. J. Oper. Res. **87**(1), 175–187 (1995)
16. Pisinger, D.: Where are the hard knapsack problems? Comput. Oper. Res. **32**, 2271–2284 (2005)
17. Smith, J.C., Song, Y.: A survey of network interdiction models and algorithms. Eur. J. Oper. Res. **283**(3), 797–811 (2020)
18. Tahernejad, S., Ralphs, T.K., DeNegre, S.T.: A branch-and-cut algorithm for mixed integer bilevel linear optimization problems and its implementation. Math. Program. Comput. **12**(4), 529–568 (2020)
19. Tang, Y., Richard, J.P.P., Smith, J.C.: A class of algorithms for mixed-integer bilevel min-max optimization. J. Global Optim. **66**, 225–262 (2016)

Multiplicative Auction Algorithm for Approximate Maximum Weight Bipartite Matching

Da Wei Zheng[1(✉)] and Monika Henzinger[2]

[1] University of Illinois Urbana-Champaign, Urbana, IL, USA
dwzheng2@illinois.edu
[2] IST Austria, Klosterneuburg, Austria
monika.henzinger@ista.ac.at

Abstract. We present an *auction algorithm* using multiplicative instead of constant weight updates to compute a $(1 - \varepsilon)$-approximate maximum weight matching (MWM) in a bipartite graph with n vertices and m edges in time $O(m\varepsilon^{-1}\log(\varepsilon^{-1}))$, matching the running time of the linear-time approximation algorithm of Duan and Pettie [JACM '14]. Our algorithm is very simple and it can be extended to give a dynamic data structure that maintains a $(1 - \varepsilon)$-approximate maximum weight matching under (1) one-sided vertex deletions (with incident edges) and (2) one-sided vertex insertions (with incident edges sorted by weight) to the other side. The total time time used is $O(m\varepsilon^{-1}\log(\varepsilon^{-1}))$, where m is the sum of the number of initially existing and inserted edges.

1 Introduction

Let $G = (U \cup V, E)$ be an edge-weighted bipartite graph with $n = |U \cup V|$ vertices and $m = |E|$ edges where each edge $uv \in E$ with $u \in U$ and $v \in V$ has a non-negative weight $w(uv)$.

The *maximum weight matching* (MWM) problem asks for a matching $M \subseteq E$ that attains the largest possible weight $w(M) = \sum_{uv \in M} w(uv)$. This paper will focus on approximate solutions to the MWM problem. More specifically, if we let M^* denote a maximum weight matching of G, our goal is to find a matching M such that $w(M) \geq (1 - \varepsilon)w(M^*)$ for any small constant $\varepsilon > 0$.

Matchings are a very well studied problem in combinatorial optimization. Kuhn [13] in 1955 published a paper that started algorithmic work in matchings, and presented what he called the "Hungarian algorithm" which he attributed the work to Kőnig and Egerváry. Munkres [15] showed that this algorithm runs in $O(n^4)$ time. The running time for computing the exact MWM has been improved many times since then. Recently this year, Chen et al. [6] showed that it was possible to solve the more general problem of max flow in $O(m^{1+o(1)})$ time.

A. Del Pia and V. Kaibel (Eds.): IPCO 2023, LNCS 13904, pp. 453–465, 2023.
https://doi.org/10.1007/978-3-031-32726-1_32

For $(1 - \varepsilon)$-approximation algorithms for MWM in bipartite graphs, Gabow and Tarjan in 1989 showed an $O(m\sqrt{n}\log(n/\varepsilon))$ algorithm. Since then there were a number of results for different running times and different approximation ratios. The current best approximate algorithm is by Duan and Pettie [8] which computes a $(1-\varepsilon)$-approximate maximum weight matching in $O(m\varepsilon^{-1}\log(\varepsilon^{-1}))$ time with a scaling algorithm. We defer to their work for a more thorough survey of the history on the MWM problem.

We show in our work that the auction algorithm for matchings using multiplicative weights can give a $(1 - \varepsilon)$-approximate maximum weight matching with a running time of $O(m\varepsilon^{-1}\log(\varepsilon^{-1}))$ for bipartite graphs. This matches the best known running time of Duan and Pettie [8]. However, in comparison to their rather involved algorithm, our algorithm is simple and only uses elementary data structures. Furthermore, we are able to use properties of the algorithm to support two dynamic operations, namely one where vertices on one side are deleted and vertices on the other side are added with all incident edges given in sorted order of weight.

1.1   Dynamic Matching Algorithms

Dynamic Weighted Matching. There has been a large body of work on dynamic matching and many variants of the problem have been studied, e.g., the maximum, maximal, as well as α-approximate setting for a variety of values of α, both in the weighted as well as in the unweighted setting. See [10] for a survey of the current state of the art for the fully dynamic setting. We just mention here a few of the most relevant prior works. For any constant $\delta > 0$ there is a conditional lower bound based on the OMv conjecture that shows that any dynamic algorithm that returns the *exact* value of a maximum cardinality matching in a bipartite graph with polynomial preprocessing time cannot take time $O(m^{1-\delta})$ per query and $O(m^{1/2-\delta})$ per edge update operation [11]. For *general weighted* graphs Gupta and Peng [9] gave the first algorithm in the *fully dynamic* setting with edge insertions and deletions to maintain a $(1 - \varepsilon)$-approximate matching in $O(\varepsilon^{-1}\sqrt{m}\log w_{max})$ time, where the edges fall into the range $[1, w_{max}]$.

Vertex Updates. By vertex update we refer to updates that are vertex insertion (resp. deletion) that also inserts (resp. deletes) all edges incident to the vertex. There is no prior work on maintaining matchings in weighted graphs under vertex updates. However, vertex updates in the *unweighted bipartite* setting has been studied. Bosek et al. [4] gave an algorithm that maintains the $(1 - \varepsilon)$-approximate matching when vertices of one side are deleted in $O(\varepsilon^{-1})$ amortized time per changed edge. The algorithm can be adjusted to the setting where vertices of one side are inserted in the same running time, but it cannot handle both vertex insertions and deletions. Le et al. [14] gave an algorithm for maintaining a *maximal* matching under vertex updates in constant amortized time per changed edge. They also presented an $e/(e - 1) \approx 1.58$ approximate algorithm for maximum matchings in an unweighted graph when vertex updates are only allowed on one side of a bipartite graph.

We give the first algorithm to maintain a $(1 - \varepsilon)$-approximate maximum weight matching where vertices can undergo vertex insertions on one side *and* vertex deletions on the other side in $O(\varepsilon^{-1} \log(\varepsilon^{-1}))$ amortized time per edge inserted.

1.2   Linear Program for MWM

The MWM problem can be expressed as the following *linear program* (LP) where the variable x_{uv} denotes whether the edge uv is in the matching. It is well known [17] that the below LP is integral, that is the optimal solution has all variables $x_{uv} \in \{0, 1\}$.

$$\max \quad \sum_{uv \in E} w(uv)x_{uv}$$

$$s.t. \quad \sum_{v \in N(u)} x_{uv} \leq 1 \qquad \forall u \in U$$

$$\sum_{u \in N(u)} x_{uv} \leq 1 \qquad \forall v \in V$$

$$x_{uv} \geq 0 \qquad \forall uv \in E$$

We can also consider the dual problem that aims to find dual weights y_u and y_v for every vertex $u \in U$ and $v \in V$ respectively.

$$\min \quad \sum_{u \in U} y_u + \sum_{v \in V} y_v$$

$$s.t. \quad y_u + y_v \geq w(uv) \qquad \forall uv \in E$$

$$y_u \geq 0 \qquad \forall u \in U$$

$$y_v \geq 0 \qquad \forall v \in V$$

1.3   Multiplicative Weight Updates for Packing LPs

Packing LPs are LPs of the form $\max\{c^T x \mid Ax \leq b\}$ for $c \in \mathbb{R}^n_{\geq 0}$, $b \in \mathbb{R}^m_{\geq 0}$ and $A \in \mathbb{R}^{n \times m}_{\geq 0}$. The LP for MWM is a classical example of a packing LP. The *multiplicative weight update method* (MWU) has been investigated extensively to provide faster algorithms for finding approximate solutions[1] to packing LPs [1,5,12,16,18,19]. Typically the running times for solving these LPs have a dependence on ε of ε^{-2}, e.g. the algorithm of Koufogiannakis and Young [12] would obtain a running time of $O(m\varepsilon^{-2} \log n)$ when applied to the matching LP.

The fastest multiplicative weight update algorithm for solving packing LPs by Allen-Zhu and Orecchia [1] would obtain an $O(m\varepsilon^{-1} \log n)$ running time for

---

[1] By *approximate solution* we mean a possibly fractional assignments of variables that obtains an approximately good LP objective. If we find such an approximate solution to MWM, fractional solutions need to be rounded to obtain an actual matching.

MWM. Very recently, work by Battacharya, Kiss, and Saranurak [3] extended the MWU for packing LPs to the *partially dynamic setting*. When restricted to the MWM problem means the weight of edges either only increase or only decrease. However as packing LPs are more general than MWM, these algorithms are significantly more complicated and are slower by $\log n$ factors (and worse dependence on ε for [3]) when compared to our static and dynamic algorithms.

We remark that our algorithm, while it uses multiplicative weight updates, is unlike typical MWU algorithms as it has an additional monotonicity property. We only increase dual variables on one side of the matching, and only (implicitly) decrease dual variables on the other side.

1.4   Auction Algorithms

Auction algorithms are a class of primal dual algorithms for solving the MWM problem that view U as a set of *goods* to be sold, V as a set of *buyers*. The goal of the auction algorithm is to find a welfare-maximizing allocation of goods to buyers. The algorithm is commonly attributed Bertsekas [2], as well as to Demange, Gale, and Sotomayor [7].

An auction algorithm initializes the prices of all the goods $u \in U$ with a price $y_u = 0$ (our choice of y_u is intentional, as prices correspond directly to dual variables), and has buyers initially *unallocated*. For each buyer $v \in V$, the *utility* of that buyer upon being allocated $u \in U$ is $util(uv) = w(uv) - y_u$. The auction algorithm proceeds by asking an unallocated buyer $v \in V$ for the good they desire that maximizes their utility, i.e. for $u_v = \arg\max_{u \in N(v)} util(uv)$. If $util(u_v v) < 0$, the buyer remains unallocated. Otherwise the algorithm allocates u_v to v, then increases the price y_u to $y_u + \varepsilon$. The algorithm terminates when all buyers are either allocated or for every unallocated buyer v, it holds that $util(u_v v) < 0$. If the maximum weight among all the edges is w_{max}, then the auction algorithm terminates after $O(n\varepsilon^{-1}w_{max})$ rounds and outputs a matching that differs from the optimal by an additive factor of at most $n\varepsilon$.

1.5   Our Contribution

We present the following modification of the auction algorithm:

When v is allocated u, increase y_u to $y_u + \varepsilon \cdot util(uv)$ instead of $y_u + \varepsilon$.

Note that this decreases $util(v)$ by a factor of $(1 - \varepsilon)$ and, thus, we will call algorithms with this modification *multiplicative auction algorithms*. Surprisingly, we were not able to find any literature on this simple modification. Changing the constant additive weight update to a multiplicative weight update has the effect of taking much larger steps when the weights are large, and so we are able to show that the algorithm can have no dependence on the size of the weights. In fact, we are able to improve the running time to $O(m\varepsilon^{-1}\log(\varepsilon^{-1}))$, the same as the fastest known matching algorithm of Duan and Pettie [8]. While the algorithm of [8] has the advantage that it works for general graphs and ours is limited to bipartite graphs, our algorithm is simpler as it avoids the scaling algorithm framework and is easier to implement.

Theorem 1. *Let $G = (U \cup V, E)$ be a weighted biparitite graph. There is a multiplicative auction algorithm running in time $O(m\varepsilon^{-1}\log(\varepsilon^{-1}))$ that finds a $(1 - \varepsilon)$-approximate maximum weight matching of G.*

Furthermore, it is straightforward to extend our algorithm to a setting where *edges are deleted* and *vertices on one side are added with all incident edges given in sorted order of weight*. When the inserted edges are not sorted by weight, the running time per inserted edge increases by an additive term of $O(\log n)$ to sort all incident inserted edges.

Theorem 2. *Let $G = (U \cup V, E)$ be a weighted biparitite graph. There exists a dynamic data structure that maintains a $(1 - \varepsilon)$-approximate maximum weight matching of G and supports the following operations:*

(1) Deleting a vertex in U
(2) Adding a new vertex into V along with all its incident edges sorted by weight

in total time $O(m\varepsilon^{-1}\log(\varepsilon^{-1}))$, where m is sum of the number of initially existing, and inserted edges.

2   The Static Algorithm

We assume that the algorithm is given as input some fixed $0 < \varepsilon' < 1$.

Notation For sake of notation let $N(u) = \{v \in V \mid uv \in E\}$ be the set of neighbors of $u \in U$ in G, and similarly for $N(v)$ for $v \in V$.

Preprocessing of the Weights. Let $w_{max} > 0$ be the maximum weight edge of E. For our static auction algorithm we may ignore any edge $uv \in E$ of weight less than $\varepsilon' \cdot w_{max}/n$ as $w(M^*) \geq w_{max}$ as taking n of these small weight edges would not even contribute $\varepsilon' \cdot w(M^*)$ to the matching. Thus, we only consider edges of weight at least $\varepsilon' \cdot w_{max}/n$, which allows us to rescale all edge weights by dividing them by $\varepsilon' \cdot w_{max}/n$. As a result we can assume (by slight abuse of notation) in the following that the minimum edge weight is 1 and the largest edge weight w_{max} equals n/ε'. Furthermore, since we only care about approximations, we will also round down all edge weights to the nearest power of $(1 + \varepsilon)$ for some $\varepsilon < \varepsilon'/2$ and, again by slight abuse of notation, we will use w to denote these edge weights. Formally to round, we define $\text{ILOG}(x) = \lfloor \log_{1+\varepsilon}(x) \rfloor$ and $\text{ROUND}(x) = (1 + \varepsilon)^{\text{ILOG}(x)}$.

Let $k_{max} = \text{ILOG}(w_{max}) = \text{ILOG}(n/\varepsilon) = O(\varepsilon^{-1}\log(n/\varepsilon))$. Let k_{min} be the smallest integer such that $(1 + \varepsilon)^{-k_{min}} \leq \varepsilon$. Observe that as $\log(1 + \varepsilon) \leq \varepsilon$ for $0 \leq \varepsilon \leq 1$ it holds that

$$k_{min} \geq \frac{\log(\varepsilon^{-1})}{\log(1 + \varepsilon)} \geq \varepsilon^{-1}\log(\varepsilon^{-1}).$$

Thus we see that $k_{min} = \Theta(\varepsilon^{-1}\log(\varepsilon^{-1}))$.

Algorithm. The algorithm first builds for every $v \in V$ a list Q_v of pairs (i, uv) for each edge uv and each value i with $-k_{min} \leq i \leq j_{uv} = \text{ILOG}(w_{uv})$ and then sorts Q_v by decreasing value of i. After, it calls the function $\text{MATCHR}(v)$ on every $v \in V$. The function $\text{MATCHR}(v)$ matches v to the item that maximizes its utility and updates price according to our multiplicative update rule. While matching v, another vertex v' originally matched to v may become unmatched. If this happens, $\text{MATCHR}(v')$ is called immediately after $\text{MATCHR}(v)$.

Algorithm 2.1: MULTIPLICATIVEAUCTION($G = (U \cup V, E)$)

$M \leftarrow \emptyset.$
$y_u \leftarrow 0$ for all $u \in U.$
$j_v \leftarrow k_{max}$ for all $v \in V$ # This is only used in the analysis
$Q_v \leftarrow \emptyset$ for all $v \in V.$
For $v \in V$:

1. For $u \in N(v)$:
    (a) $j_{uv} \leftarrow \text{ILOG}(w(uv))$
    (b) $j_v \leftarrow \max\{j_v, j_{uv}\}$
    (c) For i from j_{uv} to $-k_{min}$:
        i. Insert the pair (i, uv) into Q_v.
2. Sort all $(i, uv) \in Q_v$ so elements are in non-increasing order of i.

For $v \in V$:

1. $\text{MATCHR}(v)$.

Return M.

MATCHR(v)

While Q_v is not empty:

1. $(j, uv) \leftarrow$ the first element of Q_v, and remove it from Q_v.
2. $j_v \leftarrow j$ # This is only used in the analysis
3. $util(uv) \leftarrow w(uv) - y_u$
4. If $util(uv) \geq (1 + \varepsilon)^j$:
    (a) $y_u \leftarrow y_u + \varepsilon \cdot (util(uv))$ # $util(uv) \leftarrow (1 - \varepsilon) \cdot util(uv)$
    (b) If u was matched to v' in M:
        – Remove (u, v') from M
        – Add (u, v) to M
        – $\text{MATCHR}(v')$
    (c) Else:
        – Add (u, v) to M
        – Return

Data Structure. We store for each vertex $v \in V$ the list Q_v as well as its currently matched edge if it exists. In the pseudocode below we keep for each vertex v a value j_v corresponding to the highest weight threshold $(1 + \varepsilon)^{j_v}$ that we will consider. This value is only needed in the analysis.

Running Time. The creation and sorting of the lists Q_v takes time $O(|N(v)|(k_{max} + k_{min}))$ if we use bucket sort as there are only $k_{max} + k_{min}$ distinct weights. The running time of all calls to MATCHR(v) is dominated by the size of Q_v, as each iteration in MATCHR(v) removes an element of Q_v and takes $O(1)$ time. Thus, the total time is $O\left(\sum_{v \in V} |N(v)|(k_{max} + k_{min})\right) = O(m(k_{max} + k_{min})) = O(m\varepsilon^{-1} \log(n/\varepsilon))$.

Invariants Maintained by the Algorithm. Consider the following invariants maintained throughout by the algorithm:

Invariant 1. *For all $v \in V$, and all $u \in N(v)$, $util(uv) < (1 + \varepsilon)^{j_v+1}$.*

Proof. This clearly is true at the beginning, since j_v is initialized to $\max_{u \in N(v)} j_{uv}$, and

$$util(uv) = w(uv) < (1 + \varepsilon)^{j_{uv}+1}.$$

As the algorithm proceeds, $util(uv)$ which equals $w(uv) - y_u$ only decreases as y_u only increases. Thus, we only have to make sure that the condition holds whenever j_v decreases. The value j_v only decreases from some value, say $j + 1$, to a new value j, in MATCHR(v) and when this happens Q_v does not contain any pairs (j', uv) with $j' > j$ anymore. Thus, there does not exist a neighbor u of v with $util(uv) \geq (1 + \varepsilon)^{j+1}$. It follows that when j_v decreases to j for all $u \in N(v)$ it holds that $util(uv) < (1 + \varepsilon)^{j_v+1}$.

Invariant 2. *If $uv \in M$, then for all other $u' \in N(v)$, $util(uv) \geq (1 - 2\varepsilon) \cdot util(u'v)$.*

Proof. When v was matched to u, right before we updated y_u, we had that $(1 + \varepsilon)^{j_v} \leq util(uv)$ and, by Invariant 1, $util(u'v) \leq (1 + \varepsilon)^{j_v+1}$. Thus, $(1 + \varepsilon)util(uv) \geq util(u'v)$. The update of y_u decreases y_u by $\varepsilon \cdot util(uv)$, which decreases $util(uv)$ by a factor of $(1 - \varepsilon)$, but does not affect $util(u'v)$. Thus we have now that:

$$util(uv) \geq (1 - \varepsilon)(1 + \varepsilon)^{-1} \cdot util(u'v) \geq (1 - 2\varepsilon) \cdot util(u'v).$$

Invariant 3. *If $u \in U$ is not matched, then $y_u = 0$. If $uv \in M$, then $y_u > 0$.*

Proof. If u is never matched, we never increment y_u, so it stays 0. The algorithm increments y_u by $\varepsilon \cdot util(uv) > 0$ when we add uv into the matching M.

Invariant 4. *For all $v \in V$ for which MATCHR(v) was called at least once, either v is matched, or Q_v is empty.*

Proof. MATCHR(v) terminates (i) after it matches v and recurses or (ii) if Q_v is empty. It is possible that for some other $v' \in V$ with $v' \neq v$, that v becomes temporarily unmatched during MATCHR(v'), but we would immediately call MATCHR(v) to rematch v.

Approximation Factor. We will show the approximation factor of the matching M found by the algorithm by primal dual analysis. We remark that it is possible to show this result purely combinatorially as well which we include in Appendix A, as it may be of independent interest. We will show that this M and a vector y satisfy the complementary slackness condition up to a $1 \pm \varepsilon$ factor, which implies the approximation guarantee. This was proved by Duan and Pettie [8] (the original lemma was for general matchings, we have specialized it here to bipartite matchings).

Lemma 1 (Lemma 2.3 of [8]). *Let M be a matching and let y be an assignment of the dual variables. Suppose y is a valid solution to the LP in the following approximate sense: For all $uv \in E, y_u + y_v \geq (1 - \varepsilon_0) \cdot w(uv)$ and for all $e \in M$, $y_u + y_v \leq (1 + \varepsilon_1) \cdot w(uv)$. If the y-values of all unmatched vertices are zero, then M is a $\left((1 + \varepsilon_1)^{-1}(1 - \varepsilon_0)\right)$-approximate maximum weight matching.*

This lemma is enough for us to prove the approximation factor of our algorithm.

Lemma 2. MULTIPLICATIVEAUCTION$(G = (U \cup V, E))$ *outputs a $(1 - \varepsilon')$-approximate maximum weight matching of the bipartite graph G.*

Proof. Let $\varepsilon > 0$ be a parameter depending on ε' that we will choose later. We begin by choosing an assignment of the dual variables y_u for $u \in U$ and y_v for $v \in V$. Let all y_u's be those obtained by the algorithm for $u \in U$. For $v \in V$, let $y_v = 0$ if v is not matched in M and $y_v = util(uv) = w(uv) - y_u$ if v is matched to u in M. By Invariant 3 all unmatched vertices $u \in U$ have $y_u = 0$.

Observe that for $uv \in M$ we have $y_u + y_v = util(uv)$. It remains to show that for $uv \notin M$ we have that $y_u + y_v \geq (1 - \varepsilon_0)w(uv)$ for some $\varepsilon_0 > 0$. First we consider if v is unmatched, so $y_v = 0$. Since v is unmatched, by Invariant 4 then for all $u \in N(v)$, we must have $util(uv) < (1 + \varepsilon)^{-k_{min}} \leq \varepsilon$. Since we rescaled weights so that $w(uv) \geq 1$, we know that $util(uv) < \varepsilon \leq \varepsilon \cdot w(uv)$. Furthermore, observe that as $y_u = w(uv) - util(uv)$ by definition of utility, it follows that:

$$y_u + y_v = y_u = w(uv) - util(uv) > (1 - \varepsilon)w(uv). \tag{1}$$

Now we need to consider if v was matched to some vertex $u' \neq u$. To do so we use Invariant 2:

$$
\begin{aligned}
y_u + y_v &= y_u + util(u'v) & &\text{By definition of } y \\
&\geq y_u + (1 - 2\varepsilon) \cdot util(uv) & &\text{By Invariant 2} \\
&= y_u + (1 - 2\varepsilon) \cdot (w(uv) - y_u) & &\text{By definition of } util \\
&\geq (1 - 2\varepsilon)w(uv) + 2\varepsilon \cdot y_u & & \\
&\geq (1 - 2\varepsilon)w(uv) & &\text{Since } y_u \geq 0
\end{aligned}
$$

Thus we have satisfied Lemma 1 with $\varepsilon_0 = 2\varepsilon$ and $\varepsilon_1 = 0$. Setting $\varepsilon = \varepsilon'/2$ gives us a $(1 - \varepsilon')$-approximate maximum weight matching.

Thus we have shown the following result that is weaker than what we have set out to prove by a factor of $\log(n\varepsilon^{-1})$ that we will show how to get rid of in the next section.

Theorem 3. *Let* $G = (U \cup V, E)$ *be a weighted biparitite graph. There exists a multiplicative auction algorithm running in time* $O(m\varepsilon^{-1}\log(n\varepsilon^{-1}))$ *that finds a* $(1 - \varepsilon)$-*approximate maximum weight matching of* G.

2.1   Improving the Running Time

To improve the running time to $O(m\varepsilon^{-1}\log(\varepsilon^{-1}))$, we observe that all we actually need for Lemma 2 in Equation (1) is that $util(uv) \leq \varepsilon \cdot w(uv)$. Recall that $j_{uv} = \text{ILOG}(w(uv))$. Thus it suffices if we change line (b) in MULTIPLICATIVEAUCTION to range from j_{uv} to $j_{uv} - k_{min}$, since:

$$(1 + \varepsilon)^{j_{uv} - k_{min}} = (1 + \varepsilon)^{-k_{min}} \cdot (1 + \varepsilon)^{j_{uv}} \leq \varepsilon \cdot w(uv).$$

This change implies that we insert $O(k_{min}|N(v)|)$ items into Q_v for every $v \in V$. However, sorting Q_v for every vertex individually, even with bucket sort, would be too slow. We will instead perform one bucket sort on all the edges, then go through the weight classes in decreasing order to insert the pairs into the corresponding Q_v. We explicitly give the pseudocode below as MULTIPLICATIVEAUCTION+.

Algorithm 2.2: MULTIPLICATIVEAUCTION+$(G = (U \cup V, E))$

$M \leftarrow \emptyset$.
$y_u \leftarrow 0$ for all $u \in U$.
$Q_v \leftarrow \emptyset$ for all $v \in V$.
$L_i \leftarrow \emptyset$ for all i from $-k_{min}$ to k_{max}.
For $uv \in E$:

1. $j_{uv} \leftarrow \text{ILOG}(w(uv))$
2. For i from j_{uv} to $j_{uv} - k_{min}$:
   (a) Insert the pair (i, uv) into L_i.

For i from k_{max} to $-k_{min}$:

1. For all $(i, uv) \in L_i$:
   (a) Insert the pair (i, uv) to the back of Q_v.

For $v \in V$:

1. MATCHR(v).

Return M.

New Runtime. Bucket sorting all mk_{min} pairs and initializing the sorted Q_v for all $v \in V$ takes total time $O(mk_{min} + (k_{max} + k_{min})) = O(m\varepsilon^{-1}\log(\varepsilon^{-1}))$. The total amount of work done in MATCHR(v) for a vertex $v \in V$ is $O(|N(v)|k_{min})$ which also sums to $O(m\varepsilon^{-1}\log(\varepsilon^{-1}))$. Thus we get our desired running time and have proven our main theorem that we restate here.

Theorem 1. *Let $G = (U \cup V, E)$ be a weighted biparitite graph. There is a multiplicative auction algorithm running in time $O(m\varepsilon^{-1}\log(\varepsilon^{-1}))$ that finds a $(1-\varepsilon)$-approximate maximum weight matching of G.*

3   Dynamic Algorithm

There are many monotonic properties of our static algorithm. For instance, for all $u \in U$, the y_u values strictly increase. As another example, for all $v \in V$, the value of j_v strictly decreases. These monotonic properties allow us to extend MULTIPLICATIVEAUCTION+ to a dynamic setting with the following operations.

Theorem 2. *Let $G = (U \cup V, E)$ be a weighted biparitite graph. There exists a dynamic data structure that maintains a $(1 - \varepsilon)$-approximate maximum weight matching of G and supports the following operations:*

(1) Deleting a vertex in U
(2) Adding a new vertex into V along with all its incident edges sorted by weight

in total time $O(m\varepsilon^{-1}\log(\varepsilon^{-1}))$, where m is sum of the number of initially existing, and inserted edges.

Type (1) operations: Deleting a vertex in U. To delete a vertex $u \in U$, we can mark u as deleted and skip all edges uv in Q_v for any $v \in V$ in all further computation. If u were matched to some vertex $v \in V$, that is if there exists an edge $uv \in M$, we need to unmatch v and remove uv from M. All our invariants hold except Invariant 4 for the unmatched v. To restore this invariant we simply call MATCHR(v).

Type (2) Operations: Adding a New Vertex to V Along with All Incident Edges. To add a new vertex v to V with ℓ incident edges to $u_1v, ..., u_\ell v$ with $w(u_1v) > \cdots > w(u_\ell v)$, we can create the queue Q_v by inserting the $O(\varepsilon^{-1}\log(\varepsilon^{-1}))$ pairs such that it is non-increasing in the first element of the pair. Afterwards we call MATCHR(v). All invariants hold after doing so.

If the edges are not given in sorted order, we can sort the ℓ edges in $O(\ell \log \ell)$ time, or in $O(\ell + \varepsilon^{-1}\log(w(u_1v)/w(u_\ell v)))$ time by bucket sort.

Acknowledgements. The first author thanks to Chandra Chekuri for useful discussions about this paper.

This project has received funding from the European Research Council (ERC) under the European Union's Horizon 2020 research and innovation programme (Grant agreement No. 101019564 "The Design of Modern Fully Dynamic Data Structures (MoDynStruct)" and from the Austrian Science Fund (FWF) project "Fast Algorithms for a Reactive Network Layer (ReactNet)", P 33775-N, with additional funding from the *netidee SCIENCE Stiftung*, 2020–2024.

A   Combinatorial proof of Lemma 2

We start with a simple lemma.

Lemma 3. *Let $G = (U \cup V, E)$ be a weighted bipartite graph. Let M be the matching found by* MULTIPLICATIVEAUCTION+(G) *for $\varepsilon > 0$, and M' be any other matching. Then for any alternating path, i.e. a set of edges of the form $u_1 v_1$, $u_2 v_1$, $u_2 v_2$, ..., $u_k v_k$, $u_{k+1} v_k$ with all edges of $u_i v_i \in M'$ and $u_{i+1} v_i \in M$, we have that:*

$$(1 - 2\varepsilon) \cdot \sum_{i=1}^{k} w(u_i v_i) \leq \sum_{i=1}^{k} w(u_{i+1} v_i) + (1 - 2\varepsilon) \cdot y_{u_1} - y_{u_{k+1}}$$

Proof. By Invariant 2, for all i from 1 to k, since M matched v_{i+1} to u_i we have that:

$$(1 - 2\varepsilon) util(u_i v_i) \leq util(u_i v_{i+1})$$

Adding all such equations together we get

$$(1 - 2\varepsilon) \cdot \sum_{i=1}^{k} util(u_i v_i) \leq \sum_{i=1}^{k} util(u_i v_{i+1})$$

$$(1 - 2\varepsilon) \cdot \sum_{i=1}^{k} \left(w(u_i v_i) - y_{u_i} \right) \leq \sum_{i=1}^{k} \left(w(u_{i+1} v_i) - y_{u_{i+1}} \right)$$

$$(1 - 2\varepsilon) \cdot \left(\sum_{i=1}^{k} w(u_i v_i) - \sum_{i=1}^{k} y_{u_i} \right) \leq \sum_{i=1}^{k} w(u_{i+1} v_i) - \sum_{i=1}^{k} y_{u_{i+1}}$$

$$(1 - 2\varepsilon) \cdot \sum_{i=1}^{k} w(u_i v_i) \leq \sum_{i=1}^{k} w(u_{i+1} v_i) + (1 - 2\varepsilon) \cdot y_{u_1} - y_{u_{k+1}}$$

Theorem 3. *Let $G = (U \cup V, E)$ be a weighted bipartite graph and $\varepsilon' > 0$ be an input parameter. Let M be the matching found by* MULTIPLICATIVEAUCTION+(G) *with $\varepsilon = \varepsilon'/2$. M is a $(1 - \varepsilon')$-approximate maximum weight matching of the bipartite graph G.*

Proof. Let M^* be a maximum weight matching of G. Consider the symmetric difference of M with M^*. It consists of paths and and even cycles. It suffices to show that the weight obtained by M on the path or even cycle is at least $(1 - \varepsilon)$ the weight of M^*. We consider the following cases:

1. Consider any even cycle $u_1 v_1$, $u_2 v_1$, $u_2 v_2$, ..., $u_k v_k$, $u_1 v_k$ with $u_i v_i \in M^*$ for all $i = 1, ..., k$ and the other edges in M. Applying Lemma 3 with $u_{k+1} = u_1$, and by Invariant 3 $y_{u_1} > 0$ as u_1 is matched gives:

$$(1 - 2\varepsilon) \cdot \sum_{i=1}^{k} w(u_i v_i) \leq \sum_{i=1}^{k} w(u_{i+1} v_i) + (1 - 2\varepsilon) y_{u_1} - y_{u_1} < \sum_{i=1}^{k} w(u_{i+1} v_i).$$

2a. Consider any even length path u_1v_1, u_2v_1, u_2v_2, ..., u_kv_k, $u_{k+1}v_k$ with $u_iv_i \in M^*$ for all $i = 1, ..., k$ and the other edges in M. By Invariant 3 u_1 is unmatched in M so $y_{u_1} = 0$, and u_{k+1} is matched so $y_{u_{k+1}} > 0$. Applying Lemma 3 gives:

$$(1 - 2\varepsilon) \cdot \sum_{i=1}^{k} w(u_iv_i) \le \sum_{i=1}^{k} w(u_{i+1}v_i) + (1 - 2\varepsilon) \cdot 0 - y_{u_{k+1}} < \sum_{i=1}^{k} w(u_{i+1}v_i).$$

2b. Consider any odd length path u_1v_1, u_2v_1, u_2v_2, ..., u_kv_k, $u_{k+1}v_k$, u_{k+1}, v_{k+1} with $u_iv_i \in M^*$ for all $i = 1, ..., k$ and the other edges in M. Since v_k is unmatched in M, we have that $w(u_{k+1}v_{k+1}) - y_{u_{k+1}} = util(u_{k+1}v_{k+1}) \le \varepsilon w(u_{k+1}v_{k+1})$. Rearranging, we get that $y_{u_{k+1}} \ge (1 - \varepsilon)w(u_{k+1}v_{k+1}) > (1 - 2\varepsilon)w(u_{k+1}v_{k+1})$. Adding this equation to Lemma 3, and by Invariant 3 we have $y_{u_1} = 0$, so:

$$(1 - 2\varepsilon) \cdot \sum_{i=1}^{k+1} w(u_iv_i) < \sum_{i=1}^{k} w(u_{i+1}v_i) + (1 - 2\varepsilon) \cdot y_{u_1} = \sum_{i=1}^{k} w(u_{i+1}v_i).$$

2c. Consider any even length path u_0v_1, u_1v_1, u_2v_1, u_2v_2, ..., u_kv_k, $u_{k+1}v_k$ with $u_iv_i \in M^*$ for all $i = 1, ..., k$ and the other edges in M. By Invariant 2, $w(u_1v_0) - y_{u_0} \ge (1 - 2\varepsilon)(w(u_1v_1) - y_{u_1})$, Adding this inequality to what we get when we apply Lemma 3 to the path starting at u_1v_1, and remarking that u_0 and u_{k+1} are matched so Invariant 3 applies gives:

$$(1 - 2\varepsilon) \cdot \sum_{i=1}^{k+1} w(u_iv_i) \le \sum_{i=0}^{k} w(u_{i+1}v_i) - y_{u_0} - y_{u_{k+1}} < \sum_{i=1}^{k} w(u_{i+1}v_i).$$

In all cases we achieve $(1 - 2\varepsilon)$ the weight of M^*. We may choose ε such that $\varepsilon = \varepsilon'/2$, then the theorem holds.

References

1. Allen-Zhu, Z., Orecchia, L.: Nearly linear-time packing and covering LP solvers - achieving width-independence and -convergence. Math. Program. **175**(1–2), 307–353 (2019)
2. Bertsekas, D.P.: A new algorithm for the assignment problem. Math. Program. **21**(1), 152–171 (1981)
3. Bhattacharya, S., Kiss, P., Saranurak, T.: Dynamic algorithms for packing-covering lPS via multiplicative weight updates. CoRR, abs/2207.07519 (2022)
4. Bosek, B., Leniowski, D., Sankowski, P., Zych, A.: Online bipartite matching in offline time. In: 55th IEEE Annual Symposium on Foundations of Computer Science, FOCS 2014, Philadelphia, PA, USA, October 18–21, 2014, pp. 384–393. IEEE Computer Society (2014)
5. Chekuri, C., Quanrud, K.: Randomized MWU for positive lPS. In: Czumaj, A., editor, Proceedings of the Twenty-Ninth Annual ACM-SIAM Symposium on Discrete Algorithms, SODA 2018, New Orleans, LA, USA, January 7–10, 2018, pp. 358–377. SIAM (2018)

6. Chen, L., Kyng, R., Liu, Y.P., Peng, R., Gutenberg, M.P., Sachdeva, S.: Maximum flow and minimum-cost flow in almost-linear time. CoRR, abs/2203.00671 (2022)
7. Demange, G., Gale, D., Sotomayor, M.: Multi-item auctions. J. Polit. Econ. **94**(4), 863–872 (1986)
8. Duan, R., Pettie, S.: Linear-time approximation for maximum weight matching. J. ACM **61**(1), 1:1-1:23 (2014)
9. Gupta, M., Peng, R.: Fully dynamic $(1 + \epsilon)$-approximate matchings. In: 54th Symposium on Foundations of Computer Science, FOCS, pp. 548–557. IEEE Computer Society (2013)
10. Hanauer, K., Henzinger, M., Schulz, C.: Recent advances in fully dynamic graph algorithms (invited talk). In: Aspnes, J., Michail, O., editors, 1st Symposium on Algorithmic Foundations of Dynamic Networks, SAND 2022, March 28–30, 2022, Virtual Conference, volume 221 of LIPIcs, pp. 1:1–1:47. Schloss Dagstuhl - Leibniz-Zentrum für Informatik (2022)
11. Henzinger, M., Krinninger, S., Nanongkai, D., Saranurak, T.: Unifying and strengthening hardness for dynamic problems via the online matrix-vector multiplication conjecture. In Proceedings of the Forty-Seventh Annual ACM Symposium on Theory of Computing, pp. 21–30 (2015)
12. Koufogiannakis, C., Young, N.E.: A nearly linear-time PTAS for explicit fractional packing and covering linear programs. Algorithmica **70**(4), 648–674 (2014)
13. Kuhn, H.W.: The hungarian method for the assignment problem. Naval Research Logistics Quarterly, 2(1–2), 83–97 (1955)
14. Le, H., Milenkovic, L., Solomon, S., Williams, V.V.: Dynamic matching algorithms under vertex updates. In: Braverman, M., editor, 13th Innovations in Theoretical Computer Science Conference, ITCS 2022, January 31 - February 3, 2022, Berkeley, CA, USA, volume 215 of LIPIcs, pp. 96:1–96:24. Schloss Dagstuhl - Leibniz-Zentrum für Informatik (2022)
15. Munkres, J.: Algorithms for the assignment and transportation problems. J. Soc. Ind. Appl. Math. **5**(1), 32–38 (1957)
16. Quanrud, K.: Nearly linear time approximations for mixed packing and covering problems without data structures or randomization. In: Farach-Colton, M., Gørtz, I.L., editors, 3rd Symposium on Simplicity in Algorithms, SOSA 2020, Salt Lake City, UT, USA, January 6–7, 2020, pp. 69–80. SIAM (2020)
17. Schrijver, A., et al.: Combinatorial Optimization: Polyhedra and Efficiency, vol. 24. Springer (2003)
18. Wang, D., Rao, S., Mahoney, M.W.: Unified acceleration method for packing and covering problems via diameter reduction. In: Chatzigiannakis, I., Mitzenmacher, M., Rabani, Y., Sangiorgi, D., editors, 43rd International Colloquium on Automata, Languages, and Programming, ICALP 2016, July 11–15, 2016, Rome, Italy, vol. 55 of LIPIcs, pp. 50:1–50:13. Schloss Dagstuhl - Leibniz-Zentrum für Informatik (2016)
19. Young, N.E.: Nearly linear-time approximation schemes for mixed packing/covering and facility-location linear programs. CoRR, abs/1407.3015 (2014)

A Linear Time Algorithm for Linearizing Quadratic and Higher-Order Shortest Path Problems

Eranda Çela[1]([✉])(iD), Bettina Klinz[1]([✉])(iD), Stefan Lendl[2]([✉])(iD),
Gerhard J. Woeginger[3](iD), and Lasse Wulf[1]([✉])(iD)

[1] Institute of Discrete Mathematics, Graz University of Technology, Graz, Austria
{cela,klinz,wulf}@math.tugraz.at
[2] Institute of Operations and Information Systems, University of Graz, Graz, Austria
stefan.lendl@uni-graz.at
[3] Department of Computer Science, RWTH Aachen, Aachen, Germany

Abstract. An instance of the NP-hard Quadratic Shortest Path Problem (QSPP) is called linearizable iff it is equivalent to an instance of the classic Shortest Path Problem (SPP) on the same input digraph. The linearization problem for the QSPP (LinQSPP) decides whether a given QSPP instance is linearizable and determines the corresponding SPP instance in the positive case. We provide a novel linear time algorithm for the LinQSPP on acyclic digraphs which runs considerably faster than the previously best algorithm. The algorithm is based on a new insight revealing that the linearizability of the QSPP for acyclic digraphs can be seen as a local property. Our approach extends to the more general higher-order shortest path problem.

Keywords: quadratic shortest path problem · higher-order shortest path problem · linearization

1 Introduction

In this paper we consider the linearization problem for nonlinear generalizations of the *Shortest Path Problem (SPP)*, a classic combinatorial optimization problem. An instance of the SPP consists of a digraph $G = (V, A)$, a source vertex $s \in V$, a sink vertex $t \in V$, and a cost function $c \colon A \to \mathbb{R}$, which maps each arc $a \in A$ to its cost $c(a)$. The cost of a simple directed s-t-path P, is given by[1]

$$\mathrm{SPP}(P, c) := \sum_{a \in P} c(a). \tag{1}$$

The goal is to find a simple directed s-t-path in G which minimizes the objective (1). In general it is assumed that there are no circuits of negative weight in G.

G. J. Woeginger—Deceased in April 2022.
[1] We use the same notation for the path P and the set of its arcs.

© The Author(s), under exclusive license to Springer Nature Switzerland AG 2023
A. Del Pia and V. Kaibel (Eds.): IPCO 2023, LNCS 13904, pp. 466–479, 2023.
https://doi.org/10.1007/978-3-031-32726-1_33

Consider now a number $d \in \mathbb{N}$. The *Order-d Shortest Path Problem (SPP$_d$)* takes as input a digraph $G = (V, A)$, a source vertex $s \in V$, a sink vertex $t \in V$, and an order-d arc interaction cost function $q_d \colon \{B \subseteq A : |B| \le d\} \to \mathbb{R}$. Thus q_d assigns a weight to every subset of arcs of cardinality at most d. The cost of a simple directed s-t-path P is given by

$$\text{SPP}_d(P, q_d) := \sum_{S \subseteq P \colon |S| \le d} q_d(S) \,. \tag{2}$$

The goal is to find a simple directed s-t-path in G which minimizes the objective function (2). For $d = 2$ we obtain the *Quadratic Shortest Path Problem (QSPP)* which has already been studied in the literature [2,10,11,18]. For notational convenience we write $\text{QSPP}(P, q)$ for $\text{SPP}_d(P, q_d)$ if $d = 2$.

The QSPP arises in network optimization problems where costs are associated with both single arcs and pairs of arcs. This includes variants of stochastic and time-dependent route planing problems [15,20,21] and network design problems [9,14]. For an overview on applications of the QSPP see [11,18]. We are not aware of any publications for the case $d > 2$.

While the SPP can be solved in polynomial time, the QSPP is an NP-hard problem even for the special case of the adjacent QSPP where the costs of all pairs of non-consecutive arcs are zero [18]. The QSPP is an extremely difficult problem also from the practical point of view. Hu and Sotirov [11] report that a state-of-the-art quadratic solver can solve QSPP instances with up to 365 arcs, while their tailor-made B&B algorithm can solve instances with up to 1300 arcs to optimality within one hour. Instances of the SPP can however be solved in a fraction of a second for graphs with millions of vertices and arcs.

Given the hardness of the QSPP, a research line on this problem has focussed on polynomially solvable special cases which arise if the input graph and/or the cost coefficients have certain specific properties. Rostami et al. [19] have presented a polynomial time algorithm for the adjacent QSPP in acyclic digraphs and in series-parallel graphs. Hu and Sotirov [10] have shown that the QSPP can be solved in polynomial time if the quadratic costs build a nonnegative symmetric product matrix, or if the quadratic costs build a sum matrix and all s-t-paths in G have the same number of arcs.

These two polynomially solvable special cases of the QSPP belong to the larger class of the *linearizable* SPP$_d$ *instances* defined as follows.

Definition 1. *An instance of the SPP$_d$ with an input digraph $G = (V, A)$, a source node s, a sink node t and a cost function q_d is called linearizable if there exists a cost function $c \colon A \to \mathbb{R}$ such that for any simple directed s-t-path P in G the equality $SPP(P, c) = SPP_d(P, q_d)$ holds. A linearizable instance of the QSPP is defined analogously, just replacing $SPP_d(P, q_d)$ by $QSPP(P, q)$.*

The recognition of linearizable QSPP (SPP$_d$) instances, also called *the linearization problem for the QSPP (SPP$_d$)*, abbreviated by LinQSPP (LinSPP$_d$) arises as a natural question. In this problem the task consists of deciding whether a given instance of the QSPP (SPP$_d$) is linearizable and in finding the linear cost function c in the positive case. The notion of linearizable special cases of hard combinatorial optimization problems goes back to Bookhold [1] who introduced it for the quadratic assignment problem (QAP). For symmetric linearizable QAP instances a full characterization has been obtained while only partial results are available for the linearizability of the general QAP, see [3,6–8,13,16,23]. The linearization problem has been studied for several other quadratic combinatorial optimization problems, see [4,22] for the quadratic minimum spanning tree problem, [17] for the quadratic TSP, [5] for the quadratic cycle cover problem and [12] for general binary quadratic programs. Linearizable instances of a quadratic problem can be used to generate lower bounds needed in B&B algorithms. For example, Hu and Sotirov introduce the family of the so-called *linearization-based bounds* [12] for the binary quadratic problem. Each specific bound of this family is based on a set of linearizable instances of the problem. The authors show that well-known bounds from the literature are special cases of the newly introduced bounds. Clearly, fast algorithm for the linearization problem are important in this context.

While LINSPP$_d$ has not been investigated in the literature so far (to the best of our knowledge), the LINQSPP has been subject of investigation in some recent papers. In [2] Çela, Klinz, Lendl, Orlin, Woeginger and Wulf proved that it is coNP-complete to decide whether a QSPP instance on an input graph containing a directed cycle is linearizable. Thus, a nice characterization of linearizable QSPP instances for such graphs seems to be unlikely. In the acyclic case, Hu and Sotirov first described a polynomial-time algorithm for the LINQSPP on directed two-dimensional grid graphs [10]. Recently, in [12] they generalized this result to all acyclic digraphs and proposed an algorithm which solves the problem in $\mathcal{O}(nm^3)$, where n and m denote the number of vertices and arcs in G.

Finally, let us mention a related concept, the so-called universal linearizability, studied in [2,10]. A digraph G is called *universally linearizable with respect to the QSPP* iff every instance of the QSPP on the input graph G is linearizable for every choice of the cost function q. In [10] it is shown that a particular class of grid graphs is universally linearizable. In [2] a characterization of universally linearizable grid graphs in terms of structural properties of the set of s-t-paths is given. Moreoever, for acyclic digraphs a forbidden subgraphs characterization of the universal linearizability is given in [2].

Contribution and Organization of the Paper. In this paper we provide a novel and simple characterization of linearizable QSPP instances on acyclic digraphs. Our characterization shows that the linearizability can be seen as a *local* property. In particular, we show that an instance of the QSPP on an acyclic digraph G is linearizable if and only if each subinstance obtained by considering a subdigraph of G consisting of two s-t-paths in G is linearizable. Our simple characterization also works for the SPP$_d$ and even for completely arbitrary cost functions, which assign some cost $f(P)$ to every s-t-path P without any further restrictions. The latter problem is referred to as the *Generic Shortest Path Problem* (GSPP) and is formally introduced in Sect. 2. Indeed, the characterization of the linearizable instances of the SPP_d follows from the characterization of the linearizable instances of the GSPP, both on acyclic digraphs.

Further, we propose a linear time algorithm which can check the local condition mentioned above for the QSPP and the SPP_d. We note that this is not straightforward, because the number of the subinstances for which the condition needs to be checked is in general exponential. As a side result our approach reveals an interesting connection between the LINQSPP and the problem of deciding whether all s-t-paths in a digraph have the same length. As a result, we obtain an algorithm which solves the LINQSPP linearization in $\mathcal{O}(m^2)$ time, thus improving the best previously known running time of $\mathcal{O}(nm^3)$ obtained in [12]. Our approach yields an $\mathcal{O}(d^2 m^d)$ time algorithm for the LINSPP_d, thus providing the first (polynomial time) algorithm for this problem. Note that the running time of the proposed algorithms is linear in the input size for both problems, LINQSPP and LINSPP_d, respectively. (The costs of all $\Omega(m^2)$ pairs of arcs, in the case of the QSPP, and the costs of all $\Omega(m^d/d!)$ subsets of arcs of cardinality d, in the case of the SPP_d, need to be encoded in the input.)

Finally, we also obtain a polynomial time algorithm that given an acyclic digraph G computes a basis of the subspace of all linearizable degree-d cost functions on G. Such a basis can be used to obtain better linearization-based bounds usable in B&B algorithms.

The paper is organized as follows. After introducing some notations and preliminaries in Sect. 2 we present the result on the characterization of the linearizable QSPP and SPP_d instances on acyclic input digraphs in Sect. 3. The algorithms for the linearization problems LINQSPP and LINSPP_d are presented in Sect. 4. Section 5 deals with computing a basis of the subspace of all linearizable d-degree cost functions on an acyclic digraph G.

2   Notations and Preliminaries

Given a digraph $G = (V, A)$, a simple directed s-t-path P in G is specified as a sequence of arcs $P = (a_1, a_2, \ldots, a_p)$ such that a_1 starts at s, a_p ends at t, nonconsecutive arcs do not share a vertex and the end vertex of a_i coincides with the start vertex of a_{i+1} for any $i \in \{1, \ldots, p-1\}$. The number p of arcs in P is called the length of the path. We sometimes use the same notation for a path P and the set of its arcs. We consider a single arc (x, y) as an x-y-path of length 1 and a single vertex x as a trivial x-x-path of length 0. Given an x-y-path P_1 and a y-z-path P_2, we denote the *concatenation* of P_1 and P_2 by $P_1 \cdot P_2$. We also consider concatenations of paths and arcs, that is, terms of the form $P \cdot a$ for some x-y-path P and some arc $a = (y, z)$.

In the linearization problem, we are concerned with acyclic digraphs $G = (V, A)$ with a source vertex s and a sink vertex t. We denote by $\mathcal{P}_{st}$ the set of all simple directed s-t-paths. We often assume that G is $\mathcal{P}_{st}$-*covered*, that is, every arc in G is traversed by at least one path in $\mathcal{P}_{st}$. It is easy to see that this assumption can be made without loss of generality.

Let $d \geq 2$ be a natural number. The *Order-d interaction costs* are given by a mapping $q_d \colon \{B \subseteq A : |B| \leq d\} \to \mathbb{R}$, assigning a (potentially negative) interaction cost to every subset of at most d arcs. The cost $\mathrm{SPP}_d(P, q_d)$ of some path P under interaction costs q_d is defined as in equation (2). If d is unambiguously clear form the context, we use the more compact notation $f_q(P) := \mathrm{SPP}_d(P, q_d)$. In this paper we explicitly allow the case $q(\emptyset) \neq 0$, because this simplifies the calculations. The *linearization problem for the Order-d Shortest Path Problem* (LinSPP$_d$) is formally defined as follows.

Problem: The linearization problem for the SPP$_d$ (LinSPP$_d$)

Instance: A $\mathcal{P}_{st}$-covered directed graph $G = (V, A)$ with $s, t \in V$, $s \neq t$; an integer $d \geq 2$; an order-d arc interaction cost function $q_d : \{B \subseteq A : |B| \leq d\} \to \mathbb{R}$.

Question: Find a *linearizing cost function* $c \colon A \to \mathbb{R}$ such that $\mathrm{SPP}_d(P, q_d) = \mathrm{SPP}(P, c)$ for all $P \in \mathcal{P}_{st}$ or decide that such a linearizing cost function does not exist.

In the special case $d = 2$, we obtain the linearization problem for the QSPP (LinQSPP).

Finally, let us consider the *Generic Shortest Path Problem* (GSPP) which takes as input a digraph $G = (V, A)$ with a source vertex s, a sink vertex t, $s \neq t$, and a generic cost function $f \colon \mathcal{P}_{st} \to \mathbb{R}$ assigning a cost $f(P)$ to every path $P \in \mathcal{P}_{st}$[2] We assume w.l.o.g. that G is $\mathcal{P}_{st}$-covered. The goal is to find an s-t-path which minimizes the objective function $f(P)$ over $\mathcal{P}_{st}$. A linearizable instance of the GSPP and the linearization problem for the GSPP (LinGSPP) are defined analogously as in the respective definitions for SPP$_d$.

3   A Characterization of Linearizable Instances of the GSPP on Acyclic Digraphs

The main result of this section is Theorem 1, our novel characterization of linearizable instances of the GSPP on acyclic digraphs defined as in Sect. 2.

Definition 2. *Let $G = (V, A)$ be a $\mathcal{P}_{st}$-covered acyclic digraph. For some vertex v, let P_1, P_2 be two s-v-paths, and let Q_1, Q_2 be two v-t-paths. The 5-tuple (v, P_1, P_2, Q_1, Q_2) is called a* two-path system *contained in G. The system is called* linearizable *with respect to the function $f : \mathcal{P}_{st} \to \mathbb{R}$, if there exists a cost function $c : A \to \mathbb{R}$ such that for all four paths $P \in \{P_1 \cdot Q_1, P_1 \cdot Q_2, P_2 \cdot Q_1, P_2 \cdot Q_2\}$ we have $f(P) = SPP(P, c)$. Such a c is called a* linearizing cost function *for (v, P_1, P_2, Q_1, Q_2) with respect to f.*

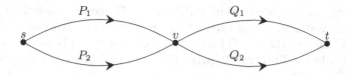

Fig. 1. A two-path system.

See Fig. 1 for an illustration of a two-path system. Note that P_1 and P_2 (as well as Q_1 and Q_2) can have common inner vertices and that the cases $P_1 = P_2$, $Q_1 = Q_2$, $v = s$ and $v = t$ are allowed. However, due to the acyclicity of G, the paths P_i and Q_j have only the vertex v in common for $i, j \in \{1, 2\}$. Further, observe that the linearizability of a two-path system is a local property, in the sense that it only depends on the four paths $P_1 \cdot Q_1, P_1 \cdot Q_2, P_2 \cdot Q_1$ and $P_2 \cdot Q_2$. Indeed, the following simple characterization holds.

Proposition 1. *A two-path system* (v, P_1, P_2, Q_1, Q_2) *is linearizable with respect to some function* $f \colon \mathcal{P}_{st} \to \mathbb{R}$ *iff*

$$f(P_1 \cdot Q_1) + f(P_2 \cdot Q_2) = f(P_1 \cdot Q_2) + f(P_2 \cdot Q_1). \tag{3}$$

Proof. First, assume that (v, P_1, P_2, Q_1, Q_2) is linearizable and let c be the corresponding linearizing cost function. Let M_1 (M_2) be the multiset resulting from the union of the sets of the arcs of the paths $P_1 \cdot Q_1$ and $P_2 \cdot Q_2$ ($P_1 \cdot Q_2$ and $P_2 \cdot Q_1$). Since M_1 and M_2 coincide we get $c(P_1 \cdot Q_1) + c(P_2 \cdot Q_2) = \sum_{a \in M_1} c(a) = \sum_{a \in M_2} c(a) = c(P_1 \cdot Q_2) + c(P_2 \cdot Q_1)$. Then, (3) follows from the definition of the linearizability of (v, P_1, P_2, Q_1, Q_2).

Assume now that Eq. (3) is true. We show the linearizability of the two-path system with respect to f by constructing a linearizing cost function c. It is easy to find a suitable c if $P_1 = P_2$ or $Q_1 = Q_2$. So let us consider the more involved case where $P_1 \neq P_2$ and $Q_1 \neq Q_2$. In this case, for each $P \in \{P_1, P_2, Q_1, Q_2\}$ there exists a (not necessarily unique) *representative* arc $a \in P$ such that a is not contained in any other path $Q \in \{P_1, P_2, Q_1, Q_2\}$, $Q \neq P$. Let a_1, a_2, e_1, e_2 be representative arcs of P_1, P_2, Q_1 and Q_2, respectively. Consider now a cost function $c : A \to \mathbb{R}$, such that $c(a) = 0$ if $a \notin \{a_1, a_2, e_1, e_2\}$, and $c(a_1)$, $c(a_2)$, $c(e_1)$, $c(e_2)$ fulfill the following linear equations:

$$
\begin{aligned}
c(a_1) &&+\ c(e_1) &&&= f(P_1 Q_1) \\
c(a_1) &&+ &\ c(e_2) &&= f(P_1 Q_2) \\
&c(a_2) + c(e_1) &&&&= f(P_2 Q_1) \\
&c(a_2) + &&\ c(e_2) &&= f(P_2 Q_2)
\end{aligned}
$$

Using basic linear algebra, one can see that this system indeed has a solution whenever Eq. (3) holds (there is even a solution with $c(e_2) = 0$). Thus, c constructed as above is a linearizing cost function for (v, P_1, P_2, Q_1, Q_2) with respect to f.   □

---

[2] We assume that f is given as an oracle.

Now, consider an instance of the GSPP with a $\mathcal{P}_{st}$-covered acyclic digraph G, with a source vertex s, a sink vertex t and a generic cost function $f\colon \mathcal{P}_{st} \to \mathbb{R}$. When is this instance (G, s, t, f) linearizable? Obviously, if G contains a two-path system which is not linearizable with respect to f, then (G, s, t, f) is not linearizable. The following theorem shows that the linearizability of each two-path system with respect to f is a sufficient condition for (G, s, t, f) being linearizable.

Theorem 1. *Let G be a $\mathcal{P}_{st}$-covered acyclic digraph with a source vertex s and a sink vertex t and let $f : \mathcal{P}_{st} \to \mathbb{R}$ be a generic cost function. Then the instance (G, s, t, f) of the GSPP is linearizable if and only if every two-path system contained in G is linearizable with respect to f.*

Before proving the theorem, we need some preparation. Let $G = (V, A)$ be a $\mathcal{P}_{st}$-covered acyclic digraph with source vertex s and sink vertex t. First we introduce a *topological arc order* as a total order $\preceq$ on A such that for any pair of arcs a, a' in A the following holds: if there exists a path P containing both a and a' such that a comes before a' in P, then $a \preceq a'$. It is easy to see that any acyclic digraph has a (in general non-unique) topological arc order. Moreover, a topological arc order can be obtained from a topological vertex order.

Further, we recall the definition of a *system of nonbasic arcs* introduced by Sotirov and Hu [12].

Definition 3. *Let G be a $\mathcal{P}_{st}$-covered acyclic digraph with a source vertex s and a sink vertex t. A set $N \subseteq A$ is called a* system of nonbasic arcs, *iff for every vertex $v \in V \setminus \{s, t\}$ exactly one of the arcs starting at v is contained in N. The latter arc is called the* nonbasic arc of v. *An arc $a \in A \setminus N$ is called* basic.

Obviously, the system of nonbasic arcs is not unique. Any such system forms an in-tree rooted at t containing all the vertices in V except for s. For some system of nonbasic arcs N and some vertex $v \in V \setminus \{s\}$, we let N_v denote the unique v-t-path consisting of nonbasic arcs (where N_t is the trivial path). A cost function $c\colon A \to \mathbb{R}$ is called *in reduced form* with respect to N, if $c(a) = 0$ for all nonbasic arcs $a \in N$. The following lemma is an easy adaption from [12], where an analogous statement was proven for the less general case of the QSPP instead of the GSPP (details are provided in the full version of this paper).

Lemma 1 (adapted from [12, Prop. 4]). *Let G be a $\mathcal{P}_{st}$-covered acyclic digraph with a source vertex s and a sink vertex t. Let $f\colon \mathcal{P}_{st} \to \mathbb{R}$ be a generic cost function and let $N \subseteq A$ be a fixed system of nonbasic arcs. If (G, s, t, f) is a linearizable instance of the GSPP, then there exists one and only one linear cost function $c\colon A \to \mathbb{R}$ which is both a linearizing cost function and in reduced form.*

Let (G, s, t, f) be a linearizable instance of the GSPP with $G = (V, A)$ and $N \subseteq A$ be a fixed system of nonbasic arcs. For a linearizing cost function $c\colon A \to \mathbb{R}$, we denote by reduced(c) the unique linearizing cost function in reduced form (which exists due to Lemma 1). It follows from the arguments in the proof of Lemma 1 that for given c one can compute reduced(c) in $\mathcal{O}(n+m)$ time. We are now ready to sketch the proof of our main theorem.

Proof (Sketch of the proof of Theorem 1).

The necessity of the conditions for linearizability is trivial. Now we prove the sufficiency. Thus we assume that every two-path system is linearizable with respect to f and show that (G, s, t, f) is linearizable. Let N be a system of nonbasic arcs. The main idea is to find a linearizing cost function which is in reduced form, i.e., which has value 0 on all nonbasic arcs. To this end we consider a topological arc order $\preceq$ on the set A of arcs in G and inductively define a linearizing cost function $c\colon A \to \mathbb{R}$ as follows. For any arc $a = (u, v)$ set

$$c(a) := \begin{cases} f(P \cdot a \cdot N_v) - \sum_{a' \in P} c(a') & a \notin N \\ 0; & a \in N \end{cases} \tag{4}$$

for some s-u-path P.

Consider now the following claim the proof of which is omitted for brevity.

Claim: If all two-path systems in G are linearizable with respect to f, then function c in Equation (4) is well-defined and independent of the concrete choice of P. Moreoever, the following equation holds for all s-u-paths P:

$$f(P \cdot a \cdot N_v) = c(a) + \sum_{a' \in P} c(a') = c(P \cdot a \cdot N_v) \tag{5}$$

Instead of a proof, we give a short intuition, why the claim is true. Due to Lemma 1, whenever we look for a linearizing function, we can w.l.o.g. look for one in reduced form. Consider now a linearizing function c such that $c(a) = 0$ holds for all nonbasic arcs. It is not hard to see that Eq. (4) is a necessary condition that must be true for every s-u-path P. Thus, we obtain a system of equations to be fulfilled by every linearizing function c as above. It turns out that the system of equations mentioned above has a solution, if every two-path system is linearizable.

Finally, observe that the claim immediately implies that (G, s, t, f) is linearizable. Indeed, let c be the cost function defined in Eq. (4) and let Q be some s-t-path. Choose $a = (x, t)$ to be the last arc on Q. Then N_t is the trivial path from t to t, so by applying Eq. (5) to the arc a, we have $f(Q) = c(Q)$.

□

Since in general a graph contains exponentially many different two-path systems, Theorem 1 does not seem to lead to an efficient algorithm for the linearization problem LINGSPP at a first glance. However, we show in the next section that this is indeed the case. The arguments are based on a more technical version of Theorem 1 and involve the concept of so-called *strongly basic arcs* and their property (π) defined below.

Definition 4. *Let $G = (V, A)$ be an acyclic $\mathcal{P}_{st}$-covered digraph with source vertex s and sink vertex t. Let $f\colon \mathcal{P}_{st} \to \mathbb{R}$ be a generic cost function and let*

$N \subseteq A$ be a system of nonbasic arcs in G. A basic arc (u, v) is called strongly basic, if it is not incident to the source vertex, that is if $u \neq s$.

A strongly basic arc $a = (u, v)$ has the property (π), if for any s-u-path P the value $\mathrm{val}(a, P) := f(P \cdot a \cdot N_v) - f(P \cdot N_u)$ does not depend on the choice of P.

Thus, if a strongly basic arc $a = (u, v)$ has the property (π), we have $\mathrm{val}(a, P) = \mathrm{val}(a, Q)$ for any two s-u-paths P, Q and this implies the existence of a value $\mathrm{val}(a) := \mathrm{val}(a, P)$ for each s-u-path P and $\mathit{val}(a)$ is well defined for each strongly basic arc. Finally, we set $\mathrm{val}(a) := f(a \cdot N_v)$ for each basic arc $a = (s, v)$.

Lemma 2. *Let $G = (V, A)$ be an acyclic $\mathcal{P}_{st}$-covered digraph with source vertex s and sink vertex t. Let $f \colon \mathcal{P}_{st} \to \mathbb{R}$ be a generic cost function and let $N \subseteq A$ be a system of nonbasic arcs in G. Then (G, s, t, f) is linearizable if and only if every strongly basic arc has the property (π). In this case, the mapping $c \colon A \to \mathbb{R}$ given by*

$$c(a) = \begin{cases} \mathrm{val}(a); & a \text{ is basic} \\ 0; & a \text{ is nonbasic} \end{cases}$$

is a linearizing cost function in reduced form.

Proof. Let $a = (u, v)$ be a strongly basic arc. We claim that a has the property (π) iff for any two s-u-paths P, Q the two-path system $(u, P, Q, N_u, a \cdot N_v)$ is linearizable with respect to f. Indeed, note that by Proposition 1, the two-path system above is linearizable with respect to f iff $f(P \cdot a \cdot N_v) + f(Q \cdot N_u) = f(P \cdot N_u) + f(Q \cdot a \cdot N_v)$. The latter equation is equivalent to $\mathrm{val}(a, Q) = \mathrm{val}(a, P)$. Recalling that the latter equality holds for every pair of P, Q iff a has the property (π) completes the proof of the claim.

Now, assume that some strongly basic arc (u, v) does not have the property (π). Then, the corresponding two-path system $(u, P, Q, N_u, a \cdot N_v)$ is not linearizable with respect to f and therefore, (G, s, t, f) is not linearizable.

Finally, assume that every strongly basic arc has the property (π). In the proof of Theorem 1 we use the linearizability assumption only for specific two-path systems of the form $(u, P, Q, N_u, a \cdot N_v)$, where $a = (u, v)$ is some strongly basic arc. Thus, if the property (π) holds for all strongly basic arcs, then each such specific two-path system is linearizable with respect to f and the linearizability of (G, s, t, f) follows. Furthermore, the value $c(a)$ of the linearizing cost function in Eq. (4) equals $\mathrm{val}(a)$ for any arc a which is either strongly basic or incident to s, while $c(a) = 0$ for any nonbasic arc a. $\square$

4   A Linear Time Algorithm for the LinSPP_d

In this section, we describe an algorithm which solves the linearization problem for SPP_d (LinSPP_d) in $\mathcal{O}(m^d)$ time which is linear compared to the input size. The algorithm uses the relationship between the LinSPP_d and the *All-Paths-Equal-Cost Problem (APECP)* which we introduce in Sect. 4.1. In Sect. 4.2 we describe the SPP_d algorithm and discuss its running time.

4.1   The All Paths Equal Cost Problem of Order-d (APECP$_d$)

The All Paths Equal Cost Problem of Order-d (APECP$_d$) is defined as follows.

Problem: ALL PATHS EQUAL COST of Order-d (APECP$_d$)

Instance: An acyclic $\mathcal{P}_{st}$-covered directed graph $G = (V, A)$ with a source vertex s and a sink vertex t, an integer $d \geq 1$; an order-d cost function $q_d \colon \{B \subseteq A : |B| \leq d\} \to \mathbb{R}$.

Question: Do all s-t-paths have the same cost, i.e. is there some $\beta \in \mathbb{R}$ such that SPP$_d(P, q_d) = \beta$ for every path P in $\mathcal{P}_{st}$?

In the following we establish a connection between the LINSPP$_d$ and the APECP$_{d-1}$ for $d \geq 2$. More precisely, we show in Lemma 3 that an instance (G, s, t, q_d) of the LINSPP$_d$ with an acyclic $\mathcal{P}_{st}$-covered digraph $G = (V, A)$ can be reduced to $\mathcal{O}(m)$ instances of APECP$_{d-1}$, each of them corresponding to exactly one strongly basic arc with respect to some fixed system of nonbasic arcs (see Definitions 3 and 4). The APECP$_{d-1}$ instance corresponding to a strongly basic arc $a = (u, v)$ is defined as follows.

Definition 5. *The instance $I^{(a)}$ of the APECP$_{d-1}$ corresponding to the strongly basic arc $a = (u, v)$ takes as input the digraph $G^{(a)} = (V_u, E_u)$ with source vertex $s' = s$, sink vertex $t' = u$, where V_u is the set of vertices in V lying on at least one s-u-path and A_u is the set of arcs in A lying on at least one s-u-path. The order-$(d-1)$ cost function $q_{d-1}^{(a)} \colon \{B \subseteq A_u : |B| \leq d - 1\} \to \mathbb{R}$ is given by*

$$q_{d-1}^{(a)}(B) := \left(\sum_{\substack{C \subseteq N_u \\ |C| \leq d - |B|}} q_d(B \cup C) \right) - \left(\sum_{\substack{C \subseteq a \cdot N_v \\ |C| \leq d - |B|}} q_d(B \cup C) \right). \tag{6}$$

Lemma 3. *Let $d \geq 2$ and let (G, s, t, q_d) be an instance of the LINSPP$_d$ with a fixed system of nonbasic arcs N. The APECP$_{d-1}$ instance $I^{(a)}$ corresponding to some strongly basic arc a is a YES-instance iff the arc a has the property (π) with respect to $f \colon \mathcal{P}_{st} \to \mathbb{R}$ given by $f(P) = SPP_d(P, q_d)$ for $P \in \mathcal{P}_{st}$. In this case, $\mathrm{val}(a) = \beta$, where β is the common cost of all paths in the APECP$_{d-1}$ instance.*

Proof (Sketch). Let $a = (u, v) \in A$ be a strongly basic arc and let P be some s-u-path in G. Then P is contained in the graph $G_a = (V_u, A_u)$. It can be shown that

$$\mathrm{val}(a, P) = f(P \cdot N_u) - f(P \cdot a \cdot N_v) = \sum_{\substack{B \subseteq P \\ |B| \leq d-1}} q_{d-1}^{(a)}(B) = f^{(a)}(P),$$

where $f^{(a)}(P) = SPP_{d-1}(P, q_{d-1}^{(a)})$ for any s-u-path P in G. We conclude that the value $\mathrm{val}(a, P)$ is independent of P, if and only if for every path the quantity $f^{(a)}(P)$ does not depend on P. The latter condition is equivalent to $I^{(a)}$ being a YES-instance. Furthermore, if this is the case, then $\mathrm{val}(a) = f^{(a)}(P)$ for any s-u-path P.   □

Lemmas 2 and 3 imply that an instance (G, s, t, q_d) of the SPP_d with an acyclic digraph G is linearizable iff each instance $I^{(a)}$ of the $APECP_{d-1}$ corresponding to some strongly basic arc a (with respect to some fixed system of nonbasic arcs) is a YES-instance. Thus, an instance of the LINSPP_d can be reduced to $\mathcal{O}(m)$ instances of the $APECP_{d-1}$. Next, in Lemma 4 we show that each instance of the $APECP_{d-1}$ can be reduced to an instance of the LINSPP_{d-1}. First, we define a specific cost function as follows.

Definition 6. *Let $G = (V, A)$ be a $\mathcal{P}_{st}$-covered acyclic digraph and $\beta \in \mathbb{R}$. The function* $\mathsf{source}_\beta : A \to \mathbb{R}$ *assigns cost β to every arc incident to the source s, and 0 to all other arcs.*

Lemma 4. *Let $G = (V, A)$ be a $\mathcal{P}_{st}$-covered acyclic digraph with source vertex s and sink vertex t and let $N \subseteq A$ a fixed system of nonbasic arcs. Let q_d be an order-d cost function. The instance (G, s, t, q_d) of the $APECP_d$ problem is a YES-instance iff the instance (G, s, t, q_d) of SPP_d is linearizable and source_β is its unique linearizing function in reduced form (with respect to N).*

Proof. Clearly, source_β is a linearizing function iff all paths have the same cost β. Furthermore, observe that all arcs incident to the source do not belong to N. Therefore source_β is in reduced form with respect to N. In fact, by Lemma 1 source_β is the unique linearizing functions in reduced form, and $\mathsf{reduced}(c') = \mathsf{source}_\beta$ for all other linearizing functions c'.   □

4.2   The Linear Time LINSPP_d algorithm

Our LINSPP_d algorithm $\mathcal{A}$ works as follows. Consider an instance (G, s, t, q_d) of the LINSPP_d with an acyclic $\mathcal{P}_{st}$-covered digraph G, with source vertex s, sink vertex t and order-d cost function q_d. We first fix some system of nonbasic arcs N and construct the instance $I^{(a)}$ of the $APECP_{d-1}$ problem given in Definition 5 for each strongly basic arc a. Then, we check each instance $I^{(a)}$ for being a YES-instance and do this by reducing $I^{(a)}$ to an instance of LINSPP_{d-1} according to Lemma 4. By iterating this process we eventually end up with APECP problems of degree 1 that can be easily solved by dynamic programming. The dynamic program is based on the fact that in a $\mathcal{P}_{st}$-covered acyclic digraph with a cost function $f : \mathcal{P}_{st} \to \mathbb{R}$ all s-t-paths have the same cost iff for every vertex v all s-v-paths have the same cost.

It is not hard to implement the algorithm described above in $\mathcal{O}(d^2 m^{d+1})$ time. With a careful implementation it is possible to achieve a better result.

Theorem 2. *The $\text{LIN}SPP_d$ on acyclic digraphs can be solved in $\mathcal{O}(d^2 m^d)$ time.*

For the sake of brevity we refer to the full version of the paper for the proof of the theorem. Here we just point out the necessity of an efficient computation of the $I^{(a)}$ instances for all strongly basic arcs as defined in Definition 5. To this end we efficiently compute the values

$$\gamma(B,x) := \sum_{\substack{C \subseteq N_x \\ |C| \leq d-|B|}} q(B \cup C).$$

for all sets $B \subseteq A$ of arcs with $|B| \leq d-1$ and all vertices $x \in V \setminus \{s\}$. These values are then used to efficiently compute the cost functions $q_{d-1}^{(a)}$ in Eq. (6). Further, with a careful management of the quantities involved in the computation of the linearizing functions (see Lemma 4) we obtain a linear time algorithm. We assume that the input q_d is stored in d-dimensional array Then the input size required to encode the cost function q_d equals $\sum_{k=0}^{d} \binom{m}{k} \geq m^d/d!$. Thus, $\mathcal{O}(d^2 m^d)$ is linear in the input size and hence optimal if d is considered a constant, like for example in the QSPP.

Finally, one could also ask about the case where q_d is sparse. In this case, our algorithm still has running time $\mathcal{O}(d^2 m^d)$, but this time is no longer linear in the input size. Hence, it is an interesting open question, whether one can also find a linear time algorithm for this case.

5   The Subspace of Linearizable Instances

Let $d \in \mathbb{N}$, $d \geq 2$, and a $\mathcal{P}_{st}$-covered acyclic digraph $G = (V,A)$ with source vertex s and sink vertex t be fixed. Let $H^{(d)} := \{B \subseteq H \mid |B| \leq d\}$ be the set of all subsets of at most d arcs in arc set $H \subseteq A$. Every order-d cost function $q_d \colon A^{(d)} \to \mathbb{R}$ can be uniquely represented by a vector $x \in \mathbb{R}^{A^{(d)}}$ with $q_d(F) = x_F$ for all $F \in A^{(d)}$, and vice-versa. Thus, each instance (G,s,t,q_d) can be identified with the corresponding vector $x \in \mathbb{R}^{A^{(d)}}$ and we will say that $x \in \mathbb{R}^{A^{(d)}}$ is an instance of the SPP$_d$. It is straightforward to see that the linearizable instances of the SPP$_d$ on the fixed digraph G form a linear subspace $\mathcal{L}_d$ of $\mathbb{R}^{A^{(d)}}$.

Methods to compute this subspace are useful in B&B algorithms for the SPP$_d$ as they can be applied to compute better lower bounds along the lines of what Hu and Sotirov [12] did for general quadratic binary programs. Hu and Sotirov showed that for $d = 2$ a basis of $\mathcal{L}_d$ can be computed in polynomial time [12, Prop. 5]. We extend their result to the case $d > 2$.

Theorem 3. *Let $G = (V,A)$ be a $\mathcal{P}_{st}$-covered, acyclic digraph with source vertex s and sink vertex t and let $d \in \mathbb{N}$ be a constant. A basis of the subspace $\mathcal{L}_d$ of the linearizable instances of the SPP$_d$ can be computed in polynomial time.*

Proof (Sketch). The proof idea is to specify a $k \in \mathbb{N}$ and a matrix M of polynomially bounded dimensions, such that for $f : \mathbb{R}^{A^{(d)}} \to \mathbb{R}^k$ with $f(x) = Mx$, we have: $f(x) = 0$ iff x is a linearizable instance of the SPP$_d$. Thus, the linearizable instances x of the SPP$_d$ form $\ker(M)$ which can be efficiently computed.

The construction of M is done iteratively and exploits the relationship between SPP_d and $APECP_{d-1}$ similarly as in the algorithm $\mathcal{A}$ from Sect. 4.2. In particular we use the following two facts:

(i) For each strongly basic arc $a = (u, v)$, the function which maps $x \in \mathbb{R}^{A^{(d)}}$ to $q^{(a)}_{d-1} : A^{(d-1)}_u \to \mathbb{R}$ is linear (see Eq. (6) and recall Definition 5 for A_u).

(ii) The function $c \mapsto \mathsf{reduced}(c)$ (defined after Lemma 1) is linear.

Using (i) and (ii) iteratively as in algorithm $\mathcal{A}$, we show by induction that for each strongly basic arc $a = (u, v)$ and each $d \geq 2$ there exist $k_a \in \mathbb{N}$ and a linear function $g_a : \mathbb{R}^{A^{(d-1)}_u} \to \mathbb{R}^{k_a}$ s.t. $g_a(x) = 0$ iff if x corresponds to a YES-instance of $APECP_{d-1}$. Then we construct the linear function g'_a on the same domain as g_a, by setting $g'_a(x) = \beta$ whenever $g_a(x) = 0$, where β is the common path cost of the corresponding instance $x$project/63605ec6eb4e243dcb2eeb7d of $APECP_{d-1}$. Next we show that for each vertex u there exists a $k_u \in \mathbb{N}$ and a linear function $f_u : \mathbb{R}^{A^{(d-1)}_u} \to \mathbb{R}^{k_u}$ such that $f_u(x) = 0$ iff x is a linearizable instance of SPP_{d-1} corresponding to $APECP_{d-1}$ (see Lemma 4). Then we construct the linear function f'_u on the same domain as f_u by setting $f'_u(x)$ equal to the linearizing cost function of the instance x of the SPP_{d-1} whenever x is linearizable (i.e. when $f_u(x) = 0$). The construction of M is done by repeating these steps iteratively until $d = 1$. One can ensure that the size of the matrix representations of all involved functions stays polynomial. $\square$

Acknowledgement. This research has been supported by the Austrian Science Fund (FWF): W1230.

References

1. Bookhold, I.: A contribution to quadratic assignment problems. Optimization **21**(6), 933–943 (1990)

2. Çela, E., Klinz, B., Lendl, S., Orlin, J.B., Woeginger, G.J., Wulf, L.: Linearizable special cases of the quadratic shortest path problem. In: Kowalik, Ł., Pilipczuk, M., Rzążewski, P. (eds.) WG 2021. LNCS, vol. 12911, pp. 245–256. Springer, Cham (2021). https://doi.org/10.1007/978-3-030-86838-3_19

3. Cela, E., Deineko, V.G., Woeginger, G.J.: Linearizable special cases of the QAP. J. Comb. Optim. **31**(3), 1269–1279 (2016)

4. Ćustić, A., Punnen, A.P.: A characterization of linearizable instances of the quadratic minimum spanning tree problem. J. Comb. Optim. **35**(2), 436–453 (2018)

5. De Meijer, F., Sotirov, R.: The quadratic cycle cover problem: special cases and efficient bounds. J. Comb. Optim. **39**(4), 1096–1128 (2020)

6. Erdoğan, G.: Quadratic assignment problem: linearizations and polynomial time solvable cases, Ph. D. thesis, Bilkent University (2006)

7. Erdoğan, G., Tansel, B.: A branch-and-cut algorithm for quadratic assignment problems based on linearizations. Comput. Oper. Res. **34**(4), 1085–1106 (2007)

8. Erdoğan, G., Tansel, B.C.: Two classes of quadratic assignment problems that are solvable as linear assignment problems. Discret. Optim. **8**(3), 446–451 (2011)

9. Gamvros, I.: Satellite network design, optimization and management. University of Maryland, College Park (2006)

10. Hu, H., Sotirov, R.: Special cases of the quadratic shortest path problem. J. Comb. Optim. **35**(3), 754–777 (2018)
11. Hu, H., Sotirov, R.: On solving the quadratic shortest path problem. INFORMS J. Comput. **32**(2), 219–233 (2020)
12. Hu, H., Sotirov, R.: The linearization problem of a binary quadratic problem and its applications. Annal. Oper. Res. **307**, 229–249 (2021)
13. Kabadi, S.N., Punnen, A.P.: An $O(n^4)$ algorithm for the QAP linearization problem. Math. Oper. Res. **36**(4), 754–761 (2011)
14. Murakami, K., Kim, H.S.: Comparative study on restoration schemes of survivable ATM networks. In: Proceedings of INFOCOM1997, vol. 1, pp. 345–352. IEEE (1997)
15. Nie, Y.M., Wu, X.: Reliable a priori shortest path problem with limited spatial and temporal dependencies. In: Lam, W., Wong, S., Lo, H. (eds.) Transportation and Traffic Theory 2009: Golden Jubilee, pp. 169–195. Springer, Boston (2009). https://doi.org/10.1007/978-1-4419-0820-9_9
16. Punnen, A.P., Kabadi, S.N.: A linear time algorithm for the Koopmans-Beckmann QAP linearization and related problems. Discret. Optim. **10**(3), 200–209 (2013)
17. Punnen, A.P., Walter, M., Woods, B.D.: A characterization of linearizable instances of the quadratic traveling salesman problem. arXiv preprint arXiv:1708.07217 (2017)
18. Rostami, B., et al.: The quadratic shortest path problem: complexity, approximability, and solution methods. Eur. J. Oper. Res. **268**(2), 473–485 (2018)
19. Rostami, B., Malucelli, F., Frey, D., Buchheim, C.: On the quadratic shortest path problem. In: Bampis, E. (ed.) SEA 2015. LNCS, vol. 9125, pp. 379–390. Springer, Cham (2015). https://doi.org/10.1007/978-3-319-20086-6_29
20. Sen, S., Pillai, R., Joshi, S., Rathi, A.K.: A mean-variance model for route guidance in advanced traveler information systems. Transp. Sci. **35**(1), 37–49 (2001)
21. Sivakumar, R.A., Batta, R.: The variance-constrained shortest path problem. Transp. Sci. **28**(4), 309–316 (1994)
22. Sotirov, R., Verchére, M.: The quadratic minimum spanning tree problem: lower bounds via extended formulations. arXiv preprint arXiv:2102.10647 (2021)
23. Waddell, L., Adams, W.: Characterizing linearizable QAPs by the level-1 reformulation-linearization technique. (2021). https://optimization-online.org/?p=17020, preprint

Author Index

Printed in the United States
by Baker & Taylor Publisher Services